PRINCIPLES OF MAGNETOHYDRODYNAMICS
With Applications to Laboratory and Astrophysical Plasmas

This textbook provides a modern and accessible introduction to magnetohydrodynamics (MHD). It describes the two main applications of plasma physics – laboratory research on thermonuclear fusion energy, and plasma-astrophysics of the solar system, stars and accretion discs – from the single viewpoint of MHD. This approach provides effective methods and insights for the interpretation of plasma phenomena on virtually all scales, from the laboratory to the Universe. It equips the reader with the necessary tools to understand the complexities of plasma dynamics in extended magnetic structures. The classical MHD model is developed in detail without omitting steps in the derivations, and problems are included at the end of each chapter. This text is ideal for senior-level undergraduate and graduate courses in plasma physics and astrophysics.

J. P. (Hans) Goedbloed is Senior Research Scientist at the FOM-Institute for Plasma Physics and Professor of Theoretical Plasma Physics in the Department of Physics and Astronomy at Utrecht University. He coordinates a large scale computational effort of the Dutch Science Organization on Fast Changes in Complex Flows involving scientists of different disciplines. He loves teaching and research in magnetohydrodynamics, both for its beauty and its clues on the coherence of the Universe.

Stefaan Poedts obtained a Ph.D. in mathematics at the Katholieke Universiteit Leuven, Belgium in 1988 and returned in 1996 as Research Associate at the Centre for Plasma-Astrophysics. He is presently Professor in the Department of Mathematics. His research is in the field of computational and nonlinear magnetohydrodynamics with applications to solar astrophysics. He teaches courses on mathematics, plasma physics of the Sun, and numerical methods to undergraduate students in engineering, mathematics, and physics.

PRINCIPLES OF MAGNETOHYDRODYNAMICS

With Applications to Laboratory and Astrophysical Plasmas

J. P. (HANS) GOEDBLOED
*FOM-Institute for Plasma Physics 'Rijnhuizen',
and Astronomical Institute, Utrecht University*

STEFAAN POEDTS
Centre for Plasma-Astrophysics, Katholieke Universiteit Leuven

PUBLISHED BY THE PRESS SYNDICATE OF THE UNIVERSITY OF CAMBRIDGE
The Pitt Building, Trumpington Street, Cambridge, United Kingdom

CAMBRIDGE UNIVERSITY PRESS
The Edinburgh Building, Cambridge CB2 2RU, UK
40 West 20th Street, New York, NY 10011–4211, USA
477 Williamstown Road, Port Melbourne, VIC 3207, Australia
Ruiz de Alarcón 13, 28014 Madrid, Spain
Dock House, The Waterfront, Cape Town 8001, South Africa

http://www.cambridge.org

© J. P. Goedbloed and S. Poedts 2004

This book is in copyright. Subject to statutory exception
and to the provisions of relevant collective licensing agreements,
no reproduction of any part may take place without
the written permission of Cambridge University Press.

First published 2004

Printed in the United Kingdom at the University Press, Cambridge

Typeface Times 11/14 pt.   *System* LATEX 2_ε [TB]

A catalogue record for this book is available from the British Library

Library of Congress Cataloging in Publication data
Goedbloed, J. P. (Hans), 1940–
Principles of magnetohydrodynamics: with applications to laboratory and
astrophysical plasmas / by J.P. Goedbloed and S. Poedts.
p.   cm.
Includes bibliographical references and index.
ISBN 0 521 62347 2   ISBN 0 521 62607 2 (paperback)
1. Magnetohydrodynamics.   I. Poedts, S. (Stefaan), 1962–   II. Title.
QC718.5.M36G64 2004
538'.6–dc22   2003070010

ISBN 0 521 62347 2 hardback
ISBN 0 521 62607 2 paperback

The publisher has used its best endeavours to ensure that the URLs for external websites referred to in this book
are correct and active at the time of going to press. However, the publisher has no responsibility for the websites
and can make no guarantee that a site will remain live or that the content is or will remain appropriate.

To Antonia and Micheline

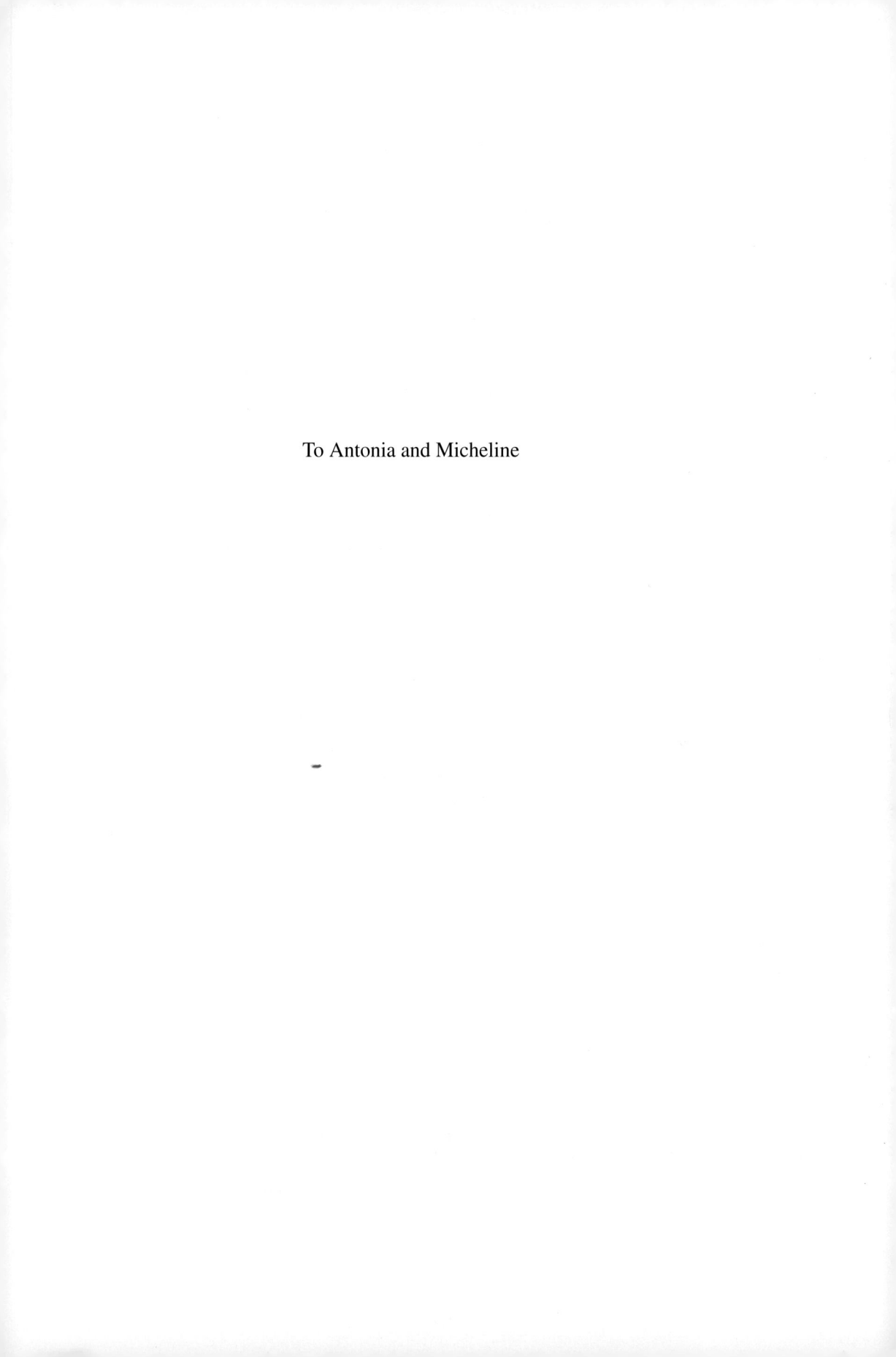

Contents

Preface				*page* xiii
Part I	**Plasma physics preliminaries**			1
1	Introduction			3
	1.1	Motivation		3
	1.2	Thermonuclear fusion and plasma confinement		4
		1.2.1	Fusion reactions	4
		1.2.2	Conditions for fusion	6
		1.2.3	Magnetic confinement and tokamaks	10
	1.3	Astrophysical plasmas		13
		1.3.1	Celestial mechanics	13
		1.3.2	Astrophysics	15
		1.3.3	Plasmas enter the stage	18
		1.3.4	The standard view of nature	21
	1.4	Definitions of the plasma state		23
		1.4.1	Microscopic definition of plasma	23
		1.4.2	Macroscopic approach to plasma	27
	1.5	Literature and exercises		29
2	Elements of plasma physics			34
	2.1	Theoretical models		34
	2.2	Single particle motion		34
		2.2.1	Cyclotron motion	34
		2.2.2	Excursion: basic equations of electrodynamics and mechanics	38
		2.2.3	Drifts, adiabatic invariants	41
	2.3	Kinetic plasma theory		47
		2.3.1	Boltzmann equation and moment reduction	48
		2.3.2	Collective phenomena: plasma oscillations	54
		2.3.3	Landau damping	58

	2.4	Fluid description	65
		2.4.1 From the two-fluid to the MHD description of plasmas	67
		2.4.2 Alfvén waves	71
		2.4.3 Equilibrium and stability	74
	2.5	In conclusion	79
	2.6	Literature and exercises	80
3	'Derivation' of the macroscopic equations★		83
	3.1	Two approaches★	83
	3.2	Kinetic equations★	84
		3.2.1 Boltzmann equation★	84
		3.2.2 Moments of the Boltzmann equation★	88
		3.2.3 Thermal fluctuations and transport★	90
		3.2.4 Collisions and closure★	94
	3.3	Two-fluid equations★	98
		3.3.1 Electron–ion plasma★	98
		3.3.2 The classical transport coefficients★	99
		3.3.3 Dissipative versus ideal fluids★	104
		3.3.4 Excursion: waves in two-fluid plasmas★	108
	3.4	One-fluid equations★	119
		3.4.1 Maximal ordering for MHD★	119
		3.4.2 Resistive and ideal MHD equations★	124
	3.5	Literature and exercises★	126

Part II Basic magnetohydrodynamics — 129

4	The MHD model		131
	4.1	The ideal MHD equations	131
		4.1.1 Postulating the basic equations	131
		4.1.2 Scale independence	138
		4.1.3 A crucial question	139
	4.2	Magnetic flux	140
		4.2.1 Flux tubes	140
		4.2.2 Global magnetic flux conservation	142
	4.3	Conservation laws	145
		4.3.1 Conservation form of the MHD equations	145
		4.3.2 Global conservation laws	148
		4.3.3 Local conservation laws – conservation of magnetic flux	152
		4.3.4 Magnetic helicity	155
	4.4	Dissipative magnetohydrodynamics	161
		4.4.1 Resistive MHD	161
		4.4.2 (Non-)conservation form of the dissipative equations★	165

	4.5	Discontinuities	167
		4.5.1 Shocks and jump conditions	167
		4.5.2 Boundary conditions for plasmas with an interface	171
	4.6	Model problems	173
		4.6.1 Laboratory plasmas (models I–III)	174
		4.6.2 Energy conservation for interface plasmas	178
		4.6.3 Astrophysical plasmas (models IV–VI)	180
	4.7	Literature and exercises	182
5	Waves and characteristics	186
	5.1	Physics and accounting	186
		5.1.1 Introduction	186
		5.1.2 Sound waves	186
	5.2	MHD waves	190
		5.2.1 Symmetric representation in primitive variables	190
		5.2.2 Entropy wave and magnetic field constraint	194
		5.2.3 Reduction to velocity representation: three waves	198
		5.2.4 Dispersion diagrams	202
	5.3	Phase and group diagrams	205
		5.3.1 Basic concepts	205
		5.3.2 Application to the MHD waves	207
		5.3.3 Asymptotic properties	212
	5.4	Characteristics★	213
		5.4.1 The method of characteristics★	213
		5.4.2 Classification of partial differential equations★	216
		5.4.3 Characteristics in ideal MHD★	219
	5.5	Literature and exercises	227
6	Spectral theory	230
	6.1	Stability: intuitive approach	230
		6.1.1 Two viewpoints	230
		6.1.2 Linearization and Lagrangian reduction	233
	6.2	Force operator formalism	237
		6.2.1 Equation of motion	237
		6.2.2 Hilbert space	242
		6.2.3 Proof of self-adjointness of the force operator	244
	6.3	Spectral alternatives★	250
		6.3.1 Mathematical intermezzo★	250
		6.3.2 Initial value problem in MHD★	253
	6.4	Quadratic forms and variational principles	256
		6.4.1 Expressions for the potential energy	256

		6.4.2	Hamilton's principle	259
		6.4.3	Rayleigh–Ritz spectral variational principle	259
		6.4.4	Energy principle	261
	6.5	Further spectral issues		263
		6.5.1	Normal modes and the energy principle★	263
		6.5.2	Proof of the energy principle★	266
		6.5.3	σ-stability	268
		6.5.4	Returning to the two viewpoints	271
	6.6	Extension to interface plasmas		274
		6.6.1	Boundary conditions at the interface	276
		6.6.2	Self-adjointness for interface plasmas	280
		6.6.3	Extended variational principles	283
		6.6.4	Application to the Rayleigh–Taylor instability	287
	6.7	Literature and exercises		296
7	Waves and instabilities of inhomogeneous plasmas			300
	7.1	Hydrodynamics of the solar interior		300
		7.1.1	Radiative equilibrium model	301
		7.1.2	Convection zone	305
	7.2	Hydrodynamic waves and instabilities of a gravitating slab		308
		7.2.1	Hydrodynamic wave equation	309
		7.2.2	Convective instabilities	312
		7.2.3	Gravito-acoustic waves	313
		7.2.4	Helioseismology and MHD spectroscopy	317
	7.3	MHD wave equation for a gravitating magnetized plasma slab		322
		7.3.1	Preliminaries	322
		7.3.2	Derivation of the MHD wave equation for a gravitating slab	327
		7.3.3	Gravito-MHD waves	335
	7.4	Continuous spectrum and spectral structure		345
		7.4.1	Singular differential equations	345
		7.4.2	Alfvén and slow continua	351
		7.4.3	Oscillation theorems	357
		7.4.4	Cluster spectra★	363
	7.5	Gravitational instabilities of plasmas with magnetic shear		365
		7.5.1	Energy principle for a gravitating plasma slab	366
		7.5.2	Interchange instabilities in sheared magnetic fields	371
		7.5.3	Interchanges in the absence of magnetic shear	376
	7.6	Literature and exercises		379
8	Magnetic structures and dynamics			384
	8.1	Plasma dynamics in laboratory and nature		384

	8.2	Solar magnetism	385
		8.2.1 The solar cycle	387
		8.2.2 Magnetic structures in the solar atmosphere	395
	8.3	Planetary magnetic fields	407
		8.3.1 The geomagnetic dynamo	409
		8.3.2 Magnetic fields of the other planets	413
	8.4	Magnetospheric plasmas	415
		8.4.1 The solar wind and the heliosphere	415
		8.4.2 Solar wind and planetary magnetospheres	419
	8.5	Perspective	426
	8.6	Literature and exercises	427
9	Cylindrical plasmas	431
	9.1	Equilibrium of cylindrical plasmas	431
		9.1.1 Diffuse plasmas	431
		9.1.2 Interface plasmas	436
	9.2	MHD wave equation for cylindrical plasmas	438
		9.2.1 Derivation of the MHD wave equation for a cylinder	438
		9.2.2 Boundary conditions for cylindrical interfaces	445
	9.3	Spectral structure	450
		9.3.1 One-dimensional inhomogeneity	450
		9.3.2 Cylindrical model problems	453
		9.3.3 Cluster spectra★	462
	9.4	Stability of cylindrical plasmas	462
		9.4.1 Oscillation theorems for stability	462
		9.4.2 Stability of plasmas with shearless magnetic fields	469
		9.4.3 Stability of force-free magnetic fields	475
		9.4.4 Stability of the 'straight tokamak'	482
	9.5	Literature and exercises	492
10	Initial value problem and wave damping★	496
	10.1	Implications of the continuous spectrum★	496
	10.2	Initial value problem★	497
		10.2.1 Reduction to a one-dimensional representation★	498
		10.2.2 Restoring the three-dimensional picture★	502
	10.3	Damping of Alfvén waves★	507
		10.3.1 Green's function★	509
		10.3.2 Spectral cuts★	512
	10.4	Quasi-modes★	516
		10.4.1 Dispersion equation★	517
		10.4.2 Exponential damping★	520
		10.4.3 Different kinds of quasi-modes★	522

Contents

	10.5	Leaky modes*		523
		10.5.1 Model equations and boundary conditions*		525
		10.5.2 Normal-mode analysis*		528
		10.5.3 Initial value problem approach*		529
	10.6	Literature and exercises*		530
11	Resonant absorption and wave heating			533
	11.1	Ideal MHD theory of resonant absorption		534
		11.1.1 Analytical solution of a simple model problem		534
		11.1.2 Role of the singularity		541
		11.1.3 Resonant 'absorption' versus resonant 'dissipation'		549
	11.2	Heating and wave damping in tokamaks and coronal magnetic loops		553
		11.2.1 Tokamaks		553
		11.2.2 Coronal loops and arcades		554
		11.2.3 Numerical analysis of resonant absorption		555
	11.3	Alternative excitation mechanisms		561
		11.3.1 Foot point driving		562
		11.3.2 Phase mixing		565
		11.3.3 Applications to solar and magnetospheric plasmas		567
	11.4	Literature and exercises		573
	Appendices			577
A	Vectors and coordinates			577
	A.1	Vector identities		577
	A.2	Vector expressions in orthogonal coordinates		578
		A.2.1 Cartesian coordinates (x, y, z)		580
		A.2.2 Cylinder coordinates (r, θ, z)		581
		A.2.3 Spherical coordinates (r, θ, ϕ)		582
B	Tables of physical quantities			585
	References			594
	Index			607

Preface

This book describes the two main applications of plasma physics, laboratory research on thermonuclear fusion energy and plasma-astrophysics of the solar system, stars, accretion discs, etc., from the single viewpoint of magnetohydrodynamics (MHD). This provides effective methods and insights for the interpretation of plasma phenomena on virtually all scales, ranging from the laboratory to the Universe. The key issue is understanding the complexities of plasma dynamics in extended magnetic structures.

The book starts with an exposition of the elements of plasma physics, followed by an in-depth derivation of the MHD model. By means of the conservation laws, different model problems for laboratory and astrophysical plasmas are formulated. The spectral theory of MHD waves and instabilities is then developed in analogy with quantum mechanics. The centrepiece is the analysis of inhomogeneous plasmas with intricate spectral structures that provide a unified view of waves and instabilities in plasmas as different as tokamaks and coronal flux tubes. This is illustrated by the magnetic structures and dynamics observed in the solar system, and analysed in detail for cylindrical flux tubes. Advanced chapters on wave damping and resonant heating expose the wonderful interplay of physics and mathematics.

In order to provide the student with all the tools that are necessary to understand plasma dynamics, the classical MHD model is developed in great detail without omitting steps in the derivations. The necessary restriction to ideal dissipationless plasmas, in static equilibrium and with inhomogeneity in one direction, is more than compensated by the insight gained in the intricacies of magnetized plasmas. With this objective the size of the original manuscript, including advanced topics of magnetohydrodynamics, became impractical so that we decided to split it into two volumes.

In the companion volume *Advanced Magnetohydrodynamics*, that will appear later, the restrictions of the classical theory are relaxed one by one: introducing stationary background flows, resistivity and reconnection, two-dimensional toroidal geometry, linear and nonlinear computational techniques, and transonic flows and shocks. These topics transform the subject into a vital new area with many applications in laboratory (thermonuclear fusion), space (space weather), and astrophysical plasmas (stellar winds, accretion discs and jets).

This book (Volume 1) and its companion (Volume 2) consist of three parts:

- *Plasma Physics Preliminaries* (Volume 1, Chapters 1–3),
- *Basic Magnetohydrodynamics* (Volume 1, Chapters 4–11),
- *Advanced Magnetohydrodynamics* (Volume 2).

Inevitably, with the chosen distinction between topics for Volume 1 (mostly ideal linear phenomena described by self-adjoint linear operators) and Volume 2 (mostly non-ideal and nonlinear phenomena), the difference between 'basic' and 'advanced' levels of magnetohydrodynamics could not be strictly maintained. The logical order required a quite advanced derivation of the MHD equations from kinetic theory (Chapter 3) at an early stage, different sections on advanced topics interspersed throughout the book, and a rather complete discussion of the initial value problem (Chapter 10) at the end. These parts are marked by a star ($\star$) and can be skipped on a first study of the book. The same applies to text put in small print, in between triangles ($\triangleright \cdots \triangleleft$), usually containing tedious derivations or advanced material. The serious student is advised though not to skip the Exercises, which are also put in small print for typographical reasons only. In particular, frequent use of the vector expressions and tables of the appendices is encouraged. The subject of magnetohydrodynamics can only be mastered through extensive practice.

An overview of the subject matter of the different chapters of this volume may help the reader to find his way:

- Chapter 1 gives an introduction to laboratory fusion and astrophysical plasmas, and formulates provisional microscopic and macroscopic definitions of the plasma state.
- Chapter 2 discusses the three complementary points of view of single particle motion, kinetic theory and fluid description. The corresponding theoretical models provide the opportunity to introduce some of the basic concepts of plasma physics.
- Chapter 3 gives the 'derivation' of the macroscopic equations from the kinetic (Boltzmann) equation. The quotation marks because a fully satisfactory derivation cannot be given at present in view of the largely unknown contribution of turbulent transport processes. The presentation given is meant to provide some idea on the limitations of the macroscopic viewpoint.
- Chapter 4 defines the MHD model and introduces the concept of scale independence. The central importance of the conservation laws is discussed at length. Based on this, the similarities and differences of laboratory and astrophysical plasmas are articulated in terms of a number of generic boundary value problems.

- Chapter 5 derives the basic MHD waves and describes their properties, with an eye on their important role in spectral analysis and computational MHD. The theory of characteristics is introduced as a vehicle for the propagation of nonlinear disturbances.
- Chapter 6 treats the subject of waves and instabilities from the unifying point of view of spectral theory. The force operator formulation and the energy principle are extensively discussed. The analogy with quantum mechanics is pointed out and exploited. The difficult extension to interface systems is treated in detail.
- Chapter 7 applies the spectral analysis developed in Chapter 6 to inhomogeneous plasmas in a plane slab. The wave equation for gravito-MHD waves is derived and solved in various limits. Here, all the intricacies of the subject enter: continuous spectra, damping of Alfvén waves, local instabilities, etc. The analogy between helioseismology and MHD spectroscopy in tokamaks is shown to hold great promise for the investigation of plasma dynamics.
- Chapter 8 introduces the enormous variety of magnetic phenomena in astrophysics, in particular the solar system (dynamo, solar wind, magnetospheres, etc.), and provides basic examples of plasma dynamics worked out in later chapters.
- Chapter 9 is the cylindrical counterpart of Chapter 7, with a wave equation describing the various waves and instabilities. It presents the stability analysis of diffuse cylindrical plasmas (classical pinches and present tokamak models) from the spectral perspective.
- Chapter 10 solves the initial value problem for one-dimensional inhomogeneous MHD and the associated damping due to the continuous spectrum.
- Chapter 11 discusses resonant absorption and phase mixing in the context of heating mechanisms of solar and stellar coronae. Anticipating Volume 2, numerical methods to solve these problems are indicated. Sunspot seismology is introduced as another example of MHD spectroscopy.

We wish to acknowledge support of our colleagues and collaborators over many years: Jeff Freidberg (his returning question starting the day, 'What is the news, Hans?', remains a source of inspiration), Paulo Sakanaka, Dan D'Ipolito, Ricardo Galvão, Jan Rem, Marcel Goossens, Wolfgang Kerner, Marnix van der Wiel (his stimulation of new scientific enterprises has significantly facilitated our research), Max Kuperus, Tony Hearn, Sasha Lifschitz (his usual remark at the end of lengthy calculations, 'And now comes the hard part, the part we want to avoid at all cost, now we have to think', has crucially motivated the writing of this book), Henk van der Vorst, Bram Achterberg, Lydia Van Driel-Gesztelyi, Brigitte Schmieder, Andro Rogava, Eric Priest, Bernard Roberts, Alan Hood, Tom Bogdan, Boon Chye Low, Herman Deconinck and, last but not least, Rony Keppens. Ph.D. students and post-docs created the essential inquisitive environment for scientific research: Guido Huysmans, Giel Halberstadt, Hanno Holties, Sander Beliën, Ronald Nijboer, Bart van der Holst, Hans De Sterck, Arpád Csík and Fabien

Casse. Rob Rutten contributed substantial improvement of Chapter 8 (remaining misconceptions are entirely ours). Numerous students contributed suggestions for improvement; Victor Land produced most of the exercises. With the gradual take-over of supporting tasks by computer programs, the original time-consuming figure drawing, type-writing, and literature search by Wim Tukker, Rosa Tenge and Hajnal Vörös for earlier versions of the manuscript is gratefully acknowledged. We thank our copy-editor, Frances Nex, for careful and efficient editing of our text.

Part I

Plasma physics preliminaries

1
Introduction

1.1 Motivation

Under ordinary circumstances, matter on Earth occurs in the three phases of solid, liquid, and gas. Here, 'ordinary' refers to the circumstances relevant for human life on this planet. This state of affairs does not extrapolate beyond earthly scales: astronomers agree that, ignoring the more speculative nature of dark matter, matter in the Universe consists more than 90% of plasma. Hence, *plasma is the ordinary state of matter in the Universe*. The consequences of this fact for our view of nature are not generally recognized yet (see Section 1.3.4). The reason may be that, since plasma is an exceptional state on Earth, the subject of plasma physics is a relative latecomer in physics.

For the time being, the following crude definition of plasma suffices. *Plasma is a completely ionized gas, consisting of freely moving positively charged ions, or nuclei, and negatively charged electrons.*[1] In the laboratory, this state of matter is obtained at high temperatures, in particular in thermonuclear fusion experiments ($T \sim 10^8$ K). In those experiments, the mobility of the plasma particles facilitates the induction of electric currents which, together with the internally or externally created magnetic fields, permits magnetic confinement of the hot plasma. In the Universe, plasmas and the associated large-scale interactions of currents and magnetic fields prevail under much wider conditions.

Hence, we will concentrate our analysis on the two mentioned broad areas of application of plasma physics, viz.

(a) *Magnetic plasma confinement for the purpose of future energy production by controlled thermonuclear reactions* (CTR); this includes the pinch experiments of the 1960s and

---

[1] In plasma physics, one can hardly avoid mentioning exceptions: in pulsar electron–positron magnetospheres, the role of positively charged particles is taken by positrons. In considerations of fusion reactions with exotic fuels like muonium, the role of negatively charged particles is taken by muons.

early 1970s, and the tokamaks and alternatives (stellarator, spheromak, etc.) developed in the 1980s and 1990s and, at present, sufficiently matured to start designing prototypes of the fusion reactors themselves;

(b) *The dynamics of magnetized astrophysical plasmas*; this includes the ever growing research field of solar magnetic activity, planetary magnetospheres, stellar winds, interstellar medium, accretion discs of compact objects, pulsar magnetospheres, etc.

The common ground of these two areas is the subject of *plasma interacting with a magnetic field*. To appreciate the power of this viewpoint, we first discuss the conditions for laboratory fusion in Section 1.2, then switch to the emergence of the subject of plasma-astrophysics in Section 1.3, and finally refine our definition(s) of plasma in Section 1.4. In the latter section, we also provisionally formulate the approach to plasmas by means of magnetohydrodynamics.

The theoretical models exploited lead to *nonlinear partial differential equations*, expressing *conservation laws*. The boundary conditions are imposed on an extended spatial domain, associated with the *complex magnetic plasma confinement geometry*, whereas the temporal dependence leads to *intricate nonlinear dynamics*. This gives theoretical plasma physics its particular, mathematical, flavour.

1.2 Thermonuclear fusion and plasma confinement

1.2.1 Fusion reactions

Both fission and fusion energy are due to nuclear processes and, ultimately, described by Einstein's celebrated formula $E = mc^2$. Hence, in nuclear reactions $A + B \rightarrow C + D$, net energy is released if there is a mass defect, i.e. if

$$(m_A + m_B)\, c^2 > (m_C + m_D)\, c^2 \,. \tag{1.1}$$

In laboratory fusion, reactions of hydrogen isotopes are considered, where the deuterium–tritium reaction (Fig. 1.1) is the most promising one for future reactors:

$$D^2 + T^3 \;\rightarrow\; He^4\,(3.5\,\text{MeV}) + n\,(14.1\,\text{MeV})\,. \tag{1.2}$$

This yields two kinds of products, viz. α-particles (He^4), which are *charged* so that they can be captured by a confining magnetic field, and neutrons, which are *electrically neutral* so that they escape from the magnetic configuration. The former contribute to the heating of the plasma (so-called α-particle heating) and the latter have to be captured in a surrounding Li^6/Li^7 blanket, which recovers the fusion energy and also breeds new T^3.

Deuterium abounds in the oceans: out of 6500 molecules of water one molecule is D_2O. Thus, in principle, 1 litre of sea water contains 10^{10} J of deuterium fusion

1.2 Thermonuclear fusion and plasma confinement

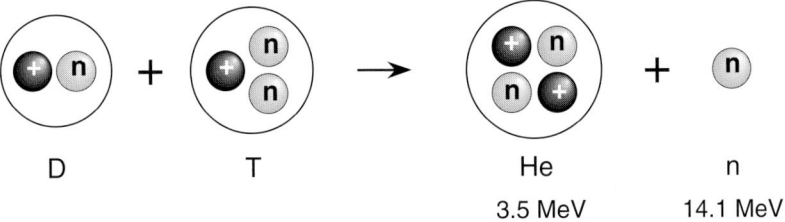

Fig. 1.1. Deuterium–tritium reactions.

energy. This is a factor of about 300 more than the combustion energy of 1 litre of gasoline, which yields 3×10^7 J.

A number of other reactions also occur, in particular reactions producing T^3 and He^3 which may be burned again. Complete burn of all available D^2 would involve the following reactions:

$$\begin{aligned} D^2 + D^2 &\to He^3 \,(0.8\,\text{MeV}) + n\,(2.5\,\text{MeV})\,, \\ D^2 + D^2 &\to T^3 \,(1.0\,\text{MeV}) + p\,(3.0\,\text{MeV})\,, \\ D^2 + T^3 &\to He^4 \,(3.5\,\text{MeV}) + n\,(14.1\,\text{MeV})\,, \\ D^2 + He^3 &\to He^4 \,(3.7\,\text{MeV}) + p\,(14.6\,\text{MeV})\,, \end{aligned} \quad (1.3)$$

so that in effect

$$6D^2 \;\to\; 2\,He^4 + 2\,p + 2\,n + 43.2\,\text{MeV}\,. \quad (1.4)$$

In the liquid Li blanket, fast neutrons are moderated, so that their kinetic energy is converted into heat, and the following reactions occur:

$$\begin{aligned} n + Li^6 &\to T^3 \,(2.1\,\text{MeV}) + He^4 \,(2.8\,\text{MeV})\,, \\ n\,(2.5\,\text{MeV}) + Li^7 &\to T^3 + He^4 + n\,. \end{aligned} \quad (1.5)$$

This provides the necessary tritium fuel for the main fusion reaction (1.3)(c) [156].

Typical numbers associated with thermonuclear fusion reactors, as presently envisaged, are:

$$\begin{aligned} &\text{temperature} \quad T \sim 10^8 \,\text{K}\,(10\,\text{keV})\,, \quad \text{power density} \sim 10\,\text{MW}\,\text{m}^{-3}\,, \\ &\text{particle density}\; n \sim 10^{21}\,\text{m}^{-3}\,, \quad\quad\;\; \text{time scale} \quad \tau \sim 100\,\text{s}\,. \end{aligned} \quad (1.6)$$

It is often said that controlled thermonuclear fusion in the laboratory is an attempt to harness the power of the stars. This is actually a quite misleading statement since the fusion reactions which take place in, e.g., *the core of the Sun* are different reactions of hydrogen isotopes, viz.

$$p + p \to D^2 + e^+ + \nu_e + 1.45\,\text{MeV} \quad (2\times),$$
$$p + D^2 \to He^3 + \gamma + 5.5\,\text{MeV} \quad (2\times), \quad (1.7)$$
$$He^3 + He^3 \to He^4 + 2p + 12.8\,\text{MeV},$$

so that complete burn of all available hydrogen amounts to

$$4p \to He^4 + 2e^+ + 2\nu_e\,(0.5\,\text{MeV}) + 2\gamma\,(26.2\,\text{MeV}). \quad (1.8)$$

The positrons annihilate with electrons, the neutrinos escape, and the gammas (carrying the bulk of the thermonuclear energy) start on a long journey to the solar surface, where they arrive millions of years later (the mean free path of a photon in the interior of the Sun is only a few centimetres) [190]. In the many processes of absorption and re-emission the wavelength of the photons gradually shifts from that of gamma radiation to that of the visible and UV light escaping from the photosphere of the Sun, and producing one of the basic conditions for life on a planet situated at the safe distance of one astronomical unit (1.5×10^{11} m) from the Sun.

At higher temperatures another chain of reactions is effective, where carbon acts as a kind of catalyst. This so-called CNO cycle involves a chain of fusion reactions where C^{12} is successively converted into N^{13}, C^{13}, N^{14}, O^{15}, N^{15}, and back into C^{12} again. However, the net result of incoming and outgoing products is the same as that of the proton–proton chain, viz. Eq. (1.8).

Typical numbers associated with thermonuclear reactions in the stars, in particular the core of the Sun, are the following ones:

$$\begin{array}{ll} \text{temperature} \quad T \sim 1.5 \times 10^7\,\text{K}, & \text{power density} \sim 3.5\,\text{W}\,\text{m}^{-3}, \\ \text{particle density} \quad n \sim 10^{32}\,\text{m}^{-3}, & \text{time scale} \quad \tau \sim 10^7\,\text{years}. \end{array} \quad (1.9)$$

Very different from the numbers (1.6) for a prospective fusion reactor on Earth!

1.2.2 Conditions for fusion

Thermonuclear fusion happens when a gas of, e.g., deuterium and tritium atoms is sufficiently heated for the thermal motion of the nuclei to become so fast that they may overcome the repulsive Coulomb barrier (Fig. 1.2) and come close enough for the attractive nuclear forces to bring about the fusion reactions discussed above. This requires particle energies of $\sim 10\,\text{keV}$, i.e. temperatures of about 10^8 K. At these temperatures the electrons are completely stripped from the atoms (the ionization energy of hydrogen is $\sim 14\,\text{eV}$) so that a plasma rather than a gas is obtained (cf. our crude definition of Section 1.1).

1.2 Thermonuclear fusion and plasma confinement

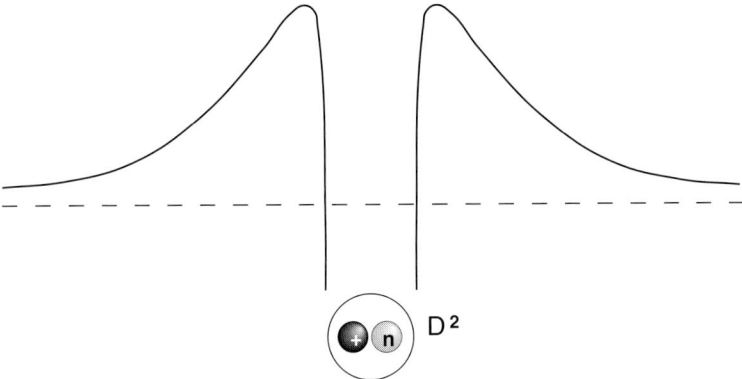

Fig. 1.2. Nuclear attraction and Coulomb barrier of a deuteron.

Because the charged particles (occurring in about equal numbers of opposite charge) are freely moving and rarely collide at these high temperatures, a plasma may be considered as a *perfectly conducting fluid* for many purposes. In such fluids, electric currents are easily induced and the associated magnetic fields in turn interact with the plasma to confine or to accelerate it. The appropriate theoretical description of this state of matter is called *magnetohydrodynamics* (MHD), i.e. the dynamics of magneto-fluids (Section 1.4.2).

Why are magnetic fields necessary? To understand this, we need to discuss the power requirements for fusion reactors (following Miyamoto [156] and Wesson [244]). This involves three contributions, viz.

(a) the thermonuclear output power per unit volume:

$$P_T = n^2 f(\tilde{T}), \qquad f(\tilde{T}) \equiv \tfrac{1}{4} \langle \sigma v \rangle E_T, \qquad E_T \approx 22.4 \text{ MeV}, \qquad (1.10)$$

where n is the particle density, σ is the cross-section of the D-T fusion reactions, v is the relative speed of the nuclei, $\langle \sigma v \rangle$ is the average nuclear reaction rate, which is a well-known function of temperature, and E_T is the average energy released in the fusion reactions (i.e. more than the 17.6 MeV of the D-T reaction (1.3)(c) but, of course, less than the 43.2 MeV released for the complete burn (1.4));

(b) the power loss by Bremsstrahlung, i.e. the radiation due to electron–ion collisions:

$$P_B = \alpha n^2 \tilde{T}^{1/2}, \qquad \alpha \approx 3.8 \times 10^{-29} \text{ J}^{1/2} \text{ m}^3 \text{ s}^{-1}; \qquad (1.11)$$

(c) the losses by heat transport through the plasma:

$$P_L = \frac{3n\tilde{T}}{\tau_E}, \qquad (1.12)$$

where $3n\tilde{T}$ is the total plasma kinetic energy density (ions + electrons), and τ_E is the energy confinement time (an empirical quantity). The latter estimates the usually anomalous (i.e. deviating from classical transport by Coulomb collisions between the charged particles) heat transport processes.

Here, we have put a tilde on the temperature to indicate that energy units of keV are exploited:

$$\tilde{T} \text{ (keV)} = 8.62 \times 10^{-8} T \text{ (K)},$$

since $\tilde{T} = 1 \text{ keV} = 1.60 \times 10^{-16}$ J corresponds with $T = 1.16 \times 10^7$ K (using Boltzmann's constant, see Appendix Table B.1).

If the three power contributions are considered to become externally available for conversion into electricity and back again into plasma heating, with efficiency η, *the Lawson criterion* [140],

$$P_B + P_L = \eta (P_T + P_B + P_L), \tag{1.13}$$

tells us that there should be power balance between the losses from the plasma (LHS) and what is obtained from plasma heating (RHS). Typically, $\eta \approx 1/3$. Inserting the explicit expressions (1.10), (1.11), and (1.12) into Eq. (1.13) leads to a condition to be imposed on the product of the plasma density and the energy confinement time:

$$n\tau_E = \frac{3\tilde{T}}{\frac{\eta}{1-\eta} f(\tilde{T}) - \alpha \tilde{T}^{1/2}}. \tag{1.14}$$

This relationship is represented by the lower curve in Fig. 1.3. Since Bremsstrahlung losses dominate at low temperatures and transport losses dominate at high temperatures, there is a minimum in the curve at about

$$n\tau_E = 0.6 \times 10^{20} \text{ m}^{-3} \text{ s}, \quad \text{for } \tilde{T} = 25 \text{ keV}. \tag{1.15}$$

This should be considered to be the threshold for a fusion reactor under the given conditions.

By a rather different, more recent, approach of fusion conditions, *ignition* occurs when the total amount of power losses is balanced by the total amount of heating power. The latter consists of α-particle heating P_α and additional heating power P_H, e.g. by radio-frequency waves or neutral beam injection. The latter heating sources are only required to bring the plasma to the ignition point, when α-particle heating may take over. Hence, at ignition we may put $P_H = 0$ and the power balance becomes

$$P_B + P_L = P_\alpha = \tfrac{1}{4}\langle\sigma v\rangle n^2 E_\alpha, \quad E_\alpha \approx 3.5 \text{ MeV}. \tag{1.16}$$

1.2 Thermonuclear fusion and plasma confinement

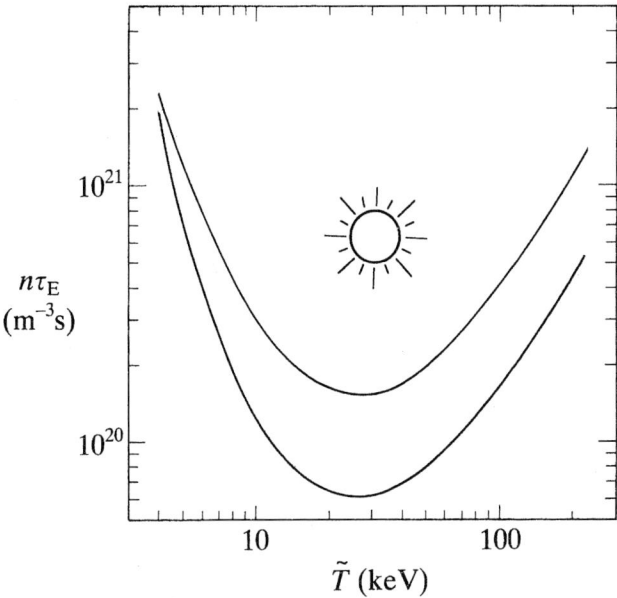

Fig. 1.3. Conditions for net fusion energy production according to the Lawson criterion (lower curve) and according to the view that power losses should be completely balanced by α-particle heating (upper curve). Adapted from Wesson [244].

Formally, this may be described by the same equation (1.14) taking now $\eta \approx 0.135$ so that a 2.5 times higher threshold for fusion is obtained:

$$n\tau_E = 1.5 \times 10^{20} \, \text{m}^{-3} \, \text{s}, \quad \text{for } \tilde{T} = 30 \, \text{keV}. \quad (1.17)$$

This relationship is represented by the upper curve of Fig. 1.3.

Roughly speaking then, products of density and energy confinement time $n\tau_E \sim 10^{20} \, \text{m}^{-3} \, \text{s}$ and temperatures $\tilde{T} \sim 25 \, \text{keV}$, or $T \sim 3 \times 10^8 \, \text{K}$, are required for controlled fusion reactions. As a figure of merit for fusion experiments one frequently constructs the product of these two quantities, which should approach

$$n\tau_E \tilde{T} \sim 3 \times 10^{21} \, \text{m}^{-3} \, \text{s} \, \text{keV} \quad (1.18)$$

for a fusion reactor. To get rid of the radioactive tritium component, one might consider pure D-D reactions in a more distant future. This would require yet another increase of the temperature by a factor of 10. Considering the kind of progress obtained over the past 40 years, though (see Fig. 1.1.1 of Wesson [244]: a steady increase of the product $n\tau_E T$ with a factor of 100 every decade!), one may hope that this difficulty eventually will turn out to be surmountable.

Returning to our question on the magnetic fields: no material containers can hold plasmas with densities of $10^{20} \, \text{m}^{-3}$ and temperatures of 100–300 million K during times in the order of minutes, or at least seconds, without immediately extinguishing the 'fire'. One way to solve this problem is to make use of the *confining properties of magnetic fields*, which may be viewed from quite different angles:

(a) the charged particles of the plasma rapidly and tightly gyrate around the magnetic field lines (they 'stick' to the field lines, see Section 2.2);
(b) fluid and magnetic field move together ('the magnetic field is frozen into the plasma', see Section 2.4), so that engineering of the geometry of the magnetic field configuration also establishes the geometry of the plasma;
(c) the thermal conductivity of plasmas is highly anisotropic with respect to the magnetic field, $\kappa_\perp \ll \kappa_\parallel$ (see Sections 2.3.1 and 3.3.2), so that heat is easily conducted along the field lines and the magnetic surfaces they map out, but impeded across.

Consequently, what one needs foremost is a *closed magnetic geometry* facilitating stable, static plasma equilibrium with roughly bell-shaped pressure and density profiles and nested magnetic surfaces. This is the subject of the next section.

1.2.3 Magnetic confinement and tokamaks

Controlled thermonuclear fusion research started in the 1950s in the weapons laboratories after the 'successful' development of the hydrogen bomb: fusion energy had been unleashed on our planet! The development of the peaceful, controlled, counterpart appeared to be a matter of a few years, as may become clear by considering the simplicity of early pinch experiments. The history of the subject is schematically illustrated in Fig. 1.4. In the upper part the two early attempts with the simple schemes of θ- and z-pinch are shown. Here, θ and z refer to the direction of the plasma current in terms of a cylindrical r, θ, z coordinate system. Since it is relatively straightforward to produce plasma by ionizing hydrogen gas in a tube, a very conductive fluid is obtained in which a strong current may be induced by discharging a capacitor bank over an external coil surrounding the gas tube. In a *z-pinch experiment*, this current is induced in the z-direction and it creates a transverse magnetic field B_θ, so that the resulting Lorentz force $(\mathbf{j} \times \mathbf{B})_r = -j_z B_\theta$ is pointing radially inward. In this manner, the confining force as well as near thermonuclear temperatures ($\sim 10^7$ K) are easily produced. There is only one problem: the curvature of the magnetic field B_θ causes the plasma to be extremely unstable, with growth rates in the order of μseconds. To avoid these instabilities, the orthogonal counterpart, the *θ-pinch experiment*, suggested itself. Here, current is induced in the θ-direction, it causes a radial decrease of the externally applied magnetic field B_z, so that the net Lorentz force $j_\theta \Delta B_z$ is again directed inward. In the θ-pinch, thermonuclear temperatures are also obtained, and the plasma is now macroscopically stable. However, pinching of the plasma column produces

1.2 Thermonuclear fusion and plasma confinement

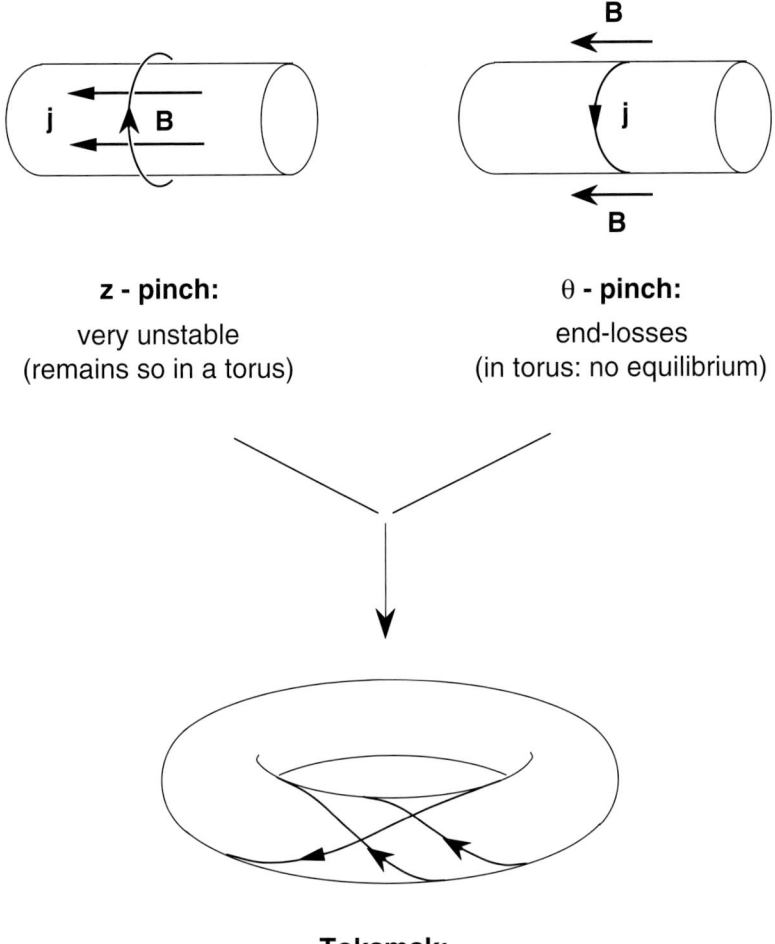

Fig. 1.4. Interaction of currents and magnetic fields: a schematic history of plasma confinement experiments.

unbalanced longitudinal forces so that the plasma is squirted out of the ends, again terminating plasma confinement on the µs time scale. In conclusion, in pinch experiments the densities and temperatures needed for thermonuclear ignition are easily produced but the confinement times fall short by a factor of a million to a billion.

With these obstacles ahead, the nations involved with thermonuclear research decided it to be opportune to declassify the subject. This fortunate decision was landmarked by the Second International UN Conference on Peaceful Uses of Atomic Energy in Geneva in 1958, where all scientific results obtained so far were

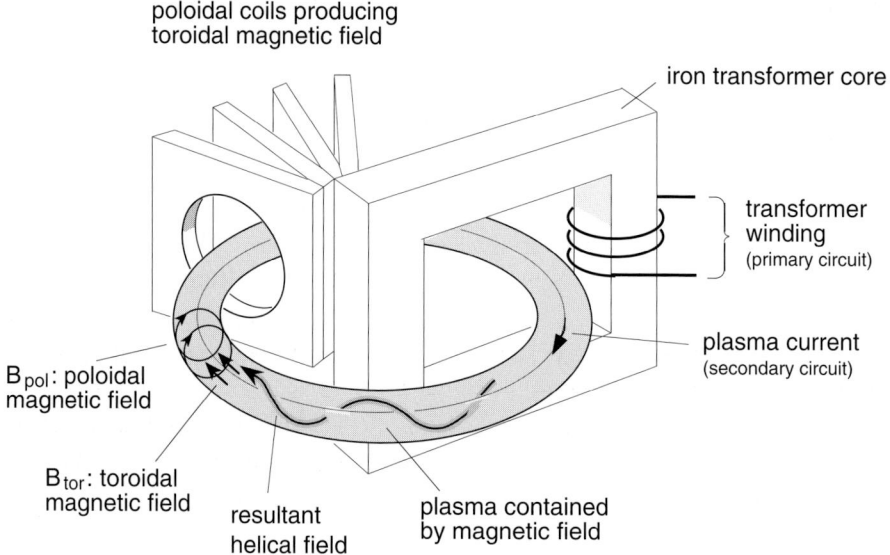

Fig. 1.5. Schematic presentation of magnetic confinement in a tokamak.

presented. Prospects then gradually became much brighter with the emergence of *the tokamak* alternative line (bottom part of Fig. 1.4 and Fig. 1.5) developed in the 1960s in the Soviet Union, and internationally accepted in the 1970s as the most promising scheme towards fusion. Crudely speaking, the tokamak configuration cures the main problems of the z-pinch (its instability due to the curvature of the transverse magnetic field) and of the θ-pinch (its end losses), both destroying the configuration on the μs time scale, by combining them into a single scheme. The vessel is now a torus rather than a straight tube and the magnetic field is helical, with a poloidal and a larger toroidal component. The latter component of the magnetic field provides the crucial longitudinal 'backbone' for stability. Whereas the toroidal geometry simply eliminates the end-loss problem of the θ-pinch, it is not quite true that the kink instability problem of the z-pinch is eliminated as well. Instead, the basic MHD problem of tokamak confinement turns out to be a delicate balance between equilibrium considerations, favouring a large toroidal current, and stability considerations, which favour a minimum current so as to eliminate the driving force of the kink instabilities. Thus, tokamak performance is an *intricate optimization problem* which makes it both interesting and impressive. Concerning the latter: to have improved upon a technological parameter by a factor of 10^8 in thirty years (from confinement times of microseconds in the sixties to minutes in the nineties) is a kind of progress which is only paralleled by developments in computer technology.

1.3 Astrophysical plasmas

For more on the history of fusion research: see Braams and Stott [37].

1.3 Astrophysical plasmas

We have sketched the efforts in controlled thermonuclear confinement experiments, where the prospect of abundant energy has driven scientists to ever deeper exploration of the plasma state. At this point an entirely different line of research should enter the presentation. This is the rapidly growing field of plasma-astrophysics, which has much older credentials than laboratory plasma research. We will introduce this topic by means of the example of the solar system, where the usual gravitational picture completely masks the dynamics of the plasmas that are present. To understand how this picture has changed in recent times, we introduce some basic astrophysical notions and recall events in space research. We will also use the opportunity to introduce numerical values of certain quantities that may not be familiar to some readers.

1.3.1 Celestial mechanics

To set the stage, recall *the traditional picture of the solar system*: the Sun is the central massive object (a thousand times more massive than Jupiter) which keeps the nine planets orbiting around it by its gravitational attraction. (See Fig. 1.6 and the numerical values summarized in Table B.7.)

The planets move according to *Kepler's laws* (1610):

(a) *The planetary orbits are ellipses* lying in or close to the ecliptic (the orbital plane of the Earth) *with the Sun in one of the focal points*. The inclination of the orbit with respect to the ecliptic is modest ($< 4°$) for most of the planets, whereas the largest values occur for the innermost planet (Mercury: $7°$) and for the outermost one (Pluto: $17°$). (Recall that this planet was only discovered in 1930.) The ellipses are characterized by the eccentricity parameter $e \equiv c/a = (1 - b^2/a^2)^{1/2}$, where c is the distance of the focal points to the origin and a and b are the lengths of the semi-axes of the ellipse. Again, the highest eccentricities occur for Mercury ($e = 0.206$) and Pluto ($e = 0.250$), whereas they are small for the other planets ($e < 0.1$). Incidentally, it is to be noted that the ellipticity as measured by the ratio of the semi-axes, $b/a = \sqrt{1 - e^2}$, is $\sqrt{0.96} \approx 0.98$ for Mercury and 0.97 for Pluto, i.e. just deviations of 2% and 3% from a circle, and much less for the other planets. The original approximation of circular motion by the ancients appears not all that stupid. The big effect is not the deviation from a circle though, but the eccentricity, i.e. *the shift c of the near-circular orbit*. This gives rise to significant variations in the distance to the Sun, as measured by the ratio

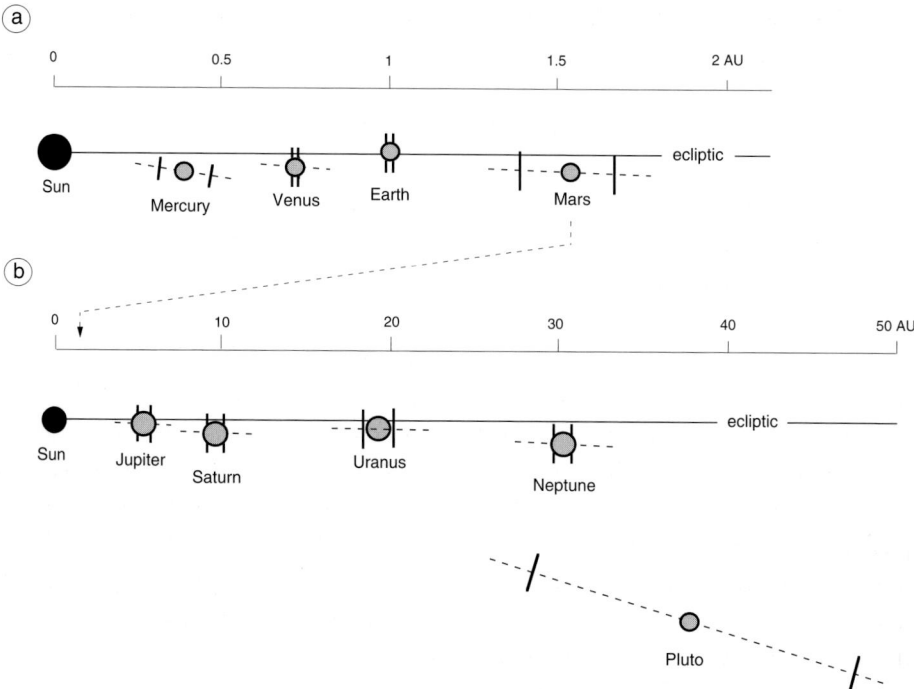

Fig. 1.6. Schematic representation of the size and inclination of the planetary orbits: (a) terrestrial planets, (b) giant planets and Pluto. The planets (not to scale) are drawn at a distance a from the Sun, where the vertical dashes indicate the extrema $a - c$ (perihelion) and $a + c$ (aphelion) of the orbital motion (of course, situated on opposite sides of the Sun).

$(a - c)/(a + c) = (1 - e)/(1 + e)$ which is 0.66 for Mercury and 0.60 for Pluto, as is evident from Fig. 1.6.

(b) *The radius vector of the Sun to the planet sweeps out equal areas in equal times.* Hence, the orbital velocity is highest in the perihelion (the orbital point closest to the Sun) and smallest in the aphelion (the point farthest from the Sun). This *law of areas* is a consequence of conservation of angular momentum.

(c) *The harmonic law: The cubes of the semi major axis a of the orbits of the planets are proportional to the squares of the orbital period τ,*

$$a^3/\tau^2 = \text{const} \approx GM_\odot/4\pi^2 = 1 \, (\text{AU})^3/\text{y}^2 = 3.38 \times 10^{18} \, \text{m}^3 \, \text{s}^{-2}. \quad (1.19)$$

Here, G is the gravitational constant, $M_\odot$ is the mass of the Sun, $1 \, \text{AU} = 1.5 \times 10^{11}$ m is the distance from the Earth to the Sun (the astronomical unit), and $1 \, \text{y} = 3.16 \times 10^7$ s is, of course, the orbital period of the Earth.

▷ **Exercise.** Use Table B.7 to check this number for the different planetary orbits. ◁

Kepler's laws were then founded on the laws of mechanics, in particular *Newton's law of gravitational attraction* (1666):

$$F = G \frac{M_1 M_2}{r^2} = -\frac{dV}{dr}, \qquad (1.20)$$

where $V = -G M_1 M_2 / r$ is the gravitational potential energy. This law implies that the planets move as point particles in the gravitational field of the Sun, whereas the whole solar system is kept together in dynamical equilibrium by gravity. All this belongs to the subject of *celestial mechanics* which is at the root of classical mechanics, which in turn constitutes the basis of physics. Thus, progress in understanding may schematically be depicted by the sequence Kepler (1609) → Newton (1687) → Lagrange (1782), Laplace (1799) → Hamilton (1845). After the work of these giants, the subject of classical mechanics (as, e.g., summarized by Goldstein [91]) has long been considered a closed subject. However, the sequence continues with the more recent names of Kolmogorov, Arnold and Moser (1964) associated with fundamental work on the stability of dynamical systems. At the present time, there is a resurgence of the subject of Hamiltonian mechanics through the development of the science of nonlinear dynamics.

So far, plasmas did not appear on the stage. Obviously, the gravitational attraction dominates everything. Gravitational and centrifugal acceleration balance perfectly in the leading order picture where the celestial bodies are treated as massive point particles. Since this is so, next order effects should be quite important (just like astronauts in an orbiting spacecraft may be accelerated by forces that are totally negligible as compared to gravity). Hence, when the internal structure of the stars (in this case, the Sun) and the planets is taken into account, the whole picture changes dramatically.

1.3.2 Astrophysics

In the nineteenth and twentieth centuries, there is a gradual shift away from exclusive interest in celestial mechanics towards the study of *the structure and evolution of stars and stellar systems*: the subject of astrophysics is born. Here, a basic postulate provides the guiding principle, viz. that *the laws of physics are valid throughout the Universe*. In historical perspective, the revolutionary character of this point of view can hardly be overestimated. Quintessence (according to Webster's Dictionary, 'the fifth and highest essence in ancient and medieval philosophy that permeates all nature and is the substance composing the heavenly bodies') is no longer essential, and 'heavenly' or 'celestial' are no longer descriptive adjectives for astronomical objects. A particularly relevant example is provided by the work of Kirchhoff and Bunsen (1859) who interpreted the observed dark lines in the spectrum of solar light, discovered by Fraunhofer (1814), as due to absorption by chemical elements in exactly the same way as spectra obtained in the laboratory. Consequently, most of our knowledge of the stars comes from *spectroscopy*, i.e. atomic physics applied to the photospheres of the stars where the spectra are determined by the temperature T of the surface and the different abundances of the chemical elements.

A quantitative measure for the relative brightness of a star is *the apparent magnitude m*:

$$m \equiv m_0 - 2.5 \times {}^{10}\log(l/l_0). \tag{1.21}$$

Here, l is the flux, i.e. the amount of electromagnetic radiation energy passing per unit time through a unit area (taken at the position of our eye, or any other observing apparatus on Earth), and the subscript 0 refers to a reference star. The value of m_0 for the reference star is fixed by convention. This definition has been chosen to conform with the ancient classification based on what the human eye can distinguish, viz. five steps in a brightness scale ranging from $m = 0$ for the brightest star to $m = 5$ for the faintest one, corresponding to a decrease by a factor of $1/100$ in the flux.

Obviously, two stars of equal apparent magnitude may have a completely different value of the luminosity L, which is the total radiation energy output per unit time, since the flux l depends on the distance d from the star according to

$$l = L/(4\pi d^2). \tag{1.22}$$

Hence, a quantity of more intrinsic physical interest is *the absolute magnitude M*, which is based on the flux $\hat{l}$ that would be produced at the position of the Earth (ignoring atmospheric extinction) if the star were moved from its actual distance d to a fictitious distance $\hat{d} = 10\,\text{pc}$ (parsec) from the Earth.[2] In other words, the absolute magnitude is defined as the apparent magnitude the star would have if positioned at $\hat{d}$, so that we obtain from Eqs. (1.21) and (1.22):

$$M \equiv m_0 - 2.5 \times {}^{10}\log(\hat{l}/l_0) = m - 2.5 \times {}^{10}\log(\hat{l}/l)$$
$$= m - 2.5 \times {}^{10}\log(d^2/100) = m + 5 - 5 \times {}^{10}\log d, \tag{1.23}$$

---

[2] Note on *distance scales*: a star at a distance $d = 1\,\text{pc}$ produces a parallax of $1'' = 4.85 \times 10^{-6}\,\text{rad}$, so that $1\,\text{pc} = (4.85 \times 10^{-6})^{-1}\,\text{AU} = 2.06 \times 10^5\,\text{AU} = 3.26\,\text{light-years} = 3.09 \times 10^{16}\,\text{m}$. The distance to the next nearest star, Alpha Centauri, is $1.3\,\text{pc}$. The size of our Galaxy (the Milky Way) is $50\,\text{kpc} = 1.6 \times 10^5$ light-years $= 1.3 \times 10^8 \times$ the size of the solar system (which is taken to be the diameter of the orbit of Pluto, i.e. $2 \times 40\,\text{AU}$).

Light also provides useful estimates for *time scales*: a photon would take 2 s to travel from the centre of the Sun to the surface if the Sun were optically thin. (In reality, because of the innumerable absorptions and re-emissions it takes about 10^7 years, as we have already noted in Section 1.2.1.) It then takes 8.3 min to reach Earth, 5.6 hours to reach Pluto, and 4.2 years to reach Alpha Centauri.

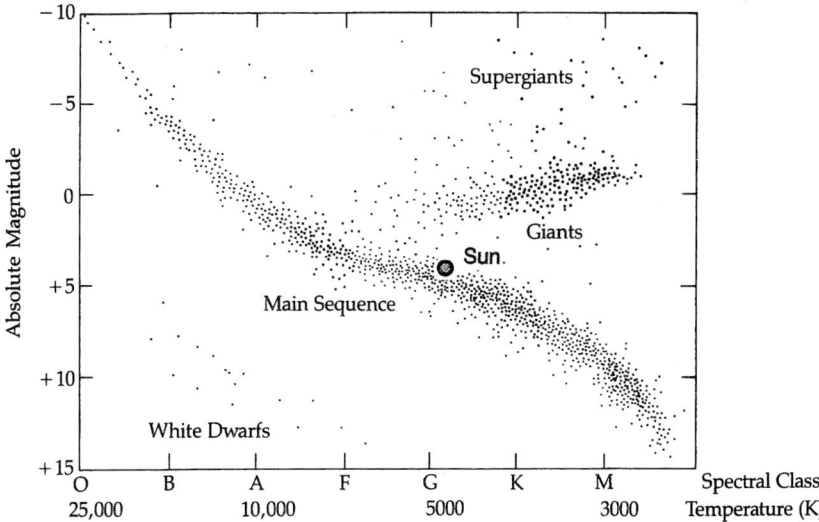

Fig. 1.7. Hertzsprung–Russell diagram: the Sun is an ordinary, main sequence, star. Adapted from Zeilik and Smith [248].

where d is measured in pc. For the Sun, with $d_\odot = 1\,\text{AU} = 1.5 \times 10^8\,\text{km} = 5 \times 10^{-6}\,\text{pc}$, we get a huge difference between the absolute and the apparent magnitude: $M_\odot = m_\odot + 31.5$.[3] The reason is clear: the apparent magnitude is based on night-time observation and, hence, totally out of range for the Sun. On the other hand, for the absolute magnitude of the Sun the very ordinary value $M_\odot = 4.7$ is obtained: apart from its proximity, the Sun is just an ordinary star.

A particularly effective way of representing the absolute magnitudes of a large number of stars is the celebrated *Hertzsprung–Russell diagram*, where the absolute magnitude M is plotted versus the effective surface temperature T_{eff}, or the associated spectral class indicated by the letters O, B, etc. (Fig. 1.7). A crude estimate of the curve for the main sequence stars may be obtained by using the Stefan–Boltzmann black-body radiation law for the luminosity,

$$L = 4\pi R^2 \sigma\, T_{\text{eff}}^4. \tag{1.24}$$

Here, R is the radius of the star, $\sigma = 5.67 \times 10^{-8}\,\text{W m}^{-2}\,\text{K}^{-4}$ is the constant of Stefan–Boltzmann, and T_{eff} is the effective surface temperature of the star. For stars of equal size, we obtain from the first line of Eq. (1.23) the following difference in

---

[3] Note that we have used the same symbol $M_\odot$ already in Section 1.3.1 to indicate the solar mass. Every now and then, we will not be able to avoid context-dependent notation.

their absolute magnitudes:

$$\Delta M = -2.5 \times \Delta(^{10}\log \hat{l}) = -2.5 \times \Delta(^{10}\log L) = -10 \times \Delta(^{10}\log T_{\text{eff}}). \tag{1.25}$$

This roughly checks with the overall slope of the Hertzsprung–Russell diagram.

For the sun, $R_\odot = 700\,000$ km and $T_{\text{eff},\odot} = 5777$ K (i.e., spectral class G) so that $L_\odot = 3.89 \times 10^{26}$ W. Incidentally, the flux $l = L_\odot/(4\pi d^2)$ at the position of the Earth is called the solar constant. Since $d = 1$ AU, its value turns out to be $l = 1.38\,\text{kW m}^{-2}$: just the right value for human and other life. However, at this point of our exposition, we have turned away from Earth-centred considerations to the intrinsic properties of the stars. The central position of the point representing the Sun in the Hertzsprung–Russell diagram is then just another way of expressing that the Sun is but an ordinary main sequence star. Yet, as far as distance is concerned, we should consider ourselves lucky to have a typical star close enough to permit *spatially resolved* observations! We will have plenty of opportunities to appreciate that this is crucial for our understanding of plasma dynamics in the Universe as a whole.

Not only is the solar system kept together by gravity, but the individual celestial bodies of the Sun and the planets are also kept together by gravity and, as a result, they *contract*. Stars with masses like that of the Sun (Jupiter is just too small to qualify as a star) contract so much that in the centre *densities* and *temperatures* are reached that are *high enough* for *thermonuclear burn by fusion* reactions of hydrogen, viz. $T_c = 1.5 \times 10^7$ K, and $\rho_c = 1.5 \times 10^5\,\text{kg m}^{-3}$. We have already encountered these fusion reactions and their conditions in Eqs. (1.8) and (1.9). Recall that, under these conditions, matter is ionized so that we encounter the plasma state again in the core of the Sun and the other stars. It appears that we have now closed the circle and that the announcement of the theme for this book will simply be: laboratory fusion and astrophysical fusion reactions require the study of plasma physics. However, this is not the case. Reality is much more interesting (and subtle) than this.

1.3.3 Plasmas enter the stage

With the discovery of Bethe and von Weiszäcker in 1939 that thermonuclear fusion reactions take place in the centre of the Sun and the other stars, we know the ultimate source of the enormous amounts of power emitted in the form of visible and ultraviolet light. However, there is quite some distance in space and time between this source and the starlight it eventually produces. In the intermediate stages, this gigantic thermonuclear energy source indirectly excites a wide variety of

1.3 Astrophysical plasmas

additional, plasma dynamical, phenomena. Could we 'see' that? If only the blinding splendour of the solar disc were blocked for a few minutes we would be able to tell. Fortunately, provision has been made for that: the relative sizes of the Moon and the Sun, and their distances to the Earth, are precisely of the right magnitude to permit occultation of the Sun every now and then to exhibit an extremely beautiful phenomenon. At the moment of the eclipse, even the birds hold their breath, and a human being lucky enough to be at the right spot at the right moment can see a hot (millions of degrees) plasma with his own eyes: a diffuse light due to scattering of sunlight by the coronal plasma and stretching out over several solar diameters. Even the magnetic structures supporting it are visible to the (admittedly prejudiced) physicist in the form of streamers of plasma tracing out magnetic field lines and helmet structures associated with magnetic cusps (see Chapter 8). Hence, at a solar eclipse, one catches a wonderful glimpse of a huge magnetized plasma structure which engulfs the whole solar system.

This structure is the solar corona expanding into the *solar wind*, which forms magnetospheres when encountering the magnetic fields of the planets and which is a giant magnetosphere by itself, called the heliosphere, terminating only at distances beyond the solar system. The solar wind carries the wave-like signals of its creation, but it also carries the intermittent radiation and high-energy particle signatures of violent outbursts of magnetic energy releases by *flares* and *coronal mass ejections* (CMEs) at the solar surface (see Chapter 8). This highly unsteady plasma dynamical state creates the critical conditions for magnetic storms in the magnetosphere[4] and forms a threat for safety of personnel and proper functioning of equipment aboard spacecrafts. This aspect of solar wind dynamics is called *space weather*. (Running ahead of our argument: its prediction involves Advanced Magnetohydrodynamics, which is the subject of the companion Volume 2 of this book.)

Receding now to the interior of the star, closer to the thermonuclear energy source, we encounter the phenomena responsible for all this: radiation transport and convection which, together with the differential rotation of the star, create the conditions required for a dynamo. This dynamo produces magnetic fields that do not stay inside the star but are expelled, together with the plasma, to form the extremely hot coronae and stellar winds that are the characteristics of X-ray emitting stars. Incidentally, the creation of magnetic fields in the interior of stars and the high temperatures of coronal plasmas are two plasma physical problems that are far from being solved at present. While we do not pretend to solve them here,

---

[4] *The* magnetosphere is always shorthand for the magnetosphere of the Earth, or *our* magnetosphere, like *the* Galaxy always stands for the Milky Way, or *our* Galaxy.

we do believe that for progress one needs to delve deeply into basic magnetohydrodynamics, which is the subject of this book. Hence, the connection between laboratory and astrophysical plasmas is not the thermonuclear fusion reactions but their indirect result far away: *magnetized plasmas are present anywhere in the Universe!*

How do we know? High-resolution astronomical observations over the whole range of the electromagnetic frequency spectrum by means of 'telescopes', ground-based or from space vehicles, have produced irrefutable evidence for that. Whereas Sputnik (1957) and the Apollo flight to the Moon (1969) have spoken to the imagination of a large public, the less-known observations of the Sun and stars by means of X-ray telescopes on board rockets and the risky (manned) Skylab missions of 1973 and 1974 may have produced a more lasting revision of our scientific picture of the cosmos. It revealed the tremendously dynamic magnetic structure of the solar atmosphere and corona with myriads of closed magnetic flux tubes, containing hot plasma, bordering open magnetic regions, so-called coronal holes where the cooler plasma is associated with reduced X-ray emission. These early observations were finally superseded by the higher resolution images obtained from the Japanese satellite Yohkoh and the NASA-ESA Solar and Heliospheric Observatory SOHO, launched in 1992 and 1995, respectively. In the meanwhile, the plasma physics picture of the solar system has been augmented considerably by planetary missions like Voyager 2 (launched 1989) travelling to the outer edges of the solar system and also measuring the magnetic fields of the giant planets (see Table B.8), or the flight of Ulysses (launched 1990) which was slung in an orbit over the magnetic poles of the Sun (i.e. out of the ecliptic) by means of a swing-by of Jupiter, whereas Cluster II (launched in 2000) will provide many more details of the three-dimensional structures of the magnetospheres. In the same period, the picture of the structure of the Galaxy and the Universe, essentially including galactic and cosmic magnetic fields, has also evolved explosively due to the ever improved resolution of the traditional telescopes, the radio telescopes in large and very large arrays, and the numerous space missions, culminating in the launch of the Hubble Space Telescope in 1990.

We summarize by making a few sweeping statements, obviously not meant to present final scientific truths:

– By means of X-ray observations, the few exciting minutes of a solar eclipse have been extended almost indefinitely to provide a picture of the corona as a high-temperature plasma with *extremely complex dynamical magnetic structures* (Priest [190]).
– The interaction of the solar wind with the planetary magnetospheres is one of the most interesting plasma laboratories in space, offering the possibility of studying *spatially resolved plasma dynamics*.

- Finally, since the Sun is an ordinary star, what has been learned there may be extrapolated to other stars, of course with due modifications (Schrijver and Zwaan [204]). Going one step further, including neutron stars and pulsars (Mestel [154]), accretion discs about compact objects, etc.: what has been learned from magnetic plasma structures in the solar system may be extrapolated, again with due modifications, to the more exotic astrophysical objects that cannot be observed with spatial resolution but that do provide *intricate temporal signatures*.

Thus, a secondary layer (considering gravity and nuclear fusion as the primary layer) of phenomena has been revealed in the solar system that is present everywhere in the Universe. This brings us to our next subject.

1.3.4 The standard view of nature

Consider the standard view of nature, as developed in the twentieth century and widely held to provide the correct scientific representation of the Universe (Fig. 1.8). The four fundamental forces govern phenomena at immensely separate length scales, at least at times beyond 'The First Three Minutes' (Steven Weinberg, 1978) after the big bang. At the risk of caricaturing the wonderful achievements of elementary particle physics, on a scale of increasing dimensions, the weak and strong nuclear forces in the end just produce the different kinds of nuclei and electrons, which constitute the main building blocks of matter. In a sense, these forces are exhausted beyond the length scale of 10^{-15} m. Since nuclei are positively charged and electrons negatively, the much longer range electric forces then take over, giving rise to the next stage of the hierarchy, viz. that of 'ordinary' matter consisting of atoms and molecules with sizes of the order of 10^{-9} m. Since these particles are electrically neutral, all there appears to remain is the gravitational force which requires the collective effect of huge amounts of matter over length scales beyond 10^9 m in order to become sizeable. This gives rise to the different astrophysical systems of stars, galaxies, clusters of galaxies, etc. Since the gravitational force is a long-range force which is solely attractive (there is no screening by repulsive negative mass particles), this force is only 'exhausted' at the scale of the Universe itself.

It will be noticed that the 'picture' of Fig. 1.8 jumps the eighteen orders of magnitude from atoms to stars (indicated by the dots) under the assumption that nothing of fundamental interest happens there. One could remark that we just happen to live on the least interesting level of the physical Universe, or one could dwell on the disproportion of man between the infinities of the small and the large (Pascal), or one could join the recent chorus of holistic criticism on the reductionism of physics. So much appears to be correct in the latter viewpoint that the given picture does not have any place for the complexities of solid state physics, fluid dynamics

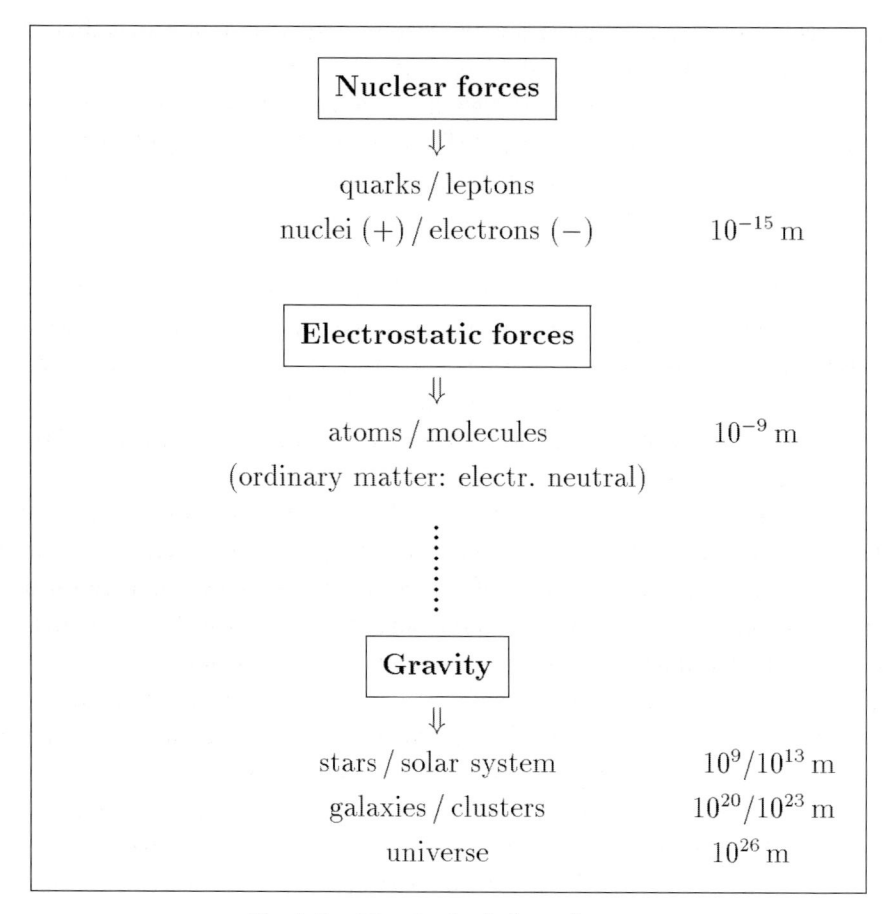

```
┌─────────────────────────────────────────────────────┐
│                 ┌─────────────────┐                 │
│                 │ Nuclear forces  │                 │
│                 └─────────────────┘                 │
│                          ⇓                          │
│              quarks / leptons                       │
│          nuclei (+) / electrons (−)      $10^{-15}$ m │
│                                                     │
│               ┌─────────────────────┐               │
│               │ Electrostatic forces│               │
│               └─────────────────────┘               │
│                          ⇓                          │
│               atoms / molecules          $10^{-9}$ m │
│         (ordinary matter: electr. neutral)          │
│                                                     │
│                         ⋮                           │
│                                                     │
│                    ┌─────────┐                      │
│                    │ Gravity │                      │
│                    └─────────┘                      │
│                          ⇓                          │
│              stars / solar system    $10^{9}/10^{13}$ m │
│              galaxies / clusters    $10^{20}/10^{23}$ m │
│                    universe              $10^{26}$ m │
└─────────────────────────────────────────────────────┘
```

Fig. 1.8. The standard view of nature.

or biological systems, to name just a few. It should come as a big disappointment that nature would hang together from elementary particles to cosmology without really involving the intermediate stages.

For our subject, however, another misrepresentation is implicit in Fig. 1.8. We have started our discussion in Section 1.1 by noting that more than 90% of matter in the Universe is plasma so that *the Universe does not consist of ordinary matter (in the usual sense) but most of it is plasma*! It is true that plasma is usually also almost electrically neutral, like atoms and molecules, but the important difference is that the ions and electrons are not tied together in atoms but *move about freely as fluids*. The large scale result of this dynamics is the formation of *magnetic fields* which in turn determine the plasma dynamics: a highly nonlinear situation. These magnetic fields not only bridge the gap between microscopic and macroscopic physics, but they also reach far into the astrophysical realm at all scales.

The subject of plasma-astrophysics is of basic importance for understanding phenomena occurring everywhere in the Universe.

It will have been noted that, in our presentation, we have ignored the unification of electric and magnetic forces brought about by Maxwell's theory of electromagnetism. There is a good reason for this since, in the domain of plasma dynamics, electric and magnetic forces are associated with quite different effects operating on immensely different length scales with the magnetic forces dominating on the longer length scales. Consequently, most of plasma dynamics is well described by exploiting the so-called pre-Maxwell equations, i.e. Maxwell's equations without the displacement term. We will see in later chapters that the dynamics of magnetic fields is so interwoven with the dynamics of the plasma itself that its proper description takes precedence over the one where electric and magnetic fields are treated on an equal footing.

The most important law for magnetic fields is $\nabla \cdot \mathbf{B} = 0$, which implies that there are no sources or sinks. This law is incompatible with spherical symmetry so that the simplest basic geometries of magnetized plasmas are completely different from the ones prevailing on the atomic and gravitational scales. In particular, large scale tubular magnetic structures occur which move with the plasma so that magnetic forces are transmitted with the fluid. One could hardly imagine a bigger contrast with central electrostatic and gravitational forces decaying in vacuum with distance as r^{-2} ! Striking examples are solar flares, the X-ray emitting corona of the Sun, and coronal mass ejections (plasma expelled from the main body of the Sun against the gravitational pull), the interaction of the solar wind with the planetary magnetic fields, waves and flows in neutron star magnetospheres, extragalactic jets, spiral arm instabilities, etc.

In conclusion: the standard view of nature fails over a wide range of scales because it does not recognize the presence of magnetized plasmas all over the Universe. Magnetic fields are an important aspect of modern astrophysics. Hence, *the nonlinear interaction of plasma and complex magnetic structures* presents itself as an important common theme of laboratory and astrophysical plasma research.

1.4 Definitions of the plasma state

1.4.1 Microscopic definition of plasma

Turning now to the subject of plasmas proper, we need to refine the crude definition given in Section 1.1. This involves a closer study of the microscopic properties required for the plasma state. To that end, we follow the exposition given by F. F. Chen in the first chapter of his book on Plasma Physics [53].

First, we need to relax the condition of complete ionization given in our crude definition since plasma behaviour is already encountered when the ionization is only partial. A simpler definition of plasma would then be: *a plasma is an ionized gas*. However, how much ionization is required? An estimate may be obtained from the *Saha equation* which gives an expression for the amount of ionization of a gas in thermal equilibrium:

$$\frac{n_i}{n_n} = \left(\frac{2\pi m_e k}{h^2}\right)^{3/2} \frac{T^{3/2}}{n_i} e^{-U_i/kT} \approx 2.4 \times 10^{21} \frac{T^{3/2}}{n_i} e^{-15.8 \times 10^4 / T}. \quad (1.26)$$

Here, n_i and n_n are the particle densities of ions and neutrals (in m^{-3}), U_i is the ionization energy (in J), T is the temperature (in K), and the other symbols have their usual meaning. The numbers on the RHS are obtained by exploiting Table B.1, $(2\pi m_e k/h^2)^{3/2} = 2.4 \times 10^{21}$ K$^{-3/2}$ m^{-3}, and inserting the ionization energy of hydrogen, $U_i = 13.6$ eV. (Ionization energies are usually given in eV, where 1 eV $= 1.6 \times 10^{-19}$ J, which corresponds with 1.16×10^4 K if one divides by the Boltzmann factor k.)

For *air* at room temperature, where $n_n = 3 \times 10^{25}$ m^{-3}, $T = 300$ K, $U_i = 14.5$ eV (ionization potential for nitrogen), one finds a huge negative factor (-560) in the exponent of Eq. (1.26) so that the final ratio of the densities of ions and neutrals is extremely small: $n_i/n_n \approx 2 \times 10^{-122} \ll 1$. As expected, the degree of ionization of air at room temperature is totally negligible: air is not a plasma. For hydrogen in *a tokamak machine* with $T = 10^8$ K and $n \equiv n_e = n_i = 10^{20}$ m^{-3}, one finds that the expression in the exponent $U_i \ll kT$ so that $\exp(U_i/kT) \approx 1$ and $n_i/n_n \approx 2.4 \times 10^{13} \gg 1$: in such machines genuine plasmas are obtained. However, for *the core of the Sun* with $T = 1.6 \times 10^7$ K and $n = 10^{32}$ m^{-3}, one finds that $n_i/n_n \approx 1.5$. Surprisingly, although thermonuclear reactions take place, ionization is not complete in the core of the Sun and plasma behaviour is not completely dominant! This is due to the extremely high densities there. On the other hand, *in the corona of the Sun*, with typical values of $T = 10^6$ K (not thermonuclear, but anomalously high: a subject which will occupy us in later chapters) and $n = 10^{12}$ m^{-3}, we have $n_i/n_n = 2.4 \times 10^{18}$: matter in the corona is an excellent plasma!

Even though we now have a measure for the degree of ionization required for plasmas, we still do not have a criterion for plasma behaviour. A much more precise definition, as given by Chen, reads: *a plasma is a quasi-neutral gas of charged and neutral particles which exhibits collective behaviour.*

In an ordinary gas neutral molecules move about freely (there are no net electromagnetic forces) until a collision occurs. This is a *short-range, binary*, event in which two particles hit each other. In a hard-sphere model of the molecules, the cross-section for such a collision is just the cross-section of the particles. In a plasma, on the other hand, the charged particles are subject to *long-range*,

1.4 Definitions of the plasma state

collective, Coulomb interactions with many distant encounters (so-called pitch angle scattering). Although the electrostatic force between two charged particles decays with the mutual distance ($\sim 1/r^2$), the combined effect of all charged particles may not even decay, since the interacting volume increases ($\sim r^3$). This is a typical collective effect, the result of the statistics of many particles, each moving in the average electrostatic field created by all the other particles.

We now discuss this electrostatic collective aspect quantitatively. For collective plasma behaviour, again according to Chen [53], three conditions should be satisfied.

(a) The long-range Coulomb interaction between charged particles should dominate over the short-range binary collisions with neutrals. Indicating typical *time scales* of collective oscillatory motion by τ ($\sim 1/\omega$ when ω is the angular frequency of the oscillations), this implies that

$$\tau \ll \tau_n \equiv \frac{1}{n_n \sigma v_{\text{th}}} \approx \frac{10^{17}}{n_n \sqrt{T}}, \quad (1.27)$$

where τ_n is the mean time between collisions of charged plasma particles with neutrals. The estimate on the RHS is obtained by writing $\tau_n \approx \lambda_{\text{mfp}}/v_{\text{th}}$, where λ_{mfp} is the mean free path and v_{th} is the thermal speed of the particles. With $\lambda_{\text{mfp}} = (n_n \sigma)^{-1}$, where the cross-section $\sigma = \pi a^2 \approx 10^{-19}$ m^2 is obtained by taking the radius $a \approx 2 \times 10^{-10}$ m of a neutral H atom, and $v_{\text{th}} \approx \sqrt{kT/m_p} \approx 100\sqrt{T}$, we obtain the expression (1.27) for τ_n. Since we are interested in plasma conditions, we should convert this expression from neutral density n_n to ion density n_i by means of the Saha equation (1.26). For solar coronal plasma with $T = 10^6$ K and $n_i = 10^{12}$ m^{-3}, so that $n_n = 4 \times 10^{-7}$ m^{-3}, this implies $\tau \ll \tau_n \approx 2 \times 10^{20}$ s (!). For tokamaks with $T = 10^8$ K and $n_i = 10^{20}$ m^{-3}, the condition becomes $\tau \ll \tau_n \approx 2.4 \times 10^6$ s. Clearly, the condition (1.27) represents very mild restrictions on the time scales for plasma behaviour.

(b) The *length scale* of plasma dynamics should be much larger than the minimum size over which the condition of *quasi-neutrality* holds. Production of overall charge imbalance creates huge electric fields which in turn produce huge accelerations, so that such an imbalance is neutralized almost instantaneously and the plasma maintains charge neutrality to a high degree of accuracy. However, local charge imbalances may be produced by thermal fluctuations. To estimate their size, one should compare the thermal energy kT of the particles with their electrostatic energy $e\Phi$. The latter can be estimated through Poisson's law, $dE/dx = -d^2\Phi/dx^2 = -(1/\epsilon_0) en$, so that $kT \approx e\Phi \approx (1/\epsilon_0) e^2 n \lambda_D^2$. Here, the gradient length has been equated to the *Debye length*, which is the typical size of a region over which charge imbalance due to thermal fluctuations may occur. Hence, length scales for a quasi-neutral plasma should satisfy

$$\lambda \gg \lambda_D \equiv \sqrt{\frac{\epsilon_0 kT}{e^2 n}} \approx 70\sqrt{\frac{T}{n}}, \qquad (1.28)$$

where $n \equiv n_e \approx Zn_i$ (with Z indicating the ion charge number). Inserting the numbers for coronal plasma again, we find $\lambda_D = 0.07\,\text{m}$. Considering typical transverse length scales of coronal loops, $\lambda \sim 10\,000\,\text{km} = 10^7\,\text{m}$, this condition is easily satisfied.

▷ **Exercise.** Exploit the tables of Appendix B to also find out what this condition means for other cases, like tokamak plasmas. ◁

Note that the concept of Debye length alleviates our original statement about long-range electrostatic forces considerably: sizeable regions with charge accumulation do not form through thermal fluctuations alone. A free charge, which in vacuum would have a potential $\Phi = q/r$, in a plasma is surrounded by a cloud of particles of opposite charge, which effectively shields the Coulomb potential for distances much larger than the Debye length: $\Phi_{\text{eff}} = (q/r)\exp(-r/\lambda_D)$ (called *Debye shielding*). This just implies that $Zn_i \approx n_e$, i.e. *quasi* charge-neutrality holds. It does not mean that electric fields do not arise in plasmas. Actually, quite the opposite: electric fields arise almost automatically when plasmas move in a magnetic field. However, charge imbalances are extremely small when measured in terms of the total charge of the separate species:

$$\frac{|Zn_i - n_e|}{n_e} \ll 1. \qquad (1.29)$$

Hence $Zn_i \approx n_e$ holds to a high degree of accuracy in plasmas.

(c) Finally, in order for statistical considerations to be valid, sufficiently many particles should be present in a *Debye sphere*, i.e. a sphere of radius λ_D:

$$N_D \equiv \tfrac{4}{3}\pi \lambda_D^3 n \approx 1.4 \times 10^6 \sqrt{\frac{T^3}{n}} \gg 1. \qquad (1.30)$$

For our example of a coronal plasma, this yields $N_D = 1.4 \times 10^9 \gg 1$, which is again easily satisfied. Note that both $\lambda_D \sim n^{-1/2}$ and $N_D \sim n^{-1/2}$ so that very high density plasmas are OK with respect to condition (1.28), but not with respect to condition (1.30). For example, for the core of the Sun, $\lambda_D = 3 \times 10^{-11}\,\text{m}\,(!)$, but $N_D \approx 9$: not so good for the application of statistical mechanics.

In conclusion: collective plasma behaviour is encountered when the time scales are sufficiently short with respect to collision times with neutrals, $\tau \ll \tau_n$, the length scales are much larger than the Debye length, $\lambda \gg \lambda_D$, and there are many particles in a Debye sphere, $N_D \gg 1$. These conditions can be translated in terms

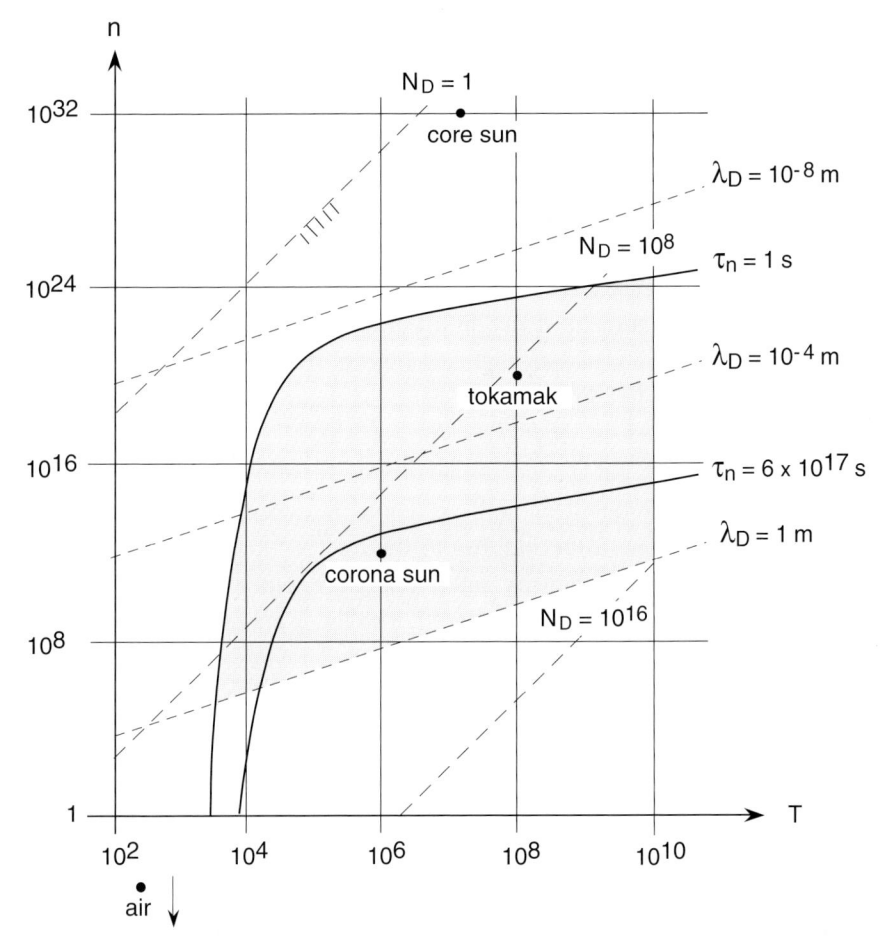

Fig. 1.9. Conditions for collective plasma behaviour, in terms of the density $n \equiv n_e \approx Zn_i$ and temperature $T \sim T_e \sim T_i$, are satisfied in the shaded area for time scales $\tau < \tau_n = 1\,\text{s}$ and length scales $\lambda > \lambda_D = 1\,\text{m}$, where $N_D \gg 1$ is also satisfied. The restrictions on the upper time limit of low density astrophysical plasmas quickly approach the age of the Universe, whereas the restrictions on the lower length limit for high density laboratory fusion experiments approach microscopic dimensions.

of conditions on the density and the temperature, which are satisfied under a wide variety of conditions, as shown in Fig. 1.9. This picture confirms our statement of Section 1.1: plasma is a very normal state of matter in the Universe.

1.4.2 Macroscopic approach to plasma

So far, the most important physical variable in laboratory and astrophysical plasmas, viz. the magnetic field, has been conspicuously absent from our definition of the plasma state. The reason is that we have followed the traditional exposition of basic plasma theory, which starts with the microscopic point of view and stresses the collective phenomena involving electric fields. Whereas the length and time scales appropriate for these phenomena may be discussed in terms of the *local* values of the plasma density n and the temperature T, the magnetic field **B** brings in entirely different, *global*, considerations. (Incidentally, here one may detect one of the ways in which reductionism fails to recognize the emergence of new levels in the description of nature.) We have already observed the central importance of magnetic fields in confinement of fusion plasmas (Section 1.2) and in the dynamics of an enormous variety of astrophysical objects (Section 1.3), where we have stressed their basic non-locality. We now have to quantify these observations.

The macroscopic point of view does not set aside the microscopic conditions derived in Section 1.4.1 but it incorporates them as follows. A macroscopic description requires (1) frequent enough collisions between electrons and ions to establish fluid behaviour, (2) in addition to the microscopic conditions of length and time scales involving the density and temperature, global conditions on length and time scales involving the magnetic field. The latter quantities have to be large in order to permit *averaging over the microscopic dynamics*. To quantify this step requires the consideration of the cyclotron (or gyro) motion of the electrons and ions, which will only be discussed in the next chapter. Anticipating that discussion, the cyclotron radii $R_{e,i}$ and the inverse cyclotron frequencies $\Omega_{e,i}^{-1}$ of the electrons and ions will be shown to be inversely proportional to the magnetic field strength, $R_{e,i} \sim B^{-1}$ and $\Omega_{e,i}^{-1} \sim B^{-1}$, where the ion expressions provide the most limiting conditions on macroscopic length and time scales. Consequently, 'large enough' means that macroscopic length and time scales should be much larger than R_i and Ω_i^{-1}, respectively. This is possible when the magnetic field is large enough for the plasma volume under consideration to contain many ion gyro radii and when the dynamic phenomena last many ion gyro periods.

Summarizing: *for a valid macroscopic model of a particular magnetized plasma dynamical configuration, size, duration, density and magnetic field strength should be large enough to establish fluid behaviour and to average out the microscopic phenomena (i.e. collective plasma oscillations and cyclotron motions of the electrons and ions)*.

The distinguishing feature for macroscopic plasma dynamics is the interaction of plasma motion and magnetic field geometry. This fluid aspect of plasmas concerns the motion of the plasma as a whole, without considering the separate electrons and ions, under the influence of magnetic fields. These fields are, in turn, generated by the plasma motion itself: a highly nonlinear situation. *The theoretical tool to describe this global interplay of plasma and magnetic field is called MHD ≡ magnetohydrodynamics*. The objective of this book is to demonstrate how this theory provides the common basis for the description of laboratory and astrophysical plasma dynamics.

The (surprisingly many) different aspects of the given definition of a macroscopic plasma model will be discussed one by one in the following chapters. In particular, Chapters 2 and 3 will provide the missing quantitative elements of microscopic plasma physics needed for the foundation of macroscopic plasma dynamics. The basic Chapter 2 may be skipped by readers who are already familiar with basic plasma physics. The advanced Chapter 3 may be skipped as well by readers who wish to start with magnetohydrodynamics proper as soon as possible.

One question, answered in detail in the next chapters, must be addressed at least provisionally here, viz.: why is the electric field not even mentioned in the above discussion of macroscopic plasma dynamics? The reason is that the electric field becomes, in fact, a secondary quantity in MHD. Large electrostatic fields due to charge imbalances only occur over Debye length scales, which are averaged out, and electromagnetic waves are absent in non-relativistic MHD since the displacement current is negligible. The electric field is then determined from the primary variables of the velocity $\mathbf{v}$ and the magnetic field $\mathbf{B}$ by means of 'Ohm's law' for a nearly perfectly conducting plasma: $\mathbf{E} + \mathbf{v} \times \mathbf{B} \approx 0$, i.e. the electric field in a frame moving with the plasma vanishes.

1.5 Literature and exercises

Notes on literature

Some general references for the whole book are given under the different headings below. The complete information on the references is given at the end of the book.

Introductory plasma physics:

- Boyd & Sanderson, *Plasma Dynamics* [35] is one of the older textbooks on plasma physics that is still quite useful. It has been revised completely in *The Physics of Plasmas* [36].
- Chen, *Introduction to Plasma Physics and Controlled Fusion* [53] is the most readable, and probably most widely used, basic textbook on plasma physics. An older edition also contained material on controlled fusion, which will appear in a separate second volume.
- Bittencourt, *Fundamentals of Plasma Physics* [31] is a basic theoretical course on plasma physics with detailed calculations.
- Sturrock, *Plasma Physics* [221] is a basic text on plasma physics written for graduate students from astrophysics, space science, physics, and engineering departments.
- Goldston & Rutherford, *Introduction to Plasma Physics* [92] is a basic text on plasma physics based on teaching at Princeton University by two experts in tokamak physics.

- Nishikawa & Wakatani, *Plasma Physics* [168] is a text on basic plasma theory with applications to magnetic as well as inertial confinement of fusion plasmas.

Topics in advanced plasma physics:

- Leontovich (ed.), *Reviews of Plasma Physics,* Vols. 1–5 [142] contain the unsurpassed Russian expositions of the basics of plasma theory after the declassification of the 1958 Geneva Conference. The different chapters will be quoted by separate references.
- Akhiezer, Akhiezer, Polovin, Sitenko & Stepanov, *Plasma Electrodynamics* [4] is another classic from one of the Soviet plasma theory schools, systematically building up plasma physics by kinetic and hydrodynamic methods and progressing to the diverse linear and nonlinear manifestations of the plasma state.
- Dendy (ed.), *Plasma Physics: an Introductory Course* [65] contains the material taught at the yearly Culham summerschools on plasma physics.
- Donné, Rogister, Koch & Soltwisch (eds.), *Proc. Second Carolus Magnus Summer School on Plasma Physics* [68] contains the material taught at the mentioned summerschool on plasma physics held every other year.

Magnetohydrodynamics:

- Freidberg, *Ideal Magnetohydrodynamics* [72] is a textbook on ideal MHD, based on lectures at MIT for graduate students and researchers, which puts perfect conductivity and the applications to fusion research centre stage.
- Lifschitz, *Magnetohydrodynamics and Spectral Theory* [146] is an advanced text on MHD stressing the unity of physics and mathematics through spectral theory. This kind of complex, yet faultless, calculation is a rare commodity in plasma physics.
- Polovin & Demutskii, *Fundamentals of Magnetohydrodynamics* [187] is an introduction of the various aspects of MHD, written in the lucid style of the great Russian theoreticians.
- Biskamp, *Nonlinear Magnetohydrodynamics* [29] is a monograph on nonlinear MHD processes like evolution of large amplitude instabilities, reconnection, turbulence, disruptions, field reversals, and flares.

Tokamaks:

- Wesson, *Tokamaks* [244] is a veritable encyclopaedia of the plasma physics involved in nuclear fusion research in tokamaks.
- White, *Theory of Toroidally Confined Plasmas* [245] contains the material of a graduate course at Princeton University on fundamental plasma theory of tokamaks.
- Hazeltine & Meiss, *Plasma Confinement* [107] provides the advanced theory of magnetic plasma confinement with stress on derivations from first principles.
- Braams & Stott, *Nuclear Fusion: Half a Century of Magnetic Confinement Fusion Research* [37] gives the history of nuclear fusion research up to the present, leading up to the citation of Artsimovich 'Fusion will be there when society needs it'.

The Sun:

- Priest, *Solar Magnetohydrodynamics* [190] is the classical introduction of magnetohydrodynamics of the Sun, in particular the solar corona.

- Stix, *The Sun* [217] is a textbook on the physics of the Sun with due account of innumerable observational facts.
- Foukal, *Solar Astrophysics* [69] aims at making the advances in understanding of the Sun accessible to students and non-specialists by means of simple physical concepts and observations.

Space physics:

- Hasegawa & Sato, *Space Plasma Physics* [106] is a monograph on the physics of stationary plasmas, small amplitude waves, and the stationary magnetosphere.
- Kivelson & Russell (eds.), *Introduction to Space Physics* [127] is an introduction of all aspects of space and solar plasmas for senior undergraduate and graduate students, written by experts in the various fields. The different chapters will be quoted by separate references.
- Baumjohann & Treumann, *Basic Space Plasma Physics*, and (same authors in reverse order) *Advanced Space Plasma Physics* [19] are the basic material presented in a space plasma physics course at the University of Munich, and the advanced nonlinear aspects of the various waves and instabilities.

Plasma astrophysics:

- Battaner, *Astrophysical Fluid Dynamics* [18] is a systematic theoretical treatise of the dynamics of classical, relativistic, photon and plasma fluids, progressing from stars to the Universe at large.
- Choudhuri, *The Physics of Fluids and Plasmas* [55] is an introduction to fluid dynamics, plasma physics and stellar dynamics for graduate students of astrophysics.
- Mestel, *Stellar Magnetism* [154] is a monograph on MHD applied to the magnetism of stars, including stellar dynamos, star formation and pulsar electrodynamics.

Exercises

The exercises are meant to increase understanding of the principles of plasma dynamics, where estimating orders of magnitude is an essential part. Frequent use of the numerical tables of the appendices is recommended. Difficult problems are marked with a star.

[1.1] *Fusion reactions*

We know two methods of energy production by nuclear processes, namely nuclear fission and nuclear fusion. For both, the net energy released is described by the same formula.

- What is expressed by that formula? What is the major difference between fission and fusion? What is actually explained by the mentioned formula for the most likely reaction in future fusion reactors, $D^2 + T^3 \rightarrow He^4$ (3.5 MeV) + n (14.1 MeV) ?

[1.2] *Fusion power*

If we want a tokamak reactor to produce energy and sustain itself, we need balance between thermonuclear power output and power losses. This leads to a condition on required particle density n, energy confinement time τ_E, and temperature $\tilde{T}$.

32   *Introduction*

- If the power output is given by $P_T = n^2 f(\tilde{T})$, where f is a known function of temperature, and the power losses consist of Bremsstrahlung, $P_B = \alpha n^2 \tilde{T}^{1/2}$, and heat transport, $P_L = 3n\tilde{T}/\tau_E$, derive the criterion for fusion energy production. Assume that all three power contributions can be converted in plasma heating with efficiency η.
- A more recent approach states that for ignition of a fusion reactor, α-particle heating of the plasma, P_α, should make up for the power losses. Express this criterion by a similar equation above.
- For such a reactor, the product of the three mentioned quantities will have to be $n\tau_E \tilde{T} \approx 3 \times 10^{21}$ m^{-3} s keV. Give some realistic values for n, τ_E and $\tilde{T}$. Using Table B.1, convert the temperature $\tilde{T}$ in keV to T in degrees K.
- Why do magnetic fields play such an important role in thermonuclear fusion?

[1.3] *Astrophysical observations*

When observing objects in the night sky, we detect a certain amount of light coming from those objects. However, objects that appear dimmer do not really emit less light. They just have a smaller apparent magnitude, defined by $m \equiv m_0 - 2.5 \times {}^{10}\!\log(l/l_0)$.

- Give the meaning of the different variables in this equation and explain how the dependence on distance is incorporated.
- The absolute magnitude is the magnitude based on the amount of flux that would be collected from the object if it were located at a distance of 10 parsecs from the observer. From this, derive the formula for the absolute magnitude.
- Does the same relationship hold for observations with a spectrographic filter that selects a specific frequency band?

[1.4] *Solar plasmas*

All the light we receive is the result of specific nuclear reactions which release energy. For instance, nuclear reactions in the centre of the Sun are the ultimate cause of light escaping at positions where the Sun becomes optically thin for this radiation.

- Which nuclear reactions take place in the centre of the Sun? What kind of radiation is produced? Why doesn't this extremely energetic light escape right away? Why is the light we collect on Earth mostly in the visible range? Why is it the less energetic light that escapes from the Sun? Explain what this says about the density profile of the Sun.
- Normally we cannot see the solar atmosphere further outwards since we are blinded by the escaping light. A beautiful exception occurs during a solar eclipse. Explain why we can observe such a phenomenon at all on Earth.
- The coronal structures revealed during a solar eclipse show the footprints of a giant engine that produces the solar activity. What is the mechanism of that engine and how is that connected to the kind of structures observed?

[1.5] *Plasma definitions and applications*

Putting plasmas in a wide perspective, discuss the following aspects.

- What is a plasma? How is it different from ordinary gases and fluids?
- Name some applications of plasma physics.
- How can plasmas be confined?

[1.6] *Terminology*

Explain the following terms and their connection with plasma or fusion physics:
- Coulomb interaction,
- Saha equation,
- quasi-neutrality, Debye length,
- θ-pinch, z-pinch, tokamak,
- Ohm's law.

[1.7] *Forces in nature*

Explain the major forces present in nature, together with their relative strength and decay distance and, thus, the scale at which they are dominant. Explain why gravity is such a special force. What forces dominate the plasma regime?

2
Elements of plasma physics

2.1 Theoretical models

Plasma processes are described by quite different theoretical models. Which one is to be chosen depends on the kind of phenomenon one is interested in. Broadly speaking, three kinds of theoretical description are used:

(a) the theory of the motion of *individual charged particles* in given magnetic and electric fields (Section 2.2);
(b) the *kinetic theory* of a collection of such particles, describing plasmas microscopically by means of particle distribution functions $f_{e,i}(\mathbf{r}, \mathbf{v}, t)$ (Section 2.3);
(c) the *fluid theory* (magnetohydrodynamics), describing plasmas in terms of averaged macroscopic functions of $\mathbf{r}$ and t (Section 2.4).

By way of introduction, within each of these descriptions, we will give a simple example which illustrates the plasma property that is relevant for our subject, viz. plasma confinement by magnetic fields.

2.2 Single particle motion

2.2.1 Cyclotron motion

The motion of a charged, non-relativistic particle in an electric and magnetic field is described by the well-known equation of motion

$$m\frac{d\mathbf{v}}{dt} = q(\mathbf{E} + \mathbf{v} \times \mathbf{B}), \qquad (2.1)$$

where $\mathbf{E}(\mathbf{r}, t)$ and $\mathbf{B}(\mathbf{r}, t)$ are considered to be given (of course, in agreement with Maxwell's equations) and one has to solve for the particle velocity $\mathbf{v}(\mathbf{r}, t)$. For the moment, we do not specify the mass m and the charge q of the particles. They will be fixed later to correspond to either electrons ($m = m_e$, $q = -e$) or ions

2.2 Single particle motion

with mass number A and charge number Z (i.e. multiples of the proton mass and charge: $m = m_i = A m_p$, $q = Ze$).

Consider a charged particle in a constant magnetic field, taken in the z-direction, in the absence of an electric field: $\mathbf{B} = B\mathbf{e}_z$, $\mathbf{E} = 0$. Performing two simple vector operations on Eq. (2.1) provides some preliminary insight: (1) projecting on $\mathbf{B}$ (exploiting the vector identity (A.1) of Appendix A) gives

$$m \frac{dv_\parallel}{dt} = 0, \quad \text{so that} \quad v_\parallel = \text{const}, \tag{2.2}$$

and (2) projecting on $\mathbf{v}$ gives

$$\frac{d}{dt}(\tfrac{1}{2} m v^2) = 0 \;\rightarrow\; \tfrac{1}{2} m v^2 = \text{const} \;\stackrel{(2.2)}{\longrightarrow}\; \tfrac{1}{2} m v_\perp^2 = \text{const} \;\rightarrow\; v_\perp = \text{const}. \tag{2.3}$$

The expressions (2.2) and (2.3) already suggest the kind of orbits to be expected.

We now solve Eq. (2.1) systematically. With $\mathbf{v} = d\mathbf{r}/dt = (\dot{x}, \dot{y}, \dot{z})$ we obtain two coupled differential equations for the motion in the perpendicular plane:

$$\ddot{x} - (qB/m)\, \dot{y} = 0,$$
$$\ddot{y} + (qB/m)\, \dot{x} = 0. \tag{2.4}$$

Defining *the gyro-* or *cyclotron frequency*,

$$\Omega \equiv \frac{|q|B}{m}, \tag{2.5}$$

the solution to these equations represents a periodic circular motion about a point $x = x_c$, $y = y_c$ (*the guiding centre*):

$$x(t) = x_c + (\dot{x}_0/\Omega) \sin \Omega t - (\dot{y}_0/\Omega) \cos \Omega t,$$
$$y(t) = y_c + (\dot{y}_0/\Omega) \sin \Omega t + (\dot{x}_0/\Omega) \cos \Omega t. \tag{2.6}$$

Since the coupled differential equations constitute a fourth order system, there are four free constants. They are fixed by the choice of the initial positions x_0, y_0 and the initial velocities $\dot{x}_0$, $\dot{y}_0$, so that $x_c = x_0 + \dot{y}_0/\Omega$, $y_c = y_0 - \dot{x}_0/\Omega$. Check the property (2.3):

$$v_\perp = \sqrt{\dot{x}^2(t) + \dot{y}^2(t)} = \sqrt{\dot{x}_0^2 + \dot{y}_0^2} = \text{const}. \tag{2.7}$$

Also,

$$\sqrt{[x(t) - x_c]^2 + [y(t) - y_c]^2} = \sqrt{\dot{x}_0^2 + \dot{y}_0^2}/\Omega = v_\perp/\Omega = \text{const},$$

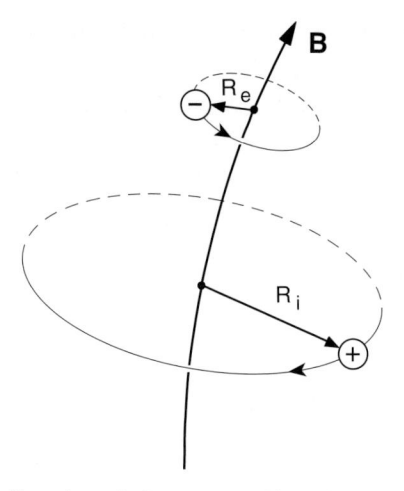

Fig. 2.1. Gyration of electrons and ions in a magnetic field.

so that *the gyro-* or *cyclotron radius*[1]

$$R \equiv \frac{v_\perp}{\Omega} = \text{const}. \tag{2.8}$$

Hence, the complete orbit consists of gyration $\perp$ **B** and inertial motion $\parallel$ **B**:

$$\begin{aligned} x(t) &= x_c + R\cos(\Omega t - \theta_0), \\ y(t) &= y_c - R\sin(\Omega t - \theta_0), \\ z(t) &= z_c(t) = z_0 + v_\parallel t, \end{aligned} \tag{2.9}$$

where θ_0 is the initial value of the polar angle in the perpendicular plane. This helical motion already constitutes an important confining feature of a magnetic field: *charged particles stick to the field lines*. In other words: the magnetic field **B** determines *the geometry of the dynamics* of both kinds of particles and, hence, of the plasma.

Electrons and ions gyrate in opposite directions (Fig. 2.1) with quite different gyro-frequencies and gyro-radii because of the smallness of the mass ratio m_e/m_i:

$$\begin{aligned} \Omega_e &\equiv \frac{eB}{m_e} \gg \Omega_i \equiv \frac{ZeB}{m_i}, \\ R_e &\equiv \frac{v_{\perp,e}}{\Omega_e} \ll R_i \equiv \frac{v_{\perp,i}}{\Omega_i} \quad (\text{assuming } T_e \sim T_i). \end{aligned} \tag{2.10}$$

---

[1] The expressions (2.5) and (2.8) are usually called the Larmor frequency and the Larmor radius. Those are misnomers: the Larmor motion proper refers to the precession of a magnetic dipole in an applied magnetic field, as in the classical theory of nuclear magnetic resonance. This precession frequency is an entirely different physical effect, as evidenced by the fact that its frequency is half the value given by Eq. (2.5).

Inserting a typical value for the magnetic field in tokamaks (see Table B.3), viz. $B = 3\,\text{T}$ (= 30 kgauss), and inserting the values for e, m_e, and m_p of Table B.1, we find for the angular frequencies of protons and electrons

$$\Omega_e = 5.3 \times 10^{11} \text{ rad s}^{-1} \quad \text{(i.e., a frequency of 84 GHz)},$$
$$\Omega_i = 2.9 \times 10^{8} \text{ rad s}^{-1} \quad \text{(i.e., a frequency of 46 MHz)}. \quad (2.11)$$

To estimate the gyro-radii, we consider particles with thermal speed[2] so that $v_\perp = v_{\text{th}} \equiv \sqrt{2kT/m}$. For electrons and protons at $\widetilde{T} = 10\,\text{keV}$, i.e. $T_e = T_i = 1.16 \times 10^8\,\text{K}$, this implies

$$v_{\text{th},e} = 5.9 \times 10^7 \text{ m s}^{-1} \quad \Rightarrow \quad R_e = 1.1 \times 10^{-4} \text{ m} \approx 0.1\,\text{mm},$$
$$v_{\text{th},i} = 1.4 \times 10^6 \text{ m s}^{-1} \quad \Rightarrow \quad R_i = 4.9 \times 10^{-3} \text{ m} \approx 5\,\text{mm}. \quad (2.12)$$

This gives an impression of the time and length scales for gyro-motion in a tokamak. Clearly, these time scales are very small compared to the typical time scales needed for thermonuclear fusion and the length scales are small compared to the macroscopic dimensions of a tokamak machine. This permits averaging over the gyro-motion in the macroscopic description of plasma dynamics.

▷ **Exercise.** Consult Table B.3 for the values of the gyration parameters of a solar plasma. Also complete the empty column of Table B.4: it is important to familiarize yourself with the different orders of magnitude! ◁

One significant feature of the gyro-frequency is its dependence on the value of the magnetic field alone. Except for the fundamental constants of nature, no other quantities appear in the expressions (2.10)(a). Hence, detection of oscillatory motion at the cyclotron frequencies provides an important means for the determination of the value of the magnetic field in plasmas.

Since the equation of motion for the perpendicular motion is a fourth order differential equation (two coupled second order equations), quite complicated drift motions can occur if the electromagnetic fields **B** and **E** are not constant in time or if they are inhomogeneous in space. One then introduces expansions exploiting the smallness of the gyro-radius of the particles as compared to the length scale of the field inhomogeneities. This is the subject of individual *orbit theory*. For many applications, time and length scales of phenomena are large enough to permit averaging over the rapid gyration of the particles so that only the motion of the guiding centre needs to be considered. This is called the *guiding centre approximation*. We return to this subject in Section 2.2.3.

---

[2] The convention for the definition of the thermal speed v_{th} differs for different authors: some have the factor 2, like we do, others do not have it.

2.2.2 Excursion: basic equations of electrodynamics and mechanics

We have treated the simple non-relativistic motion of a charged particle in a constant, prescribed, magnetic field. Before we proceed to motion in more complex fields, it is useful to recall some of the basic equations from electrodynamics and particle mechanics.

In plasma theory, where we consider the dynamics of a collection of charged particles in electromagnetic fields, the appropriate form of *Maxwell's equations* in mksA units is given by:

$$\begin{cases} \nabla \times \mathbf{E} = -\dfrac{\partial \mathbf{B}}{\partial t} & \text{(Faraday)}, \\[6pt] \nabla \times \mathbf{B} = \mu_0 \mathbf{j} + \dfrac{1}{c^2} \dfrac{\partial \mathbf{E}}{\partial t} & \text{('Ampère')}, \qquad c^2 = (\epsilon_0 \mu_0)^{-1}, \\[6pt] \nabla \cdot \mathbf{E} = \dfrac{\tau}{\epsilon_0} & \text{(Poisson)}, \\[6pt] \nabla \cdot \mathbf{B} = 0 & \text{(no magnetic monopoles)}. \end{cases} \quad (2.13)$$

We have ignored polarization and magnetization effects, i.e. $\epsilon = \epsilon_0$ and $\mu = \mu_0$ so that $\mathbf{D} = \epsilon_0 \mathbf{E}$ and $\mathbf{H} = (\mu_0)^{-1} \mathbf{B}$, since these effects are absorbed in the definitions of charge and current density:

$$\begin{cases} \tau = \sum_\alpha q_\alpha n_\alpha \\ \mathbf{j} = \sum_\alpha q_\alpha n_\alpha \mathbf{u}_\alpha \end{cases} \quad (\alpha = e, i). \quad (2.14)$$

Here, n_α and $\mathbf{u}_\alpha$ are the particle density and the macroscopic velocity of particles of type α. This implies that *the plasma is viewed as existing of point charges moving in the electromagnetic fields, which they partially create themselves*. Obviously, adding an equation of motion of the form of Eq. (2.1) for every particle of the plasma would constitute a complete dynamical problem, but it would be foolish to proceed from that point of view since one would have to solve, say, 10^{20} equations of motion. In Section 2.3, we will see how this problem may be reformulated by means of a statistical approach.

In Section 2.4, we will introduce the next level of description, viz. the macroscopic approach of plasmas. For the majority of macroscopic plasma phenomena, the displacement current $\epsilon_0 \partial \mathbf{E}/\partial t$ and Poisson's equation are unimportant and may be dropped from Eqs. (2.13). In that case, Faraday's law expresses the dynamics of the magnetic field, 'Ampère's' law will become Ampère's law again, expressing the relation between the current and the magnetic field, and $\nabla \cdot \mathbf{B} = 0$ is then the usual restriction of the possible initial conditions for $\mathbf{B}$. In this description in

2.2 Single particle motion

terms of the so-called *pre-Maxwell's equations*, the magnetic field is a more basic quantity than the electric field.

Of course, to get *electromagnetic waves* the displacement current and Poisson's equation are essential. Keeping these terms, for a *vacuum* ($\tau = 0$, $\mathbf{j} = 0$) we get from the original set of Maxwell's equations (2.13) two identical wave equations for $\mathbf{E}$ and $\mathbf{B}$:

$$\nabla^2 \mathbf{E} - \frac{1}{c^2}\frac{\partial^2 \mathbf{E}}{\partial t^2} = 0, \qquad \nabla^2 \mathbf{B} - \frac{1}{c^2}\frac{\partial^2 \mathbf{B}}{\partial t^2} = 0, \tag{2.15}$$

where $c = (\epsilon_0 \mu_0)^{-\frac{1}{2}}$ is the velocity of light in vacuum. Considering plane wave solutions

$$\mathbf{E}(\mathbf{r},t) = \hat{\mathbf{E}}\, e^{i(\mathbf{k}\cdot\mathbf{r}-\omega t)}, \qquad \mathbf{B}(\mathbf{r},t) = \hat{\mathbf{B}}\, e^{i(\mathbf{k}\cdot\mathbf{r}-\omega t)}, \tag{2.16}$$

where $\hat{\mathbf{E}}$ and $\hat{\mathbf{B}}$ are complex (the physical quantities correspond to the real parts of $\mathbf{E}$ and $\mathbf{B}$), we just obtain the familiar relation between angular frequency ω and wave vector $\mathbf{k}$:

$$\omega^2 = k^2 c^2. \tag{2.17}$$

Inserting these relations back into Eqs. (2.13) we find

$$\omega\hat{\mathbf{B}} = \mathbf{k} \times \hat{\mathbf{E}}, \qquad \omega\hat{\mathbf{E}} = -c^2\,\mathbf{k} \times \hat{\mathbf{B}},$$
$$\mathbf{k}\cdot\hat{\mathbf{B}} = 0, \qquad \mathbf{k}\cdot\hat{\mathbf{E}} = 0, \tag{2.18}$$

so that $|\hat{\mathbf{E}}| = c|\hat{\mathbf{B}}|$ and the vectors $\{\hat{\mathbf{E}}, \hat{\mathbf{B}}, \mathbf{k}\}$ form an orthogonal triad: electromagnetic waves are *transverse waves* where $\mathbf{E}$ and $\mathbf{B}$ oscillate in a plane perpendicular to the direction of propagation given by $\mathbf{k}$.

For the sake of reference, we present *the Lorentz transformation* of two inertial frames [117] moving with relative velocity $\mathbf{v}$:

$$\begin{cases} \mathbf{x}' = \mathbf{x} + \dfrac{\gamma - 1}{v^2}\mathbf{v}\mathbf{v}\cdot\mathbf{x} - \gamma\mathbf{v}t, \\ t' = \gamma\left(t - \dfrac{1}{c^2}\mathbf{v}\cdot\mathbf{x}\right), \qquad \gamma \equiv \dfrac{1}{\sqrt{1 - v^2/c^2}}. \end{cases} \tag{2.19}$$

The electromagnetic fields transform according to

$$\begin{cases} \mathbf{E}' = \gamma\left(\mathbf{E} + \mathbf{v}\times\mathbf{B}\right) - \dfrac{\gamma^2}{(\gamma+1)c^2}\mathbf{v}\mathbf{v}\cdot\mathbf{E}, \\ \mathbf{B}' = \gamma\left(\mathbf{B} - \dfrac{1}{c^2}\mathbf{v}\times\mathbf{E}\right) - \dfrac{\gamma^2}{(\gamma+1)c^2}\mathbf{v}\mathbf{v}\cdot\mathbf{B}. \end{cases} \tag{2.20}$$

Notice that the symmetry between **E** and **B** is lost here since we exploit mksA units.

A consistent treatment of particles in electromagnetic waves would require the replacement of the classical equation of motion (2.1) by the relativistic one:

$$\frac{d\mathbf{p}}{dt} = q(\mathbf{E} + \mathbf{v} \times \mathbf{B}), \quad (2.21)$$

where **p** is the relativistic momentum of the particle. *The relativistic expressions for the energy E* (not to be confused here with the electric field **E**) *and the momentum* **p** *of particles moving with the velocity* **v** *are given by:*

$$E = \gamma mc^2 \quad (\approx mc^2 + \tfrac{1}{2}mv^2 \text{ for } v \ll c),$$
$$\mathbf{p} = \gamma m\mathbf{v} \quad (\approx m\mathbf{v} \text{ for } v \ll c). \quad (2.22)$$

Hence, $E = \sqrt{p^2c^2 + m^2c^4}$, where E includes the rest mass energy mc^2. For photons, the rest mass $m \to 0$ and $v \to c$, so that the expressions (2.22) for E and **p** become undetermined but their relation is still given by $E = pc$.

Quantum mechanical expressions for the energy and momentum of particles, including photons, involve Planck's constant h:

$$E = h\nu, \qquad p = h/\lambda,$$

where ν and λ are the de Broglie frequency and wavelength of the wave functions associated with the particles. It is convenient to include the direction of the momentum vector **p** in these relations. This is done by introducing the wave vector **k**, where $|\mathbf{k}| \equiv 2\pi/\lambda$, and defining the angular frequency $\omega \equiv 2\pi\nu$ and $\hbar \equiv h/2\pi$:

$$E = \hbar\omega, \qquad \mathbf{p} = \hbar\mathbf{k}. \quad (2.23)$$

These expressions are also valid for photons, for which frequency and wavelength are related by

$$\nu = c/\lambda, \quad \text{or} \quad \omega = kc, \quad (2.24)$$

according to the dispersion equation (2.17), so that we recover the relation $E = pc$ again.

In this book, quantum mechanical and relativistic effects will not play an important role. It is of some interest though to extend the analysis of Section 2.2.1 to the *cyclotron motion of relativistic particles* since high energy non-thermal electrons frequently occur in plasmas (e.g., 60 keV electrons require a relativistic correction $v^2/c^2 \sim 0.2$). Consider again a charged particle moving in a constant magnetic field **B**, but now exploit Eq. (2.21) rather than Eq. (2.1):

$$\frac{d\mathbf{p}}{dt} = \frac{|q|}{\gamma m} \mathbf{p} \times \mathbf{B}. \tag{2.25}$$

Performing the operations of projection onto **B** and **p** again, we find that $p_\parallel =$ const and $|\mathbf{p}| = $ const so that, according to Eq. (2.22), $v = $ const and the relativistic factor $\gamma = $ const. Hence, particles turn out to gyrate around the magnetic field with the modified gyro-frequency

$$\Omega = \frac{|q|B}{\gamma m}, \tag{2.26}$$

whereas the corresponding gyro-radius becomes

$$R = \frac{p_\perp}{|q|B} = \frac{v_\perp}{\Omega}. \tag{2.27}$$

The ratio $p_\perp/q$ ($= RB$), which depends on the particle properties only, is a measure of how little a particle can be deflected by a magnetic field. It is called *the magnetic rigidity* of a particle. Energies of cosmic ray particles are often expressed in terms of this quantity.

After this excursion, we *return to classical plasmas* in the double sense of non-relativistic (i.e. neglect of v/c terms and displacement current in 'Ampère's' equation) and non-quantum mechanical (i.e. neglect of the discreteness of the energy expressed by Eq. (2.23)).

2.2.3 Drifts, adiabatic invariants

In Section 2.2.1, we considered the gyro-motion of charged particles in a constant magnetic field **B**. Let us now add a constant electric field **E** to the problem. Projecting Eq. (2.1) on **B** then gives

$$m\frac{dv_\parallel}{dt} = qE_\parallel, \tag{2.28}$$

which represents a constant acceleration along the magnetic field. In hot plasmas, such accelerations may lead to a high energy tail of so-called *runaway electrons* when the electric field exceeds a certain critical value which depends on the electron–ion collision frequency.

For the present purpose, it is more relevant to consider the effect of a perpendicular electric field: $\mathbf{E}$ ($= E\mathbf{e}_y$) $\perp \mathbf{B}$ ($= B\mathbf{e}_z$). The transverse motion is then described by the differential equations

$$\begin{aligned} \ddot{x} - \frac{qB}{m}\dot{y} &= 0, \\ \ddot{y} + \frac{qB}{m}(\dot{x} - E/B) &= 0, \end{aligned} \tag{2.29}$$

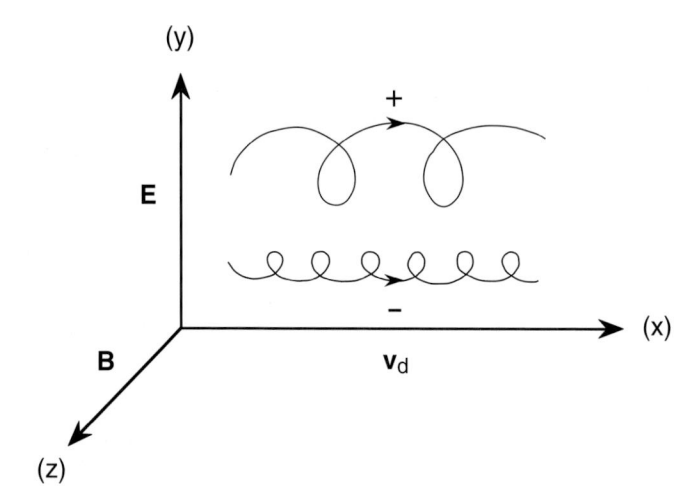

Fig. 2.2. Drift of ions and electrons in crossed electric and magnetic fields.

which just differ from the previous ones (2.4) by a transformation $\dot{x} \to \dot{x} - E/B$. Hence, the motion is virtually the same as before, except that it is superposed on a constant drift in the x-direction. This drift, which is called the $\mathbf{E} \times \mathbf{B}$ *drift*, may be written as

$$\mathbf{v}_d = \frac{\mathbf{E} \times \mathbf{B}}{B^2}. \tag{2.30}$$

Notice that it is independent of the charge so that electrons and ions both drift in the same direction. The reason is the periodic increase and decrease of the perpendicular velocity due to the acceleration and deceleration of the charge as it moves in the electric field (Fig. 2.2). This results in a periodic change of the size of the gyro-radius. Electrons are decelerated and ions are accelerated but, since the two orbits have an opposite sense of circulation, the net effect is a drift of ions and electrons in the same direction.

The $\mathbf{E} \times \mathbf{B}$ drift (2.30) permits a more general interpretation. Consider an inertial frame moving with a velocity $\mathbf{v}$ in the direction of $\mathbf{E} \times \mathbf{B}$ (i.e., in the x-direction). According to Eq. (2.20), since $\mathbf{E}$, $\mathbf{B}$, and $\mathbf{v}$ are mutually orthogonal, the fields in the moving frame are given by the Lorentz transformation $\mathbf{E}' = \gamma(\mathbf{E} + \mathbf{v} \times \mathbf{B})$, $\mathbf{B}' = \gamma(\mathbf{B} - c^{-2}\mathbf{v} \times \mathbf{E})$. Choosing $\mathbf{v} = \mathbf{v}_d$ according to Eq. (2.30), we get $\mathbf{E} + \mathbf{v} \times \mathbf{B} = \mathbf{E} + (\mathbf{E} \times \mathbf{B}) \times \mathbf{B}/B^2 = 0$, so that $\mathbf{E}' = 0$: the particles move in such a way that the electric field in the moving frame vanishes! This is in precise agreement with the motion of a perfectly conducting plasma consisting of a huge number of oppositely charged particles: $\mathbf{E} + \mathbf{v} \times \mathbf{B} = 0$ is one of the fundamental equations for such a plasma (see Section 2.4.1). Note that the argument does not

2.2 Single particle motion

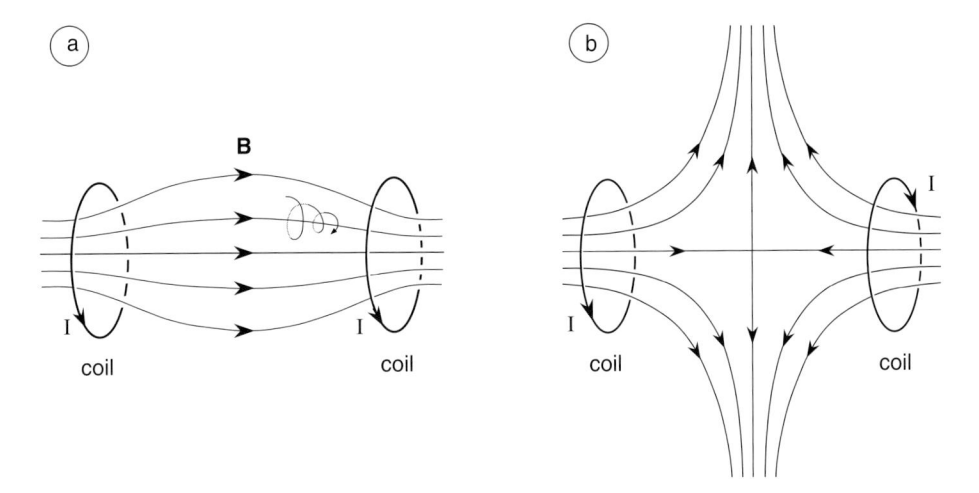

Fig. 2.3. (a) Mirror and (b) cusp magnetic confinement schemes.

require relativistic velocities: whereas $\mathbf{B}' \approx \mathbf{B}$ for $v \ll c$, the electric field $\mathbf{E}'$ in the moving frame is certainly very different from $\mathbf{E}$ for $v \ll c$.

If we replace the electric force $q\mathbf{E}$ by some other force $\mathbf{F}$, e.g. the gravitational force $m\mathbf{g}$, we find a similar expression for the resulting drift:

$$\mathbf{v}_d = \frac{\mathbf{F} \times \mathbf{B}}{qB^2} \,. \tag{2.31}$$

If the force $\mathbf{F}$ is charge-independent, like gravity, the drift itself becomes charge-dependent so that electrons and ions drift in opposite directions (neglecting interactions between the particles). This implies the flow of an electric current.

In inhomogeneous magnetic fields many more drifts occur, e.g. one due to the gradient of $\mathbf{B}$. This can again be understood, like the $\mathbf{E} \times \mathbf{B}$ drift, as caused by the periodic variation of the size of the gyro-radius. This so-called $\mathbf{B} \times \nabla \mathbf{B}$ drift is again charge-dependent, so that it is associated with a current flow. Another drift is due to the curvature of the magnetic field lines, resulting in a centrifugal force for particles moving along the field lines. According to Eq. (2.31) this force gives rise to an additional drift velocity.

An important application of orbit theory is the *mirror effect*: particles entering the regions of higher magnetic field strength (created by so-called magnetic mirrors, see Fig. 2.3(a)) are reflected back into the region of smaller magnetic field strength where the gyro-radius is larger and the perpendicular velocity smaller. A proper treatment requires the consideration of the magnetic moment of the gyro-motion, which may be shown to be an adiabatic invariant (see below). Like many other confinement schemes considered in research on controlled thermonuclear reactions, the magnetic mirror (also called magnetic bottle) has been investigated extensively and subsequently abandoned as a candidate for fusion reactors. However, it remains an extremely important concept in plasma dynamics since all magnetic confinement schemes involve highly inhomogeneous magnetic fields with associated *trapping of particles* in between the maxima of the field. Also, the magnetic mirror concept plays a prominent role in space and plasma astrophysics. Examples are the Van Allen belts in the magnetosphere of the Earth and acceleration mechanisms for cosmic rays.

A definite disadvantage of the mirror confinement scheme is the curvature of the field lines, which is convex with respect to the confined plasma and, hence, subject to *interchange instability*. It was realized early in controlled fusion research that this instability is eliminated in the cusp confinement scheme (Fig. 2.3(b)) where the magnetic field geometry consists of two mirrors connected with a cusped structure which is produced by simply reversing the direction of the current in one of the coils. The plasma is now stable up to very high values of $\beta \equiv 2\mu_0 p/B^2$ (an important parameter in plasma confinement, which we will encounter extensively later on) since the field lines are concave with respect to the plasma. However, confinement of particles has even more severe limitations than in the mirror scheme.

Periodic motion in inhomogeneous magnetic fields calls for a more systematic treatment exploiting the just-mentioned *adiabatic invariants*. This is based on the notion that there are distinct spatial scales in the problem, viz. one scale associated with the gyro-motion and another, much larger, scale associated with the field inhomogeneities. Action variables $J \equiv \oint P \, dQ$ (Goldstein [91]) are exploited, where P is the generalized momentum conjugate to a periodic coordinate Q. For non-relativistic particles in an electromagnetic field the generalized momentum is defined by $\mathbf{P} \equiv m\mathbf{v} + q\mathbf{A}$ (Jackson [117]), where $\mathbf{A}$ is the vector potential corresponding to the magnetic field $\mathbf{B} = \nabla \times \mathbf{A}$.

We now consider the gyro-motion in local cylindrical coordinates r, θ, z about the magnetic field, with $\mathbf{B} = B \, \mathbf{e}_z$ and $\mathbf{A} = \frac{1}{2} Br \, \mathbf{e}_\theta$, where the particle moves on an orbit $r = R = v_\perp/\Omega$ in the direction of *decreasing* angle θ (as the ion in Fig. 2.1). In the first adiabatic invariant, the transverse momentum corresponding to the gyro-motion enters:

$$J_1 \equiv \oint \mathbf{P}_\perp \cdot d\mathbf{l} = \oint (mv_\perp - \tfrac{1}{2} q B R) R \, d\theta = \frac{\pi m v_\perp^2}{\Omega}. \tag{2.32}$$

This invariant may be expressed in terms of *the magnetic moment* $\mu \equiv \pi R^2 I$ of the gyro-motion, where $I = (2\pi)^{-1} q\Omega$ is the current due to the circulating charge. Inserting the expressions (2.5) and (2.8) for Ω and R gives the required result:

$$\mu \equiv \pi R^2 I = \tfrac{1}{2} q R^2 \Omega = \frac{\tfrac{1}{2} m v_\perp^2}{B} \quad \Rightarrow \quad J_1 = \frac{2\pi m}{q} \mu. \tag{2.33}$$

2.2 Single particle motion

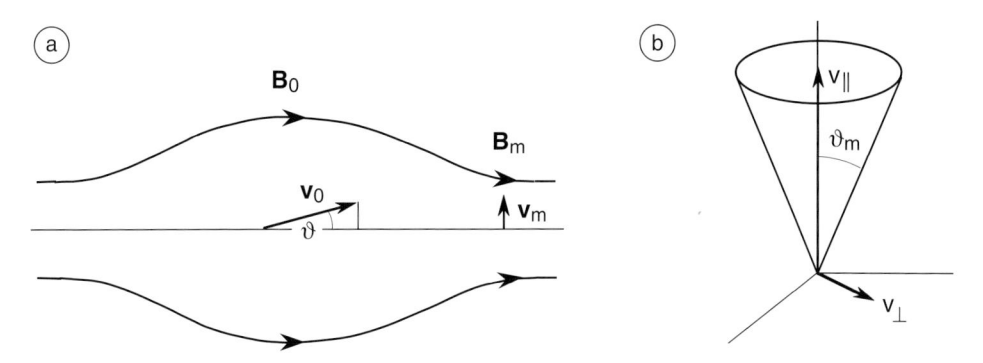

Fig. 2.4. (a) Magnetic mirror geometry and (b) loss cone in velocity space.

According to convention, the corresponding vector of the magnetic moment, μ, is defined as pointing opposite to **B**. The first adiabatic invariant may also be expressed in terms of the magnetic flux Ψ_R enclosed by the gyro-orbit: $\Psi_R = \pi R^2 B = (2\pi m/q^2)\mu$, so that $J_1 = q\Psi_R$. The adiabatic invariance implies that J_1 is constant when the external parameters only vary slowly. Consequently, for rapid gyro-motion in a slowly varying magnetic field, the magnetic moment μ and the contained magnetic flux Ψ_R are also constant.

The use of such adiabatic invariants can be demonstrated with the motion in a mirror field (Fig. 2.4). Because $\mu = $ const, if the particle moves into the mirror, $v_\perp$ must increase since B increases. Because of energy conservation, this can happen only if $v_\parallel$ decreases. As a result, particles are reflected by the magnetic mirror. (Hence, its name.) Clearly, for particles with a very small value of $v_\perp/v_\parallel$, the mirror does not work since these particles are lost along the axis. This loss is determined by *the mirror ratio* B_m/B_0, where the subscript m refers to the mirror throat, where B has a maximum, and the subscript 0 refers to the mid-plane, where B has a minimum. One can express the pitch angle of *the loss cone* in velocity space (Fig. 2.4(b)) in terms of this mirror ratio by exploiting the constancy of μ,

$$v_{\perp,0}^2/B_0 = v_{\perp,m}^2/B_m, \qquad (2.34)$$

and energy conservation for particles with velocities at the transition from trapped to untrapped ($v_{\parallel,m} = 0$),

$$v_{\parallel,0}^2 + v_{\perp,0}^2 = v_{\perp,m}^2, \qquad (2.35)$$

so that

$$\left(1 + v_{\parallel,0}^2/v_{\perp,0}^2\right) B_0 = B_m .$$

The pitch angle is defined by $\vartheta \equiv \arctan(v_\perp/v_\parallel)$. Hence, particles with large enough parallel velocity $v_\parallel$, such that

$$\vartheta < \vartheta_m \equiv \arctan\sqrt{\frac{B_0}{B_m - B_0}}, \qquad (2.36)$$

are lost.

The particles outside the loss cone are trapped. They bounce back and forth between the mirrors. With this motion a second adiabatic invariant is associated, viz. the *longitudinal invariant*

$$J_2 \equiv \oint P_\parallel \, dl \approx \oint m v_z \, dz = \oint m v_z^2 \, dt = \frac{\pi m \hat{v}_z^2}{\omega_b}, \qquad (2.37)$$

where the motion along the symmetry axis of the configuration (the z-direction) has been assumed to be a harmonic oscillation with velocity $v_z = \hat{v}_z \cos \omega_b t$, where ω_b is the bounce frequency.

A *third adiabatic invariant* is associated with the slow drift of the guiding centres of the particles across the field lines. This drift is caused by curvature and gradients of the magnetic field, with the associated variation of the size of the gyro-radius. For example, electrons gyrating about magnetic field lines of the Earth's dipole drift eastward and ions drift westward, creating a huge *ring current* system around the Earth. The guiding centres drift across the magnetic field lines while they stay on the 'magnetic surface' (called drift shell) mapped out by those field lines. Accordingly, the third adiabatic invariant is characterized by the value of the magnetic flux Ψ_d enclosed by the surface to which the drift is confined. This flux may be calculated by exploiting another cylindrical coordinate system r, ϕ, z with the magnetic field in the r, z-plane and ϕ the ignorable azimuthal coordinate in the direction of the drift (see Fig. 2.5). For the example of the Earth's dipole field, it is easiest to evaluate the flux at the equatorial plane. From $\mathbf{B} = \nabla \times \mathbf{A}$, we obtain $B_z = (1/r)\partial(rA_\phi)/\partial r$, so that the flux enclosed between the drift shell and a reference shell through $r = r_0, z = 0$, where we chose $A_\phi(r_0, 0) = 0$, may be written as

$$\Psi_d = 2\pi \int_{r_0}^{r} B_z r \, dr = 2\pi r A_\phi(r, 0).$$

Hence, the expression for the third adiabatic invariant becomes

$$J_3 \equiv \oint P_\phi \, r \, d\phi = \oint (m v_d + q A_\phi) \, r \, d\phi \approx 2\pi q \, r A_\phi = q \, \Psi_d. \qquad (2.38)$$

Here, the term with the drift velocity has been neglected since it is much smaller than the contribution of the vector potential A_ϕ (see Hasegawa and Sato [106], Chapter 1).

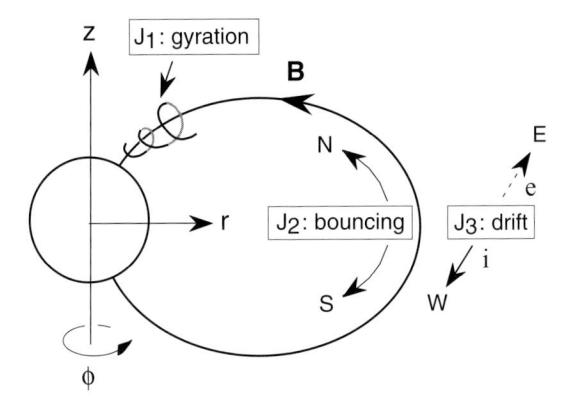

Fig. 2.5. Adiabatic invariants for particles in the magnetosphere.

Summarizing the use of adiabatic invariants describing the motion of charged particles in an inhomogeneous magnetic field, e.g. that of the Earth, associated with the Van Allen belts (Fig. 2.5):

(a) electrons and ions execute a fast gyration in opposite directions about the magnetic field lines conserving the first adiabatic invariant J_1, i.e. the magnetic moment of the guiding centres;
(b) they bounce back and forth between the mirrors on the northern and southern hemispheres on a slower time scale, conserving J_2;
(c) they drift on a slower time scale yet in opposite longitudinal directions conserving the third adiabatic invariant J_3, i.e. the magnetic flux inside the drift shell.

Obviously, the three adiabatic invariants are conserved in decreasing order of robustness. The fluctuating interaction of the solar wind with the magnetosphere will not invalidate the assumptions underlying the adiabatic invariance of J_1, since this invariant concerns very fast motion, but it may easily invalidate the invariance of the third adiabatic invariant J_3.

Thus, a very effective description of charged particle motion in inhomogeneous magnetic fields has been *sketched*. The precise formulation of the conditions of validity and *proofs* of adiabatic invariance would be another matter. This would lead from the early seminal work by Northrop [169] and by Kruskal to modern developments in Hamiltonian mechanics (see, e.g., Goldston and Rutherford [92], Chapter 4, and Balescu [14], Chapter 1).

2.3 Kinetic plasma theory

The motion of a *single* (non-relativistic) charged particle in electric and magnetic fields is described by the equation of motion (2.1). Single particle orbit theory is only valid when the density of charged particles is so low that the interactions between the particles can be ignored. However, a plasma consists of a very large number of *interacting* particles and, hence, it is appropriate to use a *statistical approach* for its analysis. It is the task of *kinetic plasma theory* to derive equations describing the *collective behaviour* of the many charged particles that constitute a plasma by applying the methods of statistical mechanics. One should be aware of the formidable amount of theoretical analysis involved in even a partial performance of this task. (See, e.g. the basic papers by Trubnikov [231] and Braginskii [41] in the first volume of the excellent series 'Reviews of Plasma Physics' for early contributions, and the more recent comprehensive treatise by Balescu [14].) Here, we exploit one of the end results of this program, viz. *the Boltzmann equation*, which may be derived by heuristic arguments as long as no specific expression for the collision term is needed. Single particle orbit theory (Section 2.2) ignores collective effects, and the fluid description of plasmas (Section 2.4) averages out microscopic fluctuations. Kinetic theory includes these important aspects of plasma dynamics and it is, therefore, more comprehensive than both orbit theory and the fluid description of plasmas, but it is also much more complicated.

First, we just give an introduction to some of the basic kinetic concepts (Section 2.3.1). A more detailed exposition is relegated to Chapter 3, which may be consulted for details on the derivations of the equations. Next, we consider a simple example of collective behaviour (Section 2.3.2), viz. electron plasma oscillations, associated with a fundamental plasma parameter, *the plasma frequency*. Section 2.3.3 deals with the damping of these oscillations through kinetic effects.

2.3.1 Boltzmann equation and moment reduction

Consider a plasma consisting of electrons and one kind of ion. In the statistical description, the information on the individuality of the particles is lost but the relevant physical information on the plasma as a whole is retained and expressed in terms of time-dependent *distribution functions* $f_\alpha(\mathbf{r}, \mathbf{v}, t)$ for the electrons and ions ($\alpha = e, i$). These are defined as the density of the representative points of particles of type α in a six-dimensional *phase space* formed by the three position coordinates (x, y, z) and the three velocity coordinates (v_x, v_y, v_z) (see, e.g., Bittencourt [31]). The probable number of particles of type α in the six-dimensional volume element $d^3r\, d^3v$ centred at $(\mathbf{r}, \mathbf{v})$ is then given by $f_\alpha(\mathbf{r}, \mathbf{v}, t)\, d^3r\, d^3v$. The total number of particles, $N_\alpha \equiv \iint f_\alpha d^3r\, d^3v$, will be assumed constant. Clearly, to describe more general plasmas like thermonuclear ones, where fusion reactions create and annihilate particles, more than two distribution functions are needed and the respective total number of particles will not be constant.

2.3 Kinetic plasma theory

The motion of the swarm of representative points in phase space is described by the total time derivative of the distribution function $f_\alpha(\mathbf{r}, \mathbf{v}, t)$:

$$\frac{df_\alpha}{dt} \equiv \frac{\partial f_\alpha}{\partial t} + \frac{\partial f_\alpha}{\partial \mathbf{r}} \cdot \frac{d\mathbf{r}}{dt} + \frac{\partial f_\alpha}{\partial \mathbf{v}} \cdot \frac{d\mathbf{v}}{dt}$$
$$= \frac{\partial f_\alpha}{\partial t} + \mathbf{v} \cdot \frac{\partial f_\alpha}{\partial \mathbf{r}} + \frac{q_\alpha}{m_\alpha}(\mathbf{E} + \mathbf{v} \times \mathbf{B}) \cdot \frac{\partial f_\alpha}{\partial \mathbf{v}}, \quad (2.39)$$

where the expression (2.1) for the acceleration $d\mathbf{v}/dt$ of the particles has been inserted in the second line. The notation of inner products involving derivatives with respect to the vectors $\mathbf{r}$ and $\mathbf{v}$ just indicates that the sum over the products of the three vector components is to be taken: $\mathbf{v} \cdot \partial/\partial \mathbf{r} \equiv v_x \partial/\partial x + v_y \partial/\partial y + v_z \partial/\partial z$, and similarly for the term with $\partial/\partial \mathbf{v}$. Also note a subtle, but important, difference in notation: d/dt for the total time derivative and d/dt for ordinary time derivatives. In the absence of binary interactions between particles, the density of representative points in phase space remains constant in time so that $df_\alpha/dt = 0$ (Liouville's theorem, Goldstein [91]).

Of course, the interesting part of kinetic theory comes with the introduction of interactions or, rather, collisions between the particles. The variation in time of the distribution functions of both electrons and ions is then found from a kinetic equation, known as *the Boltzmann equation*:

$$\frac{\partial f_\alpha}{\partial t} + \mathbf{v} \cdot \frac{\partial f_\alpha}{\partial \mathbf{r}} + \frac{q_\alpha}{m_\alpha}(\mathbf{E} + \mathbf{v} \times \mathbf{B}) \cdot \frac{\partial f_\alpha}{\partial \mathbf{v}} = C_\alpha \equiv \left(\frac{\partial f_\alpha}{\partial t}\right)_{\text{coll}}. \quad (2.40)$$

Now, $\mathbf{E}(\mathbf{r}, t)$ and $\mathbf{B}(\mathbf{r}, t)$ consist of the contributions of the external fields *and* of the averaged internal fields originating from the long-range inter-particle interactions. The symbolic expression on the RHS of Eq. (2.40) represents the rate of change of the distribution function due to the short-range inter-particle interactions, which are somewhat arbitrarily called *collisions*. In a plasma, these may be considered as the cumulative effect of many small-angle velocity changes effectively resulting in large-angle scattering, described by a Fokker–Planck type collision operator. It is the first objective of kinetic theory to justify the distinction between long-range interactions and binary collisions, to determine the ranges of its validity, and to derive suitable expressions for the collision term. One such expression, discussed in Section 3.2, is the Landau collision integral (1936) [135]. On the other hand, neglect of the collisions leads to the *Vlasov equation* (1938) [239]:

$$\frac{\partial f_\alpha}{\partial t} + \mathbf{v} \cdot \frac{\partial f_\alpha}{\partial \mathbf{r}} + \frac{q_\alpha}{m_\alpha}(\mathbf{E} + \mathbf{v} \times \mathbf{B}) \cdot \frac{\partial f_\alpha}{\partial \mathbf{v}} = 0, \quad (2.41)$$

where it is to be realized that the particles still interact through the long-range interactions represented by the averaged internal parts of the $\mathbf{E}$ and $\mathbf{B}$ fields.

A closed system of equations is now obtained by combining either the Boltzmann equation (2.40) or the Vlasov equation (2.41), determining the distribution functions $f_\alpha(\mathbf{r}, \mathbf{v}, t)$, with Maxwell's equations (2.13), determining the electric and magnetic fields $\mathbf{E}(\mathbf{r}, t)$ and $\mathbf{B}(\mathbf{r}, t)$, and the expressions (2.14) for the charge and current density source terms $\tau(\mathbf{r}, t)$ and $\mathbf{j}(\mathbf{r}, t)$. The latter are related to the particle densities and the average velocities:

$$n_\alpha(\mathbf{r}, t) \equiv \int f_\alpha(\mathbf{r}, \mathbf{v}, t)\, d^3v, \qquad \tau(\mathbf{r}, t) \equiv \sum q_\alpha n_\alpha, \quad (2.42)$$

$$\mathbf{u}_\alpha(\mathbf{r}, t) \equiv \frac{1}{n_\alpha(\mathbf{r}, t)} \int \mathbf{v} f_\alpha(\mathbf{r}, \mathbf{v}, t)\, d^3v, \qquad \mathbf{j}(\mathbf{r}, t) \equiv \sum q_\alpha n_\alpha \mathbf{u}_\alpha. \quad (2.43)$$

This completes the microscopic equations.

A systematic procedure to obtain macroscopic equations, not involving details of velocity space any more, is to expand in *a finite number of moments of the Boltzmann equation* (2.40), obtained by first multiplying the expressions with powers of $\mathbf{v}$ and then integrating over velocity space:

$$\int d^3v \cdots, \quad \int d^3v\, \mathbf{v} \cdots, \quad \int d^3v\, v^2 \cdots \bigg|_{\text{truncate}}. \quad (2.44)$$

This, in turn, involves the moments of the distribution function itself, like the zeroth moment associated with the particle density $n_\alpha(\mathbf{r}, t)$ and the first moment associated with the average velocity $\langle \mathbf{v} \rangle_\alpha \equiv \mathbf{u}_\alpha(\mathbf{r}, t)$, just defined. In order for this expansion to be practical, it needs to be truncated at a very limited number of terms, like the five (one scalar + one vector + one scalar) indicated in Eq. (2.44). The justification of this truncation is part of transport theory, to be discussed below and more extensively in Chapter 3. In general, macroscopic variables $\langle g \rangle_\alpha(\mathbf{r}, t)$ will appear as the average of some phase space function $g(\mathbf{r}, \mathbf{v}, t)$ over velocity space:

$$\langle g \rangle_\alpha(\mathbf{r}, t) \equiv \frac{1}{n_\alpha(\mathbf{r}, t)} \int g(\mathbf{r}, \mathbf{v}, t)\, f_\alpha(\mathbf{r}, \mathbf{v}, t)\, d^3v. \quad (2.45)$$

This definition obviously requires that the distribution functions f_α fall off rapidly enough for $v \to \infty$ to yield a finite answer.

The different moments of the collision term in the RHS of the Boltzmann equation should also be determined. Without specifying the particular form of the collision operator, important conclusions can be drawn from general physical principles. To that end, the collision term on the RHS of Eq. (2.40) is decomposed in contributions $C_{\alpha\beta}$ due to collisions of particles α (e.g. electrons) with particles β (i.e. electrons as well as ions):

$$C_\alpha = \sum_\beta C_{\alpha\beta}. \quad (2.46)$$

2.3 Kinetic plasma theory

In the absence of fusion reactions, the total number of particles α at a certain position does not change by collisions with particles β, so that

$$\int C_{\alpha\beta}\, d^3v = 0. \tag{2.47}$$

Similarly, momentum and energy conservation lead to corresponding expressions. Details of these and other manipulations are worked out in Chapter 3. Here, we only present the derivation of the lowest moment equation (describing mass conservation) to give an impression of the procedure.

The *zeroth moment* of Eq. (2.40), obtained by integrating over velocity space, yields the following terms:

$$\int \frac{\partial f_\alpha}{\partial t} d^3v = \frac{\partial n_\alpha}{\partial t} \qquad \text{(definition (2.42))},$$

$$\int \mathbf{v} \cdot \frac{\partial f_\alpha}{\partial \mathbf{r}} d^3v = \nabla \cdot (n_\alpha \mathbf{u}_\alpha) \qquad \text{(definition (2.43))},$$

$$\int \frac{q_\alpha}{m_\alpha} (\mathbf{E} + \mathbf{v} \times \mathbf{B}) \cdot \frac{\partial f_\alpha}{\partial \mathbf{v}} d^3v = 0 \qquad \text{(integrating by parts)},$$

$$\int C_\alpha\, d^3v = 0 \qquad \text{(summing Eq. (2.47))}.$$

Adding these expressions gives *the continuity equation* for particles of species α:

$$\frac{\partial n_\alpha}{\partial t} + \nabla \cdot (n_\alpha \mathbf{u}_\alpha) = 0. \tag{2.48}$$

In the same vein, the *first moment* of Eq. (2.40), obtained by multiplying with $m_\alpha \mathbf{v}$ and integrating over the velocities, yields *the momentum equation*:

$$\frac{\partial}{\partial t}(n_\alpha m_\alpha \mathbf{u}_\alpha) + \nabla \cdot \left(n_\alpha m_\alpha \langle \mathbf{vv}\rangle_\alpha\right) - q_\alpha n_\alpha (\mathbf{E} + \mathbf{u}_\alpha \times \mathbf{B}) = \int C_{\alpha\beta}\, m_\alpha \mathbf{v}\, d^3v. \tag{2.49}$$

Finally, *the scalar second moment* of Eq. (2.40), obtained by multiplying with $\frac{1}{2} m_\alpha v^2$ and integrating over velocity space, yields *the energy equation*:

$$\frac{\partial}{\partial t}\left(n_\alpha \tfrac{1}{2} m_\alpha \langle v^2\rangle_\alpha\right) + \nabla \cdot \left(n_\alpha \tfrac{1}{2} m_\alpha \langle v^2 \mathbf{v}\rangle_\alpha\right) - q_\alpha n_\alpha \mathbf{E} \cdot \mathbf{u}_\alpha = \int C_{\alpha\beta} \tfrac{1}{2} m_\alpha v^2\, d^3v. \tag{2.50}$$

See Section 3.2.2 for the explicit steps in the derivation of these equations.

Clearly, this sequence can be continued indefinitely, but Eqs. (2.48)–(2.50) can be turned into a closed set by making additional assumptions. In broad outlines, the procedure is as follows:

(a) Split the particle velocity $\mathbf{v}$ into an average part $\mathbf{u}_\alpha$ and a random part $\tilde{\mathbf{v}}_\alpha$ defined as

$$\tilde{\mathbf{v}}_\alpha \equiv \mathbf{v} - \mathbf{u}_\alpha, \quad \text{where } \langle \tilde{\mathbf{v}}_\alpha \rangle = 0. \tag{2.51}$$

This permits the definition of *thermal quantities*:

$$T_\alpha(\mathbf{r}, t) \equiv \frac{m_\alpha}{3k} \langle \tilde{v}_\alpha^2 \rangle \qquad \text{(temperature)}, \tag{2.52}$$

$$\mathbf{P}_\alpha(\mathbf{r}, t) \equiv n_\alpha m_\alpha \langle \tilde{\mathbf{v}}_\alpha \tilde{\mathbf{v}}_\alpha \rangle = p_\alpha \mathbf{I} + \boldsymbol{\pi}_\alpha,$$
$$p_\alpha \equiv n_\alpha k T_\alpha \qquad \text{(stress tensor)}, \tag{2.53}$$

$$\mathbf{h}_\alpha(\mathbf{r}, t) \equiv \tfrac{1}{2} n_\alpha m_\alpha \langle \tilde{v}_\alpha^2 \tilde{\mathbf{v}}_\alpha \rangle \qquad \text{(heat flow)}, \tag{2.54}$$

$$\mathbf{R}_\alpha(\mathbf{r}, t) \equiv m_\alpha \int C_{\alpha\beta} \, \tilde{\mathbf{v}}_\alpha \, d^3v \qquad \text{(momentum transfer)}, \tag{2.55}$$

$$Q_\alpha(\mathbf{r}, t) \equiv \tfrac{1}{2} m_\alpha \int C_{\alpha\beta} \, \tilde{v}_\alpha^2 \, d^3v \qquad \text{(heat transfer)}. \tag{2.56}$$

Here, $\mathbf{I}$ is the unit tensor, so that $\boldsymbol{\pi}_\alpha$ represents the off-diagonal terms of the pressure tensor $\mathbf{P}$. A particular example, consistent with these definitions, is *the Maxwell distribution* for thermal equilibrium:

$$f_\alpha^0(\mathbf{r}, \mathbf{v}, t) = n_\alpha \left(\frac{m_\alpha}{2\pi k T_\alpha} \right)^{3/2} \exp\left(-\frac{m_\alpha \tilde{v}_\alpha^2}{2k T_\alpha} \right). \tag{2.57}$$

For this distribution, the LHS of the Boltzmann equation (2.40) vanishes so that the collision term on the RHS should vanish as well, i.e. when the two distributions have equal average velocities ($\mathbf{u}_e = \mathbf{u}_i$) and temperatures ($T_e = T_i$). Plasma kinetic theory is concerned with deviations from this thermal equilibrium and the way in which collisions cause relaxation to thermal equilibrium in the course of time (Braginskii [41]).

(b) The equations of continuity, momentum, and heat balance then take the form:

$$\frac{\partial n_\alpha}{\partial t} + \nabla \cdot (n_\alpha \mathbf{u}_\alpha) = 0, \tag{2.58}$$

$$n_\alpha m_\alpha \left(\frac{\partial \mathbf{u}_\alpha}{\partial t} + \mathbf{u}_\alpha \cdot \nabla \mathbf{u}_\alpha \right) + \nabla \cdot \mathbf{P}_\alpha - n_\alpha q_\alpha (\mathbf{E} + \mathbf{u}_\alpha \times \mathbf{B}) = \mathbf{R}_\alpha, \tag{2.59}$$

$$\tfrac{3}{2} n_\alpha k \left(\frac{\partial T_\alpha}{\partial t} + \mathbf{u}_\alpha \cdot \nabla T_\alpha \right) + \mathbf{P}_\alpha : \nabla \mathbf{u}_\alpha + \nabla \cdot \mathbf{h}_\alpha = Q_\alpha. \tag{2.60}$$

In Eq. (2.59), the divergence of the stress tensor may be decomposed into an isotropic part, involving the scalar pressure p_α, and an anisotropic part, involving

2.3 Kinetic plasma theory

the off-diagonal pressure tensor π_α:

$$\nabla \cdot \mathbf{P}_\alpha = \nabla p_\alpha + \nabla \cdot \pi_\alpha. \tag{2.61}$$

The double dot in Eq. (2.60) indicates that a double sum over the Cartesian components is to be taken, $\mathbf{P} : \nabla \mathbf{u} \equiv \sum_i \sum_j P_{ij} \partial u_i / \partial x_j$, so that decomposition into diagonal and off-diagonal contributions gives

$$\mathbf{P}_\alpha : \nabla \mathbf{u}_\alpha = p_\alpha \nabla \cdot \mathbf{u}_\alpha + \pi_\alpha : \nabla \mathbf{u}_\alpha. \tag{2.62}$$

The logical next step is to transform the temperature evolution equation (2.60) into a pressure evolution equation by exploiting Eqs. (2.53) and (2.58):

$$\frac{\partial p_\alpha}{\partial t} + \mathbf{u}_\alpha \cdot \nabla p_\alpha + \gamma p_\alpha \nabla \cdot \mathbf{u}_\alpha + (\gamma - 1)(\pi_\alpha : \nabla \mathbf{u}_\alpha + \nabla \cdot \mathbf{h}_\alpha) = (\gamma - 1) Q_\alpha. \tag{2.63}$$

Here, we have introduced the ratio of specific heats, $\gamma \equiv C_p/C_v = 5/3$, to demonstrate the connection with gas dynamics. The equations (2.58) for n_α, (2.59) with (2.61) for $\mathbf{u}_\alpha$, and (2.63) for p_α now appear rather macroscopic, but they hide the unsolved kinetic dependence in the variables π_α and $\mathbf{h}_\alpha$, which involve higher order moments, and the variables $\mathbf{R}_\alpha$ and Q_α, which involve the unspecified collision operator.

(c) The truncated set of moment equations is closed by exploiting the transport coefficients derived by *transport theory* (Braginskii [41], Balescu [14]). This theory concerns deviations from local thermodynamic equilibrium, expressed by Eq. (2.57), where the distribution functions are developed in powers of a small parameter measuring that deviation. This results in relationships, involving *transport coefficients*, between the thermal quantities defined in Eqs. (2.52)–(2.56) and the gradients of the macroscopic quantities. It is the second objective of kinetic theory (the first one being the derivation of the kinetic equation with a collision operator) to provide these coefficients: another formidable task. Just exploiting the final outcome, i.e. the explicit expressions of the transport coefficients, the closing relationships schematically take the form:

$$\pi_\alpha \sim \mu_\alpha \nabla \mathbf{u}_\alpha \qquad \text{(viscosity)},$$

$$\mathbf{h}_\alpha \sim -\kappa_\alpha \nabla (kT_\alpha) \qquad \text{(heat conductivity)},$$

$$\mathbf{R}_\alpha \approx -q_\alpha n_\alpha \eta \mathbf{j}, \qquad \sum Q_\alpha \approx \eta |\mathbf{j}|^2 \quad \text{(resistivity)}. \tag{2.64}$$

Here, we have just indicated the form of the expressions, omitting many terms, suppressing the anisotropic tensor structure of the transport coefficients, and leaving their dependencies on the densities, temperatures, and magnetic field unspecified. (See Section 3.3.2 for the explicit expressions.)

54   *Elements of plasma physics*

With respect to the anisotropy of the transport coefficients, an example is the huge difference between the electron heat conductivities parallel and perpendicular to the magnetic field: $\kappa_\perp^e/\kappa_\parallel^e \sim (\Omega_e \tau_e)^{-2} \ll 1$, where Ω_e is the electron gyro-frequency and τ_e is the electron collision time. This anisotropy is crucial for the possibility of magnetic confinement in fusion machines. On the other hand, the perpendicular resistivity only differs by a factor 2 from the classical value of the parallel resistivity, the so-called Spitzer resistivity:

$$\eta_\parallel \approx \frac{m_e}{2e^2 n_e \tau_e} = \frac{e^2 \sqrt{m_e} Z \ln \Lambda}{6\epsilon_0^2 (2\pi kT)^{3/2}} \approx 1.63 \times 10^{-9} \frac{Z \ln \Lambda}{\tilde{T}^{3/2}}, \qquad \eta_\perp \approx 2\eta_\parallel, \quad (2.65)$$

where $\ln \Lambda$ (~ 20) is the Coulomb logarithm. (A crude isotropic resistivity model that is frequently exploited actually assumes $\eta = 2\eta_\parallel = \eta_\perp$; see Section 3.3.3.) These two loose remarks suffice to illustrate the intricacies of the subject of classical transport. All of this will be discussed more fully in Chapter 3. Here, we will not further dwell on this but just remark that in the derivation of the macroscopic equations most of the transport is neglected, i.e. assumed to operate on time scales which are much longer than those of interest for macroscopic dynamics. The moment equations (2.58)–(2.60), together with Maxwell's equations (2.13), then transform into the closed set of *two-fluid* and *one-fluid plasma equations*. This subject will be continued in Section 2.4.

We will now present a highly simplified application of the two-fluid description (Section 2.3.2). In that application, most of the complicated terms discussed do not occur. Nevertheless, it illustrates an important basic physical mechanism at work, viz. collective electrostatic oscillations. Next, we return to the kinetic description in terms of distribution functions and show how velocity space effects lead to the surprising kinetic phenomenon called Landau damping (Section 2.3.3).

2.3.2 Collective phenomena: plasma oscillations

We have encountered the concepts of quasi-neutrality and Debye length in Section 1.4. We extend these electric field concepts in two steps. First, we study perturbations of quasi-neutrality in a cold plasma by *plasma oscillations*, also called Langmuir waves (1929) after the name of the author who also introduced the term 'plasma' in 1923. Next, we study the thermal effects on these oscillations in terms of the Debye length.

Consider the highly simplified case of a cold plasma in the absence of a magnetic field ($\mathbf{B} = 0$). This implies that all thermal effects are neglected ($\mathbf{P}_\alpha$, $\mathbf{h}_\alpha$, $\mathbf{R}_\alpha$, and Q_α vanish), so that all complicated terms in the equations of motion (2.59) disappear and the energy equations (2.60) may be dropped. We then just need to exploit the continuity equations (2.58),

$$\frac{\partial n_\alpha}{\partial t} + \nabla \cdot (n_\alpha \mathbf{u}_\alpha) = 0 \qquad (\alpha = e, i), \qquad (2.66)$$

and the simplified momentum equations (2.59),

$$m_\alpha \left(\frac{\partial \mathbf{u}_\alpha}{\partial t} + \mathbf{u}_\alpha \cdot \nabla \mathbf{u}_\alpha \right) = q_\alpha \mathbf{E} \qquad (\alpha = e, i). \qquad (2.67)$$

The electric field can be determined self-consistently from *Poisson's equation* (2.13)(c), where *the charge density* is obtained from Eq. (2.14)(a):

$$\nabla \cdot \mathbf{E} = \frac{\tau}{\epsilon_0} = \frac{e}{\epsilon_0} (Zn_i - n_e). \qquad (2.68)$$

These equations constitute a complete set for the variables $n_{e,i}(\mathbf{r}, \mathbf{v}, t)$, $\mathbf{u}_{e,i}(\mathbf{r}, \mathbf{v}, t)$, and $\mathbf{E}(\mathbf{r}, t)$ describing the problem of electrostatic oscillations.

We have already encountered one of the most fundamental properties of plasmas, viz. that *plasmas maintain approximate charge neutrality*. Indeed, charge imbalances on a macroscopic scale L would create huge electric fields ($E \sim \tau L/\epsilon_0$) which would neutralize these imbalances extremely fast by accelerating the electrons, so that the plasma maintains charge neutrality to a high degree of accuracy.

Considered on a finer time and length scale, however, charge imbalances do occur in the form of oscillations which are very typical for plasmas. For these oscillations, the heavy ions ($m_i \gg m_e$) may be considered as a fixed ($\mathbf{u}_i = 0$) neutralizing background in which only the light electrons move ($\mathbf{u}_e \neq 0$). Perturbing a small region inside the plasma by displacing the electrons there, charge neutrality will be disturbed ($n_e \neq Zn_i$). The electron variables then determine the problem:

$$\begin{aligned} n_e &\approx n_0 + n_1(\mathbf{r}, t), \\ \mathbf{u}_e &\approx \mathbf{u}_1(\mathbf{r}, t), \end{aligned} \qquad (2.69)$$

whereas the ion variables simplify to

$$n_i \approx n_0/Z = \text{const}, \qquad \mathbf{u}_i \approx 0. \qquad (2.70)$$

Hence, the two ion equations (2.66) and (2.67) for $\alpha = i$ may be dropped.[3] The subscripts 0 and 1 refer to the constant background and the perturbations, respectively. The density perturbation $|n_1(\mathbf{r}, t)| \ll n_0$ occurs in a small region of the plasma and is zero elsewhere so that *linearization* is appropriate, i.e. terms involving products of perturbations are neglected since they are small compared to linear

---

[3] Heavy immobile ions imply taking the limits $m_i \to \infty$, $\mathbf{u}_i \to 0$ such that the LHS of Eq. (2.67) becomes undetermined. Such procedures always require justification in terms of a small parameter, in this case the mass ratio $m_e/m_i \ll 1$. The ion dynamics would enter as a higher order correction in terms of this parameter.

terms. The electric field $\mathbf{E}_1$ that is created is then proportional to n_1. This small electric field creates a small electron flow velocity $\mathbf{u}_1$, which is also proportional to n_1.

The linearized electron density equation (2.66), the momentum equation (2.67) (both with $\alpha = e$), and the Poisson equation (2.68) then yield a complete set of equations:

$$\frac{\partial n_1}{\partial t} + n_0 \nabla \cdot \mathbf{u}_1 = 0,$$

$$m_e \frac{\partial \mathbf{u}_1}{\partial t} = -e\mathbf{E}_1,$$

$$\nabla \cdot \mathbf{E}_1 = \frac{\tau_1}{\epsilon_0} = -\frac{e}{\epsilon_0} n_1. \qquad (2.71)$$

These equations may be reduced to a single wave equation for n_1:

$$\frac{\partial^2 n_1}{\partial t^2} = -n_0 \nabla \cdot \frac{\partial \mathbf{u}_1}{\partial t} = \frac{n_0 e}{m_e} \nabla \cdot \mathbf{E}_1 = -\frac{n_0 e^2}{\epsilon_0 m_e} n_1. \qquad (2.72)$$

The solutions $n_1(\mathbf{r}, t) = \hat{n}_1(\mathbf{r}) \exp(-i\omega t)$ represent electron density oscillations, called *plasma oscillations*, with a characteristic frequency, called *the electron plasma frequency*:

$$\omega = \pm \omega_{pe}, \quad \omega_{pe} \equiv \sqrt{\frac{n_0 e^2}{\epsilon_0 m_e}}. \qquad (2.73)$$

This frequency is one of the fundamental parameters of a plasma. Since it depends only on the plasma density, detection of plasma oscillations provides a diagnostic for the determination of the plasma density.

The plasma frequency is usually very high because m_e is very small. In tokamak plasmas, e.g., a typical density $n_0 = 10^{20}$ m^{-3} gives

$$\omega_{pe} = 5.7 \times 10^{11} \text{ rad s}^{-1} \quad (\text{i.e. } 91 \text{ GHz}).$$

Comparing this frequency with the gyro-frequencies of Section 2.2, we find that the electron plasma frequency is of the same order of magnitude as the electron cyclotron frequency for tokamaks with very strong magnetic fields ($B \sim 3$ T).

▷ **Exercise.** In the solar corona, the density is much lower, so that the electron plasma frequency is also much lower than for tokamaks. The representative numbers in Table B.3 have been chosen somewhat artificially such that the ratio ω_{pe}/Ω_e is the same for tokamaks and coronal loops. What is the plasma frequency for the latter if $n_0 = 10^{14}$ m^{-3} is taken? How does this change the ratio ω_{pe}/Ω_e? Also, complete the empty entries of Table B.4. ◁

Note that the spatial form of the amplitude $\hat{n}_1(\mathbf{r})$ of the plasma oscillations is not determined in cold plasma theory. This becomes different for 'warm' plasmas,

2.3 Kinetic plasma theory

where deviations from charge neutrality due to thermal fluctuations occur in small regions of a size of the order of the *Debye length*

$$\lambda_D \equiv \sqrt{\frac{\epsilon_0 k_B T_e}{n_0 e^2}} = \frac{v_{\text{th},e}}{\sqrt{2}\,\omega_{pe}}. \tag{2.74}$$

We here indicate the Boltzmann constant with a subscript, k_B, to distinguish it from the wave number k of the waves that now enters the analysis. Inserting numbers again for thermonuclear plasmas, with $\widetilde{T} = 10\,\text{keV}$, $v_{\text{th},e} = 5.9 \times 10^7\,\text{m s}^{-1}$, $\omega_{pe} = 5.7 \times 10^{11}\,\text{rad s}^{-1}$ gives

$$\lambda_D = 7.4 \times 10^{-5}\,\text{m} \approx 0.07\,\text{mm},$$

i.e. of the order of the electron gyro-radius R_e.

▷ **Exercise.** What is the Debye length in the solar corona for $\widetilde{T} = 100\,\text{eV}$ and $n_0 = 10^{14}\,\text{m}^{-3}$? For what value of the coronal magnetic field does this become of the order of the electron gyro-radius? ◁

Because of these thermal fluctuations, the frequency of the plasma oscillations becomes dependent on the wavelength. This part of the thermal contributions may be computed by means of the two-fluid equations (2.58)–(2.63) for an unmagnetized plasma ($\mathbf{B} = 0$), assuming an isotropic pressure and neglecting heat transport and collisions:

$$\frac{\partial n_\alpha}{\partial t} + \nabla \cdot (n_\alpha \mathbf{u}_\alpha) = 0, \tag{2.75}$$

$$n_\alpha m_\alpha \left(\frac{\partial \mathbf{u}_\alpha}{\partial t} + \mathbf{u}_\alpha \cdot \nabla \mathbf{u}_\alpha \right) + \nabla p_\alpha = n_\alpha q_\alpha \mathbf{E}, \tag{2.76}$$

$$\frac{\partial p_\alpha}{\partial t} + \mathbf{u}_\alpha \cdot \nabla p_\alpha + \gamma p_\alpha \nabla \cdot \mathbf{u}_\alpha = 0. \tag{2.77}$$

Again assuming immobile ions and linearizing the equations (2.75)–(2.77) for the electrons, we now get a modified eigenvalue problem where the pressure $p_0 = n_0 k_B T_0$, i.e. the temperature, of the background plasma enters:

$$\frac{\partial n_1}{\partial t} + n_0 \nabla \cdot \mathbf{u}_1 = 0, \tag{2.78}$$

$$n_0 m_e \frac{\partial \mathbf{u}_1}{\partial t} + \nabla p_1 = -e n_0 \mathbf{E}_1, \tag{2.79}$$

$$\frac{\partial p_1}{\partial t} + \gamma p_0 \nabla \cdot \mathbf{u}_1 = 0, \tag{2.80}$$

$$\nabla \cdot \mathbf{E}_1 = -\frac{e}{\epsilon_0} n_1. \tag{2.81}$$

Assuming plane waves in the x-direction, and ignoring spatial dependences in the y- and z-directions,

$$n_1(x, t) = \hat{n}_1 e^{i(kx-\omega t)} \tag{2.82}$$

(and similar expressions for $\mathbf{u}_1$, p_1, $\mathbf{E}_1$), the gradients $\nabla \to ik\mathbf{e}_x$ and the time derivatives $\partial/\partial t \to -i\omega$, so that Eqs. (2.78)–(2.82) become an algebraic system of equations for the amplitudes $\hat{n}_1$, $\hat{\mathbf{u}}_1$, $\hat{p}_1$, and $\hat{\mathbf{E}}_1$. The determinant provides the dispersion equation:

$$\omega^2 = \omega_{pe}^2 (1 + \gamma k^2 \lambda_D^2). \tag{2.83}$$

Here, since the oscillations are one-dimensional, we should exploit the value $\gamma = 3$ (see Chen [53], Chapter 4). Note that the old result (2.73) is recovered for long wavelengths, where $k^2 \lambda_D^2 \ll 1$, but there is a large effect now on the oscillations for wavelengths of the order of or smaller than the Debye length. However, this effect is not quite correctly described since the fluid description actually breaks down because of a peculiar kinetic effect that will be discussed in the next section.

2.3.3 Landau damping

A more refined analysis of longitudinal plasma oscillations for 'warm' plasmas should take velocity space effects into account, exploiting the Vlasov, or *collisionless* Boltzmann, equation (2.41) for the perturbations $f_1(\mathbf{r}, \mathbf{v}, t)$ of the electron distribution function. Taking again plane wave solutions $\sim \exp i(\mathbf{k} \cdot \mathbf{r} - \omega t)$, one immediately runs into a mathematical problem:

$$\frac{\partial f_1}{\partial t} + \mathbf{v} \cdot \frac{\partial f_1}{\partial \mathbf{r}} = -i(\omega - \mathbf{k} \cdot \mathbf{v}) f_1 = \frac{e}{m_e} \mathbf{E}_1 \cdot \frac{\partial f_0}{\partial \mathbf{v}}, \tag{2.84}$$

so that inversion of the operator $\partial/\partial t + \mathbf{v} \cdot \partial/\partial \mathbf{r}$, to express f_1 in terms of $\mathbf{E}_1$, leads to singularities when $\omega - \mathbf{k} \cdot \mathbf{v} = 0$. Incorporated in a proper treatment of the initial value problem, these singularities were shown by Landau (1946) [136] to give rise to damping of the plasma oscillations. This *Landau damping* is a surprising phenomenon since it occurs in a purely collisionless medium, i.e. there is no dissipation! Much later, experiments by Malmberg and Wharton [150] first verified the phenomenon of Landau damping (1966), and then also demonstrated that, in fact, the information contained in the initial signal is not irreversibly lost but that it may be recovered by means of plasma wave echos (1968).

2.3 Kinetic plasma theory

A complementary approach to the electrostatic plasma oscillations by means of a normal mode analysis was given by Van Kampen (1955) [237, 238]. He showed that the singularities $\omega - \mathbf{k} \cdot \mathbf{v} = 0$ lead to a continuous spectrum of singular, δ-function type, modes (the Van Kampen modes), which constitute a complete set of 'improper' eigenmodes for this system. Damping occurs because a package of those modes rapidly loses its spatial phase coherence (*phase mixing*). The occurrence of a continuous spectrum is a very intriguing aspect of the analysis of plasma oscillations, which is also encountered in the fluid description of macroscopic waves (as we will see in Chapter 6).

Let us analyse the problem in some more detail for one space dimension x and one velocity space dimension v ($\equiv v_x$). Since Landau damping is due to velocity space effects we have to redo the problem of Section 2.3.2 in terms of distribution functions, leading to the so-called Vlasov–Poisson problem:

$$\frac{\partial f_1}{\partial t} + v \frac{\partial f_1}{\partial x} = \frac{e}{m_e} \frac{\partial f_0}{\partial v} E_1 ,$$

$$\frac{\partial E_1}{\partial x} = -\frac{e}{\epsilon_0} n_1 = -\frac{e}{\epsilon_0} \int_{-\infty}^{\infty} f_1 \, dv . \tag{2.85}$$

Inserting plane wave solutions

$$f_1(x, v, t) = \hat{f}_1(v) \, e^{i(kx - \omega t)} , \qquad E_1(x, t) = \hat{E}_1 \, e^{i(kx - \omega t)} , \tag{2.86}$$

in these equations, i.e. making the replacements $\partial/\partial x \to ik$, $\partial/\partial t \to -i\omega$, they transform into

$$-i(\omega - kv) \hat{f}_1 = \frac{e}{m_e} \frac{\partial f_0}{\partial v} \hat{E}_1 ,$$

$$ik \hat{E}_1 = -\frac{e}{\epsilon_0} \int_{-\infty}^{\infty} \hat{f}_1 \, dv . \tag{2.87}$$

Just expressing $\hat{f}_1$ in terms of $\hat{E}_1$ by means of the first equation (*assuming $\omega \neq kv$!*), and inserting $\hat{f}_1$ into the equation for $\hat{E}_1$, we obtain

$$\left[1 - \frac{e^2}{\epsilon_0 m_e k^2} \int_{-\infty}^{\infty} \frac{1}{v - \omega/k} \frac{\partial f_0}{\partial v} \, dv \right] \hat{E}_1 = 0 . \tag{2.88}$$

Hence, the expression inside the square brackets should vanish, providing the

f₀ diagram with Maxwell distribution showing v_{ph} marked on the velocity axis.

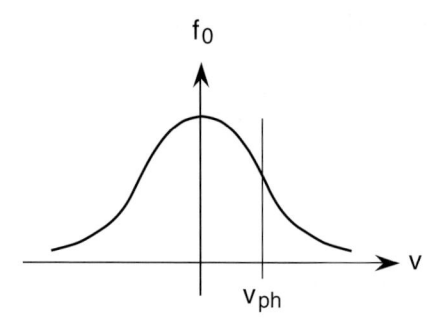

Fig. 2.6. Maxwell velocity distribution and a phase speed $v_{ph} = \omega/k$ of plasma oscillations in the thermal region.

dispersion equation, i.e. the relation between ω and k, we are looking for:

$$D_V(k,\omega) \equiv 1 - \frac{\omega_{pe}^2}{k^2 n_0} \int_{-\infty}^{\infty} \frac{1}{v - \omega/k} \frac{\partial f_0}{\partial v} dv = 0. \tag{2.89}$$

Here, we have inserted the plasma frequency ω_{pe} defined in Eq. (2.73).

For definiteness, we now have to specify the distribution function f_0 of the background. It is logical to choose the one-dimensional form of the Maxwell equilibrium distribution introduced in Eq. (2.57):

$$f_0(v) = \frac{n_0}{\sqrt{\pi} v_{th}} e^{-v^2/v_{th}^2}, \qquad v_{th} \equiv \sqrt{\frac{2 k_B T_e}{m_e}}. \tag{2.90}$$

Since we wish to concentrate on velocity space effects, we assume spatial homogeneity of the background equilibrium so that the density n_0 and the thermal speed v_{th} (i.e. the electron temperature T_e) are constant. These apparently innocent assumptions imply that a whole 'zoo' of kinetic and macroscopic instabilities is eliminated at once.

Obviously, the assumption $\omega \neq kv$ cannot be justified if the frequency ω of the plane waves is real since the integration in Eq. (2.89) is then right across the singularity. This singularity occurs for particles with speeds that are resonant with the phase velocity of the waves: $v = v_{ph} \equiv \omega/k$ (vertical line in Fig. 2.6). An *apparent* way out, proposed by Vlasov [239], is to exploit the principal value of the integral for real ω, defined as:

$$\mathcal{P} \int_{-\infty}^{\infty} dv \cdots \equiv \lim_{\delta \to 0} \left[\int_{-\infty}^{v_{ph}-\delta} dv \cdots + \int_{v_{ph}+\delta}^{\infty} dv \cdots \right]. \tag{2.91}$$

One can crudely estimate this integral for long wavelengths ($k \to 0$), when $v_{ph} \gg$

2.3 Kinetic plasma theory

$v_{\rm th}$, so that the Maxwell distribution f_0 becomes quite small at the singular point:

$$\mathcal{P}\int_{-\infty}^{\infty} \frac{1}{v-\omega/k}\frac{\partial f_0}{\partial v}\,dv = -\frac{2n_0}{\sqrt{\pi}v_{\rm th}^3}\,\mathcal{P}\int_{-\infty}^{\infty}\frac{v}{v-v_{\rm ph}}\,e^{-v^2/v_{\rm th}^2}\,dv$$

$$\approx \frac{n_0}{v_{\rm ph}^2}\left(1+\frac{3}{2}\frac{v_{\rm th}^2}{v_{\rm ph}^2}\right). \quad (2.92)$$

This gives the following approximate dispersion equation:

$$\omega^2 \approx \omega_{pe}^2 + \frac{3}{2}k^2 v_{\rm th}^2 = \omega_{pe}^2(1+3k^2\lambda_{\rm D}^2)\,, \quad (2.93)$$

i.e. one obtains the thermal correction of the frequency (2.73) of the plasma oscillations in terms of the product of the wave vector of the oscillations and the Debye length. This correction turns out to agree with Eq. (2.83) obtained from the fluid approximation.

However, this procedure is much too cavalier, as pointed out by Landau [136] in his severe criticism of the work of Vlasov [239][4]: there is no justification of the use of the principal value integral, and a more careful analysis of the singularity reveals that there is an imaginary contribution (the Landau damping) to the frequency of the waves. The more careful analysis is somewhat beyond the level of this chapter, but too fundamental to be skipped altogether. Therefore, we put it in small print here.

▷ **Landau's solution of the initial value problem.** We return to the basic equations (2.85) of the Vlasov–Poisson problem. Instead of the Ansatz (2.86) of plane wave solutions, we now keep the exp(ikx) spatial dependence but treat the time dependence through the Laplace transform:

$$\tilde{f}_1(v;\omega) \equiv \int_0^{\infty} f_1(v;t)e^{i\omega t}\,dt\,, \qquad \tilde{E}_1(\omega) \equiv \int_0^{\infty} E_1(t)e^{i\omega t}\,dt\,. \quad (2.94)$$

(We exploit the variable ω instead of the standard Laplace variable $p \equiv -i\omega$ so that convergence in the right half p-plane is replaced by convergence in the upper half ω-plane.) The Laplace transform of the time derivative $\partial f_1/\partial t$ in Eq. (2.85) produces a similar expression $-i\omega \tilde{f}_1$ as before,

$$\int_0^{\infty}\frac{\partial f_1}{\partial t}e^{i\omega t}\,dt = \left[f_1 e^{i\omega t}\right]_{t=0}^{t\to\infty} - i\omega\int_0^{\infty}f_1 e^{i\omega t}\,dt = -g(v) - i\omega\tilde{f}_1(v,\omega)\,, \quad (2.95)$$

but with an additional contribution of the initial value of the perturbation f_1 of the

---

[4] The cited two references of Vlasov both contain the important contribution of the collisionless Boltzmann equation (with the justification of the neglect of collisions for many plasma phenomena), but, unfortunately, also the incorrect wave analysis of the plasma oscillations.

distribution function:

$$g(v) \equiv f_1(v, t = 0). \qquad (2.96)$$

The contribution for $t \to \infty$ vanishes since we assume that $\operatorname{Im}\omega > 0$.

In effect, the Laplace transform of Eqs. (2.85) is virtually the same as Eqs. (2.87), except for the additional term $-g(v)$. We again express $\tilde{f}_1$ in terms of $\tilde{E}_1$,

$$\tilde{f}_1(v,\omega) = \frac{i}{\omega - kv} \left[\frac{e}{m_e} \frac{\partial f_0}{\partial v} \tilde{E}_1(\omega) + g(v) \right], \qquad (2.97)$$

and insert $\tilde{f}_1$ into the equation for $\tilde{E}_1$:

$$\tilde{E}_1(\omega) = \frac{e}{\epsilon_0 k^2} \frac{\int \frac{1}{v - \omega/k} g\, dv}{1 - \frac{\omega_{pe}^2}{k^2 n_0} \int \frac{1}{v - \omega/k} \frac{\partial f_0}{\partial v}\, dv} \equiv \frac{e}{\epsilon_0 k^2} \frac{N(\omega)}{D(\omega)}. \qquad (2.98)$$

Maybe not too surprising, the denominator *appears to be identical* to the expression D_V of the dispersion equation (2.89): the zeros of $D(\omega)$, i.e. the solutions of the dispersion equation $D(\omega) = 0$, become poles of the complex function $\tilde{E}_1(\omega)$. However, in contrast to Vlasov's approach, we now have a procedure to make sense of the singularities $v = \omega/k$ in the velocity integrals, viz. by completing the solution of the initial value problem by means of the inverse Laplace transform:

$$E_1(t) = \frac{1}{2\pi} \int_{iv_C - \infty}^{iv_C + \infty} \tilde{E}_1(\omega) e^{-i\omega t}\, dt, \text{ and similarly for } f_1(v, t) \quad (\operatorname{Im}\omega = v_C > 0).$$

$$\qquad (2.99)$$

Formally, the problem is now solved since these integrals avoid the singularities altogether by staying away from them on the path C_ω in the upper half of the ω-plane, indicated in Fig. 2.7(a).

To describe the collective plasma oscillations, the expressions (2.99) need to be evaluated with respect to their asymptotic time dependence for $t \to \infty$. To that end, it is expedient to deform the contour C_ω into the contour C'_ω in the lower half of the ω-plane. For the latter contour, the contribution of the straight pieces may be neglected ($\operatorname{Im}\omega \ll 0$) whereas the residue of the uppermost pole (indicated by ω_0) will survive the longest:

$$\tilde{E}_1(\omega) \approx A(\omega - \omega_0)^{-1} \Rightarrow E_1(t) \approx -\frac{A}{2\pi} \oint \frac{1}{\omega - \omega_0} e^{-i\omega t}\, d\omega = -iAe^{-i\omega_0 t}. \qquad (2.100)$$

Of course, the explicit value of ω_0 is to be computed yet. This requires knowledge of the complex function $\tilde{E}_1(\omega)$ for $\operatorname{Im}\omega \leq 0$, whereas it was only defined for $\operatorname{Im}\omega > 0$. The canonical way to obtain that knowledge is by means of *analytic continuation* of the functions $N(\omega)$ and $D(\omega)$. If one assumes that $g(v)$ is an entire function (i.e. an analytic function which is regular for all finite values of v; see e.g. Nehari [162], Chapter 3), then the Cauchy contour integral along the closed path C_v $(= C_{v1} + C_{v2})$ will just pick up the residue at $v = \omega/k$ when $\operatorname{Im}\omega > 0$ (Fig. 2.7(b_1)). Whereas this implies that one could relate the value of the integral along the real axis C_{v1} (plus the residue) to the value of the integral along the semi-circular path C_{v2} with $R \to \infty$, this does not provide us with the usual simplification of the algebra, since the latter integral does not vanish. However, it does show what to do when $\operatorname{Im}\omega \leq 0$ (when the singularity crosses the real axis): one should deform the contour C_{v1} to remain on the same side of the singularity, as indicated in

2.3 Kinetic plasma theory

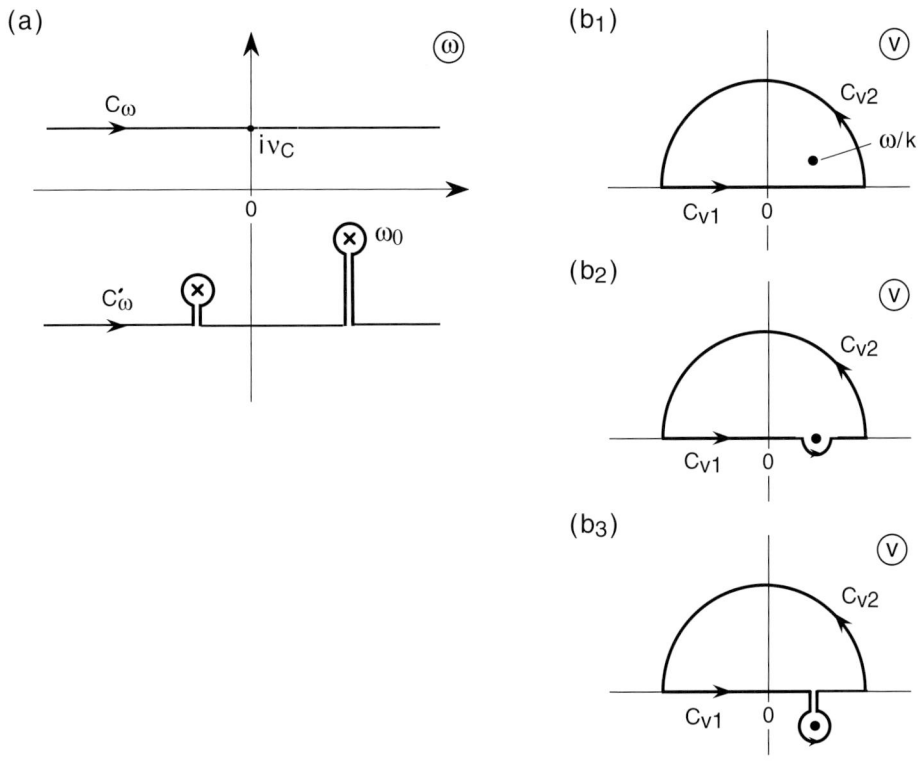

Fig. 2.7. Techniques used by Landau in his solution of the initial value problem: (a) deforming the contour C_ω in the complex ω-plane of the inverse Laplace transform to a contour C'_ω above the solutions of the dispersion equation; (b) analytic continuation of the velocity integrals, considered as functions of ω, by means of a Cauchy contour integral in the complex v-plane along a path C_v ($= C_{v1} + C_{v2}$) enclosing the singularity $v = \omega/k$ from values $\mathrm{Im}\,\omega > 0$ (b_1) to $\mathrm{Im}\,\omega = 0$ (b_2) and $\mathrm{Im}\,\omega < 0$ (b_3).

Figs. 2.7(b_2) and 2.7(b_3). Then, the appropriate analytic continuation of $N(\omega)$ is obtained by integrating along the real v-axis and adding no residue, half a residue, or the full residue depending on the value of $\mathrm{Im}\,\omega$:

$$\int_{C_{v1}} \frac{1}{v - \omega/k} g(v)\,dv = \begin{cases} \int_{-\infty}^{\infty} \frac{1}{v - \omega/k} g(v)\,dv & (\mathrm{Im}\,\omega > 0) \\ \mathcal{P}\int_{-\infty}^{\infty} \frac{1}{v - \omega/k} g(v)\,dv + i\pi\, g(\omega/k) & (\mathrm{Im}\,\omega = 0) \\ \int_{-\infty}^{\infty} \frac{1}{v - \omega/k} g(v)\,dv + i2\pi\, g(\omega/k) & (\mathrm{Im}\,\omega < 0). \end{cases}$$

(2.101)

The integral in $D(\omega)$ may be analytically continued in a similar way using the fact that, for the Maxwell distribution (2.90), $\partial f_0/\partial v$ is also an entire function.

The three expressions (2.101) may be conveniently combined in a single one, viz. the middle expression (2.101)(b), which may be considered as *valid for all values of ω* if the principal value integral is interpreted as the average of the two integrals along paths just above (C_a) and just below (C_b) the singularity:

$$\mathcal{P}\int dv \cdots \equiv \tfrac{1}{2}\int_{C_a} dv \cdots + \tfrac{1}{2}\int_{C_b} dv \cdots . \tag{2.102}$$

For $\mathrm{Im}\,\omega = 0$, the addition of two semi-circles above and below the singularity does not change anything since those contributions cancel. However, for $\mathrm{Im}\,\omega > 0$, the singularity contributes $-i\pi\, g(\omega/k)$ from the path C_a, cancelling the second term of Eq. (2.101)(b) so that the expression (2.101)(a) is obtained. For $\mathrm{Im}\,\omega < 0$, the singularity contributes $+i\pi\, g(\omega/k)$ from the path C_b, doubling the second term so that the expression (2.101)(c) is obtained.

Hence, the correct dispersion equation, valid for all values of ω, may be written as

$$D(k,\omega) \equiv 1 - \frac{\omega_{pe}^2}{k^2 n_0}\left[\mathcal{P}\int_{-\infty}^{\infty}\frac{1}{v-\omega/k}\frac{\partial f_0}{\partial v}\,dv + i\pi\left.\frac{\partial f_0}{\partial v}\right|_{\omega/k}\right]$$

$$= 1 + (k\lambda_D)^{-2}\left[1 + \frac{1}{\sqrt{\pi}}\zeta\,\mathcal{P}\int_{-\infty}^{\infty}\frac{1}{\bar{v}-\zeta}e^{-\bar{v}^2}d\bar{v} + i\sqrt{\pi}\,\zeta e^{-\zeta^2}\right] = 0, \tag{2.103}$$

where the Maxwell equilibrium distribution (2.90) with $\bar{v} \equiv v/v_{\mathrm{th}}$ and $\zeta \equiv \omega/(k v_{\mathrm{th}})$ have been inserted in the second line. Clearly, the interpretation of the integrals in Eq. (2.98) (where we purposely omitted the integration path since that was yet to be determined) is completely different from the one given by Vlasov in Eqs. (2.89) and (2.91): there is an imaginary part so that one does not obtain oscillations with a real frequency.

All ambiguity about the mathematical physics of the plasma oscillations has now been resolved. The rest is technical (though intricate) details about the evaluation of the velocity integrals for complex values of ω. That part of the problem is usually collected in the properties of the plasma dispersion function, which is a well-known tabulated complex function described in many textbooks on waves in plasmas (see, e.g., Stix [218] or Swanson [223]):

$$Z(\zeta) \equiv \frac{1}{\sqrt{\pi}}\int_{-\infty}^{\infty}\frac{1}{\bar{v}-\zeta}e^{-\bar{v}^2}d\bar{v} \qquad (\mathrm{Im}\,\zeta > 0), \tag{2.104}$$

where the analytic continuation for $\mathrm{Im}\,\zeta \le 0$ is obtained as above. In terms of $Z(\zeta)$, the dispersion equation becomes

$$D(k,\omega) \equiv 1 + (k\lambda_D)^{-2}\left[1 + \zeta Z(\zeta)\right] = 0, \qquad \zeta \equiv \frac{\omega}{k v_{\mathrm{th}}} = \frac{\omega/\omega_{pe}}{\sqrt{2}\,k\lambda_D}. \tag{2.105}$$

The power series of the plasma dispersion function $Z(\zeta)$ for $|\zeta| < 1$ reads

$$Z(\zeta) = i\sqrt{\pi}\,e^{-\zeta^2} - 2\zeta + \frac{4}{3}\zeta^3 - \frac{8}{15}\zeta^5 \cdots . \tag{2.106}$$

A more useful expression for physical applications is the asymptotic expansion for $|\zeta| \gg 1$:

$$Z(\zeta) \approx -\frac{1}{\zeta} - \frac{1}{2\zeta^3} - \frac{3}{4\zeta^5}\cdots + is\sqrt{\pi}\,e^{-\zeta^2}, \qquad s = \begin{cases} 0 & (\mathrm{Im}\,\zeta > 0) \\ 1 & (|\mathrm{Im}\,\zeta| < |\mathrm{Re}\,\zeta|^{-1} \ll 1) \\ 2 & (\mathrm{Im}\,\zeta < 0), \end{cases} \tag{2.107}$$

where the middle expression can be applied to a finite region about the real axis. (As usual in asymptotic expansions, the boundaries indicated in brackets do not imply that the expansion is valid up to that boundary but, rather, that the expression is no longer valid there because of the Stokes phenomenon; see Bender and Orszag [23], Chapter 3.) Iterating on the real and imaginary parts of the solution ζ of the dispersion equation for long wavelengths ($k\lambda_D \ll 1$) yields the expression (2.108) below for the complex frequency of the plasma oscillations. ◁

In conclusion: Landau's study of the initial value problem of electrostatic plasma oscillations shows that there is an important contribution of the singularities $v = v_{\text{ph}} \equiv \omega/k$ where the particles are in resonance with the phase velocity of the waves. For a Maxwell distribution, the solution of the dispersion equation (2.105) for long wavelengths ($k\lambda_D \ll 1$) is given by

$$\omega \approx \omega_{pe} \left\{ 1 + \frac{3}{2} k^2 \lambda_D^2 - i \sqrt{\frac{\pi}{8}} (k\lambda_D)^{-3} \exp\left[-\frac{1}{2}(k\lambda_D)^{-2} - \frac{3}{2} \right] \right\}, \qquad (2.108)$$

where the imaginary part represents *damping of the waves*. For long wavelengths, this damping is exponentially small. For short wavelengths ($k\lambda_D \sim 1$), the damping becomes very strong so that wave motion with wavelengths smaller than the Debye length becomes impossible.

2.4 Fluid description

Kinetic theory involves details of the distribution functions that evolve on very short length and time scales, like the Debye length λ_D and the plasma frequency ω_{pe}. Since the subject of this book is the macroscopic dynamics of magnetized plasmas, we now have to face the main difficulty, viz. how to bridge the enormous difference between these scales and the macroscopic ones. The development of the fluid picture of plasmas involves three major steps, illustrated in Fig. 2.8 and elaborated in the next chapter:

(a) Collisionality A first step has been taken in Section 2.3.1 with the formulation of the lowest moments (2.58)–(2.60) of the Boltzmann equation and the transport closure relations indicated in Eqs. (2.64). In this manner, a system of *two-fluid equations* is obtained describing the plasma dynamics in terms of the ten variables $n_{e,i}$, $\mathbf{u}_{e,i}$, $T_{e,i}$. To justify such a fluid description, the electrons and ions must undergo *frequent collisions* to establish the separate electron and ion fluids. Transport theory provides the quantitative criterion for the time scale τ_H on which the hydrodynamic description is valid, compared to the collisional relaxation times τ_e and τ_i of the electrons and ions:

$$\tau_H \gg \tau_i \left[\, \gg \tau_e \,\right]. \qquad (2.109)$$

```
                    Kinetic theory
                          ⇓
                 ( frequent collisions )
                          ⇓
                   Two-fluid theory
                          ⇓
                     ( large scales )
                          ⇓
        Diss. MHD  ⇒ ( slow dissipation ) ⇒  Ideal MHD
```

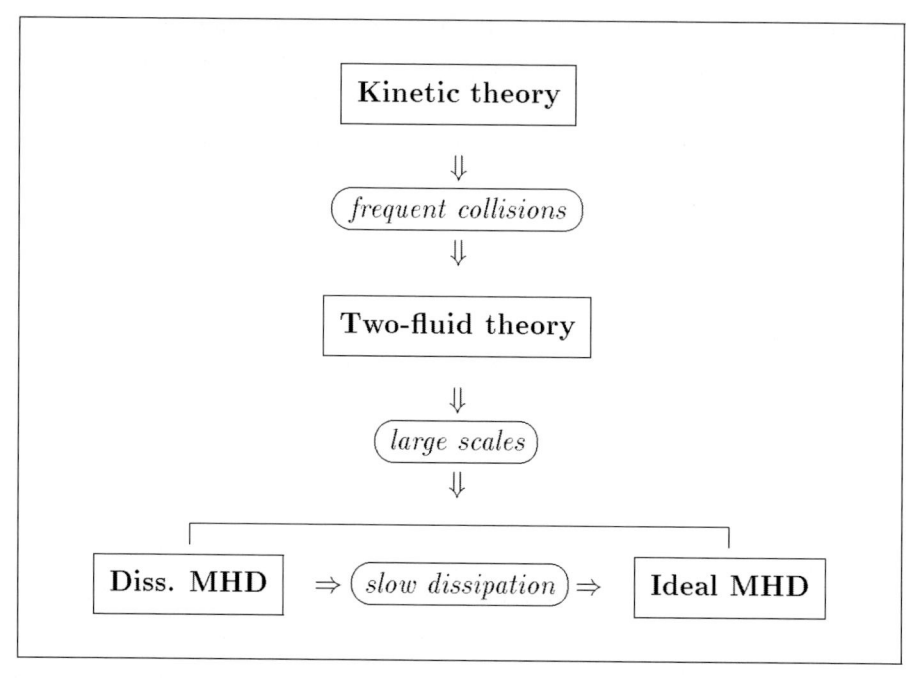

Fig. 2.8. Different theoretical plasma models and their connections.

The explicit expressions for τ_e and τ_i may be found in Section 3.2.4 (Eqs. (3.50) and (3.51), which demonstrate the faster relaxation of the electrons due to the smallness of the mass ratio: $\tau_e/\tau_i \sim (m_e/m_i)^{1/2} \ll 1$). Once more: transport theory is an enormous field of research by itself, the needed results of which have been collected in Chapter 3. Here, we just indicate the main line of thought leading to the fluid description. In particular, from the explicit expressions for τ_e and τ_i, frequent collisions imply that the plasma densities $n_{e,i}$ should be high enough for given values of the temperatures $T_{e,i}$.

(b) Macroscopic scales The plasma dynamics described by the two-fluid equations still involves the small length and time scales of the fundamental phenomena we have encountered, viz. the plasma frequency ω_{pe}, the cyclotron frequencies $\Omega_{e,i}$, the Debye length λ_D, and the cyclotron radii $R_{e,i}$ (and also a quantity not yet encountered, viz. the electron skin depth $\delta_e \equiv c/\omega_{pe}$; see Eq. (3.100)). Therefore, the essential second step towards the *magnetohydrodynamics* (MHD) description of plasmas is to consider *large length and time scales*:

$$\lambda_{\text{MHD}} \sim a \gg R_i, \qquad \tau_{\text{MHD}} \sim a/v_A \gg \Omega_i^{-1}. \tag{2.110}$$

Here, the magnetic field crucially enters: the larger the magnetic field strength, the more easily these conditions are satisfied. On these scales, the plasma is considered as a *single* conducting fluid without distinguishing its individual species: the MHD equations describe the behaviour of the plasma as a whole. Therefore, the dimension of the plasma appears in the estimate of the length scale and the Alfvén velocity v_A (a new fundamental plasma quantity, to be introduced in Section 2.4.2) appears in the time scale. The derivation of the MHD equations from the two-fluid equations will be sketched in Section 2.4.1, and presented in full detail in Section 3.4.

(c) Ideal fluids A third step is to consider the plasma dynamics on time scales *faster* than the *slow dissipation* connected with the decay of the macroscopic variables, in particular the resistive decay of the magnetic field:

$$\tau_{\text{MHD}} \ll \tau_R \sim a^2/\eta. \tag{2.111}$$

Here, the actual value of the resistivity given by Eq. (2.65) is so small that the condition may be satisfied even for the relatively small sizes of fusion machines, and very easily for the huge sizes of astrophysical plasmas. (Actually, too easily: a major problem is to find dissipation mechanisms fast enough to explain the onset of frequently observed disruptions, like stellar flares.) This condition leads to the model of *ideal MHD*: the most robust macroscopic description of magnetized plasmas. An ideal two-fluid counterpart may also be formulated since the condition (2.111) is certainly satisfied on the two-fluid time scale when it is satisfied on the MHD time scale ($\tau_H \ll \tau_{\text{MHD}}$).

2.4.1 From the two-fluid to the MHD description of plasmas

To derive the MHD equations, we continue the exposition of Section 2.3.1. Again, the derivation will omit most of the intermediate steps, which may be found in Chapter 3. We now specify the plasma to consist of electrons, $q_e = -e$, and one kind of ion, with $q_i = Ze$. Eqs. (2.58)–(2.60) then give the double set of *two-fluid moment equations*, which is closed by the specification of the transport coefficients, like in Eqs. (2.64). Since macroscopic dynamics on the MHD time scale is generally much faster than changes due to dissipative transport, as indicated in Eq. (2.111), we neglect most of the dissipative terms:

$$\pi_{e,i} \to 0, \quad \mathbf{h}_{e,i} \to 0 \quad \text{(neglect of viscosity and heat flow).} \tag{2.112}$$

However, we still keep the small terms due to momentum transfer and generated heat associated with resistivity, for reasons explained below, so that

$$\mathbf{R}_e = -\mathbf{R}_i \approx en_e\eta\mathbf{j}, \tag{2.113}$$

$$Q_e + Q_i = -(\mathbf{u}_e - \mathbf{u}_i) \cdot \mathbf{R}_e \approx \eta|\mathbf{j}|^2 \quad \text{(scalar resistivity).} \tag{2.114}$$

68 *Elements of plasma physics*

From Eqs. (2.58), (2.59) and (2.63), we then get the following set of *resistive two-fluid equations* (with $\alpha = e$ for the electrons and $\alpha = i$ for the ions):

$$\frac{\partial n_\alpha}{\partial t} + \nabla \cdot (n_\alpha \mathbf{u}_\alpha) = 0, \qquad (2.115)$$

$$n_\alpha m_\alpha \left(\frac{\partial \mathbf{u}_\alpha}{\partial t} + \mathbf{u}_\alpha \cdot \nabla \mathbf{u}_\alpha \right) + \nabla p_\alpha - n_\alpha q_\alpha (\mathbf{E} + \mathbf{u}_\alpha \times \mathbf{B}) = \mathbf{R}_\alpha, \qquad (2.116)$$

$$\frac{\partial p_\alpha}{\partial t} + \mathbf{u}_\alpha \cdot \nabla p_\alpha + \gamma p_\alpha \nabla \cdot \mathbf{u}_\alpha = (\gamma - 1) Q_\alpha. \qquad (2.117)$$

This set is completed by adding Maxwell's equations (2.13).

Next, we combine the two-fluid equations such that a set of nearly equivalent one-fluid equations is obtained. This is done by defining macroscopic one-fluid variables that are linear combinations of the two-fluid variables:

$$\rho \equiv n_e m_e + n_i m_i \qquad \text{(total mass density)}, \qquad (2.118)$$

$$\tau \equiv -e(n_e - Z n_i) \qquad \text{(charge density)}, \qquad (2.119)$$

$$\mathbf{v} \equiv (n_e m_e \mathbf{u}_e + n_i m_i \mathbf{u}_i)/\rho \qquad \text{(centre of mass velocity)}, \qquad (2.120)$$

$$\mathbf{j} \equiv -e(n_e \mathbf{u}_e - Z n_i \mathbf{u}_i) \qquad \text{(current density)}, \qquad (2.121)$$

$$p \equiv p_e + p_i \qquad \text{(pressure)}. \qquad (2.122)$$

(Notice the new meaning of $\mathbf{v}$, which can now be used without confusion with the particle velocities since distribution functions will not be used any more in this chapter.) This implies that no information of the mass and momentum equations is lost since the one-fluid variables $\rho, \tau, \mathbf{v}, \mathbf{j}$ are just linear combinations of the two-fluid variables $n_e, n_i, \mathbf{u}_e, \mathbf{u}_i$. The essential assumption of one-fluid dynamics is that the temperature equilibration time between electrons and ions is short compared to other characteristic times, so that $T_e = T_i$. (This assumption presents a significant simplification of the model, but one has to keep in mind that some plasmas in nature (e.g. the solar wind) are not adequately described this way.) This implies that the information on the separate electron and ion temperatures is annihilated so that there is one variable fewer in one-fluid theory and, hence, one equation fewer to be solved.

The one-fluid evolution equations are then obtained by operating on the two-fluid equations (2.115)–(2.117) by adding pairs, multiplied with mass and charge factors:

$m_e \, (2.115)_e + m_i \, (2.115)_i \Rightarrow \partial \rho / \partial t, \qquad -e \, (2.115)_e + Ze \, (2.115)_i \Rightarrow \partial \tau / \partial t,$

$(2.116)_e + (2.116)_i \Rightarrow \partial \mathbf{v} / \partial t, \qquad -\dfrac{e}{m_e} (2.116)_e + \dfrac{Ze}{m_i} (2.116)_i \Rightarrow \partial \mathbf{j} / \partial t,$

$(2.117)_e + (2.117)_i \Rightarrow \partial p / \partial t.$

2.4 Fluid description

In principle, this results in a set of one-fluid equations in terms of the variables ρ, τ, $\mathbf{v}$, $\mathbf{j}$, and p alone. However, to remove all two-fluid variables from the equations one needs to exploit the inverses of the relations (2.118)–(2.121). These simplify significantly (see Eqs. (3.132) of Section 3.4.1) if one exploits the approximation

$$|n_e - Zn_i| \ll n_e \qquad \text{(quasi charge-neutrality)}, \qquad (2.123)$$

which is extremely well satisfied already at the microscopic level, as we have seen in Sections 1.4.1 and 2.3.2. The resulting one-fluid equations still contain the small length and time scale phenomena of the two-fluid equations (see Eqs. (3.135)–(3.139) of Section 3.4.1). Those are removed by the use of two additional approximations:

$$|\mathbf{u}_i - \mathbf{u}_e| \ll v \qquad \text{(small relative velocity of ions and electrons)}, \qquad (2.124)$$

which implies that the electron skin depth should be small, $\delta_e \ll (m_e/m_i)^{1/2}$, and

$$v \ll c \qquad \text{(non-relativistic speeds)}. \qquad (2.125)$$

As a result, the evolution expressions for the variables τ and $\mathbf{j}$ disappear.

Combining the one-fluid moment equations thus obtained with the pre-Maxwell equations (i.e., according to Section 2.2.2, dropping the displacement current and Poisson's equation), results in *the resistive MHD equations*:

$$\frac{\partial \rho}{\partial t} + \nabla \cdot (\rho \mathbf{v}) = 0 \qquad \text{(continuity)}, \qquad (2.126)$$

$$\rho \left(\frac{\partial \mathbf{v}}{\partial t} + \mathbf{v} \cdot \nabla \mathbf{v} \right) + \nabla p - \mathbf{j} \times \mathbf{B} = 0 \qquad \text{(momentum)}, \qquad (2.127)$$

$$\frac{\partial p}{\partial t} + \mathbf{v} \cdot \nabla p + \gamma p \nabla \cdot \mathbf{v} = (\gamma - 1)\eta |\mathbf{j}|^2 \qquad \text{(internal energy)}, \qquad (2.128)$$

$$\frac{\partial \mathbf{B}}{\partial t} + \nabla \times \mathbf{E} = 0 \qquad \text{(Faraday)}, \qquad (2.129)$$

where

$$\mathbf{j} = \mu_0^{-1} \nabla \times \mathbf{B} \qquad \text{(Ampère)}, \qquad (2.130)$$

$$\mathbf{E}' \equiv \mathbf{E} + \mathbf{v} \times \mathbf{B} = \eta \mathbf{j} \qquad \text{(Ohm)}, \qquad (2.131)$$

and

$$\nabla \cdot \mathbf{B} = 0 \qquad \text{(no magnetic monopoles)} \qquad (2.132)$$

is just an initial condition on Faraday's law.

We will have plenty of opportunity to come to appreciate the power of this set of equations for the description of macroscopic plasma dynamics. For now, it suffices to make a few remarks.

(a) The momentum equation (2.127) represents balance between acceleration, on the one hand, and pressure gradient and Lorentz force (per unit volume), on the other. Additional forces, like the gravitational force $\mathbf{F}_{\text{grav}} = \rho \mathbf{g}$, may be put on the RHS of this equation. In laboratory plasmas, that force is completely negligible compared to the pressure gradient and the Lorentz force. Note that the Lorentz force is directed *perpendicular to* the magnetic field, so that acceleration *along* magnetic field lines must be produced by pressure gradients or gravity.

(b) The electric current density $\mathbf{j}$ and the electric field $\mathbf{E}$ have become secondary quantities in MHD, to be derived from Ampère's law and Ohm's law. Substituting them in Faraday's law (2.129) yields *the induction equation*:

$$\frac{\partial \mathbf{B}}{\partial t} = \nabla \times (\mathbf{v} \times \mathbf{B}) - \mu_0^{-1} \nabla \times (\eta \nabla \times \mathbf{B}). \qquad (2.133)$$

This equation couples the dynamics of the magnetic field to that of the plasma through the velocity term. When $\mathbf{v}$ was known, the induction equation could be used to determine $\mathbf{B}$, where $\nabla \cdot \mathbf{B} = 0$ should be imposed as an initial condition.

(c) The algebraic relation (2.131) between the electric field $\mathbf{E}$ and the electric current density $\mathbf{j}$ is the generalization of Ohm's law for moving conducting media. According to this law, $\mathbf{j}$ is proportional to the electric field $\mathbf{E}'$ in a frame moving with the plasma. Many laboratory and astrophysical plasmas are nearly perfectly conducting, so that

$$\mathbf{E}' \equiv \mathbf{E} + \mathbf{v} \times \mathbf{B} = 0 \qquad \text{(perfect conductivity: ideal MHD)} \qquad (2.134)$$

almost everywhere. (Also recall from Section 2.2.3 that single particles drift in such a way that $\mathbf{E}' = 0$.) Therefore, the resistive form (2.131) of Ohm's law is only needed in regions of high current concentration, which are usually extremely thin. However, such current sheets do occur and play an important role in models for the disruptive phenomena mentioned above. That is why we have kept the term $\eta \mathbf{j}$ in Ohm's law. For the same reason, we have kept the term $(\gamma - 1)\eta |\mathbf{j}|^2$, representing heating due to Ohmic dissipation, in the energy equation (2.128). This term is usually very small compared to the terms on the left hand side.

(d) The slow time scale for *resistive diffusion* of the magnetic field, introduced in Eq. (2.111), can be estimated from the induction equation (2.133):

$$\tau_R \sim \frac{\mu_0 l_0^2}{\eta}, \qquad (2.135)$$

2.4 Fluid description

where $l_0 \sim a$ is the length scale of gradients. In terms of a representative plasma speed v_0 and length scale l_0, the ratio of the magnitudes of the convective and diffusive terms on the RHS of the induction equation (2.133) is a dimensionless number,

$$R_m \equiv \frac{\mu_0 l_0 v_0}{\eta}, \qquad (2.136)$$

called the *magnetic Reynolds number*, in analogy with the Reynolds number $R \equiv v_0 l_0 / \nu$ (where $\nu \equiv \mu/\rho$ is the kinematic viscosity), measuring the relative importance of inertial and viscous effects in ordinary hydrodynamics. The magnetic Reynolds number is a measure of the strength of the coupling between the flow and the magnetic field. Since η is extremely small for most plasmas of interest (e.g. tokamaks and the solar corona), $R_m \gg 1$ so that this coupling is very strong and resistive decay is negligible on the time scales specified in Eqs. (2.111).

The neglect of resistivity, and the substitution of **j** and **E**, finally leads to *the ideal MHD equations* in their most compact form:

$$\frac{\partial \rho}{\partial t} + \nabla \cdot (\rho \mathbf{v}) = 0, \qquad (2.137)$$

$$\rho \left(\frac{\partial \mathbf{v}}{\partial t} + \mathbf{v} \cdot \nabla \mathbf{v} \right) + \nabla p - \mu_0^{-1} (\nabla \times \mathbf{B}) \times \mathbf{B} = 0, \qquad (2.138)$$

$$\frac{\partial p}{\partial t} + \mathbf{v} \cdot \nabla p + \gamma p \nabla \cdot \mathbf{v} = 0, \qquad (2.139)$$

$$\frac{\partial \mathbf{B}}{\partial t} - \nabla \times (\mathbf{v} \times \mathbf{B}) = 0, \qquad \nabla \cdot \mathbf{B} = 0. \qquad (2.140)$$

These *coupled nonlinear partial differential equations* govern the evolution of the density, the velocity, the pressure and the magnetic field. Their properties and solutions will occupy us for most of the remainder of this book.

2.4.2 Alfvén waves

Let us consider wave propagation in a *homogeneous* plasma occupying all space, with a constant magnetic field in the z-direction:

$$\rho_0 = \text{const}, \quad \mathbf{v}_0 = 0, \quad p_0 = \text{const}, \quad \mathbf{B}_0 = B_0 \, \mathbf{e}_z \Rightarrow \mathbf{j}_0 = 0. \qquad (2.141)$$

We perturb this plasma with small deviations ρ_1, $\mathbf{v}_1$, p_1, $\mathbf{B}_1$ from the background state. This permits us to *linearize* the equations (2.137)–(2.140):

$$\frac{\partial \rho_1}{\partial t} = -\rho_0 \nabla \cdot \mathbf{v}_1, \tag{2.142}$$

$$\rho_0 \frac{\partial \mathbf{v}_1}{\partial t} = -\nabla p_1 + \mu_0^{-1}(\nabla \times \mathbf{B}_1) \times \mathbf{B}_0, \tag{2.143}$$

$$\frac{\partial p_1}{\partial t} = -\gamma p_0 \nabla \cdot \mathbf{v}_1, \tag{2.144}$$

$$\frac{\partial \mathbf{B}_1}{\partial t} = \nabla \times (\mathbf{v}_1 \times \mathbf{B}_0), \tag{2.145}$$

producing a complete set of equations for the unknowns ρ_1, $\mathbf{v}_1$, p_1, and $\mathbf{B}_1$.

Note that the density perturbation ρ_1 only appears in Eq. (2.142) so that it need not be determined to solve the other equations. Moreover, in order to concentrate on the magnetic effects, let us neglect the equilibrium pressure ($p_0 = 0$ or, in terms of the parameter β introduced in Section 2.2.3, $\beta = 0$) so that Eq. (2.144) implies that $p_1 = 0$ as well. We then obtain from the remaining Eqs. (2.143) and (2.145) a wave equation for the velocity $\mathbf{v}_1$:

$$\rho_0 \frac{\partial^2 \mathbf{v}_1}{\partial t^2} = \mu_0^{-1} \left(\nabla \times \frac{\partial \mathbf{B}_1}{\partial t} \right) \times \mathbf{B}_0 = \mu_0^{-1} \mathbf{B}_0 \times \left(\nabla \times \left(\nabla \times (\mathbf{B}_0 \times \mathbf{v}_1) \right) \right). \tag{2.146}$$

Inserting plane wave solutions of the form

$$\mathbf{v}_1(\mathbf{r}, t) = \hat{\mathbf{v}}\, e^{i(\mathbf{k}\cdot\mathbf{r} - \omega t)}, \tag{2.147}$$

i.e. replacing $\partial/\partial t \to -i\omega$ and $\nabla \to i\mathbf{k}$, Eq. (2.146) transforms into an algebraic eigenvalue equation

$$-\rho_0 \omega^2 \hat{\mathbf{v}} = -\mu_0^{-1} B_0^2\, \mathbf{e}_z \times \left(\mathbf{k} \times \left(\mathbf{k} \times (\mathbf{e}_z \times \hat{\mathbf{v}}) \right) \right). \tag{2.148}$$

Because the expression on the RHS is a vector $\perp \mathbf{e}_z$ (i.e. $\perp \mathbf{B}_0$), the immediate result is that the parallel velocity vanishes: $\hat{v}_\parallel \equiv \mathbf{e}_z \cdot \hat{\mathbf{v}} = 0$. This leaves two components for $\hat{\mathbf{v}}_\perp$, which turn out to oscillate independently, each with its own characteristic frequency. The flow fields of the two modes differ with respect to their direction relative to the plane through $\mathbf{k}$ and $\mathbf{B}_0$. To distinguish them, it is convenient to fix the direction of the wave vector $\mathbf{k}$ to lie in the x-z plane, as indicated in Fig. 2.9: $\mathbf{k} = (k_\perp, 0, k_\parallel)$. Obviously, this can be done without loss of generality. Focusing on the most significant branch, viz. the one with a flow velocity perpendicular to both $\mathbf{k}$ and $\mathbf{B}_0$ (i.e. $\hat{\mathbf{v}} \sim \mathbf{k} \times \mathbf{B}_0$), and manipulating with the vector

2.4 Fluid description

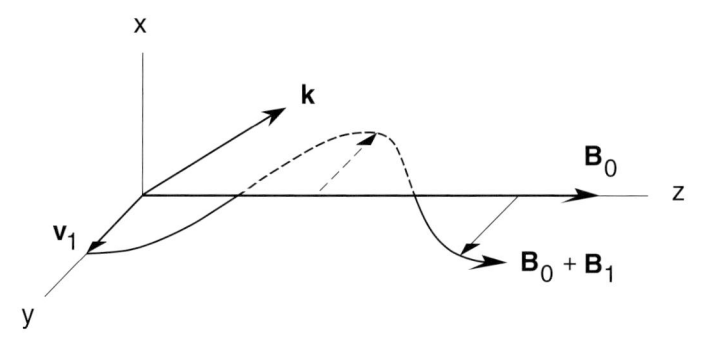

Fig. 2.9. Alfvén wave.

products (using Eq. (A.2) of Appendix A), the eigenvalue problem becomes

$$(\omega^2 - k_\parallel^2 v_A^2)\,\hat{v}_y = 0, \tag{2.149}$$

where

$$v_A \equiv \frac{B_0}{\sqrt{\mu_0 \rho_0}} \tag{2.150}$$

is the celebrated *Alfvén velocity*. Hence, one obtains the two *Alfvén waves*, discovered by Alfvén in 1942 [5], with frequency

$$\omega = \pm \omega_A, \quad \text{where} \quad \omega_A \equiv k_\parallel v_A = \mathbf{k} \cdot \mathbf{B}/\sqrt{\mu_0 \rho_0}, \tag{2.151}$$

and phase velocity v_A, corresponding to waves that run along the magnetic field to the right (+) or to the left (−).

▷ **Magneto-sonic waves.** For $\omega = 0$, an extra solution is obtained, viz. the remnant of the slow magneto-sonic mode, with $\hat{v}_\parallel \neq 0$. Another branch, $\omega^2\,\hat{v}_x = (k_\perp^2 + k_\parallel^2) v_A^2\,\hat{v}_x$, represents degenerate fast magneto-sonic waves which, accidentally, also propagate with the Alfvén speed. These waves are not well described here since the pressure has been neglected. They will be discussed more extensively later (in Chapter 5). ◁

The Alfvén waves are caused by tension of the magnetic field lines, which tends to restore the initial shape (as schematically indicated in Fig. 2.9). Another important aspect of the macroscopic MHD waves is that their frequency depends on the value of the wave vector **k**, i.e. on the wavelength, in contrast to the microscopic plasma oscillations with the plasma frequency (2.73), which is independent of the wavelength. This points to an extremely significant property of Alfvén waves, and of MHD perturbations in general: the slowest of them have the longest wavelengths and, hence, they sample the magnetic field in the large, i.e. they 'feel' the global magnetic geometry. Of course, infinite homogeneous plasmas do not have a particularly interesting magnetic geometry. However, when the magnetic field is bent into the relevant shape for plasma confinement, e.g. a torus, this property survives: *Alfvén waves carry the information of the overall magnetic geometry.*

Inserting numbers for a typical tokamak experiment, viz. $n = 10^{20}\,\mathrm{m}^{-3}$, $B = 3\,\mathrm{T}$, so that $\rho = nm_i = 1.7 \times 10^{-7}\,\mathrm{kg\,m^{-3}}$, we obtain a phase velocity

$$v_A \approx 6 \times 10^6\,\mathrm{m\,s^{-1}}$$

(rather fast, but still non-relativistic: only 2% of the velocity of light). For long wavelengths, e.g. $\lambda_\| = 20\,\mathrm{m}$ ($\approx 2\pi R$ for a torus with major radius $R \approx 3\,\mathrm{m}$), so that $k_\| \equiv 2\pi/\lambda_\| \approx 0.3\,\mathrm{m^{-1}}$, this gives an Alfvén frequency of

$$\omega_A = k_\| v_A \approx 1.8 \times 10^6\,\mathrm{s^{-1}}\,.$$

Hence, with wavelengths of the order of the size of the confinement experiment, we are now on the μsec time scale: $\tau = 2\pi R/v_A = 2\pi/\omega_A \approx 3\,\mu\mathrm{sec}$.

▷ **Exercise.** Consult Table B.5 for corresponding values of a solar plasma. What is the Alfvén transit time for a solar coronal loop of length $L = 100\,000\,\mathrm{km}$, $n = 10^{14}\,\mathrm{m^{-3}}$, $B = 0.003\,\mathrm{T}$? How does this compare to the 'times of interest' indicated in this table? The latter refer to time scales needed for fusion in tokamaks or typical life times of coronal loops. How do tokamaks compare with coronal loops as far as the number of Alfvén transit times during that time span is concerned? Also complete the empty column of Table B.6. ◁

On the Alfvén transit time scale, the overall magnetic confinement geometry becomes 'known' to the plasma, which then will exhibit the dynamics corresponding to the intrinsic stability or instability of the configuration. These possibilities necessarily involve inhomogeneous plasmas with curved magnetic fields. This brings us to our next topic.

2.4.3 Equilibrium and stability

The basic theoretical approach to plasma confinement for a given magnetic geometry typically consists of, first, fixing an *equilibrium* state, next, determining the different kinds of *waves* produced by perturbing this state, and, finally, finding out whether amongst those perturbations there are *instabilities* that would lead to destruction of the configuration. We have just encountered an example of the second topic, viz. Alfvén waves. We will now treat simple examples of the first and last topic.

Consider an *inhomogeneous* plasma in *static* equilibrium, i.e. the fluid is at rest ($\mathbf{v} = 0$) and all time derivatives in the MHD equations (2.137)–(2.140) vanish. In contrast to ordinary hydrodynamics, even this highly idealized situation is non-trivial because of the possibility of balancing pressure gradients by the Lorentz

2.4 Fluid description

force:
$$\nabla p = \mathbf{j} \times \mathbf{B}. \tag{2.152}$$

This is the basis of all *magnetic confinement* systems for controlled thermonuclear fusion experiments. (In laser fusion, or inertial confinement experiments, the inertial forces replace the magnetic ones.) The current, of course, has to satisfy Ampère's law

$$\mathbf{j} = \mu_0^{-1} \nabla \times \mathbf{B}, \tag{2.153}$$

and the magnetic field should satisfy

$$\nabla \cdot \mathbf{B} = 0. \tag{2.154}$$

The latter equation is no longer an initial condition now, but it remains a restriction on the kind of vectors that qualify as magnetic fields. These are the equations which the variables p, $\mathbf{j}$, and $\mathbf{B}$ have to satisfy in order to produce a *static equilibrium* state.

(a) Equilibrium of a z-pinch As an example of a simple equilibrium, consider a cylindrical plasma column with a strong electric current. This current creates a magnetic field, resulting in a force $\mathbf{j} \times \mathbf{B}$ directed radially inward: *the pinch effect*. On the basis of this effect, two interesting plasma confinement configurations can be distinguished, viz. the θ- and the z-pinch illustrated in Fig. 1.4. Here, the labels θ and z refer to the direction of the current in the cylindrical coordinate representation of the plasma column. As we have seen in Section 1.2.3, both types of configurations have been considered in the early days of nuclear fusion research, both without success, at least in their pure form, but the concepts remain important because they represent the basic forces in plasma confinement.

For *the z-pinch* in cylinder geometry (current in the z-direction, magnetic field in the θ-direction, gradients in the r-direction; see Fig. 2.10) the equilibrium equations (2.152)–(2.154) reduce to

$$\frac{dp}{dr} = -j_z B_\theta, \quad \text{and} \quad j_z = \frac{1}{\mu_0 r} \frac{d}{dr}(r B_\theta). \tag{2.155}$$

This gives the following relationship between the pressure profile $p(r)$ and the magnetic field profile $B_\theta(r)$:

$$\frac{dp}{dr} = -\frac{B_\theta}{\mu_0 r} \frac{d}{dr}(r B_\theta). \tag{2.156}$$

This is the only restriction on these profiles as far as the equilibrium is concerned. We are free to choose e.g., for simplicity, a constant current profile $j_z = $ const.

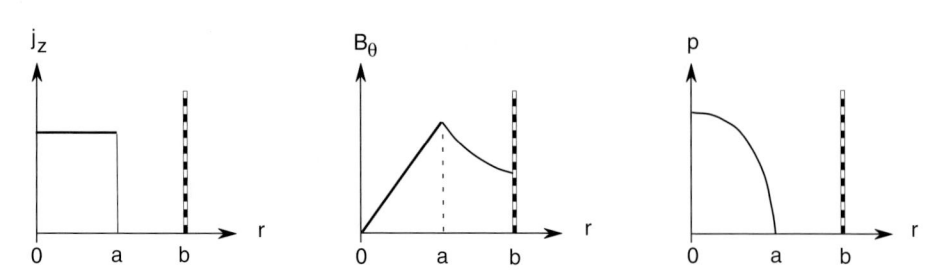

Fig. 2.10. Radial distributions of current density, magnetic field and pressure in a z-pinch.

This choice determines the other profiles:

$$B_\theta = \tfrac{1}{2}\mu_0 r j_z, \quad \text{and} \quad p = p_c(1 - r^2/a^2), \quad p_c \equiv \tfrac{1}{4}\mu_0 a^2 j_z^2, \qquad (2.157)$$

where a is the radius of the plasma cylinder.

The plasma cylinder may be surrounded by a vacuum with a magnetic field $\hat{\mathbf{B}}$, satisfying $\nabla \times \hat{\mathbf{B}} = 0$, $\nabla \cdot \hat{\mathbf{B}} = 0$, so that $\hat{j}_z = 0$ and $\hat{B}_\theta(r) = B_\theta(a)\, a/r$ there. Obviously, this radially decaying magnetic field is produced by the total current I_z flowing within the plasma interval $0 \leq r \leq a$. Finally, the configuration may be closed off by putting a perfectly conducting wall at some radius $r = b$. The central pressure p_c is related to the total current I_z:

$$I_z = \pi a^2 j_z, \quad \text{so that} \quad p_c = \frac{\mu_0 I_z^2}{4\pi^2 a^2}. \qquad (2.158)$$

Inserting typical numbers for the early laboratory pinch plasmas, viz. $n = 10^{22}$ m^{-3} (i.e. two orders higher than present-day tokamaks), $T = 10^8$ K, $a = 0.1$ m, we find

$$p_c = nkT = 1.38 \times 10^7 \, \text{N m}^{-2} \ (= 136 \, \text{atm !}),$$

$$I_z = 2\pi a (p_c/\mu_0)^{1/2} = 2.1 \times 10^6 \, \text{A},$$

$$B_\theta = \mu_0 I_z/(2\pi a) = 4.2 \, \text{T} \ (= 42 \, \text{kgauss}).$$

Although these currents and fields are quite large and exert huge pressures, they were already well within reach of the technology of the 1950s: a thermonuclear reactor by just passing a current through a linear plasma column?

(b) Kink instabilities In Section 2.4.2, we found the frequency of Alfvén waves for a homogeneous plasma with straight magnetic field lines. However, for the z-pinch equilibrium, the field lines are curved. Consequently, the analysis of the dispersion equation is significantly more complicated. We will not enter that analysis here

2.4 Fluid description

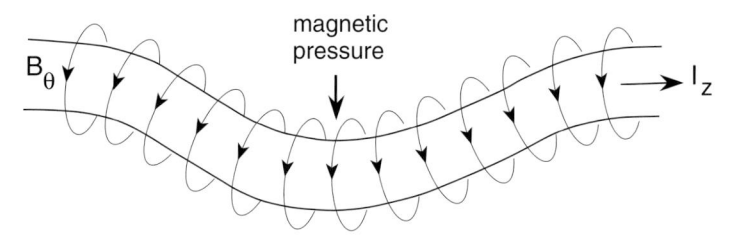

Fig. 2.11. External kink instability of a z-pinch.

since it will be the subject of Chapter 9, but just indicate a result that is relevant to the present discussion. Amongst the normal mode solutions $\sim \exp[i(m\theta + kz - \omega t)]$ of the cylindrical configuration there are some 'Alfvén' waves, with low mode number m (typically $m = 1$), that have a complex frequency ($\omega^2 < 0$) so that they are exponentially growing. These modes are called *kink instabilities* because of the associated helical deformation of the plasma column. For wavelengths $k_z^{-1} \gg a$, the expression for their growth rate demonstrates the cause of the instability, viz. the curved magnetic field component $B_\theta(a)$ at the plasma edge:

$$\omega^2 \approx -\frac{B_\theta^2(a)}{2\mu_0\rho_0 a^2}. \tag{2.159}$$

As illustrated by Fig. 2.9, the magnetic force $\mathbf{F} = \mathbf{j}_1 \times \mathbf{B}_0$ in the RHS of Eq. (2.143) is opposite to $\mathbf{v}_1$ in the case of genuine Alfvén waves. Hence, this force is restoring, i.e. stabilizing (the effect counteracts its cause). However, in the case of a z-pinch equilibrium there is an extra term $\mathbf{j}_0 \times \mathbf{B}_1$, due to the plasma current, which points in the same direction as $\mathbf{v}_1$ so that it is destabilizing (the effect enhances its cause). That term involves the detailed distribution of the magnetic field B_θ in the plasma ($r \leq a$) *and* the surrounding vacuum ($r > a$) shown in Fig. 2.10. For the latter reason, these modes are called *external* kink instabilities.

The mechanism of the external kink instability can easily be illustrated from the similar situation of a current carrying wire. Here, a helical deformation of the wire tends to be magnified due to the compression of the field lines on the inside and expansion on the outside bends of the deformation. The net result is an unbalanced increase of the *magnetic pressure* $B^2/(2\mu_0)$ on the inside as compared to the outside bends (see Fig. 2.11). This is fatal for the z-pinch since time scales for the instability are of the same order of magnitude as the time scale of Alfvén waves, viz. microseconds, whereas at least seconds are needed for thermonuclear ignition. That answers the question posed above: no simple fusion reactor is obtained this way: no free rides in science and technology!

For thermonuclear confinement, this is the first reason that something more clever had to be done than just a straight z-pinch. Here, the long history of thermonuclear research starts: the next step was the linear θ-pinch, which was equally disastrous because of the open ends. Success gradually came when these two concepts were combined in the tokamak (illustrated in Fig. 1.4).

How is a violent instability like the external kink mode remedied in a tokamak? For a change, this turned out to be quite simple. The cylinder is replaced by a torus and, since kink modes are long wavelength instabilities, the parameters may be chosen such that the range of unstable wavelengths simply does not fit in the torus. This leads to the famous *Kruskal–Shafranov condition* for external kink mode stability, which puts a limit on the total plasma current:

$$I_z(a) < \frac{2\pi a^2 B_z}{\mu_0 R_0}. \tag{2.160}$$

Notice that this requires a contribution of the θ-pinch concept: a stabilizing 'backbone' of longitudinal magnetic field B_z is necessary. As is evident from this expression, a z-pinch ($B_z = 0$) does not have a stable range of the current.

The Kruskal–Shafranov condition can also be written in terms of the '*safety factor*' $q(a)$. This concept is easily understood by considering a double periodic cylinder (periodic in θ with period 2π and in z with period $2\pi R_0$) as a mathematical model for a torus with an inverse aspect ratio $\epsilon \equiv a/R_0$. The physical justification of such a model is that it provides a first approximation to a genuine, but slender ($\epsilon \ll 1$), torus which neglects the toroidal curvature but accounts for the more important toroidal periodicity. The cylindrical helical magnetic field $\mathbf{B} = B_\theta(r)\mathbf{e}_\theta + B_z(r)\mathbf{e}_z$ then provides a model of nested magnetic surfaces (cylinders of radius $r \leq a$ and length $2\pi R_0$) with field lines having a pitch that is constant on each surface, but varies from surface to surface. This may be visualized by unrolling one of the cylinders (Fig. 2.12). The parameter $q(r)$ is defined as the inverse of the pitch of the field lines with a normalization chosen such that $q = 1$

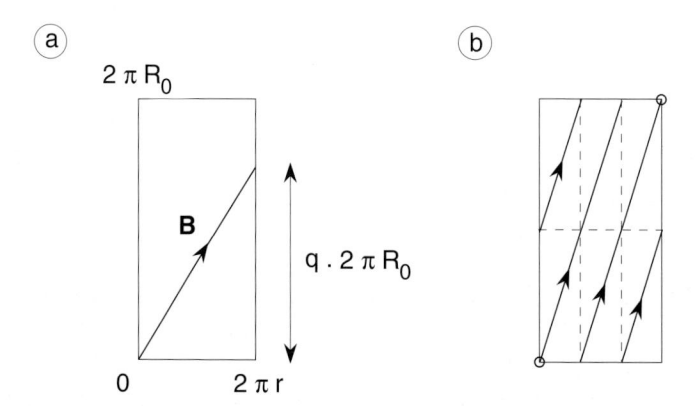

Fig. 2.12. 'Safety factor' in periodic cylinder model of a toroidal plasma: (a) field line with $q < 1$; (b) rational field line with $q = 3/2$.

corresponds to a topology where the field lines close upon themselves after one revolution the short way and one revolution the long way around the torus:

$$q(r) \equiv \frac{rB_z(r)}{R_0 B_\theta(r)} = \frac{2\pi r^2 B_z(r)}{\mu_0 R_0 I_z(r)}. \tag{2.161}$$

Hence, the Kruskal–Shafranov limit can be expressed as $q(a) > 1$. This *accidental agreement* with an integer value is rather unfortunate since it has led to many erroneous statements in the literature about this limit being connected with rational field lines 'biting' in their own tail. Actually, this has nothing to do with the external kink mode mechanism, which is driven by the total plasma current $I_z(a)$ creating a curved magnetic field $B_\theta(a)$ on the outside of the plasma column. In fact, in more general configurations, like a genuine torus with non-circular cross-section, the external kink mode is no longer associated with integer values of $q(a)$.

Curing the remaining instabilities (the internal ones, not perturbing the vacuum magnetic field) was not so simple. Here, rational values of the safety factor do play an important role since most internal instabilities are localized about rational field lines and surfaces where $q = -m/n$, when m and n indicate the mode numbers in the periodic directions (see Fig. 2.12(b)). This has led to a long period of steady experimental trial and error, eventually leading to confinement on the second and even minute time scales in the last part of the twentieth century.

2.5 In conclusion

In this chapter, we have introduced the three main theoretical approaches of plasmas, viz. the theory of single particle motion in prescribed electric and magnetic fields, the kinetic theory of collections of many such particles (with roughly equal numbers of the opposite charges), and the theory of magnetohydrodynamics pertaining to the global macroscopic aspects of the dynamics of plasmas in complex magnetic field geometries. We have also considered a sample of the many different dynamical phenomena associated with these theoretical models. Three effects were encountered which give plasmas the *coherence* that is necessary for thermonuclear confinement of laboratory plasmas and which is also characteristic for magnetized plasmas encountered in nature.

(a) Within the single particle picture, we found that *particles of either charge stick to the magnetic field lines by their gyro-motion* restraining the perpendicular motion.
(b) Because of the large electric fields that occur when electrons and ions are separated, deviations from charge neutrality can occur only in very small regions (of the size of a Debye length). Over larger regions, *ions and electrons stay together to maintain approximate charge neutrality*.

(c) In the fluid picture, it was found that *currents in the plasma create their own magnetic field, pinching the plasma,* and that *Alfvén waves act as a restoring agent on magnetic field distortions*. However, we also encountered the first effect *destroying the coherence of plasmas, viz. the global external kink instability*.

In the next chapter, we delve more deeply in the kinetic theory aspects (loosely considering two-fluid theory also under this heading, although it is strictly a fluid theory, because of the microscopic scales involved). The exposition of MHD is continued in Chapter 4.

2.6 Literature and exercises

Notes on literature

The list below is a very limited choice from the numerous books and articles on the topics introduced in this chapter, which are discussed in all textbooks on plasma physics.

Basic texts on classical mechanics and electrodynamics:

- Goldstein, *Classical Mechanics* [91].
- Jackson, *Classical Electrodynamics* [117].

Single particle motion:

- Chen, *Introduction to Plasma Physics* [53], Chapter 2 on single particle motions (elementary).
- Sturrock, *Plasma Physics* [221], Chapters 3–5 on orbit theory and adiabatic invariants (intermediate).
- Northrop, *The Adiabatic Motion of Charged Particles* [169]: a whole monograph on adiabatic motion of charged particles (advanced).
- Balescu, *Transport Processes in Plasmas* [14], Chapter 1 on modern developments in Hamiltonian mechanics of charged particles (advanced).

Kinetic theory and Landau damping:

- Bittencourt, *Fundamentals of Plasma Physics* [31], Chapters 5–8 on kinetic theory and the derivation of macroscopic equations, Chapter 18 on waves in hot plasmas with Landau damping (elementary).
- Goldston & Rutherford, *Introduction to Plasma Physics* [92], Chapters 22–24 on kinetic theory and Landau damping (elementary).
- Hasegawa & Sato, *Space Plasma Physics* [106], Chapter 1 on the physics of stationary plasmas with condensed material on motion of charged particles, adiabatic invariants, kinetic theory, and MHD equations (intermediate).

Fluid theory and MHD:

- Freidberg, *Ideal Magnetohydrodynamics* [72], Chapter 2 on the derivation of the MHD equations from the Boltzmann equation (intermediate).

- Roberts, *An Introduction to Magnetohydrodynamics* [194], Chapter 1 on the MHD equations as a part of continuum mechanics (intermediate).
- Alfvén's 1942 paper on the 'Existence of electromagnetic-hydrodynamic waves' [5] is just one column in *Nature*, with one paragraph on the discovery of Alfvén waves, that earned him the Nobel prize. The last paragraph, with an application to sunspot migration, is an unfortunate example (the waves he discovered propagate along the magnetic field, not across) at a time when solar coronal dynamics was still largely unknown.

Exercises

[2.1] *Cyclotron motion*

We start from the equation of motion for an electron or ion in an electromagnetic field, with **B** in the z-direction and **E** in the (y, z)-plane.

- Write down the component for the motion parallel to the magnetic field. What happens to electrons when collisions are negligible?
- Now, choose **E** perpendicular to **B** and solve the full equation of motion. What is different compared to a situation with $\mathbf{E} = 0$? Is there an extra velocity component and, if so, in which direction? Is there a difference between electrons and ions?
- Assume particles with thermal perpendicular speed, $v_{\text{th}} = (2kT/m)^{1/2}$, at room temperature! Calculate the gyro-frequencies and radii for electrons and protons for $B = 1\,\text{T}$. Which particles have the largest gyro-radius? How do they encircle the field lines?
- For a background density typical of tokamaks, $n = 10^{20}\,\text{m}^{-3}$, calculate the plasma frequency and the Debye length. Compare this to the gyro-radius. What do you conclude?

[2.2] *Drift velocities*

Assume we have a cyclotron device with a magnetic field strength of 1 T in the z-direction. The radius of the device is 2 metres.

- On both sides we put a plate, with a potential difference of $+300$ V between the right and left one. Calculate the drift velocities for electrons and protons.
- We remove the plates but take gravity into account. Unfortunately, the apparatus is old and the magnetic field lines are not vertical any more but slightly tilted, 2° to be specific. What are the drift velocities now? In what direction are they pointing?

[2.3] *Momentum equation*

Derive the momentum equation from the Boltzmann equation. (If you want to check your answer, the complete derivation is given in Section 3.2.2.)

[2.4] *Plasma oscillations*

We displace electrons with respect to ions in a cold, unmagnetized, plasma and study the resulting oscillations from the two-fluid equations and Poisson's law.

- Give an argument why the ion equations will not play a significant role.
- Since we are left with the electron equations now, linearize them and Poisson's law to construct the wave equation for the perturbed density. Assuming harmonic time dependence, find the frequency of the oscillations. What does it depend on?
- Try to design a device that can measure the background density profile of a plasma.

[2.5] *Landau damping*

What is Landau damping? What are the implications for possible wave motions in a plasma?

[2.6] *Alfvén waves*

We start from the ideal MHD equations (2.137)–(2.140). Assuming a static homogeneous plasma with a constant magnetic field in the z-direction, perturb all quantities and linearize.

- Consider a low β plasma. What does that mean for the pressure perturbations?
- Find the wave equation for the velocity. Insert plane wave solutions and construct the eigenvalue equation. What can you say at once about the velocity?
- Choose the **k** vector in the x-z plane and the velocity in the y-direction. Give the solutions to the new eigenvalue problem. How many are there?
- What does the Alfvén velocity depend on? Could that be useful for plasma diagnostics?

[2.7] *MHD time scales*

Estimate the longest time scale τ_{2F} of the different high frequency waves described by the two-fluid model. Also estimate the slow diffusion time scale τ_R of the resistive MHD model. Use a characteristic length scale l_0 for the gradients. Show that the window $\tau_{2F} \ll \tau_{MHD} \ll \tau_R$ easily accommodates Alfvén wave dynamics. Insert numbers for the different plasmas of Table B.5.

[2.8] *Equilibrium of a plasma cylinder*

Consider a cylindrical plasma column in static equilibrium, with a diameter of 2 metres and carrying a current of 10^6 A in the z-direction.

- Write down the equations for the pressure, the current density and the magnetic field in cylindrical coordinates. Determine the number of free profiles and parameters.
- Assuming a constant current density profile, determine the pressure profile and calculate the pressure at the centre of this cylinder. How does it compare with atmospheric pressures?

[2.9] *Kink instabilities*

How is the violent kink instability mechanism of the z-pinch eliminated in a tokamak? Assuming a 'straight tokamak' model of a finite-length periodic cylinder, determine the limit on the total current from the order of magnitude estimate $q = 1$ for the 'safety factor'.

3
'Derivation' of the macroscopic equations*

3.1 Two approaches*

There are basically two ways of introducing the equations of magnetohydrodynamics:

(a) pose them as reasonable postulates for a hypothetical medium called 'plasma';
(b) derive them by appropriate averaging of kinetic equations.

Our approach, starting with Chapter 4, is mainly along the lines of the first method, pioneered by Grad [98, 32] in a series of lecture notes, using physical arguments and mathematical criteria to justify the results. In this chapter, the main steps of the second method will be discussed and shown to be somewhat unsatisfactory since they involve a number of approximations that are often difficult to justify. The reason for going through this analysis anyway is that it provides understanding of the domain of validity of the MHD description and that it indicates what kind of modifications are in order when this description fails.

Mathematically inclined readers may skip this digression, where most results from kinetic theory are not derived but simply stated, and continue reading with Chapter 4. Also, the serious student of magnetohydrodynamics is advised to turn to a detailed study of the present chapter only after a first reading of Chapters 4–11 on basic MHD since the level of this chapter is essentially that of the advanced theory, but it has been placed here because this is where it logically belongs.

We give a 'derivation' of the MHD equations by averaging the kinetic equations for plasmas. The quotation marks are meant to remind the reader of the embarrassing assumptions that have to be made when starting from kinetic equations and then averaging them to obtain the transport coefficients and the macroscopic equations (like in the classics by Chapman and Cowling [52] and Braginskii [41]). The point is not so much that transport theory is in a poor state (an excellent monograph by Balescu [14] exists), but the justification of local classical transport in the face of the observed dissipative processes. Nature just appears to mock classical transport. For example, the very existence of the solar cycle, a magnetohydrodynamic phenomenon with a period of about 22 years, requires a turbulent resistivity which is a factor of about 10^{10} larger than the classical Spitzer value! Clearly, such order of magnitude 'anomalies' cannot be maintained indefinitely in a scientific enterprise.

3.2 Kinetic equations*

3.2.1 Boltzmann equation*

Recall our introductory exposition in Section 2.3. The plasma is considered as a collection of charged particles moving in an electromagnetic field $\mathbf{E}, \mathbf{B}$. Different species of particles, specifically ions and electrons, are distinguished by a subscript α. The electrons and ions are described by time-dependent distribution functions in six-dimensional phase space:

$$f_\alpha = f_\alpha(\mathbf{r}, \mathbf{v}, t) \qquad (\alpha = e, i). \tag{3.1}$$

The probable number of particles of species α in the six-dimensional volume element $d^3r\, d^3v$ centred at $\mathbf{r}, \mathbf{v}$ will then be $f_\alpha(\mathbf{r}, \mathbf{v}, t)\, d^3r\, d^3v$. The variation in time of these distribution functions is determined by a *Boltzmann equation* for each particle species:

$$\frac{df_\alpha}{dt} \equiv \frac{\partial f_\alpha}{\partial t} + \mathbf{v} \cdot \frac{\partial f_\alpha}{\partial \mathbf{r}} + \frac{q_\alpha}{m_\alpha}(\mathbf{E} + \mathbf{v} \times \mathbf{B}) \cdot \frac{\partial f_\alpha}{\partial \mathbf{v}} = C_\alpha. \tag{3.2}$$

Here, $\mathbf{E}$ and $\mathbf{B}$ consist of contributions of the external fields and of the averaged internal fields originating from the long-range inter-particle interactions. The RHS of Eq. (3.2) gives the rate of change of the distribution function due to short-range binary particle interactions called collisions. Neglect of these collisions leads to *the Vlasov equation*, also called the *collisionless Boltzmann equation*[1]:

$$\frac{\partial f_\alpha}{\partial t} + \mathbf{v} \cdot \frac{\partial f_\alpha}{\partial \mathbf{r}} + \frac{q_\alpha}{m_\alpha}(\mathbf{E} + \mathbf{v} \times \mathbf{B}) \cdot \frac{\partial f_\alpha}{\partial \mathbf{v}} = 0. \tag{3.3}$$

A closed system of equations is obtained by adding Maxwell's equations to determine $\mathbf{E}(\mathbf{r}, t)$ and $\mathbf{B}(\mathbf{r}, t)$.

At this point, some remarks are in order.

(a) In this chapter, we assume that the plasma consists of electrons and one kind of ion so that we just have the two distribution functions f_e and f_i to worry about (more than enough!). The temperature is assumed to be high enough to have complete ionization, so that ionization, recombination and charge exchange processes

---

[1] Therefore, punned the 'Boltzmannless' Boltzmann equation by Rosenbluth.

no longer occur. We also assume that fusion reactions do not yet occur. There will be conservation of the total number of each of the two kinds of charged particles separately. These assumptions can be relaxed, but at the cost of enormously complicating the analysis by the appearance of even more unknown distribution functions. However, even with a large fraction of neutral particles present, one still can show dominance of magnetohydrodynamic processes. Our main interest here is to exhibit those.

(b) Following standard terminology in plasma physics, the basic kinetic equation (3.2) has been called the 'Boltzmann' equation but it should be stressed that the expression C_α for the rate of change of the distribution function by collisions is completely different from the one derived by Boltzmann himself. He was concerned with collisions of neutral particles, generally resulting in large-angle scattering, described by an integral of the product of distribution functions evaluated for the different relative velocities of the colliding particles. Whereas neutral particles only collide when they hit each other, charged particles interact through the long-range Coulomb force. In high-temperature plasmas, most of those scattering events are small-angle deflections. By judiciously exploiting this feature, Landau [135] (1936) was able to convert the Boltzmann collision integral into an expression valid for charged particles: Eq. (3.13) below. The proper way of handling the statistics of many small-angle deflections is by means of a Fokker–Planck equation. This was done by Rosenbluth *et al.* [196] (see also Trubnikov [231]), resulting in an expression that is equivalent to the Landau form (Eq. (3.16) below): another example of Landau's great intuition.

(c) A fundamental assumption for the validity of the Landau collision integral (as well as the Boltzmann integral) is the dependence on the so-called *one-particle* distribution functions $f_\alpha(\mathbf{r}, \mathbf{v}, t)$ alone. These functions express the probability of finding particles of type α at a certain time at the phase space position $\mathbf{r}, \mathbf{v}$. However, binary collisions essentially depend on the two-particle distribution function, i.e. on the probability of finding one particle at $\mathbf{r}_1, \mathbf{v}_1$, simultaneous with another particle at $\mathbf{r}_2, \mathbf{v}_2$, where $\mathbf{r}_1 \approx \mathbf{r}_2$. The evolution of the one-, two-, and more-particle distribution functions leads to the BBGKY hierarchy of equations (see, e.g., [238], [114], [14]). To truncate this infinite set and to reduce the expressions to only involve the one-particle distribution function requires the formulation of a kinetic regime for weakly coupled plasmas, where the potential energy of the particles is very small compared to their mean kinetic energy [14]. For the validity of this regime, the plasma must be not too hot and not too dense (essentially the shaded region of Fig. 1.9). In conclusion: the three conditions (1.27), (1.28), and (1.30) of Section 1.4 for collective plasma behaviour should be satisfied to justify a description of plasmas by means of the Boltzmann equation (3.2) with a Landau-type collision integral (3.13).

In order to determine the charges and currents that occur in Maxwell's equations we take moments of the distribution functions. The zeroth moment gives the number of particles of species α per unit volume:

$$n_\alpha(\mathbf{r}, t) \equiv \int f_\alpha(\mathbf{r}, \mathbf{v}, t)\, d^3v \qquad \text{(particle density)}. \qquad (3.4)$$

Exploiting the definition (2.45) for averages, the first moment yields the average velocity:

$$\mathbf{u}_\alpha(\mathbf{r}, t) = \langle \mathbf{v} \rangle_\alpha \equiv \frac{1}{n_\alpha(\mathbf{r}, t)} \int \mathbf{v} f_\alpha(\mathbf{r}, \mathbf{v}, t)\, d^3v \qquad \text{(average velocity)}. \qquad (3.5)$$

The charge and current density then follow by multiplying with q_α and summing over particle species:

$$\tau(\mathbf{r}, t) = \sum_\alpha q_\alpha n_\alpha(\mathbf{r}, t) \qquad \text{(charge density)}, \qquad (3.6)$$

$$\mathbf{j}(\mathbf{r}, t) = \sum_\alpha q_\alpha n_\alpha(\mathbf{r}, t)\, \mathbf{u}_\alpha(\mathbf{r}, t) \qquad \text{(current density)}. \qquad (3.7)$$

Since all charges and currents in the plasma are supposed to be free, polarization and magnetization effects are negligible so that Maxwell's equations only involve $\mathbf{E}$ and $\mathbf{B}$. In the rationalized mks system of units we then have:

$$\nabla \times \mathbf{E} = -\frac{\partial \mathbf{B}}{\partial t}, \qquad (3.8)$$

$$\nabla \times \mathbf{B} = \mu_0 \mathbf{j} + \frac{1}{c^2} \frac{\partial \mathbf{E}}{\partial t}, \qquad (3.9)$$

$$\nabla \cdot \mathbf{E} = \frac{\tau}{\epsilon_0}, \qquad (3.10)$$

$$\nabla \cdot \mathbf{B} = 0, \qquad (3.11)$$

where $c = (\epsilon_0 \mu_0)^{-1/2}$. In the Vlasov theory of plasmas, Eqs. (3.3)–(3.11) constitute a complete set of equations for the variables $f_\alpha(\mathbf{r}, \mathbf{v}, t)$, $\mathbf{E}(\mathbf{r}, t)$, and $\mathbf{B}(\mathbf{r}, t)$.

With collisions, an explicit expression for the collision term on the RHS of Eq. (3.2) is needed. We decompose this term into contributions $C_{\alpha\beta}$ due to collisions of particles of species α with particles of species β:

$$C_\alpha = \sum_\beta C_{\alpha\beta}, \qquad \text{where} \quad C_{\alpha\beta} = C_{\alpha\beta}(f_\alpha, f_\beta). \qquad (3.12)$$

We only consider two kinds of particles, viz. electrons (e) and ions (i), so that the indices α and β just run over the two values e and i (giving collision terms C_{ee}, C_{ei} for the electron Boltzmann equation, and C_{ie}, C_{ii} for the ion Boltzmann

3.2 Kinetic equations*

equation). The dependence on f_α and f_β in brackets indicates that the collision operator is a quadratic form where f_β is integrated over velocity space.

We now give the explicit form of the collision operator in terms of the *Landau collision integral* (1936) [135]. We suppress the dependence on $\mathbf{r}$ and t in $f_\alpha(\mathbf{r}, \mathbf{v}, t)$ since only operations on the velocity variable occur. The effect on the distribution function $f_\alpha(\mathbf{v})$ of binary collisions with particles of type β is then expressed by the following integral over velocity space:

$$C_{\alpha\beta} = \frac{q_\alpha^2 q_\beta^2}{8\pi\epsilon_0^2} \ln \Lambda_{\alpha\beta} \int \frac{1}{m_\alpha} \frac{\partial}{\partial \mathbf{v}} \cdot \left[\mathbf{G}(\mathbf{v} - \mathbf{v}') \cdot \left(\frac{1}{m_\alpha} \frac{\partial}{\partial \mathbf{v}} - \frac{1}{m_\beta} \frac{\partial}{\partial \mathbf{v}'} \right) f_\alpha(\mathbf{v}) f_\beta(\mathbf{v}') \right] d^3 v', \tag{3.13}$$

where the prime indicates the integration variable. The integrand involves a double contraction (defined by $\mathbf{a} \cdot \mathbf{T} \cdot \mathbf{b} \equiv \sum_i \sum_j a_i T_{ij} b_j$ for Cartesian components) with the Landau tensor function of the relative velocity

$$\mathbf{G}(\mathbf{v} - \mathbf{v}') \equiv \frac{1}{|\mathbf{v} - \mathbf{v}'|} \left[\mathbf{I} - \frac{(\mathbf{v} - \mathbf{v}')(\mathbf{v} - \mathbf{v}')}{|\mathbf{v} - \mathbf{v}'|^2} \right]. \tag{3.14}$$

It satisfies the useful equality $(\mathbf{v} - \mathbf{v}') \cdot \mathbf{G}(\mathbf{v} - \mathbf{v}') = 0$. The Coulomb logarithm $\ln \Lambda_{\alpha\beta}$ represents the screening of the Coulomb potential in a plasma for volumes outside a Debye sphere. Since the logarithm strongly suppresses the numerical dependence on the physical quantities, so that $\ln \Lambda_{ee} \sim \ln \Lambda_{ei} \sim \ln \Lambda_{ii}$, one may as well drop the subscripts and exploit the simple expression (see Bittencourt [31]) for a plasma with singly charged ions ($Z = 1$) and equal temperatures ($T_e = T_i$):

$$\ln \Lambda = \ln(9N_D) = \ln \frac{12\pi(\epsilon_0 kT)^{3/2}}{e^3 n_e^{1/2}} \approx \ln\left(4.9 \times 10^{17} \frac{\widetilde{T}^{3/2}}{n_e^{1/2}} \right). \tag{3.15}$$

The value of $\ln \Lambda$ just ranges from 10 to 30 for plasmas of laboratory and astrophysical interest. For example, for $\widetilde{T}_e = \widetilde{T}_i = 1$ keV and $n = 10^{20}$ m^{-3}, one gets $\ln \Lambda \approx 17.7$.

By some elementary algebra, the Landau collision integral (3.13) may be transformed into the following form:

$$C_{\alpha\beta} = \frac{q_\alpha^2 q_\beta^2}{4\pi\epsilon_0^2} \frac{\ln \Lambda_{\alpha\beta}}{m_\alpha^2} \left\{ -\frac{\partial}{\partial \mathbf{v}} \cdot \left[\frac{\partial h_\beta(\mathbf{v})}{\partial \mathbf{v}} f_\alpha(\mathbf{v}) \right] + \frac{1}{2} \frac{\partial^2}{\partial \mathbf{v} \partial \mathbf{v}} : \left[\frac{\partial^2 g_\beta(\mathbf{v})}{\partial \mathbf{v} \partial \mathbf{v}} f_\alpha(\mathbf{v}) \right] \right\}, \tag{3.16}$$

where the Rosenbluth potentials g_β and h_β [196], [231] are defined as integrals of the distribution function f_β:

$$g_\beta \equiv \int |\mathbf{v} - \mathbf{v}'| f_\beta(\mathbf{v}') d^3 v', \qquad h_\beta(\mathbf{v}) \equiv \left(1 + \frac{m_\alpha}{m_\beta} \right) \int \frac{f_\beta(\mathbf{v}')}{|\mathbf{v} - \mathbf{v}'|} d^3 v'. \tag{3.17}$$

The two terms of the Fokker–Planck expression (3.16) represent dynamical friction and diffusion in velocity space, respectively [92].

Apart from its beauty, the symmetric form of the Landau collision operator also guarantees the satisfaction of the conservation properties (3.18)–(3.22) below, that are essential for the derivation of macroscopic equations. The generalization by Balescu [13] and Lenard [141] to include the collective effects of plasma oscillations leads to a similar form where the tensor **G** contains an additional summation over the wave vector **k** associated with the collective modes.

3.2.2 Moments of the Boltzmann equation*

The fact that the distribution function is a function of seven independent variables presents formidable complications in the analysis. Since we wish to study plasmas in complex magnetic geometries, we clearly have to get rid of some of the independent variables in order to make progress. The most logical approach is to remove the velocity as an independent variable by taking *moments of the Boltzmann equation*. This approach will run into the problem of producing an infinite chain of equations which somehow has to be truncated in order to make sense. At that point assumptions need to be made that imply restrictions on the domain of validity of the theory.

The different moments of the Boltzmann equation are obtained by multiplying the Boltzmann equation (3.2) with powers of **v** and integrating over velocity space. In the derivations below, integration by parts will produce surface integrals over a surface at $|\mathbf{v}| = \infty$. It is assumed that the distribution functions $f_\alpha(\mathbf{v}) \to 0$ sufficiently rapidly as $|\mathbf{v}| \to \infty$, in particular that $\lim_{|\mathbf{v}| \to \infty} [g(\mathbf{v}) f_\alpha(\mathbf{v})] = 0$ for all functions $g(\mathbf{v})$ that appear, so that these surface integrals do not contribute.

For the time being, we do not need to go into the specific form of the collision term (3.12). It suffices to list a few general properties following from conservation principles. Since the total number of particles of species α at a certain position is not changed by collisions with particles of species β (only their velocities change), we have

$$\int C_{\alpha\beta} \, d^3v = 0 \quad \text{(including } \beta = \alpha\text{)}. \tag{3.18}$$

Also, momentum and energy are conserved for collisions between like particles:

$$\int m_\alpha \mathbf{v} \, C_{\alpha\alpha} \, d^3v = 0, \tag{3.19}$$

$$\int \tfrac{1}{2} m_\alpha v^2 C_{\alpha\alpha} \, d^3v = 0, \tag{3.20}$$

3.2 Kinetic equations*

whereas for collisions between unlike particles ($\beta \neq \alpha$) the following relations hold:

$$\int m_\alpha \mathbf{v} \, C_{\alpha\beta} \, d^3v + \int m_\beta \mathbf{v} \, C_{\beta\alpha} \, d^3v = 0, \tag{3.21}$$

$$\int \tfrac{1}{2} m_\alpha v^2 C_{\alpha\beta} \, d^3v + \int \tfrac{1}{2} m_\beta v^2 C_{\beta\alpha} \, d^3v = 0. \tag{3.22}$$

Obviously, the Landau form (3.13) of the collision operator should satisfy these properties and it does (prove!). For a Maxwellian distribution function, as given by Eq. (2.57), the separate terms of Eqs. (3.21) and (3.22) also vanish. We will return to this fundamental fact in Section 3.2.3.

We now derive the zeroth moment of the Boltzmann equation (3.2) by just integrating over velocity space. As we have already seen in Section 2.3.1, this results in *the continuity equation* for particles of species α:

$$\frac{\partial n_\alpha}{\partial t} + \nabla \cdot (n_\alpha \mathbf{u}_\alpha) = 0 \qquad \text{(mass conservation)}. \tag{3.23}$$

This is the most robust macroscopic equation since its derivation does not require any restrictive assumptions on transport coefficients. In brackets we have indicated the conservation law that is expressed by this equation.

The first moment of the Boltzmann equation (3.2) is obtained by multiplying Eq. (3.2) by $\mathbf{v}$ and integrating over velocity space. This results in the following terms:

$$\int \frac{\partial f_\alpha}{\partial t} \mathbf{v} \, d^3v = \frac{\partial}{\partial t}(n_\alpha \mathbf{u}_\alpha),$$

$$\int \mathbf{v} \cdot \frac{\partial f_\alpha}{\partial \mathbf{r}} \mathbf{v} \, d^3v = \nabla \cdot \int \mathbf{v}\mathbf{v} f_\alpha \, d^3v = \nabla \cdot \left(n_\alpha \langle \mathbf{v}\mathbf{v} \rangle_\alpha\right),$$

$$\int \frac{q_\alpha}{m_\alpha}(\mathbf{E} + \mathbf{v} \times \mathbf{B}) \cdot \frac{\partial f_\alpha}{\partial \mathbf{v}} \mathbf{v} \, d^3v = -\frac{q_\alpha n_\alpha}{m_\alpha}(\mathbf{E} + \mathbf{u}_\alpha \times \mathbf{B}),$$

$$\int C_\alpha \mathbf{v} \, d^3v = \int C_{\alpha\beta} \, \mathbf{v} \, d^3v \qquad (\beta \neq \alpha).$$

Adding them, we obtain *the momentum equation* for particles of species α:

$$\frac{\partial}{\partial t}(n_\alpha m_\alpha \mathbf{u}_\alpha) + \nabla \cdot \left(n_\alpha m_\alpha \langle \mathbf{v}\mathbf{v} \rangle_\alpha\right) - n_\alpha q_\alpha (\mathbf{E} + \mathbf{u}_\alpha \times \mathbf{B}) = \int C_{\alpha\beta} \, m_\alpha \mathbf{v} \, d^3v$$
$$\text{(momentum conservation)}. \tag{3.24}$$

Here, new averaged quantities like $n_\alpha \langle \mathbf{v}\mathbf{v} \rangle_\alpha$ and the collision term appear that require further evaluation.

The final relevant equation is obtained from one of the second moment equations, viz. the scalar one obtained by multiplying Eq. (3.2) with v^2. The following terms result:

$$\int \frac{\partial f_\alpha}{\partial t} v^2 d^3v = \frac{\partial}{\partial t}\left(n_\alpha \langle v^2 \rangle_\alpha\right),$$

$$\int \mathbf{v} \cdot \frac{\partial f_\alpha}{\partial \mathbf{r}} v^2 d^3v = \nabla \cdot \left(n_\alpha \langle v^2 \mathbf{v} \rangle_\alpha\right),$$

$$\int \frac{q_\alpha}{m_\alpha}(\mathbf{E} + \mathbf{v} \times \mathbf{B}) \cdot \frac{\partial f_\alpha}{\partial \mathbf{v}} v^2 d^3v = -2\frac{n_\alpha q_\alpha}{m_\alpha} \mathbf{E} \cdot \mathbf{u}_\alpha,$$

$$\int C_\alpha v^2 d^3v = \int C_{\alpha\beta} v^2 d^3v \quad (\beta \neq \alpha).$$

Multiplying these terms by $\frac{1}{2}m_\alpha$ and adding them gives *the energy equation*:

$$\frac{\partial}{\partial t}\left(n_\alpha \tfrac{1}{2} m_\alpha \langle v^2 \rangle_\alpha\right) + \nabla \cdot \left(n_\alpha \tfrac{1}{2} m_\alpha \langle v^2 \mathbf{v} \rangle_\alpha\right) - n_\alpha q_\alpha \mathbf{E} \cdot \mathbf{u}_\alpha = \int C_{\alpha\beta} \tfrac{1}{2} m_\alpha v^2 d^3v$$

(energy conservation). (3.25)

Again, averages and a collision term appear that require further reduction to really give a self-consistent macroscopic equation.

This chain of moment equations can be continued indefinitely. As a matter of fact, each moment introduces a new unknown whose temporal evolution is described by the next order moment of the Boltzmann equation. For example, the zeroth order moment (3.23) is an evolution equation for the particle density n_α and it introduces the average velocity $\mathbf{u}_\alpha$ as a new unknown; the first order moment (3.24) then yields an evolution equation for this $\mathbf{u}_\alpha$ but it contains the unknown $\langle \mathbf{vv} \rangle_\alpha$, etc. This infinite procedure will have to be truncated at some point to get a closed set of equations. The 'art' of obtaining macroscopic equations resides in an appropriate closure of the chain at a very limited number of moments. In the usual fluid theories this number is just five: the (scalar) continuity equation (3.23), the (vector) momentum equation (3.24), and the (scalar) energy equation (3.25).

3.2.3 Thermal fluctuations and transport*

The derived moment equations (3.23)–(3.25) are the only ones needed here. In order to turn them into a closed set, a number of assumptions have to be made. Before we do this it is useful to transform the momentum and energy equation into a form that has a more macroscopic appearance. To that end, as already discussed in Section 2.3.1, we separate the effects of thermal fluctuations from the macroscopic background, by defining the random velocity $\tilde{\mathbf{v}}_\alpha$ of the particles with respect to the average velocity $\mathbf{u}_\alpha$:

$$\tilde{\mathbf{v}}_\alpha \equiv \mathbf{v} - \mathbf{u}_\alpha, \qquad \text{where} \quad \langle \tilde{\mathbf{v}}_\alpha \rangle = 0. \tag{3.26}$$

The random velocity part of the scalar $\langle v^2 \rangle_\alpha$ in the energy equation (3.25) then gives rise to a quantity measuring the mean kinetic energy of the particles in a frame moving with the velocity $\mathbf{u}_\alpha$, which is *the temperature T_α*:

$$T_\alpha(\mathbf{r}, t) \equiv \frac{m_\alpha}{3k} \langle \tilde{v}_\alpha^2 \rangle. \tag{3.27}$$

Likewise, the random velocity part of the term $\langle \mathbf{vv} \rangle_\alpha$ in the momentum equation (3.24) gives rise to *the stress tensor $\mathbf{P}_\alpha$* defined as

$$\mathbf{P}_\alpha \equiv n_\alpha m_\alpha \langle \tilde{\mathbf{v}}_\alpha \tilde{\mathbf{v}}_\alpha \rangle = p_\alpha \mathbf{I} + \boldsymbol{\pi}_\alpha, \tag{3.28}$$

where the isotropic part is directly related to the temperature,

$$p_\alpha(\mathbf{r}, t) \equiv \tfrac{1}{3} \text{Tr}(\mathbf{P}_\alpha) = \tfrac{1}{3} n_\alpha m_\alpha \langle \tilde{v}_\alpha^2 \rangle = n_\alpha k T_\alpha, \tag{3.29}$$

and the traceless tensor $\boldsymbol{\pi}_\alpha(r, t)$ is the contribution due to the anisotropy of the distribution function:

$$\boldsymbol{\pi}_\alpha(\mathbf{r}, t) \equiv n_\alpha m_\alpha \langle \tilde{\mathbf{v}}_\alpha \tilde{\mathbf{v}}_\alpha - \tfrac{1}{3} \tilde{v}_\alpha^2 \mathbf{I} \rangle. \tag{3.30}$$

Finally, the random velocity part of the vector $\langle v^2 \mathbf{v} \rangle_\alpha$ in the energy equation (3.25) gives rise to a quantity

$$\mathbf{h}_\alpha(\mathbf{r}, t) \equiv \tfrac{1}{2} n_\alpha m_\alpha \langle \tilde{v}_\alpha^2 \tilde{\mathbf{v}}_\alpha \rangle, \tag{3.31}$$

which is *the heat flow* by random motion of the particles of species α.

The collision terms may also be simplified by transforming to a frame moving with the velocity $\mathbf{u}_\alpha$. From Eq. (3.18) it follows that only the random part contributes to the RHS of the momentum equation (3.24):

$$\int C_{\alpha\beta}\, m_\alpha \mathbf{v}\, d^3v = \mathbf{R}_\alpha = \int C_{\alpha\beta}\, m_\alpha \tilde{\mathbf{v}}_\alpha\, d^3v \qquad (\beta \neq \alpha), \tag{3.32}$$

which is the friction force, i.e., *the mean momentum transfer* from particles β to particles α. Similarly, the RHS of the energy equation (3.25) may be written as

$$\int C_{\alpha\beta}\, \tfrac{1}{2} m_\alpha v^2\, d^3v = \int C_{\alpha\beta} \left(m_\alpha \mathbf{u}_\alpha \cdot \tilde{\mathbf{v}}_\alpha + \tfrac{1}{2} m_\alpha \tilde{v}_\alpha^2 \right) d^3v = \mathbf{u}_\alpha \cdot \mathbf{R}_\alpha + Q_\alpha, \tag{3.33}$$

where

$$Q_\alpha \equiv \int C_{\alpha\beta}\, \tfrac{1}{2} m_\alpha \tilde{v}_\alpha^2\, d^3v \qquad (\beta \neq \alpha) \tag{3.34}$$

is *the heat transferred* to the system of particles α due to collisions with the unlike particles β.

Whereas the mass conservation equation (3.23) already has the required macroscopic form, the momentum and energy conservation equations (3.24) and (3.25) still need further transformation. Substituting the definitions (3.27)–(3.34), they become

$$\frac{\partial}{\partial t}(n_\alpha m_\alpha \mathbf{u}_\alpha) + \nabla \cdot (n_\alpha m_\alpha \mathbf{u}_\alpha \mathbf{u}_\alpha) + \nabla \cdot \mathbf{P}_\alpha - n_\alpha q_\alpha (\mathbf{E} + \mathbf{u}_\alpha \times \mathbf{B}) = \mathbf{R}_\alpha, \tag{3.35}$$

and

$$\frac{\partial}{\partial t}\left(\tfrac{1}{2}n_\alpha m_\alpha u_\alpha^2\right) + \frac{\partial}{\partial t}\left(\tfrac{3}{2}n_\alpha k T_\alpha\right) + \nabla \cdot \left(\tfrac{1}{2}n_\alpha m_\alpha u_\alpha^2 \mathbf{u}_\alpha + \tfrac{3}{2}n_\alpha k T_\alpha \mathbf{u}_\alpha \right.$$
$$\left. + \mathbf{u}_\alpha \cdot \mathbf{P}_\alpha + \mathbf{h}_\alpha\right) - n_\alpha q_\alpha \mathbf{E} \cdot \mathbf{u}_\alpha = \mathbf{u}_\alpha \cdot \mathbf{R}_\alpha + Q_\alpha. \tag{3.36}$$

The momentum equation (3.35) may be simplified by using the continuity equation (3.23) to remove contributions $\partial n_\alpha/\partial t$, whereas the energy equation (3.36) may be simplified by removing the bulk kinetic energy part by means of Eqs. (3.23) and (3.35). The three lowest moments of the Boltzmann equation then take the compact form

$$\left(\frac{\partial}{\partial t} + \mathbf{u}_\alpha \cdot \nabla\right) n_\alpha + n_\alpha \nabla \cdot \mathbf{u}_\alpha = 0 \qquad \text{(mass)}, \tag{3.37}$$

$$n_\alpha m_\alpha \left(\frac{\partial}{\partial t} + \mathbf{u}_\alpha \cdot \nabla\right) \mathbf{u}_\alpha + \nabla(n_\alpha k T_\alpha) - n_\alpha q_\alpha (\mathbf{E} + \mathbf{u}_\alpha \times \mathbf{B})$$
$$= -\nabla \cdot \boldsymbol{\pi}_\alpha + \mathbf{R}_\alpha \qquad \text{(momentum)}, \tag{3.38}$$

$$\tfrac{3}{2} n_\alpha \left(\frac{\partial}{\partial t} + \mathbf{u}_\alpha \cdot \nabla\right) k T_\alpha + n_\alpha k T_\alpha \nabla \cdot \mathbf{u}_\alpha$$
$$= -\boldsymbol{\pi}_\alpha : \nabla \mathbf{u}_\alpha - \nabla \cdot \mathbf{h}_\alpha + Q_\alpha \qquad \text{(heat)}, \tag{3.39}$$

which we already encountered in Section 2.3.1. They constitute the equations of continuity, motion and heat balance for particles of species α. It will not have escaped the attentive reader that apparent progress has been made by just hiding the problems in simple looking variables that are abbreviations of intricate kinetic processes. Clearly, we need substantial additional information concerning the variables $\boldsymbol{\pi}_\alpha$, $\mathbf{h}_\alpha$, $\mathbf{R}_\alpha$ and Q_α to be able to express them in terms of the macroscopic variables n_α, $\mathbf{u}_\alpha$, T_α and the electromagnetic fields $\mathbf{E}$, and $\mathbf{B}$, to really close the set of macroscopic equations (3.37)–(3.39) so that they become genuine hydrodynamic equations. Such information comes from transport theory.

An important equilibrium distribution function in kinetic theory, conforming to the temperature definition (3.27), is *the Maxwell distribution* already introduced in

Eq. (2.57), which we here repeat for convenience:

$$f_\alpha^0(\mathbf{r}, \mathbf{v}, t) = n_\alpha \left(\frac{m_\alpha}{2\pi k T_\alpha} \right)^{3/2} \exp\left(-\frac{m_\alpha \tilde{v}_\alpha^2}{2k T_\alpha} \right), \qquad \tilde{v}_\alpha^2 \equiv |\mathbf{v} - \mathbf{u}_\alpha|^2. \qquad (3.40)$$

This function represents *local thermal equilibrium*: it just depends on the local values of the macroscopic variables $n_\alpha(\mathbf{r}, t)$, $\mathbf{u}_\alpha(\mathbf{r}, t)$, $T_\alpha(\mathbf{r}, t)$ which are supposed to evolve according to the hydrodynamics equations, that are still to be derived by the appropriate closure of Eqs. (3.37)–(3.39). Substitution of f_α^0 in the definitions (3.30) and (3.31) for the anisotropic pressure tensor and the heat flow gives $\boldsymbol{\pi}_\alpha[f_\alpha^0] = 0$ and $\mathbf{h}_\alpha[f_\alpha^0] = 0$. Moreover, due to the equality below Eq. (3.14), the Landau collision integral vanishes for like-particle collisions of Maxwellian particles: $C_{\alpha\alpha}(f_\alpha^0, f_\alpha^0) = 0$. If the average velocities and the temperatures were equal, i.e. $u_\alpha = u_\beta$ (no electric current!) and $T_\alpha = T_\beta$, the collision integral for unlike-particle collisions would vanish as well, $C_{\alpha\beta}(f_\alpha^0, f_\beta^0) = 0$, so that also the momentum transfer and the heat transfer would vanish: $\mathbf{R}_\alpha[f_\alpha^0, f_\beta^0] = 0$ and $Q_\alpha[f_\alpha^0, f_\beta^0] = 0$. Under these circumstances, the two sets of hydrodynamical variables evolve according to the equations (3.37)–(3.39) with vanishing RHSs, whereas initial charge neutrality and the equality of average velocities and temperatures would even remove the separate identities of the electron and ion fluids. If this state were established on a collision time scale, thermonuclear fusion would be impossible and most interesting plasma-astrophysical phenomena would not occur. However, the global boundary conditions on the hydrodynamical variables constantly drive the system away from this state of equilibrium and collisions establish a state of quasi-equilibrium with non-vanishing dissipative quantities $\boldsymbol{\pi}_\alpha$, $\mathbf{h}_\alpha$, $\mathbf{R}_\alpha$, and Q_α that are due to the systematic deviations from Maxwellian distributions f_α^0 and f_β^0. *Transport theory* is concerned with the analysis of these deviations from the state of local thermodynamic equilibrium.

The fundamentals of classical transport theory are clearly explained in the seminal paper by Braginskii [41]. Here, we follow the theoretical framework expanded by Balescu [14] since it is more general and incorporates later developments in transport theory. While it is impossible to do justice to the diversity and beauty of this field, we just collect the results needed for our present purpose, in particular the relaxation processes (Section 3.2.4) and the transport coefficients (Section 3.3.2). For full details the reader is referred to the quoted references.

To analyse the consequences of deviations from local thermodynamic equilibrium, the distribution functions are developed in powers of some small parameter (to be specified in Section 3.2.4) measuring these deviations (called the Chapman–Enskog procedure [52]):

$$f_\alpha \approx f_\alpha^0 + f_\alpha^1, \qquad \text{where} \quad |f_\alpha^1| \ll f_\alpha^0. \qquad (3.41)$$

It is expedient to extract a factor f_α^0 from f_α^1,

$$f_\alpha(\mathbf{v}; \mathbf{r}, t) \approx f_\alpha^0(\mathbf{v}; \mathbf{r}, t)[1 + \chi_\alpha(\mathbf{v}; \mathbf{r}, t)], \tag{3.42}$$

so that the deviations from the local Maxwellians are now given by the functions χ_α. These functions can be normalized such that n_α, $\mathbf{u}_\alpha$, and T_α not only measure the average with respect to the Maxwellian f_α^0 but also with respect to the full distribution function f_α. In the representation of Balescu, χ_α is systematically expanded in irreducible tensorial Hermite polynomials $H^{(\cdots)}_{\cdots}(\mathbf{w}_\alpha)$ in the dimensionless fluctuating velocity variable $\mathbf{w}_\alpha \equiv (2kT_\alpha/m_\alpha)^{-1/2} \tilde{\mathbf{v}}_\alpha$:

$$\chi_\alpha(\mathbf{w}_\alpha; \mathbf{r}, t) = \sum_{n=1}^{\infty} \left[\bar{h}_\alpha^{(2n+2)} H^{(2n+2)} + \bar{\mathbf{h}}_\alpha^{(2n+1)} \cdot \mathbf{H}^{(2n+1)} + \bar{\mathbf{h}}_\alpha^{(2n)} : \mathbf{H}^{(2n)} + \cdots \right], \tag{3.43}$$

where the three indicated terms represent the scalar, vector and traceless second rank tensor contributions, respectively. The expansion coefficients are the *unknown* dimensionless Hermitian moments $\bar{h}_{\alpha\cdots}^{(\cdots)}(\mathbf{r}, t)$. The lowest order ones correspond to the anisotropic pressure, $\boldsymbol{\pi}_\alpha \equiv 2^{1/2} n_\alpha k T_\alpha \bar{\mathbf{h}}_\alpha^{(2)}$, and the heat flow, $\mathbf{h}_\alpha \equiv (5/2)^{1/2} n_\alpha m_\alpha (kT_\alpha)^{3/2} \bar{\mathbf{h}}_\alpha^{(3)}$. The program of transport theory now consists of substituting (3.43) in the Boltzmann equation (3.2) and solving for the Hermitian moments. This results in closure of the hydrodynamic equations.

The expansion (3.43) may be truncated at different levels, where the number of moments kept will determine the accuracy. For example, the thirteen moment (13M) approximation just keeps the five plasmadynamical moments n_α, $\mathbf{u}_\alpha$, T_α, the three components of the heat flow vector $\mathbf{h}_\alpha$, and the five independent components of the anisotropic pressure tensor $\boldsymbol{\pi}_\alpha$. The 21M approximation adds the three components of the fifth order Hermitian moment $\bar{\mathbf{h}}_\alpha^{(5)}$ and the five components of the fourth order Hermitian moment $\bar{\mathbf{h}}_\alpha^{(4)}$, etc. In this manner, a general framework of transport theory is constructed that may be, and has been, used to extend the classical transport theory to incorporate the effects of particle orbits in toroidal geometry (called the neo-classical transport theory, initiated by Galeev and Sagdeev [77]) and, currently, also turbulent effects (leading to anomalous transport). Balescu [14] demonstrates that the classical transport coefficients in the 21M approximation agree to within 1% with the ones obtained much earlier by Braginskii [41] by a very different representation, but it deviates in some important aspects of the interpretation (see Section 3.3.2).

3.2.4 Collisions and closure*

So far, the Hermitian moment expansion is just a formal representation. To actually calculate the moments, the specific properties of the collision operator need

3.2 Kinetic equations*

to be exploited. At this point, the nice symmetry of the Landau collision operator with respect to electrons and ions should be taken for granted and, instead, the consequences of the smallness of the mass ratio,

$$m_e/m_i \ll 1, \tag{3.44}$$

should be utilized. This condition implies that typical ion velocities are much smaller than typical electron velocities, so that the ion distribution $f_i(\mathbf{v})$ is a much narrower function than the electron distribution $f_e(\mathbf{v})$. Consequently, e.g. in the electron–ion collision integral $C_{ei}(f_e, f_i)$, there is only interaction over a limited range of velocities, so that the Landau tensor $\mathbf{G}(\mathbf{v} - \mathbf{v}')$ can be expanded in powers of $|\mathbf{v}' - \mathbf{u}_e|/|\mathbf{v} - \mathbf{u}_e|$, giving significant contributions only for the lowest order terms. Systematically exploiting such asymmetries between electron and ions, the expressions for the linearized collision integrals may be grouped as follows.

(a) The *electron ion collision integral* splits in two contributions, where we only indicate the most important term representing pitch angle scattering of the electrons on the heavy, virtually immobile, ions in thermal equilibrium:

$$C_{ei}(f_e, f_i) \approx C_{ei}(f_e^1, f_i^0) + C_{ei}(f_e^0, f_i^1) \approx \nu_{ei}(w_e)\mathcal{L}_e(f_e^1) + \cdots. \tag{3.45}$$

The electron–ion collision frequency,

$$\nu_{ei} \equiv \frac{3\sqrt{\pi}}{4w_e^3}\tau_e^{-1}, \tag{3.46}$$

involves the relaxation time τ_e, defined below in Eq. (3.50), and the operator

$$\mathcal{L}_e \equiv \frac{1}{2}\frac{\partial}{\partial \mathbf{w}_e} \cdot \left(w_e^2\mathbf{I} - \mathbf{w}_e\mathbf{w}_e\right) \cdot \frac{\partial}{\partial \mathbf{w}_e} \tag{3.47}$$

is due to the mentioned expansion of the Landau tensor.

Like-particle collisions require a different expansion. The dominant contribution of the resulting *electron–electron collision integral* reads:

$$C_{ee}(f_e, f_e) \approx C_{ee}(f_e^1, f_e^0) + C_{ee}(f_e^0, f_e^1) \approx \nu_{ee}(w_e)\mathcal{L}_e(f_e^1) + \cdots. \tag{3.48}$$

The electron–electron collision frequency,

$$\nu_{ee} \equiv \frac{3\sqrt{\pi}}{4w_e^3} H(w_e)(Z\tau_e)^{-1}, \tag{3.49}$$

involves the same relaxation time τ_e as the electron–ion collision frequency, but it also depends on the Chandrasekhar function $H(w_e)$, which is a smoothly increasing function from $H(0) = 0$ to $H(\infty) = 1$. Hence, ν_{ee} is of the same order as ν_{ei}, but always numerically smaller. Since $C_e = C_{ee} + C_{ei}$, the expressions (3.45) and

(3.48) should be added so that the two collision processes, represented by the frequencies ν_{ei} and ν_{ee}, both contribute to the *relaxation to an electron fluid* on the time scale

$$\tau_e \equiv 6\pi\sqrt{2\pi}\epsilon_0^2 \frac{m_e^{1/2}(kT_e)^{3/2}}{\ln\Lambda Z^2 e^4 n_i} \approx \frac{1.09\times 10^{16}}{\ln\Lambda} \frac{\tilde{T}_e^{3/2}}{Zn_e}. \quad (3.50)$$

On this time scale, the electron fluid is established, i.e. the collection of electrons relaxes to its state of near local thermodynamic equilibrium. Recall that $\ln\Lambda \approx 17.7$ for laboratory plasmas with $\tilde{T}_e = 1\,\text{keV}$ and $n = 10^{20}\,\text{m}^{-3}$, so that a typical value of $\tau_e \approx 6.1\times 10^{-6}\,\text{s}$ is obtained for those plasmas.

(b) In the case of ions, $C_i = C_{ii} + C_{ie}$, but the two collision processes do not operate on the same time scale. The effect of scattering of ions on electrons, represented by C_{ie}, is negligible (i.e. an order $(m_e/m_i)^{1/2}$ smaller) on the ion relaxation time scale. Hence, the dominant process is represented by the *ion–ion collision integral* only:

$$C_{ii}(f_i, f_i) \approx C_{ii}(f_i^1, f_i^0) + C_{ii}(f_i^0, f_i^1) \approx \nu_{ii}(w_i)\mathcal{L}_i(f_i^1) + \cdots, \quad (3.51)$$

where the operator $\mathcal{L}_i$ is defined as in Eq. (3.47) with w_e replaced by w_i. The ion–ion collision frequency,

$$\nu_{ii} \equiv \frac{3\sqrt{\pi}}{4w_i^3} H(w_i)\tau_i^{-1}, \quad (3.52)$$

involves the ion relaxation time τ_i and the Chandrasekhar function $H(w_i)$. Due to the ion–ion collisions, *relaxation to an ion fluid* proceeds on the time scale

$$\tau_i \equiv 6\pi\sqrt{2\pi}\epsilon_0^2 \frac{m_i^{1/2}(kT_i)^{3/2}}{\ln\Lambda Z^4 e^4 n_i} \approx \frac{4.66\times 10^{17}}{\ln\Lambda} \frac{A^{1/2}\tilde{T}_i^{3/2}}{Z^3 n_e}. \quad (3.53)$$

On this time scale, the ion fluid is established, i.e. the collection of ions relaxes to its own state of near local thermodynamic equilibrium. For a laboratory hydrogen plasma with $\tilde{T}_i = 1\,\text{keV}$ and $n = 10^{20}\,\text{m}^{-3}$, a typical value of $\tau_i \approx 2.6\times 10^{-4}\,\text{s}$ is obtained. If $T_i \sim T_e$, due to the smallness of the mass ratio, $\tau_e/\tau_i \sim (m_e/m_i)^{1/2} \ll 1$, so that the ions relax much slower than the electrons.

(c) Finally, the most important part of the *ion–electron collision integral* is due to a non-vanishing contribution from the two Maxwellians f_i^0 and f_e^0:

$$C_{ie}(f_i^0, f_e^0) \approx \tau_{eq}^{-1}\left(\tfrac{1}{2}Z\frac{\partial}{\partial \mathbf{w}_i}\cdot\mathbf{w}_i f_i^0 + \tfrac{1}{4}\frac{m_e}{m_i}\frac{\partial}{\partial \mathbf{w}_e}\cdot\frac{\partial}{\partial \mathbf{w}_e}f_e^0\right) + \cdots. \quad (3.54)$$

3.2 Kinetic equations*

This process proceeds on the temperature equilibration time scale, which is the longest relaxation time scale:

$$\tau_{eq} \equiv \frac{m_i}{2m_e} \tau_e. \tag{3.55}$$

On this time scale a fluid *with equal electron and ion temperatures* is established. Again, for the mentioned laboratory plasma, a typical value of $\tau_{eq} \approx 5.6 \times 10^{-3}$ s is obtained.

Next, having obtained the collisional expressions, the two linearized Boltzmann equations for f_e^1 and f_i^1 are solved to produce the closure relations. Exploiting the Hermitian expansion (3.43), these relations take the form of evolution equations for the non-plasmadynamical moments $\bar{h}_{\alpha...}^{(...)}(\mathbf{r}, t)$ connecting them to the hydrodynamical sources, i.e. the gradients of $n_{e,i}$, $\mathbf{u}_{e,i}$, $T_{e,i}$ and the fields $\mathbf{E}$ and $\mathbf{B}$. The latter are obtained from the hydrodynamical equations (3.37)–(3.39), neglecting the RHSs, and Maxwell's equations (3.6)–(3.11). At this point, a crucial distinction is to be made between the relaxation time scales $\tau_{e,i}$ of the non-plasmadynamical moments and the hydrodynamical evolution time scale τ_H of the hydrodynamical variables. In order for this split to make sense at all, we should have $\tau_{e,i} \ll \tau_H$. Hence, the small parameter that was implicit in the approximations (3.42) for χ_e and χ_i can now be specified:

$$\epsilon_e \equiv \tau_e/\tau_H \ll 1, \qquad \epsilon_i \equiv \tau_i/\tau_H \ll 1. \tag{3.56}$$

Usually, $\tau_e \ll \tau_i$, so that the second restriction is the most severe one. This implies that ion relaxation should proceed faster than hydrodynamic evolution, whereas temperature equilibration might proceed on the hydrodynamical time scale, $\tau_{eq} \sim \tau_H$.

Balescu demonstrates that, under these circumstances, the initial value problem for the non-plasmadynamical moments $\bar{h}_{\alpha...}^{(...)}(\mathbf{r}, t)$ has a very simple solution. Due to the frequent collisions, the system forgets about the initial data after a few relaxation times so that the $\partial/\partial t$ terms may be neglected. In effect, the system tends towards a state where the non-plasmadynamical moments are linearly related to the hydrodynamical variables: *closure!* The resulting relations are just four (two for the electrons and the ions times two for the vectors and the tensors) uncoupled sets of algebraic equations relating the thermodynamic fluxes to the sources. They are explicitly listed in Section 3.3.2 for the simplified case of singly charged ions ($Z = 1$) and large magnetic field,

$$\Omega_e \tau_e \gg 1, \qquad \Omega_i \tau_i \gg 1, \tag{3.57}$$

in the 21M approximation. The latter conditions are not required for the transport coefficients derived by Balescu, but they greatly simplify the expressions.

Moreover, they are easily satisfied for cases of interest: $\Omega_e \tau_e \approx 3.2 \times 10^6$, $\Omega_i \tau_i \approx 7.5 \times 10^4$ for tokamak plasmas with $\widetilde{T}_i = 1\,\mathrm{keV}$, and $\Omega_e \tau_e \approx 1.0 \times 10^8$, $\Omega_i \tau_i \approx 2.4 \times 10^5$ for coronal loop plasmas with $\widetilde{T}_i = 0.1\,\mathrm{keV}$ (exploiting the values of Table B.3).

Unfortunately, the numerical values just given also reveal the tremendous restrictions imposed by the conditions (3.56), in particular for laboratory fusion plasmas (see Freidberg [72]). Exploiting the Alfvén wave crossing time $\tau_A \equiv v_A/L$ along the plasma as a measure for τ_H, these conditions become

$$\epsilon_e \equiv \frac{\tau_e v_A}{L} \ll 1, \qquad \epsilon_i \equiv \frac{\tau_i v_A}{L} \ll 1. \tag{3.58}$$

Inserting the numerical values of Table B.5, and even lowering the temperature for tokamaks to $\widetilde{T} = 1\,\mathrm{keV}$, we get $\epsilon_e = 2$ and $\epsilon_i = 87$; not small at all! On the other hand, for astrophysical plasmas, the occurrence of L in the denominator completely saves the day: $\epsilon_e = 1.3 \times 10^{-5}$ and $\epsilon_i = 5.3 \times 10^{-3}$ for coronal loops with $\widetilde{T} = 0.1\,\mathrm{keV}$. Of course, the choice of large L is not an available option in fusion research since the size of plasma confinement devices is determined by economical considerations, not by theoretical convenience.

Some comments are in order. (1) MHD fusion theory is not concerned about Alfvén wave propagation on the μsec time scale, but with the much slower residual instabilities (left after considerable experimental effort to eliminate the fastest ones) growing on the msec time scale. The description of these phenomena brings in small geometrical factors increasing the effective value of τ_H. (2) Neo-classical transport theory, incorporating the effects of particle trapping in toroidal magnetic fields [77], brings in similar geometrical factors that reduce the effective value of the relaxation times τ_e and τ_i. (3) Usually, nature produces turbulent transport processes that completely swamp the classical ones so that the concern is not so much satisfaction of the conditions (3.56), but finding regimes where anomalous transport is sufficiently reduced to make fusion possible. (4) Of course, the option always remains open to exploit one of the hybrid models, e.g. with a fluid description for the electrons but a kinetic description for the ions.

With this caveat, we just proceed on the basis of classical transport theory, observing that asymptotic results from a rigid theoretical framework frequently remain valid far outside the strict domain of validity.

3.3 Two-fluid equations★

3.3.1 Electron–ion plasma★

We now collect the two-fluid equations for a plasma consisting of electrons, $q_e = -e$, and one kind of ion with charge number Z, $q_i = Ze$. From the moment

3.3 Two-fluid equations*

equations (3.37)–(3.39) a double set of equations for the electrons and ions is obtained, where the explicit expressions for the RHS quantities $\boldsymbol{\pi}_{e,i}$, $\mathbf{h}_{e,i}$, $\mathbf{R}_{e,i}$, and $Q_{e,i}$ in terms of the macroscopic variables are listed in Section 3.3.2. The latter two pairs of quantities are not independent because of the momentum and energy conservation properties (3.21) and (3.22) of unlike-particle collisions. From momentum conservation one derives

$$\mathbf{R}_e = -\mathbf{R}_i, \tag{3.59}$$

whereas energy conservation, by the use of Eq. (3.33), yields

$$Q_e = -(\mathbf{u}_e - \mathbf{u}_i) \cdot \mathbf{R}_e - Q_i. \tag{3.60}$$

Consequently, the set of dissipative two-fluid equations becomes

$$\left(\frac{\partial}{\partial t} + \mathbf{u}_e \cdot \nabla\right) n_e + n_e \nabla \cdot \mathbf{u}_e = 0,$$

$$\left(\frac{\partial}{\partial t} + \mathbf{u}_i \cdot \nabla\right) n_i + n_i \nabla \cdot \mathbf{u}_i = 0, \tag{3.61}$$

$$n_e m_e \left(\frac{\partial}{\partial t} + \mathbf{u}_e \cdot \nabla\right) \mathbf{u}_e + \nabla p_e + e n_e (\mathbf{E} + \mathbf{u}_e \times \mathbf{B}) = -\nabla \cdot \boldsymbol{\pi}_e + \mathbf{R}_e,$$

$$n_i m_i \left(\frac{\partial}{\partial t} + \mathbf{u}_i \cdot \nabla\right) \mathbf{u}_i + \nabla p_i - Z e n_i (\mathbf{E} + \mathbf{u}_i \times \mathbf{B}) = -\nabla \cdot \boldsymbol{\pi}_i - \mathbf{R}_e, \tag{3.62}$$

$$\frac{3}{2} n_e \left(\frac{\partial}{\partial t} + \mathbf{u}_e \cdot \nabla\right) k T_e + p_e \nabla \cdot \mathbf{u}_e = -\boldsymbol{\pi}_e : \nabla \mathbf{u}_e - \nabla \cdot \mathbf{h}_e - (\mathbf{u}_e - \mathbf{u}_i) \cdot \mathbf{R}_e - Q_i,$$

$$\frac{3}{2} n_i \left(\frac{\partial}{\partial t} + \mathbf{u}_i \cdot \nabla\right) k T_i + p_i \nabla \cdot \mathbf{u}_i = -\boldsymbol{\pi}_i : \nabla \mathbf{u}_i - \nabla \cdot \mathbf{h}_i + Q_i, \tag{3.63}$$

where

$$p_e = n_e k T_e, \qquad p_i = n_i k T_i. \tag{3.64}$$

Of course, this set only becomes complete when augmented with Maxwell's equations (3.8)–(3.11) with the charge and current density,

$$\tau \equiv -e(n_e - Z n_i), \qquad \mathbf{j} \equiv -e(n_e \mathbf{u}_e - Z n_i \mathbf{u}_i), \tag{3.65}$$

acting as sources.

3.3.2 The classical transport coefficients*

Since the transport closure relations involve vector and tensor expressions with highly anisotropic coefficients (due to the presence of the magnetic field), it is convenient to introduce a notation that highlights these anisotropies. First, associated with any vector **V**, we define three auxiliary vectors

$$\mathbf{V}_\| \equiv \mathbf{bb} \cdot \mathbf{V}, \qquad \mathbf{V}_\wedge \equiv \mathbf{b} \times \mathbf{V}, \qquad \mathbf{V}_\perp \equiv (\mathbf{b} \times \mathbf{V}) \times \mathbf{b}, \qquad (3.66)$$

where $\mathbf{b} \equiv \mathbf{B}/B$ is the unit vector in the direction of the magnetic field. Hence, $\mathbf{V} = \mathbf{V}_\| + \mathbf{V}_\perp$, and $\mathbf{V}_\wedge$ is a vector of length $V_\perp$ but orthogonal to both **V** and **B**. In a projection with respect to a triad of orthogonal unit vectors $\mathbf{e}_1, \mathbf{e}_2, \mathbf{b}$, we have $\mathbf{V} = (V_1, V_2, V_\|)^T$, so that the auxiliary vectors are represented by

$$\mathbf{V}_\| = (0, 0, V_\|)^T, \qquad \mathbf{V}_\wedge = (-V_2, V_1, 0)^T, \qquad \mathbf{V}_\perp = (V_1, V_2, 0)^T. \qquad (3.67)$$

Next, we consider a typical expression $\vartheta \cdot \mathbf{V}$, where ϑ is a second rank tensor of transport coefficients. Symmetry with respect to rotations about the magnetic field implies that ϑ can only have three independent elements, which we denote by $\vartheta_\|$, $\vartheta_\wedge$ and $\vartheta_\perp$:

$$\vartheta = \begin{pmatrix} \vartheta_\perp & -\vartheta_\wedge & 0 \\ \vartheta_\wedge & \vartheta_\perp & 0 \\ 0 & 0 & \vartheta_\| \end{pmatrix} \Rightarrow \vartheta^{-1} = \begin{pmatrix} \vartheta_\perp/D & \vartheta_\wedge/D & 0 \\ -\vartheta_\wedge/D & \vartheta_\perp/D & 0 \\ 0 & 0 & 1/\vartheta_\| \end{pmatrix}, \qquad (3.68)$$

where $D \equiv \vartheta_\perp^2 + \vartheta_\wedge^2$. (The expression for the inverse tensor ϑ^{-1} will be needed below.) One then easily checks that multiplication of the left matrix (3.68) with the vector **V** is equivalent to the sum over the three auxiliary vectors (3.66) multiplied with the three transport coefficients:

$$\vartheta \cdot \mathbf{V} = \sum_\lambda \vartheta_\lambda \mathbf{V}_\lambda \qquad (\lambda = \|, \wedge, \perp). \qquad (3.69)$$

Thus, the tensor expressions are conveniently compressed by this notation.

We now list the electron vector transport coefficients under (a) and the electron tensor coefficients under (d). Recall that these coefficients are established on the time scale τ_e. The ion vector and tensor coefficients, established on the time scale τ_i, are listed under (b) and (d). The expression for the heat transfer from the electrons to the ions, established on the time scale τ_{eq}, is given under (c).

(a) Electron electrical and thermal coefficients The expressions for the electron electrical and thermal transport coefficients are the most interesting ones since they exhibit the characteristic Onsager symmetry of the transport matrix, relating thermodynamics fluxes and forces in non-equilibrium thermodynamics. In Balescu's treatment, the current density **j** and the electron heat flow $\mathbf{h}_e$ are considered

as the fluxes that are driven by the modified electric field, $\widehat{\mathbf{E}} \equiv \mathbf{E} + \mathbf{u}_i \times \mathbf{B} + (en_e)^{-1} \nabla(n_e k T_e)$, and the negative gradient of the electron temperature, $-\nabla(kT_e)$:

$$\mathbf{j} = \boldsymbol{\sigma} \cdot \widehat{\mathbf{E}} - \boldsymbol{\alpha} \cdot \nabla(kT_e), \qquad (3.70)$$

$$\mathbf{h}_e = kT_e\, \boldsymbol{\alpha} \cdot \widehat{\mathbf{E}} - \boldsymbol{\kappa}_e \cdot \nabla(kT_e). \qquad (3.71)$$

The explicit expressions for the *electrical conductivity* σ_λ, the *thermo-electric coupling* α_λ and the *electron thermal conductivity* $\kappa_{e\lambda}$ for $Z = 1$ read:

$$\begin{aligned}
\sigma_\| &\approx 1.95\,\hat{\sigma}, & \alpha_\| &\approx -1.39\,\hat{\alpha}, & \kappa_{e\|} &\approx 4.15\,\hat{\kappa}_e, \\
\sigma_\wedge &\approx 1.0\,\hat{\sigma}(\Omega_e\tau_e)^{-1}, & \alpha_\wedge &\approx -6.60\,\hat{\alpha}(\Omega_e\tau_e)^{-3}, & \kappa_{e\wedge} &\approx 2.5\,\hat{\kappa}_e(\Omega_e\tau_e)^{-1}, \\
\sigma_\perp &\approx 1.0\,\hat{\sigma}(\Omega_e\tau_e)^{-2}, & \alpha_\perp &\approx 1.5\,\hat{\alpha}(\Omega_e\tau_e)^{-2}, & \kappa_{e\perp} &\approx 4.66\,\hat{\kappa}_e(\Omega_e\tau_e)^{-2},
\end{aligned} \qquad (3.72)$$

where the dimensional factors are given by

$$\hat{\sigma} \equiv \frac{e^2 n_e \tau_e}{m_e}, \qquad \hat{\alpha} \equiv \frac{e n_e \tau_e}{m_e}, \qquad \hat{\kappa}_e \equiv \frac{n_e k T_e \tau_e}{m_e}. \qquad (3.73)$$

The inverse of the conductivity tensor is the resistivity tensor $\boldsymbol{\eta} \equiv \boldsymbol{\sigma}^{-1}$. Its coefficients are obtained by means of the right matrix (3.68):

$$\eta_\| = \frac{1}{\sigma_\|} \approx 0.51\,\hat{\sigma}^{-1}, \qquad \eta_\wedge \approx -\frac{1}{\sigma_\wedge} \approx -\hat{\sigma}^{-1}\Omega_e\tau_e, \qquad \eta_\perp \approx \frac{\sigma_\perp}{\sigma_\wedge^2} \approx \hat{\sigma}^{-1}. \qquad (3.74)$$

As stressed by Balescu, this shows the following peculiarities: (1) the perpendicular resistivity is not the inverse of the perpendicular conductivity, $\eta_\perp \neq (\sigma_\perp)^{-1}$, and the simple relationship $\eta_\perp \approx 2\eta_\|$ of the Spitzer resistivity (already introduced in Eq. (2.65)), with no dependence on the magnetic field to leading order, is essentially due to the contribution of the Hall conductivity $\sigma_\wedge$; (2) in sharp contrast to the corresponding component $\eta'_\wedge$ of the Braginskii resistivity tensor, discussed below, the component $\eta_\wedge$ becomes very large when the magnetic field is large ($\Omega_e\tau_e \gg 1$).

This does not yet provide the necessary expression for the friction force $\mathbf{R}_e$. At this point, the Hermitian moment expansion brings in an additional number of higher order moments, depending on the level of accuracy desired. In the 21M approximation,

$$\mathbf{R}_e = \frac{m_e}{e\tau_e}\left[\mathbf{j} + 0.6\,e(kT_e)^{-1}\mathbf{h}_e - 0.896\,en_e(kT_e/m_e)^{1/2}\,\bar{\mathbf{h}}_e^{(5)}\right], \qquad (3.75)$$

where the heat flow $\mathbf{h}_e$ corresponds to the dimensionless third order Hermitian moment $\bar{\mathbf{h}}_e^{(3)}$, and $\bar{\mathbf{h}}_e^{(5)}$ is the dimensionless fifth order Hermitian moment. The latter is related to the driving thermodynamic forces by

$$\bar{\mathbf{h}}_e^{(5)} = \frac{1}{en_e}\sqrt{\frac{m_e}{kT_e}}\left[\boldsymbol{\gamma}\cdot\widehat{\mathbf{E}} - \boldsymbol{\delta}\cdot\nabla(kT_e)\right], \qquad (3.76)$$

where $\boldsymbol{\gamma}$ and $\boldsymbol{\delta}$ are pseudo-transport coefficients:

$$\gamma_\parallel \approx 0.132\,\hat{\sigma}, \qquad \delta_\parallel \approx 1.23\,\hat{\alpha},$$
$$\gamma_\wedge \approx 4.64\,\hat{\sigma}(\Omega_e\tau_e)^{-3}, \qquad \delta_\wedge \approx 12.4\,\hat{\alpha}(\Omega_e\tau_e)^{-3},$$
$$\gamma_\perp \approx -0.896\,\hat{\sigma}(\Omega_e\tau_e)^{-2}, \qquad \delta_\perp \approx -2.57\,\hat{\alpha}(\Omega_e\tau_e)^{-2}.$$
$$(3.77)$$

This completes the closure of the electron fluid equations with respect to $\mathbf{R}_e$ and $\mathbf{h}_e$.

▷ **Braginskii's transport expressions** [41] originate from an entirely different approach, leading to the following relations for the momentum transfer $\mathbf{R}_e$ and the heat flow $\mathbf{h}_e$:

$$\mathbf{R}_e = en_e\,\boldsymbol{\eta}'\cdot\mathbf{j} - \boldsymbol{\beta}\cdot\nabla(kT_e), \qquad (3.78)$$

$$\mathbf{h}_e = -\frac{kT_e}{en_e}\boldsymbol{\beta}\cdot\mathbf{j} - \boldsymbol{\kappa}_e'\cdot\nabla(kT_e). \qquad (3.79)$$

Here, $\mathbf{j}$ and $-\nabla(kT_e)$ are considered as the driving thermodynamic forces and $-\mathbf{R}_e$ and $\mathbf{h}_e$ as the fluxes (a 'rather unnatural choice', according to Balescu, although Onsager symmetry is also obtained this way). The connection between Braginskii's and Balescu's expressions[2] for the electron electrical and thermal coefficients is given by

$$\boldsymbol{\eta}' = \frac{m_e}{e^2 n_e \tau_e}\left[\mathbf{I} + (0.6\,e\,\boldsymbol{\alpha} - 0.896\,\boldsymbol{\gamma})\cdot\boldsymbol{\sigma}^{-1}\right]$$
$$= \boldsymbol{\sigma}^{-1} + \frac{m_e}{e^2 n_e \tau_e}\left[\boldsymbol{\alpha} + 0.6\,e(kT_e)^{-1}\boldsymbol{\kappa}_e - 0.896\,\boldsymbol{\delta}\right]\cdot\boldsymbol{\alpha}^{-1}, \qquad (3.80)$$

$$\boldsymbol{\beta} = -en_e\,\boldsymbol{\alpha}\cdot\boldsymbol{\sigma}^{-1}, \qquad (3.81)$$

$$\boldsymbol{\kappa}_e' = \boldsymbol{\kappa}_e - kT_e\,\boldsymbol{\alpha}\cdot\boldsymbol{\sigma}^{-1}\cdot\boldsymbol{\alpha}, \qquad (3.82)$$

where the resistivity $\boldsymbol{\eta}'$ now involves the pseudo-transport coefficients $\boldsymbol{\gamma}$ and $\boldsymbol{\delta}$. This yields the following (very different from Balescu's!) explicit expressions for the *resistivity* η_λ', the

---

[2] The notations used by Balescu and Braginskii differ, where we mostly follow the first author with some exceptions dictated by consistency of the present text: $\boldsymbol{\eta} \equiv \boldsymbol{\sigma}^{-1}|_{\text{Bal}}$, $\boldsymbol{\gamma} \equiv \boldsymbol{\beta}|_{\text{Bal}}$, $\boldsymbol{\delta} \equiv \boldsymbol{\gamma}|_{\text{Bal}}$, $\boldsymbol{\eta}' \equiv \boldsymbol{\rho}|_{\text{Bal}} \equiv (en_e)^{-1}\boldsymbol{\alpha}|_{\text{Brag}}$, $\boldsymbol{\beta} \equiv -\mathbf{b}|_{\text{Bal}} \equiv \boldsymbol{\beta}|_{\text{Brag}}$, $\boldsymbol{\kappa}_e' \equiv \boldsymbol{\kappa}_e'|_{\text{Bal}} \equiv \boldsymbol{\kappa}_e|_{\text{Brag}}$. (Note that $\alpha_\wedge|_{\text{Brag}}$ and $\kappa_{i\wedge}|_{\text{Brag}}$ have been defined with opposite sign. Also note that our sign of Ω_e is positive.)

thermo-electric coupling β_λ, and the *electron thermal conductivity* $\kappa'_{e\lambda}$ for $Z = 1$:

$$\begin{aligned}
&\eta'_\| \approx 0.51\,\hat{\sigma}^{-1}, & &\beta_\| \approx 0.70\,n_e, & &\kappa'_{e\|} \approx 3.2\,\hat{\kappa}_e, \\
&\eta'_\wedge \approx -1.70\,(\hat{\sigma}\,\Omega_e\tau_e)^{-1}, & &\beta_\wedge \approx 1.5\,n_e(\Omega_e\tau_e)^{-1}, & &\kappa'_{e\wedge} \approx 2.5\,\hat{\kappa}_e(\Omega_e\tau_e)^{-1}, \\
&\eta'_\perp \approx 1.0\,\hat{\sigma}^{-1}, & &\beta_\perp \approx 5.1\,n_e(\Omega_e\tau_e)^{-2}, & &\kappa'_{e\perp} \approx 4.66\,\hat{\kappa}_e(\Omega_e\tau_e)^{-2}.
\end{aligned}$$
(3.83)

Note that Braginskii's resistivity tensor η' is not the inverse of the conductivity tensor σ. Also note the significant difference between the two thermo-electric coefficients α and β (where $\beta_\|$ does not even depend on τ_e), and between the two parallel components of the thermal conductivities κ_e and κ'_e. ◁

(b) Ion thermal coefficients Since there is only ion–ion scattering on the time scale τ_i, the expressions for the ion vector moments are much simpler than those for the electrons. Only the expression for the ion heat flow appears here, viz.

$$\mathbf{h}_i = -\boldsymbol{\kappa}_i \cdot \nabla(kT_i),$$
(3.84)

with the following *ion thermal conductivity* coefficients $\kappa_{i\lambda}$:

$$\begin{aligned}
&\kappa_{i\|} \approx 5.52\,\hat{\kappa}_i, \\
&\kappa_{i\wedge} \approx -2.5\,\hat{\kappa}_i(\Omega_i\tau_i)^{-1}, \\
&\kappa_{i\perp} \approx 1.41\,\hat{\kappa}_i(\Omega_i\tau_i)^{-2} \quad \text{with} \quad \hat{\kappa}_i \equiv \frac{n_i k T_i \tau_i}{m_i}.
\end{aligned}$$
(3.85)

(c) Heat transfer Neglecting viscous heating processes, the heat transfer functions read:

$$Q_e = \frac{1}{en_e}\mathbf{R}_e \cdot \mathbf{j} - Q_i, \qquad Q_i = \frac{3n_e k(T_e - T_i)}{2\tau_{\text{eq}}},$$
(3.86)

demonstrating that equilibration of the electron and ion temperatures proceeds on the time scale τ_{eq} defined in Eq. (3.55).

(d) Electron and ion viscosities The stress tensors $\boldsymbol{\pi}_e$ and $\boldsymbol{\pi}_i$ are related to the traceless strain tensors $\mathbf{W}_e(\mathbf{u}_e)$ and $\mathbf{W}_i(\mathbf{u}_i)$, respectively, through fourth rank viscosity tensors which contain just five different elements $\mu_{e\ell}$ and $\mu_{i\ell}$ (now: $\ell = \|, 1, 2, 3, 4$) for each species.[3] Again, this is due to symmetry with respect to rotations about the magnetic field. Suppressing the indices e and i, these

---

[3] Again for consistency of the notation, we exploit the symbol μ for the viscosity with Balescu's numbering of the components: $(\mu_\|, \mu_1, \mu_2, \mu_3, \mu_4) \equiv (\eta_\|, \eta_1, \eta_2, \eta_3, \eta_4)|_{\text{Bal}} \equiv (\eta_0, -\eta_4, \eta_2, -\eta_3, \eta_1)|_{\text{Brag}}$.

relationships are given by

$$\pi_{11} = -\tfrac{1}{2}\mu_\|(W_{11} + W_{22}) + \mu_3 W_{12} - \tfrac{1}{2}\mu_4(W_{11} - W_{22}),$$

$$\pi_{12} = \pi_{21} = -\tfrac{1}{2}\mu_3(W_{11} - W_{22}) - \mu_4 W_{12},$$

$$\pi_{13} = \pi_{31} = \mu_1 W_{23} - \mu_2 W_{13},$$

$$\pi_{22} = -\tfrac{1}{2}\mu_\|(W_{11} + W_{22}) - \mu_3 W_{12} + \tfrac{1}{2}\mu_4(W_{11} - W_{22}),$$

$$\pi_{23} = \pi_{32} = -\mu_1 W_{13} - \mu_2 W_{23},$$

$$\pi_{33} = -\mu_\| W_{33}, \tag{3.87}$$

where the strain tensors are defined by

$$W_{ij} \equiv \frac{\partial u_i}{\partial x_j} + \frac{\partial u_j}{\partial x_i} - \tfrac{2}{3}\delta_{ij} \nabla \cdot \mathbf{u}. \tag{3.88}$$

The *electron viscosity* coefficients $\mu_{e\ell}$ for $Z = 1$ read:

$$\mu_{e\|} \approx 0.73\,\hat{\mu}_e,$$
$$\mu_{e1} \approx 1.0\,\hat{\mu}_e(\Omega_e \tau_e)^{-1}, \quad \mu_{e3} \approx \tfrac{1}{2}\mu_{e1},$$
$$\mu_{e2} \approx 2.05\,\hat{\mu}_e(\Omega_e \tau_e)^{-2}, \quad \mu_{e4} \approx \tfrac{1}{4}\mu_{e2}, \quad \text{with} \quad \hat{\mu}_e \equiv n_e k T_e \tau_e. \tag{3.89}$$

The *ion viscosity* coefficients $\mu_{i\ell}$ read:

$$\mu_{i\|} \approx 1.36\,\hat{\mu}_i,$$
$$\mu_{i1} \approx -1.0\,\hat{\mu}_i(\Omega_i \tau_i)^{-1}, \quad \mu_{i3} \approx \tfrac{1}{2}\mu_{i1}, \tag{3.90}$$
$$\mu_{i2} \approx 0.85\,\hat{\mu}_i(\Omega_i \tau_i)^{-2}, \quad \mu_{i4} \approx \tfrac{1}{4}\mu_{i2}, \quad \text{with} \quad \hat{\mu}_i \equiv n_i k T_i \tau_i.$$

With the expressions listed under (a)–(d), the two-fluid equations (3.61)–(3.63) have become a closed set.

3.3.3 Dissipative versus ideal fluids*

We now come to a peculiar point in the exposition: in the end, the extensive discussion of the transport coefficients just serves to neglect most of them! This will be justified on the basis of the time scales τ_H for the hydrodynamic and τ_D for the dissipative phenomena. Paradoxically, whereas the relaxation times $\tau_{e,i}$ measure the very *short* time scales needed to establish the electron and ion fluids, the associated dissipative diffusion (decay) of the macroscopic quantities takes place on the very *long* time scale τ_D. Hence, in the restricted range

$$\tau_{e,i} \ll \tau_H \ll \tau_D, \tag{3.91}$$

3.3 Two-fluid equations*

the macroscopic fluid dynamics may be considered as dissipationless, or *ideal*. We will demonstrate that this restriction not only permits a very significant reduction of the number of variables, but also creates the essential window on the great majority of macroscopic plasma dynamical phenomena.

As an example, consider the transport of ion momentum and energy by viscosity and heat conduction. This is described by the momentum equation (3.62)(b), with the ion viscosities $\mu_{i\ell}$ given in Eq. (3.90), and the energy equation (3.63)(b), with the ion heat conductivities $\kappa_{i\lambda}$ given in Eq. (3.85). To estimate orders of magnitude, we just single out the two terms that give a diffusion equation:

$$n_i m_i \frac{\partial \mathbf{u}_i}{\partial t} \approx -\nabla \cdot \boldsymbol{\pi}_i \sim \mu_{i\ell} \frac{\partial^2 \mathbf{u}_i}{\partial x_\ell^2}, \tag{3.92}$$

$$\frac{3}{2} n_i \frac{\partial (kT_i)}{\partial t} \approx -\nabla \cdot \mathbf{h}_i \sim \kappa_{i\lambda} \frac{\partial^2 (kT_i)}{\partial x_\lambda^2}. \tag{3.93}$$

For simplicity, we consider the indices ℓ and λ to just take the values $\parallel$ and $\perp$. Since both $\mu_{i\parallel}$ and $\kappa_{i\parallel} \sim kT_i \tau_i$, we obtain the following estimates for the parallel and perpendicular ion diffusion time scales:

$$\tau_{D,i\parallel} \sim \frac{L^2}{v_{th,i}^2 \tau_i}, \quad \text{and} \quad \tau_{D,i\perp} \sim \left(\frac{a}{L}\right)^2 (\Omega_i \tau_i)^2 \tau_{D,i\parallel}. \tag{3.94}$$

This exhibits the mentioned paradox. Since the relaxation time scale τ_i is considered to be short, the parallel diffusion time $\tau_{D,i\parallel}$ will be long. Moreover, although the ratio a/L of perpendicular to parallel plasma dimensions is small, typically ~ 0.1, the factor $\Omega_i \tau_i$ is usually so large (see the numerical examples below Eq. (3.57)) that the ratio $[(a/L)\Omega_i \tau_i]^2$ between perpendicular and parallel diffusion times will be huge: classical perpendicular thermal isolation is nearly perfect! Of course, this is the very reason for using magnetic fields to confine thermonuclear plasmas in the laboratory. It is also the reason why large perpendicular temperature gradients are maintained so well in coronal magnetic flux tubes.

Let us again insert the numbers of our generic examples, discussed in Section 3.2.4, using the Tables B.3 and B.5. For a tokamak plasma with $\widetilde{T} = 1$ keV, $v_{th,i} = 0.4 \times 10^6$ m s^{-1}, $\tau_i = 2.6 \times 10^{-4}$ s, we obtain $\tau_{D,i\parallel} = 8 \times 10^{-6}$ s and $\tau_{D,i\perp} = 4.5 \times 10^4$ s (≈ 12 hrs!). For a coronal loop with $\widetilde{T} = 0.1$ keV, $v_{th,i} = 1.4 \times 10^5$ m s^{-1}, $\tau_i = 8.6 \times 10^{-3}$ s, we obtain $\tau_{D,i\parallel} = 6.2 \times 10^6$ s (≈ 72 days!) and $\tau_{D,i\perp}$ is virtually infinite. This demonstrates, once more, the easy justification of hydrodynamic models for astrophysical plasmas, due to the large length scales L. It also justifies the use of *ideal* hydrodynamic models for those plasmas. On the other hand, satisfaction of the condition (3.91) for the parallel diffusion in thermonuclear laboratory plasmas is difficult, for the same reasons as discussed in Section 3.2.4 with respect to the conditions (3.58) for hydrodynamic behaviour. The comments raised there might serve to alleviate our concerns here as well. From a practical point of view, rapid parallel diffusion is actually desirable in fusion devices since it evens out temperature gradients in the magnetic surfaces.

Electron momentum and energy transport by viscosity and heat conduction lead to similar considerations. The estimation of the numerical magnitudes is left as an exercise for the reader. Consequently, under the restrictions (3.91), the electron and ion anisotropic pressure and heat flow terms may be neglected:

$$\pi_{e,i} \to 0 \quad (\text{i.e. } \mu_{e,i\ell} \to 0), \qquad \mathbf{h}_{e,i} \to 0 \quad (\text{i.e. } \kappa_{e,i\lambda} \to 0) \qquad (3.95)$$

in the two-fluid equations (3.62)–(3.63).

The next step towards an ideal fluid description is the neglect of the ion–electron momentum transfer $\mathbf{R}_e$ in the two-fluid equations. This requires an entirely different type of argumentation since the pertinent transport coefficient is the electrical conductivity $\sigma_\|$ which, although proportional to τ_e like the other transport coefficients, should be large (nearly perfect conductivity) instead of small for the emergence of ideal fluid behaviour. We need to exploit some form of the generalized Ohm's law (see Sections 2.4.1 and 3.4.1), which is obtained by subtracting from the electron momentum equation (3.62)(a) the ion momentum equation (3.62)(b) multiplied with Zm_e/m_i and neglecting small terms in the mass ratio:

$$en_e \widehat{\mathbf{E}} = \mathbf{R}_e + \mathbf{j} \times \mathbf{B} \quad \Rightarrow \quad en_e(\mathbf{E} + \mathbf{u}_i \times \mathbf{B}) \approx \mathbf{R}_e. \qquad (3.96)$$

Here, the left expression is the one exploited by Balescu [14] to obtain his transport coefficients keeping the appropriate number of Hermitian moments (typically 21M), whereas the right expression results from neglecting the contributions of the electron pressure gradient and the Hall term. The latter approximations are necessary to obtain a consistent representation in a low number of moments. As suggested by Balescu, the above neglect of the electron heat flow ($\kappa_e \to 0$) implies that the thermo-electric coupling should be neglected as well ($\alpha \to 0$) and, hence, that all higher moments in Eq. (3.75) disappear. Effectively, we are now down to a crude 5M approximation with a very simple relationship between $\mathbf{R}_e$ and $\mathbf{j}$:

$$\mathbf{R}_e = \frac{m_e}{e\tau_e}\mathbf{j} = en_e\eta_0\mathbf{j} \quad \Rightarrow \quad \eta_0 \equiv \frac{m_e}{e^2 n_e \tau_e} = 2\eta_\| = 2\sigma_\|^{-1}. \qquad (3.97)$$

Consequently, we obtain an isotropic resistivity tensor $\eta_0\mathbf{I}$ which is off by a factor of 2 in the parallel direction and misses the off-diagonal elements.

3.3 Two-fluid equations*

Nevertheless, the resulting resistive fluid model is the most widely used one in the plasma literature.

We still have to demonstrate why and when the ion–electron momentum transfer, i.e. the resistivity, may be neglected. The crucial observation here is that resistivity is the cause of current decay, i.e. decay of magnetic field inhomogeneity. Obviously, in fusion devices, such dissipation should be small in order for the confinement configuration to be maintained. The relevant diffusion equation for magnetic field inhomogeneity is obtained by substituting in Faraday's law (3.8) the approximate electric field $\mathbf{E} \approx (\sigma)^{-1} \cdot \mathbf{j} \approx \eta_0 \mathbf{j}$ (from Eq. (3.70), neglecting the ion flow and electron temperature gradient contributions) and exploiting Ampère's law $\mathbf{j} \approx \mu_0^{-1} \nabla \times \mathbf{B}$ (from Eq. (3.9), neglecting the displacement current):

$$\frac{\partial \mathbf{B}}{\partial t} \approx -\nabla \times \mathbf{E} \approx -\nabla \times \left(\frac{\eta_0}{\mu_0} \nabla \times \mathbf{B} \right) \sim -\frac{\eta_0}{\mu_0} \frac{\partial^2 \mathbf{B}}{\partial x_\perp^2}. \tag{3.98}$$

This yields the following estimate of the resistive decay time:

$$\tau_R \sim \frac{\mu_0 a^2}{\eta_0} = \frac{\mu_0 e^2 n_e \tau_e a^2}{m_e} = a^2 \left(\frac{\omega_{pe}}{c} \right)^2 \tau_e \equiv \left(\frac{a}{\delta_e} \right)^2 \tau_e. \tag{3.99}$$

Here, a new length scale enters the discussion, viz. the *electron skin depth*,

$$\delta_e \equiv c/\omega_{pe}, \tag{3.100}$$

which measures the thickness of the layer in which high frequency electromagnetic waves can penetrate a well-conducting plasma. Using the numbers of Tables B.3 and B.5 again, we find $\tau_R = 24$ s for a 1 keV tokamak plasma and $\tau_R = 8 \times 10^{12}$ s (!) for a coronal loop plasma; sufficiently long to assume $\tau_H \ll \tau_R$ and to put

$$\mathbf{R}_e \to 0 \quad \text{(i.e. } \eta_0 \to 0\text{)} \tag{3.101}$$

in the two-fluid equations (3.62)–(3.63).

Finally, we need to insert the expression (3.86) for the electron–ion energy transfer rate Q_i into the two-fluid energy equations (3.63). These will be simplified again by putting

$$Q_i \to 0. \tag{3.102}$$

In a two-fluid plasma model, this neglect of temperature equilibration may be justified for short time scales $\tau_H \ll \tau_{eq}$, when the two temperatures T_e and T_i still evolve by themselves. In the one-fluid model, considered in Section 3.4, the opposite will be assumed, viz. $\tau_H \gg \tau_{eq}$, when temperature equilibration has already taken place so that $T_e = T_i$. In both limits, the condition (3.102) is a valid assumption. Hence, the *ideal two-fluid equations* are obtained from Eqs. (3.61)–(3.64) in the limits (3.95), (3.101), (3.102), i.e. neglecting all the RHS terms.

Even though the condition $\tau_H \ll \tau_R$ appears to be well satisfied for plasmas of interest, there are important instances when resistive effects enter the hydrodynamic time scales after all. This happens, e.g., because certain resistive instabilities evolve on a time scale faster than τ_R, or turbulence creates current sheets with much smaller length scales than a. Effectively, internal resistive boundary layers with large gradients develop, very analogous to the boundary layers of ordinary fluids. A spectacular example is the solar flare with the release of huge amounts of magnetic energy, triggered by resistive phenomena on a very small length scale (totally unrelated to the astronomical scales L and a). In those cases, it is expedient to keep the resistive terms

$$\mathbf{R}_e = en_e \eta_0 \mathbf{j}, \quad \text{and} \quad -(\mathbf{u}_e - \mathbf{u}_i) \cdot \mathbf{R}_e = (en_e)^{-1} \mathbf{j} \cdot \mathbf{R}_e = \eta_0 |\mathbf{j}|^2 \quad (3.103)$$

in the momentum and energy equations, where the right expression represents Ohmic heating. Hence, the *resistive two-fluid equations* are obtained from Eqs. (3.61)–(3.64) in the limits (3.95) and (3.102), i.e. neglecting pressure anisotropies and heat flows, but keeping the resistive terms (3.103) in the RHSs. From now on, we will drop the subscript 0 on the resistivity η_0, which we recall is just a model representation of the actual anisotropic tensor.

In conclusion: the essential, but reduced, picture of ideal two-fluid plasma dynamics is valid in the wide range of time scales (3.91), intermediate between rapid kinetics and slow transport. These conditions clearly indicate how the theory is to be modified when the ideal description fails. When the left condition is violated, a kinetic description is in order, and when the right condition is not satisfied, the dissipative transport terms should be restored in the equations. A very relevant example of the latter procedure is the resistive two-fluid model, obtained after restoring only the resistive terms in the ideal two-fluid equations.

3.3.4 Excursion: waves in two-fluid plasmas*

An instructive example of dynamics in a two-fluid plasma is the enormous variety of waves it supports. We here derive the general *dispersion equation* for a homogeneous, resistive, two-fluid plasma, jumping details to keep this subsection within reasonable bounds. The dispersion equation will be solved explicitly for the ideal case, exposing the different length and time scales of the waves in order to facilitate the discussion of Section 3.4 where only the largest scales will survive.

Our starting point is the complete set of two-fluid equations (3.61)–(3.64), neglecting the RHSs except for the resistive terms (3.103), and Maxwell's

3.3 Two-fluid equations*

equations (3.6)–(3.11). As in Section 2.3, we exploit the pressures $p_{e,i}$ instead of the temperatures as variables. We assume a homogeneous background equilibrium with the electrons and ions at rest:

$$p_{e0}, p_{i0}, \mathbf{B}_0 \text{ const}, \quad \mathbf{E}_0 = 0 \Rightarrow n_{e0} = Zn_{i0}, \quad \mathbf{u}_{e0} = \mathbf{u}_{i0} = 0 \Rightarrow \mathbf{j}_0 = 0. \quad (3.104)$$

We now perturb this equilibrium with small amplitude oscillations. Dropping the 0 on the equilibrium quantities and indicating perturbations with a tilde, the equations for the perturbations become:

$$\frac{\partial \tilde{n}_e}{\partial t} + n_e \nabla \cdot \tilde{\mathbf{u}}_e = 0,$$

$$n_e m_e \frac{\partial \tilde{\mathbf{u}}_e}{\partial t} + \nabla \tilde{p}_e + en_e(\tilde{\mathbf{E}} + \tilde{\mathbf{u}}_e \times \mathbf{B}) = \frac{m_e}{e} \nu \tilde{\mathbf{j}} = -n_e m_e \nu(\tilde{\mathbf{u}}_e - \tilde{\mathbf{u}}_i),$$

$$\frac{\partial \tilde{p}_e}{\partial t} + \gamma p_e \nabla \cdot \tilde{\mathbf{u}}_e = 0, \quad (3.105)$$

$$\frac{\partial \tilde{n}_i}{\partial t} + n_i \nabla \cdot \tilde{\mathbf{u}}_i = 0,$$

$$n_i m_i \frac{\partial \tilde{\mathbf{u}}_i}{\partial t} + \nabla \tilde{p}_i - Zen_i(\tilde{\mathbf{E}} + \tilde{\mathbf{u}}_i \times \mathbf{B}) = -\frac{m_e}{e} \nu \tilde{\mathbf{j}} = n_e m_e \nu(\tilde{\mathbf{u}}_e - \tilde{\mathbf{u}}_i),$$

$$\frac{\partial \tilde{p}_i}{\partial t} + \gamma p_i \nabla \cdot \tilde{\mathbf{u}}_i = 0, \quad (3.106)$$

$$\nabla \times \tilde{\mathbf{E}} = -\frac{\partial \tilde{\mathbf{B}}}{\partial t}, \quad \nabla \cdot \tilde{\mathbf{E}} = -\frac{e}{\epsilon_0}(\tilde{n}_e - Z\tilde{n}_i),$$

$$\nabla \times \tilde{\mathbf{B}} = \frac{1}{c^2} \frac{\partial \tilde{\mathbf{E}}}{\partial t} - \mu_0 en_e(\tilde{\mathbf{u}}_e - \tilde{\mathbf{u}}_i), \quad \nabla \cdot \tilde{\mathbf{B}} = 0. \quad (3.107)$$

Here, $\gamma \equiv 5/3$ and we have introduced an effective electron–ion collision frequency,

$$\nu \equiv \tau_e^{-1} \quad \Rightarrow \quad \eta = m_e(e^2 n_e)^{-1} \nu, \quad (3.108)$$

which turns out to be a convenient way of bookkeeping the resistive damping of the waves.

Since all equilibrium quantities are constant in space and time, we assume plane wave solutions, $n_e(\mathbf{r}, t) \sim \exp[i(\mathbf{k} \cdot \mathbf{r} - \omega t)]$, etc., so that $\nabla \to i\mathbf{k}$, $\partial/\partial t \to -i\omega$. This turns the partial differential equations into a set of algebraic equations. The determinant of this set gives the dispersion equation $\omega = \omega(\mathbf{k})$ of the waves. The manner in which the time derivatives appear in these equations dictates the number of waves to be expected: since there are five electron variables ($\tilde{n}_e$, $\tilde{\mathbf{u}}_e$, $\tilde{p}_e$), five ion variables ($\tilde{n}_i$, $\tilde{\mathbf{u}}_i$, $\tilde{p}_i$), and four EM variables (two independent components of $\tilde{\mathbf{E}}$ and two independent components of $\tilde{\mathbf{B}}$), we may expect 14 different types of waves!

To solve this system, we capitalize on all available geometrical and physical relationships between the variables. To that end, we project on unit vectors oriented with respect to $\mathbf{k}$ and $\mathbf{B}$, with $\mathbf{e}_3$ along $\mathbf{k}$:

$$\mathbf{e}_1 \equiv \mathbf{e}_2 \times \mathbf{e}_3 , \qquad \mathbf{e}_2 \equiv \mathbf{B} \times \mathbf{k}/|\mathbf{B} \times \mathbf{k}| , \qquad \mathbf{e}_3 \equiv \mathbf{k}/|\mathbf{k}| , \qquad (3.109)$$

and indicate the direction of $\mathbf{k}$ by means of the angle ϑ between $\mathbf{k}$ and $\mathbf{B}$:

$$\lambda \equiv k_\parallel/k = \cos\vartheta , \qquad \tau \equiv k_\perp/k = \sin\vartheta . \qquad (3.110)$$

In this projection, the components $\tilde{E}_3$ and $\tilde{B}_3$ ($= 0$) may be eliminated and we obtain the mentioned system of 14 variables.

Next, we isolate a peculiar class of marginal 'waves' ($\omega = 0$) consisting of a two-parameter family of modes with $\tilde{n}_e \neq 0$ and $\tilde{n}_i \neq 0$ and associated pressure perturbations that are completely out of phase,

$$\tilde{p}_e = -\tilde{p}_i = \mathrm{i}\frac{en_e}{k}\tilde{E}_3 = -\frac{e^2 n_e}{\epsilon_0 k^2}(\tilde{n}_e - Z\tilde{n}_i) , \qquad (3.111)$$

whereas the remaining variables vanish. These waves do not move, they just sit there with the pressures kept in balance by the longitudinal electric field due to the space charge clouds. Even in the absence of the latter, there still is a subclass of charge-neutral modes with $\tilde{n}_e = Z\tilde{n}_i \neq 0$, and all of the remaining variables identically zero. In the ideal MHD limit, the two-parameter marginal modes transform into the one-parameter entropy modes discussed in Section 5.2.

For the remaining 12 solutions, which represent genuine waves ($\omega \neq 0$), we can freely divide by ω to express the density and pressure variables $\tilde{n}_{e,i}$ and $\tilde{p}_{e,i}$ in terms of $\tilde{u}_{e,i3}$, and the magnetic field perturbations $\tilde{B}_{2,1}$ in terms of $\tilde{E}_{1,2}$. This yields a system of eight algebraic equations in the variables $\tilde{\mathbf{u}}_{e,i}$ and $\tilde{E}_{1,2}$. At this point, it is expedient to account for the phase differences of the variables, expressed by the factors i, and to equalize the dimensions of the different terms. This is done by introducing the following variables for the perturbations:

$$\begin{aligned}
\tilde{E}_1 &\equiv \mathrm{i}\sqrt{\epsilon_0}\,\tilde{\mathbf{E}} \cdot \mathbf{e}_1 , & \tilde{E}_2 &\equiv \sqrt{\epsilon_0}\,\tilde{\mathbf{E}} \cdot \mathbf{e}_2 , \\
\tilde{U}_{e1,3} &\equiv \sqrt{n_e m_e}\,\tilde{\mathbf{u}}_e \cdot \mathbf{e}_{1,3} , & \tilde{U}_{e2} &\equiv \mathrm{i}\sqrt{n_e m_e}\,\tilde{\mathbf{u}}_e \cdot \mathbf{e}_2 , \\
\tilde{U}_{i1,3} &\equiv \sqrt{n_i m_i}\,\tilde{\mathbf{u}}_i \cdot \mathbf{e}_{1,3} , & \tilde{U}_{i2} &\equiv \mathrm{i}\sqrt{n_i m_i}\,\tilde{\mathbf{u}}_i \cdot \mathbf{e}_2 , \qquad (3.112)
\end{aligned}$$

3.3 Two-fluid equations*

and equilibrium parameters characterizing the waves:

$$\omega_{pe} \equiv \sqrt{\frac{e^2 n_e}{\epsilon_0 m_e}}, \qquad \Omega_e \equiv \frac{eB}{m_e}, \qquad v_e \equiv \sqrt{\frac{\gamma p_e}{n_e m_e}},$$

$$\omega_{pi} \equiv \sqrt{\frac{Z^2 e^2 n_i}{\epsilon_0 m_i}} \equiv \sqrt{\mu}\,\omega_{pe}, \qquad \Omega_i \equiv \frac{ZeB}{m_i} \equiv \mu\Omega_e, \qquad v_i \equiv \sqrt{\frac{\gamma p_i}{n_i m_i}}.$$

(3.113)

Here, v_e and v_i represent the electron and ion sound speeds (differing from the thermal speeds by a factor of $(\gamma/2)^{1/2} \approx 0.913$). The ratio of masses over charges,

$$\mu \equiv \frac{Zm_e}{m_i}, \tag{3.114}$$

is not yet assumed to be small in order to profit from the symmetry of electron and ion terms in the analysis.

(a) Eigenvalue problem for a resistive two-fluid plasma The above transformations yield the following eigenvalue problem:

$$\begin{pmatrix}
\omega^2 - k^2 c^2 & 0 & \omega\omega_{pe} & 0 & 0 & -\omega\omega_{pi} & 0 & 0 \\
0 & \omega^2 - k^2 c^2 & 0 & -\omega\omega_{pe} & 0 & 0 & \omega\omega_{pi} & 0 \\
\omega_{pe} & 0 & \omega + i\nu & \lambda\Omega_e & 0 & -i\sqrt{\mu}\nu & 0 & 0 \\
0 & -\omega_{pe} & \lambda\Omega_e & \omega + i\nu & \tau\Omega_e & 0 & -i\sqrt{\mu}\nu & 0 \\
0 & 0 & 0 & \omega\tau\Omega_e & \omega^2 + i\nu\omega - k^2 v_e^2 - \omega_{pe}^2 & 0 & 0 & \omega_{pe}\omega_{pi} - i\sqrt{\mu}\nu\omega \\
-\omega_{pi} & 0 & -i\sqrt{\mu}\nu & 0 & 0 & \omega + i\mu\nu & -\lambda\Omega_i & 0 \\
0 & \omega_{pi} & 0 & -i\sqrt{\mu}\nu & 0 & -\lambda\Omega_i & \omega + i\mu\nu & -\tau\Omega_i \\
0 & 0 & 0 & 0 & \omega_{pe}\omega_{pi} - i\sqrt{\mu}\nu\omega & 0 & -\omega\tau\Omega_i & \omega^2 + i\mu\nu\omega - k^2 v_i^2 - \omega_{pi}^2
\end{pmatrix}$$

$$\cdot\,(\tilde{E}_1, \quad \tilde{E}_2, \quad \tilde{U}_{e1}, \quad \tilde{U}_{e2}, \quad \tilde{U}_{e3}, \quad \tilde{U}_{i1}, \quad \tilde{U}_{i2}, \quad \tilde{U}_{i3})^T = 0.$$

(3.115)

Because of the judicious choice of variables, most of the matrix elements are real now, except for the resistive terms ($\sim i\nu$). This implies that the waves will be damped: $\omega = \omega_0 + i\gamma_0$ with $\gamma_0 < 0$, or $k = k_0 + i\alpha_0$ with $\alpha_0 \neq 0$.

Such behaviour is easily checked for the special case of cold, unmagnetized electrons with heavy, immobile ions ($\omega_{pi} = 0$, $\Omega_e = \Omega_i = 0$, $v_e = v_i = 0$). The dispersion equation then renders two branches, one for the longitudinal plasma oscillations (discussed in Section 2.3.2) and one for the transverse electromagnetic waves:

$$\omega(\omega + i\nu) - \omega_{pe}^2 = 0 \quad \Rightarrow \quad \omega_0^2 \approx \omega_{pe}^2, \qquad \gamma_0 \approx -\tfrac{1}{2}\nu, \quad (3.116)$$

$$(\omega + i\nu)(\omega^2 - k^2 c^2) - \omega \omega_{pe}^2 = 0 \Rightarrow \omega_0^2 \approx \omega_{pe}^2 + k^2 c^2, \ \gamma_0 \approx -\tfrac{1}{2}\nu \frac{\omega_{pe}^2}{\omega_0^2}, \quad (3.117)$$

where the damping is small when $\omega_e \tau_e \gg 1$. For the transverse waves, we have assumed real k. On the other hand, if ω is imposed to be real, the wave number becomes complex, satisfying $k_0^2 - \alpha_0^2 \approx (\omega^2 - \omega_{pe}^2)/c^2$, with $\alpha_0 k_0 \approx \tfrac{1}{2}(\nu/\omega)(\omega_{pe}^2/c^2)$. Hence, the electron skin depth $\delta_e \equiv c/\omega_{pe}$, introduced in Eq. (3.100), naturally emerges as the spatial decay length of transverse EM waves.

(b) Eigenvalue problem for an ideal two-fluid plasma In the limit of vanishing resistivity ($\nu \to 0$), a significant simplification results from the elimination of the variables $\tilde{U}_{e2}$ and $\tilde{U}_{i2}$:

$$\begin{pmatrix} \omega^2 - k^2 c^2 & 0 & \omega_{pe}\omega & 0 & -\omega_{pi}\omega & 0 \\ 0 & \begin{matrix}\omega^2 - k^2 c^2 \\ -\omega_{pe}^2 - \omega_{pi}^2\end{matrix} & \lambda\omega_{pe}\Omega_e & \tau\omega_{pe}\Omega_e & \lambda\omega_{pi}\Omega_i & \tau\omega_{pi}\Omega_i \\ \omega_{pe}\omega & \lambda\omega_{pe}\Omega_e & \omega^2 - \lambda^2\Omega_e^2 & -\lambda\tau\Omega_e^2 & 0 & 0 \\ 0 & \tau\omega_{pe}\Omega_e & -\lambda\tau\Omega_e^2 & \begin{matrix}\omega^2 - k^2 v_e^2 \\ -\omega_{pe}^2 - \tau^2\Omega_e^2\end{matrix} & 0 & \omega_{pe}\omega_{pi} \\ -\omega_{pi}\omega & \lambda\omega_{pi}\Omega_i & 0 & 0 & \omega^2 - \lambda^2\Omega_i^2 & -\lambda\tau\Omega_i^2 \\ 0 & \tau\omega_{pi}\Omega_i & 0 & \omega_{pe}\omega_{pi} & -\lambda\tau\Omega_i^2 & \begin{matrix}\omega^2 - k^2 v_i^2 \\ -\omega_{pi}^2 - \tau^2\Omega_i^2\end{matrix} \end{pmatrix} \begin{pmatrix} \tilde{E}_1 \\ \tilde{E}_2 \\ \tilde{U}_{e1} \\ \tilde{U}_{e3} \\ \tilde{U}_{i1} \\ \tilde{U}_{i3} \end{pmatrix} = 0. \quad (3.118)$$

The twelfth order system is now represented by two EM variables, two electron variables, and two ion variables. These yield a two-fold degenerate dispersion equation since, due to the symmetry of the matrix, the eigenvalue parameter ω will only appear squared.

3.3 Two-fluid equations*

Since no assumption on the smallness of the mass ratio was made, the characteristic electron and ion frequencies may still be assumed to have equal orders of magnitude. Therefore, when frequencies are made dimensionless by means of the plasma frequency, the electron and ion contributions appear on an equal footing:

$$\bar{\omega} \equiv \omega/\omega_p, \quad \text{where} \quad \omega_p \equiv \sqrt{\omega_{pe}^2 + \omega_{pi}^2}. \quad (3.119)$$

Similarly, wavelengths are made dimensionless by means of the combined skin depth for electromagnetic waves,

$$\bar{k} \equiv \delta k, \quad \text{where} \quad \delta \equiv c/\omega_p, \quad (3.120)$$

and velocities by means of the speed of light in vacuum, c. This yields six dimensionless parameters, which we indicate by one-letter symbols to facilitate the formidable algebra and the numerical implementation:

$$\begin{aligned} e &\equiv \omega_{pe}/\omega_p, & E &\equiv \Omega_e/\omega_p, & v &\equiv v_e/c, \\ i &\equiv \omega_{pi}/\omega_p, & I &\equiv \Omega_i/\omega_p, & w &\equiv v_i/c. \end{aligned} \quad (3.121)$$

These parameters are not all independent, though. Charge neutrality, $n_e = Zn_i$, implies from the expressions (3.113) that e, i, and I may be eliminated in favour of the ratio of masses over charges:

$$\mu = \frac{i^2}{e^2} = \frac{I}{E} \Rightarrow e^2 = \frac{1}{1+\mu}, \quad i^2 = 1 - e^2 = \frac{\mu}{1+\mu}, \quad I = \mu E. \quad (3.122)$$

Hence, after removing dimensions with $\omega_p \sim n_e^{-1/2}$ and c, the dimensionless dispersion equation is determined by the direction λ of the wave vector and the four independent parameters μ, $E \sim B\, n_e^{-1/2}$, $v \sim T_e^{1/2}$, $w \sim T_i^{1/2}$, that are directly related to the physical variables of the background state.

The actual construction of the explicit form of the dispersion equation is one of those calculations which require, according to Stix [218], 'in small proportion, insight, and in large proportion, stamina'. Fortunately, the earlier enterprises of Braginskii [40], and in particular of Denisse and Delcroix [66], and Stringer [220], summarized by Swanson [223], permit conclusive checking of the correctness of the final expressions below. Once this is established, to avoid confusion with conventional symbols, the short-hand notation (3.121) is of course replaced again by the physical expressions.

The *dispersion equation for the waves in an ideal two-fluid plasma*, finally obtained by brute force reduction of the determinant expressions, is a polynomial of

Table 3.1. *Dispersion equation for an ideal two-fluid plasma with asymptotic limits.*

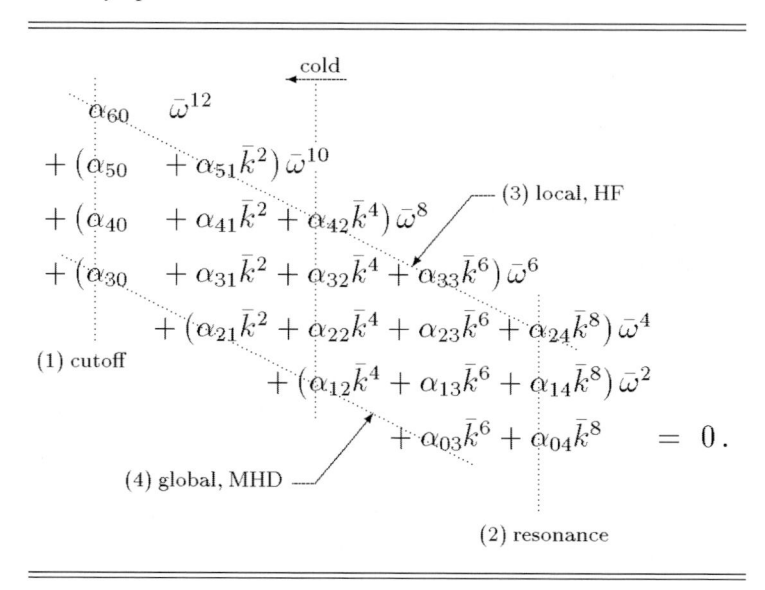

$$
\begin{aligned}
&\alpha_{60}\,\bar{\omega}^{12} \\
&+(\alpha_{50}+\alpha_{51}\bar{k}^2)\,\bar{\omega}^{10} \\
&+(\alpha_{40}+\alpha_{41}\bar{k}^2+\alpha_{42}\bar{k}^4)\,\bar{\omega}^{8} \\
&+(\alpha_{30}+\alpha_{31}\bar{k}^2+\alpha_{32}\bar{k}^4+\alpha_{33}\bar{k}^6)\,\bar{\omega}^{6} \\
&+(\alpha_{21}\bar{k}^2+\alpha_{22}\bar{k}^4+\alpha_{23}\bar{k}^6+\alpha_{24}\bar{k}^8)\,\bar{\omega}^{4} \\
&+(\alpha_{12}\bar{k}^4+\alpha_{13}\bar{k}^6+\alpha_{14}\bar{k}^8)\,\bar{\omega}^{2} \\
&+\alpha_{03}\bar{k}^6+\alpha_{04}\bar{k}^8 = 0.
\end{aligned}
$$

cold ←
(1) cutoff
(2) resonance
(3) local, HF
(4) global, MHD

the sixth degree in $\bar{\omega}^2$ and of the fourth degree in $\bar{k}^2$:

$$F(\bar{k}^2,\bar{\omega}^2) \equiv \sum_{m=0}^{6}\sum_{n=\max(0,3-m)}^{\min(4,6-m)} \alpha_{mn}\,\bar{k}^{2n}\,\bar{\omega}^{2m} = 0. \tag{3.123}$$

Arranging the terms as shown in Table 3.1 facilitates the discussion of the different asymptotic limits for which analytic solutions can be found. The explicit expressions for the 19 coefficients $\alpha_{mn} = \alpha_{mn}(\lambda^2,\mu,E^2,v^2,w^2)$ are listed in Table 3.2.

The numerical solution of this dispersion equation is shown in Fig. 3.1 for a representative choice of the parameters (the tokamak example of Table B.3). As expected, for each value of k^2, there are six two-fold degenerate waves corresponding to propagation in opposite directions ($\omega < 0$ and $\omega > 0$). Other than this degeneracy, and that of the two EM waves at high frequencies (corresponding to two different states of polarization), the six waves are non-degenerate for oblique propagation. This means that waves in two-fluid plasmas exhibit very intricate behaviour, where virtually all characteristic plasma frequency and length scales enter, as is indicated by the dotted lines in Fig. 3.1. It is beyond the scope of the present chapter to dwell on all of these aspects. The reader should consult any basic textbook on plasma physics, like Chen [53] and Goldston and Rutherford [92], or books entirely devoted to waves in plasmas, like Stix [218] and Swanson [223].

Table 3.2. *Coefficients of the dispersion equation for an ideal two-fluid plasma.*

$$\alpha_{60} = 1,$$

$$\alpha_{50} = -(3 + E^2 + I^2),$$

$$\alpha_{51} = -(2 + v^2 + w^2),$$

$$\alpha_{40} = 3 + E^2 + I^2 + 2EI + E^2 I^2,$$

$$\alpha_{41} = 4 + 2E^2 + 2I^2 + (2 + \lambda^2 E^2 + I^2)v^2 + (2 + E^2 + \lambda^2 I^2)w^2 + i^2 v^2 + e^2 w^2,$$

$$\alpha_{42} = 1 + 2v^2 + 2w^2 + v^2 w^2,$$

$$\alpha_{30} = -(1 + EI)^2,$$

$$\alpha_{31} = -\Big\{ 2(1 + EI)^2 + (1 + \lambda^2)(E^2 + I^2 - EI) \\
\qquad + [1 + I^2 + \lambda^2(3 + EI)EI]v^2 + [1 + E^2 + \lambda^2(3 + EI)EI]w^2 \\
\qquad + [2 + (1 - 3\lambda^2)EI](i^2 v^2 + e^2 w^2) \Big\},$$

$$\alpha_{32} = -\Big\{ 1 + E^2 + I^2 + 2(1 + \lambda^2 E^2 + I^2)v^2 + 2(1 + E^2 + \lambda^2 I^2)w^2 \\
\qquad + 2(i^2 v^2 + e^2 w^2) + [2 + \lambda^2(E^2 + I^2)]v^2 w^2 \Big\},$$

$$\alpha_{33} = -(v^2 + w^2 + 2v^2 w^2),$$

$$\alpha_{21} = (1 + EI)(1 + \lambda^2)EI + (1 + EI)(1 + \lambda^2 EI)(i^2 v^2 + e^2 w^2),$$

$$\alpha_{22} = (1 + EI)EI + \lambda^2(E^2 + I^2 - EI) \\
\qquad + [(1 + \lambda^2)I^2 + 2\lambda^2 EI(2 + EI)]v^2 + [(1 + \lambda^2)E^2 + 2\lambda^2 EI(2 + EI)]w^2 \\
\qquad + [2 + (1 - 5\lambda^2)EI](i^2 v^2 + e^2 w^2) + (1 + \lambda^2 EI)^2 v^2 w^2,$$

$$\alpha_{23} = (I^2 + \lambda^2 E^2)v^2 + (E^2 + \lambda^2 I^2)w^2 + i^2 v^2 + e^2 w^2 + 2[1 + \lambda^2(E^2 + I^2)]v^2 w^2,$$

$$\alpha_{24} = v^2 w^2,$$

$$\alpha_{12} = -\lambda^2 EI \big\{ EI + [2 + (1 + \lambda^2)EI](i^2 v^2 + e^2 w^2) \big\},$$

$$\alpha_{13} = -\lambda^2 \big\{ E^2 I^2 v^2 + E^2 I^2 w^2 + (E^2 + I^2)(i^2 v^2 + e^2 w^2) + 2EI(1 + \lambda^2 EI)v^2 w^2 \big\},$$

$$\alpha_{14} = -\lambda^2 (E^2 + I^2) v^2 w^2,$$

$$\alpha_{03} = \lambda^4 E^2 I^2 (i^2 v^2 + e^2 w^2),$$

$$\alpha_{04} = \lambda^4 E^2 I^2 v^2 w^2.$$

Our interest here is to demonstrate how the MHD phenomena emerge from the two-fluid ones.

To that end, we first analyse how the six waves split apart for extreme values of k^2 and ω^2. By means of the Tables 3.1 and 3.2, one may easily determine those asymptotic limits.

116 *'Derivation' of the macroscopic equations*★

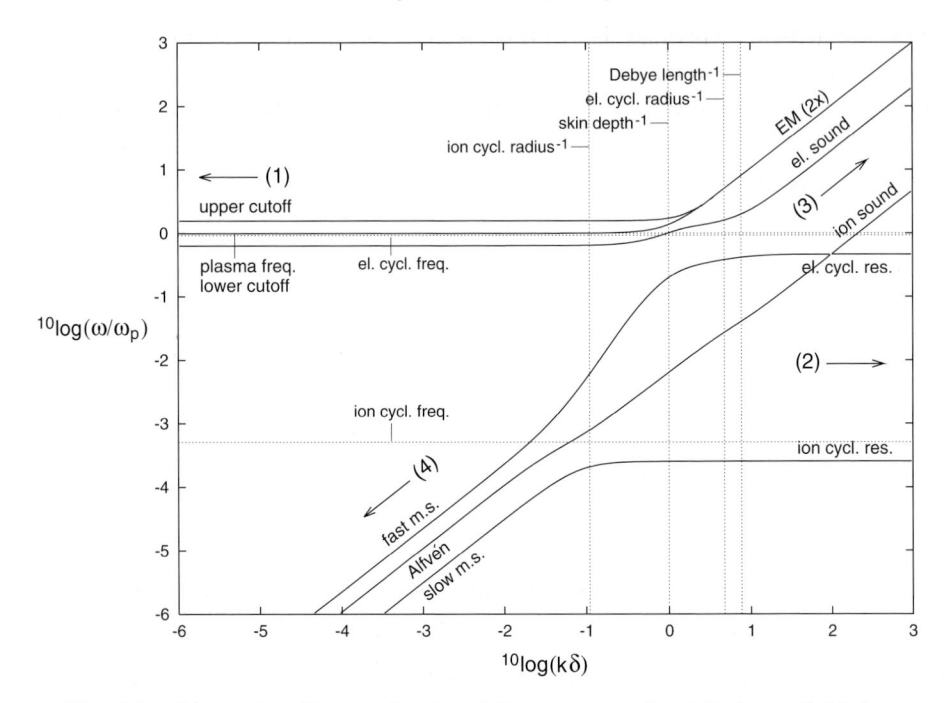

Fig. 3.1. Dispersion diagram for the oblique waves of an ideal two-fluid (hydrogen) plasma; $\lambda = 0.5$ ($\vartheta = 60°$), $\mu = 5.4 \times 10^{-4}$, $E \equiv \Omega_e/\omega_p = 0.93$, $v \equiv v_e/c = 0.2$, $w \equiv v_i/c = 4.7 \times 10^{-3}$ ($\tilde{T}_e = \tilde{T}_i = 10\,\mathrm{keV}$). The arrows refer to the asymptotic limits (1)–(4) discussed in the text.

(1) *Cutoff limits* ($k^2 \to 0$):

$$\omega^2 = \begin{cases} \omega_p^2 & \text{(plasma frequency)} \\ \omega_p^2 + \tfrac{1}{2}(\Omega_e^2 + \Omega_i^2) \pm |\Omega_e - \Omega_i|\sqrt{\omega_p^2 + \tfrac{1}{4}(\Omega_e+\Omega_i)^2} & \text{(upper \& lower cutoff).} \end{cases} \quad (3.124)$$

They represent the lower limits of the high frequency waves. Excitation at frequencies below these limits results in wave motion that is spatially evanescent ($k^2 < 0$). The huge evanescent gap in Fig. 3.1 between these waves, where the electric field dominates, and the low frequency MHD waves, where the magnetic field dominates, is the way in which the symmetry breaking of Maxwell's equations mentioned in Section 1.3.4 appears.

(2) *Resonance limits* ($k^2 \to \infty$):

$$\omega^2 = \begin{cases} \lambda^2 \Omega_e^2 & \text{(electron cyclotron resonance)} \\ \lambda^2 \Omega_i^2 & \text{(ion cyclotron resonance).} \end{cases} \quad (3.125)$$

3.3 Two-fluid equations*

These frequencies represent the asymptotic limits of spatially localized cyclotron waves. (Different resonances (with hybrid contributions of ω_p, Ω_e, and Ω_i) are obtained for cold plasmas ($v = w = 0$), as indicated by the dotted vertical line labelled 'cold' in Table 3.1 to the right of which all contributions to the dispersion equation vanish, whereas the remaining ones are also substantially simplified, as is evident from Table 3.2.)

(3) *Local, high frequency limit* ($k^2 \to \infty$, $\omega^2 \to \infty$, but ω^2/k^2 finite):

$$\omega^2 = \begin{cases} k^2 c^2 & \text{(2 degenerate EM waves)} \\ k^2 v_e^2 & \text{(electron sound wave)} \\ k^2 v_i^2 & \text{(ion sound wave)}. \end{cases} \tag{3.126}$$

In this limit, corresponding to localization in both space and time, the electromagnetic, the electron thermal, and the ion thermal contributions split apart.

(4) *Global, low frequency MHD limit* ($k^2 \to 0$, $\omega^2 \to 0$, but ω^2/k^2 finite):

$$(1 + v_A^2/c^2)^2 \left(\frac{\omega}{k}\right)^6 - (1 + v_A^2/c^2)\left[(1 + \lambda^2)v_A^2 + v_s^2 + \lambda^2 v_s^2 v_A^2/c^2\right]\left(\frac{\omega}{k}\right)^4$$

$$+ \lambda^2 v_A^2 \left[v_A^2 + 2v_s^2 + (1 + \lambda^2)v_s^2 v_A^2/c^2\right]\left(\frac{\omega}{k}\right)^2 - \lambda^4 v_A^4 v_s^2 = 0. \tag{3.127}$$

In this opposite limit, where the waves are both spatially and temporally global, all characteristic plasma parameters ω_p, Ω_e, Ω_i, v_e, and v_i contribute to produce the following *exact* expressions for the two characteristic MHD velocities:

$$EI \equiv \frac{\Omega_e \Omega_i}{\omega_p^2} \equiv \frac{B^2}{\mu_0(n_e m_e + n_i m_i)c^2} \equiv \frac{v_A^2}{c^2} \quad \text{(Alfvén speed)},$$

$$i^2 v^2 + e^2 w^2 \equiv \frac{\omega_{pi}^2 v_e^2 + \omega_{pe}^2 v_i^2}{\omega_p^2 c^2} \equiv \frac{\gamma(p_e + p_i)}{(n_e m_e + n_i m_i)c^2} \equiv \frac{v_s^2}{c^2} \quad \text{(sound speed)}. \tag{3.128}$$

In other words: in this limit, the two-fluid plasma waves become entangled to produce the three MHD waves that are central to macroscopic plasma dynamics, as will be discussed extensively in Chapter 5. Consistent with the non-relativistic form of the two-fluid equations employed, one should drop the four terms with v_A^2/c^2 ($\ll 1$) in the asymptotic dispersion equation (3.127). With this single approximation, the expressions for the three MHD waves become:

$$\omega^2 = \begin{cases} \lambda^2 k^2 v_A^2 & \text{(Alfvén wave)} \\ \frac{1}{2}k^2 \left[v_A^2 + v_s^2 \pm \sqrt{(v_A^2 + v_s^2)^2 - 4\lambda^2 v_A^2 v_s^2} \right] & \text{(fast \& slow m.s. wave).} \end{cases}$$

(3.129)

They manifest the magnetic dominance of macroscopic plasma dynamics.

The two-fluid theory represents only a very partial picture of plasma wave dynamics in the high frequency domain. This is evident from the fact that usually $\omega \tau_e \gg 1$, in violation of the left part of the condition (3.91) for fluid behaviour. In fact, in RF (radio-frequency) wave diagnostics and heating, velocity space effects determining Landau damping and the opposite phenomenon of microscopic instabilities (driven by non-monotonicities of the distribution functions) become important. Accordingly, the greater part of the mentioned books on plasma waves [218, 223] is devoted to the kinetic picture.

On the other hand, in the MHD limit of *large scales*, both in space and time, the use of fluid theory is well justified. From Fig. 3.1 it is obvious how this condition is to be quantified with respect to the different inverse time scales Ω_i, Ω_e, ω_p, and inverse length scales R_i^{-1}, δ^{-1}, R_e^{-1}, λ_D^{-1} that occur in the two-fluid model. Whereas we did not use the smallness of the mass ratio μ so far, it is expedient to exploit it now to find out that the ion scales determine the limits for the validity of the MHD model:

$$\omega \ll \Omega_i, \qquad k \ll R_i^{-1}, \tag{3.130}$$

i.e. the lower left quadrant of Fig. 3.1. Going down to larger and larger scales in this diagram, e.g. to study the transition to instability ($\omega^2 < 0$), we automatically encounter the finite size of the plasma so that we will have to drop the (enormously simplifying) assumption of plasma homogeneity that was made in this section. Hence, the counterpart of kinetic theory, paying attention to inhomogeneities in velocity space, is the large scale magnetohydrodynamic theory, paying attention to inhomogeneities in ordinary space. This will occupy us for most of the rest of this book.

3.4 One-fluid equations★

3.4.1 Maximal ordering for MHD★

We now derive the one-fluid MHD equations from the two-fluid equations (3.61)–(3.63). They are obtained as linear combinations of the pairs of the mass conservation equations for n_e and n_i, and of the momentum conservation equations for $\mathbf{u}_e$ and $\mathbf{u}_i$, and as the sum of the energy conservation equations for T_e and T_i (or, rather, p_e and p_i). At this point, we annihilate the information on the temperature difference between electrons and ions by imposing the time scale restriction $\tau_H \gg \tau_{eq}$ (temperature equilibration is assumed to have taken place already) associated with the approximation (3.102), i.e. the electron–ion energy transfer rate $Q_i \to 0$. Consequently, the number of moment equations is reduced by one. This permits the definition of the following one-fluid variables:

$$
\begin{aligned}
\rho &\equiv n_e m_e + n_i m_i & &\text{(total mass density)}, \\
\tau &\equiv -e(n_e - Z n_i) & &\text{(charge density)}, \\
\rho \mathbf{v} &\equiv n_e m_e \mathbf{u}_e + n_i m_i \mathbf{u}_i & &\text{(momentum density)}, \\
\mathbf{j} &\equiv -e(n_e \mathbf{u}_e - Z n_i \mathbf{u}_i) & &\text{(current density)}, \\
p &\equiv p_e + p_i = (n_e + n_i) kT & &\text{(pressure)},
\end{aligned} \quad (3.131)
$$

as we have already seen in Section 2.4.1. Except for the temperature difference, the full information contained in the two-fluid equations is retained by the mentioned linear combinations, when the inverses of the one-fluid variables are exploited:

$$
\begin{aligned}
n_e &= \frac{Z[\rho - (m_i/Ze)\tau]}{m_i(1+\mu)} &&\approx \frac{Z}{m_i(1+\mu)} \rho, \\
n_i &= \frac{\rho + \mu(m_i/Ze)\tau}{m_i(1+\mu)} &&\approx \frac{1}{m_i(1+\mu)} \rho, \\
\mathbf{u}_e &= \frac{\rho \mathbf{v} - (m_i/Ze)\mathbf{j}}{\rho - (m_i/Ze)\tau} &&\approx \mathbf{v} - \frac{m_i}{Ze}\frac{\mathbf{j}}{\rho}, \\
\mathbf{u}_i &= \frac{\rho \mathbf{v} + \mu(m_i/Ze)\mathbf{j}}{\rho + \mu(m_i/Ze)\tau} &&\approx \mathbf{v} + \mu\frac{m_i}{Ze}\frac{\mathbf{j}}{\rho}, \\
p_e &= n_e kT = \frac{n_e}{n_e + n_i} p &&\approx \frac{Z}{1+Z} p, \\
p_i &= n_i kT = \frac{n_i}{n_e + n_i} p &&\approx \frac{1}{1+Z} p.
\end{aligned} \quad (3.132)
$$

Here, the ratio μ of masses over charges has been defined in Eq. (3.114), and the approximations on the RHS are due to the assumption of *quasi charge-neutrality*:

$$|n_e - Zn_i| \ll n_e, \quad \text{or} \quad \frac{m_i}{Ze}|\tau| \ll \rho. \tag{3.133}$$

As we have seen in Sections 1.4.1 and 2.3.2, this approximation is extremely well satisfied for plasma phenomena with a macroscopic (hydrodynamic) length scale

$$\lambda_H \gg \lambda_D \equiv v_{\text{th},e}/(\sqrt{2}\,\omega_{pe}). \tag{3.134}$$

(For the tokamak and coronal loop examples of Tables B.3 and B.5, the values of the relevant small parameter are $\lambda_D/a = 7 \times 10^{-5}$ and 7×10^{-11}, resp.) This does not imply that space charges do not occur, but just that they do not involve a sizeable fraction of the available free charges. Hence, for the time being (i.e. until we estimate the different terms in Maxwell's equations) we will keep the electrostatic contributions when they occur by themselves (like $\tau \mathbf{E}$ in the momentum equation), but drop them in the inverse expressions (3.132).

Multiplying the pair of equations (3.61) by the masses and adding them gives the equation of *mass conservation*:

$$\frac{\partial \rho}{\partial t} + \nabla \cdot (\rho \mathbf{v}) = 0, \tag{3.135}$$

whereas multiplication by the charges and subtraction results in the equation of *charge conservation*:

$$\frac{\partial \tau}{\partial t} + \nabla \cdot \mathbf{j} = 0. \tag{3.136}$$

Likewise, adding the pair of equations (3.62), while using the approximations on the RHS of Eqs. (3.132), results in *the equation of motion*:

$$\rho \frac{\partial \mathbf{v}}{\partial t} + \rho \mathbf{v} \cdot \nabla \mathbf{v} + \mu \left(\frac{m_i}{Ze}\right)^2 \nabla \cdot \left(\frac{1}{\rho}\mathbf{j}\mathbf{j}\right) + \nabla p - \tau \mathbf{E} - \mathbf{j} \times \mathbf{B} = -\nabla \cdot (\boldsymbol{\pi}_e + \boldsymbol{\pi}_i). \tag{3.137}$$

(This equation reduces to the Navier–Stokes equation when electric and magnetic effects are absent and the stress tensor is replaced by the hydrodynamic expression.) Multiplying the pair of equations (3.62) by the charge over mass quotients and adding them results in an equation for the rate of change of the current density, which is known under the name *generalized Ohm's law*:

$$\frac{\partial \mathbf{j}}{\partial t} + \nabla \cdot \left[\mathbf{j}\mathbf{v} + \mathbf{v}\mathbf{j} - \frac{m_i}{Ze}(1-\mu)\frac{1}{\rho}\mathbf{j}\mathbf{j} \right]$$

$$+ \frac{1}{\mu}\frac{Ze}{m_i}\left[(1-\mu)\mathbf{j} \times \mathbf{B} - \frac{Z-\mu}{Z+1}\nabla p \right] - \frac{1}{\mu}\left(\frac{Ze}{m_i}\right)^2 \rho(\mathbf{E} + \mathbf{v} \times \mathbf{B})$$

$$= \frac{1}{\mu}\frac{Ze}{m_i}\left[\nabla \cdot (\pi_e - \mu\pi_i) - (1+\mu)\mathbf{R}_e \right]. \quad (3.138)$$

Finally, addition of the equations (3.63) results in *the heat balance equation*:

$$\frac{\partial p}{\partial t} + \mathbf{v} \cdot \nabla p + \gamma p \nabla \cdot \mathbf{v} + \frac{Z-\mu}{Z+1}\frac{m_i}{Ze}\left[(\gamma-1)\frac{1}{\rho}\mathbf{j} \cdot \nabla p - \gamma \mathbf{j} \cdot \nabla\left(\frac{p}{\rho}\right) \right]$$

$$= -(\gamma - 1)\left[\pi_e : \nabla \mathbf{u}_e + \pi_i : \nabla \mathbf{u}_i + \nabla \cdot (\mathbf{h}_e + \mathbf{h}_i) - (1+\mu)\frac{m_i}{Ze}\frac{1}{\rho}\mathbf{j} \cdot \mathbf{R}_e \right],$$

$$(3.139)$$

where the ratio $\gamma = 5/3$ of specific heats is introduced again.

To reduce the dissipative expressions on the RHSs of Eqs. (3.137)–(3.139), we again apply the time scale ordering (3.91) of Section 3.3.3, leading to the neglect of the electron and ion viscosities and heat conductivities:

$$\pi_{e,i} \to 0, \qquad \mathbf{h}_{e,i} \to 0, \quad (3.140)$$

and to the reduction of the ion–electron momentum transfer $\mathbf{R}_e$ in terms of the current density:

$$\mathbf{R}_e = \eta e n_e \mathbf{j} \approx \eta \frac{1}{1+\mu}\frac{Ze}{m_i}\rho \mathbf{j}. \quad (3.141)$$

Here, the factor of proportionality, *the resistivity* η, is assumed to be a scalar.

To consistently derive the *large scale* dynamics associated with the MHD description, we now apply a *maximal ordering*[4] of the one-fluid variables, starting from the expressions (3.128) for the Alfvén and sound speed:

$$v_A^2 \equiv \frac{B^2}{\mu_0 \rho} \sim v_s^2 \equiv \frac{\gamma p}{\rho} \sim v^2. \quad (3.142)$$

(This implies that we admit arbitrary values of the ratio between kinetic and magnetic pressures, $\beta \equiv 2\mu_0 p/B^2 \sim 1$, and of the Mach number of the flow, $M \equiv v/v_s \sim 1$.) The hydrodynamic length and time scales will now (in contrast

---

[4] In the sense of the principles of asymptotology, i.e. 'the art of dealing with applied mathematical systems in limiting cases', formulated by Kruskal [132].

to Section 3.3.3 !) be chosen to correspond to the size of the plasma:

$$\lambda_{MHD} \equiv |\nabla|^{-1} \sim a \gg R_i \left[\gg \delta \gg R_e \gg \lambda_D \right],$$

$$\tau_{MHD} \equiv |\partial/\partial t|^{-1} \sim a/v_A \gg \Omega_i^{-1} \left[\gg \omega_p^{-1} \sim \Omega_e^{-1} \right]. \tag{3.143}$$

(Again, admitting arbitrary values of the ratio between the transverse and longitudinal dimensions of the plasma geometry, $a/L \sim 1$.) The magnitudes of the remaining, electrodynamic, variables $\mathbf{j}$, $\mathbf{E}$, and τ are chosen to be set by Ampère's law, Ohm's law, and Poisson's law, resp.:

$$j \sim \frac{B}{\mu_0 a}, \quad E \sim vB, \quad \tau \sim \frac{\epsilon_0 E}{a} \sim \frac{\epsilon_0 v B}{a}. \tag{3.144}$$

This suffices to bring the large scale MHD phenomena to the fore, as we will see.

Let us now get rid of the remaining local two-fluid effects in Eqs. (3.137)–(3.139). To that end, we first check the assumption (3.133) of quasi-neutrality. In terms of the maximal ordering (3.142)–(3.144), the relevant small parameter becomes

$$\frac{m_i}{Ze} \frac{|\tau|}{\rho} \sim \frac{v_{th,i}}{c} \frac{\lambda_D}{a} \left[\sim \left(\frac{v_{th,i}}{c}\right)^2 \frac{1}{\sqrt{\mu}} \frac{\delta_e}{a} \right] \ll 1. \tag{3.145}$$

In fact, with the additional small factor $v_{th,i}/c$ ($\sim 5 \times 10^{-3}$ for the tokamak example again), this becomes an excellent approximation. Next, we assume the relative velocity of the ions and electrons to be small compared to the centre of mass velocity:

$$|\mathbf{u}_i - \mathbf{u}_e| \ll |\mathbf{v}|, \quad \text{or} \quad \frac{m_i}{Ze} |\mathbf{j}| \ll \rho |\mathbf{v}|. \tag{3.146}$$

The latter restriction on the magnitude of the current density is to be understood in the sense of the maximal ordering, so that there is an upper limit to the current density also for static plasmas ($\mathbf{v} = 0$), where $|\mathbf{v}|$ should be replaced by v_s or v_A in the inequality. The magnitude of the associated small parameter is

$$\frac{m_i}{Ze} \frac{|\mathbf{j}|}{\rho |\mathbf{v}|} \sim \frac{1}{\sqrt{\mu}} \frac{\delta_e}{a} \ll 1, \tag{3.147}$$

i.e. much less extreme than the condition (3.145) for quasi-neutrality (note the expression in square brackets there), but still easily satisfied for plasmas of interest (and extremely easily for astrophysical plasmas due to their huge length scales).

3.4 One-fluid equations*

With this ordering, and the condition (3.146) on the current density, the evolution equations for **v**, **j**, and p simplify to

$$\rho \frac{\partial \mathbf{v}}{\partial t} + \rho \mathbf{v} \cdot \nabla \mathbf{v} + \nabla p - \tau \mathbf{E} - \mathbf{j} \times \mathbf{B} = 0, \tag{3.148}$$

$$-\mu \left(\frac{m_i}{Ze}\right)^2 \frac{1}{\rho}\left[\frac{\partial \mathbf{j}}{\partial t} + \nabla \cdot (\mathbf{j}\mathbf{v} + \mathbf{v}\mathbf{j})\right] - \frac{m_i}{Ze}\frac{1}{\rho}\left[(1-\mu)\mathbf{j}\times\mathbf{B} - \frac{Z-\mu}{Z+1}\nabla p\right]$$

$$+ \mathbf{E} + \mathbf{v} \times \mathbf{B} = \eta \mathbf{j}, \tag{3.149}$$

$$\frac{\partial p}{\partial t} + \mathbf{v} \cdot \nabla p + \gamma p \nabla \cdot \mathbf{v} = (\gamma - 1)\eta |\mathbf{j}|^2. \tag{3.150}$$

Here, we have momentarily kept an intermediate form of the generalized Ohm's law with the time derivative term, which is small to second order in the small parameter (3.147), and the Hall current term, which is small to first order. Keeping those terms would conserve the evolutionary form of Ohm's law, but numerical computation of the current density this way would be extremely inaccurate since it would force the computation to proceed on the short time scale of the two-fluid model instead of on the much longer time scale of the MHD model. Obviously, for consistency of the MHD description, we have to drop those terms so that the usual *Ohm's law* of the second line of Eq. (3.149) remains. However, this law gives the current density $\mathbf{j} = \sigma \mathbf{E}'$ with a large (and inaccurately known) parameter $\sigma \equiv \eta^{-1}$ multiplying the electric field in the moving frame: not very accurate either! Moreover, in the ideal MHD case ($\eta = 0$), Ohm's law $\mathbf{E} + \mathbf{v} \times \mathbf{B} = 0$ completely changes character from an equation that determines $\mathbf{j}$ into one that expresses $\mathbf{E}$ in terms of $\mathbf{v}$ and $\mathbf{B}$. Apparently, something else is required to restore the peace in the system. This will be the subject of the next section.

Concluding this section, it will have been noticed that we did not exploit the obvious further reduction of the expressions (3.132) for small values of the ratio μ of masses over charges. Although $\mu = 5.4 \times 10^{-4}$ for hydrogen plasmas, we have seen in Section 3.3.4 that MHD phenomena do not really depend on the assumption $\mu \ll 1$. This implies that the MHD equations will also be valid for other plasmas, like electron–positron plasmas in the classical, and non-relativistic, limit. However, this would require adaptation of the transport coefficients since the expressions of Section 3.3.2 essentially depend on the assumption $\mu \ll 1$. Also, the sequences of inequalities (3.143) in the square brackets will no longer apply since the quantities involved become of the same order in μ for those plasmas. This is not a serious problem though, since the hydrodynamic scales will then be defined by the largest one of them.

3.4.2 Resistive and ideal MHD equations*

We still need to subject Maxwell's equations to the maximal ordering (3.142)–(3.144). Before we do that, it is expedient to collect all equations derived so far, even though not quite consistent at this point:

$$\frac{\partial \rho}{\partial t} + \nabla \cdot (\rho \mathbf{v}) = 0 \qquad \text{(continuity)}, \qquad (3.151)$$

$$\left[\frac{\partial \tau}{\partial t} + \nabla \cdot \mathbf{j} = 0 \qquad \text{(charge)} \right], \qquad (3.152)$$

$$\rho \frac{\partial \mathbf{v}}{\partial t} + \rho \mathbf{v} \cdot \nabla \mathbf{v} + \nabla p \left[- \tau \mathbf{E} \right] - \mathbf{j} \times \mathbf{B} = 0 \quad \text{(momentum)}, \qquad (3.153)$$

$$\frac{\partial p}{\partial t} + \mathbf{v} \cdot \nabla p + \gamma p \nabla \cdot \mathbf{v} = (1-\gamma)\eta j^2 \qquad \text{(internal energy)}, \qquad (3.154)$$

$$\frac{\partial \mathbf{B}}{\partial t} + \nabla \times \mathbf{E} = 0 \qquad \text{(Faraday)}, \qquad (3.155)$$

$$\left[\frac{1}{c^2} \frac{\partial \mathbf{E}}{\partial t} + \right] \mu_0 \mathbf{j} - \nabla \times \mathbf{B} = 0 \qquad \text{('Ampère')}, \qquad (3.156)$$

where

$$\eta \mathbf{j} = \mathbf{E} + \mathbf{v} \times \mathbf{B} \qquad \text{(Ohm)}, \qquad (3.157)$$

and initially the following conditions need to be satisfied:

$$\left[\nabla \cdot \mathbf{E} = \tau/\epsilon_0 \qquad \text{(Poisson)} \right], \qquad (3.158)$$

$$\nabla \cdot \mathbf{B} = 0 \qquad \text{(no magnetic monopoles)}. \qquad (3.159)$$

These equations would represent evolution equations for the variables ρ, τ, $\mathbf{v}$, p, $\mathbf{B}$ and $\mathbf{E}$, with Ohm's law determining $\mathbf{j}$ and the last two equations to be considered as initial conditions on the differential equations for $\mathbf{E}$ and $\mathbf{B}$.

First notice that the charge conservation equation (3.152) is really redundant since it follows from Eqs. (3.156) and (3.158). Therefore, we should drop it and determine τ from Poisson's law (3.158). Consequently, Poisson's law can no longer be considered as an initial condition on 'Ampère's' law, indicating that something should be changed there as well. This becomes immediately clear by estimating the displacement current in the ordering (3.142)–(3.144):

$$\frac{1}{c^2} \left| \frac{\partial \mathbf{E}}{\partial t} \right| \sim \frac{v^2}{c^2} \frac{B}{a} \ll |\nabla \times \mathbf{B}| \sim \frac{B}{a}. \qquad (3.160)$$

Hence, for the non-relativistic flows we are considering, where

$$v^2/c^2 \ll 1, \qquad (3.161)$$

the displacement current is negligible and we return to the *pre-Maxwell equations* [32]. These are characterized by the fact that Eq. (3.156) is replaced by Ampère's law proper (without the quotes, since this is the form in which Ampère posed it):

$$\mathbf{j} = \mu_0^{-1} \nabla \times \mathbf{B}. \qquad (3.162)$$

We have now lost the evolution equation for $\mathbf{E}$ and obtained instead the simple expression for $\mathbf{j}$ as determined by the curl of $\mathbf{B}$. This appears to be consistent with the result of Section 3.4.1: Ohm's law should be read from right to left, i.e. it determines $\mathbf{E}$ rather than $\mathbf{j}$.

So far so good. But how about the charge conservation equation (3.152), that we have already dropped? Equation (3.162) implies that $\nabla \cdot \mathbf{j} = 0$ so that charge conservation now appears to tell us that $\partial \tau / \partial t = 0$. This would be in conflict with Eqs. (3.156) and (3.158), which imply that

$$\frac{\partial \tau}{\partial t} = \epsilon_0 \nabla \cdot \frac{\partial \mathbf{E}}{\partial t} = -\epsilon_0 \frac{\partial}{\partial t} \nabla \cdot (\mathbf{v} \times \mathbf{B}) \neq 0,$$

in general. The way out is to consistently apply an ordering in the small parameter v^2/c^2. One then finds that the electrostatic force $\tau \mathbf{E}$ in the momentum equation is one order smaller than the macroscopic Lorentz force $\mathbf{j} \times \mathbf{B}$, so that it should be dropped:

$$|\tau \mathbf{E}| \sim \frac{\epsilon_0 E^2}{a} \sim \frac{v^2}{c^2} \frac{B^2}{\mu_0 a} \ll |\mathbf{j} \times \mathbf{B}| \sim \frac{B^2}{\mu_0 a}. \qquad (3.163)$$

After this, all equations are expressions of the same order, except for the charge conservation equation (3.152) which is one order in v^2/c^2 smaller. This justifies its elimination from the system. Poisson's equation may still be used to calculate the charge density, but, since τ does not occur in any of the other equations, that equation should be dropped as well. The resulting set of equations is a mathematically consistent set which, amongst the many other attractions exposed in this book, enjoys the property of being *Galilean invariant*.

Clearly, the approximations of the present section are of a different kind than those of the previous one. In the latter section, the conditions given are mandatory for the description of macroscopic dynamics, whereas the restriction (3.161) of the present section can be, and has been, lifted to construct a consistent relativistic MHD theory. This has been done for special as well as general relativity, see e.g. Lichnerowicz [145], Achterberg [2], and Anile [8].

In conclusion, all expressions in square brackets in Eqs. (3.151)–(3.159) should be dropped and we obtain the following set of *resistive or ideal MHD equations* determining the evolution of the macroscopic variables ρ, $\mathbf{v}$, p, and $\mathbf{B}$ (already introduced in Section 2.4.1):

$$\frac{\partial \rho}{\partial t} + \nabla \cdot (\rho \mathbf{v}) = 0 \qquad \text{(mass)}, \qquad (3.164)$$

$$\rho \frac{\partial \mathbf{v}}{\partial t} + \rho \mathbf{v} \cdot \nabla \mathbf{v} + \nabla p - \mathbf{j} \times \mathbf{B} = 0 \qquad \text{(momentum)}, \qquad (3.165)$$

$$\frac{\partial p}{\partial t} + \mathbf{v} \cdot \nabla p + \gamma p \nabla \cdot \mathbf{v} = \begin{cases} (1-\gamma) \eta j^2 & \text{(resistive)} \\ 0 & \text{(ideal)} \end{cases} \text{(internal energy)}, \quad (3.166)$$

$$\frac{\partial \mathbf{B}}{\partial t} + \nabla \times \mathbf{E} = 0 \qquad \text{(Faraday)}, \qquad (3.167)$$

where the variables $\mathbf{j}$ and $\mathbf{E}$ may be eliminated by means of the algebraic equations

$$\mathbf{j} = \mu_0^{-1} \nabla \times \mathbf{B} \qquad \text{(Ampère)}, \qquad (3.168)$$

$$\mathbf{E} + \mathbf{v} \times \mathbf{B} = \begin{cases} \eta \mathbf{j} & \text{(resistive)} \\ 0 & \text{(ideal)} \end{cases} \text{(Ohm)}, \qquad (3.169)$$

and the magnetic field needs to satisfy the initial condition

$$\nabla \cdot \mathbf{B} = 0. \qquad (3.170)$$

This is how, in the end, the double set of five moment equations of kinetic theory, combined with Maxwell's equations, lead to a very powerful description of large scale plasma dynamics, where collisions appear to have served no other purpose (i.e. in the ideal case) than to establish the coherence of the single fluid.

3.5 Literature and exercises*

Notes on literature

Kinetic theory and transport:

– Braginskii's paper [41] on 'Transport processes in a plasma' in the first volume of the Russian *Reviews of Plasma Physics* is a masterpiece on the subject that has been used by most plasma physicists in one form or the other ever since its appearance.
– Balescu's two volumes [14] on *Transport Processes in Plasmas* are fruits of a life-long dedication to the science of statistical mechanics and non-equilibrium thermo-dynamics of charged particles. His love for the subject shines through every page of these books. The first volume is devoted to 'Classical Transport' and the second to

'Neo-classical transport'. The announced future volume on 'Anomalous Transport' will be a most welcome guide through the jungle of turbulent transport processes.

- Neo-classical transport in toroidal plasmas is discussed in the review paper by Galeev & Sagdeev [77].
- Chapman & Cowling, *The Mathematical Theory of Non-uniform Gases* [52] is the classical treatise of obtaining hydrodynamic equations by expansion of the distribution functions in a small parameter.
- The papers by Rosenbluth *et al.* [196] and Trubnikov [231] contain the derivation of the Fokker–Planck form of the collision operator for charged particles.
- Montgomery & Tidman, *Plasma Kinetic Theory* [158] is a critical exposition of the principles of kinetic theory of plasmas.

Derivation of fluid equations:

- Grad, and Blank *et al.*, *Notes on Magnetohydrodynamics* [98, 32] present the pioneering lecture notes on the foundations of MHD.
- Most basic textbooks on plasma physics contain one or more chapters on the derivation and conditions of a macroscopic description, e.g.:
  Spitzer, *Physics of Fully Ionized Gases* [213], Chapter 2 and Appendix,
  Clemmov & Dougherty, *Electrodynamics of Particles and Plasmas* [58], Chapter 11,
  Schmidt, *Physics of High Temperature Plasmas* [203], Chapter 3,
  Boyd & Sanderson, *Plasma Dynamics* [35], Chapter 3,
  Krall & Trivelpiece, *Principles of Plasma Physics* [129], Chapter 3,
  Akhiezer et al., *Plasma Electrodynamics* [4], Chapter 1.
- Hazeltine & Meiss, *Plasma Confinement* [107] contains a fundamental discussion of MHD and alternative closures of the moment equations.

Exercises

[3.1] *Collision integral*
Prove that the Landau collision operator (3.13) satisfies the conservation properties (3.18)–(3.22).

[3.2] *Transport coefficients*
Estimate the numerical magnitudes of the electron momentum and energy transport by viscosity and heat conduction for tokamak and coronal loop plasmas. What conclusions can be drawn from these numbers?

[3.3]★ *Waves in two-fluid plasmas*
Write a numerical program, in the computer language of your choice, for the numerical solution of the dispersion equation, derived in Section 3.3.4, for the waves in two-fluid plasmas. Carefully copy the explicit expressions of Table 3.2 in a separate subroutine! (As always in computations, you save time by spending enough time when coding up the algebra.) The graphical representation of the solution of the dispersion equation is most easily obtained by just contour plotting the implicit dispersion function $F(\bar{k}^2, \bar{\omega}^2) = 0$. For simplicity, you may restrict the analysis to cold plasmas by dropping the thermal terms. Now, enjoy the powerful tool you have obtained to study:

- Ordinary and extra-ordinary cutoffs, upper and lower hybrid resonances, whistler waves, etc. Compare the numerical results with the analytical asymptotic results.

- Compute the dispersion diagrams for perpendicular ($\lambda = 0$) and parallel ($\lambda = 1$) propagation and compare them with Fig. 3.1. Comment on the degeneracies obtained.
- Which of the MHD waves do you obtain in the large scale limit?

[3.4] *Derivation of the one-fluid equations*

Complete the derivation of the one-fluid equations by constructing the sixth equation, for the temperature difference between electron and ions. Check for the consistency of omitting this equation under the conditions formulated.

Part II

Basic magnetohydrodynamics

Part II

Basic Diagnostic Procedures

4
The MHD model

4.1 The ideal MHD equations

The dynamics of magnetically confined plasmas, as exploited in laboratory nuclear fusion research and observed in astrophysical systems, is essentially of a macroscopic nature so that it can be studied in the fluid (MHD) model introduced in Chapter 2. The 'derivation' of the MHD equations in Chapter 3 provided indications about the range of validity and the limitations of the equations. In the present chapter, we will develop the MHD model for the interaction of plasma and magnetic field in detail and, thus, obtain a powerful 'picture' for the dynamics of the mentioned plasmas.

Recall from the introduction of Chapter 3 that the equations of magnetohydrodynamics can be introduced either by just posing them as postulates for a hypothetical medium called 'plasma' or by the much more involved procedure of averaging the kinetic equations. Whereas Chapter 3 was mainly concerned with the second method, in the present chapter we exploit the first method: we simply pose the equations and use physical arguments and mathematical criteria to justify the result. We continue the exposition of Section 2.4.1, where we already encountered the relevant equations.

4.1.1 Postulating the basic equations

The ideal MHD equations describe *the motion of a perfectly conducting fluid interacting with a magnetic field*. Hence, we need to *combine Maxwell's equations with the equations of gas dynamics* and provide *equations describing the interaction*.

First, consider *Maxwell's equations*, already encountered in Chapters 2 and 3. They describe the evolution of the electric field $\mathbf{E}(\mathbf{r}, t)$ and the magnetic field

$\mathbf{B}(\mathbf{r}, t)$ in response to the current density $\mathbf{j}(\mathbf{r}, t)$ and the space charge $\tau(\mathbf{r}, t)$:

$$\nabla \times \mathbf{E} = -\frac{\partial \mathbf{B}}{\partial t}, \tag{4.1}$$

$$\nabla \times \mathbf{B} = \mu_0 \mathbf{j} + \frac{1}{c^2}\frac{\partial \mathbf{E}}{\partial t}, \qquad c \equiv (\epsilon_0 \mu_0)^{-1/2}, \tag{4.2}$$

$$\nabla \cdot \mathbf{E} = \frac{\tau}{\epsilon_0}, \tag{4.3}$$

$$\nabla \cdot \mathbf{B} = 0. \tag{4.4}$$

Next, consider *the equations of gas dynamics* for the evolution of the density $\rho(\mathbf{r}, t)$ and the pressure $p(\mathbf{r}, t)$:

$$\frac{D\rho}{Dt} + \rho \nabla \cdot \mathbf{v} \equiv \frac{\partial \rho}{\partial t} + \nabla \cdot (\rho \mathbf{v}) = 0, \tag{4.5}$$

$$\frac{Dp}{Dt} + \gamma p \nabla \cdot \mathbf{v} \equiv \frac{\partial p}{\partial t} + \mathbf{v} \cdot \nabla p + \gamma p \nabla \cdot \mathbf{v} = 0. \tag{4.6}$$

We will see later that these equations actually express mass conservation and conservation of entropy. Note that we have used the occasion to introduce the notation

$$\frac{D}{Dt} \equiv \frac{\partial}{\partial t} + \mathbf{v} \cdot \nabla$$

for the *Lagrangian time-derivative*, evaluated while moving with the fluid, in contrast to the *Eulerian time-derivative* $\partial/\partial t$, which is evaluated at a fixed position.

So far, the two systems described by the variables $\mathbf{E}$, $\mathbf{B}$, and ρ, p do not appear to interact. Such interaction is introduced through the equations involving the velocity $\mathbf{v}(\mathbf{r}, t)$ of the fluid. First, 'Newton's' *equation of motion for a fluid element*,

$$\rho \frac{D\mathbf{v}}{Dt} = \mathbf{F} \equiv -\nabla p + \rho \mathbf{g} + \mathbf{j} \times \mathbf{B} + \tau \mathbf{E}, \tag{4.7}$$

expresses the acceleration of a fluid element (LHS) caused by the force $\mathbf{F}$ consisting of pressure gradient, gravity, and electromagnetic contributions (RHS).[1] Next, one of the most characteristic equations describing the plasma state is the equation for the electric field in a perfectly conducting moving fluid,

$$\mathbf{E}' \equiv \mathbf{E} + \mathbf{v} \times \mathbf{B} = 0, \tag{4.8}$$

which expresses that the electric field $\mathbf{E}'$ in a co-moving frame should vanish.

The system of equations (4.1)–(4.8) is now complete, but not yet in a form suitable for self-consistent calculations for the majority of plasmas occurring in the laboratory and in nature. We need to make one additional assumption (repeating

---

[1] Actually, these expressions have the dimension of *force density*.

4.1 The ideal MHD equations

some of the discussion and expressions of Section 3.4.2). For most plasma phenomena it is sufficient to restrict the analysis to *non-relativistic velocities*:

$$v \ll c. \tag{4.9}$$

In that case, we can make the following estimates for the orders of magnitude of the different terms in Eq. (4.2):

$$\frac{1}{c^2}\left|\frac{\partial \mathbf{E}}{\partial t}\right| \sim \frac{v^2}{c^2}\frac{B}{l_0} \ll |\nabla \times \mathbf{B}| \sim \frac{B}{l_0} \quad \text{(using Eq. (4.8))},$$

where we have indicated the length scale of gradients by l_0 and the time scale by t_0, so that $v \sim l_0/t_0$. Hence, the displacement current (Maxwell's great contribution to electrodynamics) is small, $\mathcal{O}(v^2/c^2)$, and can be removed again from Eq. (4.2), so that the current $\mathbf{j}$ may be expressed directly in terms of $\mathbf{B}$ and the original form of Ampère's law is recovered:

$$\mathbf{j} = \frac{1}{\mu_0}\nabla \times \mathbf{B}. \tag{4.10}$$

Furthermore, the non-relativistic approximation implies a remarkable simplification of the equation of motion (4.7) as well since the electrostatic acceleration,

$$\tau|\mathbf{E}| \sim \frac{v^2}{c^2}\frac{B^2}{\mu_0 l_0} \ll |\mathbf{j} \times \mathbf{B}| \sim \frac{B^2}{\mu_0 l_0} \quad \text{(using Eqs. (4.3), (4.8), and (4.10))},$$

is also $\mathcal{O}(v^2/c^2)$. Consequently, space charge effects can be neglected, and Poisson's law (4.3) may be dropped since it is no longer needed. The electric field then becomes a secondary quantity, to be determined from Eq. (4.8):

$$\mathbf{E} = -\mathbf{v} \times \mathbf{B}. \tag{4.11}$$

This shows that, for non-relativistic MHD motions, the order of magnitude of the electric field as compared to the magnetic field is given by $|\mathbf{E}| \sim |\mathbf{v}||\mathbf{B}|$, i.e. an order $\mathcal{O}(v/c)$ smaller than for electromagnetic waves, where $|\mathbf{E}| \sim c|\mathbf{B}|$.

Exploiting the mentioned approximations and eliminating $\mathbf{E}$ and $\mathbf{j}$ from the equations by means of Eqs. (4.10) and (4.11), *the basic equations of ideal MHD of Sections 2.4.1 and 3.4.2 are recovered*:

$$\frac{\partial \rho}{\partial t} + \nabla \cdot (\rho\mathbf{v}) = 0, \tag{4.12}$$

$$\rho\left(\frac{\partial \mathbf{v}}{\partial t} + \mathbf{v} \cdot \nabla\mathbf{v}\right) + \nabla p - \rho\mathbf{g} - \frac{1}{\mu_0}(\nabla \times \mathbf{B}) \times \mathbf{B} = 0, \tag{4.13}$$

$$\frac{\partial p}{\partial t} + \mathbf{v} \cdot \nabla p + \gamma p \nabla \cdot \mathbf{v} = 0, \tag{4.14}$$

$$\frac{\partial \mathbf{B}}{\partial t} - \nabla \times (\mathbf{v} \times \mathbf{B}) = 0, \quad \nabla \cdot \mathbf{B} = 0. \tag{4.15}$$

This is a set of eight nonlinear partial differential equations for the eight variables $\rho(\mathbf{r}, t)$, $\mathbf{v}(\mathbf{r}, t)$, $p(\mathbf{r}, t)$, and $\mathbf{B}(\mathbf{r}, t)$. Here, the magnetic field equation (4.15)(b) is to be considered as a condition on the initial values: once satisfied, it remains satisfied for all later times by virtue of Eq. (4.15)(a). Depending on the physical problem considered, the value of the gravitational acceleration $\mathbf{g}(\mathbf{r}, t)$ is either an externally fixed quantity or partially determined by the plasma itself. This aspect is discussed below.

For *static equilibria* ($\partial/\partial t = 0$ and $\mathbf{v} = 0$) in the absence of gravity ($\mathbf{g} = 0$), the MHD equations reduce to

$$\nabla p = \frac{1}{\mu_0} (\nabla \times \mathbf{B}) \times \mathbf{B}, \qquad \nabla \cdot \mathbf{B} = 0. \tag{4.16}$$

Now, the magnetic field equation (4.16)(b) fully counts, in contrast to the dynamical problem, where it is just an initial condition. Thus, we appear to obtain four equations for the determination of $p(\mathbf{r})$ and the three components of $\mathbf{B}(\mathbf{r})$. However, we have already seen in Section 2.4.3, for the specific example of a z-pinch, that the equilibrium equations leave much more freedom in the determination of these quantities than this remark suggests. This is due to the symmetry that is usually assumed for equilibria.

(a) Thermodynamic variables The above formulation of the ideal MHD equations exploits ρ, $\mathbf{v}$, p, $\mathbf{B}$ as the basic variables. It is of interest to also work out the evolution equations for the *other thermodynamical variables*, which could replace ρ and p, viz.: e – the internal energy per unit mass (which is equivalent to T – the temperature) and s – the entropy per unit mass. These are defined by the ideal gas relations, with $p = (n_e + n_i)kT$:

$$e \equiv \frac{1}{\gamma - 1} \frac{p}{\rho} \approx C_v T, \qquad C_v \approx \frac{(1 + Z)k}{(\gamma - 1)m_i},$$

$$s \equiv C_v \ln S + \text{const}, \qquad S \equiv p\rho^{-\gamma}, \tag{4.17}$$

where m_i is the mass of the ions, k is the Boltzmann constant, and $\gamma \equiv C_p/C_v$ is the ratio of specific heats at constant pressure and volume, respectively. We have introduced the variable $S \equiv \rho^{-\gamma} p$ here since it is a slightly more convenient measure for entropy than the variable s itself because it does not contain awkward constants any more.

Neglecting thermal conduction and heat flow, i.e. considering adiabatic processes, *the entropy convected by the fluid is constant*:

$$\frac{Ds}{Dt} = 0, \quad \text{or} \quad \frac{DS}{Dt} \equiv \frac{D}{Dt}(p\rho^{-\gamma}) = 0. \tag{4.18}$$

Eliminating the density ρ from the latter equation by means of Eq. (4.12) yields again the pressure evolution equation (4.14). This substantiates our claim that Eq. (4.14) actually expresses entropy conservation. On the other hand, we could also exploit the internal energy e (or T) as a basic variable. From Eqs. (4.12) and (4.14) we then have:

$$\frac{De}{Dt} + (\gamma - 1)e\nabla \cdot \mathbf{v} = 0. \tag{4.19}$$

Which pair of the four state variables ρ, p, s, e one chooses to complement the description by means of $\mathbf{v}$ and $\mathbf{B}$ is usually a matter of convenience. However, for the expression of conservation laws and certain symmetry properties this choice matters, as we will see later (Sections 4.3.1 and 5.2.1).

▷ **Exercise.** Anticipating the discussion of the conservation properties in Section 4.3.1: why is the internal energy convected by the fluid not constant? ◁

(b) Gravitation In the equation of motion (4.13), the gravitational acceleration $\mathbf{g}$ due to the masses of the plasma within the region under consideration may be derived from an internal gravitational potential Φ_{gr}^{in}. It satisfies the Poisson equation

$$\nabla^2 \Phi_{gr}^{in} = 4\pi G \rho(\mathbf{r}), \tag{4.20}$$

having the solution

$$\Phi_{gr}^{in}(\mathbf{r}) = -G \int \frac{\rho(\mathbf{r}')}{|\mathbf{r} - \mathbf{r}'|} d^3 r', \tag{4.21}$$

so that

$$\mathbf{g}^{in}(\mathbf{r}) = -\nabla \Phi_{gr}^{in}(\mathbf{r}) = -G \int \rho(\mathbf{r}') \frac{\mathbf{r} - \mathbf{r}'}{|\mathbf{r} - \mathbf{r}'|^3} d^3 r'. \tag{4.22}$$

To check the solution (4.21), use the property of the Dirac delta function in three dimensions,

$$\nabla^2 \left(\frac{1}{|\mathbf{r} - \mathbf{r}'|} \right) = -4\pi \delta(\mathbf{r} - \mathbf{r}'), \tag{4.23}$$

that is also exploited in the analogous problem of electrostatics (see Jackson [117]).

In many astrophysical systems, the internal gravitational force $\mathbf{F}_g^{in} \equiv \rho \mathbf{g}^{in}$ is completely negligible compared to the Lorentz force $\mathbf{F}_B \equiv \mathbf{j} \times \mathbf{B}$, but also compared to the gravitational force $\mathbf{F}_g^{ex} \equiv \rho \mathbf{g}^{ex}$ due to an external compact object. We will represent such an external gravitational field by a point mass M_* situated at a position $\mathbf{r} = \mathbf{r}_*$ far outside the plasma region. In that case, we have a Poisson

equation
$$\nabla^2 \Phi_{\text{gr}} = 4\pi G \Big[M_* \delta(\mathbf{r} - \mathbf{r}_*) + \rho(\mathbf{r}) \Big], \quad (4.24)$$

having the solution
$$\Phi_{\text{gr}}(\mathbf{r}) = -G \frac{M_*}{|\mathbf{r} - \mathbf{r}_*|} + \Phi_{\text{gr}}^{\text{in}}(\mathbf{r}), \quad (4.25)$$

so that
$$\mathbf{g}(\mathbf{r}) = -\nabla \Phi_{\text{gr}}(\mathbf{r}) = -G M_* \frac{\mathbf{r} - \mathbf{r}_*}{|\mathbf{r} - \mathbf{r}_*|^3} + \mathbf{g}^{\text{in}}(\mathbf{r}), \quad (4.26)$$

where the internal field is given by Eqs. (4.21) and (4.22).

Let us now estimate the relative magnitudes of the Lorentz and the gravitational forces. For a tokamak, with typical parameters as given in Table B.5, we get:

$$\begin{aligned}
|\mathbf{F}_B| &\equiv |\mathbf{j} \times \mathbf{B}| \sim \frac{B^2}{\mu_0 a} &&= 7.2 \times 10^6 \, \text{kg m}^{-2} \, \text{s}^{-2}, \\
|\mathbf{F}_g^{\text{ex}}| &\equiv |\rho \mathbf{g}^{\text{ex}}| \sim \rho G \frac{M_*}{R_*^2} &&= 1.7 \times 10^{-6} \, \text{kg m}^{-2} \, \text{s}^{-2}, \\
|\mathbf{F}_g^{\text{in}}| &\equiv |\rho \mathbf{g}^{\text{in}}| \sim \rho^2 G a &&= 1.9 \times 10^{-24} \, \text{kg m}^{-2} \, \text{s}^{-2}, \quad (4.27)
\end{aligned}$$

where a is the width of the plasma tube and M_* and R_* here, of course, refer to the Earth ($GM_E/R_E^2 = 10 \, \text{m s}^{-2}$). Clearly, the gravitational contributions are completely negligible for tokamak plasmas.

▷ **Exercise.** Compute the corresponding numbers for solar coronal flux tubes from Table B.5 and also for 'your favourite plasma' (filling out the last column of Table B.6) by finding the relevant orders of magnitude for the different quantities from whatever source you can find. (This may turn out to be rather difficult!) What do you conclude from these numbers? ◁

Let us also estimate the relative magnitudes of the forces for a typical astrophysical plasma, viz. an accretion disc surrounding a compact object (see Frank, King and Raine [70], and Balbus and Hawley [12]). A schematic picture of the system is shown in Fig. 4.1. The compact object is located in the origin and matter is accreting in a thin rotating disc of size $\sim R_d$ and height $\sim H_d$ (at a distance $\sim 0.1 R_d$ from the centre). The compact object carries its own magnetic field, which is mainly dipolar, and the disc features a protruding poloidal magnetic field B_p. One important aspect is the frequent appearance of powerful jets ejected from the centre in the direction perpendicular to the disc (along the Z-axis). To estimate forces, we will exploit the expressions (4.27) with $a = H_d$ but replacing R_* by a reasonable distance away from the central object, viz. $R = 0.1 R_d$. This object may be a black hole, but our gravitational equations refer to a Newtonian

4.1 The ideal MHD equations

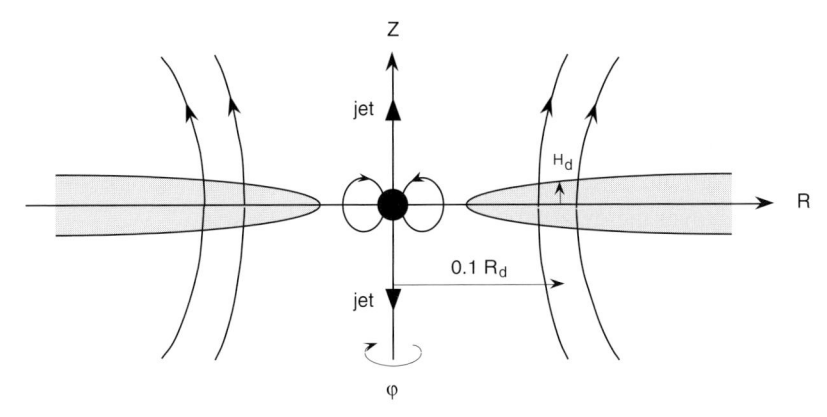

Fig. 4.1. Schematic presentation of accretion disc around compact object with ejected jets.

potential, excluding general relativistic effects, so that we have to stay far outside the Schwarzschild radius, i.e.

$$R \gg R_{\text{Schw}} = \frac{2GM_*}{c^2}. \tag{4.28}$$

This condition is easily met by our choice of the distance since R_{Schw} is just a multiple of 3 km for every solar mass unit in the black hole.

Fig. 4.1 may be used to illustrate astrophysical plasmas on entirely different scales. Let us consider the following two examples.

(a) Accretion disc around a young stellar object (YSO), a protostellar cradle for stars and planets: $R_d \sim 1$ AU ($= 1.5 \times 10^{11}$ m), $H_d \sim 0.01$ AU ($= 1.5 \times 10^9$ m). The central object has a mass of the order of the solar mass, $M_* \sim 1 M_\odot$ ($= 2.0 \times 10^{30}$ kg), and the accretion rate is of the order $\dot{M} \sim 10^{-7} M_\odot \, \text{y}^{-1}$ ($= 6 \times 10^{15}$ kg s^{-1}). Representative values of the other physical parameters are: $B = 10^{-4}$ T ($= 1$ G), $n = 10^{18}$ m^{-3}, $T = 10^4$ K, so that $\rho \approx 1.7 \times 10^{-9}$ kg m^{-3}, and $v_A = 2.2 \times 10^3$ m s^{-1}. This gives the following estimate for the three forces:

$$|\mathbf{F}_B| = 5.3 \times 10^{-12}, \quad |\mathbf{F}_g^{\text{ex}}| = 1.0 \times 10^{-9}, \quad |\mathbf{F}_g^{\text{in}}| = 2.9 \times 10^{-19} \, [\text{kg m}^{-2} \, \text{s}^{-2}]. \tag{4.29}$$

In this case, the contribution of the external gravitational field dominates.

(b) Accretion disc around an active galactic nucleus (AGN): $R_d \sim 50$ kpc ($= 1.6 \times 10^5$ ly $= 1.5 \times 10^{21}$ m), $H_d \sim 120$ pc ($= 3.3 \times 10^{18}$ m). The central object has a mass of the order of a hundred million solar masses, $M_* \sim 10^8 M_\odot$ ($= 2.0 \times 10^{38}$ kg), and the accretion rate is of the order $\dot{M} \sim 0.1 M_\odot \, \text{y}^{-1}$ ($= 6 \times 10^{21}$ kg s^{-1}). Representative values of the other physical parameters are: $B = 10^{-4}$ T ($= 1$ G), $n = 10^{12}$ m^{-3}, $T = 10^8$ K, so that $\rho \approx 1.7 \times 10^{-15}$ kg m^{-3}, and $v_A = 2.2 \times 10^6 \,\mathrm{m\,s^{-1}}$. This gives:

$$|\mathbf{F}_B| = 2.2 \times 10^{-21}, \quad |\mathbf{F}_g^{\mathrm{ex}}| = 1.0 \times 10^{-27}, \quad |\mathbf{F}_g^{\mathrm{in}}| = 6.4 \times 10^{-22} [\,\mathrm{kg\,m^{-2}\,s^{-2}}]. \tag{4.30}$$

Now, even though there may be a huge black hole in the centre, the internal gravity field dominates over the external field *at the position chosen*.

▷ **Exercise.** Complete the discussion of the different forces at work in the two types of accretion disc by also estimating the rotational and pressure terms. Do these estimates permit stationary equilibrium (i.e. solutions with $\partial/\partial t = 0$, but $\mathbf{v} \neq 0$)? What is the ratio of the plasma pressure p compared to the magnetic pressure $B^2/(2\mu_0)$? How is vertical equilibrium established? ◁

4.1.2 Scale independence

The MHD equations (4.12)–(4.15) can be made dimensionless by means of three quantities expressing a choice for the *units of length, mass, and time*. For that purpose, a typical length scale l_0 is chosen (e.g. the transverse confinement length scale of the plasma), and values for the magnitude B_0 of the magnetic field and for the plasma density ρ_0 are chosen at some representative position (e.g. on the magnetic axis in a tokamak). The unit of time then follows by exploiting the basic speed of macroscopic plasma dynamics, viz. *the Alfvén speed* introduced in Section 2.4.2:

$$v_0 \equiv v_{A,0} \equiv \frac{B_0}{\sqrt{\mu_0 \rho_0}} \quad \Rightarrow \quad t_0 \equiv \frac{l_0}{v_0}. \tag{4.31}$$

Thus, the relevant triplet of basic parameters becomes l_0, B_0, t_0.

By means of this triplet, and the derived quantities ρ_0 and v_0, we now create the dimensionless independent variables and their associated differential operators,

$$\bar{l} \equiv l/l_0, \quad \bar{t} \equiv t/t_0 \quad \Rightarrow \quad \bar{\nabla} \equiv l_0 \nabla, \quad \partial/\partial\bar{t} \equiv t_0\, \partial/\partial t, \tag{4.32}$$

and the dimensionless dependent variables,

$$\bar{\rho} \equiv \rho/\rho_0, \quad \bar{\mathbf{v}} \equiv \mathbf{v}/v_0, \quad \bar{p} \equiv p/(\rho_0 v_0^2), \quad \bar{\mathbf{B}} \equiv \mathbf{B}/B_0, \quad \bar{\mathbf{g}} \equiv (l_0/v_0^2)\,\mathbf{g}. \tag{4.33}$$

With these transformations, the equations (4.12)–(4.15) remain unchanged, except that all quantities are equipped with a bar and that the awkward quantity μ_0 (associated with the mks system of units) disappears. Obviously, we will drop the bars again and just enjoy the fact that dimensions need no longer worry us.

The important result of the exercise is the realization that the equations do not depend on the size of the plasma (l_0), on the magnitude of the magnetic field (B_0), and on the density (ρ_0), i.e. on the time scale (t_0): *the ideal MHD equations are*

Table 4.1. *Scales of different plasmas.*

	l_0 (m)	B_0 (T)	t_0 (s)
tokamak	20	3	3×10^{-6}
magnetosphere Earth	4×10^7	3×10^{-5}	6
solar coronal loop	10^8	3×10^{-2}	15
magnetosphere neutron star	10^6	10^8 *	10^{-2}
accretion disc YSO	1.5×10^9	10^{-4}	7×10^5
accretion disc AGN	4×10^{18}	10^{-4}	2×10^{12}
galactic plasma	10^{21}	10^{-8}	10^{15}
	($= 10^5$ ly)		($= 3 \times 10^7$ y)

* Some recently discovered pulsars, called magnetars, have record magnetic fields of 10^{11} T: the plasma Universe is ever expanding!

scale independent with respect to changes of these quantities. Notwithstanding the huge differences in magnitude of the parameters l_0, B_0, and t_0 encountered in nature and in the laboratory (see Table 4.1), yet the same equations of magnetohydrodynamics apply! This provides the basis for the description of macroscopic dynamics of 90% of matter in the Universe and, hence, for effective cross-fertilization between laboratory and astrophysical plasma physics (Section 1.1).

After the scaling with l_0, B_0, t_0, the value $\bar{p}_0$ of the dimensionless pressure at the reference point automatically becomes a quantity of intrinsic importance. It is directly related to the ratio of the plasma kinetic pressure to the magnetic pressure, commonly indicated by the symbol β:

$$\beta \equiv \frac{2\mu_0 p_0}{B_0^2} = 2\bar{p}_0. \qquad (4.34)$$

Since $\beta \ll 1$ for many plasmas of interest, pressure terms, labelled by the parameter β, are frequently neglected or only computed as a higher order correction in the dynamics.

▷ **Exercise.** Compute the value of β for 'your favourite plasma' from the numbers you have collected in the last column of Table B.6 and compare it with the other plasmas. Can you now estimate the relative importance of the different terms in the momentum equation? ◁

4.1.3 A crucial question

Having obtained the complete set of partial differential equations (4.12)–(4.15) for the plasma variables, and realizing their huge potential from the previous section, we now ask a crucial question: do these equations provide us with *a complete model* for plasma dynamics?

This question is to be answered with an emphatic NO! The reason is that the two most essential elements of a scientific model are still missing, viz.

(1) What is actually the physical problem we want to solve?
(2) How is this problem mathematically translated in conditions to be imposed on the solutions of the partial differential equations?

For example, one might want to study the stability of a laboratory fusion plasma or the evolution of an expanding flux tube in the solar corona. This brings in the specific space and time constraints embodied in the *boundary conditions* and *initial data*. The consideration of initial data will be our concern in the next chapter (Section 5.2). For now, this part is nearly trivial since it just amounts to prescribing arbitrary functions

$$\rho_i(\mathbf{r}) \equiv \rho(\mathbf{r}, t=0), \qquad \mathbf{v}_i(\mathbf{r}) \equiv \mathbf{v}(\mathbf{r}, t=0),$$
$$p_i(\mathbf{r}) \equiv p(\mathbf{r}, t=0), \qquad \mathbf{B}_i(\mathbf{r}) \equiv \mathbf{B}(\mathbf{r}, t=0), \tag{4.35}$$

on the domain of interest. For these initial data, there is no restriction on the kind of functions permitted other than $\nabla \cdot \mathbf{B}_i(\mathbf{r}) = 0$. However, the consideration of the appropriate boundary conditions is a much more involved one since it implies *the specification of a geometry* associated with a particular *magnetic confinement* scheme.

We will extensively dwell on the specific boundary conditions for the different magnetic confinement geometries encountered in the laboratory and nature in Section 4.6. However, before we can effectively do that, we first have to introduce the central concepts of magnetohydrodynamics underlying it all, viz. magnetic flux tubes and flux conservation (Section 4.2), the general conservation properties of the MHD equations (Sections 4.3 and 4.4), and the discontinuities permitted (Section 4.5).

4.2 Magnetic flux

4.2.1 Flux tubes

Magnetic flux tubes occur in different kinds, e.g. closed onto themselves, like in thermonuclear *tokamak* confinement machines, or connecting onto a medium of so vastly different physical characteristics that one may consider the flux tube to be finite and separated from the other medium by suitable jump conditions. The latter kind is the appropriate model for *coronal flux tubes*. These two generic plasma configurations are shown schematically in Fig. 4.2. The main point of this illustration is to show the global geometry, which is superficially similar but essentially different.

4.2 Magnetic flux

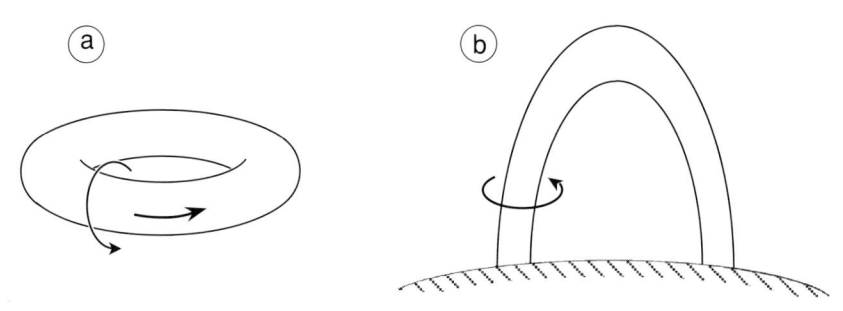

Fig. 4.2. Generic plasma confinement structures: (a) tokamak and (b) coronal magnetic loop (arrows indicate periodic directions).

▷ **Exercise.** It is frequently said that solar coronal flux loops that are 'standing' for weeks on the Sun appear to mock the stability problems encountered in laboratory plasmas. Consult Table 4.1 and compare a tokamak with a plasma that is routinely confined for half a minute (a typical operational feature at the end of the twentieth century) with a coronal loop that stands for two weeks on the Sun before erupting. Is the statement true or false? (Hint: stability is to be measured in terms of the number of Alfvén wave crossing times of the configuration.) ◁

The tokamak basically consists of closed and nested magnetic surfaces which have a toroidal shape, i.e. they are periodic in two directions (poloidal and toroidal). The typical solar coronal magnetic loop, on the other hand, is a finite flux tube bound by the photosphere, which is considered to be infinitely inert, so that there is no motion at the ends (so-called line-tying). Whereas the same MHD equations may be used to describe the macroscopic plasma dynamics in both configurations, *the boundary conditions* are obviously quite different and, consequently, the exact form of the resulting dynamics will be different as well. We will return to this point in Section 4.6.

The above examples demonstrate that magnetic fields confining plasmas, man made as well as naturally occurring, are basically *tubular structures*. This is the result of the magnetic field equation (4.15)(b), repeated here for convenience,

$$\nabla \cdot \mathbf{B} = 0, \qquad (4.36)$$

which does not permit, for example, spherically symmetric solutions. Instead, magnetic flux tubes become the essential constituents of a magnetically confined plasma configuration. Following Newcomb [163, 167], we consider a surface element $d\sigma_1$ and all magnetic field lines puncturing it (Fig. 4.3). The magnetic flux through an arbitrary other surface element $d\sigma_2$ intersecting that field line bundle is the same. This follows from the application of Gauss' theorem (A.14):

$$\iiint_V \nabla \cdot \mathbf{B} \, d\tau = \oiint \mathbf{B} \cdot \mathbf{n} \, d\sigma = -\iint_{S_1} \mathbf{B}_1 \cdot \mathbf{n}_1 \, d\sigma_1 + \iint_{S_2} \mathbf{B}_2 \cdot \mathbf{n}_2 \, d\sigma_2 = 0,$$

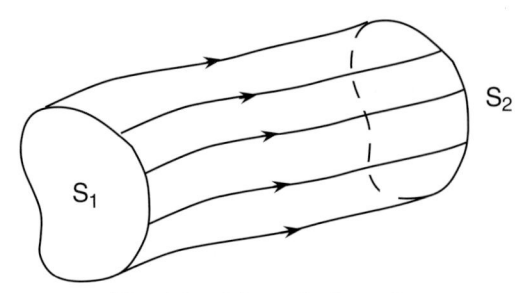

Fig. 4.3. Magnetic flux tube.

where $\mathbf{n}_1$ and $\mathbf{n}_2$ are the unit normals to the surfaces S_1 and S_2. Hence, the magnetic flux

$$\Psi \equiv \iint_S \mathbf{B} \cdot \mathbf{n}\, d\sigma \qquad (4.37)$$

through an arbitrary cross-section of the magnetic flux tube is a well-defined quantity, i.e. it does not depend on how S is taken. This holds for the flux tube as a whole (like the ones shown in Fig. 4.2), but also for the infinity of smaller flux tubes that one can imagine to be obtained by subdividing the cross-section S. Obviously, the important question to be asked next is: how do these flux tubes move when the plasma moves?

4.2.2 Global magnetic flux conservation

Whereas the concept of flux tube comes from the electromagnetic field equation (4.36), the dynamics of flux tubes requires the consideration of the two other electromagnetic field equations (4.1) and (4.8) absorbed in the induction equation (4.15)(a), which is repeated again for convenience:

$$\frac{\partial \mathbf{B}}{\partial t} = \nabla \times (\mathbf{v} \times \mathbf{B}). \qquad (4.38)$$

Exploiting this equation, one should always remember that it is really a contraction of Faraday's law, $\partial \mathbf{B}/\partial t = -\nabla \times \mathbf{E}$, describing the time dependence of the magnetic field, and 'Ohm's' law with perfect conductivity, $\mathbf{E} + \mathbf{v} \times \mathbf{B} = 0$, relating the electric field to the plasma flow. These two aspects also guide the following analysis.

Consider the complete magnetic flux tube inside the toroidal vessel of a tokamak (Fig. 4.4). The toroidal plasma volume is indicated by the letter V, the surrounding conducting wall by W, and the normal vector on the wall by $\mathbf{n}_w$. We will investigate the implications of the electromagnetic field equations for the dynamics of that flux tube inside the vessel.

4.2 Magnetic flux

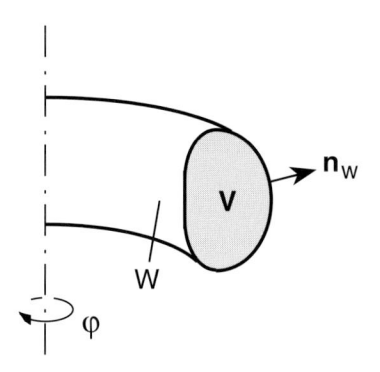

Fig. 4.4. Tokamak geometry.

First of all, we need *boundary conditions at the wall*. An obvious one is the requirement that the wall is 'perfect', i.e. it neither absorbs plasma nor emits gas (impurities). This translates into the boundary condition

$$\mathbf{n}_w \cdot \mathbf{v} = 0 \quad \text{(on } W\text{)}. \tag{4.39}$$

A second boundary condition is the requirement that the wall acts as a perfect conductor, i.e. it short-circuits tangential electric fields $\mathbf{E}_t$. Interestingly enough, this condition is not a restriction on the kind of materials a wall could be made of, but rather on the resistivity of the plasma itself: if the wall is an isolator, a thin perfectly conducting plasma layer in front of the wall serves the same purpose of short-circuiting the electric field. Hence, 'Ohm's' law applied to this layer yields:

$$\mathbf{n}_w \times [\mathbf{E} + (\mathbf{v} \times \mathbf{B})] \stackrel{(A.2)}{=} \mathbf{n}_w \times \mathbf{E}_t + \mathbf{n}_w \cdot \mathbf{B}\,\mathbf{v} - \mathbf{n}_w \cdot \mathbf{v}\,\mathbf{B} = 0. \tag{4.40}$$

Since the tangential electric field vanishes, $\mathbf{E}_t = 0$, and there is no flow across the wall, $\mathbf{n}_w \cdot \mathbf{v} = 0$, the other contribution has to vanish as well:

$$\mathbf{n}_w \cdot \mathbf{B} = 0 \quad \text{(on } W\text{)}. \tag{4.41}$$

Hence, the magnetic field lines do not intersect the wall so that this boundary condition prevents plasma flowing along the field lines hitting the wall and being lost.

Let us now study the consequences of the magnetic field evolution equation (4.38) for the total magnetic flux inside a tokamak. Since there are two magnetic field components (poloidal and toroidal), there are also two magnetic fluxes, associated with the two different surfaces obtained by a cut along or across the torus. The toroidal component is the simplest. Starting from Eq. (4.37), with $S \equiv S_{\text{pol}}$ indicating a poloidal cross-section of the torus (Fig. 4.5(a)), we find by the application of Stokes' law that the time derivative of the total toroidal flux

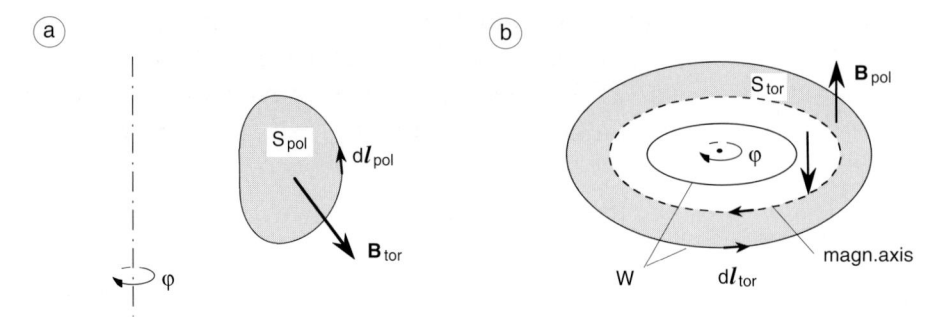

Fig. 4.5. Surfaces for (a) toroidal magnetic flux, (b) poloidal magnetic flux in a tokamak.

vanishes:

$$\frac{\partial \Psi_{\text{tor}}}{\partial t} \equiv \iint_{S_{\text{pol}}} \frac{\partial \mathbf{B}_{\text{tor}}}{\partial t} \cdot \mathbf{n}_{\text{tor}} \, d\sigma = \iint \nabla \times (\mathbf{v} \times \mathbf{B}_{\text{tor}}) \cdot \mathbf{n}_{\text{tor}} \, d\sigma$$

$$\stackrel{\text{(A.18)}}{=} \oint \mathbf{v} \mathbf{B}_{\text{tor}} \cdot d\mathbf{l}_{\text{pol}} = 0, \quad (4.42)$$

because $\mathbf{v}$, $\mathbf{B}_{\text{tor}}$, and $\mathbf{l}_{\text{pol}}$ are tangential to the wall. In other words, the two boundary conditions (4.39) and (4.41) suffice to guarantee that the total toroidal magnetic flux inside the tokamak is conserved: $\Psi_{\text{tor}} = \text{const}$.

A similar story holds for the poloidal flux:

$$\frac{\partial \Psi_{\text{pol}}}{\partial t} \equiv \iint_{S_{\text{tor}}} \frac{\partial \mathbf{B}_{\text{pol}}}{\partial t} \cdot \mathbf{n}_{\text{pol}} \, d\sigma = 0 \quad \Rightarrow \quad \Psi_{\text{pol}} = \text{const}, \quad (4.43)$$

where we exploit a surface S_{tor} bounded by a toroidal circle lying in the wall and by the magnetic axis (Fig. 4.5(b)). For the time being, this proof requires that the magnetic axis should be kept fixed. Relaxing the latter constraint requires the consideration of the local version of magnetic flux conservation. This is a slightly more complicated, but more powerful, property that will be dealt with in Section 4.3.3. Anticipating the outcome of that analysis, viz. that the magnetic flux through any surface moving with the plasma is constant, it becomes clear that *magnetic flux conservation is the central issue in magnetohydrodynamics*. To put this into proper perspective, we now turn to a systematic exposition of the conservation properties of the MHD equations.

4.3 Conservation laws

4.3.1 Conservation form of the MHD equations

We have already stated several times that the MHD equations express conservation of the main macroscopic quantities of a plasma, viz. mass, momentum, energy, and magnetic flux. We will now substantiate this claim.

A system of quasi-linear partial differential equations is said to be in conservation form if all terms can be written as a generalized divergence (incorporating the time and space derivatives on an equal footing) of the dependent variables, or simple functions of them:

$$\frac{\partial}{\partial t}(\cdots) + \nabla \cdot (\cdots) = 0. \tag{4.44}$$

The use of such a form of the equations is that one can obtain *local and global conservation laws* and *jump conditions* from them. Moreover, powerful numerical algorithms exist for the solution of such equations. We will return to this point in a later chapter on numerical magnetohydrodynamics in the companion Volume 2.

Consider again the set of nonlinear ideal MHD equations discussed in Section 4.1, exploiting *the internal energy e* rather than the pressure p as a variable, according to Eq. (4.19), and dropping the constant μ_0 for convenience:[2]

$$\frac{\partial \rho}{\partial t} + \nabla \cdot (\rho \mathbf{v}) = 0, \tag{4.45}$$

$$\rho \frac{\partial \mathbf{v}}{\partial t} + \rho \mathbf{v} \cdot \nabla \mathbf{v} + \nabla p - \mathbf{j} \times \mathbf{B} = -\rho \nabla \Phi, \quad \mathbf{j} = \nabla \times \mathbf{B},$$

$$p = (\gamma - 1)\rho e, \tag{4.46}$$

$$\frac{\partial e}{\partial t} + \mathbf{v} \cdot \nabla e + (\gamma - 1)e \nabla \cdot \mathbf{v} = 0, \tag{4.47}$$

$$\frac{\partial \mathbf{B}}{\partial t} + \nabla \times \mathbf{E} = 0, \quad \mathbf{E} = -\mathbf{v} \times \mathbf{B}, \quad \nabla \cdot \mathbf{B} = 0. \tag{4.48}$$

Evidently, only the mass conservation equation (4.45) and the magnetic field equation (4.48)(c) have the required conservation form. We only consider the effect of an *external* gravitational potential Φ in the momentum equation (4.46), where $\mathbf{g} = -\nabla \Phi$, since it is instructive to demonstrate how such a field spoils the strict conservation property of the equations. This is why this term is put on the right hand side of the equality sign.

The following text, in small print, consists of some rather intricate vector algebra that may be skipped on first reading because we will just exploit the end

---

[2] This will be done consistently from now on. To restore mks units one should make the replacements $\mathbf{B} \to \mathbf{B}/\sqrt{\mu_0}$, $\mathbf{E} \to \mathbf{E}/\sqrt{\mu_0}$, and $\mathbf{j} \to \sqrt{\mu_0}\mathbf{j}$ in the formulas.

result, viz. the conservation form of the ideal MHD equations summarized in Eqs. (4.59)–(4.62).

▷ **Transformation to conservation form.** The mass conservation form, already obtained, is repeated in Eq. (4.59) below. In order to bring the other equations into conservation form one makes use of the following vector identities (where equation numbers above the equal signs refer to auxiliary equations like those of Appendix A):

$$\nabla \cdot (a\mathbf{b}) \stackrel{(A.11)}{=} \mathbf{a} \cdot \nabla b + b \nabla \cdot \mathbf{a}, \tag{4.49}$$

$$\mathbf{a} \times (\nabla \times \mathbf{b}) \stackrel{(A.8),(A.11)}{=} (\nabla \mathbf{b}) \cdot \mathbf{a} - \nabla \cdot (\mathbf{a}\mathbf{b}) - \mathbf{b} \nabla \cdot \mathbf{a}, \tag{4.50}$$

$$\nabla \times (\mathbf{a} \times \mathbf{b}) \stackrel{(A.13)(a)}{=} \nabla \cdot (\mathbf{b}\mathbf{a} - \mathbf{a}\mathbf{b}), \tag{4.51}$$

$$\nabla(\mathbf{a} \cdot \mathbf{b}) \stackrel{(A.10)(a)}{=} (\nabla \mathbf{a}) \cdot \mathbf{b} + (\nabla \mathbf{b}) \cdot \mathbf{a}. \tag{4.52}$$

The first two terms of the momentum equation, (4.46) may then be transformed to

$$\rho \frac{\partial \mathbf{v}}{\partial t} + \rho \mathbf{v} \cdot \nabla \mathbf{v} \stackrel{(4.45)}{=} \frac{\partial}{\partial t}(\rho \mathbf{v}) + \mathbf{v} \nabla \cdot (\rho \mathbf{v}) + \rho \mathbf{v} \cdot \nabla \mathbf{v} \stackrel{(4.49)}{=} \frac{\partial}{\partial t}(\rho \mathbf{v}) + \nabla \cdot (\rho \mathbf{v}\mathbf{v}), \tag{4.53}$$

and the last term to

$$-\mathbf{j} \times \mathbf{B} = \mathbf{B} \times (\nabla \times \mathbf{B}) \stackrel{(4.50),(4.52)}{=} \nabla(\tfrac{1}{2}B^2) - \nabla \cdot (\mathbf{B}\mathbf{B}), \tag{4.54}$$

so that we obtain the conservation form of the momentum equation, Eq. (4.60) below. Similarly, the second term of Faraday's law (4.48) becomes

$$\nabla \times \mathbf{E} = -\nabla \times (\mathbf{v} \times \mathbf{B}) \stackrel{(4.51)}{=} \nabla \cdot (\mathbf{v}\mathbf{B} - \mathbf{B}\mathbf{v}), \tag{4.55}$$

so that we obtain the conservation form of the magnetic flux equation, Eq. (4.62) below.

Finally, the internal energy equation (4.47) cannot be brought into conservation form for the obvious reason that it contains only part of the energy, which can be converted into other forms of energy. We therefore need a conservation equation for the total energy density. This is obtained by adding the separate contributions of the kinetic, the internal, and the magnetic energy:

$$\mathbf{v} \cdot \text{Eq. (4.46)} \implies \rho \mathbf{v} \cdot (\frac{\partial \mathbf{v}}{\partial t} + \mathbf{v} \cdot \nabla \mathbf{v}) + \mathbf{v} \cdot \nabla p - \mathbf{v} \cdot \mathbf{j} \times \mathbf{B} = -\rho \mathbf{v} \cdot \nabla \Phi$$

$$\longrightarrow \frac{\partial}{\partial t}(\tfrac{1}{2}\rho v^2) - \tfrac{1}{2}v^2 \frac{\partial \rho}{\partial t} + \tfrac{1}{2}\rho \mathbf{v} \cdot \nabla v^2 + \mathbf{v} \cdot \nabla p - \mathbf{v} \cdot \mathbf{j} \times \mathbf{B} = -\rho \mathbf{v} \cdot \nabla \Phi$$

$$\stackrel{(4.45)}{\longrightarrow} \frac{\partial}{\partial t}(\tfrac{1}{2}\rho v^2) + \nabla \cdot (\tfrac{1}{2}\rho v^2 \mathbf{v}) + \mathbf{v} \cdot \nabla p - \mathbf{v} \cdot \mathbf{j} \times \mathbf{B} = -\rho \mathbf{v} \cdot \nabla \Phi, \tag{4.56}$$

$$\rho \text{ Eq. (4.47)} \implies \rho \frac{\partial e}{\partial t} + \rho \mathbf{v} \cdot \nabla e + (\gamma - 1)\rho e \nabla \cdot \mathbf{v} = 0$$

$$\longrightarrow \frac{\partial}{\partial t}(\rho e) - e \frac{\partial \rho}{\partial t} + \rho \mathbf{v} \cdot \nabla e + p \nabla \cdot \mathbf{v} = 0$$

$$\stackrel{(4.45)}{\longrightarrow} \frac{\partial}{\partial t}(\rho e) + \nabla \cdot (\rho e \mathbf{v}) + p \nabla \cdot \mathbf{v} = 0, \tag{4.57}$$

$$\mathbf{B} \cdot \text{Eq. (4.48)} \implies \mathbf{B} \cdot \frac{\partial \mathbf{B}}{\partial t} - \mathbf{B} \cdot \nabla \times (\mathbf{v} \times \mathbf{B}) = 0$$

$$\stackrel{(A.12)}{\longrightarrow} \quad \frac{\partial}{\partial t}(\tfrac{1}{2}B^2) + \nabla \cdot [\,\mathbf{B} \times (\mathbf{v} \times \mathbf{B})\,] - (\mathbf{v} \times \mathbf{B}) \cdot \nabla \times \mathbf{B} = 0$$

$$\stackrel{(A.2)}{\longrightarrow} \quad \frac{\partial}{\partial t}(\tfrac{1}{2}B^2) + \nabla \cdot [\,\mathbf{B} \cdot \mathbf{B}\mathbf{v} - \mathbf{v} \cdot \mathbf{B}\mathbf{B}\,] + \mathbf{v} \cdot \mathbf{j} \times \mathbf{B} = 0. \tag{4.58}$$

Adding Eqs. (4.56), (4.57), and (4.58) gives the conservation form of the energy equation, Eq. (4.61) below. ◁

Recapitulating the preceding analysis, *the conservation form of the ideal MHD equations* reads:

$$\frac{\partial \rho}{\partial t} + \nabla \cdot (\rho \mathbf{v}) = 0, \tag{4.59}$$

$$\frac{\partial}{\partial t}(\rho \mathbf{v}) + \nabla \cdot \left[\rho \mathbf{v}\mathbf{v} + (p + \tfrac{1}{2}B^2)\mathbf{I} - \mathbf{B}\mathbf{B}\right] = -\rho \nabla \Phi,$$

$$p = (\gamma - 1)\rho e, \tag{4.60}$$

$$\frac{\partial}{\partial t}(\tfrac{1}{2}\rho v^2 + \rho e + \tfrac{1}{2}B^2) + \nabla \cdot \left[(\tfrac{1}{2}\rho v^2 + \rho e + p + B^2)\mathbf{v} - \mathbf{v} \cdot \mathbf{B}\mathbf{B}\right]$$
$$= -\rho \mathbf{v} \cdot \nabla \Phi, \tag{4.61}$$

$$\frac{\partial \mathbf{B}}{\partial t} + \nabla \cdot (\mathbf{v}\mathbf{B} - \mathbf{B}\mathbf{v}) = 0, \qquad \nabla \cdot \mathbf{B} = 0. \tag{4.62}$$

Note that the energy conservation form (4.61) for the total energy, which replaces the evolution equation (4.47) for the internal energy e, required the most extensive transformation since e cannot be conserved by itself.

From the preceding analysis we conclude that the best representation of the evolution equations is in terms of the variables ρ, $\mathbf{v}$, e, and $\mathbf{B}$, as expressed by the Eqs. (4.45)–(4.48). A peculiar additional variable is *the specific entropy s* (the entropy per unit mass), introduced in Section 4.1.1. For adiabatic processes of ideal gases, which is applicable here, we have

$$S \equiv p\rho^{-\gamma} = f(s), \qquad \text{or} \quad s = C_v \ln(p\rho^{-\gamma}) + \text{const}. \tag{4.63}$$

Hence, from Eqs. (4.45) and (4.47),

$$\frac{DS}{Dt} \equiv \frac{\partial S}{\partial t} + \mathbf{v} \cdot \nabla S = 0, \tag{4.64}$$

which is not in conservation form, but expresses the conservation of specific entropy co-moving with the fluid. A genuine conservation form is obtained by transforming to the variable ρS, associated with *the entropy per unit volume*. In that variable we get

$$\frac{\partial}{\partial t}(\rho S) + \nabla \cdot (\rho S \mathbf{v}) = 0, \tag{4.65}$$

4.3.2 Global conservation laws

To understand the physical meaning of the different terms in the conservation equations (4.59)–(4.62) of ideal MHD, we define the following quantities:

- momentum density $\quad\quad \boldsymbol{\pi} \equiv \rho \mathbf{v}$, $\hfill (4.66)$

- stress tensor $\quad\quad \mathbf{T} \equiv \rho \mathbf{vv} + (p + \tfrac{1}{2}B^2)\mathbf{I} - \mathbf{BB}$, $\hfill (4.67)$

- total energy density $\quad\quad \mathcal{H} \equiv \tfrac{1}{2}\rho v^2 + \dfrac{p}{\gamma - 1} + \tfrac{1}{2}B^2$, $\hfill (4.68)$

- energy flow $\quad\quad \mathbf{U} \equiv \left(\tfrac{1}{2}\rho v^2 + \dfrac{\gamma}{\gamma - 1}p\right)\mathbf{v} + B^2 \mathbf{v} - \mathbf{v} \cdot \mathbf{BB}$, $\hfill (4.69)$

- (no name) $\quad\quad \mathbf{Y} \equiv \mathbf{vB} - \mathbf{Bv}$. $\hfill (4.70)$

Neglecting gravity, the conservation equations (4.59)–(4.62) may then be written as

$$\frac{\partial \rho}{\partial t} + \nabla \cdot \boldsymbol{\pi} = 0 \quad\quad \textit{(conservation of mass)}, \tag{4.71}$$

$$\frac{\partial \boldsymbol{\pi}}{\partial t} + \nabla \cdot \mathbf{T} = 0 \quad\quad \textit{(conservation of momentum)}, \tag{4.72}$$

$$\frac{\partial \mathcal{H}}{\partial t} + \nabla \cdot \mathbf{U} = 0 \quad\quad \textit{(conservation of energy)}, \tag{4.73}$$

$$\frac{\partial \mathbf{B}}{\partial t} + \nabla \cdot \mathbf{Y} = 0 \quad\quad \textit{(conservation of magnetic flux)}. \tag{4.74}$$

These are the evolution equations for the variables ρ, $\boldsymbol{\pi}$, $\mathcal{H}$, and $\mathbf{B}$ in conservation form. Note that the quantities appearing in the divergence terms can all be expressed in terms of these four variables so that they constitute yet another basic set of variables to describe ideal MHD.

▷ **Computation of primitive variables from conserved ones.** The conservation equations (4.71)–(4.74) are used in many numerical schemes to compute the time-advance of the conserved variables ρ, $\boldsymbol{\pi}$, $\mathcal{H}$, $\mathbf{B}$. The calculation of the primitive (original) variables ρ, $\mathbf{v}$, e or p, $\mathbf{B}$ from them is straightforward. However, calculation of the pressure,

$$p = (\gamma - 1)\left[\mathcal{H} - \tfrac{1}{2}(\pi^2/\rho + B^2)\right], \tag{4.75}$$

presents a particular numerical problem. It involves subtraction of the sum of kinetic energy $\tfrac{1}{2}\pi^2/\rho$ and magnetic energy $\tfrac{1}{2}B^2$ from the total energy $\mathcal{H}$, which may not result in

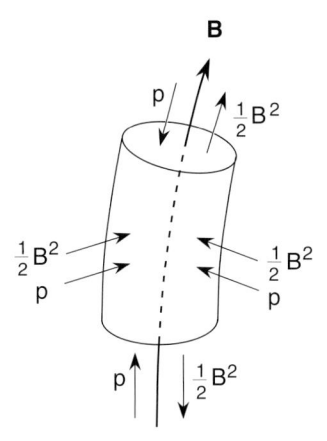

Fig. 4.6. Magnetic stress and tension in a flux tube.

a positive numerical value for p. It is clear that this requires additional care in the construction of acceptable numerical schemes since $\beta \equiv 2p/B^2 \ll 1$ for many plasmas of interest. ◁

The stress tensor **T** is composed of the *Reynolds stress tensor* $\rho\mathbf{vv}$, the *isotropic pressure* $p\mathbf{I}$, and the *magnetic part* $\frac{1}{2}B^2\mathbf{I} - \mathbf{BB}$ of the Maxwell stress tensor. In a projection based on the velocity **v**, the only non-vanishing contribution to the Reynolds stress is a positive stress ('pressure') ρv^2 along **v**. The remaining part of the stress tensor is more clearly represented in a projection based on the magnetic field **B**:

$$\begin{pmatrix} p+\frac{1}{2}B^2 & 0 & 0 \\ 0 & p+\frac{1}{2}B^2 & 0 \\ 0 & 0 & p-\frac{1}{2}B^2 \end{pmatrix} \begin{matrix} \perp \\ \perp \\ \parallel \end{matrix}.$$

Hence, the magnetic field provides positive stress, *magnetic pressure*, in directions perpendicular to **B** and negative stress, *magnetic tension*, parallel to **B** (Fig. 4.6).

The different terms of the total energy density $\mathcal{H}$ may be grouped into two parts:

$$\mathcal{H} = \mathcal{K} + \mathcal{W}, \tag{4.76}$$

where $\mathcal{K}$ is *the kinetic energy density*,

$$\mathcal{K} \equiv \tfrac{1}{2}\rho v^2, \tag{4.77}$$

and $\mathcal{W}$ is *the potential energy density*,

$$\mathcal{W} \equiv \rho e + \tfrac{1}{2}B^2 = \frac{p}{\gamma - 1} + \tfrac{1}{2}B^2. \tag{4.78}$$

In many plasmas of interest, the magnetic energy density $\frac{1}{2}B^2$ represents a huge energy reservoir that may be released suddenly, e.g. in violent disruptions in tokamaks, in solar flares, in coronal mass ejections, and in many other explosive events in astrophysics. The energy flow vector $\mathbf{U}$ is again composed of a hydrodynamic part (the term with the bracket in Eq. (4.69)) and a magnetic part. The latter part may be transformed into the usual *Poynting vector*:

$$\mathbf{U}_{\text{magn}} \equiv B^2 \mathbf{v} - \mathbf{v} \cdot \mathbf{B}\mathbf{B} \stackrel{(A.2)}{=} -(\mathbf{v} \times \mathbf{B}) \times \mathbf{B} \stackrel{(4.48)}{=} \mathbf{E} \times \mathbf{B} \equiv \mathbf{S}, \qquad (4.79)$$

which represents the flow of electromagnetic energy.

On purely formal grounds, we have introduced the tensor $\mathbf{Y} \equiv \mathbf{v}\mathbf{B} - \mathbf{B}\mathbf{v}$ in the evolution equation (4.74) for $\mathbf{B}$. We have not given a name to this symbol because it appears to have no direct intuitive meaning. The reason we wrote the flux equation in this way is that one obtains the jump conditions most easily from it (see Section 4.5) and, also, numerical algorithms based on finite volume discretization exploit this formulation. However, whereas global conservation laws for mass, momentum, and energy are obtained by the application of *Gauss' theorem* on the equations for ρ, $\boldsymbol{\pi}$ and $\mathcal{H}$, to get a global conservation law for the magnetic flux one should apply *Stokes' theorem*. For that reason, the previously exploited form of Faraday's law (4.15), with the curl operator, is far to be preferred over that of Eq. (4.74), with the divergence.

Consider now a plasma surrounded by a perfectly conducting wall so that both $\mathbf{v} \cdot \mathbf{n} = 0$ and $\mathbf{n} \cdot \mathbf{B} = 0$ at the wall (which are the b.c.s (4.39) and (4.41) derived in Section 4.2.2). Define the following quantities:

$$\text{– total mass} \qquad M \equiv \int \rho \, d\tau, \qquad (4.80)$$

$$\text{– total momentum} \qquad \boldsymbol{\Pi} \equiv \int \boldsymbol{\pi} \, d\tau, \qquad (4.81)$$

$$\text{– total energy} \qquad H \equiv \int \mathcal{H} \, d\tau, \qquad (4.82)$$

$$\text{– total magnetic flux} \qquad \Psi \equiv \int \mathbf{B} \cdot \tilde{\mathbf{n}} \, d\tilde{\sigma}, \qquad (4.83)$$

where we now abbreviate the previously used triple and double integral signs for volume and surface integration by just a single integral. In the definitions of the total mass, momentum, and energy, $\int d\tau$ is the total plasma volume enclosed by the surface $\oint d\sigma$ of the wall. However, in the definition (4.83) of the total magnetic flux, we have put a tilde on the surface element $d\tilde{\sigma}$ and the normal vector $\tilde{\mathbf{n}}$ to indicate that they refer to a cross-section of the plasma enclosed by a boundary curve $\oint d\mathbf{l}$ lying in the wall (like in Fig. 4.5 for the fluxes in a tokamak).

By applying Gauss' theorem (A.14) to the local mass conservation equation (4.71) we find the time derivative of the total mass,

$$\dot M = \int \dot\rho\, d\tau = -\int \nabla\cdot\boldsymbol{\pi}\, d\tau = -\oint \boldsymbol{\pi}\cdot\mathbf{n}\, d\sigma = 0, \qquad (4.84)$$

which vanishes by virtue of the b.c. (4.39). Hence, *the total mass is conserved*. Applying Gauss' theorem, and the b.c.s (4.39) and (4.41), to the local momentum conservation equation (4.72) gives

$$\mathbf{F} = \dot{\boldsymbol{\Pi}} = \int \dot{\boldsymbol{\pi}}\, d\tau = -\int \nabla\cdot\mathbf{T}\, d\tau = -\oint (p+\tfrac{1}{2}B^2)\mathbf{n}\, d\sigma, \qquad (4.85)$$

so that *the total momentum is conserved*, however, only *if the total force exerted by the wall vanishes*. If this were not the case, there would be an imbalance of the last term involving the total pressure. This is a logical proviso if the configuration is to remain in place (we exclude disasters like earthquakes or disruptions of the vessel). Applying Gauss' theorem, and the mentioned b.c.s, to the local energy conservation equation (4.73) gives

$$\dot H = \int \dot{\mathcal{H}}\, d\tau = -\int \nabla\cdot\mathbf{U}\, d\tau = -\oint \mathbf{U}\cdot\mathbf{n}\, d\sigma = 0, \qquad (4.86)$$

which states that *the total energy is conserved*.

On the other hand, applying Stokes' theorem (A.18), and the b.c.s (4.39) and (4.41), to the induction equation (4.48), we obtain

$$\dot\Psi = \int \dot{\mathbf{B}}\cdot\tilde{\mathbf{n}}\, d\tilde\sigma = \int \nabla\times(\mathbf{v}\times\mathbf{B})\cdot\tilde{\mathbf{n}}\, d\tilde\sigma = \oint \mathbf{v}\times\mathbf{B}\cdot d\boldsymbol{l} = 0, \qquad (4.87)$$

since $\mathbf{v}$, $\mathbf{B}$, and $d\boldsymbol{l}$ are tangential to the wall (as discussed already in Section 4.2.2). Hence, *magnetic flux is conserved* as well: it cannot leave or enter the vessel.

Consequently, in a plasma enclosed by a rigid shell (this will be called model I in Section 4.6), *the boundary conditions* $\mathbf{v}\cdot\mathbf{n} = 0$ *and* $\mathbf{n}\cdot\mathbf{B} = 0$ *guarantee that all physical quantities of interest are conserved*, so that *the system is closed*. This remains true for a plasma surrounded by vacuum (model II), as will be proved in Section 4.6 by appropriate modifications of these boundary conditions. It is no longer true if the wall is replaced by a system of external coils with time-dependent currents (model III): magnetic flux and Poynting flux may be pumped into the system. Similarly, coronal plasmas bounded by an immobile photosphere (models V and VI of Section 4.6), can be extended to incorporate coronal changes of mass, momentum, energy and magnetic flux by photospheric boundary motions. We conclude that the conservation equations (4.71)–(4.74) are an extremely powerful representation of the nonlinear macroscopic dynamics of plasmas.

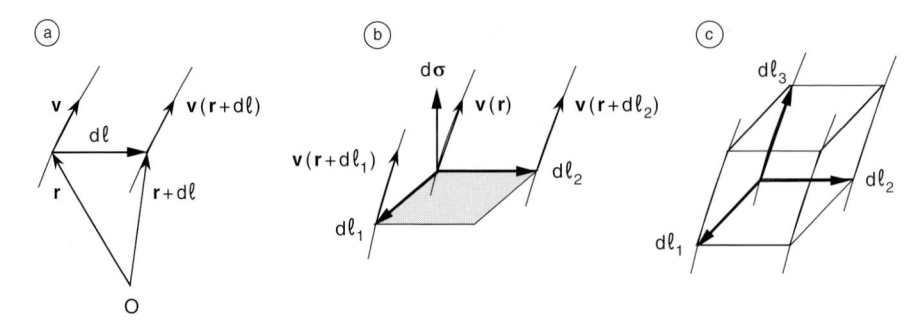

Fig. 4.7. Kinematics of (a) line element dl, (b) surface element $d\sigma$, (c) volume element $d\tau$.

4.3.3 Local conservation laws – conservation of magnetic flux

We now wish to describe the local conservation properties, in particular flux conservation, in terms of the dynamics of fluid elements. To that end, it is helpful to derive the *kinematic expressions* for the change in time of the geometric quantities.

First, consider the fluid flow $\mathbf{v}(\mathbf{r})$ at the positions $\mathbf{r}$ and $\mathbf{r} + d\mathbf{l}$ and let us find the equations describing the motion of the *line element $d\mathbf{l}$* connecting those positions (Fig. 4.7(a)). This is given by the Lagrangian derivative of $d\mathbf{l}$,

$$\frac{D}{Dt}(d\mathbf{l}) = \frac{D(\mathbf{r}+d\mathbf{l})}{Dt} - \frac{D\mathbf{r}}{Dt} = \mathbf{v}(\mathbf{r}+d\mathbf{l}) - \mathbf{v}(\mathbf{r}) = d\mathbf{l} \cdot (\nabla \mathbf{v}), \qquad (4.88)$$

which describes the kinematics of a line element.

Next, we define the *surface element* by the vector product $d\boldsymbol{\sigma} \equiv d\mathbf{l}_1 \times d\mathbf{l}_2$ (Fig. 4.7(b)). Its Lagrangian derivative is derived by applying Eq. (4.88) twice:

$$\frac{D}{Dt}(d\boldsymbol{\sigma}) = \frac{D(d\mathbf{l}_1)}{Dt} \times d\mathbf{l}_2 + d\mathbf{l}_1 \times \frac{D(d\mathbf{l}_2)}{Dt} = d\mathbf{l}_1 \cdot (\nabla \mathbf{v}) \times d\mathbf{l}_2 - d\mathbf{l}_2 \cdot (\nabla \mathbf{v}) \times d\mathbf{l}_1 .$$

Note the brackets around the expression $(\nabla \mathbf{v})$, which are introduced here to indicate that the gradient is meant to operate only on the quantity inside. To further reduce the kinematic relation obtained, we need some vector identities that are rather involved because of the tensorial character of $(\nabla \mathbf{v})$ and, therefore, put in small print.

▷ **Horrible derivation of a useful expression.** Use vector identity (A.2) of Appendix A:

$$(\mathbf{a} \times \mathbf{b}) \times \mathbf{c} = \mathbf{a} \cdot \mathbf{c}\,\mathbf{b} - \mathbf{b} \cdot \mathbf{c}\,\mathbf{a} = \mathbf{b}\mathbf{a} \cdot \mathbf{c} - \mathbf{a}\mathbf{b} \cdot \mathbf{c} ,$$

where $\mathbf{c}$ has been moved to the utmost right since we will replace it by ∇:

$$(\mathbf{a} \times \mathbf{b}) \times \nabla = \mathbf{b}\mathbf{a} \cdot \nabla - \mathbf{a}\mathbf{b} \cdot \nabla .$$

Applying this operator on $\mathbf{v}$ gives

$$[(\mathbf{a} \times \mathbf{b}) \times \nabla] \times \mathbf{v} = [\mathbf{b}\mathbf{a} \cdot \nabla] \times \mathbf{v} - [\mathbf{a}\mathbf{b} \cdot \nabla] \times \mathbf{v} = -\mathbf{a} \cdot (\nabla \mathbf{v}) \times \mathbf{b} + \mathbf{b} \cdot (\nabla \mathbf{v}) \times \mathbf{a} .$$

Setting $\mathbf{a} \equiv d\mathbf{l}_1$ and $\mathbf{b} \equiv d\mathbf{l}_2$, so that $\mathbf{a} \times \mathbf{b} \equiv d\boldsymbol{\sigma}$, we obtain the required expression:
$$-d\mathbf{l}_1 \cdot (\nabla \mathbf{v}) \times d\mathbf{l}_2 + d\mathbf{l}_2 \cdot (\nabla \mathbf{v}) \times d\mathbf{l}_1 = (d\boldsymbol{\sigma} \times \nabla) \times \mathbf{v} = (\nabla \mathbf{v}) \cdot d\boldsymbol{\sigma} - \nabla \cdot \mathbf{v} \, d\boldsymbol{\sigma},$$
where Eq. (A.9) has been used in the last step. ◁

Hence,
$$\frac{D}{Dt}(d\boldsymbol{\sigma}) = -(d\boldsymbol{\sigma} \times \nabla) \times \mathbf{v} = -(\nabla \mathbf{v}) \cdot d\boldsymbol{\sigma} + \nabla \cdot \mathbf{v} \, d\boldsymbol{\sigma}, \tag{4.89}$$
which is the kinematic expression for the motion of a surface element.

For the derivation of local magnetic flux conservation the expression (4.89) would be sufficient. However, to derive the local conservation laws for the other variables, we also need the kinematics of a *volume element*. Its definition, $d\tau \equiv d\boldsymbol{\sigma} \cdot d\mathbf{l}_3 = (d\mathbf{l}_1 \times d\mathbf{l}_2) \cdot d\mathbf{l}_3$ (Fig. 4.7(c)), gives by straightforward application of Eqs. (4.89) and (4.88):
$$\frac{D}{Dt}(d\tau) = \frac{D(d\boldsymbol{\sigma})}{Dt} \cdot d\mathbf{l}_3 + d\boldsymbol{\sigma} \cdot \frac{D(d\mathbf{l}_3)}{Dt}$$
$$= -d\mathbf{l}_3 \cdot (\nabla \mathbf{v}) \cdot d\boldsymbol{\sigma} + \nabla \cdot \mathbf{v} \, (d\boldsymbol{\sigma} \cdot d\mathbf{l}_3) + d\mathbf{l}_3 \cdot (\nabla \mathbf{v}) \cdot d\boldsymbol{\sigma}$$
$$= \nabla \cdot \mathbf{v} \, d\tau. \tag{4.90}$$

This is the kinematic expression for the motion of a volume element.

Combining now the dynamic equation (4.71) with the kinematic relation (4.90) for the motion of *the mass of a fluid element*, $dM \equiv \rho d\tau$, gives the following expression for its rate of change:
$$\frac{D}{Dt}(dM) = \frac{D}{Dt}(\rho d\tau) = \frac{D\rho}{Dt} d\tau + \rho \frac{D}{Dt}(d\tau)$$
$$= -\rho \nabla \cdot \mathbf{v} \, d\tau + \rho \nabla \cdot \mathbf{v} \, d\tau = 0. \tag{4.91}$$

Hence, the mass of a moving fluid element is constant.

Similarly, the rate of change of *the momentum of a fluid element*, $d\boldsymbol{\Pi} \equiv \boldsymbol{\pi} d\tau$, is found from Eqs. (4.72) and (4.90):
$$\frac{D}{Dt}(d\boldsymbol{\Pi}) = \frac{D\boldsymbol{\pi}}{Dt} d\tau + \boldsymbol{\pi} \frac{D}{Dt}(d\tau)$$
$$= -\nabla \cdot \mathbf{T} \, d\tau + (\mathbf{v} \cdot \nabla \boldsymbol{\pi}) d\tau + \boldsymbol{\pi} \nabla \cdot \mathbf{v} \, d\tau$$
$$= -\nabla \cdot \left[\rho \mathbf{vv} + (p + \tfrac{1}{2}B^2)\mathbf{I} - \mathbf{BB} - \mathbf{v}\boldsymbol{\pi} \right] d\tau$$
$$= \left[-\nabla(p + \tfrac{1}{2}B^2) + \nabla \cdot (\mathbf{BB}) \right] d\tau$$
$$= (-\nabla p + \mathbf{j} \times \mathbf{B}) d\tau \neq 0, \tag{4.92}$$

where the equality (4.54) has been used in the last step. Hence, the momentum of a moving fluid element is *not* constant: it changes through the force density $-\nabla p + \mathbf{j} \times \mathbf{B}$ acting upon it.

For the change of *the energy of a fluid element*, $dH \equiv \mathcal{H} d\tau$, one finds from Eqs. (4.73) and (4.90):

$$\frac{D}{Dt}(dH) = \frac{D\mathcal{H}}{Dt} d\tau + \mathcal{H} \frac{D}{Dt}(d\tau)$$

$$= (-\nabla \cdot \mathbf{U} + \mathbf{v} \cdot \nabla \mathcal{H} + \mathcal{H} \nabla \cdot \mathbf{v}) d\tau = -\nabla \cdot (\mathbf{U} - \mathbf{v}\mathcal{H}) d\tau$$

$$= -\nabla \cdot \left[\left(\tfrac{1}{2}\rho v^2 + \frac{p}{\gamma - 1} + p + B^2 \right) \mathbf{v} - \mathbf{v} \cdot \mathbf{BB} \right.$$

$$\left. - \mathbf{v} \left(\tfrac{1}{2}\rho v^2 + \frac{p}{\gamma - 1} + \tfrac{1}{2} B^2 \right) \right] d\tau$$

$$= -\nabla \cdot \left[(p + \tfrac{1}{2}B^2) \mathbf{v} - \mathbf{v} \cdot \mathbf{BB} \right] d\tau$$

$$= -\nabla \cdot \left\{ \left[(p + \tfrac{1}{2}B^2) \mathbf{I} - \mathbf{BB} \right] \cdot \mathbf{v} \right\} d\tau \neq 0 . \qquad (4.93)$$

Hence, the energy of a moving fluid element changes through the work performed by the total pressure (isotropic as well as anisotropic) on the element. So far, in the Lagrangian (co-moving) picture, we have only obtained the triviality of local mass conservation. Momentum and energy are not conserved.

There is one local Lagrangian conservation law though which is non-trivial and truly important. This does not follow from the motion of a volume element, but from that of a surface element. Consider *the magnetic flux through a surface element*, $d\Psi \equiv \mathbf{B} \cdot d\boldsymbol{\sigma}$. Departing now from the original evolution equation (4.48) for $\mathbf{B}$, rather than the artificial one (4.74), we have

$$\frac{\partial \mathbf{B}}{\partial t} = \nabla \times (\mathbf{v} \times \mathbf{B}) \stackrel{(A.13)}{=} \mathbf{B} \cdot \nabla \mathbf{v} - \mathbf{B} \nabla \cdot \mathbf{v} - \mathbf{v} \cdot \nabla \mathbf{B} , \qquad (4.94)$$

so that

$$\frac{D\mathbf{B}}{Dt} \equiv \frac{\partial \mathbf{B}}{\partial t} + \mathbf{v} \cdot \nabla \mathbf{B} = \mathbf{B} \cdot \nabla \mathbf{v} - \mathbf{B} \nabla \cdot \mathbf{v} . \qquad (4.95)$$

From this dynamic equation for $\mathbf{B}$ and the kinematic relation (4.89) for the surface element, we obtain:

$$\frac{D}{Dt}(d\Psi) = \frac{D}{Dt}(\mathbf{B} \cdot d\boldsymbol{\sigma}) = \frac{D\mathbf{B}}{Dt} \cdot d\boldsymbol{\sigma} + \mathbf{B} \cdot \frac{D}{Dt}(d\boldsymbol{\sigma})$$

$$= (\mathbf{B} \cdot \nabla \mathbf{v} - \mathbf{B} \nabla \cdot \mathbf{v}) \cdot d\boldsymbol{\sigma} + \mathbf{B} \cdot \left(-(\nabla \mathbf{v}) \cdot d\boldsymbol{\sigma} + \nabla \cdot \mathbf{v} d\boldsymbol{\sigma} \right) = 0 . \quad (4.96)$$

Hence, *the magnetic flux through a co-moving surface element is constant*. But, since this holds for any surface element, *the flux through any surface bounded by a contour C moving with the fluid is conserved*:

$$\Psi = \int_C \mathbf{B} \cdot \mathbf{n}\, d\sigma = \text{const}. \tag{4.97}$$

This completes the discussion started in Sections 4.2.1 and 4.2.2, and centred around the magnetic field equation (4.36) and the induction equation (4.38). To summarize:

(1) the magnetic flux Ψ of an arbitrary flux tube is a well-defined quantity;
(2) this flux remains constant as the flux tube moves.

Since we may shrink the cross-section of the flux tube to an arbitrarily small size, the dynamics of single magnetic field lines now also comes into view. However, in this limit the magnetic flux vanishes so that we have to divide by another quantity, that also vanishes in this limit, to get a finite result. For this we may take the mass of a segment of the flux tube. To that end, combine the induction equation (4.95) in Lagrangian form with the mass conservation equation in Lagrangian form,

$$\frac{D\rho}{Dt} = -\rho \nabla \cdot \mathbf{v}, \tag{4.98}$$

so that

$$\frac{D}{Dt}\left(\frac{\mathbf{B}}{\rho}\right) = \frac{1}{\rho}(\mathbf{B} \cdot \nabla \mathbf{v} - \mathbf{B}\nabla \cdot \mathbf{v}) + \frac{\mathbf{B}}{\rho}\nabla \cdot \mathbf{v} = \left(\frac{\mathbf{B}}{\rho}\right) \cdot \nabla \mathbf{v}. \tag{4.99}$$

Compare this expression with the kinematic relation (4.88) and note that a line element $dl \parallel \mathbf{B}$ moves in exactly the same fashion as the quantity $\mathbf{B}/\rho$. Plasma on this line element and magnetic field line move together. In the words of Alfvén (1950) [6], p. 82: *the lines of force are thus 'frozen' in the body*. Indeed, in ideal MHD (perfect conductivity!), the concept of magnetic field lines obtains more physical reality than it even had in Faraday's times.

Thus, field lines and magnetic flux moving with the plasma manifest the principal conservation property of plasmas, viz. *conservation of magnetic flux*.

4.3.4 Magnetic helicity

After the discussion of the local and global conservation properties of the basic MHD equations, the exposition of conservation laws appears to be complete. However, even though the four partial differential equations for ρ, $\mathbf{v}$, p and $\mathbf{B}$ are a complete set expressing these conservation properties, this is not so. Magnetohydrodynamics permits many different approaches which shed light on this fascinating subject, and which all start afresh from the basic equations. In particular, in studies of the dynamo mechanism for magnetic field generation, the development of MHD turbulence, and the resistive reconnection of magnetic field lines, the discussion of the *magnetic topology* properties of global plasma configurations is greatly facilitated by a representation of the magnetic field in terms of the *vector potential*.

Let us, therefore, concentrate on the magnetic field equations, taking the determination of the density, velocity, and pressure from the first three MHD equations for granted (assuming some magic black box taking care of those variables). We decompose the induction equation (4.15)(a) again in Faraday's law and the ideal MHD version of Ohm's law and, of course, take into account that **B** should be divergence free:

$$\frac{\partial \mathbf{B}}{\partial t} = -\nabla \times \mathbf{E}, \qquad \mathbf{E} = -\mathbf{v} \times \mathbf{B}, \qquad \nabla \cdot \mathbf{B} = 0. \tag{4.100}$$

The latter constraint is also called the *solenoidal condition* for the magnetic field. It may be satisfied once for all by means of the vector potential **A**:

$$\mathbf{B} = \nabla \times \mathbf{A}. \tag{4.101}$$

Ohm's law then becomes

$$\mathbf{E} = -\mathbf{v} \times (\nabla \times \mathbf{A}), \tag{4.102}$$

and Faraday's law may be integrated once to provide the induction equation in terms of **A**:

$$\frac{\partial \mathbf{A}}{\partial t} = \mathbf{v} \times (\nabla \times \mathbf{A}) - \nabla \Phi, \tag{4.103}$$

where Φ is a scalar potential. Since the equations (4.101)–(4.103) are invariant under a gauge transformation $\mathbf{A} \rightarrow \mathbf{A} + \nabla \chi$, $\Phi \rightarrow \Phi - \partial \chi / \partial t$, with arbitrary scalar function χ, there is no loss in generality if we choose $\Phi = 0$ as a gauge condition; see Jackson [117].

Next, we introduce the *magnetic helicity*

$$K(V) \equiv \int_V \mathbf{A} \cdot \mathbf{B} \, d\tau, \tag{4.104}$$

where the integration is over the volume V of some flux tube, as introduced in Section 4.2.1. The Lagrangian rate of change of this quantity is easily determined

4.3 Conservation laws

from the expressions we have derived:

$$\frac{DK}{Dt} = \int \left[\frac{D\mathbf{A}}{Dt} \cdot \mathbf{B} + \mathbf{A} \cdot \frac{D\mathbf{B}}{Dt} \right] d\tau + \int \mathbf{A} \cdot \mathbf{B} \frac{D}{Dt}(d\tau)$$

$$\stackrel{(4.90)}{=} \int \left[\frac{\partial \mathbf{A}}{\partial t} \cdot \mathbf{B} + \mathbf{v} \cdot (\nabla \mathbf{A}) \cdot \mathbf{B} + \mathbf{A} \cdot \frac{\partial \mathbf{B}}{\partial t} + \mathbf{v} \cdot (\nabla \mathbf{B}) \cdot \mathbf{A} + \mathbf{A} \cdot \mathbf{B} \nabla \cdot \mathbf{v} \right] d\tau$$

$$\stackrel{(A.6)}{=} \int \left[(\mathbf{v} \times \mathbf{B}) \cdot \mathbf{B} + \mathbf{A} \cdot \nabla \times (\mathbf{v} \times \mathbf{B}) + \nabla \cdot (\mathbf{A} \cdot \mathbf{B} \mathbf{v}) \right] d\tau$$

$$\stackrel{(A.12)}{=} \int \nabla \cdot \left[(\mathbf{v} \times \mathbf{B}) \times \mathbf{A} + \mathbf{A} \cdot \mathbf{B} \mathbf{v} \right] d\tau$$

$$\stackrel{(A.14)}{=} \int \left[\mathbf{A} \cdot \mathbf{v} \mathbf{B} - \mathbf{A} \cdot \mathbf{B} \mathbf{v} + \mathbf{A} \cdot \mathbf{B} \mathbf{v} \right] \cdot \mathbf{n} \, d\sigma = 0, \quad (4.105)$$

because $\mathbf{B} \cdot \mathbf{n} = 0$ on the boundary of the flux tube. Hence, as first shown by Woltjer [247], *the magnetic helicity of any flux tube is conserved* in ideal MHD. Clearly, since the plasma can be decomposed in infinitely many different ways in flux tubes, from infinitesimally small to globally large, there is an infinity of different magnetic helicities which are all conserved.

To appreciate the subtleties of the magnetic helicity concept, let us compute it for some magnetic field distributions in an infinitely long plasma cylinder with circular cross-section:

$$\mathbf{B} = B_\theta(r) \, \mathbf{e}_\theta + B_z(r) \, \mathbf{e}_z, \quad (4.106)$$

where r, θ, z are cylindrical coordinates. This helical magnetic field may be characterized by the distribution of the inverse pitch $\mu(r)$ of the field lines:

$$\mu(r) \equiv \frac{B_\theta(r)}{r B_z(r)}. \quad (4.107)$$

(For the finite length ($L = 2\pi R_0$) periodic cylinder model of a torus, this quantity is related to the safety factor $q(r)$, defined in Eq. (2.161) of Section 2.4.3, by $q = (\mu R_0)^{-1}$.) For the θ-pinch ($\mu = 0$) and z-pinch ($\mu \to \infty$) examples of Section 2.4.3,

$$B_\theta = 0 \Rightarrow A_z = -\int_0^r B_\theta \, dr = 0 \quad (\theta\text{-pinch}),$$
$$B_z = 0 \Rightarrow A_\theta = \frac{1}{r} \int_0^r r B_z \, dr = 0 \quad (z\text{-pinch}), \quad (4.108)$$

so that the integrand $I \equiv A_\theta B_\theta + A_z B_z$ of the helicity integral (4.104) vanishes, and, hence, the helicity vanishes as well. Apparently, the magnetic field should be

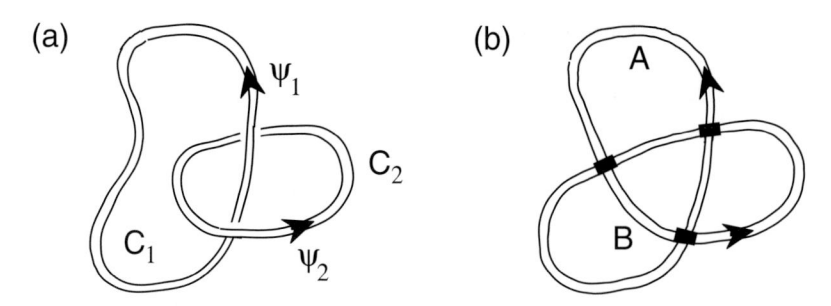

Fig. 4.8. Thin flux tube topologies: (a) two linked magnetic loops; (b) one knotted loop. (Adapted from Moffatt [157].)

at least helical to have $K \neq 0$. However, for a general helical magnetic field in cylindrical geometry, the expression for the integrand may be written as

$$I(r) = \frac{B_\theta}{r} \int_0^r r B_z \, dr - B_z \int_0^r B_\theta \, dr = \mu B_z \int_0^r \frac{B_\theta}{\mu} \, dr - B_z \int_0^r B_\theta \, dr. \tag{4.109}$$

Hence, also for a constant pitch ($\mu = $ const) helical magnetic field in a cylinder, the helicity vanishes: $I(r) = 0 \Rightarrow K(r) = 0$. In other words, magnetic helicity is *not* just the property that field lines are helical. In an infinite cylinder, the field lines should at least have different values of μ at different radii, i.e. they should have *magnetic shear*.

More important than these local considerations are the global magnetic topology implications for the magnetic helicity. This may be demonstrated from the example of *linked magnetic loops* constructed by Moffatt [157] (see Fig. 4.8). Two infinitesimally thin flux tubes C_1 and C_2 with longitudinal magnetic fluxes Ψ_1 and Ψ_2 are linked as shown in Fig. 4.8(a). Outside these loops, $\mathbf{B} = 0$. The helicity of the first loop then becomes

$$K_1 = \int_{V_1} \mathbf{A} \cdot \mathbf{B} \, d\tau = \oint_{C_1} \mathbf{A} \cdot d\mathbf{l} \int_{S_1} \mathbf{B} \cdot \mathbf{n} \, d\sigma = \Psi_1 \oint_{C_1} \mathbf{A} \cdot d\mathbf{l}. \tag{4.110}$$

Since the magnetic field vanishes in the intermediate region, the contour C_1 of the latter line integral over $\mathbf{A}$ may be shrunk to a small contour bounding the cross-section of the second loop, so that

$$\Psi_2 = \int_{S_2} \mathbf{B} \cdot \mathbf{n} \, d\sigma = \int_{S_2} (\nabla \times \mathbf{A}) \cdot \mathbf{n} \, d\sigma \stackrel{(A.18)}{=} \oint_{C_1} \mathbf{A} \cdot d\mathbf{l}. \tag{4.111}$$

A similar argument may be applied to the second loop, so that we obtain the following significant expression for the total helicity:

$$K_1 = K_2 = \Psi_1 \Psi_2 \quad \Rightarrow \quad K = K_1 + K_2 = 2\Psi_1 \Psi_2. \tag{4.112}$$

4.3 Conservation laws

When the tubes wind n times around each other, we get

$$K_1 = K_2 = \pm n \Psi_1 \Psi_2 \quad \Rightarrow \quad K = K_1 + K_2 = \pm 2n \Psi_1 \Psi_2, \qquad (4.113)$$

where the $+$ or $-$ refers to right- or left-handed orientation.

Magnetic flux tubes may also be *knotted*, with non-vanishing helicity in general. A single right-handed trefoil knot is shown in Fig. 4.8(b). Since the magnetic field has opposite directions at the points A and B, pinching the loop by moving these points towards each other produces a linking equivalent to Fig. 4.8(a). Hence, $\Psi_1 = \Psi_2 = \Psi$, so that $K = 2\Psi^2$ for this knot.

Returning now to our infinite cylindrical flux tubes, an example of sufficient complexity in the radial direction is obtained for *force-free magnetic fields*. Such fields, where $\mathbf{j} \parallel \mathbf{B}$ so that the Lorentz force vanishes, are solutions of the differential equation

$$\mathbf{j} = \nabla \times \mathbf{B} = \alpha \mathbf{B}, \qquad (4.114)$$

where α is an arbitrary function of $\mathbf{r}$. A particularly useful example is the cylindrical solution for constant α constructed by Lundquist [149]:

$$B_z = C J_0(\alpha r), \qquad B_\theta = C J_1(\alpha r), \qquad (4.115)$$

where $C \equiv B_z(0)$ is the amplitude of the magnetic field on axis, and J_0 and J_1 are the zeroth and first order Bessel functions of real argument; see Abramowitz and Stegun [1], Chapter 9. Because the Bessel functions oscillate with increasing values of r, the direction of this helical magnetic field constantly changes for the different annularly nested cylinders, which all may be considered as separate flux tubes; see Fig. 4.9(a). Using the differential properties of the Bessel functions,

$$J_1 = -J_0', \qquad x J_0 = (x J_1)', \qquad (4.116)$$

where $x \equiv \alpha r$, the expressions for the vector potential become

$$A_z = -C \int_0^r J_1(\alpha r) \, dr = \frac{C}{\alpha} \left[J_0(\alpha r) - 1 \right],$$

$$A_\theta = \frac{C}{r} \int_0^r r J_0(\alpha r) \, dr = \frac{C}{\alpha} J_1(\alpha r). \qquad (4.117)$$

Hence,

$$\bar{I}(r) \equiv \alpha I(r) \equiv \alpha \left(A_\theta B_\theta + A_z B_z \right) = C^2 \left[J_0^2(\alpha r) + J_1^2(\alpha r) - J_0(\alpha r) \right], \qquad (4.118)$$

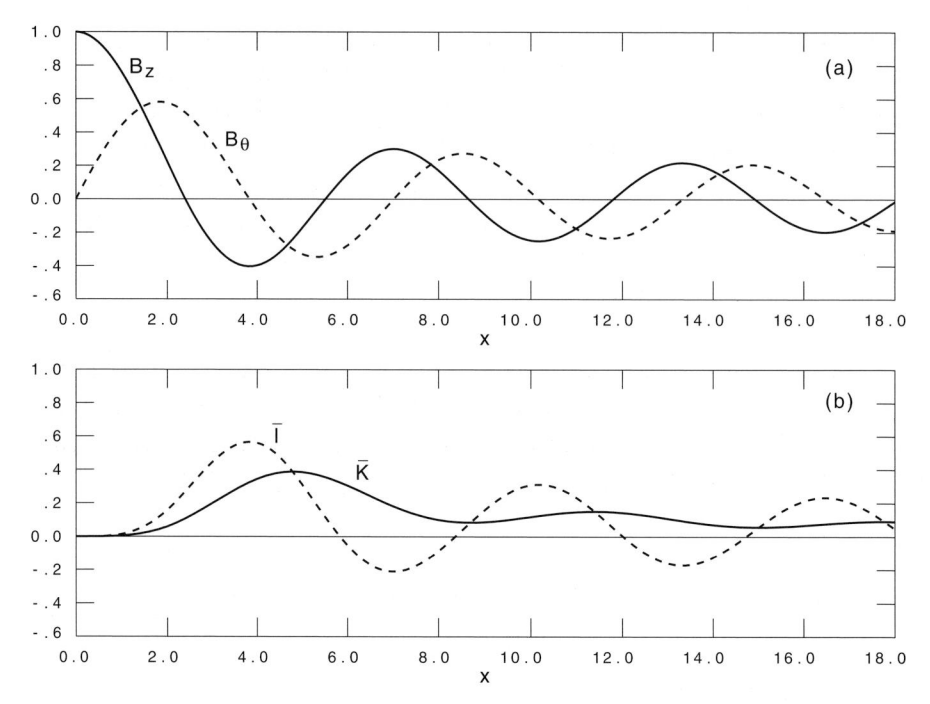

Fig. 4.9. Lundquist force-free magnetic field: (a) components; (b) normalized helicities.

and the renormalized helicity per volume of unit length L of the cylinder becomes

$$\bar{K}(r) \equiv \frac{\alpha}{\pi L r^2} K(r) = \frac{2}{r^2} \int_0^r \bar{I}(r) r \, dr$$

$$= \frac{2C^2}{\alpha^2 r^2} \int_0^{\alpha r} \left[J_0^2(x) + J_1^2(x) - J_0(x) \right] x \, dx,$$

$$= \frac{C^2}{\alpha^2 r^2} \left[\left(\alpha r J_0(\alpha r)\right)^2 + \left(\alpha r J_1(\alpha r) - 1\right)^2 - 1 + 4 \sum_{n=1}^{\infty} 2n J_{2n}^2(\alpha r) \right]. \tag{4.119}$$

The integrals of the Bessel functions are obtained from standard formulas; see Abramowitz and Stegun [1], Chapter 11. The essential point is that, whereas the integrand $I(r)$ oscillates, the integral $K(r)$ remains positive definite (for $\alpha > 0$) or negative definite (for $\alpha < 0$); see Fig. 4.9(b). Hence, the total helicity of a force-free magnetic flux tube of constant α is non-zero, in contrast to the constant pitch magnetic field (which could be force-free as well) considered above. Moreover, since the integrand $I(r)$ oscillates, the helicities of the constituent annular flux tubes have consecutively opposite signs. All these flux tubes are linked in the sense of Fig. 4.8.

Consequently, the dynamics of force-free magnetic fields, conserving all these helicities, is an intricate subject. In particular, considering the finite length periodic cylinder as a model for toroidal configurations, one should enforce the helicity to remain single-valued on the domain (which is multiply-connected in a torus). This may be done by means of a gauge-invariant generalization of the helicity; see Reiman [192]. Similarly, when the solar corona is considered as a dynamic magnetized plasma bounded by the photosphere, one should pay particular attention to the formulation of gauge-invariant boundary conditions at the photosphere; see Berger [24].

For further discussions of the helicity concept: see Moffatt [157] for applications to dynamo theory, see Biskamp [29] for the theory of Taylor [227] on the attainment of minimum energy states by self reversal of magnetic fields, and see Priest and Forbes [191] for applications to coronal magnetic fields.

4.4 Dissipative magnetohydrodynamics

4.4.1 Resistive MHD

Conservation of magnetic flux is directly connected with perfect conductivity, i.e. zero resistivity. Like in ordinary fluid mechanics, ideal fluids do not provide the full story but have to be complemented with a consideration of dissipative effects in *boundary layers* that generally occur, even though the dissipation coefficient may be extremely small; see e.g. Batchelor [17]. This happens because this coefficient multiplies the gradient of a physical quantity that is usually bounded but becomes very large in the boundary layer, producing a finite (and important) effect there. In contrast to ordinary fluid mechanics, in resistive MHD such 'boundary' layers are not associated with the physical boundaries of the system but with internal boundaries beyond which the gradient of the magnetic field (i.e. the current density) becomes very large.

The ideal MHD equations (4.12)–(4.15) for the variables ρ, $\mathbf{v}$, p and $\mathbf{B}$ permit a straightforward generalization to include one form of dissipation, viz. Ohmic dissipation through the plasma resistivity η. The resistive MHD equations[3] were introduced in Section 2.4.1. We recall the main points of the discussion: Ampère's law (4.10),

$$\mathbf{j} = \frac{1}{\mu_0} \nabla \times \mathbf{B}, \qquad (4.120)$$

---

[3] In this section, we temporarily reintroduce the constant μ_0 to establish the dimensional expressions for the dissipative parameters. This requires the replacements $\mathbf{B} \to \mathbf{B}/\sqrt{\mu_0}$, $\mathbf{E} \to \mathbf{E}/\sqrt{\mu_0}$, $\mathbf{j} \to \sqrt{\mu_0}\mathbf{j}$, and $\eta \to \eta/\mu_0$ in the previously obtained dimensionless equations.

and Ohm's law (4.8) in the moving frame, extended with the resistive term,

$$\mathbf{E}' \equiv \mathbf{E} + \mathbf{v} \times \mathbf{B} = \eta \mathbf{j}, \tag{4.121}$$

are substituted in Faraday's equation (4.1) and combined with the classical fluid equations. This gives the following set of evolution equations for resistive MHD, expressing

– *conservation of mass*:

$$\frac{\partial \rho}{\partial t} = -\nabla \cdot (\rho \mathbf{v}), \tag{4.122}$$

– *conservation of momentum*:

$$\rho \left(\frac{\partial}{\partial t} + \mathbf{v} \cdot \nabla \right) \mathbf{v} = -\nabla p + \rho \mathbf{g} + \frac{1}{\mu_0} (\nabla \times \mathbf{B}) \times \mathbf{B}, \tag{4.123}$$

– *(near) conservation of entropy*:

$$\left(\frac{\partial}{\partial t} + \mathbf{v} \cdot \nabla \right) p = -\gamma p \nabla \cdot \mathbf{v} + (\gamma - 1) \frac{\eta}{\mu_0^2} (\nabla \times \mathbf{B})^2, \tag{4.124}$$

– *(near) conservation of magnetic flux*:

$$\frac{\partial \mathbf{B}}{\partial t} = \nabla \times (\mathbf{v} \times \mathbf{B}) - \frac{1}{\mu_0} \nabla \times (\eta \nabla \times \mathbf{B}), \qquad \nabla \cdot \mathbf{B} = 0. \tag{4.125}$$

We have seen in Section 4.3.1, for the ideal MHD case, that the equations need substantial reworking to bring them in conservation form and to demonstrate the actual conservation of the indicated quantities. The resistive counterpart, i.e. demonstration of non-conservation (or, rather, conservation excepting small dissipative contributions) will be given in Section 4.4.2.

The equations have been written as evolution equations, i.e. in a form that is suitable for numerical integration. In this context, the constraint $\nabla \cdot \mathbf{B} = \mathbf{0}$ on the magnetic field complicates the structure of the evolution problem (4.122)–(4.125) significantly. Since the dynamics of plasmas is associated with an evolving magnetic geometry, its satisfaction is an important issue in computational MHD.

(a) Dimensionless parameters The ideal MHD equations (4.12)–(4.15) are obtained from Eqs. (4.122)–(4.125) by just dropping the terms with η, i.e., the Joule heating term in Eq. (4.124) and the magnetic field dissipation term in Eq. (4.125), so that exact conservation is obtained. This is justified if the *magnetic Reynolds number*, introduced in Section 2.4.1, is large:

$$R_m \equiv \frac{\mu_0 l_0 v_0}{\eta} \gg 1. \tag{4.126}$$

Here, l_0 and v_0 are characteristic length and velocity scales of the plasma flow. In Section 4.1.2, we have introduced another characteristic velocity for the plasma, viz. the Alfvén speed v_A. Using that characteristic velocity, we may define another dimensionless parameter that is large for small resistivity, viz. the *Lundquist number*:

$$L_u \equiv \frac{\mu_0 l_0 v_A}{\eta} \gg 1. \quad (4.127)$$

(In laboratory plasma literature, this quantity is usually indicated by the symbol S. Since we already have another use for this symbol, we prefer the symbol L_u here.) Which of the two dimensionless numbers is to be preferred depends on the resistive problem considered. In turbulence problems, where flow dominates, the magnetic Reynolds number is the more significant one. In resistive instabilities, also present in the absence of flow, the Lundquist number is the more relevant one (Biskamp [29]). Their ratio is determined by a third dimensionless parameter, characteristic for ideal MHD flow problems, viz. the *Alfvén Mach number*:

$$\frac{R_m}{L_u} = \frac{v_0}{v_A} \equiv M_A. \quad (4.128)$$

Note that we have not distinguished v_0 and v_A in the numerical tables of Appendix B. (In other words, $M_A = 1$ has been assumed there.)

For fusion plasmas, the value of the resistivity is usually quite small, typically, $R_m \sim 10^9$. For astrophysical plasmas, the resistivity may be somewhat larger, but then the length scales are very much larger too, so that huge values of the magnetic Reynolds number are obtained, e.g. $R_m \sim 10^{13}$ for the solar corona. It would appear that the approximation (4.126) is an extremely good one so that there is no need for the consideration of dissipative effects. However, heating and reconnection (connected with the non-conservation of magnetic flux) are generally observed in plasmas, so that resistive processes must be operating. The crucial point is that η occurs in the equations in combination with gradients of the magnetic field. Hence, if very small-scale perturbations occur, e.g. due to specific resistive modes (like tearing modes, see below) or turbulence, the associated resistive terms may become sizeable. In addition, turbulence may increase the resistive coefficient itself, which then becomes an *anomalous* transport coefficient. For numerical calculations, all this implies extreme requirements on the necessary spatial resolution.

(b) Tearing and reconnection of magnetic field lines To appreciate the global impact of small-scale resistive effects, consider the generic configuration illustrated in Fig. 4.10(a). A sheet of surface currents pointing into the plane of drawing creates a magnetic field with opposite directions in the upper and lower halves of the

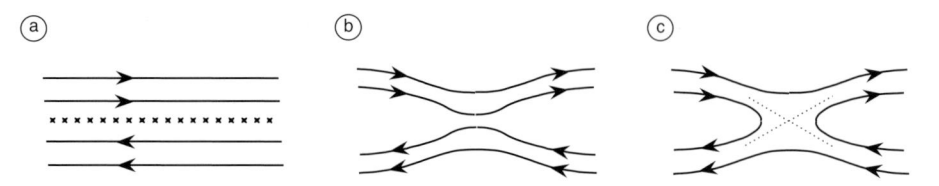

Fig. 4.10. Tearing and reconnection of magnetic field lines: (a) magnetic field of opposite directions created by sheet current (pointing into the plane); (b) ideal MHD perturbation; (c) reconfiguration of the magnetic geometry by a resistive perturbation.

space. (The concept of surface current is precisely defined in Section 4.5.2. For now, it suffices to consider it as a thin plasma layer where the current density becomes very large.) Magnetic configurations of this kind frequently occur, e.g. on the day-side of the magnetosphere, when the solar wind impinges upon it with an embedded *IMF* (interplanetary magnetic field) that may have a direction opposite to that of the magnetosphere itself, and also on the night-side of the magnetosphere, where the solar wind drags the planetary magnetic field lines stretching them out to create the *magnetotail* over a distance of many planetary radii. These situations are inherently dynamic where the two magnetic fields may be pushed together by plasma motions. In ideal MHD perturbations, as shown in Fig. 4.10(b), such dynamics is essentially flux-conserving so that the magnetic topology cannot be changed. In resistive MHD, however, the two parts of the magnetic structure may be reconfigured to form the structure shown in Fig. 4.10(c): the magnetic field lines have been broken and rejoined to form an entirely new magnetic topology with an x-point separatrix (dotted line).

The driving force leading to reconnection is the fact that the magnetic configuration depicted in Fig. 4.10(c) represents a lower energy state than that of Fig. 4.10(a). The ideal MHD perturbation of Fig. 4.10(b) is not able to create that state because of flux conservation. Hence, the important remaining question to be answered is on what time scale the constraint of flux conservation can be broken so that reconnection can take place. In the absence of current concentration, this time scale is just determined by resistive diffusion, which is extremely slow, as we have seen in Sections 2.4 and 3.3.3. However, when external forcing creates current sheets this time scale apparently becomes short enough to permit all the violent plasma phenomena observed in astrophysical plasmas. On the other hand, in laboratory fusion research efforts have been successful to avoid such disruptions.

We will return to resistive plasma dynamics in the companion Volume 2 on *Advanced Magnetohydrodynamics*, when we have obtained sufficient preparation from ideal and one-dimensional analysis. The subject of reconnection is extensively treated in the monographs by Biskamp [30] and by Priest and Forbes [191].

4.4.2 (Non-)conservation form of the dissipative equations★

It is instructive to consider how the introduction of resistivity and other dissipative effects spoils the conservation form of the MHD equations. This section is put in small print since it may be skipped on first reading of Chapter 4.

▷ **(a) Resistive effects** Introducing resistivity, the original equations (4.45) for ρ and (4.46) for $\mathbf{v}$ are unchanged, but Eq. (4.47) for the internal energy e is modified by the Ohmic dissipation term:

$$\frac{\partial e}{\partial t} + \mathbf{v} \cdot \nabla e + (\gamma - 1)e \nabla \cdot \mathbf{v} = \frac{1}{\rho} \eta j^2, \qquad e \equiv \frac{1}{\gamma - 1} \frac{p}{\rho}, \qquad (4.129)$$

whereas Eq. (4.48) for the magnetic field $\mathbf{B}$ is changed by the modification of the electric field in Ohm's law:

$$\frac{\partial \mathbf{B}}{\partial t} + \nabla \times \mathbf{E} = 0, \qquad \mathbf{E} = -\mathbf{v} \times \mathbf{B} + \eta \mathbf{j}, \qquad \nabla \cdot \mathbf{B} = 0. \qquad (4.130)$$

Consequently, the conservation equations (4.59) for ρ and (4.60) for $\rho \mathbf{v}$ remain unchanged, but the energy conservation (4.61) and the flux conservation equation (4.62) have to be modified.

In the derivation of the energy conservation equation, the contribution (4.56) remains the same, but the contributions (4.57) and (4.58) are to be replaced by

$$\frac{\partial}{\partial t}(\rho e) + \nabla \cdot (\rho e \mathbf{v}) + p \nabla \cdot \mathbf{v} = \eta j^2, \qquad (4.131)$$

$$\frac{\partial}{\partial t}\left(\frac{B^2}{2\mu_0}\right) + \frac{1}{\mu_0} \nabla \cdot \left[\mathbf{B} \times (\mathbf{v} \times \mathbf{B})\right] + \mathbf{v} \cdot \mathbf{j} \times \mathbf{B} = -\frac{1}{\mu_0} \mathbf{B} \cdot \nabla \times (\eta \mathbf{j})$$

$$\stackrel{(A.12)}{=} -\frac{1}{\mu_0} \nabla \cdot (\eta \mathbf{j} \times \mathbf{B}) - \eta j^2. \qquad (4.132)$$

Adding the contributions (4.56), (4.131), and (4.132) yields the resistive version of the energy conservation equation:

$$\frac{\partial}{\partial t}\left(\tfrac{1}{2}\rho v^2 + \rho e + \frac{B^2}{2\mu_0}\right) + \nabla \cdot \left[\left(\tfrac{1}{2}\rho v^2 + \rho e + p\right)\mathbf{v}\right.$$

$$\left. + \frac{1}{\mu_0}\left(-\mathbf{v} \times \mathbf{B} + \eta \mathbf{j}\right) \times \mathbf{B}\right] = -\rho \mathbf{v} \cdot \nabla \Phi. \qquad (4.133)$$

Clearly, this equation remains an energy conservation equation: magnetic energy may be converted into internal energy, but the sum is constant. Notice that resistivity here enters in the divergence term as $\mu_0^{-1} \mathbf{E} \times \mathbf{B}$, which is the Poynting flux discussed in Section 4.3.2.

In contrast, the magnetic flux equation becomes essentially non-conservative, due to the magnetic diffusivity coefficient η/μ_0 (with dimension $[\eta/\mu_0] = $ m^2 s^{-1}) on the RHS:

$$\frac{\partial \mathbf{B}}{\partial t} + \nabla \cdot (\mathbf{v}\mathbf{B} - \mathbf{B}\mathbf{v}) = -\nabla \times \left(\frac{\eta}{\mu_0} \nabla \times \mathbf{B}\right) \stackrel{(A.7),(A.5)}{=} \frac{\eta}{\mu_0} \nabla^2 \mathbf{B} + \mathbf{j} \times \nabla \eta, \qquad \nabla \cdot \mathbf{B} = 0. \qquad (4.134)$$

Finally, the entropy conservation equations (4.64) and (4.65) change into equations that clearly exhibit non-conservation due to Ohmic dissipation:

$$\frac{DS}{Dt} \equiv \frac{\partial S}{\partial t} + \mathbf{v} \cdot \nabla S = (\gamma - 1)\rho^{-\gamma} \eta j^2,$$

$$\frac{\partial}{\partial t}(\rho S) + \nabla \cdot (\rho S \mathbf{v}) = (\gamma - 1)\rho^{-\gamma+1} \eta j^2. \tag{4.135}$$

The equations (4.59), (4.60), (4.133), (4.134) constitute the complete set of (non-)conservation equations for resistive MHD, whereas the equations (4.135) are a mere consequence. ◁

▷ **(b) Other dissipative effects** It is also instructive to generalize the equation of motion and the internal energy equation with contributions of viscosity, heat conduction, radiative losses, etc., to indicate how the MHD equations are related to the ordinary hydrodynamics equations (like the Navier–Stokes equation); see e.g. Roberts [194]. For the sake of that argument, we ignore the tremendous complexity of the plasma transport coefficients in the presence of a magnetic field, as summarized in Section 3.3.2, and just exploit scalar one-fluid transport coefficients.

The inclusion of viscous effects turns Eq. (4.123) into an equation expressing *(near) conservation of momentum*:

$$\rho\left(\frac{\partial}{\partial t} + \mathbf{v} \cdot \nabla\right)\mathbf{v} = -\nabla p + \rho \mathbf{g} + \frac{1}{\mu_0}(\nabla \times \mathbf{B}) \times \mathbf{B} + \mathbf{F}_{\text{visc}}, \tag{4.136}$$

$$\mathbf{F}_{\text{visc}} \approx \rho \nu (\nabla^2 \mathbf{v} + \tfrac{1}{3}\nabla\nabla \cdot \mathbf{v}), \tag{4.137}$$

where $\nu \equiv \mu/\rho$ is the kinematic viscosity coefficient (with dimension $[\nu] = \text{m}^2\,\text{s}^{-1}$).

The internal energy equation (4.19) is modified by the effects of heat generation and heat flow, associated with the different dissipation mechanisms, as follows:

$$\rho \frac{De}{Dt} + (\gamma - 1)\rho e \nabla \cdot \mathbf{v} = -\nabla \cdot \mathbf{h} + Q, \tag{4.138}$$

where $\mathbf{h}$ is the heat flow and Q is the generated heat per unit volume. The heat flow is given by the expression

$$\mathbf{h} \approx -\kappa \nabla(kT) = -\rho \lambda \nabla e, \tag{4.139}$$

where κ is the coefficient of thermal conductivity and $\lambda \equiv k\kappa/(C_v \rho)$ is the coefficient of thermal diffusivity (with dimension $[\lambda] = \text{m}^2\,\text{s}^{-1}$). The total generated heat can be written as the difference between heating proper, H, due to resistivity, viscosity, thermonuclear fusion energy production, etc., and the thermal losses, L, e.g. due to radiation:

$$Q \equiv H - L, \qquad H = H_{\text{res}} + H_{\text{visc}} + H_{\text{fus}} + \cdots, \qquad L = L_{\text{rad}} + \cdots. \tag{4.140}$$

The resistive and viscous heating terms may be estimated as

$$H_{\text{res}} \approx \eta j^2 = \eta \mu_0^{-2} |\nabla \times \mathbf{B}|^2, \qquad H_{\text{visc}} \sim \rho \nu |\nabla \mathbf{v}|^2, \tag{4.141}$$

where we stress once more that the anisotropies of the transport coefficients parallel and perpendicular to the magnetic field have been ignored.

This indicates how the kinematic diffusivity ν governs diffusion of the velocity, the thermal diffusivity λ governs diffusion of the internal energy, and the magnetic diffusivity η/μ_0 governs diffusion of the magnetic field. These effects are completely negligible when the MHD description of plasmas applies, except when small-scale structures are present. ◁

4.5 Discontinuities

4.5.1 Shocks and jump conditions

An important application of the MHD conservation equations (4.59)–(4.62), or (4.71)–(4.74), is the derivation of shock conditions. Here, it is crucial that the energy conservation equation (4.61), rather than the entropy conservation equation (4.65), is exploited because *a shock is an irreversible (entropy-increasing) transition*; see Burgess [46], Courant and Friedrichs [59], or Landau and Lifshitz, *Fluid Mechanics* [137]. In ordinary gas dynamics, this transition is associated with supersonic flow upstream of the shock and subsonic flow downstream. Given the upstream state of the flow, the question there arises: what is the state on the downstream, subsonic, side of the shock? The Rankine–Hugoniot relations, i.e. the shock conditions, provide the answer to this question for ordinary gas dynamics. In that case, the characteristic speed of propagation of disturbances is the sound speed, c, and the so-called Mach number, $M \equiv v/c$, determines whether the flow is supersonic ($M > 1$) or subsonic ($M < 1$). In MHD, there are three characteristic speeds (as we will see in Chapter 5). Consequently, the subject of MHD shocks is much richer than that of gas dynamic shocks. For two-fluid (or multi-fluid) plasmas, the consideration of the separate particle effects of the electrons and ions even leads to more characteristic speeds, corresponding to the different electron and ion wave motions. In the kinetic theory of plasma dynamics, the notion of collisionless shocks occurs, which is important, e.g., for the description of magnetospheric plasmas; see Burgess [46]. Here, we will restrict the analysis to MHD shocks.

At this point, our interest is even more restricted since we have not yet developed the dynamical tools (e.g. the MHD waves to be discussed in Chapter 5) that are necessary to properly analyse the different shocks permitted. Hence, the discussion of genuine shocks is relegated to a later chapter, devoted to transonic MHD flows and shocks, in the companion Volume 2. Here, we will just use the mechanism of shock formation to derive the appropriate *jump conditions for plasmas with an internal boundary*. Our immediate aim is to generalize the boundary conditions (4.39) and (4.41), which are valid for laboratory plasmas that are completely isolated from the outside world by a rigid wall, to boundary conditions describing more compound magnetic confinement structures.

First, consider the one-dimensional flow of gas in which sound waves are excited. Local perturbations travel with the sound speed $c \equiv \sqrt{\gamma p/\rho}$. Their trajectories in the x–t plane, called *characteristics* (see Section 5.4), are two sets of parallel straight lines with derivatives $dx/dt = \pm c$. Suppose now that we suddenly increase the pressure, so that the sound speed increases (Fig. 4.11). In the x–t plane this means that the slopes of the characteristics decrease. Therefore, we may arrive at the situation where the characteristics would cross (Fig. 4.12(a)),

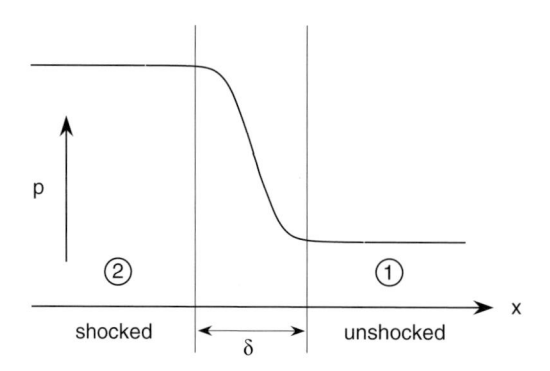

Fig. 4.11. Shock formation.

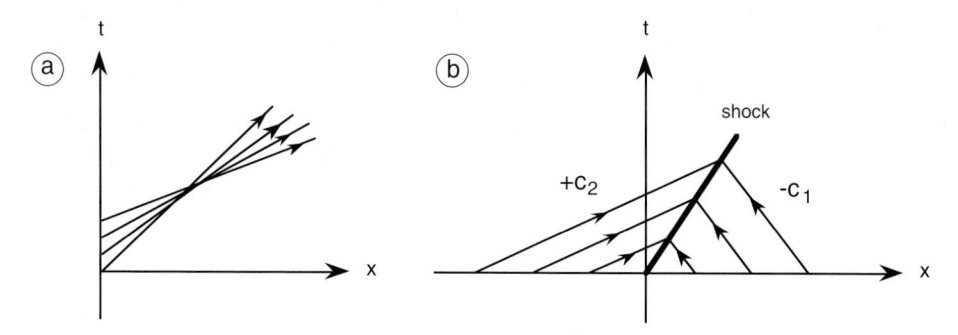

Fig. 4.12. (a) 'Crossing characteristics' due to a sudden change of the background variables. (b) In the ideal model, the characteristics meet at the shock discontinuity.

if some other mechanism did not interfere. The picture suggests what will happen: information originating from different space-time points accumulates. Consequently, gradients in the macroscopic variables build up until the point that the idealized model breaks down and dissipative effects due to the large gradients have to be taken into account. Eventually, a steady state will be reached where nonlinear and dissipative effects counterbalance: *a shock-wave* has been created. Neglecting the thickness δ of the shock, the steady state will consist of two regions with different sound speed, separated by the moving shock front. In the $x-t$ plane, this front is located at the position where the forward characteristics ($+c_2$) of the shocked part meet the backward characteristics ($-c_1$) of the unshocked part (Fig. 4.12(b)).

Without specifying the kind of dissipation, one may arrive at the so-called *shock relations* that relate variables on the two sides of the propagating shock front. The idea is that the ideal model breaks down inside a layer of infinitesimal thickness δ (i.e., a thickness proportional to some power of the dissipation coefficient, which is assumed to be vanishingly small), but it holds on either side of the layer. In the limit

4.5 Discontinuities

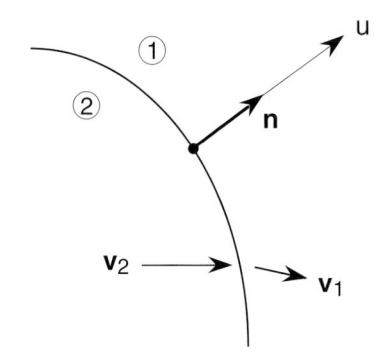

Fig. 4.13. Shock front.

$\delta \to 0$ the variables will jump across the layer, and the magnitude of the jumps is determined from the condition that mass, momentum, energy and magnetic flux should be conserved. Thus, one integrates the conservation equations (4.59)–(4.62) across the shock and keeps the leading order contributions arising from the gradients normal to the shock front only, since these gradients are infinitely large in the limit: $\partial f/\partial l \to \infty$, where f indicates any of the physical variables. Defining the jump in f by

$$[\![f]\!] \equiv f_1 - f_2, \qquad (4.142)$$

these contributions give

$$\lim_{\delta \to 0} \int_1^2 \nabla f \, dl = -\lim_{\delta \to 0} \mathbf{n} \int_1^2 \frac{\partial f}{\partial l} \, dl = \mathbf{n} [\![f]\!], \qquad (4.143)$$

where $\mathbf{n}$ is the normal to the shock front, chosen to point in the direction of the undisturbed fluid ahead of the shock (Fig. 4.13). By convention, the integration across the shock, along l, is chosen in just the opposite direction, viz. from ① (undisturbed fluid) to ② (shocked part of the fluid). The time-derivatives $\partial f/\partial t$ also contribute to the shock conditions, as may be seen by transforming to a frame moving with the normal speed u of the shock front:

$$\left(\frac{Df}{Dt} \right)_{\text{shock}} = \frac{\partial f}{\partial t} - u \frac{\partial f}{\partial l},$$

where $(Df/Dt)_{\text{shock}}$ denotes the rate of change in a frame moving with the shock. Since this quantity remains finite and $\partial f/\partial l \to \infty$, for balance we must have $\partial f/\partial t \to \infty$ as well. Hence,

$$\lim_{\delta \to 0} \int_1^2 \frac{\partial f}{\partial t} \, dl = u \lim_{\delta \to 0} \int_1^2 \frac{\partial f}{\partial l} \, dl = -u [\![f]\!]. \qquad (4.144)$$

In conclusion: from Eqs. (4.143) and (4.144) it follows that the shock relations are obtained by simply making the substitutions

$$\partial f/\partial t \to -u [\![f]\!], \qquad \nabla f \to \mathbf{n} [\![f]\!], \tag{4.145}$$

in the conservation equations.

For the ideal MHD conservation equations (4.59)–(4.62), the substitutions (4.145) result in the following general jump conditions:

$$-u [\![\rho]\!] + \mathbf{n} \cdot [\![\rho \mathbf{v}]\!] = 0, \tag{4.146}$$

$$-u [\![\rho \mathbf{v}]\!] + \mathbf{n} \cdot [\![\rho \mathbf{v}\mathbf{v} + (p + \tfrac{1}{2}B^2)\mathbf{I} - \mathbf{B}\mathbf{B}]\!] = 0, \tag{4.147}$$

$$-u \left[\!\!\left[\tfrac{1}{2}\rho v^2 + \frac{p}{\gamma - 1} + \tfrac{1}{2}B^2 \right]\!\!\right]$$
$$+ \mathbf{n} \cdot \left[\!\!\left[\left(\tfrac{1}{2}\rho v^2 + \frac{\gamma}{\gamma - 1} p + B^2 \right) \mathbf{v} - \mathbf{v} \cdot \mathbf{B}\mathbf{B} \right]\!\!\right] = 0, \tag{4.148}$$

$$-u [\![\mathbf{B}]\!] + \mathbf{n} \cdot [\![\mathbf{v}\mathbf{B} - \mathbf{B}\mathbf{v}]\!] = 0, \qquad \mathbf{n} \cdot [\![\mathbf{B}]\!] = 0, \tag{4.149}$$

where we have eliminated the internal energy variable e in favour of the pressure p. Recall that the original entropy conservation equation had to be replaced by the energy conservation equation since the latter remains valid in the presence of dissipation. In the limit $\delta \to 0$, the dissipative boundary layer contributions vanish and the variables ρ, $\mathbf{v}$, p, and $\mathbf{B}$ may become discontinuous, according to the jump conditions (4.146)–(4.149). However, the second law of thermodynamics demands that the entropy should have increased (or, rather, should not have decreased) when the shock has passed. Hence, Eqs. (4.146)–(4.149) have to be supplemented with the condition that *entropy increases across the shock*:

$$[\![s]\!] \leq 0, \quad \text{or} \quad [\![S]\!] \equiv [\![\rho^{-\gamma} p]\!] \leq 0 \qquad (entropy). \tag{4.150}$$

This miraculous condition is all that remains from the dissipative processes in the limit of infinitesimal thickness of the boundary layer.

For the steady shocks that we will now consider, it is convenient to transform to *the shock frame*, in which the shock is stationary, and the fluid velocities $\mathbf{v}' \equiv \mathbf{v} - u\mathbf{n}$ are evaluated with respect to this frame. The jump conditions (4.146)–(4.149) may then be written as

$$[\![\rho v'_n]\!] = 0 \qquad (mass), \tag{4.151}$$

$$[\![\rho v'^2_n + p + \tfrac{1}{2}B_t^2]\!] = 0 \qquad (normal\ momentum), \tag{4.152}$$

$$\rho v'_n [\![\mathbf{v}'_t]\!] = B_n [\![\mathbf{B}_t]\!] \qquad \text{(tangential momentum)}, \quad (4.153)$$

$$\rho v'_n \left[\!\left[\tfrac{1}{2}\left(v_n'^2 + v_t'^2\right) + \frac{1}{\rho}\left(\frac{\gamma}{\gamma-1}p + B_t^2\right) \right]\!\right] = B_n [\![\mathbf{v}'_t \cdot \mathbf{B}_t]\!] \quad \text{(energy)}, \quad (4.154)$$

$$[\![B_n]\!] = 0 \qquad \text{(normal flux)}, \quad (4.155)$$

$$\rho v'_n \left[\!\left[\frac{\mathbf{B}_t}{\rho} \right]\!\right] = B_n [\![\mathbf{v}'_t]\!] \qquad \text{(tangential flux)}, \quad (4.156)$$

where the momentum equation (4.147) and the magnetic flux equation (4.149) have been projected in the directions normal and tangential to the shock front. Of course, these shock conditions also have to be supplemented with the entropy condition (4.150).

We have now obtained six algebraic equations for the six jumps $[\![\rho]\!]$, $[\![v_n]\!]$, $[\![\mathbf{v}_t]\!]$, $[\![p]\!]$, $[\![B_n]\!]$, $[\![\mathbf{B}_t]\!]$, so that we may compute the values of all variables on the downstream side of the shock if their upstream values are known from the solution of the PDEs themselves. The values of the variables on the downstream boundary thus computed provide the *boundary conditions* to be imposed on the solution of the PDEs in the downstream region. The entropy condition just forbids solutions that do not correspond to an increase of the entropy on the shocked side.

4.5.2 Boundary conditions for plasmas with an interface

The jump conditions (4.150)–(4.156) provide the necessary tools to describe two quite different physical phenomena (see Landau and Lifshitz, *Electrodynamics of Continuous Media* [138]), viz.:

(1) boundary conditions for *moving plasma–plasma interfaces*, where there is *no flow across the discontinuity* ($v'_n = 0$);
(2) jump conditions for *genuine shocks*, where flow across the discontinuity ($v'_n \neq 0$) is an essential feature.

We will not enter the discussion of the second class of discontinuous phenomena at this point, except to mention that our definition of shocks is chosen such that it contains rotational discontinuities (usually not considered to be proper shocks) as well as magneto-sonic shocks. Their dynamics will be treated in a later chapter, in Volume 2.

For the first class of discontinuous phenomena (co-moving interfaces, $v'_n = 0$), the jump conditions reduce to:

$$[\![p + \tfrac{1}{2}B_t^2]\!] = 0 \qquad \text{(normal momentum)}, \quad (4.157)$$

$$B_n [\![\mathbf{B}_t]\!] = 0 \qquad \text{(tangential momentum)}, \qquad (4.158)$$

$$B_n [\![\mathbf{v}'_t \cdot \mathbf{B}_t]\!] = 0 \qquad \text{(energy)}, \qquad (4.159)$$

$$[\![B_n]\!] = 0 \qquad \text{(normal flux)}, \qquad (4.160)$$

$$B_n [\![\mathbf{v}'_t]\!] = 0 \qquad \text{(tangential flux)}. \qquad (4.161)$$

This permits two distinct possibilities for jumps, viz.

(a) *contact discontinuities*, if the magnetic field intersects the interface ($B_n \neq 0$), where the variables are alternatively

– jumping: $[\![\rho]\!] \neq 0$,
– continuous: $v'_n = 0$, $[\![\mathbf{v}'_t]\!] = 0$, $[\![p]\!] = 0$, $[\![B_n]\!] = 0$, $[\![\mathbf{B}_t]\!] = 0$;
$$\qquad (4.162)$$

(b) *tangential discontinuities*, if the magnetic field is parallel to the interface ($B_n = 0$), where the variables are alternatively

– jumping: $[\![\rho]\!] \neq 0$, $[\![\mathbf{v}'_t]\!] \neq 0$, $[\![p]\!] \neq 0$, $[\![\mathbf{B}_t]\!] \neq 0$,
– continuous: $v'_n = 0$, $B_n = 0$, $[\![p + \tfrac{1}{2} B_t^2]\!] = 0$. $\qquad (4.163)$

Note that the latter discontinuities are not a special case of the former ones.

On the basis of these two kinds of discontinuity, we may distinguish two types of magnetic configuration with an interface, viz. *astrophysical plasmas* where the magnetic fields typically originate in a planet, star, or other rotating object, with a dynamo operating inside, but intersect the surface (in the case of a star, the photosphere) where the plasma density may jump to the much lower values pertinent for a corona. Such jumps may be characterized as contact discontinuities. All variables should be continuous there, except for the density ρ (or the temperature T, or the entropy S). In *laboratory plasmas* aimed at thermonuclear energy production, on the other hand, the discontinuities of interest really serve to confine a high density plasma by a lower density one, that may even effectively qualify as a vacuum, in order to isolate it thermally from an outer wall. Such jumps are typically tangential discontinuities at a magnetic surface that is nested within the other ones. From Eq. (4.163) it is clear that these discontinuities permit much more freedom in the choice of the values of the variables. Except for the density, also the tangential velocity, the pressure, and the tangential magnetic field may jump, as long as the magnetic field stays tangential and the total pressure remains balanced.

It should be stressed that this classification is not at all strictly one-to-one corresponding with that of astrophysical and laboratory plasmas. In many astrophysical plasmas of interest (e.g. at magnetospheric boundaries) tangential discontinuities occur, whereas in laboratory plasmas magnetic fields intersecting the boundary have become an important issue (e.g. in divertor plasmas). We will treat the two kinds of discontinuity, and their associated physical models, in reverse order since tangential discontinuities (laboratory plasmas) admit a much cleaner picture of confinement (Section 4.6.1). In this respect, the apparent simplicity of contact discontinuities is deceiving because these discontinuities separate regions with widely different physical properties so that the formulation of realistic boundary conditions usually becomes a much more involved problem. In Section 4.6.3, we will just sketch the kind of astrophysical models possible.

We finish our exposition of discontinuities with the formulation of the boundary conditions for plasma–plasma interfaces in the laboratory frame. In that frame, $v_n = v'_n + u$. Eliminating the normal speed u of the interface, the interface discontinuities are characterized by $[\![v_n]\!] = 0$. The *boundary conditions for a tangential plasma–plasma interface* then become:

$$\mathbf{n} \cdot \mathbf{B} = 0 \qquad \text{(at the interface)}, \qquad (4.164)$$

$$\mathbf{n} \cdot [\![\mathbf{v}]\!] = 0 \qquad \text{(at the interface)}, \qquad (4.165)$$

$$[\![p + \tfrac{1}{2} B^2]\!] = 0 \qquad \text{(at the interface)}. \qquad (4.166)$$

The jump of the tangential magnetic field implies that there should be a *surface, or skin, current* flowing at the plasma–plasma interface. Such a current is obtained in the limit of a surface layer of thickness δ with large current density $\mathbf{j}$ when the limits $\delta \to 0$ and $|\mathbf{j}| \to \infty$ are taken in such a way that $\mathbf{j}^\star \equiv \lim_{\delta \to 0, \, |\mathbf{j}| \to \infty} (\delta \mathbf{j})$ remains finite. Note that the dimension of $\mathbf{j}^\star$ is that of current density times length, i.e. current per unit length. Application of our jump recipe (4.145) to Ampère's law, $\mathbf{j} = \nabla \times \mathbf{B}$, provides an expression for the magnitude of this surface current density:

$$\mathbf{j}^\star = \mathbf{n} \times [\![\mathbf{B}]\!]. \qquad (4.167)$$

When the plasma is rotating, so that there is a finite vorticity $\boldsymbol{\omega} \equiv \nabla \times \mathbf{v}$, there may be a *surface vorticity* as well.

$$\boldsymbol{\omega}^\star = \mathbf{n} \times [\![\mathbf{v}]\!]. \qquad (4.168)$$

For the present purpose, these expressions (representing singularities of the current density and the vorticity) are to be considered as mere consequences of the application of the interface b.c.s (4.164)–(4.166).

4.6 Model problems

We are now in a position to formulate the proper boundary conditions to distinguish two broad classes of magnetic confined configurations, of which the tokamak and the coronal loop shown in Fig. 4.2 are a particular example. We will identify six models, consisting of the MHD equations (4.12)–(4.15) + specification of a particular magnetic geometry with associated b.c.s. Models I–III refer to laboratory plasmas with tangential discontinuities, and models IV–VI refer to astrophysical plasmas with contact discontinuities.

4.6.1 Laboratory plasmas (models I–III)

The three models for confined laboratory plasmas are shown in Fig. 4.14. These configurations refer to toroidally symmetric tokamaks, i.e. the toroidal angle is an ignorable coordinate ($\partial/\partial\varphi = 0$) for all physical variables. Such problems involve the solution of partial differential equations (PDEs) in the two spatial coordinates associated with the poloidal cross-section (shaded in the upper part of Fig. 4.14). Therefore, they are called two-dimensional (2D) problems. The solution of 2D PDEs requires numerical analysis, which has been carried out extensively

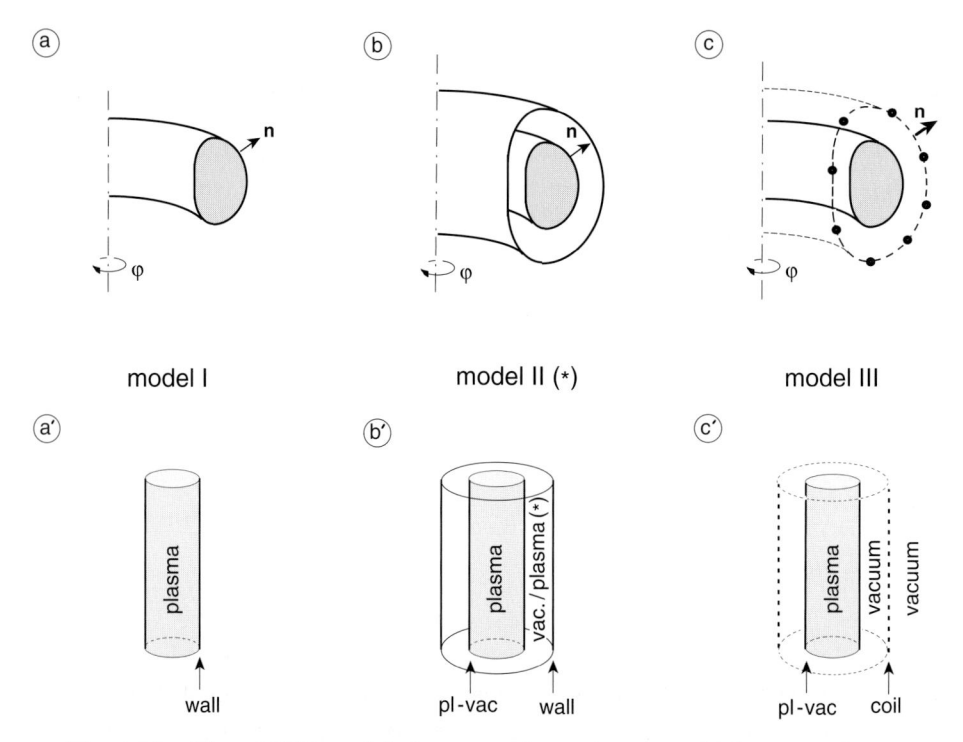

Fig. 4.14. Three MHD models for magnetic confinement of laboratory plasma (tokamak). (a) Model I: plasma surrounded by a wall; (b) model II (*): plasma isolated from the wall by a vacuum (*: or another plasma); (c) model III: plasma excited by currents in external coils. Bottom part: 1D (cylindrical) versions of the three models. Surfaces where boundary conditions have to be imposed are indicated by arrows.

for tokamaks. One step more complicated is the stellarator, also a toroidal plasma confinement system but axi-symmetry is lost there by the introduction of asymmetric external coils and cross-sectional shaping in order to eliminate the toroidal plasma current (which is essential for tokamak confinement, but associated with dissipative decay and current-driven instabilities). Such configurations are called 3D. They require even more sophisticated numerical analysis.

However, in order to build up physical understanding and to develop mathematical techniques, the approximation of the toroidal geometry by that of an infinite or 'periodic' cylinder (bottom part of Fig. 4.14) is extremely useful. Since a circular cylinder has two directions of symmetry, the resulting equations will be non-trivial in only one of the three spatial coordinates (the radial one) and reduce to ordinary differential equations (ODEs) in that coordinate. For this reason, this geometry is considered to be one-dimensional (1D). The important point for the present discussion is that this reduction does not change the b.c.s discussed, but merely facilitates the solution of the differential equations. We will extensively exploit this simplification.

(a) Model I: plasma confined inside rigid wall In this model (Fig. 4.14(a)), *the plasma is closed off from the outside world by a perfectly conducting wall*. This is appropriate for the study of equilibrium, waves, and instabilities of confined plasmas in closed vessels, as used in thermonuclear research. We already considered the appropriate boundary conditions for this model in Section 4.2.2. At the wall, both the normal magnetic field and the normal velocity have to vanish:

$$\mathbf{n} \cdot \mathbf{B} = 0 \quad \text{(at the wall)}, \qquad (4.169)$$

$$\mathbf{n} \cdot \mathbf{v} = 0 \quad \text{(at the wall)}. \qquad (4.170)$$

Recall from Section 4.3.2 that these b.c.s are sufficient to guarantee that the system conserves all important physical quantities (mass, momentum, energy, magnetic flux). Hence, this model is the simplest, and most relevant, one to describe confined plasmas.

Amazingly, only two b.c.s need to be satisfied for eight variables. This is due to the intrinsic anisotropy introduced in the system by the magnetic field. Of course, the model is a restriction to facilitate analysis of confined plasmas. In reality, one may encounter $\mathbf{n} \cdot \mathbf{v} \neq 0$, e.g. when plasma is injected or ejected, but this is a complication to be introduced only when the more basic dynamics of this model is fully understood.

(b) Model II: plasma–vacuum system confined inside rigid wall In this model (Fig. 4.14(b)), *the plasma is confined inside a rigid wall and isolated from it by a region of low enough density to be treated as a 'vacuum'*. This model again

describes confined plasmas in a closed vessel, but separated from the wall by a vacuum region. The dynamics of the vacuum field variables $\hat{\mathbf{E}}$ and $\hat{\mathbf{B}}$ should correspond to the plasma dynamics described by the non-relativistic MHD equations (4.12)–(4.15), where time and length scales were assumed to satisfy $|\partial/\partial t|/|\nabla| \sim v \ll c$. This implies neglect of the displacement current also in Maxwell's equations for the vacuum:

$$\nabla \times \hat{\mathbf{B}} = 0, \qquad \nabla \cdot \hat{\mathbf{B}} = 0, \qquad (4.171)$$

$$\nabla \times \hat{\mathbf{E}} = -\frac{\partial \hat{\mathbf{B}}}{\partial t}, \qquad \nabla \cdot \hat{\mathbf{E}} = 0. \qquad (4.172)$$

Actually, this again degrades the electric field to a secondary variable that may be computed from the magnetic field $\hat{\mathbf{B}}$. Hence, in the vacuum only one basic variable is needed, viz. $\hat{\mathbf{B}}$, satisfying Eq. (4.171) and the boundary condition

$$\mathbf{n} \cdot \hat{\mathbf{B}} = 0 \qquad \textit{(at the conducting wall)}. \qquad (4.173)$$

This boundary condition is consistent with the assumption of vanishing tangential electric field,

$$\mathbf{n} \times \hat{\mathbf{E}} = 0 \qquad \textit{(at the conducting wall)}, \qquad (4.174)$$

since this also implies that $\mathbf{n} \cdot (\nabla \times \hat{\mathbf{E}}) = 0$ there, so that Eq. (4.172) then yields the boundary condition (4.173) on the normal magnetic field.

Of course, we also need b.c.s connecting the plasma variables with the vacuum magnetic field across the plasma–vacuum interface. In order to establish those, we consider the closely related *Model II*: plasma–plasma system confined inside rigid wall*. This model is also illustrated by Fig. 4.14(b) and indicated by the asterisk. Now, instead of a vacuum with $\hat{\mathbf{B}}$ satisfying Eq. (4.171), we have another plasma with variables that satisfy the ideal MHD equations. These variables are subject to the b.c.s (4.169) and (4.170) at the wall and the plasma–plasma interface b.c.s (4.164)–(4.166) derived in Section 4.5.2. Hence, this model is complete as well. Its significance is twofold: (1) it provides a useful alternative for the description of the outer region of tokamaks; (2) it is more widely applicable to astrophysical plasmas, e.g. a solar coronal magnetic loop where an external plasma may interact with the inner one to excite waves or to provide confinement. (In that case, confinement by the outer wall is tacitly dropped and replaced by some condition on behaviour at infinity.)

We now simply extrapolate model II* to obtain the b.c.s for model II. If the outer plasma is replaced by a vacuum, the b.c. (4.165) is no longer needed since the plasma velocity just determines the velocity of the plasma–vacuum interface

4.6 Model problems

and nothing else. Hence, the b.c.s for model II proper become:

$$\mathbf{n} \cdot \mathbf{B} = \mathbf{n} \cdot \hat{\mathbf{B}} = 0 \quad \text{(at the plasma–vacuum interface)}, \quad (4.175)$$

$$[\![p + \tfrac{1}{2} B^2]\!] = 0 \quad \text{(at the plasma–vacuum interface)}. \quad (4.176)$$

These boundary conditions are quite reasonable. If a normal magnetic field component were sticking through the plasma–vacuum boundary, i.e. if Eq. (4.175) did not hold, the vacuum region simply could not exist since plasma would freely flow along the magnetic field lines into the vacuum region. Next, if there were no balance of the total pressure, i.e. if Eq. (4.176) were not valid, the plasma–vacuum interface would simply be blown apart by the huge pressure imbalance. Finally, recall from the discussion in Section 4.5.2 that this pressure balance condition permits jumps in the pressure and in the two tangential components of the magnetic field, associated with a surface current density

$$\mathbf{j}^\star = \mathbf{n} \times [\![\mathbf{B}]\!] \quad \text{(at the plasma–vacuum interface)}. \quad (4.177)$$

This is not a separate b.c. but just a consequence of the b.c.s (4.175) and (4.176).

For later applications, it is of interest to also derive the boundary condition on the electric field at the plasma–vacuum interface. In Section 4.5, we derived the jump conditions starting from the MHD equations (4.12)–(4.15), where the electric field was already eliminated. Applying the substitutions (4.145) to Faraday's law (4.1) before this elimination, with the normal interface velocity determined by the plasma velocity, $u = \mathbf{n} \cdot \mathbf{v}$, we obtain the following jump condition for the electric field at model II interfaces: $\mathbf{n} \times [\![\mathbf{E}]\!] = \mathbf{n} \cdot \mathbf{v} [\![\mathbf{B}]\!]$. On the plasma side of the interface, Ohm's law (4.8) yields $\mathbf{n} \times \mathbf{E} = -\mathbf{n} \times (\mathbf{v} \times \mathbf{B}) = \mathbf{n} \cdot \mathbf{v} \mathbf{B}$. Hence, we obtain a basic relationship between the vacuum field variables $\hat{\mathbf{E}}$ and $\hat{\mathbf{B}}$ and the normal plasma velocity:

$$\mathbf{n} \times \hat{\mathbf{E}} = \mathbf{n} \cdot \mathbf{v} \hat{\mathbf{B}} \quad \text{(at the plasma–vacuum interface)}. \quad (4.178)$$

This b.c. is actually redundant, just like the b.c. (4.174) at the conducting wall.

(c) Model III: Plasma–vacuum system excited by external currents In this model (Fig. 4.14(c)), instead of a wall we consider *an open plasma–vacuum system excited by time-dependent magnetic fields $\hat{\mathbf{B}}(t)$ that are externally created*. In laboratory plasmas, this external excitation may be caused by a system of coils. Such a system may be modelled by replacing the wall of model II by an auxiliary surface on which a time-dependent surface current $\mathbf{j}_c^\star(\mathbf{r}, t)$ forces oscillations onto the plasma–vacuum system. The effect of such an outer boundary is that the system is not isolated from the outside world: energy flows into the system. The appropriate

boundary conditions at the coil surface are again obtained by applying our jump recipe (4.145) to the vacuum magnetic field equations $\nabla \cdot \hat{\mathbf{B}} = 0$ and $\nabla \times \hat{\mathbf{B}} = \hat{\mathbf{j}}$, giving

$$\mathbf{n} \cdot [\![\hat{\mathbf{B}}]\!] = 0 \qquad \text{(at the coil surface)}, \qquad (4.179)$$

$$\mathbf{n} \times [\![\hat{\mathbf{B}}]\!] = \mathbf{j}_c^\star(\mathbf{r}, t) \qquad \text{(at the coil surface)}. \qquad (4.180)$$

Note that the surface current $\mathbf{j}_c^\star(\mathbf{r}, t)$ is now cause, not effect as in model II, since it is prescribed. Also, the magnetic field outside the coils, in principle all the way up to ∞, is needed to solve this problem. Clearly, this case is of importance for laboratory plasma confinement because this always involves external magnetic fields that have to be created somehow. External excitation of MHD waves also gives rise to this time-dependent problem.

Model III can also be exploited for the analysis of waves in astrophysical plasmas, e.g. by mimicking the effects of excitation of MHD waves by an external plasma by means of a localized set of 'coils' when the response of the internal plasma is the main issue (e.g. in the problem of sunspot oscillations excited by sound waves in the photosphere; see Section 11.3.3.)

4.6.2 Energy conservation for interface plasmas

In Section 4.3.2 we proved energy conservation of the nonlinear system of ideal MHD equations for model I (plasma enclosed by a wall). In a later chapter (Section 6.6.3) we will need the law of conservation of the total energy for a plasma–vacuum system (model II). In such interface systems, the separate energies of plasma and vacuum are not conserved but the total energy is. Also, it is instructive to consider energy conservation for model III, where the outer boundary is replaced by a current-carrying coil which transfers energy to the system.

The generalization to *model II* is straightforward. The total energy for plasma and vacuum is

$$H = \int \mathcal{H}^p \, d\tau^p + \int \mathcal{H}^v \, d\tau^v, \qquad (4.181)$$

where

$$\mathcal{H}^p \equiv \tfrac{1}{2}\rho v^2 + \frac{p}{\gamma - 1} + \tfrac{1}{2}B^2, \qquad \mathcal{H}^v \equiv \tfrac{1}{2}\hat{B}^2. \qquad (4.182)$$

In the time dependence of these energies one needs to account for the rate of change of the volume elements as given by Eq. (4.90) of Section 4.3.3. Hence,

$$\frac{D}{Dt}\int \mathcal{H}^p\, d\tau = \int \frac{D\mathcal{H}^p}{Dt}\, d\tau + \int \mathcal{H}^p \frac{D}{Dt}(d\tau) = \int \left(\frac{\partial \mathcal{H}^p}{\partial t} + \mathbf{v}\cdot\nabla \mathcal{H}^p + \mathcal{H}^p \nabla \cdot \mathbf{v}\right) d\tau$$
$$= \int \left(\frac{\partial \mathcal{H}^p}{\partial t} + \nabla \cdot (\mathcal{H}^p \mathbf{v})\right) d\tau = \int \frac{\partial \mathcal{H}^p}{\partial t}\, d\tau + \int \mathcal{H}^p \mathbf{v}\cdot \mathbf{n}\, d\sigma, \qquad (4.183)$$

where Gauss' theorem (A.14) has been applied in the last step. Although $\mathbf{v}$ is only defined in the plasma, so that Eq. (4.90) is only valid there, Eq. (4.183) obviously applies to the vacuum as well (with $\mathcal{H}^p$ replaced by $\mathcal{H}^v$) as it merely tells us that the rate of change of the energy is due to the changes of the energy density and of the total volume.

According to the energy conservation equation (4.73), with the expression (4.69) for the energy flow inserted, we may integrate the plasma contribution by parts to get:

$$\frac{\partial H^p}{\partial t} = -\int \left(\tfrac{1}{2}\rho v^2 + \frac{p}{\gamma - 1} + p + B^2\right) \mathbf{v}\cdot \mathbf{n}\, d\sigma$$
$$= -\int \mathcal{H}^p \mathbf{v}\cdot \mathbf{n}\, d\sigma - \int (p + \tfrac{1}{2}B^2)\, \mathbf{v}\cdot \mathbf{n}\, d\sigma,$$

so that

$$\frac{DH^p}{Dt} = -\int (p + \tfrac{1}{2}B^2)\, \mathbf{v}\cdot \mathbf{n}\, d\sigma. \qquad (4.184)$$

For the vacuum contribution, we exploit Faraday's law (4.1) to introduce the electric field $\hat{\mathbf{E}}$ and the Poynting vector $\hat{\mathbf{S}} \equiv \hat{\mathbf{E}} \times \hat{\mathbf{B}}$:

$$\frac{\partial \mathcal{H}^v}{\partial t} = \hat{\mathbf{B}}\cdot\frac{\partial \hat{\mathbf{B}}}{\partial t} = -\hat{\mathbf{B}}\cdot \nabla \times \hat{\mathbf{E}} \stackrel{(A.12)}{=} -\nabla \cdot (\hat{\mathbf{E}} \times \hat{\mathbf{B}}) - \hat{\mathbf{E}}\cdot \nabla \times \hat{\mathbf{B}} = -\nabla \cdot (\hat{\mathbf{E}} \times \hat{\mathbf{B}}),$$

so that

$$\frac{DH^v}{Dt} = \int \frac{\partial \mathcal{H}^v}{\partial t}\, d\tau^v - \int \mathcal{H}^v \mathbf{v}\cdot \mathbf{n}\, d\sigma = \int \hat{\mathbf{E}}\times\hat{\mathbf{B}}\cdot \mathbf{n}\, d\sigma - \int \tfrac{1}{2}\hat{B}^2 \mathbf{v}\cdot\mathbf{n}\, d\sigma. \qquad (4.185)$$

To remove the electric field from this expression again, we exploit the boundary condition (4.178) for model II interfaces derived above:

$$\frac{DH^v}{Dt} = \int \hat{B}^2 \mathbf{v}\cdot \mathbf{n}\, d\sigma - \int \tfrac{1}{2}\hat{B}^2 \mathbf{v}\cdot \mathbf{n}\, d\sigma = \int \tfrac{1}{2}\hat{B}^2 \mathbf{v}\cdot \mathbf{n}\, d\sigma. \qquad (4.186)$$

Adding Eqs. (4.184) and (4.186) for the energies, and applying the jump condition (4.176) for the total pressure, yields the desired result:

$$\frac{DH}{Dt} = \int [\![p + \tfrac{1}{2}B^2]\!]\, \mathbf{v}\cdot \mathbf{n}\, d\sigma = 0; \qquad (4.187)$$

QED.

For *model III*, where the vacuum is enclosed by coils with surface currents $\mathbf{j}_c^\star$, there is no conservation of energy for the interior region because the surface currents pump energy into the system. If we assume that these currents are arranged in such a way that no magnetic energy is lost external to these coils, i.e. $\hat{\mathbf{B}}^{\text{ext}} = 0$, the rate of change of the energy is given by:

$$\frac{DH^{\text{int}}}{Dt} = \frac{\partial H^{\text{int}}}{\partial t} = -\int \hat{\mathbf{S}} \cdot \mathbf{n}\, d\sigma_c$$

$$= -\int \hat{\mathbf{E}} \times \hat{\mathbf{B}} \cdot \mathbf{n}\, d\sigma_c = \int \mathbf{n} \times \hat{\mathbf{B}} \cdot \hat{\mathbf{E}}\, d\sigma_c = -\int \hat{\mathbf{E}} \cdot \mathbf{j}_c^\star\, d\sigma_c, \quad (4.188)$$

where we have exploited the jump condition (4.180) for the surface currents. (Note the change of sign of the Poynting flux and of the magnetic field jump since $\mathbf{n}$ points into the vacuum in Eq. (4.185) but out of it in the present case.) Hence, the rate of change of the energy internal to the coils is given by the Poynting flux across the coils, which equals the power transferred by the coils (see Jackson [117]).

4.6.3 Astrophysical plasmas (models IV–VI)

The three models for astrophysical plasmas are shown in Fig. 4.15. These configurations refer, respectively, to closed and open coronal magnetic loops, and to solar or stellar wind outflow. All three examples are 2D exhibiting only azimuthal symmetry. This is indicated by the angle θ in the loop examples (which may be described by cylinder coordinates r, θ, z) and by the angle ϕ in the wind example (which may be described by spherical coordinates r, θ, ϕ). The actual physical problems really involve genuine 3D geometries because, in general, the loops are not straight and the stellar wind outflows are not axi-symmetric. Nevertheless, it is useful for the analysis to consider the simpler quasi-1D cylindrical versions of the loops (shown in the bottom part of Fig. 4.15) as well. It is to be noted though that this simplification does not lead to genuine 1D problems because of the photospheric b.c.s to be imposed. Also, although 1D versions of the stellar outflow problem are sometimes considered, we have not indicated this possibility in the figure because it cannot be turned into a self-consistent model.

(a) Model IV: 'closed' coronal magnetic loop In this model (Fig. 4.15(a)), *the magnetic field lines of a finite plasma column are line-tied on both sides to a plasma of so much higher density that it may be considered as effectively immobile.* Of course, the magnetic field lines do not end at the bounding planes, but the tubular domain is simply closed-off by the so-called *line-tying boundary conditions*

$$\mathbf{v} = 0 \quad \text{(at the photospheric end planes)}. \quad (4.189)$$

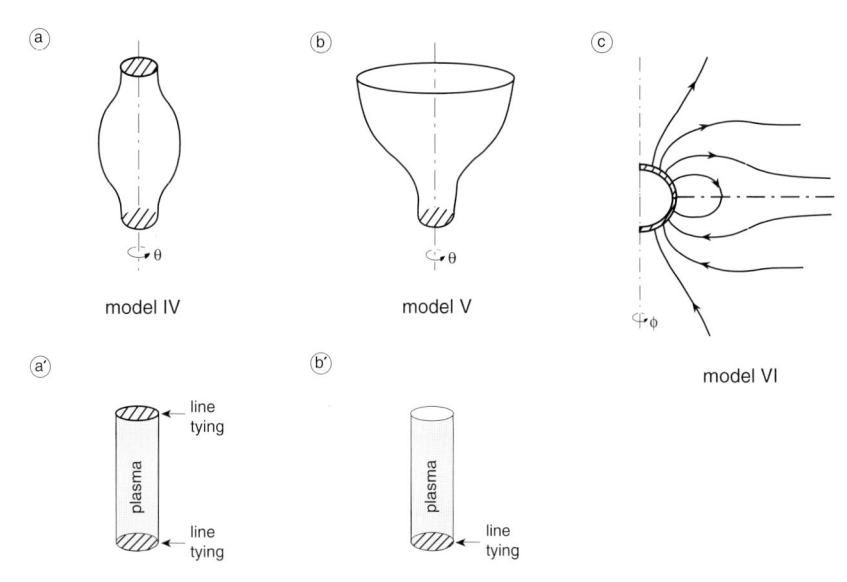

Fig. 4.15. Three MHD models for magnetic confinement in astrophysical plasmas. (a) Model IV: closed coronal magnetic loop, line-tied at both ends; (b) model V: open coronal magnetic loop, line-tied at one end and flaring on the other; (c) model VI: stellar wind outflow. Bottom part: quasi-1D (cylindrical) versions of two of the three models. Surfaces where boundary conditions have to be imposed are hatched.

These conditions represent an idealization of plasma–plasma interfaces with large density differences, where 'line-tying' refers to the fact that the magnetic field sticks through the interface and that it is effectively tied to the much heavier plasma below.

This model is appropriate for the study of *waves in solar coronal flux tubes emanating from the photosphere* because the density of the latter is of the order of a factor of at least 10^9 higher than in the corona. Clearly, in the consideration of the dynamics of a tenuous coronal flux tube, the back reaction on the photosphere may be neglected: the photosphere is too massive to be set in motion by the corona. Of course, the reverse problem is also meaningful and even quite important physically: the dynamics of the photosphere forces motion onto the coronal flux tubes. This problem may be represented by replacing the right hand side of the b.c. (4.189) by a prescribed velocity field at the photospheric boundary. In that case, Poynting flux enters the loop so that it is not closed at all then (which is why we have put quotation marks on 'closed').

(b) Model V: open coronal magnetic loop In this model (Fig. 4.15(b)), *the magnetic field lines of a semi-infinite plasma column are line-tied on one side to a massive plasma.* This model is appropriate for the description of open magnetic field

lines, emanating from the so-called *coronal holes* on the Sun, and *associated with the solar wind* which escapes along them into interplanetary space. The boundary condition (4.189) now only applies to the end that is fixed into the photosphere underneath the coronal hole. Again, one can consider either the passive problem of waves in open field lines by themselves, as represented by the b.c. (4.189), or the driven problem of wave generation by photospheric motion, which requires a non-zero right hand side of Eq. (4.189). Other applications of models IV and V are the dynamics of closed and open field lines in the magnetospheres of planets and pulsars.

(c) Model VI: stellar wind In this model (Fig. 4.15(c)), *a plasma is ejected from the photosphere of a star and accelerated mainly along the open magnetic field lines into outer space.* Clearly, this model is a composite of models IV and V, with the stress now on the outflow rather than on the waves.

This suffices for our purpose, which was the introduction of self-consistent boundary value problems that are relevant for magnetically confined laboratory and astrophysical plasmas. Clearly, the construction of the last two models has not been complete, since we have skipped over the most difficult problems there, viz. the appropriate conditions at infinity. It would be wrong to suggest that these problems have been solved to the same level of satisfaction as obtained for the models I–III. It makes no sense to discuss them separately from the dynamics. It is time to close this chapter and to move to the time domain: Chapter 5 is devoted to the simplest of all geometries possible, infinite homogeneous space, but replete with genuine plasma dynamics.

4.7 Literature and exercises

Notes on literature

Ideal MHD model and conservation laws:

– Goedbloed, *Lecture Notes on Ideal Magnetohydrodynamics* [83], containing the material of a course taught in Brazil 25 years ago, is the origin of this book that has been elaborated ever since. A chapter on the dynamics of the screw pinch, an illuminating illustration of the coupling of nonlinear plasma dynamics and external circuits has been eliminated here since we considered it outdated; the main ideas can be found in Ref. [90]. Exercise 4.12 is devoted to some intriguing aspects of this problem.

Flux tube dynamics:

– The classical paper by Newcomb [163] on 'Motion of magnetic lines of force' is still very worth studying.

Magnetic helicity:

– Moffatt, *Magnetic Field Generation in Electrically Conducting Fluids* [157], Chapter 2 on magnetokinematics with a discussion of the topology implications of magnetic helicity.

Resistive MHD and reconnection:

- Biskamp, *Nonlinear Magnetohydrodynamics* [29], Chapter 6 on the central importance of magnetic reconnection in nonlinear MHD.
- Priest and Forbes, *Magnetic Reconnection* [191], Chapter 4 on the classical solutions of steady reconnection.

Discontinuities and shocks:

- Courant and Friedrichs, *Supersonic Flow and Shock Waves* [59], Chapter IIIC on shocks in one-dimensional flows in ordinary fluids, with emphasis on irreversibility.
- Landau and Lifshitz, *Fluid Mechanics* [137], Chapter IX on shock waves in ordinary fluids.
- Landau and Lifshitz, *Electrodynamics of Continuous Media* [138], Chapter VIII on magnetic fluid dynamics with a section on MHD shock waves.

Exercises

[4.1] *Conservation laws*

Here comes the most important question on magnetohydrodynamics: what is the principal conservation law of magnetized plasmas? (If you do not know, reread Chapter 4. If you still do not know, ask your professor. If he does not know, he should not be teaching this subject.)

[4.2] *Eulerian and Lagrangian time-derivatives*

Write down and explain the two different ways of evaluating time derivatives in fluid dynamics.

[4.3] *Non-relativistic approximation*

Construct the MHD equations from Maxwell's equations and the equations for classical fluid dynamics. Show that, by assuming $v \ll c$, the displacement current and the electrostatic acceleration become negligible. Write down the resulting non-relativistic MHD equations in their most compact form.

[4.4] *Scale independence*

Why, and when, does it make sense to compare the plasma dynamics in small laboratory devices with the plasma dynamics in huge plasma-astrophysical systems?

[4.5]⋆ *Conservation form*

The MHD equations can be brought into so-called 'conservation form'.
- What is the general structure of this form and can you give an interpretation of it?
- Try to construct the conservation form of the ideal MHD equations in terms of the variables ρ, **v**, e, and **B**. If this is too much for you, try at least to indicate the basic steps involved in the construction.

[4.6] *Stress tensor*

In the conservation form of the MHD equations, one encounters the stress tensor defined by $\mathbf{T} \equiv \rho\mathbf{vv} + (p + \frac{1}{2}B^2)\mathbf{I} - \mathbf{BB}$, representing the Reynolds stress tensor, the isotropic pressure, and the magnetic part of the Maxwell stress tensor.
- Project the Reynolds stress tensor on the velocity **v** and show that the only contribution is ρv^2 along **v**. Decompose the remaining part of the stress tensor parallel and perpendicular to **B** and show how it is represented in matrix form.

– Draw a picture of a flux tube and comment on the different stresses acting on it.

[4.7] *Surface element kinematics and magnetic flux conservation*

We consider the magnetic flux $d\Psi \equiv \mathbf{B} \cdot d\boldsymbol{\sigma}$ through a surface element $d\boldsymbol{\sigma}$. When the latter element moves with the plasma, it changes in time as $D(d\boldsymbol{\sigma})/Dt = -(\nabla \mathbf{v}) \cdot d\boldsymbol{\sigma} + \nabla \cdot \mathbf{v} \, d\boldsymbol{\sigma}$.

– Use the vector identities of Appendix A to show that the Lagrangian time derivative of the magnetic flux through that surface element vanishes: $D(d\Psi)/Dt = 0$.
– What role does the assumption of perfect conductivity play here?

[4.8] *Magnetic helicity and force-free magnetic fields*

In Section 4.3.4, the magnetic helicity of a special class of cylindrical force-free magnetic fields of constant α has been computed. Let us extend these calculations in two directions.

– Calculate the components of a cylindrical force-free magnetic field with constant pitch μ. We already know that the helicity of this field vanishes. Now derive the expression for $\alpha(r)$ for this field and note that it tends to a constant value on the axis. How does that relate to the expression for the helicity of a constant α force-free field?
– Calculate the helicity for the compound configuration of a force-free magnetic field of constant α within a cylinder of radius $r = a$ surrounded by a vacuum magnetic field in the annular cylinder $a \leq r \leq b$. Does it make a difference whether that annular region is considered as a proper vacuum ($\eta = \infty$) or as just a plasma ($\eta = 0$) without current?

[4.9] *Neglect of resistivity and dimensionless parameters*

We start with the resistive induction equation for the magnetic field $\mathbf{B}$ (Eq. (4.125)).

– Use a length scale l_0 for the gradients and velocity v_0 for the flow to construct the dimensionless parameter (R_m) which justifies neglect of the resistive term when $R_m \gg 1$.
– Another dimensionless parameter (L_u) may be constructed where v_0 is replaced by the characteristic velocity of Alfvén waves. When and why would that number be relevant?

[4.10] *Jump conditions*

Boundary conditions at plasma–plasma or plasma–vacuum interfaces may be derived from jump conditions obtained from the theory of supersonic flow and shocks in gas dynamics. Integrating the partial differential equation for a variable f across a shock results in jumps of ∇f and $\partial f/\partial t$ of magnitude $\mathbf{n} [\![f]\!]$ and $-u [\![f]\!]$, resp., where $\mathbf{n}$ is the normal to the shock front, the jump $[\![f]\!] \equiv f_1 - f_2$, and u is the normal velocity of the shock.

– Derive the ideal MHD jump conditions from the conservation laws.
– Transform these conditions to a frame moving with the shock, so that the fluid velocities relative to this frame may be written as $\mathbf{v}' \equiv \mathbf{v} - u\mathbf{n}$.
– Find the required interface conditions by assuming absence of flow across and a magnetic field tangential to the interface.

[4.11]★ *Plasma–vacuum configuration between walls*

Consider a configuration of a plasma resting (against vertical gravity) on top of a vacuum with horizontal magnetic field confined between two horizontal, infinitely extended, conducting walls. (The stability of this configuration will be investigated in Section 6.6.4.)

- Derive all boundary conditions for the plasma variables at the top wall, the plasma–vacuum interface (assuming no flow across the boundary), and the vacuum variables at the bottom wall. Also give the equations for the magnetic field $\hat{\mathbf{B}}$ in the vacuum.

[4.12]* *Flux and energy conservation in a plasma coupled to external circuits*

Fig. 1.4 showed how the combination of the θ-pinch and z-pinch concepts led to the stable equilibrium of a tokamak requiring a large toroidal magnetic field and the vicinity of a wall. A similar configuration is the toroidal screw pinch; van der Laan *et al.* (1971) [236]. It is also a stable equilibrium, has a 100 × higher density than a tokamak, but operates in a pulsed fashion (~ 100 μs). The toroidal and poloidal magnetic field components are created by discharging two capacitor banks simultaneously, so that the plasma experiences an inward force consisting of the two components $j_\theta B_z$ and $j_z B_\theta$. (We neglect toroidal effects, except for the induction of the plasma current I_{z2} by means of the primary current I_{z1} in the z-coil, exploiting a finite-length cylindrical model with coordinates r, θ, z, where $0 \leq z \leq 2\pi R_0$.) We wish to solve the following (model III) problem. At $t = 0$ the two charged capacitors C_z and C_θ of voltage V_z and V_θ are switched to the z- and θ-coil surrounding the plasma. What is the resulting plasma motion and how are the available electrostatic energies $\frac{1}{2}C_z V_z^2$ and $\frac{1}{2}C_\theta V_\theta^2$ converted into magnetic field energies $W_\theta \equiv \int (B_\theta^2/2\mu_0) d\tau$ and $W_z \equiv \int (B_z^2/2\mu_0) d\tau$?

- Idealize the two coils to be one copper shell closely fitting the plasma vessel with a poloidal cut (the plane $z = 0$) over which the voltage V_z is applied and a toroidal cut (the half-plane $\theta = 0$) over which the voltage V_θ is applied. Compute the electric fields E_z and E_θ at the boundary $r = a$. Determine, by means of the ideal Ohm's law, the imposed radial velocity $v(a, t)$ and the pitch $\mu(a, t)$ of the magnetic field lines.

- Neglect the effects of pressure and density on the plasma dynamics, so that the plasma part of the problem is reduced to the determination of the time-dependence of a force-free magnetic field by the ideal MHD induction equation. Show that the way of switching of the capacitor banks produces the constant pitch force-free field of Exercise 4.8.

- The circuit equations determine the time evolution of $V_z(t)$ and $V_\theta(t)$. We assume that the coupling of the coils to the plasma is perfect, so that the primary poloidal current I_θ directly determines the plasma magnetic field $B_z(a, t)$ through $\dot{V}_\theta = C_\theta^{-1} I_\theta$, whereas the primary toroidal current I_{z1} is coupled to the induced plasma current I_{z2} through the self-inductance L_T of the torus: $V_z = -\dot{\psi}_T = L_T(\dot{I}_{z1} + \dot{I}_{z2})$, where $\dot{V}_z = C_z^{-1} I_{z1}$ and ψ_T is the poloidal flux through the hole of the torus. Now complete the solution of the full problem and show that the plasma dynamics couples the two circuits in such a way that a nonlinear oscillation is excited where the magnetic fluxes associated with B_z and B_θ are conserved separately, but the energies are only conserved as the sum $W_\theta + W_z$. (Consult Goedbloed & Zwart [90] for the detailed solution.)

5
Waves and characteristics

5.1 Physics and accounting

5.1.1 Introduction

In the previous chapter, we have stressed the *spatial* aspect of MHD by formulating the different boundary value problems associated with different plasma confinement geometries. In this chapter, we will stress the *temporal* aspect by neglecting all effects of the geometry by considering, first, the linear waves of an infinite homogeneous plasma (Sections 5.1–5.3) and, next, the nonlinear counterpart of these waves, viz. the characteristics and associated initial value problem (Section 5.4). The effects of plasma geometry and inhomogeneity will be the subject of the following chapters, dealing with spectral theory (Chapter 6) and the important topic of plasma stability (Chapter 7).

The theory of wave propagation in plasmas necessarily contains a large number of algebraic manipulations to reduce a particular problem with many variables to one that can be solved explicitly and that manifests the physics. Whereas extreme care is needed to avoid mistakes in this reduction, the bookkeeping should not obscure the physics. We will use the familiar example of sound waves to illustrate how this works in practice.

5.1.2 Sound waves

As a preliminary to the study of linear MHD waves, let us consider the simplest example of waves described by fluid equations, viz. sound waves. To that end, we start from the gas dynamic equations, contained in the MHD equations (4.12)–(4.14) as the special case of vanishing magnetic field ($\mathbf{B} = 0$), where we also neglect gravity ($\mathbf{g} = 0$):

$$\frac{\partial \rho}{\partial t} + \nabla \cdot (\rho \mathbf{v}) = 0, \tag{5.1}$$

$$\rho\left(\frac{\partial \mathbf{v}}{\partial t} + \mathbf{v} \cdot \nabla \mathbf{v}\right) + \nabla p = 0, \tag{5.2}$$

$$\frac{\partial p}{\partial t} + \mathbf{v} \cdot \nabla p + \gamma p \nabla \cdot \mathbf{v} = 0. \tag{5.3}$$

We now *linearize* these equations about a time-independent ($\partial/\partial t = 0$) infinite and homogeneous ($\nabla = 0$) background, characterized by arbitrary, but constant, values of ρ_0, $\mathbf{v}_0$, and p_0. The perturbations of this background may be written as

$$\begin{aligned}
\rho(\mathbf{r}, t) &= \rho_0 + \rho_1(\mathbf{r}, t) & (\text{where } |\rho_1| \ll \rho_0 = \text{const}), \\
p(\mathbf{r}, t) &= p_0 + p_1(\mathbf{r}, t) & (\text{where } |p_1| \ll p_0 = \text{const}), \\
\mathbf{v}(\mathbf{r}, t) &= \mathbf{v}_0 + \mathbf{v}_1(\mathbf{r}, t) & (\text{where } |\mathbf{v}_1| \text{ is small}).
\end{aligned} \tag{5.4}$$

Note that $|\mathbf{v}_1|$ is considered to be small, but not with respect to $|\mathbf{v}_0|$, because we wish to consider the special case of static ($\mathbf{v}_0 = 0$) background in particular and there is another quantity that can serve as a measure for large and small velocities, viz. the sound speed.

Inserting these expressions in the differential equations and neglecting nonlinear coupling through quadratic and higher order terms, since the amplitudes of the waves are assumed to be small, we obtain the linearized equations of gas dynamics:

$$\left(\frac{\partial}{\partial t} + \mathbf{v}_0 \cdot \nabla\right)\rho_1 + \rho_0 \nabla \cdot \mathbf{v}_1 = 0, \tag{5.5}$$

$$\rho_0\left(\frac{\partial}{\partial t} + \mathbf{v}_0 \cdot \nabla\right)\mathbf{v}_1 + \nabla p_1 = 0, \tag{5.6}$$

$$\left(\frac{\partial}{\partial t} + \mathbf{v}_0 \cdot \nabla\right)p_1 + \gamma p_0 \nabla \cdot \mathbf{v}_1 = 0. \tag{5.7}$$

Note that Eq. (5.5) for ρ_1 does not couple to the other equations, so that it may be dropped or solved separately after the equations for $\mathbf{v}_1$ and p_1 have been solved. The latter two equations may be combined by applying the operator $\partial/\partial t + \mathbf{v}_0 \cdot \nabla$ to Eq. (5.6) for $\mathbf{v}_1$ and eliminating p_1 from it by means of Eq. (5.7). This yields *the wave equation for sound waves*:

$$\left(\frac{\partial}{\partial t} + \mathbf{v}_0 \cdot \nabla\right)^2 \mathbf{v}_1 - c^2 \nabla \nabla \cdot \mathbf{v}_1 = 0, \tag{5.8}$$

where

$$c \equiv \sqrt{\gamma p_0 / \rho_0} \tag{5.9}$$

is *the velocity of sound* of the background medium. (We are free to exploit the symbol c for this purpose since we have removed the displacement current from Maxwell's equations in Section 4.1.1, so that the plasmas considered here will not support electromagnetic waves and the velocity of light does not appear any more in the equations.)

The wave equation has constant coefficients ($\mathbf{v}_0$ and c^2) so that the most general solution can be written as a superposition of *plane waves*:

$$\mathbf{v}_1(\mathbf{r}, t) = \sum_{\mathbf{k}} \hat{\mathbf{v}}_{\mathbf{k}} \, e^{i(\mathbf{k}\cdot\mathbf{r}-\omega t)} \tag{5.10}$$

(for simplicity represented as a sum, corresponding to waves in a finite box). These plane waves do not couple, since the problem is linear and homogeneous, so that each harmonic by itself is a solution of the problem. Hence, we will consider them separately and drop the subscript $\mathbf{k}$. For such solutions, the differential operators of the PDEs turn into multiplications by algebraic factors:

$$\nabla \to i\mathbf{k}, \qquad \partial/\partial t \to -i\omega. \tag{5.11}$$

This transforms the wave equation (5.8) into an algebraic eigenvalue equation:

$$\left[(\omega - \mathbf{k}\cdot\mathbf{v}_0)^2 \mathbf{I} - c^2 \mathbf{k}\mathbf{k} \right] \cdot \hat{\mathbf{v}} = 0. \tag{5.12}$$

This eigenvalue problem for the frequency ω, measured in the laboratory frame, may also be considered as a (simpler) eigenvalue problem for the *Doppler shifted frequency* ω', measured in the co-moving frame:

$$\omega' \equiv \omega - \mathbf{k}\cdot\mathbf{v}_0. \tag{5.13}$$

For homogeneous media ($\mathbf{v}_0 = \text{const}$) the Doppler shift is constant so that the difference between the two eigenvalue problems is trivial. However, our main concern in the following chapters will be inhomogeneous media where the influence of the background flow would significantly complicate the wave propagation problem. We will leave that subject for a later chapter (in Volume 2) and, for the time being, concentrate on wave propagation in *static* media ($\mathbf{v}_0 = 0$) where $\omega' = \omega$. For those problems, the eigenvalue ω only appears squared.

Since there is no preferred direction in the system, there is no loss in generality if we restrict the waves to propagate in the z-direction only, $\mathbf{k} = k\,\mathbf{e}_z$, so that the

5.1 Physics and accounting

eigenvalue problem reduces to

$$\omega^2 \hat{v}_x = 0,$$
$$\omega^2 \hat{v}_y = 0, \qquad (5.14)$$
$$(\omega^2 - k^2 c^2)\hat{v}_z = 0.$$

The important solutions are given by

$$\omega = \pm k\, c, \qquad \hat{v}_x = \hat{v}_y = 0, \qquad \hat{v}_z \text{ arbitrary}, \qquad (5.15)$$

representing plane *sound waves* travelling to the right ($+$) or to the left ($-$). These waves are *compressible* ($\nabla \cdot \mathbf{v}_1 \neq 0$) and *longitudinal* ($\mathbf{v}_1 \parallel \mathbf{k}$). The other solutions,

$$\omega^2 = 0, \qquad \hat{v}_x, \hat{v}_y \text{ arbitrary}, \qquad \hat{v}_z = 0, \qquad (5.16)$$

just correspond to time-independent incompressible transverse ($\mathbf{v}_1 \perp \mathbf{k}$) translations. They do not represent interesting physics, but simply establish the completeness of the velocity representation.

Recapitulating: we have transformed the original system (5.5)–(5.7) of first order PDEs for the primitive variables ρ_1, $\mathbf{v}_1$, p_1 to the second order wave equation (5.8) in terms of $\mathbf{v}_1$ alone. This transformation highlights the important physics of the problem, which is sound wave propagation, but also yields some trivial $\omega = 0$ solutions that are not interesting from the physical point of view. Of course, the two representations should be equivalent.

Now count: the first order system appears to have five degrees of freedom represented by the five primitive variables, whereas the second order system appears to have six degrees of freedom since there are three components of $\mathbf{v}_1$ and the eigenvalue is squared. One easily recognizes though that the second order system actually only has four degrees of freedom, since the quadratic dependence on ω does not double the actual number of translations (5.16). This spurious doubling of the eigenvalue $\omega = 0$ happened when we applied the operator $\partial/\partial t + \mathbf{v}_0 \cdot \nabla$ to Eq. (5.6) to eliminate p_1. One easily checks that the solutions (5.15) and (5.16), with $\omega = 0$, are solutions of the original system of equations (5.5)–(5.7). Hence, we have actually *lost one degree of freedom* in the reduction to the wave equation in terms of $\mathbf{v}_1$ only. This happened when we dropped Eq. (5.5) for ρ_1. By inserting $\mathbf{v}_1 = 0$ in the original system we find the signature of this lost mode:

$$\omega \hat{\rho} = 0 \quad \Rightarrow \quad \omega = 0, \quad \hat{\rho} \text{ arbitrary}, \quad \text{but } \hat{\mathbf{v}} = 0 \text{ and } \hat{p} = 0. \qquad (5.17)$$

This mode is called an *entropy wave*: it represents a perturbation of the density and, hence, of the entropy function $S \equiv p\rho^{-\gamma}$, since the pressure is not perturbed. Like the translations (5.16), the entropy mode does not represent important physics but it is needed to account for the degrees of freedom of the different representations. This is an important issue in computational studies. It will return in the MHD analysis below, where the system is much more involved, so that the distinction between genuine and spurious solutions is much less transparent.

5.2 MHD waves

5.2.1 Symmetric representation in primitive variables

We now perform a similar analysis for the magnetohydrodynamic system, with the magnetic terms included. In Chapter 4, the basic MHD equations were presented in two forms, viz. Eqs. (4.12)–(4.15) for the variables ρ, $\mathbf{v}$, p, $\mathbf{B}$, and Eqs. (4.45)–(4.48) for the variables ρ, $\mathbf{v}$, e, $\mathbf{B}$. Since the derivation of conservation laws was based on the latter set of equations, we will exploit that set for the present derivation. We drop the gravitational term in the momentum equation, since it is incompatible with the assumption of a homogeneous plasma, and convert the two occurring cross-products by means of the vector identities of the Appendix:

$$-\mathbf{j} \times \mathbf{B} = -(\nabla \times \mathbf{B}) \times \mathbf{B} \stackrel{(A.8)}{=} (\nabla \mathbf{B}) \cdot \mathbf{B} - \mathbf{B} \cdot \nabla \mathbf{B},$$

$$\nabla \times \mathbf{E} = -\nabla \times (\mathbf{v} \times \mathbf{B}) \stackrel{(A.13)}{=} -\mathbf{B} \cdot \nabla \mathbf{v} + \mathbf{B} \nabla \cdot \mathbf{v} + \mathbf{v} \cdot \nabla \mathbf{B}.$$

This yields the following set of equations:

$$\frac{\partial \rho}{\partial t} + \nabla \cdot (\rho \mathbf{v}) = 0, \tag{5.18}$$

$$\rho \frac{\partial \mathbf{v}}{\partial t} + \rho \mathbf{v} \cdot \nabla \mathbf{v} + (\gamma - 1)\nabla(\rho e) + (\nabla \mathbf{B}) \cdot \mathbf{B} - \mathbf{B} \cdot \nabla \mathbf{B} = 0, \tag{5.19}$$

$$\frac{\partial e}{\partial t} + \mathbf{v} \cdot \nabla e + (\gamma - 1)e\nabla \cdot \mathbf{v} = 0, \tag{5.20}$$

$$\frac{\partial \mathbf{B}}{\partial t} + \mathbf{v} \cdot \nabla \mathbf{B} + \mathbf{B} \nabla \cdot \mathbf{v} - \mathbf{B} \cdot \nabla \mathbf{v} = 0, \qquad \nabla \cdot \mathbf{B} = 0, \tag{5.21}$$

which is yet another form of the basic nonlinear MHD equations.

We again choose an *infinite homogeneous plasma at rest* ($\mathbf{v}_0 = 0$) as the background state:

$$\rho = \rho_0, \qquad e = e_0 \equiv \frac{p_0}{(\gamma - 1)\rho_0}, \qquad \mathbf{B} = \mathbf{B}_0,$$

where ρ_0, e_0, and B_0 are constants (in space and time) that characterize this state. Linearization proceeds as in Section 5.1. Since there are no gradients of the equilibrium quantities, terms like $\mathbf{v}_1 \cdot \nabla \rho_0$ disappear. Again, we only keep linear terms, like $\rho_0 \nabla \cdot \mathbf{v}_1$, but neglect nonlinear terms, like $\rho_0 \mathbf{v}_1 \cdot \nabla \mathbf{v}_1$. This results in the following set of linearized MHD equations:

$$\frac{\partial \rho_1}{\partial t} + \rho_0 \nabla \cdot \mathbf{v}_1 = 0, \tag{5.22}$$

$$\rho_0 \frac{\partial \mathbf{v}_1}{\partial t} + (\gamma - 1)(e_0 \nabla \rho_1 + \rho_0 \nabla e_1) + (\nabla \mathbf{B}_1) \cdot \mathbf{B}_0 - \mathbf{B}_0 \cdot \nabla \mathbf{B}_1 = 0, \tag{5.23}$$

$$\frac{\partial e_1}{\partial t} + (\gamma - 1) e_0 \nabla \cdot \mathbf{v}_1 = 0, \tag{5.24}$$

$$\frac{\partial \mathbf{B}_1}{\partial t} + \mathbf{B}_0 \nabla \cdot \mathbf{v}_1 - \mathbf{B}_0 \cdot \nabla \mathbf{v}_1 = 0, \qquad \nabla \cdot \mathbf{B}_1 = 0. \tag{5.25}$$

Two characteristic speeds now describe the background state, viz. *the sound speed* and *the vectorial Alfvén speed*:

$$c \equiv \sqrt{\frac{\gamma p_0}{\rho_0}}, \qquad \mathbf{b} \equiv \frac{\mathbf{B}_0}{\sqrt{\rho_0}}, \tag{5.26}$$

where $|\mathbf{b}| \equiv v_A$ is the scalar Alfvén speed, already defined in Eq. (2.150) of Section 2.4.2. Together, they govern the speed of propagation of magnetohydrodynamic waves.

By trial and error, we construct *dimensionless variables* for the perturbations,

$$\tilde{\rho} \equiv \frac{\rho_1}{\gamma \rho_0}, \qquad \tilde{\mathbf{v}} \equiv \frac{\mathbf{v}_1}{c}, \qquad \tilde{e} \equiv \frac{e_1}{\gamma e_0}, \qquad \tilde{\mathbf{B}} \equiv \frac{\mathbf{B}_1}{c \sqrt{\rho_0}}, \tag{5.27}$$

such that the linearized MHD equations only involve the coefficients c and $\mathbf{b}$ (and γ):

$$\gamma \frac{\partial \tilde{\rho}}{\partial t} + c \nabla \cdot \tilde{\mathbf{v}} = 0, \tag{5.28}$$

$$\frac{\partial \tilde{\mathbf{v}}}{\partial t} + c \nabla \tilde{\rho} + c \nabla \tilde{e} + (\nabla \tilde{\mathbf{B}}) \cdot \mathbf{b} - \mathbf{b} \cdot \nabla \tilde{\mathbf{B}} = 0, \tag{5.29}$$

$$\frac{\gamma}{\gamma - 1} \frac{\partial \tilde{e}}{\partial t} + c \nabla \cdot \tilde{\mathbf{v}} = 0, \tag{5.30}$$

$$\frac{\partial \tilde{\mathbf{B}}}{\partial t} + \mathbf{b} \nabla \cdot \tilde{\mathbf{v}} - \mathbf{b} \cdot \nabla \tilde{\mathbf{v}} = 0, \qquad \nabla \cdot \tilde{\mathbf{B}} = 0. \tag{5.31}$$

This form is most appropriate for at least one of our purposes, to be disclosed below.

Again, let us consider *plane wave solutions*,

$$\tilde{\rho} = \tilde{\rho}(\mathbf{r}, t) = \hat{\rho}\, e^{i(\mathbf{k}\cdot\mathbf{r} - \omega t)}, \quad \text{etc.} \tag{5.32}$$

We then obtain an algebraic system of eigenvalue equations:

$$c\mathbf{k}\cdot\hat{\mathbf{v}} = \gamma\omega\hat{\rho},$$

$$\mathbf{k}c\hat{\rho} + \mathbf{k}c\hat{e} + (\mathbf{k}\mathbf{b}\cdot - \mathbf{k}\cdot\mathbf{b})\hat{\mathbf{B}} = \omega\hat{\mathbf{v}},$$

$$c\mathbf{k}\cdot\hat{\mathbf{v}} = \frac{\gamma}{\gamma - 1}\omega\hat{e},$$

$$(\mathbf{b}\mathbf{k}\cdot - \mathbf{b}\cdot\mathbf{k})\hat{\mathbf{v}} = \omega\hat{\mathbf{B}}, \qquad \mathbf{k}\cdot\hat{\mathbf{B}} = 0. \tag{5.33}$$

Ignoring the constraint $\mathbf{k}\cdot\hat{\mathbf{B}} = 0$ for the moment, this system is solvable if the determinant of the 8×8 matrix vanishes. Since the medium is now anisotropic, through the presence of a preferred direction given by the background magnetic field $\mathbf{B}_0$ (or $\mathbf{b}$), we should not choose the vectors $\mathbf{b}$ and $\mathbf{k}$ parallel. However, there is no loss of generality if we choose $\mathbf{b}$ along the z-axis and $\mathbf{k}$ in the x-z plane (like in Fig. 5.1):

$$\mathbf{b} = (0, 0, b), \qquad \mathbf{k} = (k_\perp, 0, k_\parallel). \tag{5.34}$$

This leads to the following matrix representation of the eigenvalue problem:

$$\begin{pmatrix} 0 & k_\perp c & 0 & k_\parallel c & 0 & 0 & 0 & 0 \\ k_\perp c & 0 & 0 & 0 & k_\perp c & -k_\parallel b & 0 & k_\perp b \\ 0 & 0 & 0 & 0 & 0 & 0 & -k_\parallel b & 0 \\ k_\parallel c & 0 & 0 & 0 & k_\parallel c & 0 & 0 & 0 \\ 0 & k_\perp c & 0 & k_\parallel c & 0 & 0 & 0 & 0 \\ 0 & -k_\parallel b & 0 & 0 & 0 & 0 & 0 & 0 \\ 0 & 0 & -k_\parallel b & 0 & 0 & 0 & 0 & 0 \\ 0 & k_\perp b & 0 & 0 & 0 & 0 & 0 & 0 \end{pmatrix} \begin{pmatrix} \hat{\rho} \\ \hat{v}_x \\ \hat{v}_y \\ \hat{v}_z \\ \hat{e} \\ \hat{B}_x \\ \hat{B}_y \\ \hat{B}_z \end{pmatrix} = \omega \begin{pmatrix} \gamma\hat{\rho} \\ \hat{v}_x \\ \hat{v}_y \\ \hat{v}_z \\ \frac{\gamma}{\gamma-1}\hat{e} \\ \hat{B}_x \\ \hat{B}_y \\ \hat{B}_z \end{pmatrix}.$$

(5.35)

Except for a few shortcomings, to be addressed below, this expression is satisfactory since it clearly identifies the two new features of MHD waves as compared

5.2 MHD waves

to sound waves, viz. the occurrence of the Alfvén speed in addition to the sound speed and the anisotropy expressed by the appearance of two components of the wave vector. We could compute the dispersion equation from the determinant and study the associated waves, but we have good reasons to postpone this to Section 5.2.3.

In the formulation of Eq. (5.35), one purpose has been realized, viz. the demonstration of *the symmetry of the operator describing the linearized system*, which is one of the most important properties of ideal magnetohydrodynamics. This has been demonstrated now for the simplest case, that of homogeneous plasmas. Its generalization to inhomogeneous plasmas will be an important issue in the spectral theory of MHD, to be discussed in Chapter 6. The symmetry of the linearized system is closely related to an analogous property of the original nonlinear equations, viz. *the nonlinear ideal MHD equations are symmetric hyperbolic partial differential equations*. This will be proved in Section 5.4.3. These symmetry properties were behind our preference for the representation in terms of the variables ρ, $\mathbf{v}$, e, $\mathbf{B}$, corresponding to the variables of the moment expansion of the kinetic equations (Chapter 3). This formulation lends itself most naturally for extensions, e.g. with dissipative effects, so that it is frequently exploited in large scale numerical programs for the solution of the MHD equations for practical applications.

▷ **Generalized eigenvalue problem.** A minor shortcoming of the eigenvalue problem (5.35) is that it still contains the factors γ and $\gamma/(\gamma-1)$ in the vector on the RHS. Formally, this implies that we are dealing with a *generalized eigenvalue problem*:

$$\mathbf{A} \cdot \mathbf{x} = \lambda \mathbf{B} \cdot \mathbf{x}, \tag{5.36}$$

where $\mathbf{A}$ denotes the 8×8 matrix and $\mathbf{x}$ is the 8-vector of unknowns on the LHS of Eq. (5.35), $\lambda \equiv \omega$ is the eigenvalue, and $\mathbf{B}$ is a diagonal matrix which deviates from the unit matrix by the two mentioned factors multiplying the variables $\hat{\rho}$ and $\hat{e}$. An *ordinary eigenvalue problem*

$$\mathbf{A}' \cdot \mathbf{x}' = \lambda \mathbf{x}' \tag{5.37}$$

is obtained by the transformation

$$\mathbf{A}' \equiv \mathbf{B}^{-1/2} \cdot \mathbf{A} \cdot \mathbf{B}^{-1/2}, \qquad \mathbf{x}' \equiv \mathbf{B}^{1/2} \mathbf{x}, \tag{5.38}$$

so that

$$\mathbf{A}' = \begin{pmatrix} 0 & \frac{1}{\sqrt{\gamma}} \mathbf{k}^T c & 0 & 0 \\ \frac{1}{\sqrt{\gamma}} \mathbf{k} c & 0 & \sqrt{\frac{\gamma-1}{\gamma}} \mathbf{k} c & \mathbf{k} \mathbf{b}^T - (\mathbf{k} \cdot \mathbf{b}) \mathbf{I} \\ 0 & \sqrt{\frac{\gamma-1}{\gamma}} \mathbf{k}^T c & 0 & 0 \\ 0 & \mathbf{b} \mathbf{k}^T - (\mathbf{k} \cdot \mathbf{b}) \mathbf{I} & 0 & 0 \end{pmatrix}, \quad \mathbf{x}' = \begin{pmatrix} \sqrt{\gamma} \hat{\rho} \\ \hat{\mathbf{v}} \\ \sqrt{\frac{\gamma}{\gamma-1}} \hat{e} \\ \hat{\mathbf{B}} \end{pmatrix},$$

$$\tag{5.39}$$

where the superscript 'T' indicates row vectors. The 'cost' of this transformation is the appearance of odd square root factors in the matrix $\mathbf{A}'$ as well as in the vector $\mathbf{x}'$. These γ factors are really spurious in the eigenvalue problem since they will not appear in the final dispersion equation, as one can directly check from the present form (but much easier from the one derived in Section 5.2.3). ◁

5.2.2 Entropy wave and magnetic field constraint

Before we transform to the more transparent velocity representation of Section 5.2.3, in analogy with our exposition of Section 5.1.2 for sound waves, we first have to account for two subtleties related to the presence of a genuine and a spurious solution $\omega = 0$ of the eigenvalue problem. Such solutions are called *marginal* since the transition from stability to instability occurs at that value of ω, as we will see in Chapter 6. The analysis of these modes will be facilitated by means of the apparent detour of replacing the thermodynamic variables ρ and e by the entropy s and the pressure p. As a bonus, the redundant factors γ will disappear.

(a) Marginal entropy wave To that end, we linearize the expressions for the pressure, $p = (\gamma - 1)\rho e$, and the entropy function, $S \equiv p\rho^{-\gamma}$, and construct the appropriate dimensionless variables:

$$p_1 = (\gamma - 1)(\rho_0 e_1 + e_0 \rho_1) \quad \Rightarrow \quad \tilde{p} \equiv \frac{p_1}{\gamma p_0} = \tilde{e} + \tilde{\rho}, \tag{5.40}$$

$$S_1 = S_0 \left(\frac{p_1}{p_0} - \gamma \frac{\rho_1}{\rho_0} \right) \quad \Rightarrow \quad \tilde{S} \equiv \frac{S_1}{\gamma S_0} = \tilde{p} - \gamma \tilde{\rho} = \tilde{e} - (\gamma - 1)\tilde{\rho}. \tag{5.41}$$

Hence, in addition to Eqs. (5.28) and (5.33)(a) for the density $\tilde{\rho}$, and Eqs. (5.30) and (5.33)(c) for the internal energy $\tilde{e}$, we also get equations for $\tilde{p}$ and $\tilde{S}$:

$$\frac{\partial \tilde{p}}{\partial t} + c \nabla \cdot \tilde{\mathbf{v}} = 0 \quad \Rightarrow \quad -\omega \hat{p} + c \mathbf{k} \cdot \hat{\mathbf{v}} = 0, \tag{5.42}$$

$$\frac{\partial \tilde{S}}{\partial t} = 0 \quad \Rightarrow \quad -\omega \hat{S} = 0. \tag{5.43}$$

We may now choose any two of the four variables $\tilde{\rho}, \tilde{e}, \tilde{p}, \tilde{S}$ together with $\tilde{\mathbf{v}}$ and $\tilde{\mathbf{B}}$ to express the eigenvalue problem. However, not every choice leads to a symmetric eigenvalue problem!

5.2 MHD waves

It turns out to be expedient to transform to the unusual $(\tilde{S}, \tilde{\mathbf{v}}, \tilde{p}, \tilde{\mathbf{B}})$ representation for the eigenvalue problem:

$$
\begin{aligned}
&-\omega \hat{S} = 0, \\
&-\omega \hat{\mathbf{v}} + \mathbf{k}\, c\, \hat{p} + (\mathbf{k}\mathbf{b}\cdot - \mathbf{k}\cdot\mathbf{b})\, \hat{\mathbf{B}} = 0, \\
&-\omega \hat{p} + c\, \mathbf{k}\cdot \hat{\mathbf{v}} = 0, \\
&-\omega \hat{\mathbf{B}} + (\mathbf{b}\mathbf{k}\cdot - \mathbf{b}\cdot\mathbf{k})\, \hat{\mathbf{v}} = 0, \qquad \mathbf{k}\cdot\hat{\mathbf{B}} = 0.
\end{aligned} \tag{5.44}
$$

This representation is more compact than the previous one, fully symmetric, and free of redundant factors γ. More important for our present purpose is that it clearly exhibits the peculiar marginal *entropy wave*, for which most of the physical variables (velocity, pressure, and magnetic field perturbations) vanish, except for the entropy (and, hence, the internal energy and density) perturbations:

$$
\omega = 0, \qquad \hat{p} = \hat{e} + \hat{\rho} = 0, \qquad \hat{S} = \gamma \hat{e} = -\gamma \hat{\rho} \neq 0. \tag{5.45}
$$

Superposition of the different Fourier harmonics, like in Eq. (5.10) of Section 5.1.2, gives a mode with a completely arbitrary spatial distribution of the entropy. This may be realized, for example, by a local increase $\hat{e}$ of the internal energy (or the temperature) and a decrease $\hat{\rho}$ of the density such that there is no net pressure change $\hat{p}$. *This marginal mode is genuine but does not represent important physics*, at least not in the context of the ideal MHD model. Since it does not propagate and since the entropy does not couple to the other variables, not much were lost if this part of the problem were dropped altogether so that, actually, a 7×7 representation would be obtained. This is effected by the transformation to the velocity representation, to be discussed in the next section. In that transformation, also a spurious solution is removed, so that, effectively, a 6×6 representation is obtained. This brings us to our next topic.

(b) Magnetic field constraint So far, we have ignored accounting for the constraint $\mathbf{k}\cdot\hat{\mathbf{B}} = 0$, originating from the general condition $\nabla\cdot\mathbf{B} = 0$ on magnetic fields. If one were to just solve the eigenvalue problem as it is, one would get eight solutions with *one spurious eigenvalue* $\omega = 0$ (to be distinguished from the genuine eigenvalue $\omega = 0$ of the entropy wave). This may be seen by operating with the projector $\mathbf{k}\cdot$ onto Eq. (5.33)(d), or (5.44)(d), which gives

$$
\omega\, \mathbf{k}\cdot\hat{\mathbf{B}} = 0,
$$

suggesting that $\omega = 0$ is an eigenvalue with eigenvectors satisfying $\mathbf{k}\cdot\hat{\mathbf{B}} \neq 0$, in glaring conflict with the constraint. The problem of spurious marginal eigenvalues is a more serious one than it appears at this stage. For example, in numerical MHD programs calculating eigenvalues to determine whether a certain plasma configuration is stable or not, due to truncation errors, a spurious eigenvalue $\omega = 0$ may not be distinguishable from a genuine transition from stability to instability.

To remove this spurious marginal eigenvalue, the constraint has to be incorporated in the eigenvalue problem by eliminating one of the magnetic field variables to obtain a 7×7 matrix representation. For example, one could eliminate $\hat{B}_z$, by simply substituting $\hat{B}_z = -(k_\perp/k_\parallel) \hat{B}_x$ in the second row and dropping the last row of Eq. (5.35). However, the symmetry of the matrix would then be lost. A more satisfactory approach to eliminate the spurious eigenvalue is obtained by exploiting the wave vector projection, discussed below.

▷ **Wave vector projection.** In this approach we define *new variables reflecting the physics of the problem*, e.g. by projecting the Fourier components of the vectors $\hat{\mathbf{v}}$ and $\hat{\mathbf{B}}$ on the three directions associated with the wave vector $\mathbf{k}$. This is equivalent to exploiting the two components of *the vorticity* $\nabla \times \hat{\mathbf{v}}$ and *the compressibility* $\nabla \cdot \hat{\mathbf{v}}$ as velocity field variables, and restricting the magnetic field variables to two components of *the current density* $\nabla \times \hat{\mathbf{B}}$ (since $\nabla \cdot \hat{\mathbf{B}} = 0$):

$$\hat{v}_{1,2} \equiv [(\mathbf{k}/k) \times \hat{\mathbf{v}}]_{x,y}, \quad \hat{v}_3 \equiv (\mathbf{k}/k) \cdot \hat{\mathbf{v}}, \quad \hat{B}_{1,2} \equiv [(\mathbf{k}/k) \times \hat{\mathbf{B}}]_{x,y}. \tag{5.46}$$

Operating with the projectors $(\mathbf{k}/k)\cdot$ and $(\mathbf{k}/k)\times$ on Eqs. (5.44)(b) and (5.44)(d) yields the following 7×7 representation in terms of the variables $\hat{S}$, $\hat{v}_1$, $\hat{v}_2$, $\hat{v}_3$, $\hat{p}$, $\hat{B}_1$ and $\hat{B}_2$:

$$\begin{pmatrix} 0 & 0 & 0 & 0 & 0 & 0 & 0 \\ 0 & 0 & 0 & 0 & 0 & -k_\parallel b & 0 \\ 0 & 0 & 0 & 0 & 0 & 0 & -k_\parallel b \\ 0 & 0 & 0 & 0 & kc & -k_\perp b & 0 \\ 0 & 0 & 0 & kc & 0 & 0 & 0 \\ 0 & -k_\parallel b & 0 & -k_\perp b & 0 & 0 & 0 \\ 0 & 0 & -k_\parallel b & 0 & 0 & 0 & 0 \end{pmatrix} \begin{pmatrix} \hat{S} \\ \hat{v}_1 \\ \hat{v}_2 \\ \hat{v}_3 \\ \hat{p} \\ \hat{B}_1 \\ \hat{B}_2 \end{pmatrix} = \omega \begin{pmatrix} \hat{S} \\ \hat{v}_1 \\ \hat{v}_2 \\ \hat{v}_3 \\ \hat{p} \\ \hat{B}_1 \\ \hat{B}_2 \end{pmatrix}. \tag{5.47}$$

Again, the representation is symmetric and even simpler than that of Eq. (5.35). Discarding the entropy wave, by dropping the first row and first column, results in the 6×6 representation for the main MHD waves discussed in the next section. ◁

▷ **Numerical 8-wave scheme with $\nabla \cdot \mathbf{B}$ wave.** In numerical studies of the nonlinear MHD equations, the constraint $\nabla \cdot \mathbf{B} = 0$ also poses a problem since it does not correspond to a proper evolution equation. This problem has been addressed by Powell [188, 189], following an older idea of Godunov [80], by artificially extending the MHD system with a '$\nabla \cdot \mathbf{B}$ wave'. To that end, the RHSs of the conservation equations (4.59)–(4.62), or rather (4.71)–(4.74) without gravity, are replaced by source terms proportional to $\nabla \cdot \mathbf{B}$ (which should become vanishingly small in the limit of vanishing step size):

$$\frac{\partial \rho}{\partial t} + \nabla \cdot \boldsymbol{\pi} = 0,$$

$$\frac{\partial \boldsymbol{\pi}}{\partial t} + \nabla \cdot \left[\boldsymbol{\pi}\mathbf{v} + (p + \tfrac{1}{2}B^2)\mathbf{I} - \mathbf{BB} \right] = -\mathbf{B}\nabla \cdot \mathbf{B} \qquad (\text{RHS} \Rightarrow 0 \text{ in Ref. [118]}),$$

$$\frac{\partial \mathcal{H}}{\partial t} + \nabla \cdot \left[\mathcal{H} + (p + \tfrac{1}{2}B^2)\mathbf{v} - \mathbf{v} \cdot \mathbf{BB} \right] = -\mathbf{v} \cdot \mathbf{B}\nabla \cdot \mathbf{B} \qquad (\text{RHS} \Rightarrow 0 \text{ in Ref. [118]}),$$

$$\frac{\partial \mathbf{B}}{\partial t} + \nabla \cdot (\mathbf{vB} - \mathbf{Bv}) = -\mathbf{v}\nabla \cdot \mathbf{B}. \tag{5.48}$$

This provides a numerical scheme of (near) conservation equations that is employed, for example, for the construction of solutions of the interaction of the solar wind with the magnetosphere of the Earth.

Of course, the addition of $\nabla \cdot \mathbf{B}$ terms in these equations implies that unphysical, numerically created, magnetic monopoles are permitted [38]. It was noticed by Janhunen [118] that the problem of possible non-positive numerical values of the pressure, calculated by using Eq. (4.75), is enlarged this way since the above modified momentum and energy 'conservation' equations involve unbalanced forces on those monopoles. By means of the modified Maxwell's equations involving hypothetical magnetic monopoles, as given in the textbook by Jackson [117], p. 252, he then shows that a more consistent set of modified MHD 'conservation' equations is obtained by restoring the momentum and energy equations to have a vanishing RHS, so that only the induction equation keeps the $\nabla \cdot \mathbf{B}$ source term. Later, it was shown by Dellar [63] that this set is easily extended to a modified set of relativistic, Lorentz invariant, MHD equations, which contain the earlier, Galilean invariant, set in the limit $v/c \to 0$.

Anyway, the modification of Faraday's law, Eq. (4.62)(d), is the essential one. With some algebra one obtains an evolution equation for $\nabla \cdot \mathbf{B}$ from it,

$$\frac{\partial}{\partial t}(\nabla \cdot \mathbf{B}) + \nabla \cdot (\mathbf{v}\nabla \cdot \mathbf{B}) = 0, \tag{5.49}$$

which replaces the constraint $\nabla \cdot \mathbf{B} = 0$. This equation is in genuine conservation form now: $\nabla \cdot \mathbf{B}$ is convected with the fluid in the same way as the density ρ. This implies that if $\nabla \cdot \mathbf{B}$ is small enough initially and at the boundaries, it should remain small; numerical errors $\nabla \cdot \mathbf{B} \neq 0$ are simply convected with the flow and, hopefully, out of the computational domain. Clearly, since Maxwell's equations are modified here, the physicist cannot offer assistance to the accountant. All that matters for him is whether the books finally balance, i.e. whether the numerical scheme is accurate enough to keep $\nabla \cdot \mathbf{B}$ small enough to honestly qualify as zero. For more on computational aspects of the $\nabla \cdot \mathbf{B} = 0$ constraint: see Tóth [230]. ◁

5.2.3 Reduction to velocity representation: three waves

We now transform to the velocity representation since it is the more powerful one. It gets rid of the entropy wave and absorbs the constraint $\mathbf{k} \cdot \hat{\mathbf{B}} = 0$ so that the associated spurious eigenvalue $\omega = 0$ is eliminated as well. For the reduction to the velocity variable, the perturbations ρ_1, e_1, $\mathbf{B}_1$ are expressed in terms of $\mathbf{v}_1$ by means of Eqs. (5.22), (5.24) and (5.25), and substituted into the momentum equation (5.23). This yields *the wave equation for MHD waves in a homogeneous medium*:

$$\frac{\partial^2 \mathbf{v}_1}{\partial t^2} - \left[(\mathbf{b} \cdot \nabla)^2 \mathbf{I} + (b^2 + c^2) \nabla \nabla - \mathbf{b} \cdot \nabla (\nabla \mathbf{b} + \mathbf{b} \nabla) \right] \cdot \mathbf{v}_1 = 0. \quad (5.50)$$

Note that this equation contains the sound wave equation (5.8), with $\mathbf{v}_0 = 0$, as the special case $\mathbf{b} = 0$ (as it should). Inserting plane wave solutions gives the required eigenvalue equation:

$$\left\{ \left[\omega^2 - (\mathbf{k} \cdot \mathbf{b})^2 \right] \mathbf{I} - (b^2 + c^2) \mathbf{k} \mathbf{k} + \mathbf{k} \cdot \mathbf{b} (\mathbf{k} \mathbf{b} + \mathbf{b} \mathbf{k}) \right\} \cdot \hat{\mathbf{v}} = 0, \quad (5.51)$$

or, in components:

$$\begin{pmatrix} -k_\perp^2 (b^2 + c^2) - k_\parallel^2 b^2 & 0 & -k_\perp k_\parallel c^2 \\ 0 & -k_\parallel^2 b^2 & 0 \\ -k_\perp k_\parallel c^2 & 0 & -k_\parallel^2 c^2 \end{pmatrix} \begin{pmatrix} \hat{v}_x \\ \hat{v}_y \\ \hat{v}_z \end{pmatrix} = -\omega^2 \begin{pmatrix} \hat{v}_x \\ \hat{v}_y \\ \hat{v}_z \end{pmatrix}. \quad (5.52)$$

Hence, a 3×3 symmetric matrix equation is obtained in terms of the variable $\hat{\mathbf{v}}$, with *quadratic eigenvalue* ω^2, corresponding to the original 6×6 representation with eigenvalue ω.

The reduction to a description in terms of the velocity alone may be generalized to inhomogeneous plasmas, where it leads to the very powerful force operator formalism (Section 6.2). However, it should be kept in mind that this reduction is only possible in *ideal* MHD, i.e. in the absence of dissipative effects like resistivity. In the presence of dissipation, the basic equations (5.18)–(5.21) have additional terms (see, e.g., Eq. (2.133)), spoiling the possibility of a reduction in terms of the velocity alone. However, the analysis leading to Eq. (5.44) may be extended easily. That is why we have presented both the representation in primitive variables and the velocity representation.

With the description in terms of the velocity we have lost the marginal entropy mode $\omega = 0$. For the sake of completeness, we will include this mode again. This is done by simply multiplying the determinant of Eq. (5.52) with the factor ω to

5.2 MHD waves

yield the determinant of the original system (5.44), or rather (5.47):

$$\det = \omega \left(\omega^2 - k_\parallel^2 b^2 \right) \left[\omega^4 - k^2(b^2 + c^2)\omega^2 + k_\parallel^2 k^2 b^2 c^2 \right] = 0, \quad (5.53)$$

where $k^2 \equiv k_\perp^2 + k_\parallel^2$. This algebraic expression, with its solutions $\omega = \omega_i(\mathbf{k})$ ($i = 1, \ldots, 7$), is called *the dispersion equation* for the MHD waves. Subsequently putting each of the four factors equal to zero gives the eigenfrequencies of the waves, and substituting that frequency back into Eq. (5.44) or (5.52) then gives the relationships between the amplitudes $\hat{s}$, $\hat{\mathbf{v}}$, $\hat{p}$, $\hat{\mathbf{B}}$ which characterize the eigenfunctions.

The dispersion equation (5.53) admits four kinds of solutions.

(a) Entropy 'waves' As we have already seen in Section 5.2.2(a), the eigenfrequency of these waves vanishes:

$$\omega = \omega_E \equiv 0, \quad (5.54)$$

whereas the eigenfunctions just involve the entropy:

$$\hat{\mathbf{v}} = \hat{\mathbf{B}} = 0, \quad \hat{p} = 0, \quad \text{but} \quad \hat{S} \neq 0. \quad (5.55)$$

The entropy waves are quite degenerate: they do not propagate, they do not involve flow, magnetic field, or pressure perturbations. They just constitute a perturbation of the entropy (or density and internal energy) which would be carried with the flow if there were a background velocity field. Now, they just sit there: not very exciting. The use of such degenerate solutions is usually that they serve as a reminder for the possibility of new waves if additional physics is brought into the model. For the present discussion, they do not serve any purpose. That is why the formulation in terms of $\hat{\mathbf{v}}$ is so attractive: these marginal modes are automatically eliminated from the analysis.

(b) Alfvén waves The eigenfrequency of these waves is determined by the parallel wave vector and the Alfvén velocity:

$$\omega = \pm \omega_A, \quad \omega_A \equiv \mathbf{k} \cdot \mathbf{b} = k_\parallel b = kb \cos \vartheta,$$
$$\left[\omega_A^2 = k^2 b^2 \cos^2 \vartheta, \right] \quad (5.56)$$

where ϑ is the angle between $\mathbf{k}$ and $\mathbf{b}$, i.e. $\mathbf{B}_0$. There are two solutions, one for $\omega = \omega_A$ corresponding to waves propagating in the direction of $\mathbf{B}_0$, and another one for $\omega = -\omega_A$ corresponding to waves propagating in the opposite direction.

The eigenfunctions of the Alfvén waves just involve the perpendicular velocity and magnetic field (Fig. 5.1), whereas all other components and the perturbations

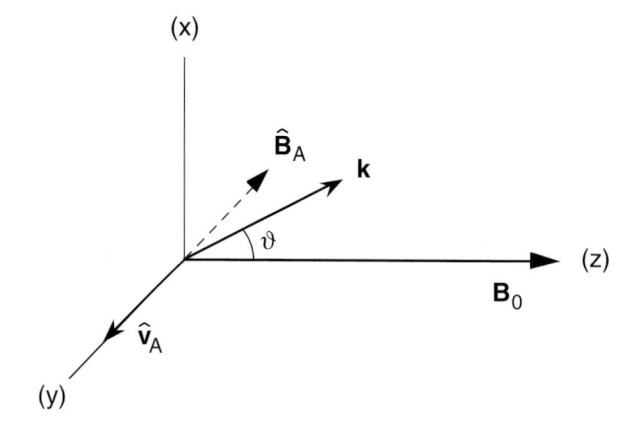

Fig. 5.1. Velocity and magnetic field perturbations for Alfvén waves.

of all thermodynamic variables vanish:

$$\hat{B}_y = -\hat{v}_y \neq 0, \qquad \hat{v}_x = \hat{v}_z = \hat{B}_x = \hat{B}_z = \hat{S} = \hat{p} = 0. \tag{5.57}$$

▷ **Alternative variables** for the wave vector projection:
$$\hat{v}_1 = -\hat{B}_1 = -(k_\parallel/k)\,\hat{v}_y \neq 0, \qquad \hat{v}_2 = \hat{v}_3 = \hat{B}_2 = 0. \qquad ◁$$

These are the waves already introduced in Section 2.4.2. They are incompressible and purely transverse, as regards both $\hat{\mathbf{v}}$ and $\hat{\mathbf{B}}$. Alfvén waves are the most important MHD waves since they are a direct result of flux conservation and magnetic field lines frozen into the fluid: a perpendicular flow velocity $\hat{\mathbf{v}}$ results in a field perturbation $\hat{\mathbf{B}}$ of opposite sign which just causes the field lines to follow the flow (as illustrated in Fig. 2.7).

(c) Fast and slow magneto-acoustic waves The eigenfrequencies of these waves are obtained from the quartic factor of the dispersion equation (5.53):

$$\omega = \pm \omega_{s,f}, \qquad \omega_{s,f} \equiv k\sqrt{\tfrac{1}{2}(b^2+c^2) \pm \tfrac{1}{2}\sqrt{(b^2+c^2)^2 - 4(k_\parallel^2/k^2)\,b^2 c^2}},$$

$$\left[\,\omega_{s,f}^2 = \tfrac{1}{2}k^2(b^2+c^2)\left(1 \pm \sqrt{1 - \sigma \cos^2\vartheta}\right),\,\right] \tag{5.58}$$

where the first $\pm$ sign refers to wave propagation to the right ($+$) and to the left ($-$), and the second $\pm$ sign (under the square root) refers to *the fast* ($+$) and *slow* ($-$) *magneto-acoustic wave*, respectively. The auxiliary parameter

$$\sigma \equiv \frac{4b^2 c^2}{(b^2+c^2)^2} = \left[\tfrac{1}{2}(b/c + c/b)\right]^{-2} \qquad (0 \leq \sigma \leq 1) \tag{5.59}$$

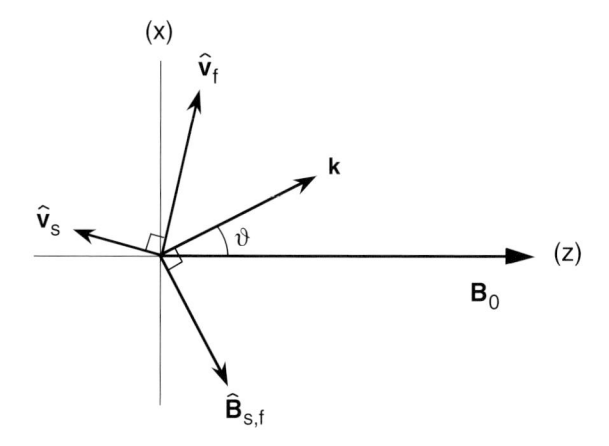

Fig. 5.2. Velocity and magnetic field perturbations for magneto-sonic waves.

is just a function of the ratio of the two characteristic speeds (b and c) in the problem, where $\sigma = 0$ when $b = 0$ or $c = 0$, and $\sigma = 1$ when $b = c$.

The eigenfunctions of the magneto-sonic waves exhibit a complicated dependence on almost all of the variables (Fig. 5.2):

$$\hat{v}_z = \alpha_{s,f} (k_\parallel/k_\perp) \hat{v}_x \neq 0, \qquad \hat{v}_y = 0,$$
$$\hat{B}_z = -(k_\perp/k_\parallel) \hat{B}_x = (k_\perp b/\omega_{s,f}) \hat{v}_x \neq 0, \qquad \hat{B}_y = 0,$$
$$\hat{p} = [\alpha_{s,f} \omega_{s,f}/(k_\perp c)] \hat{v}_x \neq 0, \qquad \hat{S} = 0, \qquad (5.60)$$

where the factor $\alpha_{s,f}$ distinguishes between the fast and slow eigenfunctions:

$$\alpha_{s,f} \equiv 1 - \frac{k^2 b^2}{\omega_{s,f}^2}, \qquad \text{so that} \quad \alpha_s \leq 0 \quad \text{and} \quad \alpha_f \geq 0. \qquad (5.61)$$

▷ **Alternative variables** for the wave vector projection:

$$\hat{v}_2 = [k_\parallel k_\perp b^2/(\omega_{s,f}^2 - k_\parallel^2 b^2)] \hat{v}_3 = (k_\parallel k b^2/\omega_{s,f}^2) \hat{v}_x \neq 0, \qquad \hat{v}_1 = 0,$$
$$\hat{B}_2 = -(kb/\omega_{s,f})\hat{v}_x \neq 0, \qquad \hat{B}_1 = 0. \qquad ◁$$

The magneto-acoustic waves are composed of both magnetic (b) and acoustic (c) constituents. The perturbations $\hat{\mathbf{v}}$ and $\hat{\mathbf{B}}$ are lying in the plane through $\mathbf{k}$ and $\mathbf{B}_0$, so that the total perturbed field ($\mathbf{B}_0 + \mathbf{B}_1$) also lies in that plane. The fast and slow velocity fields are orthogonal to each other, whereas the magnetic field perturbations are both orthogonal to $\mathbf{k}$ (as a consequence of $\mathbf{k} \cdot \hat{\mathbf{B}} = 0$). Since the normalization of the waves is arbitrary, we have chosen $\hat{\mathbf{B}}_s = \hat{\mathbf{B}}_f$ in the picture for simplicity.

5.2.4 Dispersion diagrams

From now on, we will ignore the entropy waves. We then notice two quite general properties of the eigenfrequencies and eigenfunctions of the remaining three (slow, Alfvén, and fast) MHD waves, viz.:

- *the eigenfrequencies are well ordered,*

$$0 \leq \omega_s^2 \leq \omega_A^2 \leq \omega_f^2 < \infty ; \tag{5.62}$$

- *the eigenfunctions are mutually orthogonal,*

$$\hat{\mathbf{v}}_s \perp \hat{\mathbf{v}}_A \perp \hat{\mathbf{v}}_f . \tag{5.63}$$

The first property will turn out to play a crucial role in the spectral theory of MHD waves, to be discussed in Chapter 6. The second property guarantees that an arbitrary velocity field may be decomposed at all times (e.g. at $t = 0$) in the three MHD waves. This implies that *the initial value problem is a well-posed problem.*

The three MHD waves exhibit a strong anisotropy depending on the direction of the wave vector $\mathbf{k}$ with respect to the magnetic field $\mathbf{B}_0$. This is most completely expressed by the two dispersion diagrams $\omega^2 = \omega^2(k_\parallel)$ and $\omega^2 = \omega^2(k_\perp)$ shown in Fig. 5.3. They are obtained from the dispersion equations (5.56) for the Alfvén waves and (5.58) for the magneto-sonic waves, where $k_\perp$ is kept fixed in the upper diagram and $k_\parallel$ is kept fixed in the lower diagram. They clearly illustrate the ordering principle (5.62) for the three MHD waves. Note that the diagram for $\omega^2 = \omega^2(k_\perp)$, in particular for large values of $k_\perp$, provides the most distinctive representation of the three dispersion curves. We return to this point in Section 5.3.3.

The behaviour of the magneto-sonic eigenfrequencies at small $k_\parallel$ or small $k_\perp$ is obtained by expanding the square root factor in the second dispersion expression (5.58). For *approximately perpendicular propagation* ($\vartheta \approx \pi/2$, i.e. $k_\parallel \ll k$), this yields

$$\sqrt{1 - \sigma \cos^2 \vartheta} \approx 1 - \tfrac{1}{2}\sigma \, (k_\parallel/k)^2, \tag{5.64}$$

so that

$$\omega_s^2 \approx k_\parallel^2 \, \frac{b^2 c^2}{b^2 + c^2} \; \leq \; \omega_A^2 = k_\parallel^2 b^2 \; \ll \; \omega_f^2 \approx k_\perp^2 (b^2 + c^2). \tag{5.65}$$

Hence, the slow and the Alfvén frequencies behave similarly for small $k_\parallel$ and even vanish when $k_\parallel \to 0$. The importance of the latter property for stability can

5.2 MHD waves

(a) Dispersion diagrams for $\bar{k}_\perp = 1.0$

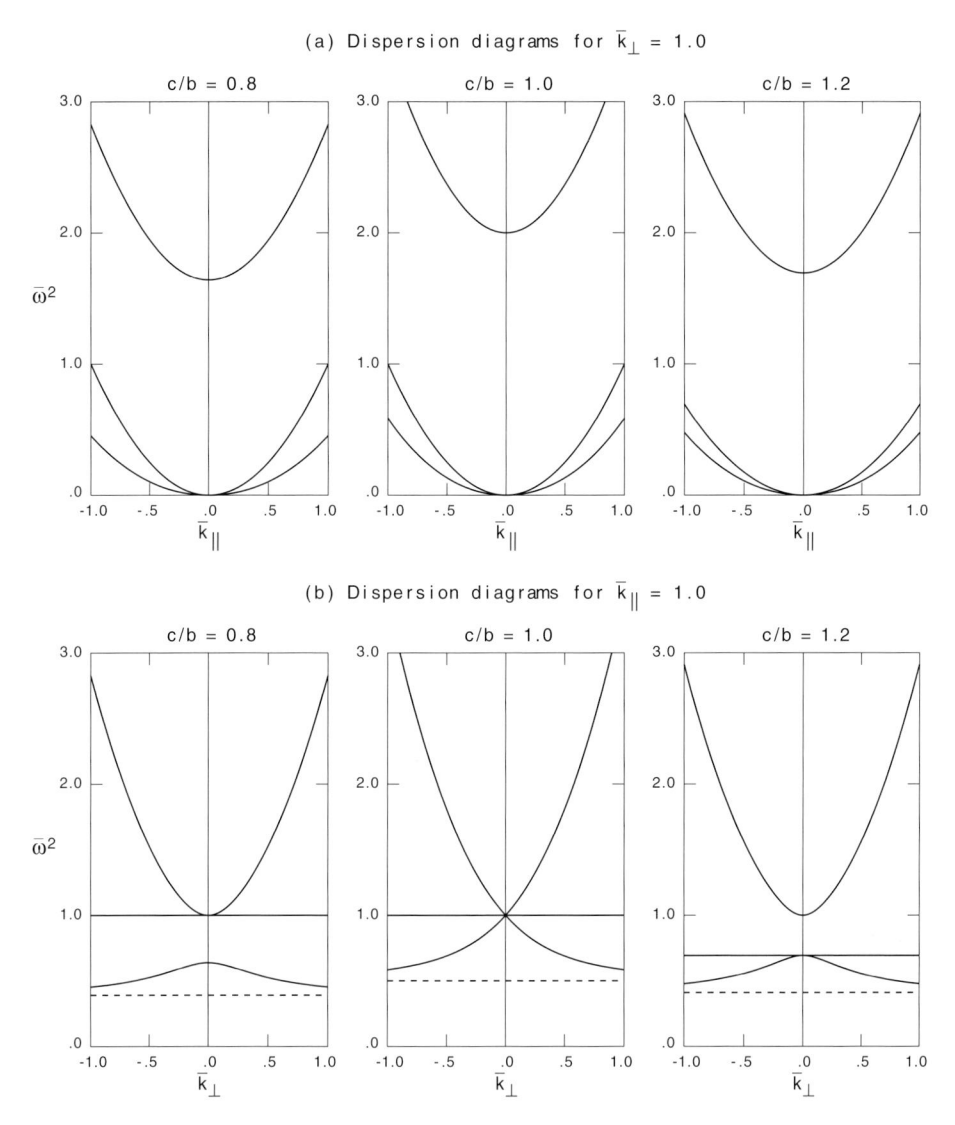

(b) Dispersion diagrams for $\bar{k}_\parallel = 1.0$

Fig. 5.3. Dispersion diagrams for three values of the ratio c/b of sound to Alfvén speed: (a) $\omega^2 = \omega^2(k_\parallel)$ keeping $k_\perp$ fixed, (b) $\omega^2 = \omega^2(k_\perp)$ keeping $k_\parallel$ fixed. Eigenvalues are normalized as $\bar{\omega}^2 \equiv \omega^2 \ell^2 / \max(b^2, c^2)$ and wave vectors as $\bar{k} \equiv k\ell$, where ℓ is a unit length.

hardly be overestimated. We will have plenty of opportunity to return to this in later chapters. For *purely parallel propagation* ($\vartheta = 0$, i.e. $k_\perp = 0$), the square root factor becomes

$$\sqrt{1 - \sigma} = \frac{|b^2 - c^2|}{b^2 + c^2}, \tag{5.66}$$

so that

$$\omega_s^2 = k_\parallel^2 \times \min(b^2, c^2) \quad \leq \quad \omega_A^2 = k_\parallel^2 b^2 \quad \leq \quad \omega_f^2 = k_\parallel^2 \times \max(b^2, c^2). \tag{5.67}$$

Hence, either the slow frequency coincides with the Alfvén frequency (if $b^2 < c^2$) or the fast frequency coincides with ω_A^2 (if $b^2 > c^2$). Complete degeneracy of the three MHD waves is obtained for $k_\perp = 0$ and $b = c$ (middle frame of Fig. 5.3(b)).

The dispersion diagrams shown in Fig. 5.3 depend qualitatively on the parameter c/b, measuring the relative contributions of hydrodynamic and magnetic effects. For later reference, we notice that the important parameter β which measures the ratio of the plasma pressure and the magnetic pressure (defined in Eq. (4.34)), is related to the square of this ratio of the sound and the Alfvén velocity:

$$\beta \equiv \frac{2\mu_0 p_0}{B_0^2} = \frac{2}{\gamma} \frac{c^2}{b^2}. \tag{5.68}$$

For $\gamma = 5/3$, this implies the simple relation $\beta = 1.2\,(c^2/b^2)$. The assumption $\beta \ll 1$, or $c^2 \ll b^2$, is valid for quite a number of relevant plasmas (like those in tokamaks and in the solar corona) so that we may exploit this approximation for those cases.

We will have many occasions to return to the dispersion diagrams of Fig. 5.3 when discussing wave propagation in inhomogeneous media. Already now, we wish to draw the attention to some generic features.

(a) The Alfvén and slow frequencies ω_A and ω_s vanish for $k_\parallel = 0$. In that case, $\mathbf{k} \perp \mathbf{B}_0$, which implies that the waves do not bend the background magnetic field $\mathbf{B}_0$. This condition has important consequences for *stability*, as we will see.
(b) The eigenfrequencies ω^2 of the different waves depend monotonically on the value of either one of the two components of the wave vector $\mathbf{k}$. (In particular, notice the unusual decreasing dependence $\omega^2 = \omega^2(k_\perp)$ for the slow magneto-sonic waves.) This property determines the *general structure of the spectrum of MHD waves* in plasmas.
(c) For large values of the wave vector the eigenfrequencies asymptotically tend to either ∞ or some finite value, which turns out to be very robust in the sense of being rather independent of the assumption of homogeneity of the plasma. This has extremely important consequences for *local wave propagation*.

In one form or another, all of these properties survive in the more general inhomogeneous plasmas which we will investigate in later chapters.

▷ **Exercise.** Derive asymptotic expressions for the frequencies of the three MHD waves in the limits $k_\parallel \to 0$, $k_\parallel \to \infty$, $k_\perp \to 0$, $k_\perp \to \infty$, and convince yourself that they correspond with the diagrams shown in Fig. 5.3. ◁

5.3 Phase and group diagrams

5.3.1 Basic concepts

We have discussed the peculiar properties of the MHD dispersion diagrams in Section 5.2.4. We now analyse the more general implications of the *dispersion equation* $\omega = \omega(\mathbf{k})$ of wave phenomena, relating the angular frequency ω to the wave vector $\mathbf{k}$. For the ideal MHD waves this relation is implicitly expressed by Eq. (5.53). From the dispersion equation two important quantities may be derived. The first one is *the phase velocity*

$$\mathbf{v}_{\text{ph}} \equiv (\omega/k)\,\mathbf{n}, \qquad \mathbf{n} \equiv \mathbf{k}/k, \tag{5.69}$$

which gives the speed of propagation of a single plane wave in the direction of $\mathbf{k}$. The magnitude of the phase velocity of the three MHD waves depends on the angle ϑ between $\mathbf{k}$ and the background magnetic field $\mathbf{B}_0$, so that $v_{\text{ph}} \equiv \omega/k = f(\vartheta)$, but it does not depend on the magnitude k of the wave vector. Such waves are called *non-dispersive* because a wave packet, consisting of many components with different wave numbers, may propagate without distortion (at least in one direction). Such a packet propagates with *the group velocity*

$$\mathbf{v}_{\text{gr}} \equiv \frac{\partial \omega}{\partial \mathbf{k}} \left[\equiv \frac{\partial \omega}{\partial k_x}\mathbf{e}_x + \frac{\partial \omega}{\partial k_y}\mathbf{e}_y + \frac{\partial \omega}{\partial k_z}\mathbf{e}_z \right], \tag{5.70}$$

which also gives the direction of the flow of the energy carried by the wave packet; see Braddick [39], Section 5.2.

The concept of group velocity deserves some further amplification; see Bittencourt [31]. Consider a wave packet consisting of a superposition of plane waves obeying a dispersion equation $\omega = \omega(\mathbf{k})$:

$$\Psi_i(\mathbf{r}, t) = \frac{1}{(2\pi)^{3/2}} \int_{-\infty}^{\infty} A_i(\mathbf{k})\, e^{i(\mathbf{k}\cdot\mathbf{r} - \omega(\mathbf{k})t)}\, d^3k. \tag{5.71}$$

This packet evolves from an initial shape, given by the Fourier synthesis

$$\Psi_i(\mathbf{r}, 0) = \frac{1}{(2\pi)^{3/2}} \int_{-\infty}^{\infty} A_i(\mathbf{k})\, e^{i\mathbf{k}\cdot\mathbf{r}}\, d^3k. \tag{5.72}$$

Vice versa, the amplitudes $A_i(\mathbf{k})$ are related to the initial values $\Psi_i(\mathbf{r}, 0)$ by Fourier analysis:

$$A_i(\mathbf{k}) = \frac{1}{(2\pi)^{3/2}} \int_{-\infty}^{\infty} \Psi_i(\mathbf{r}, 0)\, e^{-i\mathbf{k}\cdot\mathbf{r}}\, d^3r. \tag{5.73}$$

In our case, Ψ_i and A_i represent the perturbations ρ_1, $\mathbf{v}_1$, e_1, $\mathbf{B}_1$ and their Fourier amplitudes, respectively, so that the index $i = 1, 2, \ldots, 8$. (Alternatively, one could consider the representation (5.47), where $i = 1, 2, \ldots, 7$, or the velocity representation (5.51), where $i = 1, 2, 3$.) Furthermore, the pertinent dispersion equation $\omega = \omega_A(\mathbf{k})$ of the Alfvén waves, given by Eq. (5.56), or $\omega = \omega_{s,f}(\mathbf{k})$ of the magneto-sonic waves, given by Eq. (5.58), should be inserted. It is to be noted that only one of the amplitudes $A_i(\mathbf{k})$ can be chosen freely as a normalization, whereas the others should be chosen in agreement with the eigenfunction relations (5.57) for the Alfvén waves, or (5.60) for the magneto-sonic waves.

Let us assume that the wave packet consists of harmonics with wave vectors centred about some central value $\mathbf{k}_0$. A typical example is the Gaussian distribution,

$$A_i(\mathbf{k}) = \hat{A}_i \, e^{-\frac{1}{2}|(\mathbf{k}-\mathbf{k}_0)\cdot \mathbf{a}|^2}, \tag{5.74}$$

where the components of the auxiliary vector $\mathbf{a}$ measure the width of the distribution in each of the three directions. This corresponds to an initial wave packet consisting of a main harmonic, with wave vector $\mathbf{k}_0$, and a modulated amplitude centred at the origin $\mathbf{r} = 0$:

$$\Psi_i(\mathbf{r}, 0) = e^{i\mathbf{k}_0 \cdot \mathbf{r}} \times \frac{\hat{A}_i}{a_x a_y a_z} e^{-\frac{1}{2}[(x/a_x)^2 + (y/a_y)^2 + (z/a_z)^2]}. \tag{5.75}$$

An extreme example is the δ-function distribution, $A_i(\mathbf{k}) = (2\pi)^{3/2} \hat{A}_i \, \delta(\mathbf{k} - \mathbf{k}_0)$, corresponding to a plane wave $\Psi_i(\mathbf{r}, 0) = \hat{A}_i \exp(i\mathbf{k}_0 \cdot \mathbf{r})$ with a single wave vector $\mathbf{k}_0$.

For an arbitrary wave packet with a reasonably localized range of wave vectors (i.e. not infinitely narrow or infinitely wide), we may expand the dispersion equation about the central value $\mathbf{k}_0$:

$$\omega(\mathbf{k}) \approx \omega_0 + (\mathbf{k} - \mathbf{k}_0) \cdot \left(\frac{\partial \omega}{\partial \mathbf{k}}\right)_{\mathbf{k}_0}, \qquad \omega_0 \equiv \omega(\mathbf{k}_0). \tag{5.76}$$

Inserting this approximation in the expression (5.71) for the wave packet gives

$$\Psi_i(\mathbf{r}, t) \approx e^{i(\mathbf{k}_0 \cdot \mathbf{r} - \omega_0 t)} \times \frac{1}{(2\pi)^{3/2}} \int_{-\infty}^{\infty} A_i(\mathbf{k}) \, e^{i(\mathbf{k} - \mathbf{k}_0) \cdot (\mathbf{r} - (\partial \omega / \partial \mathbf{k})_{\mathbf{k}_0} t)} \, d^3k, \tag{5.77}$$

representing a carrier wave $\exp i(\mathbf{k}_0 \cdot \mathbf{r} - \omega_0 t)$ with an amplitude-modulated envelope. Through constructive interference of the plane wave harmonics, the envelope maintains its shape during an extended interval of time, whereas the surfaces of constant phase of this envelope move precisely with the group velocity,

$$\mathbf{v}_{\text{gr}} = \left(\frac{d\mathbf{r}}{dt}\right)_{\text{const phase}} = \left(\frac{\partial \omega}{\partial \mathbf{k}}\right)_{\mathbf{k}_0}, \tag{5.78}$$

in agreement with the definition (5.70).

5.3.2 Application to the MHD waves

The relation with the geometric construction of wave fronts by Huygens' principle is well known. For MHD waves, this is most strikingly illustrated for the case of Alfvén waves. From Eq. (5.56), the phase velocity for Alfvén waves is given by

$$(\mathbf{v}_{\text{ph}})_A \equiv (v_{\text{ph}})_A \, \mathbf{n}, \qquad (v_{\text{ph}})_A = b \cos \vartheta, \qquad (5.79)$$

so that the locus of the endpoints of the vector $(\mathbf{v}_{\text{ph}})_A$, i.e. the *phase diagram*, consists of two circles touching the origin (one of which is shown in Fig. 5.4(a)). The anisotropy of Alfvén wave propagation is even more strongly manifested by a localized wave packet resulting from a point perturbation in the origin at $t = 0$. This gives rise to the *group diagram*, which is the envelope at unit time of the wave fronts of a superposition of plane waves having passed in all directions through the origin at $t = 0$. Because the endpoints of the phase velocity vectors $(\mathbf{v}_{\text{ph}})_A$ of plane Alfvén waves lie on a circle, these wave fronts all go through a single point (Fig. 5.4(b)). Hence, the group diagram (or *caustic*) for Alfvén waves just consists of the two points $\pm b$ along $\mathbf{B}_0$ (Fig. 5.4(c)), so that the group velocity of Alfvén waves is given by

$$(\mathbf{v}_{\text{gr}})_A = \mathbf{b}. \qquad (5.80)$$

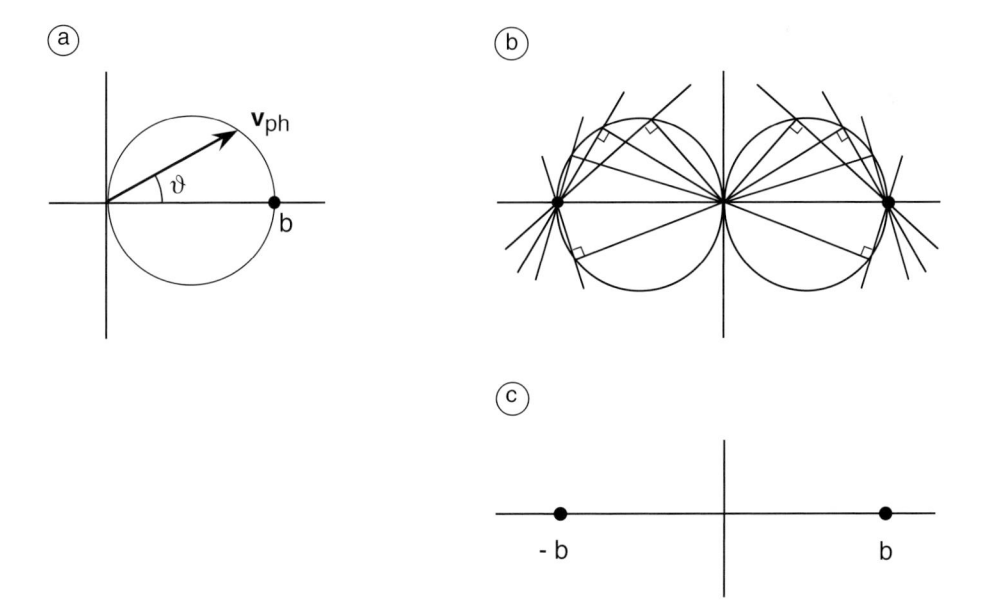

Fig. 5.4. Construction of the phase and group diagrams for Alfvén waves.

This provides the most extreme example of anisotropy of plasma waves guided by a magnetic field: *Alfvén wave point disturbances, and their associated energy flow, just propagate along single magnetic field lines.* Of course, Eq. (5.80) is also obtained algebraically from the dispersion equation (5.56).

For the magneto-acoustic waves similar geometrical constructions can be made, but, in this case, it is easier to exploit the algebraic expressions for the phase and group velocities following from the dispersion equation (5.58). It is expedient to introduce unit vectors in the x-z plane, i.e. the plane of $\mathbf{k}$ and $\mathbf{b}$:

$$\mathbf{n} \equiv \mathbf{k}/k = (\sin\vartheta, 0, \cos\vartheta),$$
$$\mathbf{t} \equiv \left[(\mathbf{b}/b) \times \mathbf{n}\right] \times \mathbf{n} = (\cos\vartheta, 0, -\sin\vartheta). \tag{5.81}$$

The expression for the phase velocity of the magneto-acoustic waves then reads

$$(\mathbf{v}_{\text{ph}})_{s,f} \equiv (v_{\text{ph}})_{s,f}\,\mathbf{n}, \qquad (v_{\text{ph}})_{s,f} = \sqrt{\tfrac{1}{2}(b^2 + c^2)}\sqrt{1 \pm \sqrt{1 - \sigma\cos^2\vartheta}}, \tag{5.82}$$

whereas the expression for the group velocity becomes

$$(\mathbf{v}_{\text{gr}})_{s,f} = (v_{\text{ph}})_{s,f}\left[\mathbf{n} \pm \frac{\sigma \sin\vartheta \cos\vartheta}{2\sqrt{1 - \sigma\cos^2\vartheta}\,\left[1 \pm \sqrt{1 - \sigma\cos^2\vartheta}\right]}\,\mathbf{t}\right]. \tag{5.83}$$

The derivation of the latter expression requires some straightforward algebra which we leave as an exercise.

▷ **Exercise.** Carry this out. Hint: compute $(v_{\text{gr}})_\perp = \partial\omega/\partial k_\perp = (\sin\vartheta/v_{\text{ph}})\,\partial\omega^2/\partial k_\perp^2$, and similarly for $(v_{\text{gr}})_\parallel$, and project on the unit vectors $\mathbf{n}$ and $\mathbf{t}$. ◁

All of the above expressions are brought together in the phase and group diagrams shown in Fig. 5.5, which depict $\mathbf{v}_{\text{ph}}(\vartheta)$ and $\mathbf{v}_{\text{gr}}(\vartheta)$ for the three MHD waves for three different values of c/b. Pictures of this kind were first constructed by K. O. Friedrichs [76] and, therefore, rightly named *Friedrichs diagrams*. The strong anisotropy of the MHD waves is manifest. In particular, notice the very different manner of fast and slow wave propagation: *the fast magneto-sonic waves may be considered as generalized sound waves with significant contributions of the magnetic pressure.* Not surprisingly, these contributions increase the perpendicular speed of propagation. In accordance with the orthogonality condition (5.63), expressing a deep duality of the waves, the waves behave in exactly the opposite manner: *the slow magneto-sonic waves may be considered as sound waves with strong magnetic guidance.* Like the Alfvén waves, internal focusing by the magnetic field produces wave packages that propagate dominantly along the magnetic field. This leads to the interesting cusp-shaped caustics exhibited by the slow wave group diagrams in the bottom part of Fig. 5.5.

Clearly, all MHD wave properties depend strongly on the value of the parameter c/b, which we showed to be related to the usual parameter β in Eq. (5.68). In Fig. 5.5, we have depicted three curves for values of c/b deviating little from 1, to show the interesting degeneracy at $c = b$, but also because the slow wave

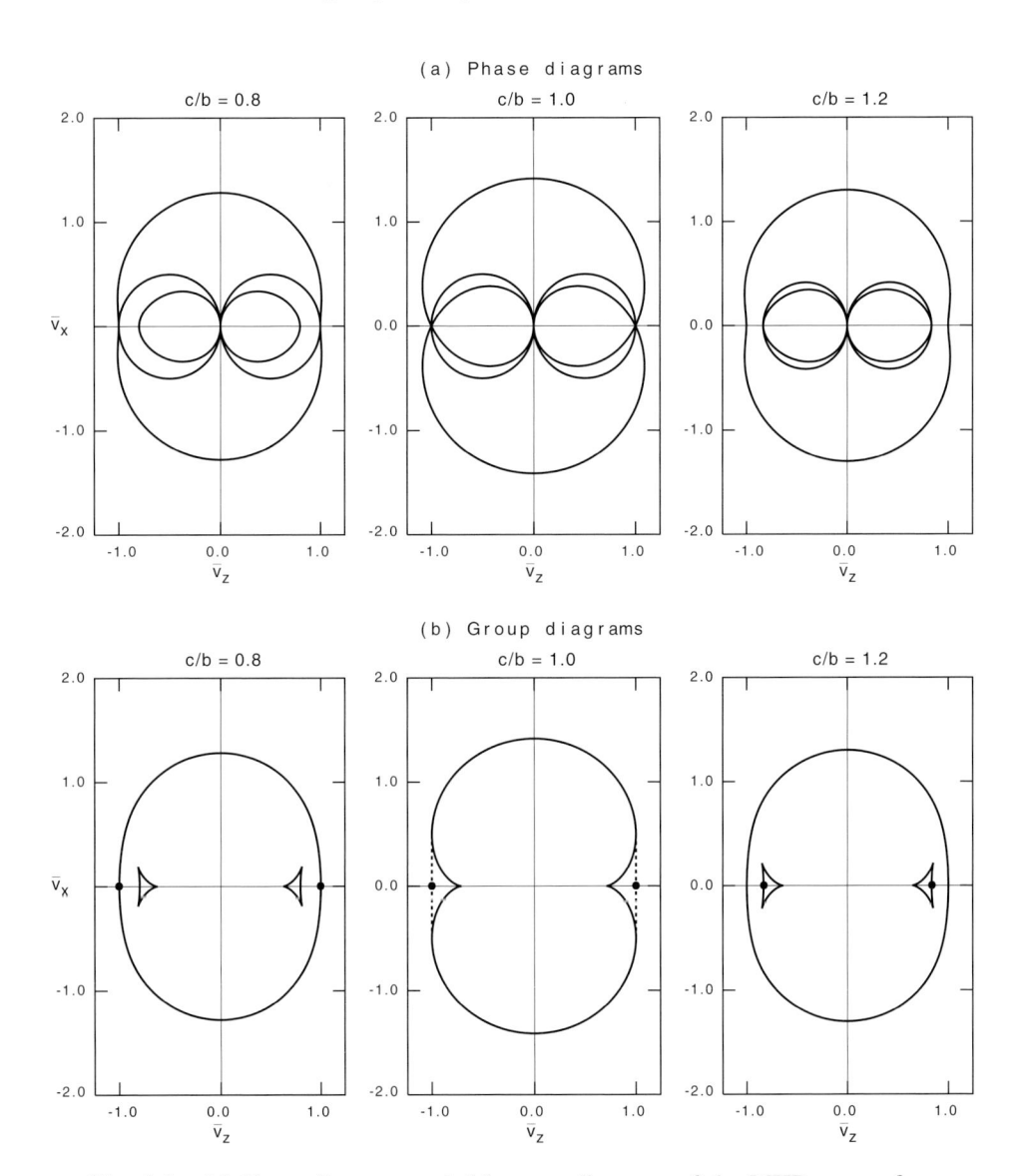

Fig. 5.5. (a) Phase diagrams and (b) group diagrams of the MHD waves for three values of the ratio c/b of the sound speed to the Alfvén speed. The phase and group velocities are normalized as $\bar{v} \equiv v/\max(b, c)$.

curves are still visible for this choice. For the important case of low β plasmas (e.g. $c/b = 0.1$, so that $\beta = 0.012$ when $\gamma = 5/3$), the slow branches virtually disappear in the origin. This wide separation from the Alfvén and fast branches is an important property that is frequently exploited to simplify wave and stability studies of, e.g., tokamaks. The theoretical framework is called *the low β expansion scheme*.

▷ **Question.** What happens in the opposite limit ($\beta \gg 1$ or $c/b \gg 1$)? Also discuss the limiting case of pure gas dynamics ($\beta \to \infty$ or $b = 0$). ◁

The construction of the phase diagram is much easier than that of the group diagram since $\mathbf{v}_{\text{ph}}$ and $\mathbf{n}$ have the same direction. However, the group diagram has a wider validity since it represents the response to local disturbances, which remains valid for inhomogeneous plasmas. The complicating factor, not visible in Fig. 5.5, is that the direction of the group velocity $\mathbf{v}_{\text{gr}}(\vartheta)$ deviates significantly from the direction $\mathbf{n}(\vartheta)$ of the central wave vector. This is also evident from the expressions (5.80) for the Alfvén waves and (5.83) for the magneto-sonic waves. In particular, slow magneto-sonic wave packages behave quite oddly (see below in the discussion of Fig. 5.6) so that one has to be very careful with the limiting values of parallel and perpendicular propagation for these waves. From the dispersion equations (5.56) and (5.58) one obtains the following limiting values for the phase and group velocities:

$\vartheta = 0$:
$$(v_{\text{ph/gr}})_s = \min(b, c) \quad \leq \quad (v_{\text{ph/gr}})_A = b \quad \leq \quad (v_{\text{ph/gr}})_f = \max(b, c),$$
$\vartheta = \pi/2$:
$$(v_{\text{ph}})_s = 0 \qquad\qquad = \quad (v_{\text{ph}})_A = 0 \quad < \quad (v_{\text{ph}})_f = \sqrt{b^2 + c^2},$$
$$(v_{\text{gr}})_s = \frac{bc}{\sqrt{b^2+c^2}}\text{(cusp)} \quad \leq \quad (v_{\text{gr}})_A = b \quad \leq \quad (v_{\text{gr}})_f = \sqrt{b^2 + c^2},$$

(5.84)

in agreement with the expressions (5.67) and (5.65), respectively.

Finally, we wish to present the peculiar differences of the relationship between the directions of $\mathbf{n}(\vartheta)$ and $\mathbf{v}_{\text{gr}}(\vartheta)$ for the three MHD waves. In Fig. 5.6 one of the group diagrams is depicted once more with the central wave vector $\mathbf{n}$ lying in the first quadrant and the three group velocities exhibiting their mutually exclusive directions of propagation: when the direction of $\mathbf{n}$ changes from $\vartheta = 0$ (parallel to $\mathbf{B}$) to $\vartheta = \pi/2$ (perpendicular to $\mathbf{B}$), the fast group velocity also changes from parallel to perpendicular (though it does not remain parallel to $\mathbf{n}$), the Alfvén group velocity just remains purely parallel, but the slow group velocity initially changes *clockwise* from parallel to some negative angle at $\vartheta = \vartheta_m$ (computed

5.3 Phase and group diagrams 211

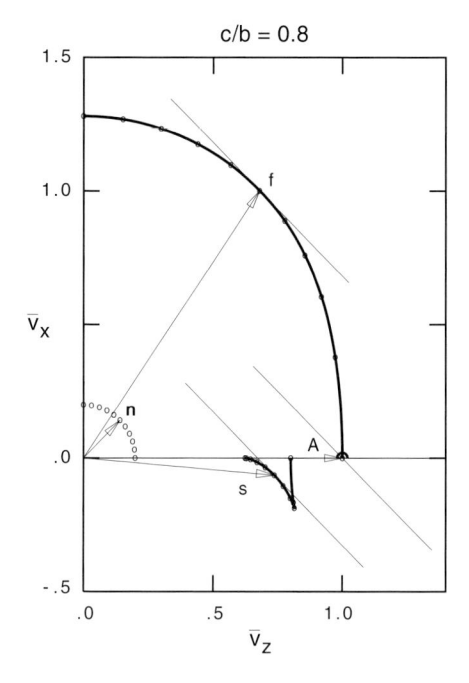

Fig. 5.6. Group diagrams with the group velocity vectors relative to the normal **n** of the three MHD waves in the first quadrant.

below) and then back again (i.e. anti-clockwise) to purely parallel. For the latter direction, the value of the slow group velocity, called the *cusp velocity*, is indicated on the last line of Eq. (5.84). It plays an important role in the analysis of local waves in inhomogeneous plasmas. Notice that slow wave packages propagate in the perpendicular direction opposite to the direction of **n**! Somehow, magnetic focusing is perfect for Alfvén waves, but slightly overshoots for slow waves.

▷ **Computation of the return angle in the slow wave group diagram.** The angle ϑ_m between **k** and $\mathbf{B}_0$ where the slow wave group vector $(\mathbf{v}_{\text{gr}})_s$ returns may be computed from Eq. (5.83) by defining the square root expression $R(\vartheta) \equiv \sqrt{1 - \sigma \cos^2 \vartheta}$. This quantity increases monotonically with ϑ so that it can be used as a parameter measuring the angle instead of ϑ itself. In terms of R, the magnitude of the slow group velocity is given by

$$(v_{\text{gr}}^2)_s = \frac{1}{8}(b^2 + c^2) \frac{-3R^3 + 5R^2 - (1-\sigma)R - (1-\sigma)}{R^2}. \tag{5.85}$$

Its maximum is reached for $R_m \equiv R(\theta_m)$ satisfying the cubic equation

$$3R_m^3 - (1-\sigma)R_m - 2(1-\sigma) = 0, \tag{5.86}$$

which has only one physically acceptable solution:

$$R_m = 3\tau^{1/3} \left[(1 + \sqrt{1-\tau})^{1/3} + (1 - \sqrt{1-\tau})^{1/3} \right], \qquad \tau \equiv \frac{1}{81}(1-\sigma), \qquad (5.87)$$

from which the required angle $\vartheta_m = \arccos(\sqrt{(1-R_m^2)/\sigma})$ is obtained. The corresponding magnitude of $(\mathbf{v}_{\text{gr}})_s$, and its angle with $\mathbf{B}_0$ (not to be confused with ϑ_m!), is obtained from Eq. (5.83) by substituting ϑ_m. Check that $(v_{\text{gr}})_\perp < 0$. ◁

5.3.3 Asymptotic properties

Clearly, the three MHD waves exhibit a distinct difference with respect to their propagation properties in the different directions. This was already discussed in Section 5.2.4 and illustrated by the dispersion diagrams of Fig. 5.3, plotting ω^2 as a function of $k_\parallel$ with $k_\perp$ fixed, and vice versa. In Fig. 5.7, we show the dispersion diagrams for large values of $k_\parallel$ and $k_\perp$ once more, in a schematic fashion, in order to highlight their asymptotic properties. In the left diagram ($\omega^2 = \omega^2(k_\parallel)$ for fixed $k_\perp$), the fast branch starts at $\omega_1^2 \equiv k_\perp^2(b^2 + c^2)$ and tends to $\omega_2^2 \equiv k_\parallel^2 b^2 \equiv \omega_A^2$ as $k_\parallel \to \infty$, whereas the slow branch starts at 0 and tends to $\omega_3^2 \equiv k_\parallel^2 c^2$. In the right diagram ($\omega^2 = \omega^2(k_\perp)$ for fixed $k_\parallel$), the fast branch ranges from $\omega_4^2 \equiv k_\parallel^2 b^2 \equiv \omega_A^2$ at $k_\perp = 0$ to $\omega_7^2 \equiv k_\perp^2(b^2 + c^2) \to \infty$, and the slow one ranges from $\omega_5^2 \equiv k_\parallel^2 c^2$ to $\omega_6^2 \equiv k_\parallel^2 b^2 c^2/(b^2 + c^2)$.

While the group velocity in the parallel direction is positive, $\partial \omega/\partial k_\parallel > 0$, for all three kinds of waves, the group velocity $\partial \omega/\partial k_\perp$ in the perpendicular direction and the asymptotic value of the frequency for $k_\perp \to \infty$ display a very

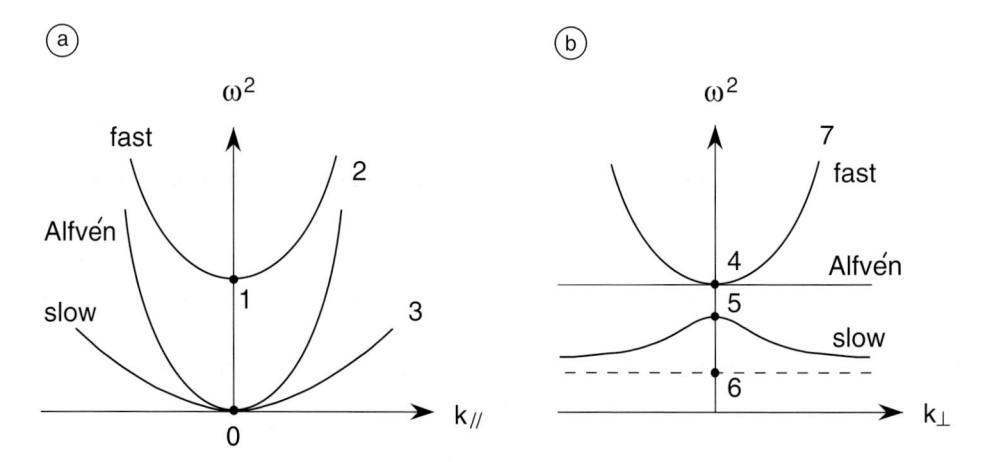

Fig. 5.7. Schematic dispersion diagrams and asymptotics for large wave numbers ($b > c$): (a) $\omega^2 = \omega^2(k_\parallel)$ for fixed $k_\perp$; (b) $\omega^2 = \omega^2(k_\perp)$ for fixed $k_\parallel$.

characteristic difference for the three waves:

$$
\begin{cases}
\partial\omega/\partial k_\perp > 0, & \omega_f^2 \to \infty & \text{for the fast waves,} \\
\partial\omega/\partial k_\perp = 0, & \omega_A^2 \to k_\parallel^2 b^2 & \text{for the Alfvén waves,} \\
\partial\omega/\partial k_\perp < 0, & \omega_s^2 \to k_\parallel^2 \dfrac{b^2 c^2}{b^2 + c^2} & \text{for the slow waves.}
\end{cases} \quad (5.88)
$$

Note that this implies that the energy propagation of slow wave packets is opposite to that of fast wave packets in the perpendicular direction! This is caused by the peculiar behaviour of the slow group velocity depicted in Fig. 5.6.

The group diagram has a much wider applicability than just wave propagation in infinite homogeneous plasmas. This is so because the construction of a wave packet involves the contributions of large **k**-vectors (small wavelengths) so that *the concept of group velocity is essentially a local one.* This is one of the reasons why it returns in the context of nonlinear MHD of inhomogeneous plasmas, where the associated *characteristics* become the carriers of the information of disturbances of the plasma. This is our next subject.

5.4 Characteristics*

5.4.1 The method of characteristics*

Consider the simple example of the linear advection equation in one spatial dimension,

$$\frac{\partial \Psi}{\partial t} + u \frac{\partial \Psi}{\partial x} = 0, \quad (5.89)$$

where $\Psi(x, t)$ is the unknown and the advection velocity u is considered to be given. If $u = \text{const}$, the solution is trivial:

$$\Psi = f(x - ut), \quad \text{where} \quad f = \Psi_0 \equiv \Psi(x, t = 0). \quad (5.90)$$

Hence, the initial data Ψ_0 are simply propagated along the set of parallel straight lines $dx/dt = u$, which are the *characteristics* for this case.

For our subject (the study of the MHD equations (5.18)–(5.21)), it is important that this procedure may be generalized to systems of first order partial differential equations (PDEs) in three spatial dimensions and time, so that Ψ becomes a vector and u a matrix. For example, the linearized equations for one-dimensional sound waves, following from Eqs. (5.5)–(5.7), read:

$$\frac{\partial}{\partial t}\begin{pmatrix} v_1 \\ p_1 \end{pmatrix} + \begin{pmatrix} 0 & 1/\rho_0 \\ \gamma p_0 & 0 \end{pmatrix} \frac{\partial}{\partial x}\begin{pmatrix} v_1 \\ p_1 \end{pmatrix} = 0, \quad (5.91)$$

where we have omitted the redundant density equation. As we have seen in Section 5.1.2, for constant values of ρ_0 and p_0, the solutions (5.10) represent plane sound waves with frequencies $\omega = \pm kc$, where $c \equiv \sqrt{\gamma p_0/\rho_0}$ is the sound speed. Exploiting real notation, the general solution may be written as

$$v_1(x,t) = \sum_k \left[\alpha_k \sin k(x \pm ct) + \beta_k \cos k(x \pm ct) \right], \qquad (5.92)$$

and a similar expression for p_1. Here, the coefficients α_k and β_k follow from Fourier decomposition of the initial data,

$$v_1(x,0) = \sum_k (\alpha_k \sin kx + \beta_k \cos kx), \qquad (5.93)$$

demonstrating that the initial data do, in fact, propagate along the two sets of straight-line characteristics $dx/dt = \pm c$.

Returning to the linear advection equation (5.89), but now assuming that u is not constant, the characteristics become solutions of the ordinary differential equations (ODEs)

$$\frac{dx}{dt} = u(x,t). \qquad (5.94)$$

Along these curves, the solution $\Psi(x,t)$ is constant,

$$\frac{d\Psi}{dt} \equiv \frac{\partial \Psi}{\partial t} + \frac{\partial \Psi}{\partial x}\frac{dx}{dt} = 0, \qquad (5.95)$$

as is evident by comparison with Eq. (5.89). As a result, for given initial data $\Psi_0(x)$, the solution can be determined at any time $t_1 > 0$ by constructing the characteristics through a suitable set of points $\{\ldots, x_i, x_{i+1}, \ldots\}$ so that, e.g., $\Psi(x_i', t_1) = \Psi_0(x_i)$, where x_i' lies on the characteristic through $x = x_i$. This is illustrated in Fig. 5.8 for the case that a weak discontinuity (a 'tent' function with a discontinuous derivative) is applied at $t = 0$. The characteristic through $x = x_i$ propagates this discontinuity forward in space-time.

Most important for the study of gas dynamics and magnetohydrodynamics, in particular for modern developments in computational fluid dynamics (CFD) and computational magneto-fluid dynamics (CMFD), is the fact that *the method of characteristics* generalizes to *nonlinear* partial differential equations. Since it lends itself naturally to numerical implementation, it is at the basis of many methods in CFD and CMFD. Following the clear exposition of LeVeque [143, 144] on this topic, this may be illustrated again with the advection equation which becomes

5.4 Characteristics*

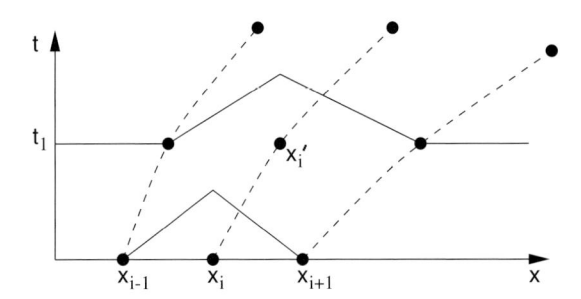

Fig. 5.8. Pointwise propagation of a solution along the characteristics.

quasi-linear when u is also a function of the unknown Ψ itself. A particularly relevant example is the case $u = \Psi$ which leads to Burgers' equation:

$$\frac{\partial \Psi}{\partial t} + \Psi \frac{\partial \Psi}{\partial x} = \nu \frac{\partial^2 \Psi}{\partial x^2}, \tag{5.96}$$

where a viscous term is added on the RHS to model the balance between nonlinear and dissipative processes when gradients build up in the solutions. (This equation may be obtained from Eqs. (4.136) and (4.137) for plane incompressible fluid flow in one dimension, where $\Psi \equiv v_x$.) At first neglecting this small term, the characteristics are the solutions of the ODE

$$\frac{dx}{dt} = \Psi(x(t), t), \tag{5.97}$$

which are just a set of straight lines with slopes determined by the initial data, like in the first step of Fig. 5.8. For large times, the characteristics will cross, in general, where build-up of large gradients of Ψ is counteracted by smoothing through the dissipative term on the RHS of Burgers' equation. This will occur in a very narrow region, so that effectively a valid solution with a *shock* is obtained in the limit $\nu \to 0$, where the condition of increasing entropy across the shock is to be applied to the ideal model to eliminate unphysical solutions, as we have already seen in Section 4.5.1 and Fig. 4.12.

We will return to computational methods in the chapters on this subject in Volume 2. For the present purpose, the essential issue connected with characteristics is the fact that the equations of MHD are actually *hyperbolic* partial differential equations. This means that they possess a complete set of real characteristics related to the eigenvalues of the linearized system. This will be demonstrated in Section 5.4.3.

5.4.2 Classification of partial differential equations*

Long before computational methods for solving PDEs became common tools, the theory of second order partial differential equations was already a central subject of mathematical physics, where the method of characteristics was applied extensively; see Courant and Hilbert [60], Garabedian [78], or Morse and Feshbach [159]. Therefore, it is useful to recall some of the concepts developed there before we return to the study of the MHD equations proper. Consider the following second order partial differential equation in two dimensions:

$$A\Phi_{xx} + 2B\Phi_{xy} + C\Phi_{yy} = D(\Phi_x, \Phi_y, x, y), \tag{5.98}$$

where A, B, and C are functions of only x and y for the time being. These independent variables may indicate two spatial dimensions as well as one space and one time coordinate, or linear combinations. Subscripts x and y indicate differentiation with respect to those variables: $\Phi_x \equiv \partial \Phi/\partial x$, etc. The *Cauchy* problem consists in finding the solution Φ away from a boundary C (Fig. 5.9), when, e.g., both Φ and its normal derivative $\Phi_n \equiv \mathbf{n} \cdot \nabla \Phi$ are specified on it. Introducing new variables

$$\Psi_1 \equiv \Phi_x, \qquad \Psi_2 \equiv \Phi_y,$$

Eq. (5.98) is transformed into an equivalent system of first order equations

$$\begin{aligned} A\Psi_{1x} + B\Psi_{1y} + B\Psi_{2x} + C\Psi_{2y} &= D(\Psi_1, \Psi_2, x, y), \\ \Psi_{1y} - \Psi_{2x} &= 0. \end{aligned} \tag{5.99}$$

The pertinent Cauchy problem is now to determine Ψ_1 and Ψ_2 away from the boundary, when they are given on C.

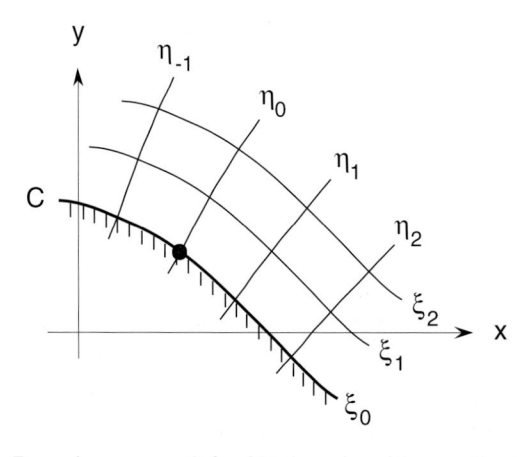

Fig. 5.9. Boundary curve C for 2D domain with coordinates ξ and η.

5.4 Characteristics*

To facilitate the solution of the Cauchy problem, let us replace the Cartesian coordinates x, y by boundary fitted coordinates ξ, η, where the boundary curve C is given by $\xi(x, y) = \xi_0$. These coordinates may be chosen orthogonal, $\nabla\xi \cdot \nabla\eta = 0$, but this is not essential for the discussion. The boundary data can then be expressed as

$$\Psi_1(\xi_0, \eta) = f_1(\eta), \qquad \Psi_2(\xi_0, \eta) = f_2(\eta). \tag{5.100}$$

We wish to investigate under which conditions $\Psi_1(\xi, \eta)$ and $\Psi_2(\xi, \eta)$ may be obtained by means of a power series solution about a particular point (ξ_0, η_0) on the boundary (the dot in Fig. 5.9):

$$\Psi_1(\xi, \eta) = \Psi_1(\xi_0, \eta_0) + (\xi - \xi_0)\left(\frac{\partial \Psi_1}{\partial \xi}\right)_0 + (\eta - \eta_0)\left(\frac{\partial \Psi_1}{\partial \eta}\right)_0 + \cdots,$$

$$\Psi_2(\xi, \eta) = \Psi_2(\xi_0, \eta_0) + (\xi - \xi_0)\left(\frac{\partial \Psi_2}{\partial \xi}\right)_0 + (\eta - \eta_0)\left(\frac{\partial \Psi_2}{\partial \eta}\right)_0 + \cdots. \tag{5.101}$$

Here, the expressions $\Psi_i(\xi_0, \eta_0) \equiv f_i(\eta_0)$ and $(\partial \Psi_i/\partial \eta)_0 \equiv df_i/d\eta\,(\eta_0)$ ($i = 1, 2$) are known from the boundary conditions (5.100), so that we need to investigate under which circumstances the remaining expressions $(\partial \Psi_i/\partial \xi)_0$ ($i = 1, 2$) can be calculated. Once the latter two derivatives are known, the higher derivatives in the expansion (5.101) may be found by successive differentiations of the original equations (5.99) so that the problem may be considered to be solved, i.e. in the neighbourhood of $\xi = \xi_0$. The process is then repeated by moving C to $\xi = \xi_1$, etc., until the solution is known everywhere.

We transform the partial differential equation (5.99) to ξ-η coordinates by writing

$$\Psi_{1x} = \frac{\partial \Psi_1}{\partial \xi}\xi_x + \frac{\partial \Psi_1}{\partial \eta}\eta_x, \quad \text{etc.}$$

This yields

$$(A\xi_x + B\xi_y)\frac{\partial \Psi_1}{\partial \xi} + (B\xi_x + C\xi_y)\frac{\partial \Psi_2}{\partial \xi} = D - (A\eta_x + B\eta_y)\frac{\partial \Psi_1}{\partial \eta}$$
$$- (B\eta_x + C\eta_y)\frac{\partial \Psi_2}{\partial \eta},$$

$$\xi_y\frac{\partial \Psi_1}{\partial \xi} - \xi_x\frac{\partial \Psi_2}{\partial \xi} = -\eta_y\frac{\partial \Psi_1}{\partial \eta} + \eta_x\frac{\partial \Psi_2}{\partial \eta}, \tag{5.102}$$

where the LHS contains the unknown derivatives and the RHS the known ones. The unknown derivatives $\partial \Psi_1/\partial \xi$ and $\partial \Psi_2/\partial \xi$ may then be determined from Eqs. (5.103) *if the determinant of the coefficients on the left hand side does not vanish*. On the other hand, the condition that the determinant vanishes,

$$\begin{vmatrix} A\xi_x + B\xi_y & B\xi_x + C\xi_y \\ \xi_y & -\xi_x \end{vmatrix} = -A\xi_x^2 - 2B\xi_x\xi_y - C\xi_y^2 = 0, \quad (5.103)$$

defines two directions in every point of the plane, *the characteristic directions*, along which posing Cauchy boundary conditions does not determine the solution. The *characteristics* are the curves in the x-y plane that are everywhere tangent to these characteristic directions. Since $\xi(x, y) = \xi_0 \Rightarrow d\xi = \xi_x dx + \xi_y dy = 0$ along C, those characteristic directions (to be avoided for a proper Cauchy boundary) are given by

$$\left.\frac{dy}{dx}\right|_{\text{char}} = -\frac{\xi_x}{\xi_y} = \frac{B \pm \sqrt{B^2 - AC}}{A}. \quad (5.104)$$

Three cases may be distinguished:

(a) $B^2 > AC$ – the characteristics are real and the equation is called *hyperbolic* (example: the wave equation $\Phi_{xx} - (1/c^2)\Phi_{tt} = 0$);
(b) $B^2 = AC$ – the characteristics are real but coincide and the equation is *parabolic* (example: the heat equation $\Phi_{xx} - (1/\lambda)\Phi_t = 0$);
(c) $B^2 < AC$ – the characteristics are complex and the equation is *elliptic* (example: Laplace's equation $\Phi_{xx} + \Phi_{yy} = 0$).

In the following, we will be concerned mainly with hyperbolic equations. Cauchy *initial* conditions (the variable y then becomes t) may be considered appropriate if the boundary (in x-t now) is not a characteristic. For the example of the wave equation

$$\Phi_{xx} - \frac{1}{c^2}\Phi_{tt} = 0,$$

the characteristic directions are given by $dx/dt = \pm c$ (Fig. 5.10). The initial data propagate along those characteristics. In spaces of higher dimension than 2, for the Cauchy problem to be well posed, it is not sufficient that the boundaries are not coincident with a characteristic. One has to demand in addition that they are space-like, as in Fig. 5.10. The reason is that the spatial part by itself is generally elliptic, so that Cauchy's problem is ill-posed if we consider time-independent solutions. In physical problems initial data are usually given along space-directions,[1] so that this does not really present a restriction.

---

[1] An exception is the excitation of waves by time-dependent forcing terms at the boundary of the plasma. In that case data are given on time-like boundaries.

5.4 Characteristics*

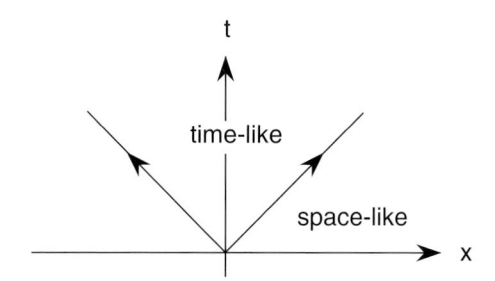

Fig. 5.10. Characteristic directions for the wave equation.

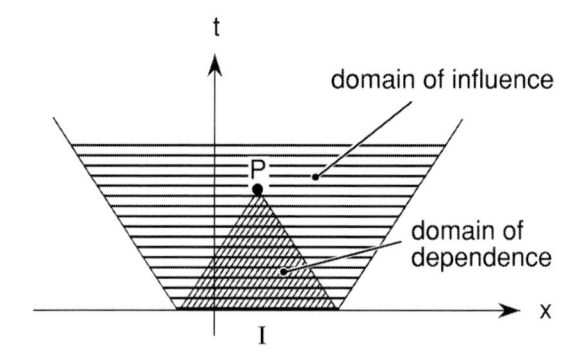

Fig. 5.11. Domains for hyperbolic equations.

Finally, it is expedient to distinguish two useful concepts for hyperbolic partial differential equations, viz. *the domain of influence* of *I*, which is the region in the *x-t* plane where the influence of the initial data *I* is felt, and *the domain of dependence* of the space–time point *P*, which is the region which influences the behaviour at *P*. These concepts are illustrated in Fig. 5.11.

Notice that the analysis above is unchanged when the coefficients *A*, *B*, and *C* depend on Ψ_1 and Ψ_2 as well, so that the method of characteristics also works for nonlinear equations, specifically *quasi-linear partial differential equations*.

5.4.3 Characteristics in ideal MHD*

We generalize the preceding discussion to partial differential equations in more than two independent variables and also more than two dependent variables, in particular the ideal MHD equations (5.18)–(5.21) for the variables ρ, **v**, e, **B** as a function of **r** and t. Now, instead of a 2-vector (Ψ_1, Ψ_2), the unknowns will be represented by an 8-vector Ψ_i ($i = 1, \ldots, 8$). We will show that the MHD equations are *symmetric hyperbolic partial differential equations*, where the nonlinearity is only of a *quasi-linear* nature.

The equations of ideal MHD are partial differential equations with respect to the independent variables $\mathbf{r}, t$. Consequently, characteristics will be three-dimensional manifolds

$$\xi(\mathbf{r}, t) = \xi_0, \tag{5.105}$$

in four-dimensional space–time $\mathbf{r}, t$. These manifolds may be visualized as being swept out by the motion of surfaces in ordinary three-dimensional space $(\mathbf{r})$ when time t progresses. We apply the same techniques as in the previous section to determine when $\xi(\mathbf{r}, t) = \xi_0$ is a characteristic manifold.

Assume that boundary data for $\rho(\mathbf{r}, t)$, $\mathbf{v}(\mathbf{r}, t)$, $e(\mathbf{r}, t)$, $\mathbf{B}(\mathbf{r}, t)$ are given on the manifold $\xi(\mathbf{r}, t) = \xi_0$. (Notice that *the initial value problem* corresponds to giving $\rho(\mathbf{r}, 0)$, $\mathbf{v}(\mathbf{r}, 0)$, $e(\mathbf{r}, 0)$, $\mathbf{B}(\mathbf{r}, 0)$ on the domain of interest in ordinary three-space. In order for this problem to be well posed, ordinary three-space should not be a characteristic. Here, we consider the opposite case that data are given on a characteristic, so that the Cauchy problem is not well posed.) As in Section 5.4.2, we consider ξ as a coordinate and introduce additional coordinates η, ζ, and τ, such that four-space $(\mathbf{r}, t)$ is covered by the coordinates ξ, η, ζ, and τ. The boundary data may then be written as

$$\rho(\xi_0, \eta, \zeta, \tau) = \rho_0(\eta, \zeta, \tau), \quad \text{etc.,} \tag{5.106}$$

where η, ζ, and τ parameterize the boundary manifold $\xi(\mathbf{r}, t) = \xi_0$. Since $\rho_0(\eta, \zeta, \tau)$ is a known function, the derivatives $\partial \rho_0/\partial \eta$, $\partial \rho_0/\partial \zeta$, and $\partial \rho_0/\partial \tau$ may also be considered to be known. Similarly, for the other variables $\mathbf{v}_0$, e_0 and $\mathbf{B}_0$.

We want to find out under which conditions the solutions $\rho(\xi, \eta, \zeta, \tau)$, $\mathbf{v}(\xi, \eta, \zeta, \tau)$, $e(\xi, \eta, \zeta, \tau)$, and $\mathbf{B}(\xi, \eta, \zeta, \tau)$ may be obtained away from the boundary $\xi = \xi_0$ or, rather, may not be obtained since then $\xi = \xi_0$ is a characteristic. To that end, we write the variables in terms of a power series:

$$\rho(\xi, \eta, \zeta, \tau) = \rho_0(\eta_0, \zeta_0, \tau_0) + (\xi - \xi_0)\left(\frac{\partial \rho}{\partial \xi}\right)_0$$
$$+ (\eta - \eta_0)\left(\frac{\partial \rho}{\partial \eta}\right)_0 + (\zeta - \zeta_0)\left(\frac{\partial \rho}{\partial \zeta}\right)_0 + (\tau - \tau_0)\left(\frac{\partial \rho}{\partial \tau}\right)_0 + \cdots, \tag{5.107}$$

and likewise for $\mathbf{v}$, e, and $\mathbf{B}$. As in the previous section, we may consider the problem to be solvable if $(\partial \rho/\partial \xi)_0$, $(\partial \mathbf{v}/\partial \xi)_0$, $(\partial e/\partial \xi)_0$ and $(\partial \mathbf{B}/\partial \xi)_0$ can be constructed, since the other first order derivatives are found from the boundary data (5.106), whereas the higher order ones may be obtained by subsequent differentiations of the original partial differential equations.

5.4 Characteristics*

For convenience, we denote the unknown derivatives with respect to ξ with a prime:

$$\rho' \equiv \frac{\partial \rho}{\partial \xi}, \quad \mathbf{v}' \equiv \frac{\partial \mathbf{v}}{\partial \xi}, \quad e' \equiv \frac{\partial e}{\partial \xi}, \quad \mathbf{B}' \equiv \frac{\partial \mathbf{B}}{\partial \xi}. \tag{5.108}$$

The derivatives in the MHD equations (5.18)–(5.20) may then be written as:

$$\nabla \rho = \nabla \xi \, \rho' + \nabla \eta \frac{\partial \rho}{\partial \eta} + \nabla \zeta \frac{\partial \rho}{\partial \zeta} + \nabla \tau \frac{\partial \rho}{\partial \tau},$$

$$\frac{D\rho}{Dt} = (\xi_t + \mathbf{v} \cdot \nabla \xi) \, \rho' + (\eta_t + \mathbf{v} \cdot \nabla \eta) \frac{\partial \rho}{\partial \eta} + (\zeta_t + \mathbf{v} \cdot \nabla \zeta) \frac{\partial \rho}{\partial \zeta}$$

$$+ (\tau_t + \mathbf{v} \cdot \nabla \tau) \frac{\partial \rho}{\partial \tau}, \tag{5.109}$$

and similarly for the other variables. Hence, with respect to the primed variables, the coordinate transformation amounts to the replacements

$$\nabla f \rightarrow \mathbf{n} f' + \cdots, \qquad \mathbf{n} \equiv \nabla \xi,$$

$$\frac{Df}{Dt} \rightarrow -u f' + \cdots, \qquad -u \equiv \xi_t + \mathbf{v} \cdot \nabla \xi. \tag{5.110}$$

Here, $\mathbf{n}$ is the *normal to the space-part of the characteristic* (where ξ has been chosen such that $|\nabla \xi| = 1$, so that $\mathbf{n}$ has unit length), and u is the *characteristic speed*, i.e. the normal velocity of the characteristic ξ measured with respect to the fluid velocity $\mathbf{v}$. (We just note the correspondence between these transformations for the characteristics (weak discontinuities) and the transformations (4.145) for the shocks (strong discontinuities). Further discussion of this intriguing correspondence has to await the analysis of transonic flows in Volume 2.)

Inserting the expressions (5.109) in Eqs. (5.18)–(5.21), and keeping the primed (unknown) variables on the left hand side but moving the known variables to the right hand side of the equation, we obtain the following set of algebraic equations:

$$-u\rho' + \rho \, \mathbf{n} \cdot \mathbf{v}' = \cdots,$$

$$-\rho u \mathbf{v}' + (\gamma - 1) \mathbf{n} \, (e\rho' + \rho e') + (\mathbf{n} \, \mathbf{B} \cdot - \mathbf{n} \cdot \mathbf{B}) \, \mathbf{B}' = \cdots,$$

$$-u e' + (\gamma - 1) e \, \mathbf{n} \cdot \mathbf{v}' = \cdots,$$

$$-u \mathbf{B}' + (\mathbf{B} \, \mathbf{n} \cdot - \mathbf{n} \cdot \mathbf{B}) \, \mathbf{v}' = \cdots, \qquad \mathbf{n} \cdot \mathbf{B}' = \cdots. \tag{5.111}$$

Here, the dots on the RHS indicate the known derivatives with respect to η, ζ, and τ. In order to get equations of the same dimension we multiply the four lines of Eq. (5.111) by c/ρ, $1/\rho$, γ/c, and $1/\sqrt{\rho}$, resp., where we recall that

$c^2 = \gamma(\gamma - 1)e$. By now, the reader will have noted that we have virtually reduced the algebra to that of Section 5.2, Eqs. (5.33), where $\mathbf{n}$ has taken the place of the wave vector $\mathbf{k}$ and u that of the frequency ω (or, rather, the Doppler shifted frequency $\omega - \mathbf{k} \cdot \mathbf{v}$). Hence, we again choose $\mathbf{B}$ along the z-axis and $\mathbf{n}$ in the x-z plane,

$$\mathbf{B} = (0, 0, B), \qquad \mathbf{n} = (n_x, 0, n_z), \tag{5.112}$$

and introduce *the Alfvén speed* and *the sound speed*,

$$b \equiv B/\sqrt{\rho}, \qquad c \equiv \sqrt{\gamma p/\rho}. \tag{5.113}$$

The system of equations (5.111) then becomes

$$\begin{pmatrix} -\gamma u & n_x c & 0 & n_z c & 0 & 0 & 0 & 0 \\ n_x c & -u & 0 & 0 & n_x c & -n_z b & 0 & n_x b \\ 0 & 0 & -u & 0 & 0 & 0 & -n_z b & 0 \\ n_z c & 0 & 0 & -u & n_z c & 0 & 0 & 0 \\ 0 & n_x c & 0 & n_z c & -\frac{\gamma}{\gamma-1} u & 0 & 0 & 0 \\ 0 & -n_z b & 0 & 0 & 0 & -u & 0 & 0 \\ 0 & 0 & -n_z b & 0 & 0 & 0 & -u & 0 \\ 0 & n_x b & 0 & 0 & 0 & 0 & 0 & -u \end{pmatrix} \begin{pmatrix} \frac{c}{\gamma \rho} \rho' \\ v'_x \\ v'_y \\ v'_z \\ \frac{\gamma-1}{c} e' \\ \frac{1}{\sqrt{\rho}} B'_x \\ \frac{1}{\sqrt{\rho}} B'_y \\ \frac{1}{\sqrt{\rho}} B'_z \end{pmatrix} = \cdots, \tag{5.114}$$

where $n_z = B_n/B$, and $n_x = [1 - (B_n/B)^2]^{1/2}$, and the constraint on $\mathbf{n} \cdot \mathbf{B}'$ is ignored for the time being. Not surprising any more, the homogeneous (LHS) part of this equation corresponds to the eigenvalue problem (5.35) for the MHD waves. And, if the algebra is the same, the physics probably is also the same. We will now show this to be the case.

The characteristics are obtained when the determinant of the LHS of Eq. (5.114) vanishes so that the full inhomogeneous problem cannot be solved. The solutions cannot be propagated away from the manifold $\xi = \xi_0$ in that case. This condition may be written as

$$\Delta = \frac{\gamma^2}{\gamma - 1} u^2 (u^2 - b_n^2) \left[u^4 - (b^2 + c^2)u^2 + b_n^2 c^2 \right] = 0, \tag{5.115}$$

where b_n is the normal Alfvén speed, $b_n \equiv \mathbf{n} \cdot \mathbf{B}/\sqrt{\rho}$. Comparison with Eq. (5.53) shows that we have recovered the dispersion equation for the linear MHD waves in the disguise of an equation for the characteristic speeds u. From the analysis of Sections 5.2.1 and 5.2.2, we immediately conclude that a spurious root $u = 0$ has been introduced from the condition $u\, \mathbf{n} \cdot \mathbf{B}' = 0$, following from the homogeneous part of the first equation (5.111)(d), which should be eliminated on account of the second equation. Hence, *seven real characteristics* are obtained, corresponding to the eight variables minus the redundant one needed to describe the system. The matrix on the LHS of Eq. (5.114) is real, symmetric, and has only real eigenvalues. Consequently, *the equations of ideal MHD are symmetric hyperbolic equations* and *the initial value problem*, where values are assigned to the variables $\mathbf{v}$, $\mathbf{B}$, e, and ρ in ordinary three-dimensional space at $t = 0$, *is well posed*. This important result is due to Friedrichs [76].

Disregarding the redundant root, Eq. (5.115) yields seven characteristic speeds:

$$u = u_E \equiv 0, \tag{5.116}$$

$$u = u_A \equiv \pm b_n, \tag{5.117}$$

$$u = u_s \equiv \pm \left[\tfrac{1}{2}(b^2 + c^2) - \tfrac{1}{2}\sqrt{(b^2 + c^2)^2 - 4b_n^2 c^2}\, \right]^{1/2}, \tag{5.118}$$

$$u = u_f \equiv \pm \left[\tfrac{1}{2}(b^2 + c^2) + \tfrac{1}{2}\sqrt{(b^2 + c^2)^2 - 4b_n^2 c^2}\, \right]^{1/2}. \tag{5.119}$$

The solution (5.116) corresponds to *entropy disturbances* that just follow the stream-lines of the flow. The pair of solutions (5.117) corresponds to *Alfvén disturbances* moving forward (+) or backward (−) with respect to the flow. The pair of solutions (5.118) are forward and backward *slow magneto-acoustic disturbances*, whereas the solutions (5.119) constitute forward and backward *fast magneto-acoustic disturbances*. The characteristic speeds are ordered according to the sequence of inequalities

$$0 = |u_E| \le |u_s| \le |u_A| \le |u_f| < \infty. \tag{5.120}$$

Degeneracies occur for $\mathbf{n} \parallel \mathbf{B}$:

$$|u_s| = \min(b, c), \qquad |u_A| = b, \qquad |u_f| = \max(b, c), \tag{5.121}$$

and for $\mathbf{n} \perp \mathbf{B}$:

$$|u_s| = |u_A| = 0, \qquad |u_f| = (b^2 + c^2)^{1/2}. \tag{5.122}$$

The equations of gas dynamics are obtained in the limit $b \to 0$, when the slow and Alfvén waves disappear in the origin ($u \to 0$) and the fast magneto-acoustic waves degenerate into ordinary sound waves ($u_f \to \pm c$). Another limit of interest is the case of incompressible plasma ($c \to \infty$). In that case the speed of the fast magneto-acoustic wave disappears at infinity (instantaneous propagation), whereas the slow magneto-acoustic speed and the Alfvén speed coincide. The

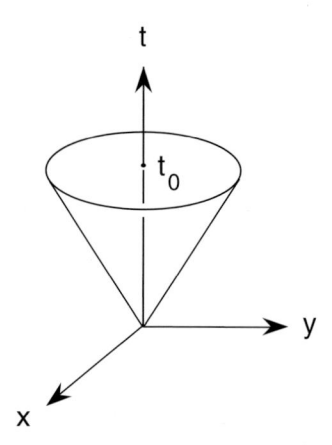

Fig. 5.12. Characteristic manifold for the 2D wave equation.

waves themselves do not coincide, of course, because their physical properties (e.g. polarization) are different.

Let us now consider the spatial part of a characteristic manifold at a certain time $t = t_0$. This is called the *ray surface*. It may be considered as a wave front, i.e. a surface across which weak discontinuities (as illustrated in Fig. 5.8) occur, emitted at time $t = 0$ from the origin $x = y = z = 0$. For example, in the absence of a magnetic field, a characteristic manifold would just be the spherical sound front $x^2 + y^2 + z^2 = c^2 t^2$ so that the ray surface would be the sphere with radius ct_0. Dropping the z-dependence, the characteristic in x, y, t space then becomes a cone through the point $x = y = t = 0$ (Fig. 5.12), whereas the circular intersection of this cone with the plane $t = t_0$ constitutes the ray surface. Of course, MHD gives much more complicated figures because the medium is anisotropic and the coefficients of the partial differential equations are not constant.

To get the ray surface we first of all compute from Eqs. (5.116)–(5.119) the distance ut_0 which a plane wave front travels along $\mathbf{n}$ after having passed the origin at $t = 0$. The collection of these points gives Fig. 5.13(a) (which is a schematic rendering of the computed phase diagram of Fig. 5.5(a)). However, this is *not* the ray surface, but the so-called *reciprocal normal surface*. (Of course, everything is symmetric around the direction of $\mathbf{B}$ so that the three-dimensional pictures are obtained by just rotating the figure around the $\mathbf{B}$-axis.) To get the ray surface depicted in Fig. 5.13(b) (which is again a schematic rendering, but now of the group diagram of Fig. 5.5(b)) we have to take the envelope of the plane wave fronts since the ray surface corresponds to a wave front due to a point disturbance at the origin at $t = 0$. Taking the envelope of the fronts indicated by s, A and f in Fig. 5.13(a) results in the completely different and, in particular, more singular picture of Fig. 5.13(b). As we have seen in Section 5.3.2, the reciprocal normal

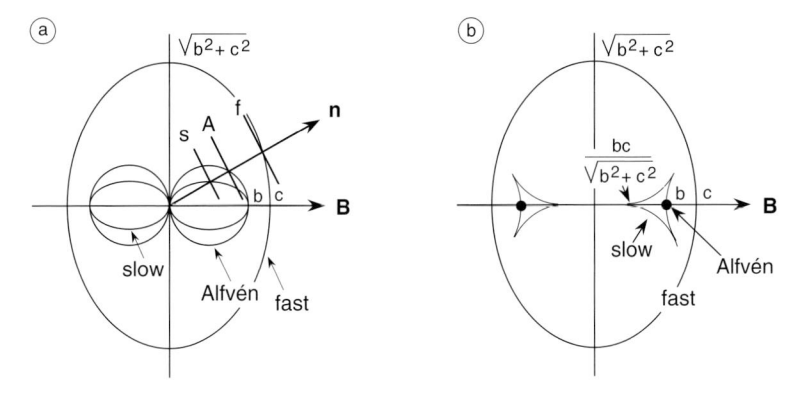

Fig. 5.13. Friedrichs diagrams: schematic representation of (a) reciprocal normal surface (or phase diagram) and (b) ray surface (or group diagram) of the MHD waves for $b < c$.

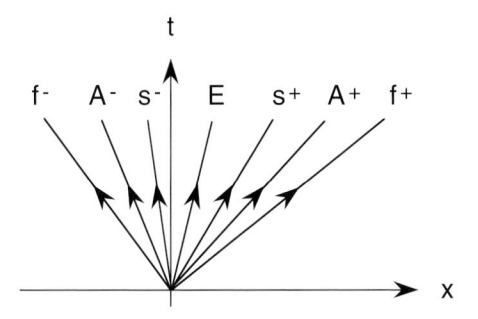

Fig. 5.14. The seven characteristic directions for MHD (when the x-axis is not perpendicular or parallel to $\mathbf{B}$).

surface for the Alfvén wave consists of two spheres touching the origin. Correspondingly, *the Alfvén ray surface* just consists of the two points $z = \pm b$, so that Alfvén waves travel as point disturbances along the magnetic field. The ray surface for the slow magneto-acoustic wave also exhibits a quite anisotropic character: it consists of two cusped figures. The fast magneto-acoustic waves exhibit the least degree of anisotropy. In that respect, they resemble ordinary sound waves most.

A simple representation of the MHD characteristics is obtained by dropping one more spatial dimension and exploiting the expressions (5.116)–(5.119) for the seven speeds to compute the x-t cross-sections of the characteristics. This yields Fig. 5.14. Recall that the characteristic speed u is measured with respect to the background velocity $\mathbf{v}$, $d\mathbf{r}/dt|_{\text{char}} = u + \mathbf{v} \cdot \mathbf{n}$, so that the inclination of the entropy characteristic indicates that the plasma is assumed to stream to the right here.

On a characteristic manifold, $\Delta = 0$, so that the homogeneous counterpart of Eq. (5.114) has a solution with fixed relations between the values ρ', $\mathbf{v}'$, e'

and **B′** on that manifold. A physical interpretation of the characteristics is to consider those primed variables as discontinuities of the derivative across the characteristic, so that $\rho' \equiv \partial\rho/\partial\xi \gg \partial\rho/\partial\eta$, etc., justifying the neglect of the RHS of Eq. (5.114). The meaning of this is that we may consider these quantities as *weak discontinuities* of the flow that are propagated along with the characteristics. We then find the following relationships for the different characteristics which, of course, reproduce the relationships found for the waves in Section 5.2.3:

(a) *Entropy characteristics* ($u = 0$):

$$S'/S = -\gamma\rho'/\rho = \gamma e'/e \neq 0, \qquad \mathbf{v}' = 0, \qquad \mathbf{B}' = 0. \tag{5.123}$$

Only the thermodynamic variables are perturbed, in particular the entropy.

(b) *Alfvén characteristics* ($u = u_A$):

$$B'_y = -\sqrt{\rho}\, v'_y \neq 0, \qquad \rho' = 0, \qquad e' = 0, \qquad v'_x = v'_z = 0, \qquad B'_x = B'_z = 0. \tag{5.124}$$

These are purely transverse disturbances where **v′** and **B′** are perpendicular to the plane through **n** and **B** (Fig. 5.1). The thermodynamic variables are not perturbed.

(c) *Magneto-acoustic characteristics* ($u = u_{s,f}$):

$$v'_x = \frac{n_x}{n_z}\frac{u^2}{u^2 - b^2} v'_z \neq 0, \qquad B'_z = -\frac{n_x}{n_z} B'_x = \frac{n_x b}{u}\sqrt{\rho}\, v'_x \neq 0,$$

$$\rho' = \frac{\gamma\rho}{c^2} e' = \frac{\rho u}{n_z c^2} v'_z \neq 0, \qquad v'_y = 0, \qquad B'_y = 0. \tag{5.125}$$

These disturbances are polarized in the plane through **n** and **B**, with the fast and slow polarizations perpendicular to each other (Fig. 5.2). This difference arises through the factor $u^2 - b^2$ which is positive for the fast characteristic and negative for the slow one. The thermodynamic variables are perturbed, except for the entropy.

For characteristic directions $\mathbf{n} \cdot \mathbf{B} = 0$ (i.e. the magnetic field is tangential to the space-part of the characteristic), the root $u = 0$ is fivefold degenerate. The homogeneous version of Eqs. (5.111) then reduces to the following two conditions:

$$\mathbf{n} \cdot \mathbf{v}' = 0, \tag{5.126}$$

$$p' + \mathbf{B} \cdot \mathbf{B}' = (p + \tfrac{1}{2}B^2)' = 0, \tag{5.127}$$

whereas we now also have to include the separate condition

$$\mathbf{n} \cdot \mathbf{B}' = 0. \tag{5.128}$$

All other components of the variables are arbitrary. These disturbances are called *tangential discontinuities*. An example would be an equilibrium of two adjacent plasmas with different pressure, density, tangential magnetic field, tangential velocity, but satisfying the relations (5.126)–(5.128). At the interface, a surface current $\mathbf{j}^{\star\prime} = \mathbf{n} \times \mathbf{B}'$ and a surface vorticity $\omega^{\star\prime} = \mathbf{n} \times \mathbf{v}'$ produce disturbances of the tangential field and velocity. Hence, quite unexpectedly, we recover the interface conditions of Section 4.5.2 and the circle is closed: boundary value problem and initial value problem are interwoven in MHD.

The fivefold degeneracy for vanishing characteristic speed ($u = 0$ for $\mathbf{n} \cdot \mathbf{B} = 0$) corresponds to the degeneracy of marginal entropy, Alfvén, and slow magnetosonic waves ($\omega = 0$ for $\mathbf{k} \cdot \mathbf{B} = 0$). In general, degeneracies like that signal the need to extend the theory with more sophisticated assumptions on the model. The present degeneracies originate from the assumption of *locality* on the group diagrams and the characteristics, which neglect the global plasma properties associated with *inhomogeneity*. Hence, in order to make progress, we now have to enter the vast territory of waves and instabilities in inhomogeneous plasmas. This is the subject where MHD acquires its particular strength and beauty. It will occupy us for the rest of this book.

5.5 Literature and exercises

Notes on literature

Phase and group diagrams:

– Bittencourt, *Fundamentals of Plasma Physics* [31], Section 14.6 and Chapter 15.

Characteristics:

– Courant & Hilbert, *Methods of Mathematical Physics II* [60], Chapter VI.
– Garabedian, *Partial Differential Equations* [78], Chapters 2, 3, 4, 6, 14.

Computational fluid dynamics:

– LeVeque, *Numerical Methods for Conservation Laws* [143], and his contributions to LeVeque, Mihalas, Dorfi & Müller, *Computational Methods for Astrophysical Fluid Flow* [144].

Characteristics in MHD:

– The demonstration that the MHD equations are symmetric hyperbolic equations is due to Friedrichs. It probably appeared for the first time in print in the Courant Institute lecture notes [76], which is somewhat hard to find nowadays but still very well worth reading.
– Akhiezer *et al.*, *Plasma Electrodynamics*, Vol. 1 [4], Chapter 3.

Exercises

[5.1] *Sound waves*

We start with a plasma without magnetic field and neglect gravity.

– Which three variables are involved? Write down the equations governing their evolution.
– We assume a time-independent, infinite and homogeneous background and perturb it. Derive the equations for the linear perturbations. Do all three variables still couple?
– Derive the dispersion equation for plane waves from the wave equation for the perturbed velocity, using the sound speed $c \equiv (\gamma p_0/\rho_0)^{1/2}$. Since there is no preferred direction in the plasma (why not?), we may choose a direction of wave propagation. Do that and solve the dispersion equation. How many solutions are there? What do they represent?

[5.2] *Towards MHD waves*

We now introduce a magnetic field into the problem and rewrite the vector products in terms of inner products by means of the vector identities of Appendix A.

– Express the MHD equations in terms of the variables ρ, $\mathbf{v}$, $e = p/[(\gamma - 1)\rho]$ and $\mathbf{B}$.
– Again perturb the quantities and derive the linearized MHD equations in terms of dimensionless variables $\tilde{\rho}$, $\tilde{\mathbf{v}}$, $\tilde{e}$, $\tilde{\mathbf{B}}$, using γ, c, and the vectorial Alfvén speed $\mathbf{b} \equiv \mathbf{B}_0/\sqrt{\rho_0}$.
– Is there a preferred direction in the plasma now? Choose general directions for $\mathbf{B}_0$ and $\mathbf{k}$ (can one choose them parallel?), and derive the matrix representation of the eigenvalue problem. What are the dimensions of this matrix? Is it symmetric, or should it be?

[5.3] *The marginal entropy wave*

To isolate this wave, we exploit the entropy function $S \equiv p\rho^{-\gamma}$ and the pressure p as variables.

– Perturb these two quantities, construct dimensionless variables $\tilde{S}$ and $\tilde{p}$, and derive the evolution equations for them. Combine these equations with those of problem [5.2] to obtain the linearized MHD equations in terms of $\tilde{S}$, $\tilde{\mathbf{v}}$, $\tilde{p}$ and $\tilde{\mathbf{B}}$.
– Show that these equations have a special solution, called the marginal entropy wave, where $\omega = 0$ and most of the variables, except $\tilde{S}$, vanish. What is the physical interpretation of this solution?

[5.4] *The velocity representation*

When the constraint $\nabla \cdot \tilde{\mathbf{B}} = 0$ is not observed, a spurious marginal 'solution' is introduced. To get rid of this solution, we reduce the MHD equations to the velocity representation.

– From the linearized MHD equations for $\tilde{\rho}$, $\tilde{\mathbf{v}}$, $\tilde{e}$ and $\tilde{\mathbf{B}}$, derive the second order wave equation for the velocity $\tilde{\mathbf{v}}$. Are the sound waves included in this equation?
– Again insert plane wave solutions and construct the matrix representation of the eigenvalue problem. What is the dimension of this matrix? Explain the counting of unknowns!
– What is the general condition for an equation like $\mathbf{A} \cdot \mathbf{x} = \lambda \mathbf{x}$ to have a solution where $\mathbf{x} \neq 0$? Give this equation for the matrix representation you derived.

[5.5] The three MHD waves

The dispersion equation just derived has three genuine ($\omega \neq 0$) wave solutions: Alfvén waves (with frequency $\pm\omega_A$), slow and fast magneto-sonic waves (with frequencies $\pm\omega_{s,f}$).

- Derive ω_A of the Alfvén waves. Are these waves longitudinal or transverse? In what direction do they propagate? Could they exist in a plasma without magnetic field?
- Derive the frequencies $\omega_{s,f}$ of the magneto-sonic waves. Which of these waves can be present in a plasma without a magnetic field? What are their frequencies in that case?
- Show that $0 \leq \omega_s^2 \leq \omega_A^2 \leq \omega_f^2 < \infty$.

[5.6]★ Phase and group velocities

There are two important velocities in the propagation of wave trains. One is the phase velocity, $\mathbf{v}_{\text{ph}} \equiv (\omega/k)\,\mathbf{k}/k$, and the other is the group velocity, $\mathbf{v}_{\text{gr}} \equiv \partial\omega/\partial\mathbf{k}$.

- Derive the equations for these velocities for the three kinds of MHD waves. Give explicit expressions for propagation parallel and perpendicular to the magnetic field.
- Show that a wave packet with wave vectors localized about a central value $\mathbf{k}_0$ propagates with the group velocity $\mathbf{v}_{\text{gr}}(\mathbf{k}_0)$.

[5.7] Characteristics in alternative representations

In the text, the characteristics were obtained from the matrix representation Eq. (5.114) in terms of the variables $\rho, \mathbf{v}, e, \mathbf{B}$. It is instructive to see what happens if $s, \mathbf{v}, p, \mathbf{B}$ are chosen as basic variables. Show that the representative matrix is again symmetric. Of course, the same characteristics should be obtained from this matrix. Show that this is also true for the system (4.12)–(4.15) for $\rho, \mathbf{v}, p, \mathbf{B}$, or the one obtained from it for the variables $\rho, \mathbf{v}, s, \mathbf{B}$. Notice that in these representations the matrix is no longer symmetric, so that the first two representations should be considered as the more adequate ones. (Friedrichs' analysis [76] makes use of the $\rho, \mathbf{v}, s, \mathbf{B}$ representation. His conclusion that this system is symmetric is based on the fact that he considers isentropic processes, where $\nabla s = 0$.)

6
Spectral theory

6.1 Stability: intuitive approach
6.1.1 Two viewpoints

How does one know whether a dynamical system is stable or not? Consider the well-known example of a ball at rest at the bottom of a trough or on the top of a hill (Fig. 6.1). There is a position (indicated by the full circle) where the potential energy W due to gravity has an extremum W_0. Displacing the ball slightly to a neighbouring position (at the open circle) results in either a higher or a lower potential energy W_1. This corresponds to a *stable* system in the first case ($W_1 > W_0$) and an *unstable* system in the second case ($W_1 < W_0$).

Already at this stage some important observations can be made, viz.:

(a) We have tacitly assumed that the constraining surface is curved, i.e. either convex or concave, so that there is a position of rest, which is called the *equilibrium* position. In this case, one may rescale the potential energy such that the equilibrium state corresponds to $W_0 = 0$, and W_1 becomes the potential energy of the displacement, which is called the *perturbation*.
(b) If the constraining surface is flat and inclined, the system is not in equilibrium and the ball simply rolls along the plane. This *lack of equilibrium*, when W has no extremum, should be well distinguished from *neutral* or *marginal stability*, when $W_1 = W_0$. The latter situation occurs when the surface is horizontal, so that the value $W = 0$ may be assigned to both W_0 and W_1.

This simple example illustrates the general theoretical approach to linear stability, where the study of the original nonlinear equations is simplified by means of a split in equilibrium and perturbations. First, a time-independent equilibrium state is to be found. This still involves the solution of the nonlinear equations, but with the simplifying condition that the time dependence of the variables vanishes and, usually, that the equilibrium is (translationally and/or rotationally) symmetric.

6.1 Stability: intuitive approach

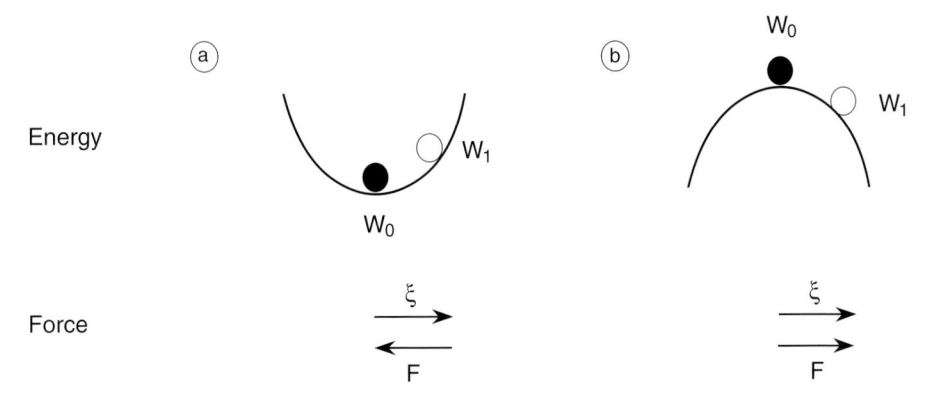

Fig. 6.1. Two viewpoints: energy and force in (a) linearly stable and (b) unstable situations.

An additional simplification, which is usually made for MHD and which is quite pertinent for fusion machines (not for astrophysical plasmas!), is the assumption of *static equilibrium*, not involving flow ($\mathbf{v} = 0$). Next, this equilibrium is subjected to small, i.e. linear, perturbations. This involves the study of the linearized, time-dependent equations.

Such a study may be conducted by means of two broad classes of methods, viz. by using variational principles involving *quadratic forms* (like the energy) or by solving *the (partial) differential equations* themselves. These methods are just a generalization of the two intuitive approaches illustrated in Fig. 6.1. The upper part illustrates the investigation of stability by itself by means of the so-called *energy principle*, i.e. a study of the sign of the potential energy W_1 of the perturbations ($W_1 > 0$: stable, $W_1 = 0$: marginally stable, $W_1 < 0$: unstable). The full dynamics of the system may be obtained from a variational principle which not only involves the potential energy but also the kinetic energy of the perturbations. The more usual approach is the solution of differential equations, in particular the equation of motion, which involves a study of the *forces* acting on the system. With respect to stability, this method is illustrated in the bottom part of Fig. 6.1. If a displacement ξ creates a force F in the opposite direction, the state of equilibrium tends to be restored and the system is stable. On the other hand, if the resulting force is in the same direction as the displacement, the motion will be away from equilibrium and the system is unstable.

These intuitive notions on displacements, forces, and energies may be generalized to continuous media, in particular magneto-fluids. This leads to the two alternative representations by means of *the equation of motion*, involving the plasma displacement vector field $\boldsymbol{\xi}(\mathbf{r}, t)$ and the linear force operator $\mathbf{F}(\boldsymbol{\xi})$ acting on that field, on the one hand (Section 6.2), and by means of *variational quadratic forms*,

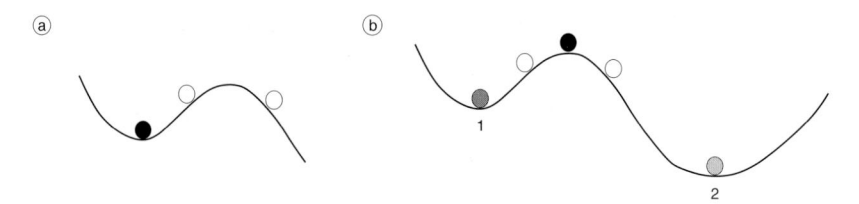

Fig. 6.2. (a) Linearly stable, nonlinearly unstable, (b) linearly unstable, nonlinearly 'stable'.

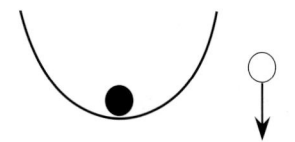

Fig. 6.3. Violating constraints.

involving the potential energy functional $W[\xi]$ and the kinetic energy functional $K[\dot{\xi}]$, on the other hand (Section 6.4).

▷ **Question.** When a glass of water is turned upside down the contents will drop out, as is generally known (first situation). However, if the glass is filled to the rim and covered by a piece of paper the water will not drop out when the glass is turned over (second situation). Discuss these facts in terms of equilibrium and stability properties of the configuration. Is the first situation lack of equilibrium or instability? How about the second one? What is the function of the piece of paper?   ◁

Nonlinear stability or instability concerns the behaviour of dynamical systems with respect to finite (non-infinitesimal) amplitudes of the displacements. Some examples are shown in Fig. 6.2. On the left a system is shown which is stable when subjected to small perturbations, but which becomes unstable when the amplitude of the perturbation is big enough. Vice versa, a system may be unstable with respect to small perturbations (Fig. 6.2(b)) but may possess neighbouring equilibrium states (indicated by the labels 1 and 2) which are stable. If those states are accessible, the system may turn out to be nonlinearly 'stable'. (Quotation marks since the original system is unstable but it evolves towards another state, which is stable.) This subject is quite relevant for the properties of confined plasmas, but too complicated to be covered in this chapter.

▷ **Subtlety:** Consider the situation of Fig. 6.3. The potential energy of the ball has been lowered by taking it outside the cup. Is this situation unstable? Of course, this is cheating: we had tacitly agreed upon the *constraint* that the ball should stay on the surface of the cup. However, such subtleties turn out to be relevant for our study of plasma instabilities, as we will see later (end of Section 6.1.2) for the constraint of 'frozen' field lines. For the moment it suffices to say that it is important to make constraints explicit, and to observe them.   ◁

6.1.2 Linearization and Lagrangian reduction

Let us now carry out the mentioned program on force and energy for plasmas. The starting point is again the ideal MHD equations, which we write in the following peculiar order (which will become clear in a moment) in terms of the variables $\mathbf{v}$, p, $\mathbf{B}$ and ρ:

$$\rho\left(\frac{\partial \mathbf{v}}{\partial t} + \mathbf{v} \cdot \nabla \mathbf{v}\right) = -\nabla p + \mathbf{j} \times \mathbf{B} - \rho \nabla \Phi, \qquad \mathbf{j} = \nabla \times \mathbf{B}, \qquad (6.1)$$

$$\frac{\partial p}{\partial t} = -\mathbf{v} \cdot \nabla p - \gamma p \nabla \cdot \mathbf{v}, \qquad (6.2)$$

$$\frac{\partial \mathbf{B}}{\partial t} = \nabla \times (\mathbf{v} \times \mathbf{B}), \qquad \nabla \cdot \mathbf{B} = 0, \qquad (6.3)$$

$$\frac{\partial \rho}{\partial t} = -\nabla \cdot (\rho \mathbf{v}). \qquad (6.4)$$

For the time being, we restrict the analysis to model I, i.e. application of the b.c.s

$$\mathbf{n} \cdot \mathbf{v} = 0, \qquad (6.5)$$

$$\mathbf{n} \cdot \mathbf{B} = 0, \qquad (6.6)$$

at the wall. These b.c.s do not need to be linearized since they are already linear. (That is so because $\mathbf{n}$ is fixed; the complications arising for interface plasmas, when $\mathbf{n}$ is not fixed, will be discussed in Section 6.6.1.)

Linearization proceeds by first defining a background equilibrium state about which the dynamics is supposed to take place. The simplest and most relevant choice is that of a *static equilibrium* ($\mathbf{v}_0 = 0$):

$$\mathbf{j}_0 \times \mathbf{B}_0 = \nabla p_0 + \rho_0 \nabla \Phi, \qquad \mathbf{j}_0 = \nabla \times \mathbf{B}_0, \qquad \nabla \cdot \mathbf{B}_0 = 0, \qquad (6.7)$$

$$\mathbf{n} \cdot \mathbf{B}_0 = 0 \quad (\text{at the wall}). \qquad (6.8)$$

These equations only partly determine the equilibrium functions $\rho_0(\mathbf{r})$, $p_0(\mathbf{r})$, $\mathbf{j}_0(\mathbf{r})$ and $\mathbf{B}_0(\mathbf{r})$. Clearly, a lot of freedom is left in the choice of the equilibrium profiles. In particular, in the absence of gravity, the density $\rho_0(\mathbf{r})$ is completely arbitrary.

▷ **Example: freedom in circular cylinder equilibria.** We drop the gravity term and momentarily omit the subscript 0 on the equilibrium quantities. Introducing cylinder coordinates r, θ, z, with $\partial/\partial \theta = 0$ and $\partial/\partial z = 0$, Eq. (6.7)(b) yields $\nabla \cdot \mathbf{j} = (1/r) \, d(r j_r)/dr = 0$, so that $j_r = 0$. Similarly, Eq. (6.7)(c) yields $B_r = 0$. Hence, the physical variables $p(r)$, $\mathbf{B} = B_\theta(r) \mathbf{e}_\theta + B_z(r) \mathbf{e}_z$, and $\mathbf{j} = j_\theta(r) \mathbf{e}_\theta + j_z(r) \mathbf{e}_z$ involve five scalar functions which depend only on r. In that case, the *magnetic surfaces*, which are tangent to $\mathbf{j}$ and $\mathbf{B}$ and orthogonal to ∇p, are nested circular cylinders. The five radial equilibrium functions are further restricted by the remaining relations of Eqs. (6.7), viz.

$$p' = j_\theta B_z - j_z B_\theta, \qquad j_\theta = -B'_z, \qquad j_z = \frac{1}{r}(rB_\theta)', \qquad (6.9)$$

so that

$$(p + \tfrac{1}{2}B^2)' = -B_\theta^2/r, \qquad (6.10)$$

where the primes denote derivatives with respect to r. Consequently, Eq. (6.10) is the only restriction on the three basic functions $p(r)$, $B_\theta(r)$, $B_z(r)$, so that two of them may be chosen arbitrarily. Note that this freedom comes in addition to the arbitrariness of the density profile $\rho(r)$. For example, choosing $p = p_0(1 - r^2)$ and $B_z = B_0$, we get $B_\theta = \sqrt{p_0}\,r$, $j_\theta = 0$ and $j_z = 2\sqrt{p_0}$, which is a slight generalization (permitting a constant B_z field) of the z-pinch equilibrium considered in Section 2.4.3. ◁

We return to the general discussion. Time enters into the problem with the *perturbation* of the equilibrium:

$$\begin{aligned} \mathbf{v}(\mathbf{r}, t) &= \mathbf{v}_1(\mathbf{r}, t), \\ p(\mathbf{r}, t) &= p_0(\mathbf{r}) + p_1(\mathbf{r}, t), \\ \mathbf{B}(\mathbf{r}, t) &= \mathbf{B}_0(\mathbf{r}) + \mathbf{B}_1(\mathbf{r}, t), \\ \rho(\mathbf{r}, t) &= \rho_0(\mathbf{r}) + \rho_1(\mathbf{r}, t), \end{aligned} \qquad (6.11)$$

where p_0, $\mathbf{B}_0$, and ρ_0 now correspond to an *inhomogeneous* equilibrium, satisfying Eqs. (6.7) and (6.8), and the perturbations should satisfy $|f_1(\mathbf{r}, t)| \ll |f_0(\mathbf{r})|$ (except for the velocity $\mathbf{v}$). The resulting first order equations for the perturbations of Eqs. (6.1)–(6.4) then read:

$$\rho_0 \frac{\partial \mathbf{v}_1}{\partial t} = -\nabla p_1 + \mathbf{j}_1 \times \mathbf{B}_0 + \mathbf{j}_0 \times \mathbf{B}_1 - \rho_1 \nabla \Phi, \qquad \mathbf{j}_1 = \nabla \times \mathbf{B}_1, \qquad (6.12)$$

$$\frac{\partial p_1}{\partial t} = -\mathbf{v}_1 \cdot \nabla p_0 - \gamma p_0 \nabla \cdot \mathbf{v}_1, \qquad (6.13)$$

$$\frac{\partial \mathbf{B}_1}{\partial t} = \nabla \times (\mathbf{v}_1 \times \mathbf{B}_0), \qquad \nabla \cdot \mathbf{B}_1 = 0, \qquad (6.14)$$

$$\frac{\partial \rho_1}{\partial t} = -\nabla \cdot (\rho_0 \mathbf{v}_1), \qquad (6.15)$$

whereas the b.c.s (6.5) and (6.6) for the perturbations require

$$\mathbf{n} \cdot \mathbf{v}_1 = 0, \qquad \mathbf{n} \cdot \mathbf{B}_1 = 0 \qquad \text{(at the wall)}. \qquad (6.16)$$

At this stage, p_0, $\mathbf{j}_0$, $\mathbf{B}_0$, and ρ_0 are considered to be known from the zeroth order equilibrium equations (the unperturbed system) and $\mathbf{v}_1$, p_1, $\mathbf{B}_1$, and ρ_1 are unknown. Note that a lot of nasty terms have been eliminated by the assumption $\mathbf{v}_0 = 0$.

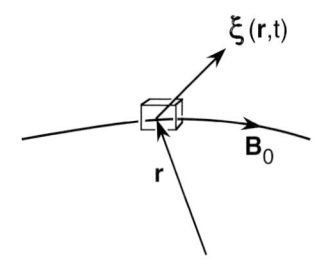

Fig. 6.4. Plasma displacement vector field.

We are now ready for an extremely useful further reduction, due to Bernstein *et al.* (1958) [26]. We have promoted the momentum conservation equation (6.12) to the first place since it describes the evolution of the velocity $\mathbf{v}_1$, whereas the perturbations p_1, $\mathbf{B}_1$ and ρ_1 may be considered as secondary, i.e. determined by $\mathbf{v}_1$ as expressed by the right hand sides of Eqs. (6.13)–(6.15). Elimination of the latter equations is straightforward if the momentum equation is differentiated with respect to time. (This procedure was followed in Section 5.2.3 to derive the wave equation (5.50) for homogeneous plasmas.) However, the opposite procedure of integrating the equations (6.13)–(6.15) is more powerful. This is effected by means of a new variable, viz. *the Lagrangian displacement vector field* $\boldsymbol{\xi}(\mathbf{r}, t)$ of a plasma element away from the equilibrium state (see Fig. 6.4). The velocity is just the Lagrangian time derivative (i.e. the variation in time experienced in a local coordinate system co-moving with the fluid) of this variable:

$$\mathbf{v} = \frac{D\boldsymbol{\xi}}{Dt} \equiv \frac{\partial \boldsymbol{\xi}}{\partial t} + \mathbf{v} \cdot \nabla \boldsymbol{\xi}, \tag{6.17}$$

which is an exact, but highly nonlinear, expression. However, for the linearized problem, we only need the first order, Eulerian, part of the time derivative:

$$\mathbf{v} \approx \mathbf{v}_1 = \frac{\partial \boldsymbol{\xi}}{\partial t}. \tag{6.18}$$

Inserting this expression, the equations (6.13)–(6.15) for p_1, $\mathbf{B}_1$ and ρ_1 can be integrated immediately:

$$p_1 = -\boldsymbol{\xi} \cdot \nabla p_0 - \gamma p_0 \nabla \cdot \boldsymbol{\xi}, \tag{6.19}$$

$$\mathbf{B}_1 = \nabla \times (\boldsymbol{\xi} \times \mathbf{B}_0), \tag{6.20}$$

$$\rho_1 = -\nabla \cdot (\rho_0 \boldsymbol{\xi}). \tag{6.21}$$

Note that the equation $\nabla \cdot \mathbf{B}_1 = 0$ can be dropped now since it is automatically satisfied by Eq. (6.20). Inserting these expressions into Eq. (6.12) leads to the desired result, viz. an equation of motion of the form

$$\rho_0 \frac{\partial^2 \boldsymbol{\xi}}{\partial t^2} = \mathbf{F}\Big(p_1(\boldsymbol{\xi}), \mathbf{B}_1(\boldsymbol{\xi}), \rho_1(\boldsymbol{\xi})\Big), \qquad (6.22)$$

where p_1, $\mathbf{B}_1$ and ρ_1 are determined by Eqs. (6.19)–(6.21). We will continue with this formulation in Section 6.2, where we will give the explicit form of the operator $\mathbf{F}$.

In the absence of gravity, the expression (6.21) for ρ_1 is actually not needed since it does not occur in the equation of motion. Hence, we could forget about this relation in that case. This changes when gravitational forces are taken into account, as needed in astrophysical plasmas. We will restrict the analysis to gravitational potentials that are external, i.e. not caused by self-gravitation of the plasma but by some external massive object.

▷ **Subtleties on the initial data.** So far, we have not mentioned initial conditions. In particular, we have omitted explicit terms for $p_1(\mathbf{r}, 0)$, $\mathbf{B}_1(\mathbf{r}, 0)$ and $\rho_1(\mathbf{r}, 0)$ in Eqs. (6.19)–(6.21). This implies that we have tacitly assumed that finite initial values for these quantities only appear as a result of a finite initial displacement $\boldsymbol{\xi}(\mathbf{r}, 0)$. Since the ultimate equation (6.22) is of second order in time, we should consider both $\boldsymbol{\xi}(\mathbf{r}, 0)$ and $\dot{\boldsymbol{\xi}}(\mathbf{r}, 0)$ as independent initial data. The following possibilities arise: (1) $\boldsymbol{\xi}(\mathbf{r}, 0) = 0$, $\dot{\boldsymbol{\xi}}(\mathbf{r}, 0) \neq 0$; (2) $\boldsymbol{\xi}(\mathbf{r}, 0) \neq 0$, $\dot{\boldsymbol{\xi}}(\mathbf{r}, 0) = 0$; and (3) combinations.

Consider case (2). Should we consider $p_1(\mathbf{r}, 0)$, $\mathbf{B}_1(\mathbf{r}, 0)$, $\rho_1(\mathbf{r}, 0)$ as additional independent initial data, together with $\boldsymbol{\xi}(\mathbf{r}, 0)$? *NO!* If we start off with a finite displacement of the plasma, we had better make sure that we got there by means of a motion consistent with the ideal MHD conservation laws. Otherwise, something could happen as illustrated in Fig. 6.5. At $t = 0$ the fluid has been displaced, but simultaneously a perturbation in $\mathbf{B}$ has been allowed that corresponds to broken field lines: not a meaningful operation in ideal MHD. Such a perturbation, *not satisfying the constraints*, should be considered as the analogue of taking the ball out of the cup (Fig. 6.3).

A similar story holds for case (1), where we should assume that $p_1(\mathbf{r}, 0) = 0$, $\mathbf{B}_1(\mathbf{r}, 0) = 0$, $\rho_1(\mathbf{r}, 0) = 0$ because $\boldsymbol{\xi}(\mathbf{r}, 0) = 0$. Hence, on the basis of physical arguments, we tacitly impose an additional condition, viz. that the initial data on p_1, $\mathbf{B}_1$ and ρ_1 should be consistent with ideal MHD. This implies that the equations (6.19)–(6.21) are considered to be valid at all times. Consequently, all perturbations are expressed in terms of $\boldsymbol{\xi}(\mathbf{r}, t)$ alone and the possibility of Fig. 6.5 is eliminated. ◁

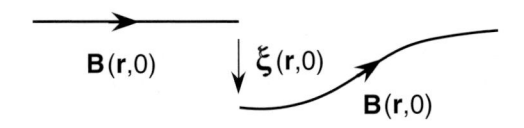

Fig. 6.5. Cheating on the initial data.

6.2 Force operator formalism

6.2.1 Equation of motion

We complete the discussion of the equation of motion (6.22). The right hand side of this equation defines the ideal MHD *force operator* $\mathbf{F}(\boldsymbol{\xi})$. Exploiting the expressions of Section 6.1.2, the explicit form of the linearized equation of motion may be written as

$$\mathbf{F}(\boldsymbol{\xi}) \equiv -\nabla \pi - \mathbf{B} \times (\nabla \times \mathbf{Q}) + (\nabla \times \mathbf{B}) \times \mathbf{Q} + (\nabla \Phi) \nabla \cdot (\rho \boldsymbol{\xi}) = \rho \frac{\partial^2 \boldsymbol{\xi}}{\partial t^2},$$
(6.23)

where

$$\pi \equiv p_1 = -\gamma p \nabla \cdot \boldsymbol{\xi} - \boldsymbol{\xi} \cdot \nabla p, \qquad (6.24)$$
$$\mathbf{Q} \equiv \mathbf{B}_1 = \nabla \times (\boldsymbol{\xi} \times \mathbf{B}). \qquad (6.25)$$

The symbols π (not to be confused with the earlier use of π in Chapter 3) and $\mathbf{Q}$ are introduced here for no other reason than to be able to drop the subscripts 0 and 1 on the equilibrium and perturbation variables. This we will do from now on. For model I (wall on the plasma), the only boundary condition needed to close the system is obtained from Eq. (6.16)(a) for $\mathbf{v}_1$:

$$\mathbf{n} \cdot \boldsymbol{\xi} = 0 \qquad \text{(at the wall)}. \qquad (6.26)$$

Once the geometry of the plasma has been specified (see Section 4.6), the equation of motion (6.23) and the boundary condition (6.26) completely determine the linearized dynamics of the plasma. Consequently, introducing the plasma displacement vector $\boldsymbol{\xi}$ has enabled us to cast the linear equations of ideal MHD in the compact form (6.23), which may be considered to be *Newton's law for a plasma element*. This is the physical basis of the very powerful spectral methods that can be applied to these problems.

▷ **Magnetic field boundary condition.** The boundary condition (6.16)(b) for $\mathbf{B}_1$, leading to $\mathbf{n} \cdot \nabla \times (\boldsymbol{\xi} \times \mathbf{B}) = 0$, is automatically satisfied since $\mathbf{n} \cdot \boldsymbol{\xi} = 0$ and $\mathbf{n} \cdot \mathbf{B} = 0$. This is proved by noting that the latter two conditions imply that $\boldsymbol{\xi} \times \mathbf{B} = f \mathbf{n}$, with f some scalar function. Hence, $\mathbf{n} \cdot \nabla \times (\boldsymbol{\xi} \times \mathbf{B}) = \mathbf{n} \cdot [(\nabla f) \times \mathbf{n} + f \nabla \times \mathbf{n}] = f \mathbf{n} \cdot \nabla \times \mathbf{n} = 0$. The last equality is easily obtained from the general expression (A.34) for the curl of a vector in curvilinear coordinates by choosing one of the coordinates along $\mathbf{n}$, so that $n_1 = 1$ and $n_2 = n_3 = 0$. ◁

It is always important to count unknowns and equations. In linearized ideal MHD only one vector field $\boldsymbol{\xi}(\mathbf{r}, t)$ appears, whereas in the nonlinear theory the variables $\mathbf{v}$, $\mathbf{B}$, p and ρ are needed. This is a substantial simplification (in addition to the linearization). However, the order of the system should be independent of the choice of the variables exploited. Equation (6.23), with the definitions (6.24) and (6.25) substituted, constitutes a sixth order system[1] of three second order PDEs for the three components of $\boldsymbol{\xi}$. On the other hand, the equations (6.12)–(6.14) also constitute a sixth order system, viz. of six first order PDEs for the three components of $\mathbf{v}$, the scalar p, and two of the three components of $\mathbf{B}_1$. The third component of $\mathbf{B}_1$ is redundant because of the condition $\nabla \cdot \mathbf{B}_1 = 0$, which is an initial condition for Eq. (6.14), but automatically satisfied in Eq. (6.25). In the absence of gravity, the equation for ρ_1 is redundant in both representations since it does not couple to the rest of the equations. The presence of a gravity field does not raise the order of the system either since gradients of the variable ρ_1 do not appear in the equations (6.12) and (6.15).

Recall the similar reduction from representation in primitive variables to velocity representation in Section 5.2.3, which we showed to be equivalent if the marginal entropy mode given by $\omega_E S_1 = 0$, or $\omega_E \rho_1 = 0$, was ignored. We conclude that the $\boldsymbol{\xi}$-representation, also called the *Lagrangian representation*, is only equivalent to the original *Eulerian representation* in primitive variables if the Eulerian entropy mode $\omega_E = 0$ is neglected. This is why the subscript E has been put on this mode.

Since the equilibrium quantities $p(\mathbf{r})$, $\mathbf{B}(\mathbf{r})$ and $\rho(\mathbf{r})$ appearing in Eq. (6.23) do not depend on time we may consider solutions in the form of *normal modes*:

$$\boldsymbol{\xi}(\mathbf{r}, t) = \hat{\boldsymbol{\xi}}(\mathbf{r}) e^{-i\omega t} . \tag{6.27}$$

This transforms Eq. (6.23) into the eigenvalue problem

$$\mathbf{F}(\hat{\boldsymbol{\xi}}) = -\rho \omega^2 \hat{\boldsymbol{\xi}}, \tag{6.28}$$

involving the linear operator $\mathbf{F}$ (or rather $\rho^{-1}\mathbf{F}$) and its eigenvalues ω^2. As well as discrete eigenvalues, ideal MHD also allows for continuous (or 'improper') eigenvalues, as we will see in Chapter 7. The collection $\{\omega^2\}$ of these two kinds of eigenvalues is called *the spectrum of ideal MHD*. The central property, which puts this subject on an equal mathematical footing to quantum mechanics, is that *the operator $\rho^{-1}\mathbf{F}$ is self-adjoint* so that *the eigenvalues ω^2 are real*. This will be proved in Section 6.2.3. Consequently, the eigenvalues ω themselves are either real or purely imaginary.

---

[1] The attentive reader may be puzzled by the appearance of only one boundary condition, viz. Eq. (6.26). Actually, in order not to slow down the exposition by too much detail, we had to be a little cavalier here. This will be remedied in later explicit calculations where additional regularity and periodicity conditions appear, dictated by the specific geometry which is chosen. Yet, it should be stressed already at this point that the general statement about the sixth order character of the problem only holds with respect to two of the three spatial coordinates. Due to the extreme anisotropy of magnetically confined systems, the dependence on the coordinate normal to the magnetic surfaces turns out to reduce to a second order system in terms of $\mathbf{n} \cdot \boldsymbol{\xi}$ alone. Hence, regularity at the magnetic axis and the condition (6.26) suffice to fix the solutions.

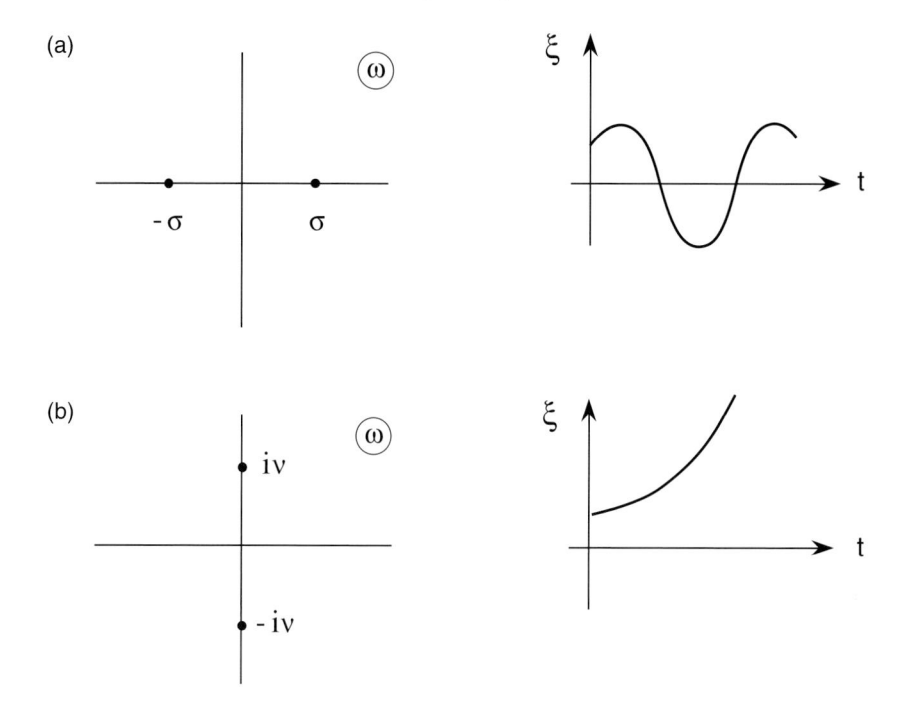

Fig. 6.6. (a) Stable waves and (b) instabilities in ideal MHD.

Writing $\omega = \sigma + i\nu$, this implies that two quite different classes of solution occur, viz. *stable waves* for $\omega^2 > 0$, so that $\nu = 0$ and the temporal dependence is given by $\exp(\mp i\sigma t)$ (Fig. 6.6(a)), and *instabilities* for $\omega^2 < 0$, so that $\sigma = 0$ and the temporal dependence $\exp(\pm \nu t)$ represents exponential growth (Fig. 6.6(b)). For eigenmodes, we already see the connection with the bottom part of the pictures of Fig. 6.1 on the relationship between displacements and forces. According to Eq. (6.28), $\mathbf{F}(\boldsymbol{\xi}) \sim -\hat{\boldsymbol{\xi}}$ for stable waves with $\omega^2 > 0$ and $\mathbf{F}(\boldsymbol{\xi}) \sim \hat{\boldsymbol{\xi}}$ for instabilities with $\omega^2 < 0$. For general motions, consisting of a superposition of eigenmodes, such a simple relationship does not hold. In that case, the sign of the potential energy (corresponding to the upper parts of Fig. 6.1) will provide the test for stability (see Section 6.4.4).

In dissipative (e.g. resistive) MHD, different eigenmodes are possible. In particular, since ω^2 need not be real then, complex values of ω may occur. This may give rise to the kind of modes depicted in Fig. 6.7: stable, but damped, modes if $\text{Im}(\omega) < 0$ and so-called 'overstable' modes if $\text{Im}(\omega) > 0$. The term 'overstable' expresses the fact that the direction of the restoring force is opposite to the displacement, as in Fig. 6.1(a), but it is too big so that the resulting motion overshoots and the net result is again an instability. These additional possibilities are

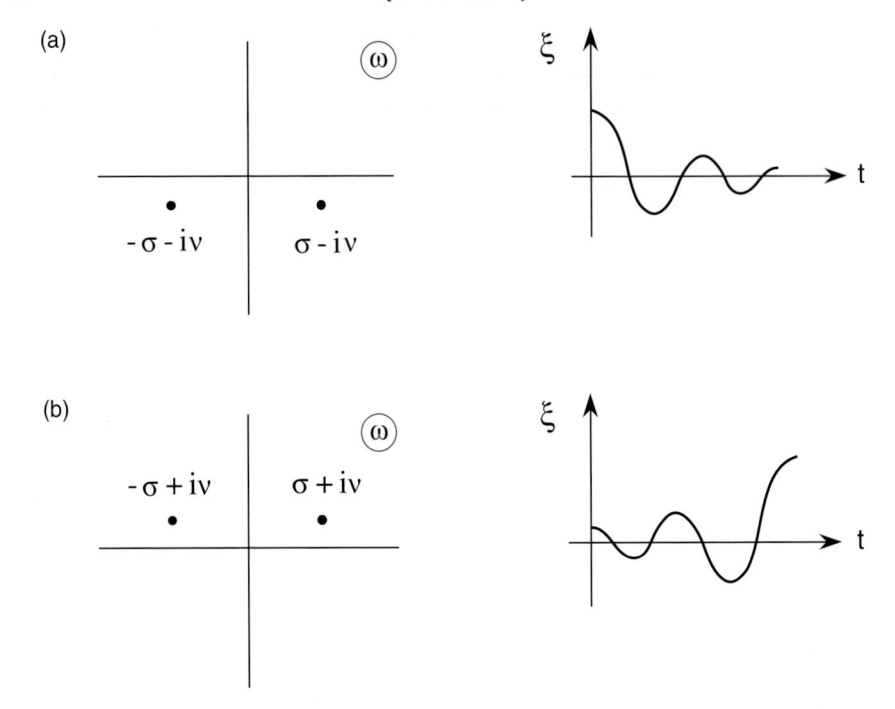

Fig. 6.7. (a) Damped stable and (b) overstable waves in dissipative MHD.

associated with the fact that the waves are non-conservative in dissipative MHD: energy may be dissipated or accumulated.

Clearly, ideal, conservative MHD of static equilibria presents a significant simplification for stability problems. Since ω^2 is real, the transition from stability to instability occurs only through the value $\omega^2 = 0$, or $\omega = 0$, i.e. through *marginal stability*. Consequently, to study the problem of stability one could study *the marginal equation of motion*

$$\mathbf{F}(\hat{\boldsymbol{\xi}}) = 0, \tag{6.29}$$

subject to the boundary condition (6.26). In general, this equation does not have a solution because it is obtained from the eigenvalue problem (6.28) and $\omega^2 = 0$ does not have to be an eigenvalue. In order to get genuine solutions one should arrange the equilibrium parameters such that $\omega^2 = 0$ becomes an eigenvalue. For example, a typical tokamak stability study would involve the variation of global equilibrium parameters like the value of $\beta \equiv 2\mu_0 p/B^2$ and the 'safety factor' at the plasma boundary, $q_1 \sim 1/I_p$, while keeping other variables fixed. For a particular value of the plasma current I_p one would push the value of β until the marginal equation of motion Eq. (6.29) is satisfied, subject to the boundary condition (6.26). In this manner one would calculate one value β_{crit} where marginal stability is obtained. By varying the value of I_p one would trace out marginal

6.2 Force operator formalism 241

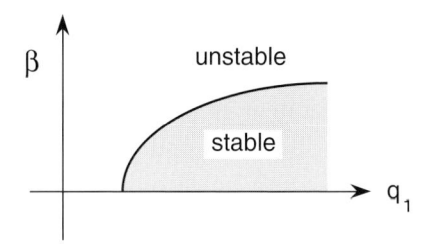

Fig. 6.8. Schematic stability diagram for tokamaks.

stability curves in the β–q_1 diagram, as shown schematically in Fig. 6.8. Physical arguments usually indicate on which side of the curve the stable states are to be found. (For example, in Fig. 6.8, this would be on the low β side.) This is the most general, though not the easiest, method of studying stability problems.

Since **F** now contains all information on MHD waves and instabilities, let us rearrange the terms slightly to see what we have:

$$\mathbf{F}(\boldsymbol{\xi}) = \nabla(\gamma p \nabla \cdot \boldsymbol{\xi}) - \mathbf{B} \times (\nabla \times \mathbf{Q}) + \nabla(\boldsymbol{\xi} \cdot \nabla p) + \mathbf{j} \times \mathbf{Q} + \nabla \Phi \nabla \cdot (\rho \boldsymbol{\xi}). \tag{6.30}$$

One recognizes, successively, an isotropic force due to plasma *compressibility*, associated with sound waves, and an anisotropic force ($\perp \mathbf{B}$) due to *bending of field lines*, which is responsible for Alfvén waves. These terms are always present, even in an infinitely homogeneous plasma ($\nabla p = 0$, $\mathbf{j} = 0$), which has been shown (Chapter 5) to give rise to stable waves only. The third and fourth terms only occur in inhomogeneous plasmas, e.g. those encountered in confined plasmas for thermonuclear research or magnetic flux tubes in astrophysical plasmas. In those plasmas, the effect of equilibrium *pressure gradients* and *currents*, which are intrinsically associated with confinement, may give rise to instabilities. Of course, the art of designing thermonuclear confinement machines is to find regions in parameter space (as exemplified by Fig. 6.8) where these potentially unstable effects are counterbalanced by the stabilizing contributions. Finally, the gravitational term is also associated with inhomogeneity. It gives rise to confinement as well as gravitational instabilities in astrophysical plasmas.

For *homogeneous* plasmas, the spectral equation (6.28) may be written as

$$\rho^{-1} \mathbf{F}(\hat{\boldsymbol{\xi}}) = c^2 \nabla \nabla \cdot \hat{\boldsymbol{\xi}} + \mathbf{b} \times (\nabla \times (\nabla \times (\mathbf{b} \times \hat{\boldsymbol{\xi}}))) = -\omega^2 \hat{\boldsymbol{\xi}}, \tag{6.31}$$

where $c \equiv \sqrt{\gamma p/\rho}$ and $\mathbf{b} \equiv \mathbf{B}/\sqrt{\rho}$ are constants. In that case, plane wave solutions $\hat{\boldsymbol{\xi}} \sim \exp(i\mathbf{k} \cdot \mathbf{r})$ give rise to the algebraic eigenvalue problem

$$\rho^{-1} \mathbf{F}(\hat{\boldsymbol{\xi}}) = -c^2 \mathbf{k} \mathbf{k} \cdot \hat{\boldsymbol{\xi}} - \mathbf{b} \times (\mathbf{k} \times (\mathbf{k} \times (\mathbf{b} \times \hat{\boldsymbol{\xi}})))$$
$$= \left[-(\mathbf{k} \cdot \mathbf{b})^2 \mathbf{I} - (b^2 + c^2) \mathbf{k}\mathbf{k} + \mathbf{k} \cdot \mathbf{b} \, (\mathbf{k}\mathbf{b} + \mathbf{b}\mathbf{k}) \right] \cdot \hat{\boldsymbol{\xi}} = -\omega^2 \hat{\boldsymbol{\xi}}. \tag{6.32}$$

This equation for $\hat{\boldsymbol{\xi}}$ is the same as Eq. (5.51) for $\hat{\mathbf{v}}$, which produced the dispersion equation (5.53) (without the marginal entropy wave), so that the three stable MHD waves of Section 5.2.3 are recovered. Hence, analogous to Eq. (5.63), the slow, Alfvén, and fast eigenvectors $\hat{\boldsymbol{\xi}}_s$, $\hat{\boldsymbol{\xi}}_A$, and $\hat{\boldsymbol{\xi}}_f$ form an orthogonal triad. This guarantees that arbitrary vectors $\hat{\boldsymbol{\xi}}$ may be decomposed into the three MHD eigenvectors so that the eigenspaces of the force operator span the whole space. We will show that this space is a Hilbert space.

Finally, before turning to mathematical techniques, it is important to note that the eigenvalue problem (6.28), with $\mathbf{F}$ given by either Eq. (6.30) or (6.31), contains Alfvén waves as the dynamical centrepiece. Thus, Eq. (6.32), with $\mathbf{B}$ and $\mathbf{k}$ in the z-direction (see Fig. 5.1) and transverse incompressible motion ($\mathbf{k} \cdot \boldsymbol{\xi} = 0$), yields the Alfvén wave equation

$$\rho^{-1}\hat{F}_y = b^2 \frac{\partial^2 \hat{\xi}_y}{\partial z^2} = -k_z^2 b^2 \hat{\xi}_y = \frac{\partial^2 \hat{\xi}_y}{\partial t^2} = -\omega^2 \hat{\xi}_y, \qquad (6.33)$$

with $\omega^2 = \omega_A^2 \equiv k_z^2 b^2$ as eigenvalue. The formidable analysis that follows is, in a sense, just the machinery needed to trace this wave in inhomogeneous plasmas.

6.2.2 Hilbert space

We have cast the ideal MHD equations in a form which is general enough to benefit from comparison with another area of physics where *linear operators* play an important role, viz. quantum mechanics. Consider a plasma volume V enclosed by a wall W. Define two plasma displacement fields:

$$\begin{aligned}\boldsymbol{\xi} &= \boldsymbol{\xi}(\mathbf{r}, t) \quad (\text{on } V), & \text{where} \quad \mathbf{n} \cdot \boldsymbol{\xi} &= 0 \quad (\text{at } W), \\ \boldsymbol{\eta} &= \boldsymbol{\eta}(\mathbf{r}, t) \quad (\text{on } V), & \text{where} \quad \mathbf{n} \cdot \boldsymbol{\eta} &= 0 \quad (\text{at } W).\end{aligned} \qquad (6.34)$$

An *inner product* may be defined with the plasma equilibrium density $\rho(\mathbf{r})$ acting as a weight function:

$$\langle \boldsymbol{\xi}, \boldsymbol{\eta} \rangle \equiv \tfrac{1}{2} \int \rho \, \boldsymbol{\xi}^* \cdot \boldsymbol{\eta} \, dV. \qquad (6.35)$$

By means of this definition, one may also define *the norm* of the vector field $\boldsymbol{\xi}(\mathbf{r}, t)$:

$$\|\boldsymbol{\xi}\| \equiv \langle \boldsymbol{\xi}, \boldsymbol{\xi} \rangle^{1/2}. \qquad (6.36)$$

For functions $\boldsymbol{\xi}$ with *a finite norm*, $\|\boldsymbol{\xi}\| < \infty$, a special linear function space is obtained, viz. a *Hilbert space*. At this stage, certain additional properties of this space with respect to completeness and separability are assumed to be true. For a more complete discussion of the mathematical aspects of MHD spectral theory see Lifschitz [146].

6.2 Force operator formalism

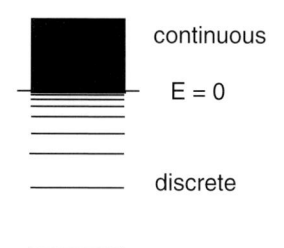

Fig. 6.9. Schematic spectrum of a quantum mechanical system.

The analogy with quantum mechanics is obvious. The Schrödinger equation for the wave function ψ,

$$H\psi = E\psi, \tag{6.37}$$

is an eigenvalue equation for the self-adjoint linear operator H (the Hamiltonian) with eigenvalues E (the energy levels). This eigenvalue problem leads to a spectrum of eigenvalues (Fig. 6.9), which may be discrete (for bound states with $E < 0$) or continuous (for free particles with $E > 0$). Ideal MHD also has a spectrum of eigenvalues ω^2, which may be discrete or continuous, with very special properties depending on the magnitude of ω^2, as we will see. It is useful to learn from another field what the relevant methods are.

The inner product in quantum mechanics,

$$\langle \psi_1, \psi_2 \rangle \equiv \tfrac{1}{2} \int \psi_1^* \psi_2 \, dV, \tag{6.38}$$

permits one to define the norm

$$\|\psi\| \equiv \langle \psi, \psi \rangle^{1/2}, \tag{6.39}$$

which defines the probability of finding the particle in the volume considered. In general, one chooses $\|\psi\| = 1$ to express the certainty that the particle is located somewhere. Clearly, normalization and Hilbert space have a very clear-cut physical background in quantum mechanics.

What is the physical background for considering vector fields $\boldsymbol{\xi}(\mathbf{r}, t)$ with a finite norm $\|\boldsymbol{\xi}\|$ in ideal MHD? To answer that question, consider *the kinetic energy of the plasma*:

$$K \equiv \tfrac{1}{2} \int \rho \mathbf{v}^2 \, dV \approx \tfrac{1}{2} \int \rho \dot{\boldsymbol{\xi}}^2 \, dV = \langle \dot{\boldsymbol{\xi}}, \dot{\boldsymbol{\xi}} \rangle \equiv \|\dot{\boldsymbol{\xi}}\|^2. \tag{6.40}$$

As far as the spatial dependence is concerned, $\dot{\boldsymbol{\xi}}(\mathbf{r}, t)$ belongs to the same class of functions as $\boldsymbol{\xi}(\mathbf{r}, t)$ (t is simply a parameter). Hence, a bounded norm implies finite kinetic energy of the perturbations. Because the total energy is conserved, this also implies finite potential energy: a very reasonable assumption, which justifies the use of Hilbert space as the mathematical device for our investigations.

Hence, in quantum mechanics, we encounter linear operators acting on the wave function ψ, and in magnetohydrodynamics, we encounter linear operators acting on the displacement vector field $\boldsymbol{\xi}$. We also have defined an inner product and a norm for the MHD displacements. Finally, the analogy would be perfect if we could also show that there is a counterpart in MHD for the important role played by self-adjoint operators L, defined by

$$\langle \psi_1, L\psi_2 \rangle = \langle L\psi_1, \psi_2 \rangle \tag{6.41}$$

in quantum mechanics. In fact, one of the central results of linearized ideal MHD turns out to be that *the force operator* $\mathbf{F}$, *or rather* $\rho^{-1}\mathbf{F}$, *is a self-adjoint linear operator in the Hilbert space of plasma displacement vectors*:

$$\langle \boldsymbol{\eta}, \rho^{-1}\mathbf{F}(\boldsymbol{\xi}) \rangle \equiv \tfrac{1}{2} \int \boldsymbol{\eta}^* \cdot \mathbf{F}(\boldsymbol{\xi}) \, dV = \tfrac{1}{2} \int \boldsymbol{\xi} \cdot \mathbf{F}(\boldsymbol{\eta}^*) \, dV \equiv \langle \rho^{-1}\mathbf{F}(\boldsymbol{\eta}), \boldsymbol{\xi} \rangle. \tag{6.42}$$

This provides linear MHD with a mathematical foundation that is the same as that of quantum mechanics, viz. linear operators in Hilbert space: a very solid ground indeed! On this basis, many analogies between MHD and quantum mechanical spectral theory should be expected.

6.2.3 Proof of self-adjointness of the force operator

Of course, after the excitement about a beautiful analogy, the dirty job of *proving the self-adjointness property* (6.42) remains; see Bernstein et al. [27]. This consists of a lot of cumbersome vector manipulations with little beauty, but it does belong to the craft of MHD. Also, we need some of the intermediate steps in later sections. Hence, we just reproduce the proof here, putting some of the more technical details in small print.

For convenience, we repeat the expression (6.30) for the force operator:

$$\mathbf{F}(\boldsymbol{\xi}) = \nabla(\gamma p \nabla \cdot \boldsymbol{\xi}) - \mathbf{B} \times (\nabla \times \mathbf{Q}) + \nabla(\boldsymbol{\xi} \cdot \nabla p) + \mathbf{j} \times \mathbf{Q} + \nabla \Phi \nabla \cdot (\rho \boldsymbol{\xi}). \tag{6.43}$$

Here, p and B should satisfy the equilibrium relations $\mathbf{j} \times \mathbf{B} = \nabla p + \rho \nabla \Phi$, $\mathbf{j} = \nabla \times \mathbf{B}$, $\nabla \cdot \mathbf{B} = 0$. The first two terms of $\mathbf{F}(\boldsymbol{\xi})$ are already present in homogeneous plasmas, but the last three terms are associated with inhomogeneity. We will first rework these three inhomogeneity terms to show that, in the absence of gravity, they give rise to *a force that is purely perpendicular to* $\mathbf{B}$ (like the second term of Eq. (6.43)).

▷ **Transformation 1 of the inhomogeneity terms.** The transformation of the last three terms of the expression (6.43) for the force operator involves extensive use of the vector identities of Appendix A.1. They are indicated above the equal signs:

$$\nabla(\boldsymbol{\xi} \cdot \nabla p) \stackrel{(A.10)}{=} (\nabla \boldsymbol{\xi}) \cdot \nabla p + \boldsymbol{\xi} \cdot \nabla \nabla p$$

$$\stackrel{(A.9)}{=} (\nabla p \times \nabla) \times \boldsymbol{\xi} + \nabla p \, \nabla \cdot \boldsymbol{\xi} + \boldsymbol{\xi} \cdot \nabla \nabla p$$

$$\stackrel{\text{equilibrium}}{=} ((\mathbf{j} \times \mathbf{B}) \times \nabla) \times \boldsymbol{\xi} - \rho(\nabla \Phi \times \nabla) \times \boldsymbol{\xi} + \nabla p \, \nabla \cdot \boldsymbol{\xi} + \boldsymbol{\xi} \cdot \nabla \nabla p$$

$$\stackrel{(A.2) \text{ on } \mathbf{j}, \mathbf{B}, \nabla}{=} (\mathbf{B}\mathbf{j} \cdot \nabla - \mathbf{j}\mathbf{B} \cdot \nabla) \times \boldsymbol{\xi} - \rho(\nabla \Phi \times \nabla) \times \boldsymbol{\xi} + \nabla p \, \nabla \cdot \boldsymbol{\xi} + \boldsymbol{\xi} \cdot \nabla \nabla p$$

$$\stackrel{(A.9) \text{ on } \nabla\Phi, \nabla, \boldsymbol{\xi}}{=} \mathbf{B} \times (\mathbf{j} \cdot \nabla \boldsymbol{\xi}) - \mathbf{j} \times (\mathbf{B} \cdot \nabla \boldsymbol{\xi}) - \rho(\nabla \boldsymbol{\xi}) \cdot \nabla \Phi + \rho \nabla \Phi \, \nabla \cdot \boldsymbol{\xi}$$

$$+ \nabla p \, \nabla \cdot \boldsymbol{\xi} + \boldsymbol{\xi} \cdot \nabla \nabla p, \quad (6.44)$$

$$\mathbf{j} \times \mathbf{Q} \stackrel{(A.13)}{=} \mathbf{j} \times (\mathbf{B} \cdot \nabla \boldsymbol{\xi} - \mathbf{B} \nabla \cdot \boldsymbol{\xi} - \boldsymbol{\xi} \cdot \nabla \mathbf{B})$$

$$= \mathbf{j} \times (\mathbf{B} \cdot \nabla \boldsymbol{\xi}) - \mathbf{j} \times \mathbf{B} \nabla \cdot \boldsymbol{\xi} - \boldsymbol{\xi} \cdot \nabla (\mathbf{j} \times \mathbf{B}) - \mathbf{B} \times (\boldsymbol{\xi} \cdot \nabla \mathbf{j}), \quad (6.45)$$

$$\nabla \Phi \, \nabla \cdot (\rho \boldsymbol{\xi}) \stackrel{(A.6)}{=} \rho \nabla \Phi \, \nabla \cdot \boldsymbol{\xi} + \nabla \Phi \, \nabla \rho \cdot \boldsymbol{\xi}, \quad (6.46)$$

where the equality $\nabla(\mathbf{a} \times \mathbf{b}) = (\nabla \mathbf{a}) \times \mathbf{b} - (\nabla \mathbf{b}) \times \mathbf{a}$ has been used in the last step of the derivation of Eq. (6.45).[2] Hence, the three inhomogeneity terms may be written as

$$\nabla(\boldsymbol{\xi} \cdot \nabla p) + \mathbf{j} \times \mathbf{Q} + \nabla \Phi \, \nabla \cdot (\rho \boldsymbol{\xi})$$

$$= \mathbf{B} \times (\mathbf{j} \cdot \nabla \boldsymbol{\xi} - \boldsymbol{\xi} \cdot \nabla \mathbf{j}) + (\nabla p - \mathbf{j} \times \mathbf{B} + 2\rho \nabla \Phi) \nabla \cdot \boldsymbol{\xi}$$
$$- \rho(\nabla \boldsymbol{\xi}) \cdot \nabla \Phi + \boldsymbol{\xi} \cdot \nabla \nabla p - \boldsymbol{\xi} \cdot \nabla(\mathbf{j} \times \mathbf{B}) + \nabla \Phi \nabla \rho \cdot \boldsymbol{\xi}$$

$$\stackrel{(A.13), \, \nabla \cdot \mathbf{j}=0}{=} -\mathbf{B} \times (\nabla \times (\mathbf{j} \times \boldsymbol{\xi})) - \mathbf{j} \times \mathbf{B} \nabla \cdot \boldsymbol{\xi} + (\nabla p - \mathbf{j} \times \mathbf{B} + 2\rho \nabla \Phi) \nabla \cdot \boldsymbol{\xi}$$
$$- \rho(\nabla \boldsymbol{\xi}) \cdot \nabla \Phi + \boldsymbol{\xi} \cdot \nabla \nabla p - \boldsymbol{\xi} \cdot \nabla(\mathbf{j} \times \mathbf{B}) + \nabla \Phi \nabla \rho \cdot \boldsymbol{\xi}$$

$$\stackrel{\text{equilibrium, (A.10)}}{=} -\mathbf{B} \times (\nabla \times (\mathbf{j} \times \boldsymbol{\xi})) - \nabla p \, \nabla \cdot \boldsymbol{\xi} - \rho \nabla(\boldsymbol{\xi} \cdot \nabla \Phi). \quad (6.47)$$

This proves our assertion: in the absence of gravity, when the last term vanishes and ∇p may be replaced by $\mathbf{j} \times \mathbf{B}$, the remaining two terms represent a force $\perp \mathbf{B}$. ◁

Using this expression, Eq. (6.43) may be rewritten in the form

$$\mathbf{F}(\boldsymbol{\xi}) = \nabla(\gamma p \nabla \cdot \boldsymbol{\xi}) - \mathbf{B} \times [\nabla \times \mathbf{Q} + \nabla \times (\mathbf{j} \times \boldsymbol{\xi})] - \nabla p \, \nabla \cdot \boldsymbol{\xi} - \rho \nabla(\boldsymbol{\xi} \cdot \nabla \Phi). \quad (6.48)$$

---

[2] The brackets are used here to delimit the range of the gradient operator. For example, the expression $(\nabla \boldsymbol{\xi}) \cdot \nabla p$ means that ∇ acts on $\boldsymbol{\xi}$ only, and contraction occurs between the components of $\boldsymbol{\xi}$ and ∇p, so that we get $\Sigma_j (\partial_i \xi_j) \partial_j p$ in Cartesian coordinates. Note that, in curvilinear coordinates, $\boldsymbol{\xi} = \Sigma_j \xi_j \mathbf{a}^j$ so that the derivatives of the basis vectors $\mathbf{a}^j$ also have to be taken into account.

In the absence of gravity (using the equilibrium relation $\nabla p = \mathbf{j} \times \mathbf{B}$), the force operator reduces to

$$\mathbf{F}_{\mathbf{g}=0}(\boldsymbol{\xi}) = \nabla(\gamma p \nabla \cdot \boldsymbol{\xi}) - \mathbf{B} \times [\nabla \times \mathbf{Q} + \nabla \times (\mathbf{j} \times \boldsymbol{\xi}) - \mathbf{j} \nabla \cdot \boldsymbol{\xi}], \quad (6.49)$$

and, in the absence of a magnetic field (using the equilibrium relation $\nabla p = \rho \mathbf{g}$), to

$$\mathbf{F}_{\mathbf{B}=0}(\boldsymbol{\xi}) = \nabla(\gamma p \nabla \cdot \boldsymbol{\xi}) - \rho \mathbf{g} \nabla \cdot \boldsymbol{\xi} + \rho \nabla(\mathbf{g} \cdot \boldsymbol{\xi}). \quad (6.50)$$

These very useful intermediate results in our derivation give an exposition, alternative to that of Eq. (6.43), of the different forces at work in an inhomogeneous plasma: isotropic (sound) contribution from the pressure and anisotropic (magnetic) contributions orthogonal to the background equilibrium magnetic field $\mathbf{B}$. Gravity, $\mathbf{g} = -\nabla \Phi$, introduces a direction additional to $\mathbf{B}$, $\mathbf{j}$, and ∇p. This contribution lifts the degeneracy implicit in the identification of ∇p and $\mathbf{j} \times \mathbf{B}$ (i.e., the identification of pressure and magnetic surfaces), which is at the heart of calculations on stability of thermonuclear plasmas. Hence, astrophysical plasmas exhibit a richer variety of waves and instabilities. It belongs to the intricacies of linear MHD theory that so many equivalent forms of the force operator (and the associated potential energy, see Section 6.4) may be given. This betrays the nonlinear origin which remains present in the equilibrium equations coupling the three inhomogeneity ingredients $\mathbf{j}$, ∇p and $\nabla \Phi$.

With respect to the proof of self-adjointness, we still have quite a way to go. We first establish two notational conveniences.
(1) Except for the two vector fields $\boldsymbol{\xi}$ and $\boldsymbol{\eta}$ defined in Eq. (6.34), we define their associated magnetic field perturbations:

$$\begin{aligned} \mathbf{Q}(\mathbf{r}) &\equiv \nabla \times (\boldsymbol{\xi} \times \mathbf{B}) \quad &(\text{on } V), \\ \mathbf{R}(\mathbf{r}) &\equiv \nabla \times (\boldsymbol{\eta} \times \mathbf{B}) \quad &(\text{on } V). \end{aligned} \quad (6.51)$$

(2) Considering the integrand $\boldsymbol{\eta}^* \cdot \mathbf{F}(\boldsymbol{\xi})$ of the quadratic expression on the left hand side of Eq. (6.42), we will drop the asterisk on the variables and temporarily return to a real type inner product, so that we do not have to write '$\boldsymbol{\eta}^* \cdot \mathbf{F}(\boldsymbol{\xi})$ + complex conjugate' all the time.

The proof of self-adjointness of the force operator basically consists of rewriting $\boldsymbol{\eta} \cdot \mathbf{F}(\boldsymbol{\xi})$, with $\mathbf{F}$ given by the expression (6.48), into a symmetric part and a remainder which should be cast in the form of a divergence. The latter part then disappears upon integration over the volume and application of the b.c.s (6.34). For example, by straightforward application of the vector identity (A.6) for the divergence, the first term of Eq. (6.48) produces the following expression:

$$\boldsymbol{\eta} \cdot \nabla(\gamma p \nabla \cdot \boldsymbol{\xi}) \stackrel{(A.6)}{=} -\gamma p \nabla \cdot \boldsymbol{\xi} \nabla \cdot \boldsymbol{\eta} + \nabla \cdot (\boldsymbol{\eta} \gamma p \nabla \cdot \boldsymbol{\xi}), \quad (6.52)$$

6.2 Force operator formalism

which obviously has the required form. Similarly, the second term may be reduced by means of the vector identities (A.1) and (A.12) and the definition (6.51) for the magnetic perturbation **R** associated with the displacement vector field $\boldsymbol{\eta}$:

$$
\begin{aligned}
-\boldsymbol{\eta} \cdot \mathbf{B} \times (\nabla \times \mathbf{Q}) &\stackrel{(A.1)}{=} -(\nabla \times \mathbf{Q}) \cdot (\boldsymbol{\eta} \times \mathbf{B}) \\
&\stackrel{(A.12)}{=} -\mathbf{Q} \cdot \nabla \times (\boldsymbol{\eta} \times \mathbf{B}) + \nabla \cdot [\,(\boldsymbol{\eta} \times \mathbf{B}) \times \mathbf{Q}\,] \\
&\stackrel{(6.51)}{=} -\mathbf{Q} \cdot \mathbf{R} + \nabla \cdot [\,(\boldsymbol{\eta} \times \mathbf{B}) \times \mathbf{Q}\,] \\
&\stackrel{(A.2)}{=} -\mathbf{Q} \cdot \mathbf{R} + \nabla \cdot [\,\mathbf{B}\,\boldsymbol{\eta} \cdot \mathbf{Q} - \boldsymbol{\eta}\,\mathbf{B} \cdot \mathbf{Q}\,] .
\end{aligned} \qquad (6.53)
$$

Again, the required form of a symmetric expression and a divergence is obtained. These terms, which correspond to the homogeneous plasma part, are still relatively simple. The transformation of the next three terms, which are due to plasma inhomogeneity, is much more complicated.

▷ **Transformation 2 of the inhomogeneity terms.** We take the three inhomogeneity terms together, exploiting the expression (6.47) (for reasons that will soon become clear):

$$
\boldsymbol{\eta} \cdot [\,\nabla(\boldsymbol{\xi} \cdot \nabla p) + \mathbf{j} \times \mathbf{Q} + \nabla\Phi \nabla \cdot (\rho \boldsymbol{\xi})\,]
$$

$$
\stackrel{(6.47)}{=} \underbrace{-\boldsymbol{\eta} \cdot \mathbf{B} \times [\,\nabla \times (\mathbf{j} \times \boldsymbol{\xi})\,]}_{\substack{\text{reduce as (6.53) with} \\ \mathbf{Q} \text{ replaced by } \mathbf{j} \times \boldsymbol{\xi}}} - (\boldsymbol{\eta} \cdot \nabla p) \nabla \cdot \boldsymbol{\xi} - \rho \boldsymbol{\eta} \cdot \nabla(\boldsymbol{\xi} \cdot \nabla\Phi)
$$

$$
= \overbrace{-\mathbf{j} \times \boldsymbol{\xi} \cdot \mathbf{R} + \nabla \cdot [\,(\boldsymbol{\eta} \times \mathbf{B}) \times (\mathbf{j} \times \boldsymbol{\xi})\,]}^{} - (\boldsymbol{\eta} \cdot \nabla p) \nabla \cdot \boldsymbol{\xi} - \rho \boldsymbol{\eta} \cdot \nabla(\boldsymbol{\xi} \cdot \nabla\Phi)
$$

$$
\stackrel{(A.2), (A.1)}{=} \boldsymbol{\xi} \cdot \mathbf{j} \times \mathbf{R} + \nabla \cdot [\,\mathbf{j}\,\mathbf{B} \cdot (\boldsymbol{\xi} \times \boldsymbol{\eta}) + \boldsymbol{\xi}\,\boldsymbol{\eta} \cdot (\mathbf{j} \times \mathbf{B})\,]
$$
$$
\qquad - (\boldsymbol{\eta} \cdot \nabla p) \nabla \cdot \boldsymbol{\xi} - \rho \boldsymbol{\eta} \cdot \nabla(\boldsymbol{\xi} \cdot \nabla\Phi)
$$

$$
\stackrel{\text{equil., (A.6)}}{=} \boldsymbol{\xi} \cdot \mathbf{j} \times \mathbf{R} + \nabla \cdot [\,\mathbf{j}\,\mathbf{B} \cdot (\boldsymbol{\xi} \times \boldsymbol{\eta})\,] + \boldsymbol{\xi} \cdot \nabla(\boldsymbol{\eta} \cdot \nabla p)
$$
$$
\qquad + \underbrace{\nabla \cdot [\,\boldsymbol{\xi}\boldsymbol{\eta} \cdot \rho \nabla\Phi\,] - \rho \boldsymbol{\eta} \cdot \nabla(\boldsymbol{\xi} \cdot \nabla\Phi)}_{\Downarrow (A.6)}
$$
$$
\qquad \overbrace{\nabla \cdot [\,\rho \nabla\Phi \cdot (\boldsymbol{\eta}\boldsymbol{\xi} - \boldsymbol{\xi}\boldsymbol{\eta})\,] + (\boldsymbol{\xi} \cdot \nabla\Phi) \nabla \cdot (\rho \boldsymbol{\eta})}^{}
$$

$$
\stackrel{\text{rearrange}}{=} \boldsymbol{\xi} \cdot [\,\nabla(\boldsymbol{\eta} \cdot \nabla p) + \mathbf{j} \times \mathbf{R} + \nabla\Phi \nabla \cdot (\rho \boldsymbol{\eta})\,]
$$
$$
\qquad + \nabla \cdot [\,\mathbf{j}\,\mathbf{B} \cdot (\boldsymbol{\xi} \times \boldsymbol{\eta}) - \rho \nabla\Phi \cdot (\boldsymbol{\xi}\boldsymbol{\eta} - \boldsymbol{\eta}\boldsymbol{\xi})\,] .
$$

Now, the first term on the right hand side is not symmetric in $\boldsymbol{\xi}$ and $\boldsymbol{\eta}$. However, this term is the exact mirror image of the left hand side. Hence, it is very easy to symmetrize the latter expression:

$$
\boldsymbol{\eta} \cdot [\,\nabla(\boldsymbol{\xi} \cdot \nabla p) + \mathbf{j} \times \mathbf{Q} + \nabla\Phi \nabla \cdot (\rho \boldsymbol{\xi})\,]
$$
$$
= \tfrac{1}{2} \boldsymbol{\eta} \cdot [\,\nabla(\boldsymbol{\xi} \cdot \nabla p) + \mathbf{j} \times \mathbf{Q} + \nabla\Phi \nabla \cdot (\rho \boldsymbol{\xi})\,]
$$
$$
\quad + \tfrac{1}{2} \boldsymbol{\xi} \cdot [\,\nabla(\boldsymbol{\eta} \cdot \nabla p) + \mathbf{j} \times \mathbf{R} + \nabla\Phi \nabla \cdot (\rho \boldsymbol{\eta})\,]
$$
$$
\quad + \tfrac{1}{2} \nabla \cdot [\,\mathbf{j}\,\mathbf{B} \cdot (\boldsymbol{\xi} \times \boldsymbol{\eta}) - \rho \nabla\Phi \cdot (\boldsymbol{\xi}\boldsymbol{\eta} - \boldsymbol{\eta}\boldsymbol{\xi})\,]
$$

$$\stackrel{(A.6),\text{ equil.}}{=} \tfrac{1}{2}\nabla\cdot[\nabla p\cdot(\boldsymbol{\xi}\boldsymbol{\eta}+\boldsymbol{\eta}\boldsymbol{\xi})] - \tfrac{1}{2}\nabla p\cdot(\boldsymbol{\xi}\nabla\cdot\boldsymbol{\eta}+\boldsymbol{\eta}\nabla\cdot\boldsymbol{\xi})$$
$$-\tfrac{1}{2}\nabla\cdot[(\mathbf{j}\times\mathbf{B}-\nabla p)\cdot(\boldsymbol{\xi}\boldsymbol{\eta}-\boldsymbol{\eta}\boldsymbol{\xi})]$$
$$-\tfrac{1}{2}\mathbf{j}\cdot(\boldsymbol{\xi}\times\mathbf{R}+\boldsymbol{\eta}\times\mathbf{Q}) + \tfrac{1}{2}\nabla\Phi\cdot[\boldsymbol{\xi}\nabla\cdot(\rho\boldsymbol{\eta})+\boldsymbol{\eta}\nabla\cdot(\rho\boldsymbol{\xi})]$$
$$+\nabla\cdot[\,\tfrac{1}{2}\mathbf{j}\mathbf{B}\cdot(\boldsymbol{\xi}\times\boldsymbol{\eta})\,]$$
$$= -\tfrac{1}{2}\nabla p\cdot(\boldsymbol{\xi}\nabla\cdot\boldsymbol{\eta}+\boldsymbol{\eta}\nabla\cdot\boldsymbol{\xi}) - \tfrac{1}{2}\mathbf{j}\cdot(\boldsymbol{\xi}\times\mathbf{R}+\boldsymbol{\eta}\times\mathbf{Q})$$
$$+\tfrac{1}{2}\nabla\Phi\cdot[\boldsymbol{\xi}\nabla\cdot(\rho\boldsymbol{\eta})+\boldsymbol{\eta}\nabla\cdot(\rho\boldsymbol{\xi})]$$
$$+\nabla\cdot[\,\boldsymbol{\eta}(\boldsymbol{\xi}\cdot\nabla p) - \tfrac{1}{2}(\mathbf{j}\times\mathbf{B})\cdot(\boldsymbol{\xi}\boldsymbol{\eta}-\boldsymbol{\eta}\boldsymbol{\xi}) + \tfrac{1}{2}\mathbf{j}\mathbf{B}\cdot(\boldsymbol{\xi}\times\boldsymbol{\eta})\,]. \quad (6.54)$$

Again, a symmetric expression and a divergence is obtained! ◁

Finally, adding up the expressions (6.52), (6.53), and (6.54) gives the required symmetric form

$$\boldsymbol{\eta}\cdot\mathbf{F}(\boldsymbol{\xi}) = -\gamma p\,\nabla\cdot\boldsymbol{\xi}\,\nabla\cdot\boldsymbol{\eta} - \mathbf{Q}\cdot\mathbf{R} - \tfrac{1}{2}\nabla p\cdot(\boldsymbol{\xi}\nabla\cdot\boldsymbol{\eta}+\boldsymbol{\eta}\nabla\cdot\boldsymbol{\xi})$$
$$-\tfrac{1}{2}\mathbf{j}\cdot(\boldsymbol{\xi}\times\mathbf{R}+\boldsymbol{\eta}\times\mathbf{Q}) + \tfrac{1}{2}\nabla\Phi\cdot[\boldsymbol{\xi}\nabla\cdot(\rho\boldsymbol{\eta})+\boldsymbol{\eta}\nabla\cdot(\rho\boldsymbol{\xi})]$$
$$+\nabla\cdot\Big[\boldsymbol{\eta}(\gamma p\,\nabla\cdot\boldsymbol{\xi}+\boldsymbol{\xi}\cdot\nabla p - \mathbf{B}\cdot\mathbf{Q})$$
$$+\mathbf{B}\boldsymbol{\eta}\cdot\mathbf{Q} + \tfrac{1}{2}\mathbf{j}\mathbf{B}\cdot(\boldsymbol{\xi}\times\boldsymbol{\eta}) - \tfrac{1}{2}(\mathbf{j}\times\mathbf{B})\cdot(\boldsymbol{\xi}\boldsymbol{\eta}-\boldsymbol{\eta}\boldsymbol{\xi})\Big]. \quad (6.55)$$

Integration over the plasma volume and application of Gauss' theorem (A.14) on the divergence term then yields the following very important result:

$$\int \boldsymbol{\eta}\cdot\mathbf{F}(\boldsymbol{\xi})\,dV = -\int\Big\{\gamma p\,\nabla\cdot\boldsymbol{\xi}\,\nabla\cdot\boldsymbol{\eta} + \mathbf{Q}\cdot\mathbf{R} + \tfrac{1}{2}\nabla p\cdot(\boldsymbol{\xi}\nabla\cdot\boldsymbol{\eta}+\boldsymbol{\eta}\nabla\cdot\boldsymbol{\xi})$$
$$+\tfrac{1}{2}\mathbf{j}\cdot(\boldsymbol{\xi}\times\mathbf{R}+\boldsymbol{\eta}\times\mathbf{Q}) - \tfrac{1}{2}\nabla\Phi\cdot[\boldsymbol{\xi}\nabla\cdot(\rho\boldsymbol{\eta})+\boldsymbol{\eta}\nabla\cdot(\rho\boldsymbol{\xi})]\Big\}dV$$
$$+\int \mathbf{n}\cdot\boldsymbol{\eta}\,[\gamma p\,\nabla\cdot\boldsymbol{\xi}+\boldsymbol{\xi}\cdot\nabla p - \mathbf{B}\cdot\mathbf{Q}]\,dS. \quad (6.56)$$

Here, the three last terms of Eq. (6.55) did not give a contribution to the surface integral since $\mathbf{n}\cdot\mathbf{B}=0$ and $\mathbf{n}\cdot\mathbf{j}=0$ on the plasma surface, whereas $\mathbf{j}\times\mathbf{B}\sim\mathbf{n}$, for the plasma-wall and plasma interface models I–III of Section 4.6. For the line-tying magnetic loop models IV and V, there is also a boundary intersecting the magnetic field and current lines so that the given argument does not apply. However, in that case, the line-tying boundary condition (4.189) translates into the conditions $\boldsymbol{\eta}=\boldsymbol{\xi}=0$ at the photospheric boundary so that the three terms do not contribute there either. We conclude that the expression (6.56) is a very general

6.2 Force operator formalism

one, applying to all of the laboratory and astrophysical problems I–V formulated in Section 4.6.

The surface integral in Eq. (6.56) is just the perturbation of the total pressure (apart from the sign). In plasmas with a free boundary (models II and III), this term couples to the external variables $\hat{\mathbf{Q}}$ and $\hat{\mathbf{R}}$ through the boundary conditions at the plasma interface so that a further analysis is needed to properly represent the contributions of the outer regions. This we leave for later (Section 6.6.2). For our present concern (wall on the plasma: model I), the term does not contribute. Consequently,

$$\int \left\{ \boldsymbol{\eta} \cdot \mathbf{F}(\boldsymbol{\xi}) - \boldsymbol{\xi} \cdot \mathbf{F}(\boldsymbol{\eta}) \right\} dV = \int \left\{ \mathbf{n} \cdot \boldsymbol{\eta} \left[\gamma p \nabla \cdot \boldsymbol{\xi} + \boldsymbol{\xi} \cdot \nabla p - \mathbf{B} \cdot \mathbf{Q} \right] \right.$$
$$\left. - \mathbf{n} \cdot \boldsymbol{\xi} \left[\gamma p \nabla \cdot \boldsymbol{\eta} + \boldsymbol{\eta} \cdot \nabla p - \mathbf{B} \cdot \mathbf{R} \right] \right\} dS = 0,$$
(6.57)

by virtue of the b.c.s $\mathbf{n} \cdot \boldsymbol{\eta} = \mathbf{n} \cdot \boldsymbol{\xi} = 0$ on the normal components of the displacement vector. $\Rightarrow$ **F** is a self-adjoint operator in model I; QED.

▷ **Challenge.** More than 40 years after the original derivation of the force operator for linear ideal MHD we still do not have a more elegant proof of the self-adjointness. Find one that is more in line with its basic character! ◁

At this point we will revert to the original complex type scalar product (6.35). We will prove that the eigenvalues of the operator $\rho^{-1}\mathbf{F}$ are real. This follows directly from the self-adjointness property (6.57). Let $\boldsymbol{\xi}_n$ be the eigenfunction belonging to an eigenvalue $-\omega_n^2$,

$$\rho^{-1}\mathbf{F}(\boldsymbol{\xi}_n) = -\omega_n^2 \boldsymbol{\xi}_n \,.$$

Then, the complex conjugate equation reads

$$\rho^{-1}\mathbf{F}^*(\boldsymbol{\xi}_n) = \rho^{-1}\mathbf{F}(\boldsymbol{\xi}_n^*) = -\omega_n^{2*} \boldsymbol{\xi}_n^* \,.$$

Multiplying the first equation with $\boldsymbol{\xi}_n^*$ and the second with $\boldsymbol{\xi}_n$, subtracting and integrating over the plasma volume yields

$$0 = (\omega_n^2 - \omega_n^{2*}) \|\boldsymbol{\xi}\|^2 \,,$$

by virtue of the self-adjointness property, so that

$$\omega_n^2 = \omega_n^{2*} \,. \tag{6.58}$$

Hence, *the eigenvalues ω_n^2 of the operator $\rho^{-1}\mathbf{F}$ are real*, so that the spectrum is confined to the real and imaginary axes of the complex ω-plane; QED. Recall from the discussion of Section 6.2.1 that this implies that the eigenvalues correspond to either waves ($\omega^2 > 0$) or exponential instabilities ($\omega^2 < 0$).

6.3 Spectral alternatives★

6.3.1 Mathematical intermezzo★

Recapitulating: we have seen in Section 6.2 that the spectral problem of linear ideal MHD arises from a study of the *dynamics*, described by the equation of motion

$$\rho^{-1}\mathbf{F}(\boldsymbol{\xi}) = \frac{\partial^2 \boldsymbol{\xi}}{\partial t^2}, \tag{6.59}$$

where normal mode solutions with an exponential time-dependence lead to the *spectral equation*

$$\rho^{-1}\mathbf{F}(\hat{\boldsymbol{\xi}}) = -\omega^2 \hat{\boldsymbol{\xi}}, \tag{6.60}$$

whereas the *stability problem* reduces to a study of the marginal equation

$$\rho^{-1}\mathbf{F}(\hat{\boldsymbol{\xi}}) = 0. \tag{6.61}$$

Before we proceed to analyse these three main physical problems again from the second point of view, that of quadratic forms and variational principles (Section 6.4), it is useful to put them into a mathematical context since other alternative formulations are encountered there.

The spectral problem of partial differential equations like Eq. (6.60) is just a generalization of the methods used in linear algebra of finite-dimensional vector spaces. In that context, three alternatives appear.

(1) *The eigenvalue problem* arises in the study of finite $N \times N$ matrices L_{ij} ($i, j = 1, 2, \ldots, N$),

$$\sum_{j=1}^{N} L_{ij} x_j = \lambda x_i, \quad \text{or} \quad \mathbf{L} \cdot \mathbf{x} = \lambda \mathbf{x}, \tag{6.62}$$

where the eigenvalues are found from the condition

$$\det(L_{ij} - \lambda \delta_{ij}) = 0, \tag{6.63}$$

and substitution back into Eq. (6.62) yields the eigenvectors $\mathbf{x}_n$.

(2) Another formulation of the same problem is obtained by constructing *the quadratic forms*, as e.g. entering the Rayleigh quotient

$$\Lambda \equiv \sum_{i=1}^{N}\sum_{j=1}^{N} x_i^* L_{ij} x_j \bigg/ \sum_{i=1}^{N} x_i^2, \tag{6.64}$$

where the eigenvalues λ are the stationary values of Λ.

(3) A third formulation arises from the consideration of *the inhomogeneous equation*

$$(\mathbf{L} - \lambda \mathbf{I}) \cdot \mathbf{x} = \mathbf{a}, \tag{6.65}$$

where **a** is a known vector. Here, *the Fredholm alternative* states that either the homogeneous equation (6.62) has a solution, so that λ coincides with one of the eigenvalues, or the inhomogeneous equation (6.65) has a solution, so that λ is outside the spectrum of eigenvalues of the matrix **L**.

Clearly, whereas our two physical viewpoints correspond to the first two spectral alternatives, the third alternative admits yet another class of physical applications. See Section 6.3.2.

In the generalization of these ideas to the infinite-dimensional Hilbert space associated with the operator $\rho^{-1}\mathbf{F}$, two kinds of mathematical problem are encountered. The first one is the fact that the operator $\rho^{-1}\mathbf{F}$ is a differential operator and, therefore, *unbounded*. In contrast, bounded operators B have the property that

$$\|Bx\| \leq M\|x\| \quad \text{for all } x \in \mathcal{H}, \tag{6.66}$$

where $\mathcal{H}$ is the Hilbert space and M is some constant. Differential operators do not have this property. Operating on a bounded (square integrable) sequence of functions in Hilbert space they may produce a sequence that is unbounded and, therefore, leads outside Hilbert space. (For example, the differential operator d/dx transforms the bounded sequence $\sin(n\pi x)$ into the diverging sequence $n\pi \sin(n\pi x)$.) One usually tries to avoid this problem by transforming it into one that involves *completely continuous* or *compact operators*. These operators have the opposite property: they transform a sequence of bounded functions into one that converges in the mean. For these operators, the theory of infinite-dimensional Hilbert space is completely analogous to that of the finite-dimensional vector spaces of linear algebra. In the case of differential operators, this implies that one tries to invert them, which leads to the study of integral operators involving Green's functions. Those operators frequently do have the required property of compactness.

Another, more serious, problem is the existence of a third class of operators where the above trick does not work, viz. that of bounded operators that are not compact. (Example: the operator of multiplication by x.) Those operators may give rise to a *continuous spectrum*, which is roughly speaking the collection of 'improper eigenvalues', for which the eigenvalue equation is solved, but not by functions that belong to Hilbert space. In the mathematical description, one then has the option of either sticking to the notion of Hilbert space by introducing the concept of *approximate spectrum*, where sequences are considered that do not converge (this is the approach of von Neumann [240] in his treatment of spectral theory for quantum mechanics), or one may consider wider classes of elements than those that belong to Hilbert space, viz. *distributions* (this is the approach of Dirac [67], perfected by Schwartz [205]). Loosely speaking, one could say that the diverging sequences of functions, that are considered in the first approach, converge to elements outside Hilbert space, which are the distributions considered in the second approach.

Using this terminology, the following generalization of the ideas of linear algebra expressed in the equations (6.62)–(6.65) may be formulated. The spectrum of a linear operator L is obtained from the study of the inhomogeneous equation

$$(L - \lambda) x = a, \qquad (6.67)$$

where a is a given element in Hilbert space and we look for solutions

$$x = (L - \lambda)^{-1} a. \qquad (6.68)$$

For complex values of λ, three possibilities arise (see Friedman [74], p. 125):

(a) $(L - \lambda)^{-1}$ does not exist because $(L - \lambda) x = 0$ has a solution
  $\Rightarrow \lambda$ belongs to *the point* or *discrete spectrum* of L ;
(b) $(L - \lambda)^{-1}$ exists but is unbounded
  $\Rightarrow \lambda$ belongs to *the continuous spectrum* of L ;
(c) $(L - \lambda)^{-1}$ exists and is bounded
  $\Rightarrow \lambda$ belongs to *the resolvent set* of L .

Thus, a complex value of λ either belongs to the spectrum or to the resolvent set, so that one may say that the spectrum of L consists of the collection of λs where the so-called *resolvent operator* $R_\lambda \equiv (L - \lambda)^{-1}$ misbehaves.

Finally, it is useful to anticipate the exposition on computational MHD in the companion Volume 2 on *Advanced Magnetohydrodynamics* where the circle will be closed by the consideration of the discretized problem, which leads back to finite-dimensional vector spaces. It will be shown that extremely powerful numerical methods exist which provide the construction of solutions of the three basic linear problems of physical interest, viz.

– determining *the temporal evolution*,

$$\mathbf{L} \cdot \mathbf{x} = \frac{\partial \mathbf{x}}{\partial t}, \qquad (6.69)$$

– determining *the spectrum*,

$$\mathbf{L} \cdot \mathbf{x} = \lambda \mathbf{x}, \tag{6.70}$$

– and determining *the stationary state of a driven plasma*,

$$(\mathbf{L} - \lambda_d \mathbf{I}) \cdot \mathbf{x} = \mathbf{f}, \tag{6.71}$$

where $\mathbf{f}$ represents the driving force with fixed frequency λ_d.

Clearly, the last problem corresponds again with the third spectral problem (6.65). It should be mentioned that the relevant computational methods are not restricted to ideal MHD, with the associated displacement vector description, but they may be generalized to dissipative MHD with an arbitrary number of unknowns.

6.3.2 Initial value problem in MHD*

To give the reader some feeling of what is yet in store, we digress on an advanced topic. According to the exposition given in the previous subsection, in particular in connection with Eqs. (6.67) and (6.68), the third, and most general, approach to the spectrum of a linear operator is to consider the inhomogeneous problem, i.e. the generalization of Eq. (6.65) to infinite-dimensional spaces. For the MHD operator $\rho^{-1}\mathbf{F}$, this approach leads to the following inhomogeneous problem:

$$(\rho^{-1}\mathbf{F} + \omega^2 \mathbf{I})\, \hat{\boldsymbol{\xi}} = i\omega \mathbf{X}, \tag{6.72}$$

where $\mathbf{I}$ is the unit operator, $\mathbf{I}(\hat{\boldsymbol{\xi}}) \equiv \hat{\boldsymbol{\xi}}$, and $\mathbf{X}$ is a known vector. Our task is then to construct the resolvent operator $(\rho^{-1}\mathbf{F} + \omega^2 \mathbf{I})^{-1}$, and to study its behaviour for complex values of ω^2.

In order to see how this is connected with physics, we consider the initial value problem. (Also notice that Eq. (6.71), obtained from model III excitation, closely corresponds to this problem.) We define *the Laplace transform* of $\boldsymbol{\xi}(\mathbf{r}; t)$ in the complex ω-plane:

$$\hat{\boldsymbol{\xi}}(\mathbf{r}; \omega) \equiv \int_0^\infty \boldsymbol{\xi}(\mathbf{r}; t)\, e^{i\omega t}\, dt, \tag{6.73}$$

so that the equation of motion (6.59) takes the form

$$\rho^{-1}\mathbf{F}(\hat{\boldsymbol{\xi}}) = \int_0^\infty \frac{\partial^2 \boldsymbol{\xi}}{\partial t^2}\, e^{i\omega t}\, dt$$

$$= -\omega^2 \hat{\boldsymbol{\xi}} + \left[\left(\frac{\partial \boldsymbol{\xi}}{\partial t} - i\omega \boldsymbol{\xi}\right) e^{i\omega t}\right]_{t=0}^{t \to \infty}. \tag{6.74}$$

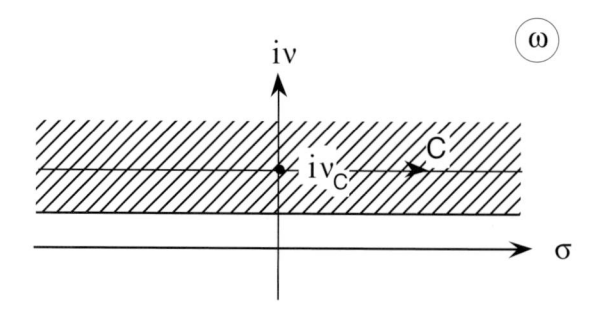

Fig. 6.10. Strip of convergence for the inverse Laplace transformation.

Writing $\omega = \sigma + i\nu$, we then get for $\nu > 0$:

$$(\rho^{-1}\mathbf{F} + \omega^2 \mathbf{I})\,\hat{\boldsymbol{\xi}}(\mathbf{r};\omega) = i\omega\boldsymbol{\xi}_i(\mathbf{r}) - \dot{\boldsymbol{\xi}}_i(\mathbf{r}) \equiv i\omega\mathbf{X}, \qquad (6.75)$$

where the vector $\mathbf{X}$ of Eq. (6.72) turns out to be the function of initial displacement $\boldsymbol{\xi}_i(\mathbf{r})$ and initial velocity $\dot{\boldsymbol{\xi}}_i(\mathbf{r})$ defined in the RHS of Eq. (6.75). In order to find the response $\boldsymbol{\xi}(\mathbf{r};t)$ to a certain initial perturbation $\mathbf{X}$, one then first has to invert Eq. (6.75) to find the Laplace transformed variable $\hat{\boldsymbol{\xi}}$ in terms of $\mathbf{X}$,

$$\hat{\boldsymbol{\xi}}(\mathbf{r};\omega) = (\rho^{-1}\mathbf{F} + \omega^2 \mathbf{I})^{-1} i\omega \mathbf{X}(\mathbf{r};\omega), \qquad (6.76)$$

and next perform *the inverse Laplace transformation*:

$$\boldsymbol{\xi}(\mathbf{r};t) = \frac{1}{2\pi}\int_{i\nu_C - \infty}^{i\nu_C + \infty} \hat{\boldsymbol{\xi}}(\mathbf{r};\omega)\,e^{-i\omega t}\,d\omega$$

$$= \frac{1}{2\pi}\int_C (\rho^{-1}\mathbf{F} + \omega^2 \mathbf{I})^{-1}\left[i\omega\boldsymbol{\xi}_i(\mathbf{r}) - \dot{\boldsymbol{\xi}}_i(\mathbf{r})\right]e^{-i\omega t}\,d\omega. \qquad (6.77)$$

Here, according to the theory of Laplace transforms, the contour C has to be placed in the strip of convergence (Fig. 6.10). The essential point is that, for the inverse transform, more is needed than just $\nu_C > 0$ because $\hat{\boldsymbol{\xi}}(\mathbf{r};\omega)$ may not exist for certain values of ω, or may be singular.

According to the discussion above, it is precisely when ω belongs to the spectrum of the operator $\rho^{-1}\mathbf{F}$ that we may expect trouble with Eq. (6.76). If ω is a point eigenvalue the operator $(\rho^{-1}\mathbf{F} + \omega^2 \mathbf{I})^{-1}$ simply does not exist, whereas for improper eigenvalues (i.e. ω in the continuum) the operator $(\rho^{-1}\mathbf{F} + \omega^2 \mathbf{I})^{-1}$ is unbounded. Before we know where to place the integration contour C for the inverse Laplace transform we, therefore, have to know the spectrum. Here, we get substantial help from the fact that $\rho^{-1}\mathbf{F}$ is Hermitian (self-adjoint) so that the eigenvalues (including the improper ones) have to be real, so that the spectrum is confined to the real and imaginary axes of the complex ω-plane. In fact, we would

6.3 Spectral alternatives*

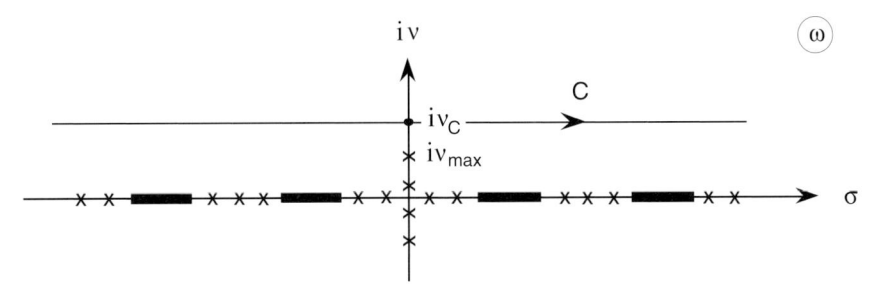

Fig. 6.11. Contour for the inverse Laplace transformation.

be in severe trouble if the operator **F** were not self-adjoint because a general theory of non-Hermitian operators does not exist.

We then conclude that the integration contour must be placed above the largest (most unstable) point eigenvalue ν_{max} of $\rho^{-1}\mathbf{F}$: $\nu_C > \nu_{max}$ (Fig. 6.11). (Because Landau's solution of the initial value problem of the Vlasov equation was also obtained by means of Laplace transforms, the similarity with the discussion of Landau damping in Section 2.3.3 (Fig. 2.7) is obvious.) In other words, the class of permissible functions $\boldsymbol{\xi}(\mathbf{r}; t)$ is restricted to functions of exponential order $\exp(\nu_C t)$ where ν_C is larger than the largest growth rate of the system. In Fig. 6.11 we have schematically indicated our knowledge so far of the spectrum of ideal MHD, which will be analysed in more detail in a later chapter. One finds two pairs of continua on the real axis, whereas point eigenvalues can occur almost everywhere on the real σ-axis and also on a bounded part $-\nu_{max} \leq \nu \leq \nu_{max}$ of the imaginary ν-axis.

Of course, it is extremely difficult to obtain the explicit time-dependence of $\boldsymbol{\xi}(\mathbf{r}; t)$ in situations of practical interest, so that one usually restricts the study to time-asymptotic solutions. It is clear that for $t \to \infty$ one wishes to deform the integration contour in the inverse Laplace transform to the lower half of the ω-plane in order to exploit the smallness of the exponential factor $\exp(-i\omega t)$ in Eq. (6.77). For this advantage one must pay in the form of a study of the analytic continuation of $\hat{\boldsymbol{\xi}}$ about the occurring *poles* (point eigenvalues) and *branch points* of $\hat{\boldsymbol{\xi}}$ (associated with the continuous spectrum). The branch point singularities lead to different branches of the complex function $\hat{\boldsymbol{\xi}}(\mathbf{r}; \omega)$ so that the inverse Laplace transform contour may be moved to another Riemann sheet, where it could pick up poles. Such poles could not correspond to point eigenvalues since these are confined to the real and imaginary axes of the principal branch of $\hat{\boldsymbol{\xi}}$, but they may have physical significance anyway.

Thus, we have sketched some of the intricacies of spectral theory even before having demonstrated that they actually occur in MHD. We will continue the analysis of the initial value problem in Chapter 10, where we consider the explicit example of an inhomogeneous slab. To appreciate the portent of the present section, the reader is advised to reread it after he has studied that example.

6.4 Quadratic forms and variational principles

6.4.1 Expressions for the potential energy

In Section 6.2, we have completed the description of linearized ideal MHD in terms of differential equations by deriving the equation of motion (6.23) with the force operator $\mathbf{F}(\boldsymbol{\xi})$, which we have proven to be self-adjoint. This property expresses the basic fact of *energy conservation*: $H = W + K = \text{const}$, where W is the potential energy and K is the kinetic energy of the perturbations (see Section 4.3.2). Accordingly, we now turn to the alternative description, viz. the one exploiting these quadratic forms. First of all, we need to derive expressions for these energies for the linearized dynamics, corresponding to the first order expression for $\mathbf{F}$. The direct derivation, discussed first, is put in small print since it turns out to be complicated, whereas the indirect one, that we discuss next, is much simpler.

▷ **Derivation from the nonlinear expression for the potential energy.** The most obvious procedure is to start from the nonlinear expressions for W and K derived in Section 4.3.2, and to exploit the energy conservation law (4.86):

$$H = \int \Big(\underbrace{\tfrac{1}{2}\rho v^2}_{K} + \underbrace{\tfrac{p}{\gamma-1} + \tfrac{1}{2}B^2}_{W} \Big) dV = \text{const}. \tag{6.78}$$

We have already derived in Eq. (6.40) the linearized expression for the kinetic energy, corresponding to the first part of Eq. (6.78), which is a second order quantity in $\dot{\boldsymbol{\xi}}$:

$$K = \tfrac{1}{2} \int \rho \dot{\boldsymbol{\xi}} \cdot \dot{\boldsymbol{\xi}} \, dV + \text{third and higher order terms}. \tag{6.79}$$

Similarly, we could work out the expression for the potential energy W, starting from the second part of Eq. (6.78):

$$W = \int \Big[\underbrace{\tfrac{p}{\gamma-1} + \tfrac{1}{2}B^2}_{W_0} + \underbrace{\tfrac{\pi}{\gamma-1} + \mathbf{B}\cdot\mathbf{Q}}_{W_1} + \underbrace{\cdots\cdots}_{W_2} \Big] dV. \tag{6.80}$$

Here, the zeroth order term corresponds to the energy of the background equilibrium, described by the pressure p and the magnetic field $\mathbf{B}$. (Note that the interpretation of p and $\mathbf{B}$ changes from Eq. (6.78), where they indicate the full nonlinear expressions, to Eq. (6.80), where they only indicate the equilibrium parts.) The first order term corresponds with the energy change produced by the plasma displacement $\boldsymbol{\xi}$, resulting in the pressure and magnetic field perturbations π and $\mathbf{Q}$ given by Eqs. (6.24) and (6.25). However, how could this term be balanced in the energy conservation law by the expression for the kinetic

6.4 Quadratic forms and variational principles

energy K, which is a second order quantity only? To see what goes on, work out the perturbations:

$$\frac{\pi}{\gamma - 1} = \frac{1}{\gamma - 1}(-\boldsymbol{\xi} \cdot \nabla p - \gamma p \nabla \cdot \boldsymbol{\xi}) = \boldsymbol{\xi} \cdot \nabla p - \frac{\gamma}{\gamma - 1} \nabla \cdot (p\boldsymbol{\xi}),$$

$$\mathbf{B} \cdot \mathbf{Q} = \mathbf{B} \cdot \nabla \times (\boldsymbol{\xi} \times \mathbf{B}) \stackrel{(A.12)}{=} (\boldsymbol{\xi} \times \mathbf{B}) \cdot \nabla \times \mathbf{B} + \nabla \cdot [(\boldsymbol{\xi} \times \mathbf{B}) \times \mathbf{B}]$$
$$\stackrel{(A.1), (A.2)}{=} -\boldsymbol{\xi} \cdot [\mathbf{j} \times \mathbf{B}] + \nabla \cdot [\boldsymbol{\xi} \cdot \mathbf{B}\mathbf{B} - B^2 \boldsymbol{\xi}],$$

so that

$$W_1 = \underbrace{\int \boldsymbol{\xi} \cdot (\nabla p - \mathbf{j} \times \mathbf{B}) \, dV}_{= 0 \text{ (equil.)}} + \underbrace{\int \nabla \cdot \left[-\frac{\gamma}{\gamma - 1} p \boldsymbol{\xi} + \boldsymbol{\xi} \cdot \mathbf{B}\mathbf{B} - B^2 \boldsymbol{\xi} \right] dV}_{= 0 \text{ (Gauss + b.c.s)}} = 0.$$

(6.81)

(Note that we have neglected gravity in this paragraph.) Consequently, in order to get relevant, non-vanishing, expressions for the potential energy W one should compute the next order contribution W_2 (also indicated as δW in the literature, although $\delta^2 W$ would be a better notation):

$$W_2 = \int \left(\frac{p_2}{\gamma - 1} + \tfrac{1}{2} Q^2 + \mathbf{B} \cdot \mathbf{B}_2 \right) dV. \tag{6.82}$$

Hence, we should compute p and $\mathbf{B}$ to second order in $\boldsymbol{\xi}$. This can be done but it is a complicated procedure, which we will not pursue further since there is a much easier method to derive the quadratic expression for the potential energy. ◁

The easier method to derive the expression for the potential energy is to exploit Eq. (6.40) for the linearized kinetic energy and to construct the linearized potential energy from energy conservation. (We now drop again the subscripts 1 and 2 on first and second order expressions since the interpretation is obvious from the context.) This is done by taking the inner product of $\dot{\boldsymbol{\xi}}^*$ with the equation of motion (6.23) and integrating over the plasma volume:

$$\int \dot{\boldsymbol{\xi}}^* \cdot \mathbf{F}(\boldsymbol{\xi}) \, dV = \int \rho \dot{\boldsymbol{\xi}}^* \cdot \ddot{\boldsymbol{\xi}} \, dV = \frac{d}{dt} \left[\tfrac{1}{2} \int \rho |\dot{\boldsymbol{\xi}}|^2 \, dV \right] = \frac{dK}{dt}. \tag{6.83}$$

From energy conservation and the self-adjointness of $\mathbf{F}$ one then easily derives that

$$\frac{dW}{dt} = -\frac{dK}{dt} = -\tfrac{1}{2} \int \dot{\boldsymbol{\xi}}^* \cdot \mathbf{F}(\boldsymbol{\xi}) \, dV - \tfrac{1}{2} \int \boldsymbol{\xi}^* \cdot \mathbf{F}(\dot{\boldsymbol{\xi}}) \, dV$$
$$= \frac{d}{dt} \left[-\tfrac{1}{2} \int \boldsymbol{\xi}^* \cdot \mathbf{F}(\boldsymbol{\xi}) \, dV \right],$$

so that

$$W = -\tfrac{1}{2} \int \boldsymbol{\xi}^* \cdot \mathbf{F}(\boldsymbol{\xi}) \, dV, \tag{6.84}$$

which is the expression for *the linearized potential energy* we were looking for. The intuitive meaning of Eq. (6.84) is clear. The increase of the potential energy due to the perturbation is just the work done against the force **F** to displace the plasma by an amount $\boldsymbol{\xi}$. The factor $\frac{1}{2}$ represents the averaging involved as the work builds up when the plasma is displaced from 0 to its actual value $\boldsymbol{\xi}$.

We can now specify the loose remark made in Section 6.1.1 that a perturbation is stable ($W > 0$), or unstable ($W < 0$), if $\boldsymbol{\xi}$ and **F** are pointing in the opposite, or in the same direction. Clearly, this has to be interpreted in the integrated sense given by the definition of the potential energy in Eq. (6.84). We will prove that W is a variational expression. This implies that a test function $\boldsymbol{\xi}$ can be substituted in W to check for the sign of that expression only. If some trial function $\boldsymbol{\xi}$ is found such that $W[\boldsymbol{\xi}] < 0$, the system is certainly unstable. In many cases, this is a much quicker way to establish instability of a particular configuration than to study the marginal equation of motion $\mathbf{F}(\boldsymbol{\xi}) = 0$. We will return to this topic in Section 6.4.4.

A bonus of the awful proof in Section 6.2.3 of the self-adjointness of **F** is the intermediate expression (6.56) which, upon identification of $\boldsymbol{\xi}$ and $\boldsymbol{\eta}$, immediately yields a more useful form of W than that of Eq. (6.84):

$$W = \tfrac{1}{2} \int \Big[\gamma p \, |\nabla \cdot \boldsymbol{\xi}|^2 + |\mathbf{Q}|^2 + (\boldsymbol{\xi}^* \cdot \nabla p) \nabla \cdot \boldsymbol{\xi} + \mathbf{j} \cdot \boldsymbol{\xi}^* \mathbf{Q}$$
$$- (\boldsymbol{\xi}^* \cdot \nabla \Phi) \nabla \cdot (\rho \boldsymbol{\xi}) \Big] \, dV . \tag{6.85}$$

The five terms represent, successively, the acoustic and magnetic energy, which are positive definite so that *homogeneous plasmas are always stable*, and the additional energies due to the pressure gradient, current density and gravity, which may have either sign so that *inhomogeneous plasmas may be unstable*.

The expression (6.85) for W is only valid for *model I* (wall on the plasma), since we have discarded the divergence term appearing in Eq. (6.56) by applying the boundary condition

$$\mathbf{n} \cdot \boldsymbol{\xi} = 0 \qquad \text{(at the wall)} . \tag{6.86}$$

The wall may be put at ∞ if one wishes to consider the limit of an infinitely extended plasma. If one wishes to incorporate the influence of an external vacuum, *model II* (plasma–vacuum system) should be studied and the divergence term in Eq. (6.56) then gives a further contribution to W. This contribution represents the energy of the moving boundary and of the external vacuum magnetic field region. The proof of self-adjointness and the derivation of the extended expression for W for that case is left for later (Section 6.6.2).

6.4.2 Hamilton's principle

We continue with our program to derive the variational counterparts of the differential equations (6.23), (6.28) and (6.29), describing the dynamics, spectrum and stability. The most general way of describing the motion of a dynamical system is through Hamilton's principle, which may be generalized to continuous systems; see Goldstein [91]. The Lagrangian and Hamiltonian formulation of nonlinear MHD has been given in a classical paper by Newcomb [166], discussed in the companion Volume 2. In the same manner as above, we could linearize the expressions given by him to obtain the linearized Lagrangian. However, this labour is superfluous since we have the ingredients already: the linearized kinetic energy K is given by Eq. (6.40) and the linearized potential energy W by Eq. (6.84).

Hence, we may state *the linearized version of Hamilton's principle* at once: *The evolution of the system from time t_1 to time t_2 through the perturbation $\boldsymbol{\xi}(\mathbf{r}, t)$ is such that the variation of the integral of the Lagrangian vanishes,*

$$\delta \int_{t_1}^{t_2} L \, dt = 0, \qquad (6.87)$$

where $L \equiv K - W$, with

$$K = K[\dot{\boldsymbol{\xi}}] = \tfrac{1}{2} \int \rho \, \dot{\boldsymbol{\xi}}^* \cdot \dot{\boldsymbol{\xi}} \, dV,$$

$$W = W[\boldsymbol{\xi}] = -\tfrac{1}{2} \int \boldsymbol{\xi}^* \cdot \mathbf{F}(\boldsymbol{\xi}) \, dV.$$

Carrying out the minimization of this expression (see Exercise [6.6]) directly leads to the following *Euler–Lagrange equation*:

$$\mathbf{F}(\boldsymbol{\xi}) = \rho \frac{\partial^2 \boldsymbol{\xi}}{\partial t^2}, \qquad (6.88)$$

which reproduces the equation of motion (6.23). Clearly, the variational formulation for the dynamics is fully equivalent to the differential equation formulation.

6.4.3 Rayleigh–Ritz spectral variational principle

In the differential equations approach, we obtained the spectral eigenvalue equation (6.28) by considering the normal modes (6.27). Let us also consider the quadratic forms for normal modes, obtained by inserting (6.27) into the expressions for K and W. However, since these expressions have to be real whereas ω may be either real or imaginary, the result of this substitution is obtained much more easily by dropping the time-dependence and starting from the eigenvalue equation (6.28), $\mathbf{F}(\hat{\boldsymbol{\xi}}) = -\rho\omega^2\hat{\boldsymbol{\xi}}$, with complex $\hat{\boldsymbol{\xi}}(\mathbf{r})$. Dotting this equation with $\hat{\boldsymbol{\xi}}^*$ gives

$$\underbrace{-\frac{1}{2}\int \hat{\boldsymbol{\xi}}^* \cdot \mathbf{F}(\hat{\boldsymbol{\xi}})\, dV}_{\equiv W[\hat{\boldsymbol{\xi}}]} = \omega^2 \underbrace{\frac{1}{2}\int \rho \hat{\boldsymbol{\xi}}^*\hat{\boldsymbol{\xi}}\, dV}_{\equiv I[\hat{\boldsymbol{\xi}}]},$$

where $I[\hat{\boldsymbol{\xi}}] \equiv \|\hat{\boldsymbol{\xi}}\|^2$ is the square of the norm, defined in Eq. (6.36), so that

$$\omega^2 = \frac{W[\hat{\boldsymbol{\xi}}]}{I[\hat{\boldsymbol{\xi}}]} \quad \text{for normal modes.} \tag{6.89}$$

This is a nice expression but rather useless, as it stands, since it is just a conclusion a posteriori, after the normal modes have been obtained, and it does not provide a recipe for actually finding the eigenvalues ω^2 and the eigenfunctions $\hat{\boldsymbol{\xi}}$.

Such a recipe is obtained by considering the right hand side of Eq. (6.89) to be a variational expression for arbitrary trial functions $\boldsymbol{\xi}(\mathbf{r})$. Here, we drop the hat not only for simplicity of notation (time does not enter any more) but also because these functions need not be eigenfunctions. Nevertheless, they do produce them, as we will prove now.

Spectral variational principle: Eigenfunctions $\boldsymbol{\xi}$ of the operator $\rho^{-1}\mathbf{F}$ make the Rayleigh quotient

$$\Lambda[\boldsymbol{\xi}] \equiv \frac{W[\boldsymbol{\xi}]}{I[\boldsymbol{\xi}]} \tag{6.90}$$

stationary; the eigenvalues ω^2 are the stationary values of Λ. Here, the quadratic forms

$$W[\boldsymbol{\xi}] \equiv -\langle \boldsymbol{\xi}, \rho^{-1}\mathbf{F}(\boldsymbol{\xi})\rangle \quad \text{and} \quad I[\boldsymbol{\xi}] \equiv \langle \boldsymbol{\xi}, \boldsymbol{\xi}\rangle \tag{6.91}$$

are the potential energy and the square of the norm, respectively.

▷ **Proof.** Let ω^2 be a stationary value of $\Lambda[\boldsymbol{\xi}]$, i.e. $\delta\Lambda = 0$. Then

$$\delta\Lambda = \frac{\delta W}{I} - \frac{W}{I^2}\delta I$$
$$= -2\frac{\langle \delta\boldsymbol{\xi}, \rho^{-1}\mathbf{F}(\boldsymbol{\xi})\rangle + \omega^2\langle \delta\boldsymbol{\xi}, \boldsymbol{\xi}\rangle}{\langle \boldsymbol{\xi}, \boldsymbol{\xi}\rangle} = -2\frac{\langle \delta\boldsymbol{\xi}, \rho^{-1}\mathbf{F}(\boldsymbol{\xi}) + \omega^2\boldsymbol{\xi}\rangle}{\langle \boldsymbol{\xi}, \boldsymbol{\xi}\rangle} = 0,$$

where we used the self-adjointness (giving the factor 2) and substituted the value ω^2 for Λ. Since $\delta\boldsymbol{\xi}$ is an arbitrary variation of $\boldsymbol{\xi}$, it follows that $\rho^{-1}\mathbf{F}(\boldsymbol{\xi}) = -\omega^2\boldsymbol{\xi}$, which is the eigenvalue equation; QED. ◁

The Rayleigh–Ritz principle is extremely useful for the approximation of eigenvalues by means of finite-dimensional subspaces of the complete, infinite-dimensional, Hilbert space. Here, one selects a suitable class of functions $(\eta_1, \eta_2, \ldots, \eta_N)$, with finite norm $\|\eta_n\|$, which are used as *trial functions* in the Rayleigh quotient (6.90). The linear combination of these functions that minimizes the functional Λ constitutes an approximation for the lowest eigenvalue ω_0^2, where the minimum value of Λ is always larger than the actual value of ω_0^2.

An approximation to the N lowest eigenvalues may be obtained by choosing the η_ns to be orthonormal,

$$\langle \eta_m, \eta_n \rangle = \delta_{mn}. \tag{6.92}$$

Since these functions are supposed to be known, one may compute the matrix elements

$$W_{mn} = \langle \eta_m, \rho^{-1} \mathbf{F}(\eta_n) \rangle. \tag{6.93}$$

Writing

$$\eta = \sum_{n=1}^{N} a_n \eta_n, \tag{6.94}$$

one then obtains the following approximation:

$$\Lambda[\xi] \approx \Lambda[\eta] = \frac{\sum_{m=1}^{N} \sum_{n=1}^{N} a_m^* W_{mn} a_n}{\sum_{n=1}^{N} |a_n|^2}. \tag{6.95}$$

Hence, the problem boils down to Eq. (6.64) of Section 6.3.1, i.e. the simultaneous diagonalization of the two finite-dimensional quadratic forms $W[\eta]$ and $I[\eta]$. In this case, since the ηs have been chosen to be orthonormal, only $W[\eta]$ needs to be diagonalized. Consequently, the eigenvalues $\omega_\eta^{2(i)}$ and eigenfunctions $\eta^{(i)}$ of the matrix W_{mn} are approximations to the lowest N eigenvalues $\omega^{2(i)}$ and corresponding eigenfunctions $\xi^{(i)}$ of the operator $\rho^{-1}\mathbf{F}$. Of course, the accuracy of this approximation depends on the choice of the basis functions $\{\eta_n\}$.

6.4.4 Energy principle

The above proof demonstrates the equivalence of the Rayleigh–Ritz variational principle with the eigenvalue equation. It also provides a formulation for stability problems that is one step more useful than the force operator equation. Since $I[\xi] \equiv \|\xi\|^2 \geq 0$, one may insert *trial functions* in W. If $W[\xi] > 0$ for all possible trial functions ξ, one may conclude that eigenvalues $\omega^2 < 0$ do not exist and that the system is *stable*. On the other hand, if one can find a single ξ for which $W[\xi] < 0$, at least one eigenvalue $\omega^2 < 0$ exists and the system is *unstable*. This is summarized in the following powerful statement, due to Bernstein, Frieman, Kruskal and Kulsrud [26] and, independently, Hain, Lüst and Schlüter [101].

Energy principle for stability An equilibrium is stable if (sufficient) and only if (necessary)

$$W[\boldsymbol{\xi}] > 0 \tag{6.96}$$

for all displacements $\boldsymbol{\xi}(\mathbf{r})$ *that are bound in norm and satisfy the boundary conditions.* Here, $W[\boldsymbol{\xi}]$ is defined by Eq. (6.85) and the appropriate boundary condition is given in Eq. (6.86).

In conclusion, the variational approach offers three different methods for determining the question of stability:

(1) By physical intuition one may *guess* a trial function $\boldsymbol{\xi}(\mathbf{r})$ that picks up the unstable part of the potential energy, so that $W[\boldsymbol{\xi}] < 0$. This provides a direct demonstration of the instability of a certain system: *necessary stability* ($\equiv$ *sufficient instability*) *criteria* are obtained this way.
(2) More systematically, one may investigate the sign of W by exploiting *a complete set of trial functions that are normalized in any convenient way* (which need not be by means of I itself): *necessary + sufficient criteria* for stability are obtained.
(3) Finally, by considering *a complete set of trial functions that are properly normalized with the correct physical norm* $I[\boldsymbol{\xi}]$, *corresponding to the kinetic energy, the complete spectrum* of eigenvalues $\{\omega^2\}$ is obtained.

Comparing these different variational methods with their differential equation counterparts, we notice that:

– Method (1) has no counterpart in the equation of motion approach (unless one is extremely clever and knows how to guess forces, i.e. complete vector fields, that should on average be parallel to the displacements associated with them). It is the most direct method to investigate stability problems. Thus, if one has a good physical intuition, one may be able to design a trial function that shows right away that the system is unstable by picking up the dominant part of the driving energy of the instability. Also, one may formalize this approach by testing with a finite class of trial functions that may be considered as a subspace of the complete Hilbert space of the system.
– Method (2) is equivalent to solving the marginal equation of motion $\mathbf{F}(\boldsymbol{\xi}) = 0$, but much simpler to apply since one may use any normalization of the trial functions in the expression (6.96) for W to test for stability. For example, one could exploit a normalization involving only the component of the perturbation perpendicular to the magnetic surfaces, $\int \rho (\mathbf{n} \cdot \boldsymbol{\xi})^2 \, dV$; see Bernstein *et al.* [27]. (Applications will be given in later chapters.) The only limitation in the choice of the normalization of the trial functions is that the original norm $\|\boldsymbol{\xi}\|$ should remain finite. Of course, in

the process of dropping the proper normalization of the Hilbert space, one loses the possibility of calculating the actual growth rates of the instabilities.
– Method (3) is equivalent to solving the spectral equation $\mathbf{F}(\boldsymbol{\xi}) = -\rho\omega^2\boldsymbol{\xi}$, i.e. to a full-blown normal mode analysis: variational method and differential equation are equivalent.

Although the above discussion of the energy principle appears to be rather solid, i.e. in no need of further proof, this turns out to be deceptive since the tacit assumption has been made that the spectrum consists of discrete eigenvalues only. In general, this is not the case in MHD, where the existence of a continuous spectrum is a rule rather than an exception, as we will see in the next chapter. For stability analysis, this presents a substantial complication. This problem will be addressed in Section 6.5, where we first discuss the lost kingdom of stability analysis with normal modes only (Section 6.5.1), and next show that the energy principle can still be proved avoiding the subtleties of the continuous spectrum (Section 6.5.2). This will also facilitate a practical modification of the energy principle (Section 6.5.3). Having settled these more advanced issues, we will return to our starting point and put the spotlight on the structural beauty and applicability of MHD spectral theory (Section 6.5.4).

6.5 Further spectral issues

6.5.1 Normal modes and the energy principle*

Consider a pair of discrete normal modes $\exp(-i\omega_n t)$ and $\exp(i\omega_n t)$ belonging to the same eigenvalue $\omega^2 = \omega_n^2$. If we neglect all other modes, e.g. by preferentially exciting this one pair of modes, the solution (6.77) of the initial value problem (with the contour C as in Fig. 6.10) easily may be completed. Since

$$\rho^{-1}\mathbf{F}(\boldsymbol{\xi}) = -\omega_n^2\,\boldsymbol{\xi}, \qquad (6.97)$$

the resolvent operator would simply be given by

$$(\rho^{-1}\mathbf{F} + \omega^2\mathbf{I})^{-1} = \left[(\omega^2 - \omega_n^2)\mathbf{I}\right]^{-1}. \qquad (6.98)$$

Hence, the discrete eigenvalue ω_n^2 gives rise to two poles $\omega = \pm\omega_n$ which, by virtue of the reality condition (6.58), are situated on either the real or the imaginary axis of the complex ω-plane. Clearly, for $\omega^2 = \omega_n^2$ the resolvent operator does not exist, but everywhere else in the complex ω-plane it is now defined (of course, when we ignore the rest of the spectrum). We may now complete the integration (6.77) by deforming the contour around the two poles $\omega^2 = \pm\omega_n$. Shifting the straight part of the contour to $\nu = -\infty$, so that $\exp(-i\omega t)$ vanishes faster than everything else, the only contribution that survives for large t will be the two

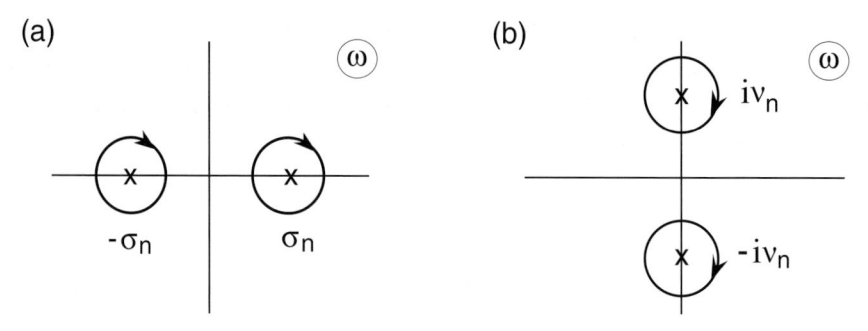

Fig. 6.12. Integration contours in the complex ω-plane for (a) stable oscillation and (b) exponential instability (corresponding to the eigenfunctions shown in Fig. 6.6).

residues picked up at the poles. As is well known (see, e.g., Churchill [57]), these residues may be computed one by one by means of Cauchy's integral formula

$$\oint \frac{f(z)}{z - z_0} dz = 2\pi i f(z_0), \tag{6.99}$$

where $f(z)$ should be an analytic function inside a closed contour encircling the point $z = z_0$. We then find for the asymptotic time-dependence of the normal modes:

$$\boldsymbol{\xi}(\mathbf{r}; t) = \frac{1}{2\pi} \int_C \frac{i\omega \boldsymbol{\xi}_i(\mathbf{r}) - \dot{\boldsymbol{\xi}}_i(\mathbf{r})}{(\omega + \omega_n)(\omega - \omega_n)} e^{-i\omega t} d\omega$$

$$= \frac{[i\omega_n \boldsymbol{\xi}_i(\mathbf{r}) + \dot{\boldsymbol{\xi}}_i(\mathbf{r})] e^{i\omega_n t} + [i\omega_n \boldsymbol{\xi}_i(\mathbf{r}) - \dot{\boldsymbol{\xi}}_i(\mathbf{r})] e^{-i\omega_n t}}{2i\omega_n} \tag{6.100}$$

(where one should notice that the contour C deformed around a pole has just the opposite sense of a Cauchy contour).

Specifically, writing $\omega_n = \sigma_n + i\nu_n$, we either have $\nu_n = 0$ or $\sigma_n = 0$. If $\nu_n = 0$, the poles are situated on the real axis (Fig. 6.12(a)) so that

$$\boldsymbol{\xi}(\mathbf{r}; t) = \boldsymbol{\xi}_i(\mathbf{r}) \cos \sigma_n t + \dot{\boldsymbol{\xi}}_i(\mathbf{r}) \sigma_n^{-1} \sin \sigma_n t, \tag{6.101}$$

which is a stable undamped oscillation excited by an initial displacement $\boldsymbol{\xi}_i(\mathbf{r})$, or by an initial velocity $\dot{\boldsymbol{\xi}}_i(\mathbf{r})$, or by a combination of both. If $\sigma_n = 0$, the poles are situated on the imaginary axis (Fig. 6.12(b)) and we have

$$\boldsymbol{\xi}(\mathbf{r}; t) = \boldsymbol{\xi}_i(\mathbf{r}) \cosh \nu_n t + \dot{\boldsymbol{\xi}}_i(\mathbf{r}) \nu_n^{-1} \sinh \nu_n t. \tag{6.102}$$

Since both $\cosh(\nu_n t)$ and $\sinh(\nu_n t)$ eventually grow as $\exp(\nu_n t)$ this is called an exponential instability, which again may be excited by initial displacements as well as initial velocities.

6.5 Further spectral issues

The important feature is that, here, true normal modes, i.e. discrete eigenvalues, are either oscillatory or exponentially growing, but never damped. This is the real simplifying feature of *ideal*, i.e. conservative, MHD perturbations of *static equilibria* expressed by the *self-adjointness* of the force operator. As a consequence, stability studies may be simplified considerably as compared to the analysis of dissipative or stationary ($\mathbf{v}_0 \neq 0$) systems, to be discussed in Volume 2. If the equilibrium is described by a set of parameters $\alpha_1, \ldots, \alpha_N$ (expressing the background equilibrium distributions), the marginal states for non-self-adjoint systems will be defined by the condition

$$\mathrm{Im}\, \omega_\kappa(\alpha_1, \ldots, \alpha_N) = 0, \quad (6.103)$$

where κ indicates parameters analogous to wave numbers labelling the different modes. This condition, called *the principle of exchange of stabilities*, determines the transition from stability to instability; see Chandrasekhar [51], reviewed by Goedbloed [81](III). Such a transition may take place at any point on the line $\mathrm{Im}\, \omega = 0$. For example, modifying an exponentially unstable static equilibrium by just adding a constant background flow $\mathbf{v}_0$ will produce a stationary equilibrium with an overstable mode, due to the Doppler shift (5.13). However, for self-adjoint systems (like in ideal MHD of static equilibria), the condition (6.103) may be replaced by the much simpler one

$$\omega_\kappa^2(\alpha_1, \ldots, \alpha_N) = 0, \quad (6.104)$$

i.e. transfer of stability to instability always takes place via the same point, viz. the origin $\omega = 0$ of the complex ω-plane. Stability may then be studied by means of either *a marginal mode analysis*, which seeks to establish the locus in parameter space $(\alpha_1, \ldots, \alpha_N)$ where the marginal equation of motion (6.29) is satisfied, or the variational counterpart expressed by the *energy principle* (6.96).

Intuitively clear as the energy principle may seem now, its proof is actually not quite straightforward, not even for the self-adjoint systems we are dealing with. If the force operator $\mathbf{F}$ would only allow for discrete eigenvalues satisfying

$$\rho^{-1}\mathbf{F}(\boldsymbol{\xi}_n) = -\omega_n^2 \boldsymbol{\xi}_n, \quad (6.105)$$

it would be reasonable to assume that the set $\{\boldsymbol{\xi}_n\}$ constitutes a complete basis for the Hilbert space. In that case, the eigenfunctions $\boldsymbol{\xi}_n$ could be chosen to be orthonormal:

$$\langle \boldsymbol{\xi}_m, \boldsymbol{\xi}_n \rangle = \delta_{mn}. \quad (6.106)$$

An arbitrary ξ could then be expanded in eigenfunctions:

$$\xi = \sum_{n=1}^{\infty} a_n \xi_n , \qquad (6.107)$$

so that

$$W = -\langle \xi, \rho^{-1} \mathbf{F}(\xi) \rangle = \sum_{n=1}^{\infty} a_n^2 \omega_n^2 . \qquad (6.108)$$

Hence, if one could find a ξ for which $W < 0$ at least one eigenvalue $\omega_n^2 < 0$ should exist. Such an eigenvalue would correspond with an exponential instability. This 'proof' was given in the original paper by Bernstein *et al.* [26], before it was known that ideal MHD systems usually have a *continuous spectrum* (see Chapter 7) *extending to the origin* $\omega^2 = 0$. This fact implies that the simplicity of the marginal stability analysis is spoiled and that more care is needed to establish necessity of the energy principle. This will be the subject of the next subsection.

6.5.2 Proof of the energy principle*

A correct proof of both necessity and sufficiency of the energy principle without invoking the assumption of a complete basis of discrete eigenvalues, but also avoiding analysis of the continuous spectrum, has been given by Laval, Mercier and Pellat in 1965 [139]. That proof is based on energy conservation,

$$\dot{H} = 0, \qquad H \equiv K + W , \qquad (6.109)$$

and the so-called virial equation

$$\ddot{I} \equiv \langle \xi, \xi \rangle'' = 2\langle \dot{\xi}, \dot{\xi} \rangle + 2\langle \xi, \ddot{\xi} \rangle = 2K - 2W . \qquad (6.110)$$

The proof of sufficiency is actually quite simple, but the proof of necessity is somewhat more involved.

Sufficiency If $W[\eta] > 0$ for all trial functions η, one cannot find a motion $\xi(t)$ such that the kinetic energy $K[\xi(t)]$ grows without bound.

Proof Since $W \equiv H - K > 0$ and H is finite, unbounded growth for K would violate energy conservation; QED. (Notice that we here exclude the class of linearly growing instabilities, where $\xi \sim t$ and K is constant.)

Necessity If a function η exists such that $W[\eta] < 0$, the system will exhibit an unbounded motion $\xi(t)$.

6.5 Further spectral issues

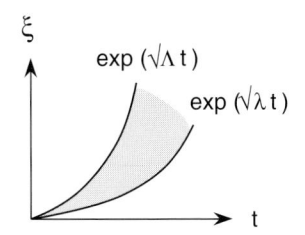

Fig. 6.13. Bounds on exponential growth.

(*First proof*) Start from $W[\eta] < 0$ and choose initial data $\xi(0) = \eta$, $\dot{\xi}(0) = 0$. From Eq. (6.109),

$$H(t) = H(0) = K(0) + W(0) = W[\eta] < 0,$$

so that

$$\ddot{I}(t) = 2K - 2W = 4K - 2H \geq -2H(t) > 0.$$

Hence, $\dot{I}$ grows without limit as $t \to \infty$ and I grows at least like $-Ht^2$. As a result, ξ grows at least linearly in t; QED. This simplified version of the proof is due to Kruskal.

▷ **Estimates of growth rates** (*Second proof*) Laval, Mercier and Pellat [139] also gave a sharper version of the proof by estimating the growth rate. Again, start from $W[\eta] < 0$ and define

$$\lambda \equiv -W[\eta]/I[\eta] > 0. \tag{6.111}$$

We then prove that there exists a $\xi(t)$ growing at least as $\exp(\sqrt{\lambda} t)$ (Fig. 6.13, lower curve). Choose as initial data $\xi(0) = \eta$, $\dot{\xi}(0) = \sqrt{\lambda}\,\eta$ (i.e., in contrast to the first case, we excite the motion with the proper relationship between ξ and $\dot{\xi}$ for an exponentially growing normal mode). Consequently,

$$H(t) = H(0) = K[\dot{\xi}(0)] + W[\xi(0)] = \lambda I[\eta] + W[\eta] = 0.$$

From Eq. (6.110) we then have

$$\ddot{I}(t) = 2K - 2W = 4K - 2H = 4K(t) > 0, \tag{6.112}$$

whereas the Schwartz inequality gives

$$\dot{I}^2(t) = 4\langle \xi, \dot{\xi}\rangle^2 \leq 4\langle \xi, \xi\rangle\langle \dot{\xi}, \dot{\xi}\rangle = 4I(t)K(t) = I(t)\ddot{I}(t). \tag{6.113}$$

Since

$$\dot{I}(0) = 2\sqrt{\lambda}\,\langle \eta, \eta\rangle = 2\sqrt{\lambda}\,I(0) > 0, \tag{6.114}$$

we have from Eq. (6.112) that $\dot{I}(t) > 0$ for $t > 0$, so that we may divide the inequality

(6.113) by $\dot{I}(t)I(t)$, giving the following sequence of inequalities:

$$\frac{\dot{I}(t)}{I(t)} \leq \frac{\ddot{I}(t)}{\dot{I}(t)} \Rightarrow \ln\frac{I(t)}{I(0)} \leq \ln\frac{\dot{I}(t)}{\dot{I}(0)} = \ln\frac{\dot{I}(t)}{2\sqrt{\lambda}I(0)} \Rightarrow \frac{I(t)}{I(0)} \leq \frac{\dot{I}(t)}{2\sqrt{\lambda}I(0)}$$

$$\Rightarrow \frac{\dot{I}(t)}{I(t)} \geq 2\sqrt{\lambda} \Rightarrow \ln\frac{I(t)}{I(0)} \geq 2\sqrt{\lambda}\,t \Rightarrow I(t) \geq I(0)\exp(2\sqrt{\lambda}\,t).$$

Consequently, ξ grows at least as $\exp(\sqrt{\lambda}\,t)$; QED.

One may also prove the following theorem, which gives an upper bound on the growth rate (Fig. 6.13, upper curve):

Theorem If the ratio $-W[\xi]/I[\xi]$ has a smallest upper bound

$$\Lambda \geq \lambda[\xi] \equiv -W[\xi]/I[\xi] \quad \text{for all } \xi,$$

then $\xi(t)$ cannot grow faster than $\exp(\sqrt{\Lambda}\,t)$.

Proof Start again from the virial expression (6.110):

$$\ddot{I}(t) = 2K(t) - 2W(t) = 2H(t) - 4W(t) \leq 2H(t) + 4\Lambda I(t).$$

Hence,

$$\ddot{I}(t) - 4\Lambda I(t) \leq 2H(t) = 2H(0).$$

Consequently, $I(t)$ grows at most like $\exp(2\sqrt{\Lambda}\,t)$, so that $\xi(t)$ cannot grow faster than $\exp(\sqrt{\Lambda}\,t)$; QED.

We have given all these proofs here because they naturally lead to the extension of the stability concept introduced in the next section. ◁

6.5.3 σ-stability

For thermonuclear confinement of plasma, the stability concept used above may be relaxed. One is not really interested in whether the plasma is stable, one is interested in whether or not one can confine it long enough to obtain fusion. For example, if the worst instability of a particular configuration were to grow as illustrated in Fig. 6.14, where a is the radial dimension of the plasma vessel and τ is the characteristic confinement time needed for fusion, one would call this configuration stable for all practical purposes. One could also choose τ to be another time scale, e.g. the time scale for which one accepts the ideal MHD model as a valid description, or one may choose τ to be the time scale of an actual experiment determined by the decay of external currents, or τ could correspond to the limit posed by the accuracy of a certain numerical stability program. For all these purposes, one may allow perturbations that grow at most like $\exp(\sigma t)$, where $\sigma \equiv 1/\tau$. We will call equilibria *σ-stable* if they do not manifest growth faster than this. This

6.5 Further spectral issues 269

Fig. 6.14. Tolerable exponential instability.

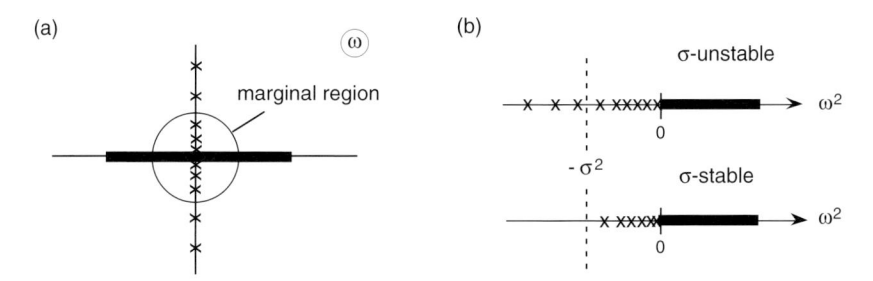

Fig. 6.15. (a) Marginal stability analysis encounters singularities at the origin of the complex ω-plane; (b) σ-stability analysis avoids them by staying away from the origin.

extension of the stability concept was introduced by Goedbloed and Sakanaka in 1974 [89], and applied to the experimental devices in use at that time [199].

As well as for practical purposes, the concept of σ-stability is also useful for analysis. We will show in Chapter 7 that the continuous spectrum nearly always reaches the origin $\omega = 0$ and that it frequently carries with it infinitely many point eigenvalues that accumulate at the edge of the continuum. Hence, the marginal point $\omega = 0$ is a highly singular point in the spectrum (Fig. 6.15(a)) so that the supposed simplicity of a marginal stability analysis, as compared to calculating actual growth rates, often turns out to be illusory. In contrast, a σ-stability analysis avoids these difficulties by staying on the unstable side of the spectrum (Fig. 6.15(b)) which consists of point eigenvalues only. At least, that is a conjecture by Grad [99] to which no exceptions have been found yet.[3] This is of particular importance for numerical stability studies where one wishes to avoid singularities as much as possible.

Since we are dealing now with point eigenvalues only, we may define an equilibrium to be σ-stable if no point eigenvalues $\omega^2 < -\sigma^2$ exist, and σ-unstable if such eigenvalues do exist. A σ-marginal stability analysis then seeks to find the

---

[3] One should actually exclude perturbations with infinitely large mode numbers since they may lead to dense sets of unstable point eigenvalues. The closure of those sets formally contains a continuous spectrum; see Spies [211] and Spies and Tataronis [212].

σ-stability boundary in parameter space, replacing the marginal stability condition (6.104) by

$$\omega_{\mathcal{K}}^2(\alpha_1, \alpha_2, \ldots, \alpha_N) = -\sigma^2. \tag{6.115}$$

This problem may be studied by means of the σ-*marginal equation of motion*:

$$\mathbf{F}^\sigma(\boldsymbol{\xi}) \equiv \mathbf{F}(\boldsymbol{\xi}) - \rho\sigma^2\boldsymbol{\xi} = 0, \tag{6.116}$$

where the force $\mathbf{F}^\sigma$ available to drive a σ-instability is reduced by the amount $\rho\sigma^2\boldsymbol{\xi}$ with respect to the force $\mathbf{F}$ for driving an instability under the original definition. The variational form of this problem is the following modification of the energy principle.

Modified energy principle for stability An equilibrium is σ-stable if and only if

$$W^\sigma[\boldsymbol{\xi}] \equiv W[\boldsymbol{\xi}] + \sigma^2 I[\boldsymbol{\xi}] > 0 \tag{6.117}$$

for all displacements $\boldsymbol{\xi}$ that are bound in norm and satisfy the boundary conditions. Hence, the amount of negative potential energy available for driving a σ-instability is reduced by the amount $\sigma^2 I[\boldsymbol{\xi}]$ as compared to that available for driving an instability under the original definition.

Comparing Eq. (6.116) with the normal mode equation (6.28), and Eq. (6.117) with the energy principle (6.96), one observes that their formal structure is the same. One might even wonder whether the whole concept of σ-stability does not boil down to a normal-mode analysis. This is not the case, the important difference being that in a normal-mode analysis the eigenvalue ω has to be determined, whereas in a σ-stability analysis σ is simply a pre-fixed parameter. Hence, the problem is of the same nature as a stability analysis by means of the energy principle, although the equations are more complicated (i.e., they have more terms). The latter complication, which is unimportant for numerical applications anyway, is more than offset by the absence of the singularities associated with the continuum at $\omega^2 = 0$.

▷ **Proof of the modified energy principle** This proof can be given in complete analogy with that of the ordinary energy principle given in the previous section. Sufficiency is proved by writing

$$W^\sigma[\boldsymbol{\eta}] = H - (K - \sigma^2 I) > 0 \quad \text{for all } \boldsymbol{\eta}, \quad H \text{ finite},$$

so that for a σ-instability, where $K - \sigma^2 I$ grows without bound, energy conservation would be violated. The necessity of the modified energy principle implies that a σ-unstable motion $\boldsymbol{\xi}(t)$ can be found if one knows a function $\boldsymbol{\eta}$ such that $W^\sigma[\boldsymbol{\eta}] < 0$. This is an immediate consequence of the proof of necessity of the ordinary energy principle. Analogous to Eq. (6.111), define

$$\mu \equiv -W^\sigma[\boldsymbol{\eta}]/I[\boldsymbol{\eta}] = -W[\boldsymbol{\eta}]/I[\boldsymbol{\eta}] - \sigma^2 \equiv \lambda - \sigma^2 > 0.$$

Then,
$$\lambda \equiv -W[\eta]/I[\eta] = \mu + \sigma^2 > \sigma^2,$$
so that $\boldsymbol{\xi}(t)$ grows at least as $\exp(\sqrt{\lambda}\,t) = \exp(\sqrt{\mu + \sigma^2}\,t)$ and the equilibrium is, therefore, σ-unstable. ◁

6.5.4 Returning to the two viewpoints

We have seen how the two intuitive viewpoints of stability, illustrated in Fig. 6.1, lead to two alternative approaches of the linearized MHD equations, viz. one in terms of differential equations (the equation of motion) and another one in terms of the quadratic forms of the potential and kinetic energy. Since this duality is also present in the formalism of quantum mechanics, expressed in the language of linear operators in Hilbert space, it is expedient to highlight the analogy.

Analogy with quantum mechanics As far as spectral theory is concerned, *the analogy between linearized MHD and quantum mechanics is complete*. In the terminology of Dirac [67], we have obtained two 'pictures' of ideal MHD spectral theory (summarized in Fig. 6.16), viz. that of *the equation of motion* in terms of $\boldsymbol{\xi}$ and that of *the variational principles* in terms of the potential and kinetic energies. They correspond to the 'Schrödinger picture' of wave mechanics (with a description in terms of the wave equation $H\psi = E\psi$), exploiting differential equations, and

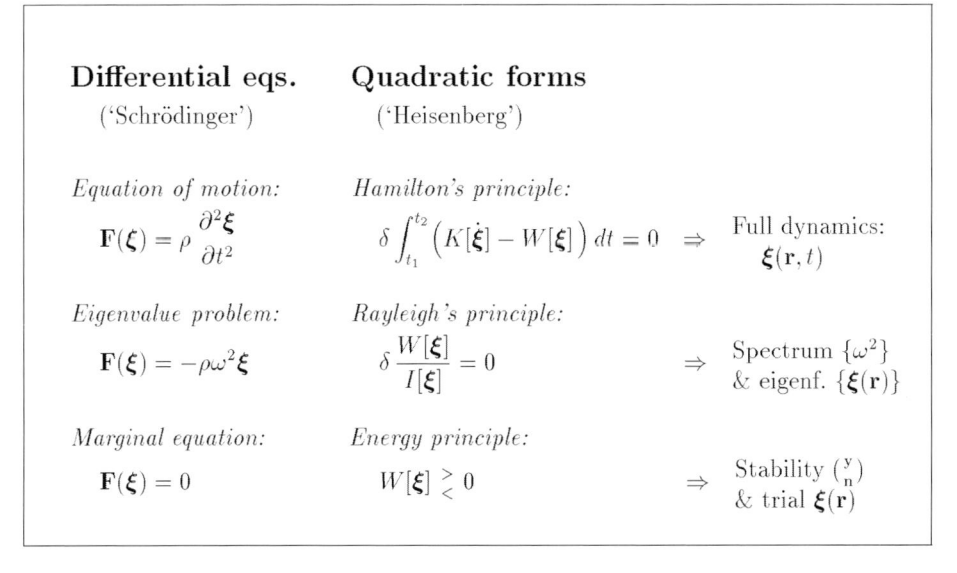

Differential eqs. ('Schrödinger')	**Quadratic forms** ('Heisenberg')	
Equation of motion: $\mathbf{F}(\boldsymbol{\xi}) = \rho\dfrac{\partial^2\boldsymbol{\xi}}{\partial t^2}$	*Hamilton's principle:* $\delta\displaystyle\int_{t_1}^{t_2}\bigl(K[\dot{\boldsymbol{\xi}}] - W[\boldsymbol{\xi}]\bigr)dt = 0 \;\Rightarrow$	Full dynamics: $\boldsymbol{\xi}(\mathbf{r},t)$
Eigenvalue problem: $\mathbf{F}(\boldsymbol{\xi}) = -\rho\omega^2\boldsymbol{\xi}$	*Rayleigh's principle:* $\delta\dfrac{W[\boldsymbol{\xi}]}{I[\boldsymbol{\xi}]} = 0 \;\Rightarrow$	Spectrum $\{\omega^2\}$ & eigenf. $\{\boldsymbol{\xi}(\mathbf{r})\}$
Marginal equation: $\mathbf{F}(\boldsymbol{\xi}) = 0$	*Energy principle:* $W[\boldsymbol{\xi}] \gtrless 0 \;\Rightarrow$	Stability $\binom{y}{n}$ & trial $\boldsymbol{\xi}(\mathbf{r})$

Fig. 6.16. The two 'pictures' of ideal MHD spectral theory.

the 'Heisenberg picture' of matrix mechanics (with a description in terms of the representative matrix elements $\langle n|H|m\rangle$ of the Hamiltonian), exploiting quadratic forms.

Of course, this analogy is mathematical, not physical. The physical systems are totally different. For example, in the spectrum, the transition from bound to free states in quantum mechanics corresponds to the completely different physical problem of transition from stable to unstable modes in magnetohydrodynamics. More important, whereas the wave function ψ in quantum mechanics is a complex quantity that requires a physical interpretation to connect it to observable properties of the atomic system, the displacement vector ξ in MHD is a real quantity that refers directly to the observable macroscopic displacement of the classical plasma fluid (using complex notation here is no more than a matter of convenience). (We do not pay much attention to the difference between a scalar ψ in quantum mechanics and a vector ξ in MHD since this appears to be of minor interest; quantum mechanical systems with wave functions with more components also occur, e.g. in quantum electrodynamics of $S = 1$ particles.)

An important reason to dwell on the analogy with quantum mechanical spectral theory is the need to disentangle useful mathematical techniques that can be transferred to another field from concepts that are essential to the physical formulation. In this respect, it should be clear (but frequently is not for physics students because of their over-exposition to quantum mechanical problems as compared to classical ones) that linear operators in Hilbert space as such have nothing to do with quantum mechanics. In fact, the mathematical formulation by Hilbert in 1912 preceded the advent of quantum mechanics by more than a decade. Essentially, the two 'pictures' illustrated in Fig. 6.16 are nothing but a translation to physics of the generalization of linear algebra concepts to infinite-dimensional vector spaces. (We are indebted to J. Moser (1973) for enlightening discussions on this topic.)

There is yet another hurdle to be taken. Whereas quantum mechanics of atomic and sub-atomic particles applies to a rich arsenal of relevant spherically symmetric systems, with the attraction of symmetry with respect to the rotation groups, MHD spectral theory refers to magnetically confined plasmas where the constraint of $\nabla \cdot \mathbf{B} = 0$ forbids spherical symmetry (as we saw in Section 1.3.4) and demands the consideration of extended magnetic structures with symmetries that are much less obvious than rotations. In fact, the application of symmetry groups to MHD spectral theory is really in its infancy, as should be evident from the more complex structure of the MHD equations compared to the quantum mechanical ones and the relatively insignificant effort in MHD compared to the impressive accomplishments of the great physicists of the twentieth century who created the quantum mechanical picture of the atomic world.

6.5 Further spectral issues

Why does the water fall out of the glass? We now show how the machinery of spectral theory and energy principle works to solve the practical problem posed in Section 6.1.1. We consider a simple fluid (no magnetic field) with a varying density in an external gravitational field. For equilibrium, the pressure and density distributions should satisfy

$$\nabla p = -\rho \nabla \Phi = \rho \mathbf{g}. \quad (6.118)$$

The expression (6.85) for the energy W then simplifies to

$$W^f = \tfrac{1}{2} \int \left[\gamma p |\nabla \cdot \boldsymbol{\xi}|^2 + (\boldsymbol{\xi}^* \cdot \nabla p) \nabla \cdot \boldsymbol{\xi} - (\boldsymbol{\xi}^* \cdot \nabla \Phi) \nabla \cdot (\rho \boldsymbol{\xi}) \right] dV$$
$$= \tfrac{1}{2} \int \left[\gamma p |\nabla \cdot \boldsymbol{\xi}|^2 + \mathbf{g} \cdot \boldsymbol{\xi}^* (2\rho \nabla \cdot \boldsymbol{\xi} + (\nabla \rho) \cdot \boldsymbol{\xi}) \right] dV. \quad (6.119)$$

Clearly, without gravity, the fluid is stable since only the positive definite first term, corresponding to compressive sound motions, remains. With gravity, the sign of W depends on the density gradient in a way that we now have to determine.

Let us specify to plane slab geometry so that pressure and density are functions of the vertical coordinate x alone: $p = p(x)$, $\rho = \rho(x)$, and gravity points in the negative x-direction: $\mathbf{g} = -g \mathbf{e}_x$. The equilibrium condition (6.118) then becomes

$$p' = -\rho g, \quad (6.120)$$

where the prime denotes differentiation with respect to x. The expression for W^f now simplifies to

$$W^f = \tfrac{1}{2} \int \left[\gamma p |\nabla \cdot \boldsymbol{\xi}|^2 - 2\rho g \xi_x^* \nabla \cdot \boldsymbol{\xi} - \rho' g |\xi_x|^2 \right] dV. \quad (6.121)$$

The energy principle according to method (1) of Section 6.4.4 is illustrated by the immediate guess suggested by this expression, viz. to exploit incompressible trial functions, $\nabla \cdot \boldsymbol{\xi} = 0$, so that

$$W^f = -\tfrac{1}{2} \int \rho' g |\xi_x|^2 dV \geq 0 \quad \Rightarrow \quad \rho' g \leq 0 \text{ (everywhere)}, \quad (6.122)$$

which is a *necessary criterion* for stability. This already more or less explains our glass of water experiment since this criterion clearly shows that lighter fluid should be on top of heavier fluid for stability. Actually, we here have derived the condition for stability inside the fluid, with respect to *internal modes*, whereas the water–air system requires an extended form of the extended principle with a two-fluid interface (model II*), permitting the description of *external modes*. This problem will be considered in Section 6.6.4. It is already clear that the physics will be the same: the density gradient then becomes a density jump that should be negative at the interface (light fluid above) for stability.

The expression (6.121) also permits the derivation of a much sharper stability condition, with respect to all possible perturbations (compressible as well as incompressible ones), according to method (2) of Section 6.4.4. In this case, it just implies rearranging terms such that two definite terms are obtained:

$$W^f = \tfrac{1}{2} \int \left[\gamma p \left| \nabla \cdot \boldsymbol{\xi} - \frac{\rho g}{\gamma p} \xi_x \right|^2 - \left(\rho' g + \frac{\rho^2 g^2}{\gamma p} \right) |\xi_x|^2 \right] dV. \qquad (6.123)$$

Since the components ξ_y and ξ_z only appear in the compressibility term $\nabla \cdot \boldsymbol{\xi}$, minimization with respect to those components is trivial:

$$\nabla \cdot \boldsymbol{\xi} = \frac{\rho g}{\gamma p} \xi_x. \qquad (6.124)$$

Consequently,

$$\rho' g + \frac{\rho^2 g^2}{\gamma p} \leq 0 \quad \text{(everywhere)} \qquad (6.125)$$

is a *necessary and sufficient criterion* for stability.

Finally, questions like 'How long does it take for the instability to develop?' or 'What do the perturbations look like?' require a complete normal mode analysis of the spectrum and eigenfunctions, according to method (3) of Section 6.4.4. This will be one of the topics covered in Chapter 7.

6.6 Extension to interface plasmas

So far, we have been concerned with spectral theory of plasmas surrounded by a rigid wall (model I). For many applications, it is useful to be able to treat plasmas with an interface by the same techniques. For example, in laboratory fusion research it is appropriate to model the very low density region close to the wall (created by a limiter) as a vacuum so that effectively a *plasma–vacuum system* is obtained (model II). In astrophysical plasmas it is frequently expedient to model plasmas with a jump in the density (e.g. to a low density force-free plasma) as a *plasma–plasma system* (model II*).

Recall the different steps of the spectral analysis of model I plasmas. The nonlinear equations (6.1)–(6.4) for the plasma variables **v**, p, **B**, and ρ were linearized about a static equilibrium, prescribed by Eqs. (6.7)–(6.8), with perturbations satisfying the differential equations (6.12)–(6.15) and the boundary conditions (6.16). We then defined the plasma displacement vector $\boldsymbol{\xi}(\mathbf{r}, t)$, and cast the dynamical problem in the form of the equation of motion (6.23) involving the force operator $\mathbf{F}(\boldsymbol{\xi})$. Next, **F** was proved to be self-adjoint in Section 6.2.3, and the quadratic form (6.85) for the associated potential energy W was derived in Section 6.4.1.

6.6 Extension to interface plasmas

These steps will now be generalized to plasmas with an interface, following the original paper on the energy principle by Bernstein *et al.* [26] and the presentation by Kadomtsev [119].

To that end, the original nonlinear equations for model II and model II* interface plasmas of Section 4.6.1 need to be subjected to the same procedure. This implies that the vacuum magnetic field splits into an equilibrium part $\hat{\mathbf{B}}$ (suppressing the subscript 0 again) and a perturbation $\hat{\mathbf{Q}}$. The *equilibrium* vacuum magnetic field $\hat{\mathbf{B}}$ satisfies

$$\nabla \times \hat{\mathbf{B}} = 0, \qquad \nabla \cdot \hat{\mathbf{B}} = 0, \tag{6.126}$$

subject to the boundary conditions

$$\mathbf{n} \cdot \mathbf{B} = \mathbf{n} \cdot \hat{\mathbf{B}} = 0, \qquad [\![p + \tfrac{1}{2} B^2]\!] = 0 \quad \text{(at the interface } S\text{)}, \tag{6.127}$$

which imply surface currents $\mathbf{j}^\star = \mathbf{n} \times [\![\mathbf{B}]\!]$, and to the boundary condition

$$\mathbf{n} \cdot \hat{\mathbf{B}} = 0 \quad \text{(at the conducting wall } \hat{W}\text{)}. \tag{6.128}$$

For model II*, the outer region is also a plasma so that the vacuum equations (6.126) should be replaced by plasma equations, like Eqs. (6.7), for $\hat{\mathbf{B}}$, $\hat{p}$, and $\hat{\rho}$, whereas the jump conditions and the outer b.c. remain the same.

The vacuum magnetic field *perturbations* $\hat{\mathbf{Q}}$ are described by

$$\nabla \times \hat{\mathbf{Q}} = 0, \qquad \nabla \cdot \hat{\mathbf{Q}} = 0, \tag{6.129}$$

subject to two entirely non-trivial boundary conditions (see below) connecting $\hat{\mathbf{Q}}$ to the plasma variable $\boldsymbol{\xi}$ at the plasma–vacuum interface, and one boundary condition

$$\mathbf{n} \cdot \hat{\mathbf{Q}} = 0 \quad \text{(at } \hat{W}\text{)}. \tag{6.130}$$

Of course, for model II* plasmas, Eqs. (6.129) should be replaced by equations analogous to Eqs. (6.23)–(6.25) for the displacement vector $\hat{\boldsymbol{\xi}}$ of the outer plasma, whereas the outer b.c. (6.130) is to be replaced by

$$\mathbf{n} \cdot \hat{\boldsymbol{\xi}} = 0 \quad \text{(at } \hat{W}\text{)}. \tag{6.131}$$

The two mentioned boundary conditions *at the interface*, connecting $\boldsymbol{\xi}$ in the plasma to $\hat{\mathbf{Q}}$ in the vacuum (model II) or to $\hat{\boldsymbol{\xi}}$ in the outer plasma (model II*), are obtained after a rather laborious derivation that deserves separate treatment. This will be given in Section 6.6.1.

Note that, in model II*, the limit of a very tenuous plasma with $\hat{p} = 0$, $\hat{\rho} = 0$, $\hat{\mathbf{j}} = 0$ still implies very different dynamics from that of a vacuum because the MHD magnetic field equation (6.3) implies the picture of frozen field lines, whereas such a condition does not hold for a vacuum magnetic field.

276   *Spectral theory*

▷ **Vector potential formulation for the vacuum field perturbations** For some of the derivations in the present section, it is useful to exploit the alternative representation of the vacuum perturbations in terms of the vector potential. To that end, recall the pre-Maxwell equations (4.171) and (4.172) given in Section 4.6.1. They yield the following expressions for the vacuum field perturbations $\hat{\mathbf{Q}}$ and $\hat{\mathbf{E}}$ in terms of the vector potential $\hat{\mathbf{A}}$ and the scalar potential $\hat{\Phi}$:

$$\hat{\mathbf{Q}} = \nabla \times \hat{\mathbf{A}}, \qquad \nabla \times \nabla \times \hat{\mathbf{A}} = 0, \qquad (6.132)$$

$$\hat{\mathbf{E}} = -\frac{\partial \hat{\mathbf{A}}}{\partial t} - \nabla \hat{\Phi}, \qquad \nabla^2 \hat{\Phi} = -\frac{\partial}{\partial t}(\nabla \cdot \hat{\mathbf{A}}). \qquad (6.133)$$

With the choice of the Coulomb gauge condition $\hat{\Phi} = 0$, which implies that $\nabla \cdot \hat{\mathbf{A}} = 0$ as well, the vector potential $\hat{\mathbf{A}}$ becomes the only variable needed to describe the vacuum. The basic equations for $\hat{\mathbf{A}}$ then become

$$\nabla^2 \hat{\mathbf{A}} = 0, \qquad \nabla \cdot \hat{\mathbf{A}} = 0, \qquad (6.134)$$

from which the vacuum field variables $\hat{\mathbf{Q}}$ and $\hat{\mathbf{E}}$ are obtained by

$$\hat{\mathbf{Q}} = \nabla \times \hat{\mathbf{A}}, \qquad \hat{\mathbf{E}} = -\frac{\partial \hat{\mathbf{A}}}{\partial t}. \qquad (6.135)$$

The appropriate boundary condition at the perfectly conducting wall is obtained from the condition that the tangential electric field has to vanish there (cf. Eq. (4.174)):

$$\mathbf{n} \times \hat{\mathbf{A}} = 0 \qquad (at\ \hat{W}). \qquad (6.136)$$

One easily demonstrates that satisfaction of this boundary condition also implies satisfaction of the boundary condition $\mathbf{n} \cdot \hat{\mathbf{Q}} = \mathbf{n} \cdot (\nabla \times \hat{\mathbf{A}}) = 0$ at the conducting wall.   ◁

6.6.1 Boundary conditions at the interface

The linearization of the two boundary conditions (4.175) and (4.176), for the normal magnetic field and the total pressure, provides the necessary connection between the plasma variable $\boldsymbol{\xi}$ and the vacuum variable $\hat{\mathbf{Q}}$ at the interface. The derivation is rather involved since we need to evaluate the physical variables *at the perturbed boundary* and we also need an expression for the *perturbation of the normal* to that boundary. Because of the importance of the interface boundary conditions for applications, we give their derivation in full and point out some pitfalls in passing. At this point, we temporarily revert to writing again the subscripts 0 and 1 for equilibrium and perturbations.

An expression for the perturbation of the normal is most easily obtained by integrating the Lagrangian time derivative of a line element (4.88), derived in Section 4.3.3, to give the perturbation of a line element moving with the fluid:

$$dl \approx dl_0 \cdot (\mathbf{I} + \nabla \boldsymbol{\xi}). \qquad (6.137)$$

6.6 Extension to interface plasmas

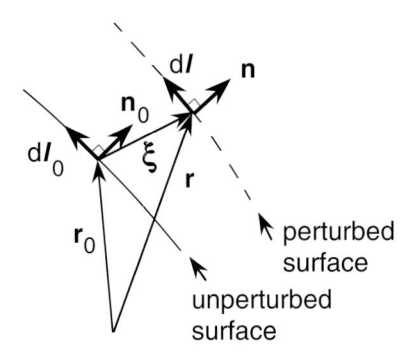

Fig. 6.17. Perturbation of the normal to the boundary.

This relation is correct to first order since the difference between the Eulerian and Lagrangian time derivative is of higher order. From this expression we obtain the following identity for a line element dl lying in the boundary surface (Fig. 6.17):

$$0 = \mathbf{n} \cdot d\mathbf{l} \approx d\mathbf{l}_0 \cdot (\mathbf{I} + \nabla \boldsymbol{\xi}) \cdot (\mathbf{n}_0 + \mathbf{n}_{1L})$$
$$\approx d\mathbf{l}_0 \cdot \mathbf{n}_0 + d\mathbf{l}_0 \cdot (\nabla \boldsymbol{\xi}) \cdot \mathbf{n}_0 + d\mathbf{l}_0 \cdot \mathbf{n}_{1L} = d\mathbf{l}_0 \cdot [\, (\nabla \boldsymbol{\xi}) \cdot \mathbf{n}_0 + \mathbf{n}_{1L} \,] ,$$

where we have put an index L on $\mathbf{n}_{1L}$ to indicate that this is a Lagrangian perturbation. Hence, $\mathbf{n}_{1L} = -(\nabla \boldsymbol{\xi}) \cdot \mathbf{n}_0 + \boldsymbol{\lambda}$, where the vector $\boldsymbol{\lambda} \perp d\mathbf{l}_0$. But $d\mathbf{l}_0$ may have any direction in the unperturbed surface so that $\boldsymbol{\lambda}$ must be parallel to $\mathbf{n}_0$: $\boldsymbol{\lambda} = \mu \mathbf{n}_0$. Since $|\mathbf{n}| = |\mathbf{n}_0| = 1$, we have $\mathbf{n}_0 \cdot \mathbf{n}_{1L} = 0$ so that $\mu = \mathbf{n}_0 \cdot (\nabla \boldsymbol{\xi}) \cdot \mathbf{n}_0$. This provides us with the required *Lagrangian perturbation of the normal*:

$$\mathbf{n}_{1L} = -(\nabla \boldsymbol{\xi}) \cdot \mathbf{n}_0 + \mathbf{n}_0 \mathbf{n}_0 \cdot (\nabla \boldsymbol{\xi}) \cdot \mathbf{n}_0 \stackrel{(A.2)}{=} \mathbf{n}_0 \times \left\{ \mathbf{n}_0 \times [\, (\nabla \boldsymbol{\xi}) \cdot \mathbf{n}_0 \,] \right\}. \quad (6.138)$$

Note that the brackets in $(\nabla \boldsymbol{\xi})$ are absolutely essential since they indicate that the derivative is to be taken on $\boldsymbol{\xi}$ alone and not on any quantity appearing after this expression. In fact, in the first investigation of the Rayleigh–Taylor and kink instabilities of plasmas with a vacuum interface by Kruskal and Schwarzschild [130], the expression for the Lagrangian perturbation of the normal was incorrect in this respect. It was corrected later by Kruskal and Tuck [131] to the above form.

(a) Model II boundary conditions The evaluation of the boundary conditions (4.175) and (4.176) for the normal magnetic field and the total pressure requires the Lagrangian expressions for the perturbed magnetic field $\mathbf{B}$ and the pressure p at the perturbed position $\mathbf{r}$ of the boundary, evaluated to first order:

$$\mathbf{B}|_\mathbf{r} \approx (\mathbf{B}_0 + \mathbf{Q} + \boldsymbol{\xi} \cdot \nabla \mathbf{B}_0)|_{\mathbf{r}_0} ,$$
$$p|_\mathbf{r} \approx (p_0 + \pi + \boldsymbol{\xi} \cdot \nabla p_0)|_{\mathbf{r}_0} = (p_0 - \gamma p_0 \nabla \cdot \boldsymbol{\xi})|_{\mathbf{r}_0} .$$
$$(6.139)$$

Here, $\mathbf{Q}$ and π are the Eulerian perturbations, defined in Eqs. (6.25) and (6.24), and the terms $\boldsymbol{\xi} \cdot \nabla \mathbf{B}_0$ and $\boldsymbol{\xi} \cdot \nabla p_0$ are due to the shift of the boundary.

Inserting Eqs. (6.138) and (6.139) into the first part of the boundary condition (4.175) for the normal magnetic field gives

$$\begin{aligned} 0 &= \mathbf{n} \cdot \mathbf{B} = [\, \mathbf{n}_0 - (\nabla \boldsymbol{\xi}) \cdot \mathbf{n}_0 + \mathbf{n}_0 \mathbf{n}_0 \cdot (\nabla \boldsymbol{\xi}) \cdot \mathbf{n}_0 \,] \cdot (\mathbf{B}_0 + \mathbf{Q} + \boldsymbol{\xi} \cdot \nabla \mathbf{B}_0) \\ &\approx -\mathbf{B}_0 \cdot (\nabla \boldsymbol{\xi}) \cdot \mathbf{n}_0 + \mathbf{n}_0 \cdot \mathbf{Q} + \boldsymbol{\xi} \cdot (\nabla \mathbf{B}_0) \cdot \mathbf{n}_0 \\ &\stackrel{(A.13)}{=} -\mathbf{n}_0 \cdot \nabla \times (\boldsymbol{\xi} \times \mathbf{B}_0) + \mathbf{n}_0 \cdot \mathbf{Q} \,. \end{aligned}$$

Interestingly, this condition is automatically satisfied by virtue of the definition (6.25) for $\mathbf{Q}$. However, exactly the same derivation applies to the second part of the boundary condition (4.175) giving the required *first interface condition* relating $\boldsymbol{\xi}$ and $\hat{\mathbf{Q}}$:

$$\mathbf{n} \cdot \nabla \times (\boldsymbol{\xi} \times \hat{\mathbf{B}}) = \mathbf{n} \cdot \hat{\mathbf{Q}} \qquad (\text{at the plasma–vacuum interface } S)\,, \qquad (6.140)$$

where we now definitively drop the 0s on the equilibrium quantities since confusion is no longer possible.

▷ **Alternative expressions for the first interface condition** That this boundary condition in fact only depends on the normal components of $\boldsymbol{\xi}$ and $\hat{\mathbf{Q}}$ may be shown by one of those tedious vector manipulations that abound in this field (see Freidberg and Haas [73]), giving:

$$\hat{\mathbf{B}} \cdot \nabla(\mathbf{n} \cdot \boldsymbol{\xi}) - \mathbf{n} \cdot (\nabla \hat{\mathbf{B}}) \cdot \mathbf{n}\, \mathbf{n} \cdot \boldsymbol{\xi} = \mathbf{n} \cdot \hat{\mathbf{Q}} \qquad (\text{at } S)\,. \qquad (6.141)$$

This form is actually to be preferred over (6.140) as it directly gives the relation between $\mathbf{n} \cdot \boldsymbol{\xi}$ and $\mathbf{n} \cdot \hat{\mathbf{Q}}$.

Another alternative expression for the first interface condition is obtained by exploiting the vector potential $\hat{\mathbf{A}}$ instead of $\hat{\mathbf{Q}}$. To that end, recall the exact model II boundary condition (4.178) in terms of the vacuum electric field $\hat{\mathbf{E}}$ and the plasma velocity $\mathbf{v}$ that was derived in Section 4.6.1. Expressing $\hat{\mathbf{E}}$ in terms of $\hat{\mathbf{A}}$ by means of Eq. (6.135), and $\mathbf{v}$ in terms of $\boldsymbol{\xi}$, this condition may be integrated to first order to yield the first interface condition in terms of the vector potential:

$$\mathbf{n} \cdot \boldsymbol{\xi}\, \hat{\mathbf{B}} = -\mathbf{n} \times \hat{\mathbf{A}} \qquad (\text{at } S)\,. \qquad (6.142)$$

This form will be exploited in Section 6.6.2. ◁

Inserting Eqs. (6.139) into the boundary condition (4.176) for the total pressure, and exploiting the equilibrium equation (6.127) to remove the equilibrium boundary contributions, leads to the *second interface condition* relating $\boldsymbol{\xi}$ and $\hat{\mathbf{Q}}$:

$$-\gamma p \nabla \cdot \boldsymbol{\xi} + \mathbf{B} \cdot \mathbf{Q} + \boldsymbol{\xi} \cdot \nabla(\tfrac{1}{2} B^2) = \hat{\mathbf{B}} \cdot \hat{\mathbf{Q}} + \boldsymbol{\xi} \cdot \nabla(\tfrac{1}{2} \hat{B}^2) \qquad (\text{at } S)\,. \qquad (6.143)$$

Note that the left hand side is just the Lagrangian perturbation of the total plasma pressure.

6.6 Extension to interface plasmas

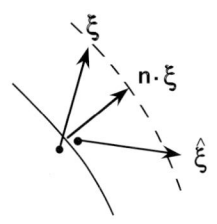

Fig. 6.18. Discontinuity of the tangential displacement.

For model II *plasma–vacuum interface systems*, the formulation of the spectral problem is now complete. The equation of motion (6.23) for $\boldsymbol{\xi}$, the Eqs. (6.129) and (6.130) for $\hat{\mathbf{Q}}$, and the boundary conditions (6.140) and (6.143) connecting $\boldsymbol{\xi}$ and $\hat{\mathbf{Q}}$ at the plasma–vacuum interface constitute a complete set of equations for the investigation of waves and stability properties of these systems.

(b) Model II* boundary conditions For *plasma–plasma interface systems*, extra care is needed in the derivation of the boundary conditions. In that case, $\hat{\boldsymbol{\xi}}$ is also defined for the outer plasma and $\hat{\mathbf{Q}} \equiv \nabla \times (\hat{\boldsymbol{\xi}} \times \hat{\mathbf{B}})$ there so that the condition (6.140) is superseded by the *first interface condition* on $\boldsymbol{\xi}$ and $\hat{\boldsymbol{\xi}}$:

$$\mathbf{n} \cdot \boldsymbol{\xi} = \mathbf{n} \cdot \hat{\boldsymbol{\xi}} \qquad \text{(at the plasma–plasma interface } S\text{)}, \qquad (6.144)$$

which is obtained directly from linearization of Eq. (4.165).

For the pressure balance equation one has to add the pressure terms of the exterior fluid to the boundary condition (6.143). One may then be tempted to infer from the continuity of the Lagrangian perturbation of the total pressure that the RHS of the boundary condition should be just the same expression as the LHS of Eq. (6.143) with $\boldsymbol{\xi}$, $\mathbf{Q}$, p, and $\mathbf{B}$ replaced by $\hat{\boldsymbol{\xi}}$, $\hat{\mathbf{Q}}$, $\hat{p}$, and $\hat{\mathbf{B}}$. In fact, such a regrettable mistake has been made in the literature (Goedbloed [81](I), corrected in [81](IV)). The point is that the two mentioned expressions do not refer to the same position on the interface since the tangential components of $\boldsymbol{\xi}$ are not continuous in general (Fig. 6.18). For the sake of symmetry between inner and exterior fluid it is, therefore, to be preferred to express the perturbation on the perturbed boundary at the position $\mathbf{r}_0 + (\mathbf{n} \cdot \boldsymbol{\xi})\mathbf{n}$, since the normal component of $\boldsymbol{\xi}$ is continuous. The expressions for the perturbations of the plasma pressure and the magnetic field pressure at that position read:

$$\Delta p = \pi + \mathbf{n} \cdot \boldsymbol{\xi} \, \mathbf{n} \cdot \nabla p,$$
$$\Delta(\tfrac{1}{2} B^2) = \mathbf{B} \cdot \mathbf{Q} + \mathbf{n} \cdot \boldsymbol{\xi} \, \mathbf{n} \cdot \nabla(\tfrac{1}{2} B^2), \qquad (6.145)$$

so that

$$\Delta(p + \tfrac{1}{2}B^2) = -\gamma p \nabla \cdot \boldsymbol{\xi} - \boldsymbol{\xi} \cdot \nabla p + \mathbf{B} \cdot \mathbf{Q} + \mathbf{n} \cdot \boldsymbol{\xi} \mathbf{n} \cdot \nabla(p + \tfrac{1}{2}B^2). \quad (6.146)$$

The *second interface condition* on $\boldsymbol{\xi}$ and $\hat{\boldsymbol{\xi}}$ then becomes nicely symmetric:

$$-\gamma p \nabla \cdot \boldsymbol{\xi} + \mathbf{B} \cdot \mathbf{Q} - \boldsymbol{\xi}_t \cdot \nabla p + \mathbf{n} \cdot \boldsymbol{\xi} \mathbf{n} \cdot \nabla(\tfrac{1}{2}B^2)$$
$$= -\gamma \hat{p} \nabla \cdot \hat{\boldsymbol{\xi}} + \hat{\mathbf{B}} \cdot \hat{\mathbf{Q}} - \hat{\boldsymbol{\xi}}_t \cdot \nabla \hat{p} + \mathbf{n} \cdot \hat{\boldsymbol{\xi}} \mathbf{n} \cdot \nabla(\tfrac{1}{2}\hat{B}^2) \quad \text{(at } S\text{),} \quad (6.147)$$

where $\boldsymbol{\xi}_t \equiv \boldsymbol{\xi} - \mathbf{n} \cdot \boldsymbol{\xi} \mathbf{n}$, and similarly for $\hat{\boldsymbol{\xi}}_t$.

Note that the difference between the correct boundary condition (6.147) and the incorrect one, obtained by assuming continuity of the Lagrangian total pressure perturbation $-\gamma p \nabla \cdot \boldsymbol{\xi} + \mathbf{B} \cdot \mathbf{Q} + \boldsymbol{\xi} \cdot \nabla(\tfrac{1}{2}B^2)$, is a jump $-[\![\boldsymbol{\xi}_t \cdot \nabla(p + \tfrac{1}{2}B^2)]\!]$. Although $[\![\boldsymbol{\xi}_t]\!] \neq 0$, this term would still vanish if $p + \tfrac{1}{2}B^2$ were constant on the interface. For the simple 1D geometries considered in the first part of this book, this is the case. However, for toroidal geometries of laboratory plasmas and oblique gravitational acceleration of astrophysical plasmas, the term does not disappear. This completes the formulation of the spectral problem for model II* interface plasmas.

6.6.2 Self-adjointness for interface plasmas

We will now extend the proof of self-adjointness of the force operator **F** to plasmas with an interface. Recall the proof of Section 6.2.3 for model I plasmas, which resulted in the expression (6.56) with a boundary term that we purposely kept for the present reduction. That term will be transformed now to a symmetric expression, both manifesting self-adjointness (this section) and immediately resulting in the necessary extensions of the energy expression (6.85) for model II interface plasmas in the next section.

As in Section 6.2.2, consider again two displacement vector fields $\boldsymbol{\xi}(\mathbf{r}, t)$ and $\boldsymbol{\eta}(\mathbf{r}, t)$ defined over the plasma volume V, not necessarily satisfying the ideal MHD equation of motion (6.23). These vector fields will be connected by means of the boundary conditions (6.140) and (6.143) to the associated magnetic perturbations $\hat{\mathbf{Q}}(\mathbf{r}, t)$ and $\hat{\mathbf{R}}(\mathbf{r}, t)$, defined over the vacuum volume $\hat{V}$, that *do* satisfy the vacuum equations (6.129) and the b.c. (6.130) on the wall W. Hence, the 'extensions' $\hat{\mathbf{Q}}$ and $\hat{\mathbf{R}}$ satisfy the following differential equations:

$$\nabla \times \hat{\mathbf{Q}} = 0, \quad \nabla \cdot \hat{\mathbf{Q}} = 0,$$
$$\nabla \times \hat{\mathbf{R}} = 0, \quad \nabla \cdot \hat{\mathbf{R}} = 0 \quad (\text{on } \hat{V}), \quad (6.148)$$

6.6 Extension to interface plasmas

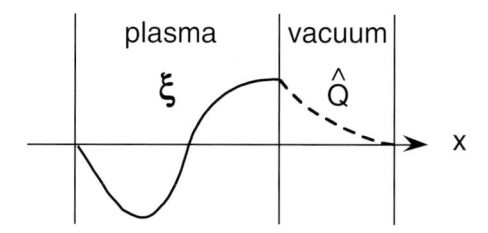

Fig. 6.19. 'Extension' of the perturbation into the vacuum.

subject to the boundary conditions

$$\mathbf{n} \cdot \nabla \times (\boldsymbol{\xi} \times \hat{\mathbf{B}}) = \mathbf{n} \cdot \hat{\mathbf{Q}},$$
$$\mathbf{n} \cdot \nabla \times (\boldsymbol{\eta} \times \hat{\mathbf{B}}) = \mathbf{n} \cdot \hat{\mathbf{R}} \qquad (on\ S), \quad (6.149)$$

$$-\gamma p \nabla \cdot \boldsymbol{\xi} + \mathbf{B} \cdot \mathbf{Q} + \boldsymbol{\xi} \cdot \nabla(\tfrac{1}{2} B^2) = \hat{\mathbf{B}} \cdot \hat{\mathbf{Q}} + \boldsymbol{\xi} \cdot \nabla(\tfrac{1}{2} \hat{B}^2),$$
$$-\gamma p \nabla \cdot \boldsymbol{\eta} + \mathbf{B} \cdot \mathbf{R} + \boldsymbol{\eta} \cdot \nabla(\tfrac{1}{2} B^2) = \hat{\mathbf{B}} \cdot \hat{\mathbf{R}} + \boldsymbol{\eta} \cdot \nabla(\tfrac{1}{2} \hat{B}^2)\ (on\ S), \quad (6.150)$$

$$\mathbf{n} \cdot \hat{\mathbf{Q}} = 0,$$
$$\mathbf{n} \cdot \hat{\mathbf{R}} = 0 \qquad (on\ \hat{W}), \quad (6.151)$$

Recall that $\mathbf{Q} \equiv \nabla \times (\boldsymbol{\xi} \times \mathbf{B})$ and $\mathbf{R} \equiv \nabla \times (\boldsymbol{\eta} \times \mathbf{B})$ in the LHSs of Eqs. (6.150).

The idea of the relations (6.148)–(6.151) is to 'extend' the function $\boldsymbol{\xi}$ into the vacuum by means of the magnetic field variable $\hat{\mathbf{Q}}$, and likewise to 'extend' $\boldsymbol{\eta}$ by means of $\hat{\mathbf{R}}$, by matching something like the function value and the normal derivative at the plasma–vacuum interface. This is schematically indicated in Fig. 6.19. It is a very remarkable property of ideal MHD that only two conditions need to be satisfied to connect two vector fields $\boldsymbol{\xi}$ and $\hat{\mathbf{Q}}$. Hence, it appears that we are dealing only with ordinary second order differential equations. The reason behind this is the extreme anisotropy of ideal MHD as regards motion inside and across the magnetic surfaces, to the study of which we will turn later.

The quadratic form (6.56) was derived in Section 6.2.3 without invoking the solid wall boundary conditions so that it remains valid for model II and model II* interface plasmas. We repeat it here for convenience:

$$\int \boldsymbol{\eta} \cdot \mathbf{F}(\boldsymbol{\xi})\, dV = -\int \Big\{ \gamma p \nabla \cdot \boldsymbol{\xi}\, \nabla \cdot \boldsymbol{\eta} + \mathbf{Q} \cdot \mathbf{R} + \tfrac{1}{2} \nabla p \cdot (\boldsymbol{\xi} \nabla \cdot \boldsymbol{\eta} + \boldsymbol{\eta} \nabla \cdot \boldsymbol{\xi})$$
$$+ \tfrac{1}{2} \mathbf{j} \cdot (\boldsymbol{\xi} \times \mathbf{R} + \boldsymbol{\eta} \times \mathbf{Q}) - \tfrac{1}{2} \nabla \Phi \cdot \big[\boldsymbol{\xi} \nabla \cdot (\rho \boldsymbol{\eta}) + \boldsymbol{\eta} \nabla \cdot (\rho \boldsymbol{\xi}) \big] \Big\} dV$$
$$+ \int \mathbf{n} \cdot \boldsymbol{\eta} \left(\gamma p \nabla \cdot \boldsymbol{\xi} + \boldsymbol{\xi} \cdot \nabla p - \mathbf{B} \cdot \mathbf{Q} \right) dS. \quad (6.152)$$

▷ **Transformation of the surface integral** By the application of the second interface condition (6.143), the surface integral may be transformed as follows:

$$\int \mathbf{n} \cdot \boldsymbol{\eta} \, (\gamma p \nabla \cdot \boldsymbol{\xi} + \boldsymbol{\xi} \cdot \nabla p - \mathbf{B} \cdot \mathbf{Q}) \, dS$$

$$= -\int \mathbf{n} \cdot \boldsymbol{\eta} \boldsymbol{\xi} \cdot [\![\nabla (p + \tfrac{1}{2} B^2)]\!] \, dS - \int \mathbf{n} \cdot \boldsymbol{\eta} \hat{\mathbf{B}} \cdot \hat{\mathbf{Q}} \, dS$$

$$= -\int \mathbf{n} \cdot \boldsymbol{\eta} \mathbf{n} \cdot \boldsymbol{\xi} \mathbf{n} \cdot [\![\nabla (p + \tfrac{1}{2} B^2)]\!] \, dS - \int \mathbf{n} \cdot \boldsymbol{\eta} \hat{\mathbf{B}} \cdot \hat{\mathbf{Q}} \, dS . \quad (6.153)$$

For the last step we used the equilibrium jump condition $[\![p + \tfrac{1}{2} B^2]\!] = 0$, which implies that the tangential derivative of the jump vanishes as well: $\mathbf{t} \cdot [\![\nabla (p + \tfrac{1}{2} B^2)]\!] = 0$, where $\mathbf{t}$ is an arbitrary unit vector tangential to the interface.

Next, we transform the term $-\int \mathbf{n} \cdot \boldsymbol{\eta} \hat{\mathbf{B}} \cdot \hat{\mathbf{Q}} \, dS$. Introducing vector potentials in the vacuum, $\hat{\mathbf{Q}} \equiv \nabla \times \hat{\mathbf{A}}$, $\hat{\mathbf{R}} \equiv \nabla \times \hat{\mathbf{C}}$, and exploiting the first interface condition (6.142) in terms of the vector potential $\hat{\mathbf{C}}$, i.e. $\mathbf{n} \cdot \boldsymbol{\eta} \hat{\mathbf{B}} = -\mathbf{n} \times \hat{\mathbf{C}}$, we get:

$$-\int \mathbf{n} \cdot \boldsymbol{\eta} \hat{\mathbf{B}} \cdot \hat{\mathbf{Q}} \, dS = \int \mathbf{n} \times \hat{\mathbf{C}} \cdot \hat{\mathbf{Q}} \, dS = \int \mathbf{n} \times \hat{\mathbf{C}} \cdot \nabla \times \hat{\mathbf{A}} \, dS$$

$$\stackrel{(A.1)}{=} -\int (\nabla \times \hat{\mathbf{A}}) \times \hat{\mathbf{C}} \cdot \mathbf{n} \, dS \stackrel{(A.14)}{=} \int \nabla \cdot \left[(\nabla \times \hat{\mathbf{A}}) \times \hat{\mathbf{C}} \right] d\hat{V}$$

$$\stackrel{(A.12)}{=} \int \left[\hat{\mathbf{C}} \cdot \nabla \times \nabla \times \hat{\mathbf{A}} - \nabla \times \hat{\mathbf{A}} \cdot \nabla \times \hat{\mathbf{C}} \right] d\hat{V}$$

$$\stackrel{(6.132)}{=} -\int \nabla \times \hat{\mathbf{A}} \cdot \nabla \times \hat{\mathbf{C}} \, d\hat{V} = -\int \hat{\mathbf{Q}} \cdot \hat{\mathbf{R}} \, d\hat{V} . \quad (6.154)$$

Here, a minus sign appears in the conversion of the surface term to the volume term because $\hat{V}$ is located outside S, and the contribution over W could be added for free since it vanishes by virtue of the boundary condition (6.136). ◁

Collecting terms yields

$$\int \boldsymbol{\eta} \cdot \mathbf{F}(\boldsymbol{\xi}) \, dV = -\int \Big\{ \gamma p \nabla \cdot \boldsymbol{\xi} \nabla \cdot \boldsymbol{\eta} + \mathbf{Q} \cdot \mathbf{R} + \tfrac{1}{2} \nabla p \cdot (\boldsymbol{\xi} \nabla \cdot \boldsymbol{\eta} + \boldsymbol{\eta} \nabla \cdot \boldsymbol{\xi})$$

$$+ \tfrac{1}{2} \mathbf{j} \cdot (\boldsymbol{\xi} \times \mathbf{R} + \boldsymbol{\eta} \times \mathbf{Q}) - \tfrac{1}{2} \nabla \Phi \cdot \left[\boldsymbol{\xi} \nabla \cdot (\rho \boldsymbol{\eta}) + \boldsymbol{\eta} \nabla \cdot (\rho \boldsymbol{\xi}) \right] \Big\} dV$$

$$- \int \mathbf{n} \cdot \boldsymbol{\eta} \mathbf{n} \cdot \boldsymbol{\xi} \mathbf{n} \cdot [\![\nabla (p + \tfrac{1}{2} B^2)]\!] \, dS - \int \hat{\mathbf{Q}} \cdot \hat{\mathbf{R}} \, d\hat{V} , \quad (6.155)$$

which is symmetric in the variables $\boldsymbol{\xi}$ and $\boldsymbol{\eta}$, and their 'extensions' $\hat{\mathbf{Q}}$ and $\hat{\mathbf{R}}$; QED.

6.6.3 Extended variational principles

Analogous to the procedure of Sections 6.2.3 and 6.4.1, and again exploiting complex-type inner products, we immediately obtain a meaningful expression for the potential energy W of interface plasmas by identifying $\boldsymbol{\xi}$ and $\boldsymbol{\eta}$, and $\hat{\mathbf{Q}}$ and $\hat{\mathbf{R}}$, in the quadratic form (6.155):

$$W[\boldsymbol{\xi}, \hat{\mathbf{Q}}] = -\tfrac{1}{2} \int \boldsymbol{\xi}^* \cdot \mathbf{F}(\boldsymbol{\xi}) \, dV = W^p[\boldsymbol{\xi}] + W^s[\xi_n] + W^v[\hat{\mathbf{Q}}], \qquad (6.156)$$

where

$$W^p[\boldsymbol{\xi}] = \tfrac{1}{2} \int \Big[\gamma p \, |\nabla \cdot \boldsymbol{\xi}|^2 + |\mathbf{Q}|^2 + (\boldsymbol{\xi}^* \cdot \nabla p) \nabla \cdot \boldsymbol{\xi} + \mathbf{j} \cdot \boldsymbol{\xi}^* \times \mathbf{Q}$$
$$- (\boldsymbol{\xi}^* \cdot \nabla \Phi) \nabla \cdot (\rho \boldsymbol{\xi}) \Big] dV, \qquad (6.157)$$

$$W^s[\xi_n] = \tfrac{1}{2} \int |\mathbf{n} \cdot \boldsymbol{\xi}|^2 \, \mathbf{n} \cdot [\![\nabla(p + \tfrac{1}{2}B^2)]\!] \, dS, \qquad (6.158)$$

$$W^v[\hat{\mathbf{Q}}] = \tfrac{1}{2} \int |\hat{\mathbf{Q}}|^2 \, d\hat{V}. \qquad (6.159)$$

This shows that the work done against the force $\mathbf{F}$ leads to an increase of the potential energy W^p of the plasma proper, the potential energy W^s of the plasma–vacuum surface, and the potential energy W^v of the vacuum: very plausible indeed.

However, how to exploit the expression for $W[\boldsymbol{\xi}, \hat{\mathbf{Q}}]$? Obviously, if we were to minimize it by exploiting trial functions $\boldsymbol{\xi}$ with vacuum 'extensions' $\hat{\mathbf{Q}}$, that would have to satisfy the differential equations (6.129) and be subject to the interface conditions (6.140) and (6.143) on S and to the boundary condition (6.130) on $\hat{W}$, we would have obtained a very awkward (asymmetrical) variational principle. Here is how the asymmetry between the use of the variable $\boldsymbol{\xi}$ and the variable $\hat{\mathbf{Q}}$ is removed and how life for a physicist becomes pleasant again.

Extended spectral variational principle Eigenfunctions $\boldsymbol{\xi}$ of the operator $\rho^{-1}\mathbf{F}$ with their vacuum 'extension' $\hat{\mathbf{Q}}$ make the Rayleigh quotient

$$\Lambda[\boldsymbol{\xi}, \hat{\mathbf{Q}}] \equiv \frac{W[\boldsymbol{\xi}, \hat{\mathbf{Q}}]}{I[\boldsymbol{\xi}]} \equiv \frac{W^p[\boldsymbol{\xi}] + W^s[\xi_n] + W^v[\hat{\mathbf{Q}}]}{\tfrac{1}{2} \int \rho |\boldsymbol{\xi}|^2 \, dV} \qquad (6.160)$$

stationary; the eigenvalues ω^2 *are the stationary values of* Λ. *Here,* $\boldsymbol{\xi}$ *and* $\hat{\mathbf{Q}}$ *have* to satisfy the following *essential boundary conditions*:

(1) $\boldsymbol{\xi}$ regular on V, [or $\mathbf{n} \cdot \boldsymbol{\xi} = 0$ (*on interior wall W*)],   (6.161)

(2) $\mathbf{n} \cdot \nabla \times (\boldsymbol{\xi} \times \hat{\mathbf{B}}) = \mathbf{n} \cdot \hat{\mathbf{Q}}$ (*on S*),   (6.162)

(3) $\mathbf{n} \cdot \hat{\mathbf{Q}} = 0$ (*on exterior wall $\hat{W}$*).   (6.163)

The regularity condition (6.161), already mentioned in the footnote of Section 6.2.1, has been added to complete the number of conditions necessary to fix solutions. For example, in cylindrical plasmas, this condition eliminates solutions that diverge at $r = 0$. Such solutions are also eliminated by removing these points from the physical domain by putting an internal wall there (like the fixed rod exploited in 'hard-core' pinches).

▷ **Proof** Let ω^2 be a stationary value of $\Lambda[\boldsymbol{\xi}, \hat{\mathbf{Q}}]$, so that

$$\delta \Lambda = \frac{\delta W^p + \delta W^s + \delta W^v - \Lambda \delta I}{I} = 0.$$

Working out the variational expressions in terms of the variations $\delta \boldsymbol{\xi}$ of $\boldsymbol{\xi}$ and $\delta \hat{\mathbf{A}}$ of the vector potential $\hat{\mathbf{A}}$ just amounts to transforming the quadratic forms in reverse order with respect to that of Section 6.6.2, imposing the essential boundary conditions (6.161)–(6.163):

$$\delta W^p - \Lambda \delta I$$
$$= - \int \delta \boldsymbol{\xi}^* \cdot \left[\mathbf{F}(\boldsymbol{\xi}) + \rho \omega^2 \boldsymbol{\xi} \right] dV - \int \delta \xi_n^* \left(\gamma p \nabla \cdot \boldsymbol{\xi} + \boldsymbol{\xi} \cdot \nabla p - \mathbf{B} \cdot \mathbf{Q} \right) dS,$$
(6.164)

$$\delta W^s = \int \delta \xi_n^* \, \xi_n \, \mathbf{n} \cdot [\![\nabla (p + \tfrac{1}{2} B^2)]\!] \, dS,$$
(6.165)

$$\delta W^v = - \int \delta \hat{\mathbf{A}}^* \cdot \nabla \times \nabla \times \hat{\mathbf{A}} \, d\hat{V} + \int \delta \xi_n^* \, \hat{\mathbf{B}} \cdot \hat{\mathbf{Q}} \, dS.$$
(6.166)

Since $\delta \boldsymbol{\xi}$ and $\delta \hat{\mathbf{A}}$ are arbitrary variations of $\boldsymbol{\xi}$ and $\hat{\mathbf{A}}$, minimization of $\Lambda[\boldsymbol{\xi}, \hat{\mathbf{Q}}]$ leads to *three Euler equations*, viz. the equation of motion from the volume part of $\delta W^p - \Lambda \delta I$, the vacuum differential equations from the volume part of δW^v, and the second interface condition from the sum of the remaining surface integrals:

$$\mathbf{F}(\boldsymbol{\xi}) = -\rho \omega^2 \boldsymbol{\xi}$$ (*on V*),   (6.167)

$$\nabla \times \hat{\mathbf{Q}} = 0, \qquad \nabla \cdot \hat{\mathbf{Q}} = 0$$ (*on $\hat{V}$*),   (6.168)

$$-\gamma p \nabla \cdot \boldsymbol{\xi} + \mathbf{B} \cdot \mathbf{Q} + \boldsymbol{\xi} \cdot \nabla (\tfrac{1}{2} B^2) = \hat{\mathbf{B}} \cdot \hat{\mathbf{Q}} + \boldsymbol{\xi} \cdot \nabla (\tfrac{1}{2} \hat{B}^2)$$ (*on S*).   (6.169)

Hence, the variational formulation (6.160) plus the essential boundary conditions (6.161)–(6.163) is fully equivalent to the spectral formulation in terms of the differential equations (6.167)–(6.168), subject to the essential boundary conditions plus the second interface condition (6.169); QED.   ◁

We have now reached the goal of symmetrizing the variational principle with respect to the use of $\boldsymbol{\xi}$ and $\hat{\mathbf{Q}}$. In this manner, the differential equations for the vacuum and the second interface condition do not need to be imposed separately; they have been absorbed in the form of $W[\boldsymbol{\xi}, \hat{\mathbf{Q}}]$ and are automatically satisfied upon minimization. For that reason the second interface condition (6.169) is called a *natural boundary condition*. The distinction between essential and natural boundary conditions is of fundamental importance since it is connected with *counting* the number of equations and unknowns. It will return when we discuss numerical methods of solving the MHD equations by means of Galerkin methods in the companion Volume 2.

For the investigation of stability, the spectral variational principle can again be simplified considerably by dropping the requirement that the functions $\boldsymbol{\xi}$ are normalized by means of I. As shown in Section 6.4.4, the central question of stability is just the determination of the sign of W so that any normalization will do. Hence, we are led to the following formulation of the stability problem.

Extended energy principle for stability A plasma–vacuum interface equilibrium is stable if (sufficient) and only if (necessary)

$$W[\boldsymbol{\xi}, \hat{\mathbf{Q}}] > 0 \tag{6.170}$$

for all trial functions $\boldsymbol{\xi}(\mathbf{r})$ in the plasma, that are bound in norm, 'extended' with trial functions $\hat{\mathbf{Q}}(\mathbf{r})$ in the vacuum, satisfying the boundary conditions (6.161)–(6.163).

The extended energy principle can be modified with the σ-stability contribution, as in Section 6.5.3: $W^\sigma[\boldsymbol{\xi}, \hat{\mathbf{Q}}] \equiv W[\boldsymbol{\xi}, \hat{\mathbf{Q}}] + \sigma^2 I[\boldsymbol{\xi}]$. This yields the *extended σ-stability principle*.

The three different methods of stability analysis discussed in Section 6.4.4 again apply with the extended expression (6.156) for $W[\boldsymbol{\xi}, \hat{\mathbf{Q}}]$. We will demonstrate them by completely solving a particular problem of gravitational stability in the next section. In order to simplify that analysis as much as possible, we will exploit an incompressible plasma model. This choice implies one more *counting* problem, associated with the incompressibility condition. We wish to settle this before becoming immersed in the detailed explicit calculations.

Incompressibility In the energy principle, one can always test for stability with respect to the restricted class of *incompressible displacements* $\nabla \cdot \boldsymbol{\xi} = 0$. That is something else than considering an *incompressible plasma model*. Since such a plasma admits only incompressible displacements, one might be tempted to simply pose $\nabla \cdot \boldsymbol{\xi} = 0$ and to drop the contribution $-\gamma p \nabla \cdot \boldsymbol{\xi}$ from the pressure perturbation π, defined in Eq. (6.24), entering the equation of motion (6.23). However, that would lead to an overdetermined system of equations for the three components of $\boldsymbol{\xi}$! The problem is that the ratio of specific heats, γ, diverges for incompressible fluids. A consistent procedure to restore the required freedom in the dynamics of incompressible plasmas is, therefore, to apply the two limits $\gamma \to \infty$ and $\nabla \cdot \boldsymbol{\xi} \to 0$ simultaneously in such a way that the Lagrangian pressure perturbation

$$\pi_L \equiv -\gamma p \nabla \cdot \boldsymbol{\xi} \quad \text{is finite} . \tag{6.171}$$

Eq. (6.24) for the Eulerian pressure perturbation π ($\equiv \pi_E$) then has to be replaced by $\pi = \pi_L - \boldsymbol{\xi} \cdot \nabla p$, so that an additional free variable π_L appears in the equation of motion. Hence, for incompressible plasmas, Eq. (6.23) is to be replaced by

$$\begin{aligned} \mathbf{F}_{\text{inc}}(\boldsymbol{\xi}) \equiv &- \nabla \pi_L + \nabla(\boldsymbol{\xi} \cdot \nabla p) - \mathbf{B} \times (\nabla \times \mathbf{Q}) \\ &+ (\nabla \times \mathbf{B})\mathbf{Q} + (\nabla \Phi) \nabla \rho \cdot \boldsymbol{\xi} = \rho \frac{\partial^2 \boldsymbol{\xi}}{\partial t^2} , \end{aligned} \tag{6.172}$$

where application of the condition

$$\nabla \cdot \boldsymbol{\xi} = 0 \tag{6.173}$$

implicitly determines the free variable π_L. For incompressible interface plasmas, there is one additional adjustment, viz. that the second interface condition (6.143) is to be replaced by

$$\pi_L + \mathbf{B} \cdot \mathbf{Q} + \boldsymbol{\xi} \cdot \nabla(\tfrac{1}{2} B^2) = \hat{\mathbf{B}} \cdot \hat{\mathbf{Q}} + \boldsymbol{\xi} \cdot \nabla(\tfrac{1}{2} \hat{B}^2) \quad \text{(at } S\text{)}, \tag{6.174}$$

where the boundary value of π_L is determined by the solutions of the plasma equations (6.172) and (6.173).

The equivalent variational principle for incompressible interface plasmas departs from an expression for the Rayleigh quotient $\Lambda[\boldsymbol{\xi}, \hat{\mathbf{Q}}]$ as in Eq. (6.160), where the expression (6.157) for the plasma energy is to be replaced by

$$W^p_{\text{inc}}[\boldsymbol{\xi}] = \tfrac{1}{2} \int \left[|\mathbf{Q}|^2 + \mathbf{j} \cdot \boldsymbol{\xi}^* \times \mathbf{Q} - (\boldsymbol{\xi}^* \cdot \nabla \Phi) \nabla \rho \cdot \boldsymbol{\xi} \right] dV , \tag{6.175}$$

and minimization of Λ should be subject to the (unchanged) boundary conditions (6.161)–(6.163) plus the incompressibility constraint (6.173). At this point, the simplification due to the incompressibility assumption is evident. However, where does the Lagrangian pressure variable π_L reside in this formulation? It enters when one minimizes Λ subject to a constraint, viz. Eq. (6.173). This brings in a Lagrange multiplier which plays the role of π_L, and which is determined by the minimization. As before, it is not necessary to subject the trial functions to the second interface condition (6.174) but functions satisfying it naturally emerge as a result of the minimization, where the boundary value of π_L is found from the Lagrange multiplier.

6.6.4 Application to the Rayleigh–Taylor instability

We now apply the extended energy principle to the problem of the gravitational instability of a magnetized plasma that is supported from below by a vacuum magnetic field. Hence, we extend the problem of the ordinary gravitational stability of an inverted glass of water (Sections 6.1.1 and 6.5.4) to a genuine plasma with an interface. The gravitational instability for interface fluids is called the *Rayleigh–Taylor instability*. We will use the same term for the generalized instability of magnetized plasmas. This presents a model problem for plasma confinement with a clear separation of inner plasma and outer vacuum, where instabilities are preferentially localized at the interface (so-called free-boundary or surface instabilities). Of course, for laboratory plasmas, the driving force of the instability, gravity, is negligible, but very similar instabilities arise in a plasma with a curved magnetic field at the interface, as shown in the pioneering investigation by Kruskal and Schwarzschild [130]. Thus, the Rayleigh–Taylor instability of magnetized plasmas allows us to discuss a number of basic concepts, like *interchange instability*, stabilization by *magnetic shear*, and *wall stabilization*. On the other hand, gravitational instabilities do arise in a wide class of astrophysical situations, notably the *Parker instability* [175] in galactic plasmas (Section 7.3.3). Hence, it pays to carefully study the different steps in the analysis of its most simple representation.

In contrast to Section 6.5.4, we now assume constant density ρ_0 in the plasma and neglect compressibility, but allow for a constant uni-directional magnetic field **B** (no current), on top of a vacuum also with a constant uni-directional magnetic field $\hat{\mathbf{B}}$, but pointing in a different direction (Fig. 6.20). The plasma occupies a

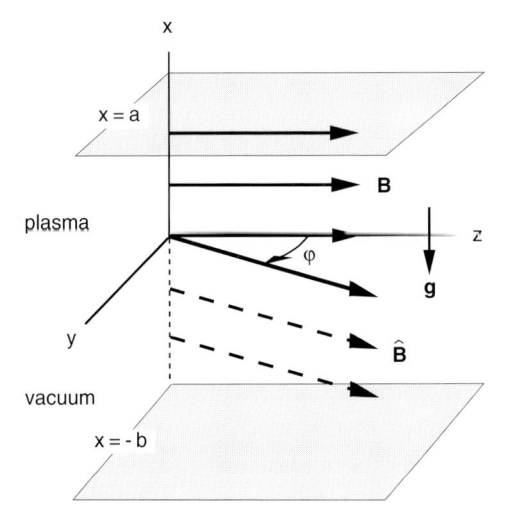

Fig. 6.20. Gravitating plasma slab supported from below by a vacuum magnetic field with conducting walls at $x = a$ and $x = -b$.

horizontal layer ($0 \leq x \leq a$) that is infinitely extended in the y and z directions, but vertically confined between a conducting wall at $x = a$ and a vacuum below. Likewise, the vacuum occupies a horizontal layer ($-b \leq x \leq 0$) confined between the plasma above and a conducting wall below at $x = -b$. The gravitational acceleration is constant and points downward in the vertical direction: $\mathbf{g} = -\nabla \Phi = -g\mathbf{e}_x$.

Gravitational equilibrium is chosen to be due to balance by the pressure gradient alone, so that we have the following quantities in the plasma volume V:

$$\rho = \rho_0, \qquad \mathbf{B} = B_0 \mathbf{e}_z, \qquad \mathbf{j} = 0,$$
$$\nabla p = -\rho \nabla \Phi \quad \Rightarrow \quad p' = -\rho_0 g \quad \Rightarrow \quad p = p_0 - \rho_0 g x \quad (\text{where } p_0 \geq \rho_0 g a);$$
(6.176)

pressure balance at the plasma–vacuum interface S yields

$$p_0 + \tfrac{1}{2} B_0^2 = \tfrac{1}{2} \hat{B}_0^2; \tag{6.177}$$

and the magnetic field in the vacuum volume $\hat{V}$ is described by

$$\hat{\mathbf{B}} = \hat{B}_0 (\sin \varphi \, \mathbf{e}_y + \cos \varphi \, \mathbf{e}_z), \tag{6.178}$$

where φ is the angle between $\hat{\mathbf{B}}$ and $\mathbf{B}$. Recall that the jump in the direction and magnitude of the magnetic field at the interface implies that a surface current flows there:

$$\mathbf{j}^\star = \mathbf{n} \times [\![\mathbf{B}]\!] = \mathbf{e}_x \times (\mathbf{B}_0 - \hat{\mathbf{B}}_0), \tag{6.179}$$

where the unit normal $\mathbf{n}$ is defined to point into the vacuum, so that $\mathbf{n} = -\mathbf{e}_x$.

(a) Reduction to a one-dimensional problem We assume the plasma to be incompressible since this permits the use of the simplified expression (6.175) for the plasma energy derived in Section 6.6.3 and to illustrate how the incompressibility constraint is handled in practice. Some of the more subtle gravitational mechanisms involving compressibility have already been encountered in Section 6.5.4, and will return in Chapter 7. It should be stressed that the assumption of incompressibility is just a convenient way to defer the difficult issue of wave propagation and instabilities in inhomogeneous plasmas, with the associated singularities, to a later stage (Chapter 7). In particular, notice that $p' \neq 0$ in the present equilibrium. For compressible plasmas, this would considerably complicate the analysis since it implies that the speed of sound varies in the layer. The simplicity of the incompressible model is really that the sound speed does not occur.

Applying the simplifications of the chosen equilibrium to the expressions W_{inc}^p, W^s, W^v, where the jump in the surface integral (6.158) exhibits the driving energy

6.6 Extension to interface plasmas

of the gravitational instability explicitly,

$$\mathbf{n} \cdot [\![\nabla(p + \tfrac{1}{2}B^2)]\!] = p' = -\rho_0 g \,, \tag{6.180}$$

we obtain the following potential energy contributions to $W[\boldsymbol{\xi}, \hat{\mathbf{Q}}]$ in the spectral variational principle (6.160):

$$W^p = \tfrac{1}{2} \int |\mathbf{Q}|^2 \, dV \,, \qquad \mathbf{Q} \equiv \nabla \times (\boldsymbol{\xi} \times \mathbf{B}) \,, \qquad \nabla \cdot \boldsymbol{\xi} = 0 \,, \tag{6.181}$$

$$W^s = -\tfrac{1}{2} \rho_0 g \int |\mathbf{n} \cdot \boldsymbol{\xi}|^2 \, dS \,, \tag{6.182}$$

$$W^v = \tfrac{1}{2} \int |\hat{\mathbf{Q}}|^2 \, d\hat{V} \,, \qquad \nabla \cdot \hat{\mathbf{Q}} = 0 \,. \tag{6.183}$$

Minimization of $\Lambda[\boldsymbol{\xi}, \hat{\mathbf{Q}}]$, or $W[\boldsymbol{\xi}, \hat{\mathbf{Q}}]$, is to be carried out for divergence-free trial functions $\boldsymbol{\xi}$ and $\hat{\mathbf{Q}}$ that satisfy the essential boundary conditions (6.161)–(6.163).

Because the slab is translation symmetric in the y and z directions, we may represent the perturbations in terms of separate Fourier modes that do not couple:

$$\begin{aligned}
\boldsymbol{\xi} &= \bigl(\xi_x(x), \xi_y(x), \xi_z(x) \bigr) \, e^{i(k_y y + k_z z)} \,, \\
\hat{\mathbf{Q}} &= \bigl(\hat{Q}_x(x), \hat{Q}_y(x), \hat{Q}_z(x) \bigr) \, e^{i(k_y y + k_z z)} \,.
\end{aligned} \tag{6.184}$$

Hence, $\nabla \to (d/dx, 0, 0)$ for equilibrium quantities, but $\nabla \to (d/dx, ik_y, ik_z)$ for perturbations. In products like $\xi_x^* \xi_y$, which occur in the integrands of the quadratic forms, the exponential y and z dependencies of the complex conjugate Fourier factors cancel out so that we just obtain one-dimensional (1D) integrations of functions of x alone. For that reason, the indices k_y and k_z that should appear on the Fourier amplitude $\xi_{x;k_y,k_z}(x)$, etc., are simply suppressed in Eqs. (6.184). Also, the exponential factor is usually omitted when working out expressions, e.g. by writing

$$\nabla \cdot \boldsymbol{\xi} = \xi_x' + ik_y \xi_y + ik_z \xi_z = 0 \,. \tag{6.185}$$

These are just shorthand notational conveniences. In case of doubt, one should restore the mentioned factors. Another matter of notation: since the slab is infinitely extended, so that the horizontal areas $A \equiv \iint dy\,dz \to \infty$, the integrals will be renormalized to correspond to the finite contribution over a unit area, $\overline{W} \equiv W/A$. However, since this is a trivial operation, the bars will not be written leaving it understood that all expressions for the Ws and I are renormalized from now on.

Eliminating the contributions of ξ_z to W^p, and of $\hat{Q}_z$ to W^v, by exploiting the divergence conditions, we get the following 1D expressions:

$$W^p = \tfrac{1}{2} B_0^2 \int_0^a \left[k_z^2 (|\xi_x|^2 + |\xi_y|^2) + |\xi_x' + ik_y \xi_y|^2 \right] dx, \qquad (6.186)$$

$$W^s = -\tfrac{1}{2} \rho_0 g |\xi_x(0)|^2, \qquad (6.187)$$

$$W^v = \tfrac{1}{2} \int_{-b}^0 \left[|\hat{Q}_x|^2 + |\hat{Q}_y|^2 + \frac{1}{k_z^2} |\hat{Q}_x' + ik_y \hat{Q}_y|^2 \right] dx. \qquad (6.188)$$

These are to be minimized subject to a normalization that may be chosen rather freely if we just wish to investigate the problem of stability. For surface instabilities, the choice

$$\xi_x(0) = \text{const} \qquad (6.189)$$

is quite effective since it stresses the contribution of the surface. (Under point (d), we will point out an important restriction though in the use of this normalization.) On the other hand, to obtain the growth rate of the instability, the full physical norm for incompressible plasmas should be exploited:

$$I = \tfrac{1}{2} \rho_0 \int_0^a \left[|\xi_x|^2 + |\xi_y|^2 + \frac{1}{k_z^2} |\xi_x' + ik_y \xi_y|^2 \right] dx. \qquad (6.190)$$

(For stability investigations, it would also be permissible to drop the second and third term and to keep a normalization involving the normal perturbation ξ_x only, as we saw in Section 6.4.4. However, in this particular case, that choice turns out not to be a clever one since it is much easier to keep the full norm, as we will see under point (c).) Finally, independent of whether the normalization (6.189) or (6.190) is used, the essential boundary conditions should be satisfied. They reduce to

$$\xi_x(a) = 0, \qquad (6.191)$$

$$\hat{Q}_x(0) = i\mathbf{k}_0 \cdot \hat{\mathbf{B}} \, \xi_x(0), \qquad (6.192)$$

$$\hat{Q}_x(-b) = 0, \qquad (6.193)$$

where $\mathbf{k}_0 \equiv (0, k_y, k_z)$ is the horizontal wave vector.

(b) Stability analysis We will investigate the stability by means of the normalization (6.189). Our first task is to get rid of the tangential components ξ_y and $\hat{Q}_y$. Since ξ_y does not occur in the reduced normalization, minimization with respect to ξ_y only involves minimization of W^p. This is accomplished algebraically by

6.6 Extension to interface plasmas

splitting off a positive definite term which is then put equal to zero:

$$W^p = \tfrac{1}{2} B_0^2 \int_0^a \left[\frac{k_z^2}{k_0^2} \xi_x'^2 + k_z^2 \xi_x^2 + \left| \frac{k_y}{k_0} \xi_x' + i k_0 \xi_y \right|^2 \right] dx$$

$$= \tfrac{1}{2} k_z^2 B_0^2 \int_0^a \left(\frac{1}{k_0^2} \xi_x'^2 + \xi_x^2 \right) dx \,. \tag{6.194}$$

Similarly, minimization of W^v with respect to $\hat{Q}_y$ is trivial:

$$W^v = \tfrac{1}{2} \int_{-b}^0 \left[|\hat{Q}_x|^2 + \frac{1}{k_0^2} |\hat{Q}_x'|^2 + \frac{1}{k_z^2} \left| \frac{k_y}{k_0} \hat{Q}_x' + i k_0 \hat{Q}_y \right|^2 \right] dx$$

$$= \tfrac{1}{2} \int_{-b}^0 \left(\frac{1}{k_0^2} |\hat{Q}_x'|^2 + |\hat{Q}_x|^2 \right) dx \,. \tag{6.195}$$

Hence, just the two unknown functions ξ_x and $\hat{Q}_x$, joined by the first interface condition (6.192), remain to be determined, whereas the other components of the vectors follow from the relations that we have already found:

$$\xi_y = i(k_y/k_0^2)\,\xi_x' \,, \qquad \xi_z = i(k_z/k_0^2)\,\xi_x' \,,$$
$$\hat{Q}_y = i(k_y/k_0^2)\,\hat{Q}_x' \,, \qquad \hat{Q}_z = i(k_z/k_0^2)\,\hat{Q}_x' \,. \tag{6.196}$$

Notice that there is no loss in generality if we assume ξ_x, $\hat{Q}_y$, $\hat{Q}_z$ to be real, and ξ_y, ξ_z, $\hat{Q}_x$ to be purely imaginary (as we tacitly did above when dropping the absolute signs in the final expression for W^p, but keeping them in W^v). This duality between the Fourier components in the direction of inhomogeneity (x) and in the homogeneous directions (y, z) will frequently be employed in the following chapters.

The stability problem has now been reduced to the minimization of

$$W = W^p[\xi_x(x)] + W^s[\xi_x(0)] + W^v[\hat{Q}_x(x)] \,, \tag{6.197}$$

subject to the constraint (6.189) and the boundary conditions (6.191)–(6.193), where W^p is given by (6.194), W^s by (6.187), and W^v by (6.195). This involves the standard variational problem of minimizing W^p with respect to real functions ξ_x on $0 \le x \le a$, and of W^v with respect to real functions $i\hat{Q}_x$ on $-b \le x \le 0$.

To carry out these minimizations, recall some general results from *variational analysis*. Minimization of the quadratic form

$$W[\xi] = \tfrac{1}{2} \int_0^a (F\xi'^2 + G\xi^2)\, dx = \tfrac{1}{2} \Big[F\xi\xi' \Big]_0^a - \tfrac{1}{2} \int_0^a \Big[(F\xi')' - G\xi \Big] \xi\, dx \,, \tag{6.198}$$

where $F \geq 0$ but, otherwise, $F(x)$ and $G(x)$ are arbitrary functions on $[0, a]$, is effected by the variation $\delta\xi(x)$ of the unknown function $\xi(x)$:

$$\delta W = \int_0^a \left(F\xi'\delta\xi' + G\xi\delta\xi \right) dx = \left[F\xi'\delta\xi \right]_0^a - \int_0^a \left[(F\xi')' - G\xi \right]\delta\xi \, dx = 0. \tag{6.199}$$

Since $\xi(0)$ and $\xi(a)$ are prescribed, $\delta\xi = 0$ at the boundaries so that the solution of the *Euler–Lagrange equation*

$$(F\xi')' - G\xi = 0 \tag{6.200}$$

minimizes $W[\xi]$. For that minimizing solution, the value of W becomes

$$W_{\min} = \tfrac{1}{2}\left[F\xi\xi' \right]_0^a = -\tfrac{1}{2}[F\xi\xi'](x=0), \tag{6.201}$$

where we imposed the boundary condition $\xi(a) = 0$ appropriate for our application.

Minimization of the integral (6.194) for W^p and of (6.195) for W^v yields the following Euler–Lagrange equations, with solutions satisfying the respective boundary conditions on the upper and lower walls:

$$\xi_x'' - k_0^2 \xi_x = 0 \quad \Rightarrow \quad \xi_x = C \sinh[k_0(a-x)],$$
$$\hat{Q}_x'' - k_0^2 \hat{Q}_x = 0 \quad \Rightarrow \quad \hat{Q}_x = i\hat{C} \sinh[k_0(x+b)]. \tag{6.202}$$

Whereas the y- and z-dependencies of the Fourier modes (6.184) correspond to *wave-like* motions (with a real wave number k_0) in the horizontal direction, the exponential x-dependencies $\exp(\pm k_0 x)$ combine into cusp-shaped vertical disturbances about the interface: the waves are *evanescent* (with a purely imaginary wave number ik_0) in the vertical direction. The constant C is related to the normalization (6.189) and $\hat{C}$ may be eliminated by means of the boundary condition (6.192):

$$\hat{C} \sinh(k_0 b) = \mathbf{k}_0 \cdot \hat{\mathbf{B}}\, \xi_x(0) = C \mathbf{k}_0 \cdot \hat{\mathbf{B}} \sinh(k_0 a). \tag{6.203}$$

Inserting these solutions of the Euler–Lagrange equations back into the integrals (6.197), using the relation (6.201), yields the final expression for the energy in terms of boundary contributions at $x = 0$ only:

$$W = -\frac{k_z^2 B_0^2}{2k_0^2} \xi_x(0)\xi_x'(0) - \tfrac{1}{2}\rho_0 g\, \xi_x^2(0) + \frac{1}{2k_0^2}\left| \hat{Q}_x(0)\hat{Q}_x'(0) \right|$$

$$= \frac{\xi_x^2(0)}{2k_0 \tanh(k_0 a)} \left[(\mathbf{k}_0 \cdot \mathbf{B})^2 - \rho_0 k_0 g \tanh(k_0 a) + (\mathbf{k}_0 \cdot \hat{\mathbf{B}})^2 \frac{\tanh(k_0 a)}{\tanh(k_0 b)} \right]. \tag{6.204}$$

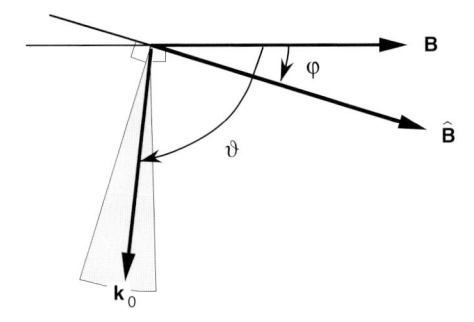

Fig. 6.21. Direction of the horizontal wave vector $\mathbf{k}_0$ for the least stable modes.

Here, the expression in square brackets has been arranged so as to correspond to the expression for the growth rate derived below.

The perturbation energy W clearly exhibits the competition between the stabilizing field line bending energies, $\sim \frac{1}{2}(\mathbf{k}_0 \cdot \mathbf{B})^2$ for the plasma and $\sim \frac{1}{2}(\mathbf{k}_0 \cdot \hat{\mathbf{B}})^2$ for the vacuum, with the destabilizing gravitational energy, $\sim -\frac{1}{2}\rho_0 k_0 g \tanh(k_0 a)$, of the Rayleigh–Taylor instability due to the motion of the interface. Since the magnetic fields $\mathbf{B}$ and $\hat{\mathbf{B}}$ do not point in the same direction, representing *magnetic shear* at the plasma–vacuum interface, no choice of $\mathbf{k}_0$ exists for which the magnetic energies vanish. However, minimum stabilization is obtained for directions of $\mathbf{k}_0$ that are, on average, perpendicular to the field lines (the shaded area in Fig. 6.21). The Rayleigh–Taylor instability may then lead to *interchange instability*, where regions of high plasma pressure and vacuum magnetic field regions are interchanged.

To discuss the dependence on the magnitude of the wave vector $\mathbf{k}_0$, we exploit approximations of the hyperbolic tangent:

$$\tanh \kappa \equiv \frac{e^\kappa - e^{-\kappa}}{e^\kappa + e^{-\kappa}} \approx \begin{cases} 1 & (\kappa \gg 1 : \text{short wavelength}) \\ \kappa & (\kappa \ll 1 : \text{long wavelength}). \end{cases} \qquad (6.205)$$

For short wavelengths of the modes or far away walls ($k_0 a \gg 1$, $k_0 b \gg 1$), the magnetic terms ($\sim k_0^2$) dominate over the gravitational term ($\sim k_0$) and the system is stable. The more dangerous regime is for long wavelengths of the modes ($k_0 a \ll 1$), when the lower wall is far away or even removed ($k_0 b \to \infty$). (To keep p_0 finite in the present equilibrium, the upper wall cannot be moved to ∞.) In that case, the plasma magnetic energy and the gravitational energy compete ($\sim k_0^2$) but, since the vacuum magnetic energy is negligible ($\sim k_0^3$), the plasma becomes Rayleigh–Taylor unstable for interchange modes ($\mathbf{k}_0 \perp \mathbf{B}$) where the plasma magnetic energy vanishes. Only when $b/a \sim 1$ is there genuine competition between the three terms ($\sim k_0^2$) so that effective *wall stabilization* may be obtained.

(c) Growth rate analysis To get the growth rate of the instabilities, when W is negative, we need to minimize the full Rayleigh quotient Λ with the complete expression (6.190) for the norm I. Exceptionally, for the present case of incompressible surface instabilities, this turns out to be a minor extension since I and W^p have the same functional dependence:

$$I = \frac{\rho_0}{k_z^2 B_0^2} W^p. \tag{6.206}$$

It should be pointed out that such blind luck is not at all typical in plasma stability calculations. Nevertheless, of course we will make use of it to get our answer quickly. Because of this relation, the expression for Λ simplifies to

$$\Lambda = \frac{k_z^2 B_0^2}{\rho_0} + \frac{W^s + W^v}{I}. \tag{6.207}$$

Minimization is now straightforward: W^s and W^v do not depend on the radial dependence of ξ so that the expressions (6.187) and (6.195) still hold, whereas, vice versa, minimization of I does not depend on the radial dependence of $\hat{Q}$ so that it proceeds along the same line as minimization of W^p. (Note that we obtain the largest growth rate by minimizing W^s and W^v (which are negative for instability) and also minimizing I (which is positive definite).) This yields

$$I = -\frac{\rho_0}{2k_0^2} \xi_x(0)\xi_x'(0) = \frac{\rho_0 \xi_x^2(0)}{2k_0 \tanh(k_0 a)}, \tag{6.208}$$

so that we obtain the following *dispersion equation*:

$$\omega^2 = \frac{W}{I} = \frac{1}{\rho_0} \left[(\mathbf{k}_0 \cdot \mathbf{B})^2 - \rho_0 k_0 g \, \tanh(k_0 a) + (\mathbf{k}_0 \cdot \hat{\mathbf{B}})^2 \frac{\tanh(k_0 a)}{\tanh(k_0 b)} \right]. \tag{6.209}$$

The expression in square brackets is identical to that of W in Eq. (6.204). In hindsight, that justifies the discussion under point (b) of the wave number dependence: Whereas the energy expression W allows a discussion of the relative importance of the three contributions, there is really no invariant measure for their absolute magnitudes, like that of Eq. (6.209), in the context of the energy principle.

To construct a *scale independent expression for the growth rate* in terms of a minimum number of essential parameters (see Section 4.1.2), we exploit the trivial parameters a, ρ_0, and $\hat{B}_0$ to get rid of the dimensions. This yields the following six dimensionless parameters:

$$\bar{k}_0 \equiv k_0 a, \quad \beta \equiv \frac{2p_0}{\hat{B}_0^2} = 1 - \frac{B_0^2}{\hat{B}_0^2}, \quad \bar{g} \equiv \frac{a\rho_0}{\hat{B}_0^2} g, \quad w \equiv \frac{b}{a}, \tag{6.210}$$

with the angle ϑ between $\mathbf{k}_0$ and $\mathbf{B}$, and the angle φ between $\hat{\mathbf{B}}$ and $\mathbf{B}$ (see Fig. 6.21). The normalized growth rate then reads:

$$\bar{\omega}^2 \equiv \frac{a^2 \rho_0}{\hat{B}_0^2} \omega^2 = (1-\beta)\bar{k}_0^2 \cos^2 \vartheta - \bar{k}_0 \bar{g} \tanh \bar{k}_0 + \bar{k}_0^2 \cos^2(\vartheta - \varphi) \frac{\tanh \bar{k}_0}{\tanh(\bar{k}_0 w)}. \tag{6.211}$$

Hence, $\bar{\omega}^2 = \bar{\omega}^2(\bar{k}_0, \vartheta; \beta, \bar{g}, \varphi, w)$, where parameters in front of the semicolon refer to the wave number of the perturbations and parameters behind it refer to the equilibrium. Their ranges are: $\bar{k}_0 \geq 0$, $0 \leq \vartheta \leq \pi$; $0 \leq \beta \leq 1$, $0 \leq \bar{g} \leq \frac{1}{2}\beta$, $0 \leq \varphi \leq \pi$, $w \geq 0$. Clearly, this is a final and complete answer that just remains to be studied in the relevant limits.

Approximating the hyperbolic tangents, we get the following expressions for the short and long wavelength limits:

$$\bar{\omega}^2 \approx \begin{cases} (1-\beta)\bar{k}_0^2 \cos^2 \vartheta - \bar{k}_0 \bar{g} + \bar{k}_0^2 \cos^2(\vartheta - \varphi) & (\bar{k}_0 \gg 1) \\ \bar{k}_0^2 \left[(1-\beta) \cos^2 \vartheta - \bar{g} + w^{-1} \cos^2(\vartheta - \varphi) \right] & (\bar{k}_0 \ll 1). \end{cases} \tag{6.212}$$

For $\beta = 1$ (no magnetic field in the plasma) and $\vartheta - \varphi = \frac{1}{2}\pi$ (horizontal wave vector $\mathbf{k}_0 \perp \hat{\mathbf{B}}$), the Rayleigh–Taylor instability survives in its pure form:

$$\bar{\omega}^2 = -\bar{k}_0 \bar{g} \tanh \bar{k}_0 \approx \begin{cases} -\bar{k}_0 \bar{g} & (\bar{k}_0 \gg 1) \\ -\bar{k}_0^2 \bar{g} & (\bar{k}_0 \ll 1). \end{cases} \tag{6.213}$$

As we saw above, the Rayleigh–Taylor instability also survives for arbitrary β and long-wavelength interchange modes in the plasma, i.e. $\vartheta = \frac{1}{2}\pi$ ($\mathbf{k}_0 \perp \mathbf{B}$), when the lower wall is removed ($w \to \infty$). This implies that the vacuum wall is essential to obtain complete stability. Further study of the different limits, and their physical significance, is left to the reader (Exercise [6.8]).

(d) Concluding remarks We have derived the stability properties of a simple gravitating plasma–vacuum system bounded by conducting walls above and below. When this system is considered as an example of confined laboratory plasmas, gravity should be viewed as mimicking magnetic curvature effects. When the interest is in astrophysical applications, the walls should either be moved to infinity or considered as a device to limit the domain of the calculation. In computational MHD, such 'walls' are nearly always present for that purpose. One should check then that the solutions found are insensitive to the positions of these 'walls'.

The growth rates found for the Rayleigh–Taylor instability depend on the wave number as $\bar{\omega}^2(\bar{k}_0^2, \vartheta)$, i.e. they just depend on the two horizontal components of

$\bar{\mathbf{k}}_0$. Why no dependence on a vertical 'wave number'? Of course, strictly speaking, there is no wave number in that direction because the plasma is inhomogeneous. However, that is just a formal answer, since it is obvious that wave-like motion must exist, also when the spatial dependence is not precisely periodic. The point is that the incompressible plane plasma slab is a highly degenerate configuration, admitting oscillatory solutions of the Alfvén and slow magneto-sonic type, but these internal modes completely decouple from the external surface modes studied in this section. It is a useful exercise, left to the reader (Exercise [6.8]), to figure out from the equation of motion how this decoupling comes about.

The merit of the present calculation is that it can be carried out completely in terms of elementary functions. The reason is that the inhomogeneities that occur are concentrated on a single surface ($x = 0$). This gives rise to surface waves with a cusped structure that is localized precisely there. For compressible plasmas with distributed inhomogeneities, the mentioned degeneracy is lifted so that all waves and instabilities are coupled and the solutions $\xi_x(x)$ are oscillatory, in general. In that case, use of the normalization (6.189) is extremely risky since $\xi_x(0)$ may become zero. From the expression (6.201) for $W_{\min}$ it is clear that such a zero marks a transition when the plasma becomes internally unstable. This demonstrates that the internal inhomogeneities, and their associated singularities, require a much more careful study than presented so far. This is the subject of Chapter 7.

6.7 Literature and exercises

Notes on literature

Stability:

– Bernstein, Frieman, Kruskal & Kulsrud [26], 'An energy principle for hydromagnetic stability problems', is probably the most quoted paper in plasma physics. The Project Matterhorn report [27] is an appendix containing useful details on the derivations.
– Hain, Lüst & Schlüter [101], 'Zur Stabilität eines Plasmas', appeared simultaneously but is less known, probably because it is written in German and uses tensor notation.
– Kadomtsev [119] is a readable presentation of the energy principle with interface extensions and applications.
– Freidberg, *Ideal Magnetohydrodynamics* [72], chapter 8, is an overview of the general theory of MHD stability.

Spectral theory:

– Friedman, *Principles and Techniques of Applied Mathematics* [74] contains the elements of linear operator theory in a form that is well adapted to use by physicists.
– Lifschitz, *Magnetohydrodynamics and Spectral Theory* [146], Chapter 2, presents the mathematical preliminaries to linear MHD from the spectral theory of operators.

Proof of the energy principle and modifications:

- Laval, Mercier & Pellat [139] presents the proofs of the 'Necessity of the energy principle for magnetostatic stability' for the first time.
- Goedbloed & Sakanaka [89] presents a 'New approach to magnetohydrodynamic stability' by introducing the σ-stability concept.

Rayleigh–Taylor instability:

- Kruskal & Schwarzschild [130] gives the first analysis of the gravitational and kink instabilities of a plasma–vacuum interface system.
- Meyer [155] shows that the gravitational instability can be stabilized by magnetic fields with different directions in the plasma and the vacuum.
- Goedbloed [81](I) demonstrates stabilization over a wider parameter range by replacing the vacuum by a tenuous plasma with a force-free magnetic field.

Exercises

[6.1] *Stability*

Clarify, with a sketch, the meaning of: stable perturbations, unstable perturbations, marginal stability, nonlinear stability, and the difference between marginal stability and lack of equilibrium.

[6.2] *The force operator*

The plasma velocity is related to the displacement vector $\boldsymbol{\xi}$ by the Lagrangian time derivative $\mathbf{v} = D\boldsymbol{\xi}/Dt$.

- Exploiting this definition, linearize the MHD equations for a plasma with a background flow. What is the difference between keeping and ignoring the flow?
- Now ignore the background flow. Insert the expression for $\mathbf{v}$ in the linearized MHD equations and find the expressions for the perturbations p_1, $\mathbf{B}_1$, and ρ_1. Insert those in the equation for $\mathbf{v}_1$, and derive the equation of motion in terms of the force operator $\mathbf{F}$.
- How is the role of the equation for ρ_1 different for laboratory and astrophysical plasmas?
- Count the number of variables and equations in this formalism and compare it with the formalism in terms of primitive variables. Does it check? Comment on the difference.

[6.3] *Stable and unstable solutions*

Express the force operator equation of motion for normal modes with $\omega = \sigma + i\nu$.

- Discuss the different classes of solutions in terms of the signs of σ and ν. Which ones are stable and which ones are unstable?
- Show that for stable solutions the force of the perturbation is restoring, while for unstable solutions it is in the same direction as the plasma displacement.
- Show that, in ideal MHD, the transition from stable to unstable solutions has to go through the marginal state where the force operator $\mathbf{F}(\boldsymbol{\xi}) = 0$.
- What happens to the perturbations when dissipation cannot be neglected?

[6.4] Force operator; meaning of the terms

The force operator is given by the expression (6.30) of Section 6.2.1.

– Give the meaning of the different terms. Which one gives rise to the Alfvén waves?
– Simplify the force operator for infinite homogeneous plasmas and show that only stable waves can occur.
– Insert plane wave solutions in the spectral equation for a homogeneous plasma and derive the algebraic eigenvalue problem. Discuss the different solutions.

[6.5]★ The resolvent operator

The spectrum of a linear operator L is obtained from the study of the inhomogeneous equation $(L - \lambda)x = a$, where a is a given element of the function space exploited, λ is one of the (complex) eigenvalues of the operator L, and x is the unknown solution.

– Invert this expression. Discuss the three possibilities for solutions dependent on λ.
– Convert the equation of motion, in terms of the force operator $\mathbf{F}$, by means of the Laplace transform $\hat{\boldsymbol{\xi}} \equiv \int_0^\infty \boldsymbol{\xi} e^{i\omega t} dt$. Show that the resulting equation is of the above type.
– Invert this expression to find the solution for the Laplace transformed variable $\hat{\boldsymbol{\xi}}$.
– Find the formal solution for $\boldsymbol{\xi}$ by performing the inverse Laplace transformation. What obstacles do you encounter in the construction of the asymptotic time-dependence?

[6.6]★ Hamilton's principle

Consider a particle with position $x_i(t)$ and velocity $\dot{x}_i(t)$, having kinetic energy T and potential energy W. The path of the particle will be such that $\delta \int_{t_0}^{t_1} L(x_i, \dot{x}_i, t) \, dt = 0$, where $L \equiv T - W$, and t_0 and t_1 are the respective times at the start and end points of the path.

– Assume that these points are fixed. Show that we get the following 'Euler–Lagrange' equations from this variational principle:

$$\frac{d}{dt}\frac{\partial L}{\partial \dot{x}_i} - \frac{\partial L}{\partial x_i} = 0.$$

For a continuum, like a plasma described by MHD, the linearized motion may be described in terms of the continuous displacement $\xi_j(x_i, t)$, so that the discrete label i is replaced by the continuous labels x_i. With the kinetic and potential energy given by

$$T = \tfrac{1}{2} \int \rho \boldsymbol{\xi}^* \cdot \dot{\boldsymbol{\xi}} \, dV, \qquad W = -\tfrac{1}{2} \int \boldsymbol{\xi}^* \cdot \mathbf{F} \, dV,$$

the Lagrangian becomes $L = \int \mathcal{L}(\partial \xi_j/\partial x_i, \dot{\xi}_j, \xi_j; x_i, t) \, dV$.

– Show that the Euler–Lagrange equations,

$$\frac{d}{dt}\frac{\partial \mathcal{L}}{\partial \dot{\xi}_j} + \sum_k \frac{d}{dx_k}\frac{\partial \mathcal{L}}{\partial (\partial \xi_j/\partial x_k)} - \frac{\partial \mathcal{L}}{\partial \xi_j} = 0,$$

result in the equation of motion. (See Goldstein [91] on continuous systems and fields.)

[6.7] The energy principle

The energy for linear plasma perturbations is given by the expression (6.85) of Section 6.4.1.

- What is the criterion for stability? Ignoring magnetic fields and gravity, is an ordinary fluid always stable?
- Specify to plane slab geometry, where the fluid is infinitely extended in the horizontal directions and the variables are functions of height only. With limited height of the fluid, and still ignoring magnetic fields but including gravity, impose the equilibrium equation and derive the expression for W. Construct a necessary stability criterion by means of incompressible trial functions and explain what it means.
- Rearrange the equation for W and obtain the Rayleigh–Taylor stability criterion by means of a complete minimization.

[6.8]* *Normal mode analysis for the Rayleigh–Taylor instability*

Work out the normal mode analysis for the Rayleigh–Taylor instability of incompressible interface plasmas by means of the equation of motion. Can you figure out where the Alfvén and slow magneto-sonic waves reside? Of course, the final expression for the growth rate should agree with the expression derived in Section 6.6.4 by means of the variational analysis. What is the role of the Lagrangian pressure perturbation?

[6.9]* *Stability criteria for plasma–vacuum configurations*

Derive stability criteria for general plasma–vacuum configurations with $\beta = 1$ and a curved interface. Discuss the qualitative difference between convex and concave curvature of the outer magnetic field. (Hint: consult Kadomtsev [119], p. 162.)

7
Waves and instabilities of inhomogeneous plasmas

7.1 Hydrodynamics of the solar interior

We have studied the MHD waves for homogeneous plasmas in Chapter 5. This theory was transformed in Chapter 6 to the higher level of spectral theory in order to facilitate the much more complicated analysis of inhomogeneous plasmas, which we want to undertake in the present chapter. Plasma inhomogeneity is not just a complication in the analysis, but also provides qualitatively new physical phenomena like wave damping, wave transformation, and, most important of all, a very wide class of global MHD instabilities of magnetically confined plasmas.

Explicit examples of inhomogeneous plasma dynamics abound in the solar system, as we will see in Chapter 8. For the Sun, a number of important phenomena may be described neglecting the magnetic field. Therefore, before we turn to magnetized plasmas in Section 7.3, we will first simplify the model to a purely hydrodynamic one and study the effects of sound and gravity separate from the three MHD waves. Since the hydrodynamic waves are clearly identified in solar observations, we will be able to clarify the potential of observing MHD wave propagation for the investigation of astrophysical objects in general (Section 7.2.4).

We summarize some basic facts of *the standard solar model* (see Priest [190], Stix [217] or Foukal [69]). The Sun is a sphere of hot material, mainly plasma, of radius $R_\odot = 7.0 \times 10^8$ m and mass $M_\odot = 2.0 \times 10^{30}$ kg. The total radiation power output of the Sun, *the solar luminosity*, is $L_\odot = 3.86 \times 10^{26}$ W. At the Earth, i.e. at a distance of 1 AU (= 1.5×10^{11} m), this results in a heat flux of $L_\odot/[4\pi \times (1 \text{ AU})^2] = 1.36 \text{ kW m}^{-2}$: *the solar constant*. This number roughly corresponds to a crude estimate which one would make on the basis of sunbathing experience. Through the work of H. A. Bethe and C. F. von Weizsäcker (1939) we know that this enormous amount of energy is not produced by gravitational contraction but by *thermonuclear reactions*.

7.1 Hydrodynamics of the solar interior

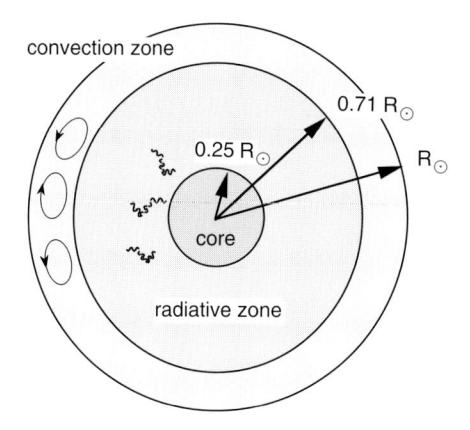

Fig. 7.1. Internal structure of the Sun: thermonuclear energy production in the core, radiation transport in the radiation zone, and convection in the outermost layer.

The thermonuclear energy source is located in *the core* ($r \leq 0.25 R_\odot$) of the Sun (Fig. 7.1), where the values of the temperature and the density are $T_0 = 1.6 \times 10^7$ K and $\rho_0 = 1.6 \times 10^5$ kg m^{-3}, respectively. This is sufficient for *the p–p fusion reaction chain* to occur, described by the equations (1.7) and (1.8) of Section 1.2.1. (Recall that these fusion reactions are completely different, with vastly different time scales, from the ones exploited in thermonuclear laboratory experiments.) The energy produced in the form of gamma radiation is transported through *the radiative zone* ($0.25 R_\odot \leq r \leq 0.713 R_\odot$) to the outer layers. As mentioned in Chapter 1, on average, this process takes millions of years per photon and the wavelengths gradually shift to those of UV and visible light. Finally, the radiative transport is exceeded by convection in the outermost layer of the Sun, *the convection zone* ($0.713 R_\odot \leq r \leq R_\odot$), for reasons that soon will become clear.

7.1.1 Radiative equilibrium model

Let us first consider the process of radiation transport of the thermonuclear power produced in the core to the outer layers of the Sun. This requires the formulation of a *radiative equilibrium model* which can predict the static equilibrium distributions of the different variables characterizing the interior of the Sun. For that purpose, the equations of magnetohydrodynamics presented in Section 4.1 should be enhanced with radiation terms (see Section 4.4.2), whereas the magnetic field contributions can be neglected.

For generality, we first present *the time-dependent hydrodynamic equations*, extending the equations (4.12)–(4.14) of MHD with internal energy production and radiation transport, but ignoring the magnetic field ($\mathbf{B} = 0$). The equation of mass conservation (4.12) is unchanged:

$$\frac{\partial \rho}{\partial t} + \nabla \cdot (\rho \mathbf{v}) = 0. \tag{7.1}$$

In the momentum conservation equation (4.13), the gravitational attraction is specified to be spherically symmetric:

$$\rho \frac{D\mathbf{v}}{Dt} + \nabla p - \rho \mathbf{g} = 0, \quad \text{where} \quad \mathbf{g} = -g\mathbf{e}_r, \quad g = g(r) = G\frac{M(r)}{r^2}. \tag{7.2}$$

Here, the gravitational constant $G = 6.67 \times 10^{-11} \, \text{N m}^2 \, \text{kg}^{-2}$ and $M(r)$ is the mass inside a sphere of radius r, which is related to the density ρ by means of the obvious differential equation

$$\frac{dM}{dr} = 4\pi r^2 \rho. \tag{7.3}$$

For the present purpose, it is convenient to replace the pressure evolution equation (4.14), which is a consequence of the entropy conservation equation, by the appropriate generalization of the internal energy equation (4.19), given by Eq. (4.138):

$$\rho \frac{De}{Dt} + p\nabla \cdot \mathbf{v} = \nabla \cdot [\lambda \nabla (kT)] + \rho \varepsilon, \quad \text{where} \quad e \equiv \frac{1}{\gamma - 1} \frac{p}{\rho}. \tag{7.4}$$

The right hand side now contains two additional terms which are due to *radiative transport*, governed by *the thermal conduction coefficient*, $\lambda(r)$,[1] and *the thermonuclear energy production per unit mass*, $\varepsilon(r)$. In general, the latter will be strongly dependent on the temperature and the concentration of hydrogen.

At this point, it becomes clear that the radiative model actually requires more than just one fluid since thermonuclear energy production and thermal conduction depend on the concentration of the different particle species. This aspect of the problem will be approximated by means of an equation of state, relating the pressure and the temperature, that is modified to take the different masses of the particles into account. For a plasma consisting of electrons and one kind of ion with equal temperatures ($T \equiv T_e = T_i$), the expression for the pressure may be converted by means of the charge neutrality relation, $n_e = Z_c n_i$, and the definition for the density, $\rho \equiv n_e m_e + n_i m_i$:

$$p = (n_e + n_i)kT = n_i(1 + Z_c)kT = \frac{1 + Z_c}{m_i(1 + Z_c m_e/m_i)} \rho kT \approx \frac{1 + Z_c}{A} \frac{\rho kT}{m_p}. \tag{7.5}$$

---

[1] To stick to conventions in the literature, we had to introduce notations conflicting with those of previous chapters: the thermal conductivity λ of the present chapter was called κ in Sections 3.3.2 and 4.4.2, but the symbol κ will here be used for the opacity (see below).

Here, A is the mass number (in multiples of the proton mass, $m_i = Am_p$) and Z_c is the charge number (in multiples of e) of the ions. The latter quantity is subscripted with the letter c since, in this context, the bare symbol Z is used for another quantity, to be introduced shortly.

So far, the equation of state (7.5) is just the usual one for a plasma consisting of electrons and one kind of ion, so that Z_c and A are constants. Now, we modify the equation of state by permitting different ion species that are radially distributed according to a function called the *mean molecular weight* μ, so that

$$p = \frac{1}{\mu}\frac{\rho kT}{m_p}, \quad \text{where } \mu \equiv \frac{A}{1+Z_c} \text{ is no longer constant}. \tag{7.6}$$

The mean molecular weight μ measures the average weight of the constituent particles in units of m_p: $\mu = 1/2$ for ionized hydrogen ($Z_c = A = 1$), $\mu = 4/3$ for ionized helium ($Z_c = 2$, $A = 4$), and $\mu \approx 2$ for the completely ionized heavier atoms ($Z_c \gg 1$, $A \approx 2Z_c$). For a plasma with a mixture of these ions, $\mu \approx (2X + \frac{3}{4}Y + \frac{1}{2}Z)^{-1}$, where X is the mass fraction of hydrogen, Y is the mass fraction of helium, and Z indicates the mass fraction of the heavier elements. Hence, $\mu = \mu(\rho(r), T(r))$ is now a function of position reflecting the abundance of the different elements in the Sun and their degrees of ionization, to be determined by Saha equations of the type discussed in Section 1.4.1. Note that the dependence of p on μ implies that the pressure decreases when hydrogen is converted into helium by the mentioned thermonuclear reactions.

To construct a *static* equilibrium model of the Sun from these equations, one sets $\mathbf{v} = 0$ and $\partial/\partial t = 0$. The mass conservation equation (7.1) is then trivially satisfied, whereas Eqs. (7.2) and (7.3) provide the equations for *hydrostatic equilibrium*:

$$\frac{dp}{dr} = -\rho g = -G\frac{\rho M}{r^2} \Rightarrow \frac{1}{r^2}\frac{d}{dr}\left(\frac{r^2}{\rho}\frac{dp}{dr}\right) = -\frac{G}{r^2}\frac{dM}{dr} = -4\pi G\rho. \tag{7.7}$$

In static equilibrium, the LHS of the internal energy equation (7.4) vanishes so that

$$\frac{1}{r^2}\frac{d}{dr}\left[r^2\lambda\frac{d(kT)}{dr}\right] = -\rho\varepsilon, \tag{7.8}$$

which expresses *radiative equilibrium*: the produced thermonuclear energy is transported outward by radiative transfer. The set of differential equations (7.7)–(7.8) is complete when supplemented with the equation of state (7.6) and equations determining the functions $\mu(r)$, $\lambda(r)$, and $\varepsilon(r)$.

Although the hydrostatic model is now complete, it is useful to relate the differential equation (7.8) for the temperature with the fundamental quantities of radiative transport, viz. luminosity, radiation pressure and opacity. Radiative equilibrium implies that the local value of *the luminosity*, $L(r)$, which is the net radiation flux per unit of time through a spherical surface, is given by

$$L(r) = \int_0^r \varepsilon(r')\,dM = 4\pi \int_0^r \rho\,\varepsilon\, r'^2\,dr'. \tag{7.9}$$

To relate this expression to the temperature gradient appearing in Eq. (7.8), we introduce *the radiation pressure*[2]

$$q \equiv p_{\text{rad}} = \left(\frac{4\sigma}{3c}\right) T^4, \tag{7.10}$$

where $\sigma \equiv 2\pi^5 k^4/(15c^2 h^3) = 5.67 \times 10^{-8}$ W m^{-2} K^{-4} is the Stefan–Boltzmann constant, and we exploit the fundamental relationship between the gradient of the radiation pressure and the absorbed radiation,

$$\frac{dq}{dr} = -\left(\frac{1}{4\pi c}\right) \frac{\kappa \rho L}{r^2}, \tag{7.11}$$

where κ is *the opacity*, i.e. the absorption coefficient of the radiation per unit mass and per unit length. Hence,

$$L(r) = -\left(\frac{16\pi\sigma}{3}\right) \frac{r^2}{\kappa\rho} \frac{dT^4}{dr}. \tag{7.12}$$

From Eqs. (7.9) and (7.12) we again obtain a differential equation for the temperature, which, of course, should be equivalent to Eq. (7.8). This implies that the thermal conductivity λ is inversely proportional to the opacity κ,

$$\lambda = \left(\frac{16\sigma}{3k}\right) \frac{T^3}{\kappa\rho}. \tag{7.13}$$

The opacity is a microscopic quantity determined by quite a number of different scattering processes between photons and atoms, ions or electrons, averaged over frequency. To compute radiative equilibrium, one considers $\kappa = \kappa(\rho, T)$ to be a known function, obtained in the form of a table from the extensive numerical calculations that have been developed for this purpose [217].

Finally, we need an explicit relation for the thermonuclear energy production rate by proton–proton reactions, as given, e.g., by Zirin [250]:

$$\varepsilon = \varepsilon(X, \rho, T) = 0.25\,\rho X^2 \left(\frac{10^6}{T}\right)^{2/3} \exp\left\{-33.8\,(10^6/T)^{1/3}\right\}, \tag{7.14}$$

---

[2] The effect of the radiation pressure is here only taken into account as far as the radiation transport is concerned, but neglected in the hydrostatic equilibrium equation. Actually, one should replace p in Eq. (7.7) by the total pressure $P \equiv p + q$, but one easily checks that this effect is negligible for relatively cool stars like the Sun. With $T_0 = 1.6 \times 10^7$ K, $n_0 = 10^{32}$ m^{-3}, and $T_{\text{ph}} = 6000$ K, $n_{\text{ph}} = 2.4 \times 10^{23}$ m^{-3}, the value of q/p ranges from 7.4×10^{-4} in the centre to 1.6×10^{-5} at the photosphere.

7.1 Hydrodynamics of the solar interior

where the dimensions of the quantities are given by $[\rho] = \text{kg m}^{-3}$, $[T] = \text{K}$, and $[\varepsilon] = \text{J kg}^{-1}\text{s}^{-1}$. It is clear that the function $\varepsilon(X, \rho, T)$ is strongly concentrated in the central regions of the Sun where the temperature is high: thermonuclear burn only occurs in *the core region* of the Sun ($r \leq 0.25 R_\odot$).

▷ **Exercise.** Estimate the fraction of the total solar mass participating in the thermonuclear energy production, using the values of $M(R_\odot)$, $L(R_\odot)$ and ρ_0 given in this section. ◁

The standard solar model consists of the two second order differential equations (7.7) and (7.8), supplemented with the equation of state (7.6) and numerically obtained values of the functions $\mu(\rho, T)$, $\kappa(\rho, T)$, and $\varepsilon(X(\rho, T), \rho, T)$. For the latter quantity one should exploit Eq. (7.14), which evidently spoils the possibility of obtaining explicit analytic solutions, even when mild assumptions are made about μ and κ.

Since two second order ODEs are equivalent to four first order ODEs, one may choose, for example, to solve for the following ODEs for the four quantities $M(r)$, $L(r)$, $p(r)$ and $T(r)$:

$$\begin{aligned}
\frac{dp}{dr} &= -G\frac{\rho M}{r^2}, \qquad \text{where } \rho = \left(\frac{m_p}{k}\right)\frac{\mu p}{T}, \\
\frac{dM}{dr} &= 4\pi r^2 \rho, \\
\frac{dT}{dr} &= -\left(\frac{3}{64\pi\sigma}\right)\frac{\kappa\rho L}{r^2 T^3}, \\
\frac{dL}{dr} &= 4\pi r^2 \rho\varepsilon.
\end{aligned} \qquad (7.15)$$

Appropriate boundary conditions for these equations are

$$p(R_\odot) \approx 0, \quad M(0) = 0, \quad T(R_\odot) \approx 0, \quad L(0) = 0. \qquad (7.16)$$

Of course, the solution obtained should reproduce the known value of the solar mass, $M(R_\odot) = M_\odot$, and of the solar luminosity, $L(R_\odot) = L_\odot$, at the known position $r = R_\odot$ of the solar radius. Also, the values p_0 and T_0 should correspond to the numbers cited for the central density and temperature to yield thermonuclear burn. This procedure actually turns out to provide quite realistic solutions (see Fig. 7.2), except for one feature, viz. the assumption of hydrostatic equilibrium *up to the solar surface*.

7.1.2 Convection zone

In the outer layers of the Sun, cooling is so strong that the absolute value of the temperature gradient exceeds a certain threshold given by the Schwarzschild

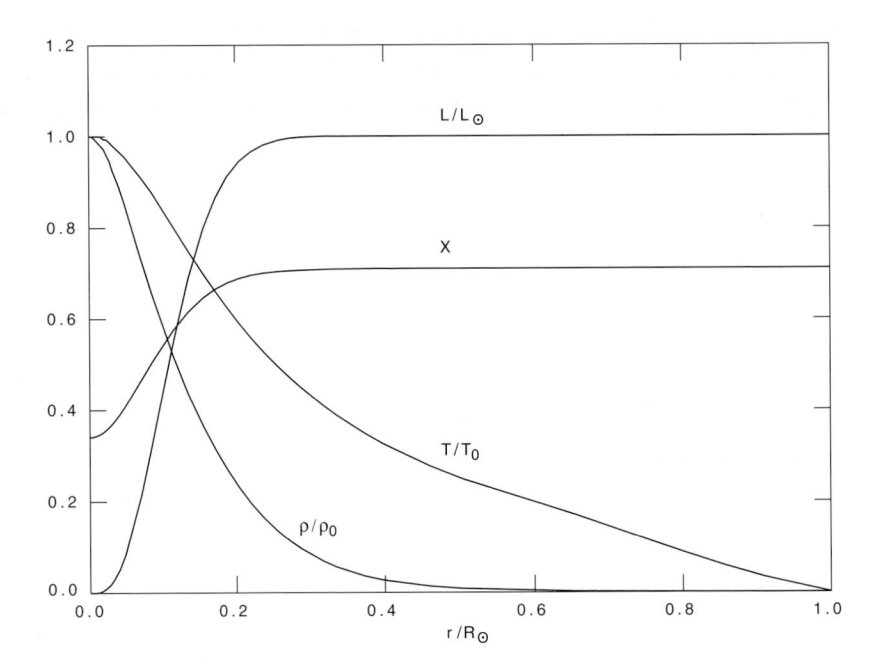

Fig. 7.2. Radial distributions of the physical quantities for a standard solar model. (From Foukal [69], using data of Bahcall and Ulrich, *Rev. Mod. Phys.* **60**, 297 (1988).)

criterion for convective stability. Since the temperature gradient is negative, that criterion is best discussed in terms of the quantity $-dT/dr$. Convective stability is achieved when $-dT/dr$ does not exceed a critical value, which is obtained as follows.

Introducing the mean particle mass, $m \equiv \mu m_p$, the ratio of the Boltzmann constant k and m may be abbreviated as a kind of gas constant, $\mathcal{R} \equiv k/m$. Anticipating the outcome of the present section, $\mathcal{R}$ is actually constant in the convection zone because of the strong mixing that occurs, so that $\mu \approx$ const there. Consequently, the equation of state (7.6) may be written as

$$p = \mathcal{R}\rho T . \qquad (7.17)$$

Hence, the temperature gradient is related to the density and pressure gradients by

$$-\frac{dT}{dr} = \frac{1}{\mathcal{R}}\left(\frac{p}{\rho^2}\frac{d\rho}{dr} - \frac{1}{\rho}\frac{dp}{dr}\right) \quad \left[= \frac{1}{\mathcal{R}}\left(\frac{p}{\rho^2}\frac{d\rho}{dr} + g\right)\right], \qquad (7.18)$$

where the equilibrium relation (7.7), $dp/dr = -\rho g$, has been inserted in the

7.1 Hydrodynamics of the solar interior

rightmost equality. Here, and below, we put square brackets around such equalities to indicate that this additional equilibrium information has been exploited.

The critical value for convective stability is obtained from the consideration that motions of the gas are neutrally stable only if the fluid is *isentropic*, i.e. if the value of the specific entropy is constant throughout the pertinent region of space:

$$S \equiv p\rho^{-\gamma} = \text{const} \Rightarrow \frac{d\rho}{dr} = \frac{\rho}{\gamma p}\frac{dp}{dr} \quad \left[= -\frac{\rho^2 g}{\gamma p} \right], \quad (7.19)$$

so that

$$-\left(\frac{dT}{dr}\right)_{\text{isentr.}} = -\frac{1}{\mathcal{R}}\frac{\gamma-1}{\gamma}\frac{1}{\rho}\frac{dp}{dr} \quad \left[= \frac{1}{\mathcal{R}}\frac{\gamma-1}{\gamma} g \right]. \quad (7.20)$$

Convective instability occurs when the actual temperature gradient $-dT/dr$ exceeds this value, i.e. when *the Schwarzschild criterion for convective stability*,

$$-\frac{dT}{dr} \leq -\left(\frac{dT}{dr}\right)_{\text{isentr.}} \Rightarrow \frac{d\rho}{dr} - \frac{\rho}{\gamma p}\frac{dp}{dr} \leq 0 \quad \left[\Rightarrow \frac{d\rho}{dr}g + \frac{\rho^2 g^2}{\gamma p} \leq 0 \right], \quad (7.21)$$

is violated. Interestingly enough, the criterion for gravitational stability (in square brackets) of Section 6.5.4 is recovered: convective and gravitational (Rayleigh–Taylor) instabilities amount to the same.

Note that we here exploit the term 'isentropic' rather than the more usual term 'adiabatic' since we wish to stress that the background equilibrium state is not isentropic in general but that the assumption of adiabatic fluid *motion*, i.e. $DS/Dt = 0$, may still be made. Recall that this assumption is underlying the whole analysis of Chapter 6, including the derivation of the gravitational stability in Section 6.5.4. Hence, adiabatic motions are stable or unstable according to whether the condition (7.21) is satisfied or violated. Only at marginal stability do the two concepts coincide: the equilibrium is then characterized by constant entropy in space and marginal motion ($\omega = 0$) does not change this distribution.

In the Sun, the Schwarzschild criterion (7.21) is violated in the region $0.713 R_\odot \leq r \leq R_\odot$, which is called *the convective region* for that reason. In this region turbulent mixing occurs, i.e. eddies of hot material are carried up and eddies of cool material are carried down so as to maintain the temperature gradient at the value given by the RHS of the first expression (7.21). Consequently, rather than applying the boundary conditions (7.16) at the solar surface ($r = R_\odot$), one obtains a more realistic solar model by separating the interior in the regions $r \leq 0.713 R_\odot$ (core and radiative zone), bounded by a surface where the boundary condition

$$-\frac{dT}{dr} = -\left(\frac{dT}{dr}\right)_{\text{isentr.}} \quad (\text{at } r = 0.713 R_\odot) \quad (7.22)$$

is imposed, and the convective region which is governed by a completely different set of equations. In particular, the assumption of static equilibrium is not appropriate for the convective region so that one should return to the time-dependent equations (7.1)–(7.6), where the flow also destroys the spherical symmetry of the problem. Unfortunately, there is little one can do about solving this problem without recourse to a large computer and, even then, there remains a large uncertainty about the nature of the solutions in the convection zone.

Ironically, there appears to be more certainty about the structure of the interior of the Sun, which is completely opaque (recall that the mean free path of photons is only a few centimetres there), than the outer regions. Also, the magnetic field, which we have ignored until now, turns out to play an important role in the outer layers of the Sun. In particular, the origin of the solar magnetic field, i.e. *the solar dynamo*, is thought to be situated at the bottom of the convection zone. The dynamo problem involves differential rotation, spherical geometry, and magnetic fields. This significantly adds to the complexity of the problem since magnetic fields cannot be fitted in a spherically symmetric geometry. This fact is very evident from the huge number of complex magnetic structures that can be observed in the atmosphere of the Sun, viz. in the photosphere, the chromosphere and the corona.

In conclusion, it appears hopeless to try to find solutions for the interior of the Sun, by means of the theoretical approach discussed so far, starting from such a complicated boundary. Fortunately, a powerful observational and analytical tool has become available that can probe the deep structure of the Sun, viz. *helioseismology*, which will be treated in the following section.

7.2 Hydrodynamic waves and instabilities of a gravitating slab

The convective layer derives its name from the convective instability which arises when the Schwarzschild criterion (7.21) is violated. Since this is associated with the transition from stability to instability of gravity waves, at this point it is instructive to consider the more general subject of solar oscillations. Here, we will consider the Sun as a whole and, again, neglect the presence of magnetic fields. The symmetry of the problem would involve the treatment of wave propagation in spherically inhomogeneous systems (background quantities depending on the radial coordinate r), leading to the occurrence of spherical harmonics in the analysis. However, we will avoid these technical complications and just consider *the gravito-acoustic waves in a planar stratification* (background quantities depending on the vertical coordinate x). Obviously, the expressions derived require a generalization to spherical stratification if one wishes to compare calculated frequencies with observed ones. Our goal is to demonstrate how spectral analysis relates to practical applications, not to treat the subject of solar and stellar oscillations in detail. (The interested reader should consult the texts by Christensen-Dalsgaard [56] or Unno *et al.* [234].)

7.2.1 Hydrodynamic wave equation

Again, we exploit the equations of gas dynamics. We now keep the adiabatic law so that the relevant equations become

$$\frac{D\rho}{Dt} + \rho \nabla \cdot \mathbf{v} = 0, \tag{7.23}$$

$$\rho \frac{D\mathbf{v}}{Dt} + \nabla p - \rho \mathbf{g} = 0, \tag{7.24}$$

$$\frac{Dp}{Dt} + \gamma p \nabla \cdot \mathbf{v} = 0. \tag{7.25}$$

Note that this implies that, for the study of waves, we neglect the RHS of Eq. (7.4), i.e. *non-adiabatic effects* caused by thermal conduction and thermonuclear energy production. Also, for simplicity, we consider the gravitational acceleration as a given constant, i.e. we neglect perturbations of the gravitational potential (which is called the *Cowling approximation*):

$$\mathbf{g} = (-g, 0, 0), \qquad g = \text{const}.$$

We will again assume a static background equilibrium, i.e. $\mathbf{v} = 0$ and $\partial/\partial t = 0$, so that the only restriction on the equilibrium comes from the momentum equation (7.24):

$$\nabla p_0 = \rho_0 \mathbf{g} \quad \Rightarrow \quad p_0'(x) = -\rho_0(x) g, \tag{7.26}$$

where the prime indicates differentiation with respect to x.

Linearization leads to the following equations:

$$\frac{\partial \rho_1}{\partial t} + \mathbf{v}_1 \cdot \nabla \rho_0 + \rho_0 \nabla \cdot \mathbf{v}_1 = 0, \tag{7.27}$$

$$\rho_0 \frac{\partial \mathbf{v}_1}{\partial t} + \nabla p_1 - \rho_1 \mathbf{g} = 0, \tag{7.28}$$

$$\frac{\partial p_1}{\partial t} + \mathbf{v}_1 \cdot \nabla p_0 + \gamma p_0 \nabla \cdot \mathbf{v}_1 = 0. \tag{7.29}$$

These equations are straightforward generalizations of Eqs. (5.5)–(5.7) with additional terms due to gravity and the associated inhomogeneity. Hence, we could derive the wave equation, corresponding to Eq. (5.8) for the homogeneous case, by differentiating Eq. (7.28) with respect to t and inserting the expressions for $\partial \rho_1/\partial t$ and $\partial p_1/\partial t$ from Eqs. (7.27) and (7.29). However, we rather exploit our newly obtained instrument of integrating by means of the displacement vector $\boldsymbol{\xi}$ (Section 6.1.2), where $\mathbf{v}_1 = \partial \boldsymbol{\xi}/\partial t$, so that $\rho_1 = -\nabla \cdot (\rho_0 \boldsymbol{\xi})$ and $p_1 = -\boldsymbol{\xi} \cdot \nabla p_0 - \gamma p_0 \nabla \cdot \boldsymbol{\xi}$. Inserting these expressions into Eq. (7.28), and cancelling some terms by exploiting the equilibrium relation (7.26), we obtain *the wave equation for gravito-acoustic waves in a plane stratified medium*:

$$\rho \frac{\partial^2 \boldsymbol{\xi}}{\partial t^2} - \nabla(\gamma p \nabla \cdot \boldsymbol{\xi}) - \rho \nabla(\mathbf{g} \cdot \boldsymbol{\xi}) + \rho \mathbf{g} \nabla \cdot \boldsymbol{\xi} = 0. \tag{7.30}$$

We have omitted the subscripts 0 and 1 for the equilibrium and perturbed quantities since there can be no confusion any more: perturbed quantities are expressed in $\boldsymbol{\xi}$, all other quantities refer to the background. The wave equation (7.30) generalizes Eq. (5.8) with two additional terms related to gravity. Of course, Eq. (7.30) is also obtained as the special case $\mathbf{B} = 0$ of the general MHD equation of motion (6.23), exploiting the appropriate expression (6.50) for the force operator.

Considering normal mode solutions of the form

$$\boldsymbol{\xi}(\mathbf{r}, t) = \hat{\boldsymbol{\xi}}(x)\, e^{i(k_y y + k_z z - \omega t)}, \tag{7.31}$$

Eq. (7.30) may be written in matrix form as

$$\begin{pmatrix} \rho\omega^2 + \dfrac{d}{dx}\gamma p \dfrac{d}{dx} & ik_y\left(\dfrac{d}{dx}\gamma p + \rho g\right) & ik_z\left(\dfrac{d}{dx}\gamma p + \rho g\right) \\ ik_y\left(\gamma p \dfrac{d}{dx} - \rho g\right) & \rho\omega^2 - k_y^2 \gamma p & -k_y k_z \gamma p \\ ik_z\left(\gamma p \dfrac{d}{dx} - \rho g\right) & -k_y k_z \gamma p & \rho\omega^2 - k_z^2 \gamma p \end{pmatrix} \begin{pmatrix} \hat{\xi}_x \\ \hat{\xi}_y \\ \hat{\xi}_z \end{pmatrix} = 0. \tag{7.32}$$

Since there is no preferred horizontal direction in the system, there is no loss in generality if we rotate the coordinate system such that $k_z = 0$, so that $\rho\omega^2 \hat{\xi}_z = 0$. Hence, ignoring marginal shifts ($\omega^2 = 0$, $\hat{\xi}_z \neq 0$), all terms with k_z and $\hat{\xi}_z$ may be neglected and $\hat{\xi}_y$ may be eliminated:

$$\hat{\xi}_y = -ik_0 \frac{\gamma p \hat{\xi}_x' - \rho g \hat{\xi}_x}{\rho\omega^2 - k_0^2 \gamma p}, \tag{7.33}$$

where $k_0 \equiv \sqrt{k_y^2 + k_z^2}$ ($= k_y$ now) is the horizontal wave number. Inserting this expression in the first component of Eq. (7.32) gives a *second order differential equation* for the unknown $\hat{\xi}_x$:

$$\frac{d}{dx}\left(\frac{\gamma p\, \rho\omega^2}{\rho\omega^2 - k_0^2\gamma p}\frac{d\hat{\xi}_x}{dx}\right) + \left[\rho\omega^2 - \frac{k_0^2\rho^2 g^2}{\rho\omega^2 - k_0^2\gamma p} - \left(\frac{k_0^2\gamma p\,\rho g}{\rho\omega^2 - k_0^2\gamma p}\right)'\right]\hat{\xi}_x = 0. \tag{7.34}$$

To solve this differential equation one should specify the vertical profiles $p(x)$ and $\rho(x)$, which should satisfy the equilibrium condition (7.26), and the boundary conditions.

▷ **Alternative expression.** The last term of Eq. (7.34) may be written as

$$-\left(\frac{k_0^2\gamma p\,\rho g}{\rho\omega^2 - k_0^2\gamma p}\right)' = \rho' g - \left(\rho g\frac{\rho\omega^2}{\rho\omega^2 - k_0^2\gamma p}\right)'.$$

This more readily exhibits the correspondence with the expressions for spherical geometry (Eq. (7.58) of Section 7.2.4) and for magnetized plasmas (Eq. (7.91) of Section 7.3.2). ◁

Formulation of the appropriate boundary conditions obviously would involve the spherical geometry of the original problem. Since we have ignored that part of the problem, the next logical step is to consider modes that do not sensitively depend on the choice of the boundary. Such modes are obtained by the assumption that the boundaries are far away. In other words: we should study modes of finite extent in the y- and z-directions (either arbitrary wave numbers k_y and k_z, or quantized if one wishes to consider the y- and z-directions as periodic), but *sufficiently localized in the direction of inhomogeneity*. Hence, imposing rigid wall boundary conditions,

$$\hat{\xi}_x(x=0) = \hat{\xi}_x(x=a) = 0, \tag{7.35}$$

where $x = 0$ would correspond to the centre and $x = a$ to the surface of the Sun, only makes sense if a is large as compared to the typical 'wavelengths' in the x-direction. (Quotation marks here since we strictly cannot define a wave vector in a direction of inhomogeneity.) For example, we might consider modes which rapidly oscillate in the x-direction:

$$\hat{\xi}_x(x) \sim e^{iqx} \quad \text{with} \quad qa \gg 1. \tag{7.36}$$

Of course, there are no walls on the Sun, and the boundary conditions (7.35) just serve to define a region of localization of the modes. In more general terms, one would consider so-called *cavity modes*, where the background pressure and density distributions create a kind of potential well in which the modes are confined.

The differential equations (7.34) with the boundary conditions (7.35) permit the study of the instabilities ($\omega^2 < 0$) and stable waves ($\omega^2 > 0$) which originate from the peculiar interaction of the sound and gravity terms. For the solar context, these two cases are exemplified by the convective instabilities of the convection zone and the solar acoustic-gravity waves (the *p*- and *g*-modes), respectively. We treat these two examples in Subsections 7.2.2 and 7.2.3.

7.2.2 Convective instabilities

Since time scales of gravitational instabilities are much longer than time scales of acoustic oscillations, we may assume $\rho |\omega^2| \ll k_0^2 \gamma p$ for the study of these instabilities. The wave equation (7.34) then simplifies to

$$\frac{d}{dx}\left(\frac{\rho\omega^2}{k_0^2}\frac{d\hat{\xi}_x}{dx}\right) - \rho\left(\omega^2 - N^2\right)\hat{\xi}_x = 0, \tag{7.37}$$

where N is *the Brunt–Väisäläa frequency*:

$$N^2 = N^2(x) \equiv -g\left(\frac{1}{\rho}\frac{d\rho}{dx} - \frac{1}{\gamma p}\frac{dp}{dx}\right) \quad \left[= -\frac{1}{\rho}\left(\rho' g + \frac{\rho^2 g^2}{\gamma p}\right)\right]. \tag{7.38}$$

Since ρ and N^2 depend on x through $\rho(x)$ and $p(x)$, of which at least one function is completely arbitrary, in general, the solution of the differential equation (7.37) can only be obtained by numerical integration. However, assuming rapid spatially oscillatory modes of the kind expressed by Eq. (7.36) with $q \gg |\nabla| \sim 1/a$, where $|\nabla|$ represents a typical gradient scale length of the equilibrium, one obtains the following estimate for the eigenfrequencies of the local instabilities:

$$\omega^2 \approx \frac{k_0^2}{k_0^2 + q^2} N^2(x). \tag{7.39}$$

Hence, the system is locally unstable in the range of x where $N^2 < 0$.

It will have been noticed that the bracket in the definition (7.38) for N^2 is just the expression entering the gravitational stability criterion derived in Section 6.5.4 by application of the energy principle. We now have additional information on this instability, viz. an estimate of the growth rates. (Note that this required application of the alternative to method (3) of Section 6.4.4, viz. the differential equation approach associated with the normal-mode analysis.)

We have already seen in Section 7.1.2 that the Schwarzschild criterion (7.21) for convective stability and the gravitational stability criterion coincide when the equilibrium conditions are taken into account. This is now immediately evident from the connection with the Brunt–Väisäläa frequency:

$$\frac{dT}{dx} - \left(\frac{dT}{dx}\right)_{\text{isentr.}} = \frac{T}{g} N^2 \geq 0 \quad \left[\Rightarrow \quad -\frac{T}{\rho g}\left(\rho' g + \frac{\rho^2 g^2}{\gamma p}\right) \geq 0\right]. \tag{7.40}$$

Hence, $N^2 \geq 0$ is nothing else than the Schwarzschild criterion.

If the Schwarzschild criterion is violated, i.e. $N^2 < 0$, Eq. (7.39) tells us that we should expect convective instabilities exponentially growing as $\exp(\sqrt{-\omega^2}\,t)$. Of course, such modes rapidly enter the nonlinear domain, where the assumptions underlying the given derivation break down. This is associated with the formation of convective cells, which may be nonlinearly stable. However, the onset of this phenomenon is correctly predicted by the linear theory. On the other hand, if $N^2 > 0$, Eq. (7.39) provides us with the expression for the frequency of slow gravitational modes (the g-modes). These will be treated in a slightly more general context, without the assumption of smallness of ω^2 made in this subsection.

7.2.3 Gravito-acoustic waves

An instructive special case of gravito-acoustic waves in an inhomogeneous medium is obtained for an *exponentially stratified medium with constant sound speed*:

$$\rho = \rho_0 e^{-\alpha x}, \quad p = p_0 e^{-\alpha x}, \quad \text{so that} \quad c^2 = \frac{\gamma p}{\rho} = \frac{\gamma p_0}{\rho_0} = \text{const.} \quad (7.41)$$

From the equilibrium equation (7.26) this implies that $p' = -\alpha p = -\rho g$ (where we recall that g is also assumed constant), so that

$$\alpha = \frac{\rho g}{p} = \frac{\rho_0 g}{p_0} = \frac{\gamma g}{c^2} = \text{const.} \quad (7.42)$$

The spectral equation (7.34) then reduces to

$$\frac{c^2 \omega^2}{\omega^2 - k_0^2 c^2} \frac{d}{dx}\left(e^{-\alpha x} \frac{d\hat{\xi}_x}{dx}\right) + \left(\omega^2 - \frac{k_0^2 g^2}{\omega^2 - k_0^2 c^2} + \alpha \frac{k_0^2 c^2 g}{\omega^2 - k_0^2 c^2}\right) e^{-\alpha x} \hat{\xi}_x = 0. \quad (7.43)$$

The expression (7.38) for the square of the Brunt–Väisälä frequency here simplifies to

$$N^2 = \alpha g - \frac{g^2}{c^2} = (\gamma - 1)\frac{g^2}{c^2} > 0, \quad (7.44)$$

so that we will only find stable waves. By means of this expression, Eq. (7.43) transforms to

$$\frac{d^2 \hat{\xi}_x}{dx^2} - \alpha \frac{d\hat{\xi}_x}{dx} + \frac{\omega^4 - k_0^2 c^2 \omega^2 + k_0^2 c^2 N^2}{c^2 \omega^2} \hat{\xi}_x = 0. \quad (7.45)$$

This is a differential equation with constant coefficients so that its solution becomes trivial.

This is another occasion of blind luck (cf. Section 6.6.4) which, we repeat, is *not at all typical* for the problem of wave propagation in inhomogeneous media. One should be aware of the limitations intrinsic to the use of special cases that are analytically solvable. Frequently, one comes across so-called 'intuition', based upon insight obtained from such cases, which may be very misleading. For example, here we have lost the important possibility of convective instability, discussed in the previous subsection, simply because $N^2 > 0$ due to the assumption of an exponential medium. In general, for arbitrary pressure and density profiles, the differential equation to be solved is just a general ODE, where numerical solution is the appropriate procedure.

The solutions of Eq. (7.45) are exponentials of the form $\hat{\xi}_x = Ce^{(\frac{1}{2}\alpha \pm iq)x}$, where

$$q \equiv \sqrt{-\frac{1}{4}\alpha^2 + \frac{\omega^4 - k_0^2 c^2 \omega^2 + k_0^2 c^2 N^2}{c^2 \omega^2}}. \tag{7.46}$$

Clearly, the expression under the square root sign has to be positive in order to obtain spatially oscillatory solutions that can satisfy the boundary conditions (7.35). The latter conditions imply that q has to be quantized:

$$qa = n\pi \quad (n = 1, 2, \ldots). \tag{7.47}$$

The *dispersion equation of the gravito-acoustic waves* is then obtained by inverting the expression (7.46) for q:

$$\omega^4 - (k_0^2 + q^2 + \tfrac{1}{4}\alpha^2)c^2\omega^2 + k_0^2 c^2 N^2 = 0, \tag{7.48}$$

having the solutions

$$\omega_{p,g}^2 = \tfrac{1}{2} k_{\text{eff}}^2 c^2 \left[1 \pm \sqrt{1 - \frac{4k_0^2 N^2}{k_{\text{eff}}^4 c^2}} \right], \quad k_{\text{eff}}^2 \equiv k_0^2 + q^2 + \tfrac{1}{4}\alpha^2, \tag{7.49}$$

where k_{eff} is the effective total 'wave number' (quotation marks since the vertical dependence of the perturbations does not correspond to a plane wave) and k_0 is the horizontal wave number. The branch with the $+$ sign refers to *the acoustic waves* or *p-modes* (pressure driven), and the branch with the $-$ sign refers to the *gravity waves* or *g-modes* (gravity driven).

In Fig. 7.3 we have drawn the curves for $\omega^2 = \omega^2(k_0^2)$ for different values of the vertical 'wave number' q. The curves for $q^2 = 0$ (dotted), which represent the so-called *turning point* frequencies ω_{p0}^2 and ω_{g0}^2, mark the boundaries between propagation (q real) and evanescence (q imaginary) of the waves. For increasing values of q^2, the eigenfrequencies ω^2 of the *p*-modes monotonically increase with an *accumulation* or *cluster point* at ∞: $\omega_p^2 \to \omega_P^2 = \infty$. This monotonicity of the eigenfrequency with the number of nodes of the eigenfunctions, so-called

7.2 Hydrodynamic waves of a gravitating slab

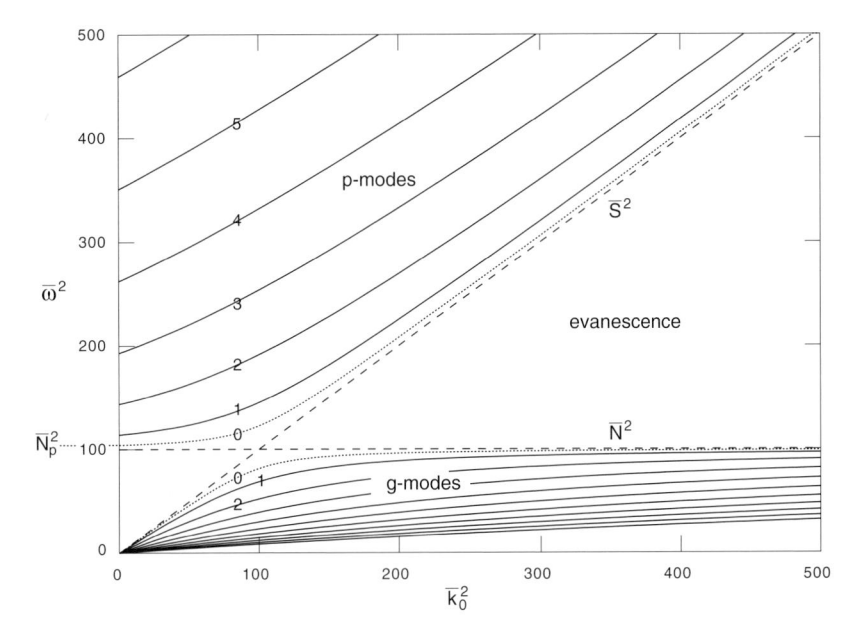

Fig. 7.3. Dispersion diagram for the p- and g-modes of a plane, exponential, atmosphere. The squared frequencies in normalized units, $\bar{\omega}^2 = \bar{\omega}^2(\bar{k}_0^2, \bar{q}^2)$, are plotted for ten values of the vertical 'wave number' $\bar{q} = n\pi$ ($n = 1, 2, \ldots, 10$). The normalized Brunt–Väisälä frequency $\bar{N} = 10$, the normalized acoustic cut-off frequency $\bar{N}_p = \frac{1}{2}\bar{\alpha} = 10.21$. Dotted curves represent the turning point frequencies ($n = 0$), dashed curves are the asymptotic values for $\bar{k}_0 \to \infty$.

Sturmian behaviour, is a well-known property of sound waves and musical instruments. On the other hand, the eigenfrequencies of the g-modes monotonically decrease (*anti-Sturmian* behaviour) with an accumulation point at 0: $\omega_g^2 \to \omega_G^2 = 0$. The lowest branch ($q^2 = 0$) of the p-modes crosses the ω^2 axis at the value N_p^2, given by

$$N_p^2 \equiv \omega_{p,\mathrm{cut}}^2 \equiv \omega_{p0}^2(k_0^2 = 0) = \tfrac{1}{4}\alpha^2 c^2 = \frac{\gamma^2 g}{4c^2} \quad (\geq N^2), \tag{7.50}$$

which is called *the acoustic cutoff frequency*. For $k_0^2 \to \infty$, the acoustic branches asymptotically tend to the ordinary sound wave frequency, which is called the Lamb frequency in this context, $\omega_p^2 \approx S^2 \equiv k_0^2 c^2$, whereas the g-modes asymptotically tend to the Brunt–Väisälä frequency, $\omega_g^2 \approx N^2$ (dashed curves).

The split in Sturmian ($\partial \omega^2/\partial q^2 > 0$) and anti-Sturmian ($\partial \omega^2/\partial q^2 < 0$) discrete modes is a quite general one for wave propagation in inhomogeneous media. We will prove in Section 7.5.1, for the more general MHD case, that this difference is associated with the sign of the coefficient in front of the highest derivative of the differential equation, i.e.

$$\frac{\rho c^2 \omega^2}{\omega^2 - k_0^2 c^2}, \tag{7.51}$$

for the gravito-acoustic waves described by Eq. (7.34). Hence, the marginal frequency ($\omega^2 = 0$) of the numerator and the Lamb frequency ($\omega^2 = k_0^2 c^2$) of the denominator delimit the frequency ranges where the modes are Sturmian or anti-Sturmian.

For the interpretation of the dispersion curves of Fig. 7.3, note that all frequencies and wave numbers have been made dimensionless by means of the sound speed c and the height a of the layer: $\bar{\omega} \equiv (a/c)\,\omega$, $\bar{k}_0 \equiv k_0 a$, $\bar{q} \equiv qa$, etc. To get estimates for the p-mode frequencies of the Sun, let us choose $a \equiv R_\odot = 7 \times 10^8$ m and a value for the sound speed c somewhere in the interior of the Sun (using data from Unno et al. [234]):

$$r = 0.5 R_\odot, \quad c = 3 \times 10^5 \text{ m s}^{-1} \quad \Rightarrow \quad f_p \equiv \frac{c}{2\pi a} = 6.8 \times 10^{-5} \text{ Hz}. \tag{7.52}$$

With the factor f_p we may convert the normalized angular frequencies $\bar{\omega}$ into the actual frequencies ν of the modes, e.g.

$$\begin{aligned}
l &= 0, & n &= 0 & &\Rightarrow & \nu_{p,\text{cut}} &\equiv \bar{N}_p \cdot f_p = 0.68 \text{ mHz}, \\
l &= 0, & n &= 20 & &\Rightarrow & \nu_p &\approx n\pi \cdot f_p = 4.3 \text{ mHz}, \\
l &= 100, & n &= 1 & &\Rightarrow & \nu_p &\approx l \cdot f_p = 6.8 \text{ mHz},
\end{aligned}$$

where $l \approx \bar{k}_0$ is the number of nodal lines on the spherical surface and $n = \bar{q}/\pi$ is the radial quantum number. These frequencies lie in the range of the observed *5 minute oscillations* ($\nu_p = 3.3$ mHz). To go beyond these order of magnitude estimates, one has to consider the spherical geometry and the actual inhomogeneities of the Sun.

▷ **Free-boundary modes.** The addition of a separate low-density layer on top of the exponential one we considered so far would permit the investigation of surface modes, analogous to the analysis of the Rayleigh–Taylor instability of Section 6.6.4. In the solar context, these oscillations are called *f-modes*, which are free-boundary oscillations of the solar surface involving both the interior and the atmosphere of the Sun. Their frequency is intermediate between the lowest n branches of the p- and g-modes so that the f-mode could be considered as the common $n = 0$ branch of these modes. However, an estimate of the short-wavelength limit of the frequency of incompressible surface waves (which remains valid for compressible fluids) shows that the mechanism is really gravitational:

$$\omega_f^2 \approx k_0 g. \tag{7.53}$$

Notice that, in contrast to the problem of Section 6.6.4, gravity is now pointing into the heavier fluid so that the waves are stable. Inserting the value of the gravitational acceleration at the solar surface, we obtain the following approximation of the frequency of the

7.2 Hydrodynamic waves of a gravitating slab

f-modes:

$$v_f \equiv \frac{\omega_f}{2\pi} \approx \sqrt{l} \cdot f_f \,, \qquad f_f \equiv \frac{1}{2\pi}\sqrt{\frac{GM_\odot}{R_\odot^3}} \approx 0.1 \text{ mHz} \,. \tag{7.54}$$

Since helioseismology actually depends on surface oscillations for detection, the extension with free-boundary modes is an important step towards reality of the model. For further discussion of surface modes, and generalization to MHD, see the review by Roberts [193]. ◁

7.2.4 Helioseismology and MHD spectroscopy

For educational purposes, we have treated the *p*- and *g*-modes in a highly idealized medium, characterized by constant values of the sound speed and the gravitational acceleration. Of course, in the Sun these conditions are not valid. Moreover, one should treat the problem in *spherical geometry*. This gives rise to a quantization condition for the horizontal wave number: $k_0 R_\odot = \sqrt{l(l+1)}$, where l is the degree of the mode, whereas the vertical wave number q and associated quantization condition (7.47) also require a modification to account for the inhomogeneities of the interior of the Sun.

▷ **Spherical geometry.** For spherical symmetry, the normal mode expansion (7.31) is to be replaced by

$$\xi_r(r,\theta,\phi,t) = \hat{\xi}(r)\, Y_l^m(\theta,\phi)\, e^{-i\omega t} \,,$$

$$\xi_\theta(r,\theta,\phi,t) = \left[\hat{\eta}(r)\frac{\partial Y_l^m}{\partial \theta} + \hat{\zeta}(r)\frac{1}{\sin\theta}\frac{\partial Y_l^m}{\partial \phi}\right] e^{-i\omega t} \,, \tag{7.55}$$

$$\xi_\phi(r,\theta,\phi,t) = \left[\hat{\eta}(r)\frac{1}{\sin\theta}\frac{\partial Y_l^m}{\partial \phi} - \hat{\zeta}(r)\frac{\partial Y_l^m}{\partial \theta}\right] e^{-i\omega t} \,,$$

where $Y_l^m(\theta,\phi)$ are spherical harmonics, well known from quantum mechanics of spherically symmetric systems. Like in quantum mechanics, the actual eigenfrequencies are determined from the radial wave equation which brings in the radial mode number n (number of nodes of the radial wave function). The degree l (total number of nodal circles on the spherical surface) takes the place of the horizontal wave number k_0, and the system is degenerate with respect to the longitudinal order m (number of nodal circles through the poles, $|m| \leq l$). This degeneracy is lifted when differential rotation is taken into account.

In the Cowling approximation, where perturbations of the gravitational potential are neglected, the original vector eigenvalue problem in terms of $\boldsymbol{\xi}$ may be reduced to the solution of just one ODE for $\hat{\xi}$, the radial wave equation. Similar to the plane case of Section 7.2.1, one of the three variables $\hat{\xi}$, $\hat{\eta}$, $\hat{\zeta}$ vanishes:

$$\omega^2 \hat{\zeta} = 0 \quad \Rightarrow \quad \hat{\zeta} = 0 \quad (\text{if } \omega^2 \neq 0) \,, \tag{7.56}$$

i.e. when marginal toroidal shifts are ignored ($\omega^2 = 0$, $\hat{\zeta} \neq 0$). Then, the component $\hat{\eta}$ may be eliminated,

$$\hat{\eta} = -\frac{(c^2/r^2)(r^2\hat{\xi})' - g\hat{\xi}}{r^2(\omega^2 - S^2)} \,, \tag{7.57}$$

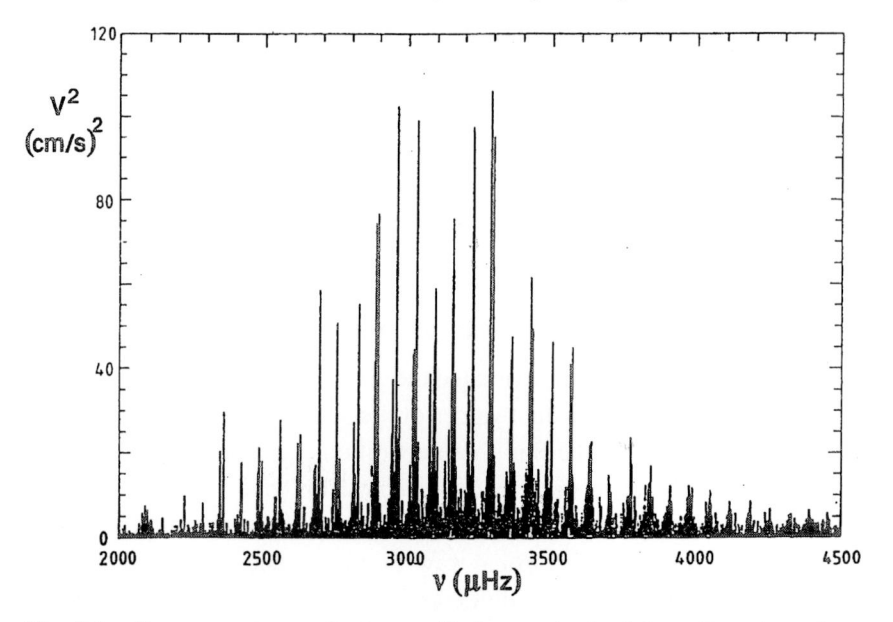

Fig. 7.4. Power spectrum of solar oscillations, obtained from Doppler velocity measurements in light integrated over the solar disc. (From Christensen-Dalsgaard [56], citing A. Claverie, G. R. Isaak, C. P. McLeod, H. B. van der Raay, P. L. Palle and T. Roca Cortes, *Mem. Soc. Astron. Ital.* **55**, 63 (1984).)

which, upon substitution, yields the radial wave equation

$$\frac{d}{dr}\left(\frac{\omega^2}{\omega^2 - S^2}\frac{\gamma p}{r^2}\frac{d}{dr}(r^2\hat{\xi})\right) + \left[\rho\omega^2 - \frac{\rho g^2}{r^2}\frac{l(l+1)}{\omega^2 - S^2} + \rho' g - r^2\left(\frac{\rho g}{r^2}\frac{\omega^2}{\omega^2 - S^2}\right)'\right]\hat{\xi} = 0, \quad (7.58)$$

where $S^2 \equiv l(l+1)c^2/r^2$ is the Lamb frequency. Note the similarity with the analogous equation for the H-atom in quantum mechanics, but also notice that there is considerably more structure to this problem (due to the presence of three background fields $\rho(r)$, $p(r)$, $g(r)$, and the reduction of a vector eigenvalue problem to a scalar wave equation, with a much more complicated dependence on the eigenvalue parameter). ◁

When the theory is modified in this manner and applied to the observations of the *p*-mode spectra, it provides a powerful tool for probing the interior of the Sun, called *helioseismology*; see Christensen-Dalsgaard [56] or Harvey [105]. Broadly speaking, this is done by 'inversion' of the observed spectra to obtain the equilibrium profiles that produced them. Fig. 7.4 shows a typical example of a *classical* spectrum of solar oscillations observed by means of a measurement of the Doppler shifts of the *quantum mechanical* spectral lines of the sunlight. At this point, it is important to realize that the Doppler shifts themselves are the object of study here since they are directly related to the radial oscillations of the photosphere. With observed radial velocities $\tilde{v}_r \sim 1$ km/s, it is quite all right to exploit linear

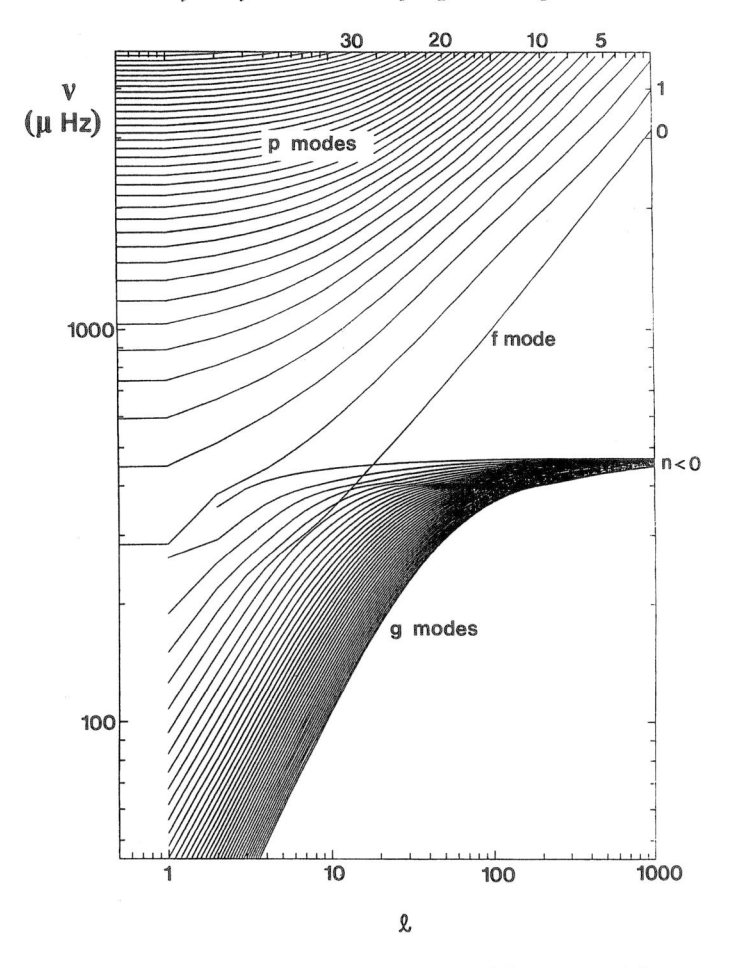

Fig. 7.5. Solar oscillations for a standard solar model: computed frequency as a function of the degree ℓ for different values of the radial mode number n. (From Christensen-Dalsgaard [56].)

theory since the amplitudes reached in typical 5 min periods are not larger than $5 \times 10^{-4} R_\odot \ll R_\odot$.

Above, 'inversion' of the solar oscillation spectrum is put in quotation marks to indicate that it is not a mechanical procedure that can be applied blindly. Whereas computing a spectrum for a particular equilibrium can be carried out numerically (see, e.g., Fig. 7.5), virtually to any degree of accuracy desired, the inverse problem requires a much larger effort. (It is not so difficult to take fingerprints at the site of the crime, but it is usually quite difficult to catch the criminal that produced them.) It consists of computing the spectrum while iterating on infinitely many possible values of the equilibrium input parameters until matching is obtained.

This procedure is ill-posed, a mathematical way of expressing that it is really a kind of art.

Also notice that a spectrum of solar oscillations, like that shown in Fig. 7.4, does not exhibit the modes on an equal footing. First, out of the infinity of modes that are possible, only a finite number are observed that have been excited by some mechanism, which, in itself, is part of the puzzle. Second, the spectral peaks observed do not correspond to the energy contents of the modes but to the amplitudes of the eigenfunctions at the position where the Doppler shifts are produced, that is in the photosphere. Hence, there is strong biasing towards modes that have a sizeable amplitude there.

One finds that p-modes of low order l penetrate the deep interior of the Sun, whereas high l modes in general are more localized towards the outside. This high l localization is most outspoken for the f-modes, which are free-boundary gravitational oscillations of the solar surface. (Note that the frequency estimate given in Eq. (7.54) checks quite well with the numbers of Fig. 7.5.) In general, g-modes are *cavity modes* trapped in the region interior to the convection zone and, hence, quite difficult to observe. The spectrum of p-modes, however, has been determined with surprising accuracy. It has been compared with computed values based on theoretical solar models. It turns out that the frequencies deduced from the Doppler shifts of spectral lines agree with the calculated ones for the p-modes to within 0.1%! This impressive agreement serves as an example for what is possible in a purely classical (as opposed to quantum mechanical) kind of spectroscopy.

▷ **Caveat on the analogy with quantum mechanical spectral theory.** In Section 6.5.4, we have stressed that the analogy of MHD (and, hence, HD) spectral theory with quantum mechanics is through the mathematics of linear operators, not through the physics of the systems. Consequently, one should be careful not to make tacit assumptions on the physical interpretation of the spectra. For example, the (accidental) degeneracy of the energy levels $E(n, l)$ of the hydrogen atom (due to the Coulomb potential), with the restriction $l \leq n$, and the association of l with the angular momentum operator are strictly limited to the quantum mechanical interpretation of the radial wave equation for $\psi(r)$. From Fig. 7.5 it is evident that the eigenfrequencies $\nu(n, l)$ of the solar oscillations do not exhibit this degeneracy and restriction on l. They are obtained from the radial wave equation (7.58) for the fluid displacement $\xi(r)$, which not only has a completely different radial structure but also a completely different physical interpretation. In particular, whereas l is associated with the number of nodal lines on the sphere in both cases, the quantum mechanical association with the angular momentum operator certainly does not carry over to fluid dynamics. ◁

On the basis of this successful example of fluid dynamical spectroscopy, we have proposed a similar activity for laboratory fusion plasmas, called *MHD spectroscopy* [88]. In order to understand the common feature, the procedure of helioseismology has been summarized in Fig. 7.6 to provide a kind of model example for other physical systems, like laboratory fusion machines. Notice that for the birth of helioseismology it was not sufficient to have collected a great number of

7.2 Hydrodynamic waves of a gravitating slab

Solar Model:
$X(r), Y(r), Z(r)$
$T(r), \rho(r), L(r)$

$\rho(r), T(r)$ →

Spectral Code:
$\hat{\xi}(r) Y_l^m(\theta, \phi) e^{i\omega t}$
(p & g modes)

→ $\{\omega_{l,n}\}_{\text{theory}}$

Extensions:
$\Omega(r, \theta)$ – diff. rotation
$\mathbf{B}(r, \theta)$ – magn. field
$f(t)$ – stellar evolution

Observations:
Doppler shifts of spectral lines

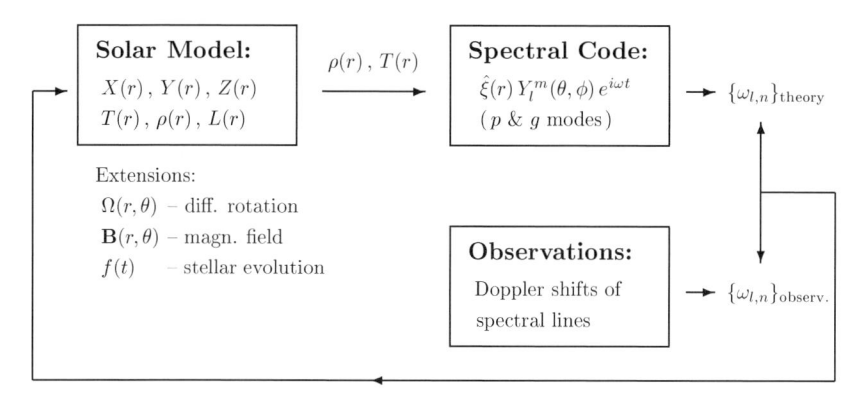

→ $\{\omega_{l,n}\}_{\text{observ.}}$

Fig. 7.6. Systematics of helioseismology.

observations of solar oscillations or to have calculated many spectra, but rather that the two activities should have matured to a sufficient degree to allow for a unique correlation between calculated and observed spectra. This interplay between theory and observations is illustrated in the figure. To leading order, the Solar Model is spherically symmetric with the equilibrium depending on the abundance of the different chemical elements H, He, and the heavier ones (indicated by the symbols X, Y, and Z), the temperature T, the density ρ, the luminosity L, etc. Perturbations of this equilibrium are studied by means of a Spectral Code which determines the collection $\{\omega_{l,n}\}$ of the frequencies of the different modes. Observations provide values of the same quantities. The discrepancies between the two then bring about improvement of the Solar Model, where the non-spherical symmetry due to the presence of differential rotation (Ω) and magnetic fields (**B**) plays an important role. Of course, the final interest is in stellar evolution, i.e. the very slow time-dependence of all the equilibrium quantities that are determined on a snapshot basis by helioseismology.

It is not difficult to recognize the counterparts of the three activities depicted in Fig. 7.6 for laboratory fusion research: transport models to describe the slow evolution of the background equilibrium, spectral calculations of the fast MHD modes, and diagnostics to determine the frequencies and other characteristics of these modes. Clearly, the success of the similar activity of MHD spectroscopy hinges on improved diagnostics and spectral calculations. Helioseismology provides the paradigm demonstrating that fluid dynamical states can be predicted with high accuracy. We will see in a later chapter that this should be realizable for tokamaks as well. Another example of this kind of HD or MHD spectroscopy has been developed in the solar context with the aim of determining the hydrodynamical and magnetohydrodynamical properties of sunspots from the absorption of sound waves propagating in the photosphere. This activity has been called *sunspot seismology* by Bogdan [34, 33]. Many other plasma-astrophysical objects may be investigated this way, e.g. there is a recent proposal to initiate *magneto-seismology of accretion discs* by Keppens *et al.* [124].

At this point in our exposition, it might appear that extension of the 1D spherical HD model to MHD with magnetic effects is the next logical step. However, recall that spherically symmetric magnetic fields are incompatible with $\nabla \cdot \mathbf{B} = 0$ (Section 1.3.4). Hence, incorporation of magnetic fields in spherical geometry inevitably involves 2D magnetic configurations (as indicated by the extensions listed outside the 1D Solar Model box in Fig. 7.6). In the following sections we will stay with 1D geometry and introduce magnetic fields in the plane slab model. This will give rise to plenty of new phenomena associated with plasma inhomogeneity. The further complications of two-dimensionality will be the subject of later chapters in Volume 2.

7.3 MHD wave equation for a gravitating magnetized plasma slab

7.3.1 Preliminaries

With the instructive examples of hydrodynamic waves and instabilities in ordinary fluids of Section 7.2, we now turn to the influence of inhomogeneity on the spectrum of waves and instabilities in *magnetized* plasmas. Macroscopic plasma dynamics inevitably involves effects of the *finite* geometry of the magnetic confinement volume. This substantially complicates the analysis since the assumption (made in Chapter 5 for homogeneous plasmas) of plane waves represented by uncoupled Fourier harmonics breaks down. Waves with different wave numbers **k** couple through plasma inhomogeneity so that *wave transformation* takes place. It is perfectly well possible to have a wave that should be characterized as a fast magneto-sonic wave in one part of the plasma volume, but which exhibits Alfvén wave properties in another part.

More important yet, from a fundamental as well as from a practical point of view, is the occurrence of two new phenomena, viz. *instabilities* and *continuous spectra*. Both are essentially due to plasma inhomogeneity, but instabilities (as the hydrodynamic convective instability encountered in Section 7.2.2) obviously occur for $\omega^2 < 0$, whereas continua are mainly located on the stable side ($\omega^2 > 0$) of the spectrum in ideal MHD. These phenomena are extremely important for a wide class of laboratory and astrophysical plasmas. This will be extensively demonstrated in the later chapters, but first we have to get acquainted with the necessary analytical tools.

We will approach the problem stepwise and, at first, restrict the analysis to inhomogeneity in one direction only (called *one-dimensional* or 1D systems) leading

7.3 MHD wave equation for a gravitating plasma slab

to a description in terms of ordinary differential equations (ODEs). A generic example of such systems is the plane gravitating plasma slab with inhomogeneities $\rho(x)$, $p(x)$ and $\mathbf{B}(x)$ of the density, pressure and magnetic field, caused by plasma currents and gravitational stratification. We have already encountered the hydrodynamic version of the basic ODE in the wave equation (7.34) of Section 7.2.1. The fundamental wave equation for the gravitating magnetized plasma slab is derived in Section 7.3.2. This equation describes a bewildering variety of waves, continua and instabilities, as will be shown in Sections 7.4 and 7.5.

(a) Transformation of the homogeneous problem As a preliminary to the study of inhomogeneous plasmas, it is expedient to restate the *homogeneous* wave problem of Section 5.2 in a slightly different manner. Recall that the equilibrium background in that case was specified by

$$\mathbf{B} = B\mathbf{e}_z, \quad \text{with } \rho, p, B = \text{const}, \tag{7.59}$$

so that the sound and Alfvén speeds, $c \equiv \sqrt{\gamma p/\rho}$ and $b \equiv B/\sqrt{\rho}$, are constant. We have already transformed the eigenvalue problem (5.51) in terms of the force operator and associated displacement vector in the spectral equation (6.31) of Section 6.2.1. Since, for the time being, all equilibrium quantities are considered constant, we may write the normal mode amplitude $\boldsymbol{\xi}(\mathbf{r})$ as a Fourier integral (or a Fourier series if one considers a finite box) of plane wave solutions which do not couple:

$$\boldsymbol{\xi}(\mathbf{r}) = (2\pi)^{-3/2} \iiint \hat{\boldsymbol{\xi}}(\mathbf{k}) \, e^{i\mathbf{k}\cdot\mathbf{r}} \, d^3k . \tag{7.60}$$

The normal modes $\hat{\boldsymbol{\xi}}(\mathbf{k}; \omega) \exp i(\mathbf{k}\cdot\mathbf{r} - \omega t)$ may then be studied separately by substituting $\nabla \rightarrow i\mathbf{k}$ in Eq. (6.31). This leads to the algebraic eigenvalue problem (6.32), which we repeat for convenience:

$$\rho^{-1}\mathbf{F}(\hat{\boldsymbol{\xi}}) = \left[-(\mathbf{k}\cdot\mathbf{b})^2 \mathbf{I} - (b^2 + c^2)\mathbf{k}\mathbf{k} + \mathbf{k}\cdot\mathbf{b}(\mathbf{k}\mathbf{b} + \mathbf{b}\mathbf{k}) \right] \cdot \hat{\boldsymbol{\xi}} = -\omega^2 \hat{\boldsymbol{\xi}}. \tag{7.61}$$

The matrix representation of this equation is identical to Eq. (5.52) with $\hat{v}_{x,y,z}$ replaced by $\hat{\xi}_{x,y,z}$. However, we will now generalize the wave vector $\mathbf{k}$ to have a non-vanishing component k_y in order to be able to distinguish between the two perpendicular directions x and y (Fig. 7.7). This will facilitate the transition to the analysis of the inhomogeneous problem, where x and y will no longer be equivalent since the eventual inhomogeneity will be in the x-direction.

The mentioned transformation amounts to a rotation of the coordinate system through an angle $-\chi \equiv -\arctan(k_y/k_x)$ about the z-axis. (The minus sign appears here because $\mathbf{k}_\perp$ is along the old x'-axis.) We indicate the old representation

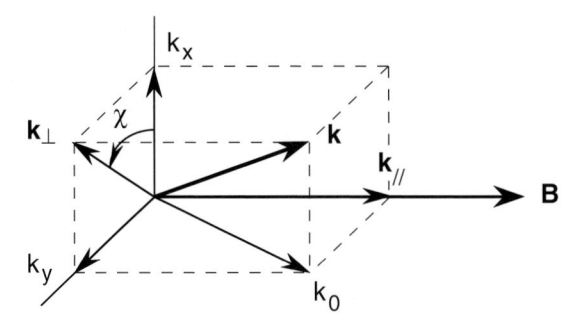

Fig. 7.7. New orientation of the **k** vector.

by primes, $\rho^{-1} \mathbf{F}' \cdot \boldsymbol{\xi}' = -\omega^2 \boldsymbol{\xi}'$, where the sans serif bold symbol $\mathbf{F}'$ is used to denote the (old) matrix representation of the operator **F**. The new representation then becomes

$$\rho^{-1}\mathbf{F} \cdot \boldsymbol{\xi} = -\omega^2 \boldsymbol{\xi}, \qquad \text{with} \quad \mathbf{F} = \mathbf{R} \cdot \mathbf{F}' \cdot \mathbf{R}^{-1} \quad \text{and} \quad \boldsymbol{\xi} = \mathbf{R} \cdot \boldsymbol{\xi}', \qquad (7.62)$$

where

$$\mathbf{R} \equiv \begin{pmatrix} \cos\chi & -\sin\chi & 0 \\ \sin\chi & \cos\chi & 0 \\ 0 & 0 & 1 \end{pmatrix}. \qquad (7.63)$$

This results in the following matrix representation of the spectral equation (7.61):

$$\begin{pmatrix} -k_x^2(b^2+c^2)-k_z^2 b^2 & -k_x k_y(b^2+c^2) & -k_x k_z c^2 \\ -k_x k_y(b^2+c^2) & -k_y^2(b^2+c^2)-k_z^2 b^2 & -k_y k_z c^2 \\ -k_x k_z c^2 & -k_y k_z c^2 & -k_z^2 c^2 \end{pmatrix} \begin{pmatrix} \xi_x \\ \xi_y \\ \xi_z \end{pmatrix} = -\omega^2 \begin{pmatrix} \xi_x \\ \xi_y \\ \xi_z \end{pmatrix}, \qquad (7.64)$$

where we have dropped the hats on $\boldsymbol{\xi}$ for notational simplicity. Note that Eq. (5.52) is recovered, as it should, when the limit $k_y \to 0$ is taken. Solutions are obtained by setting the determinant of the system to zero. Of course, this results in the same dispersion equations (5.56) for the Alfvén eigenvalues ω_A^2 and (5.58) for the magneto-sonic eigenvalues $\omega_{s,f}^2$:

$$\omega_A^2 = k_\parallel^2 b^2, \qquad \omega_{s,f}^2 = \tfrac{1}{2} k^2 (b^2+c^2) \left[1 \pm \sqrt{1 - \frac{4 k_\parallel^2 b^2 c^2}{k^2 (b^2+c^2)^2}} \right], \qquad (7.65)$$

where now

$$k^2 = k_x^2 + k_y^2 + k_z^2, \qquad k_\perp = \sqrt{k_x^2 + k_y^2}, \qquad k_\parallel = k_z.$$

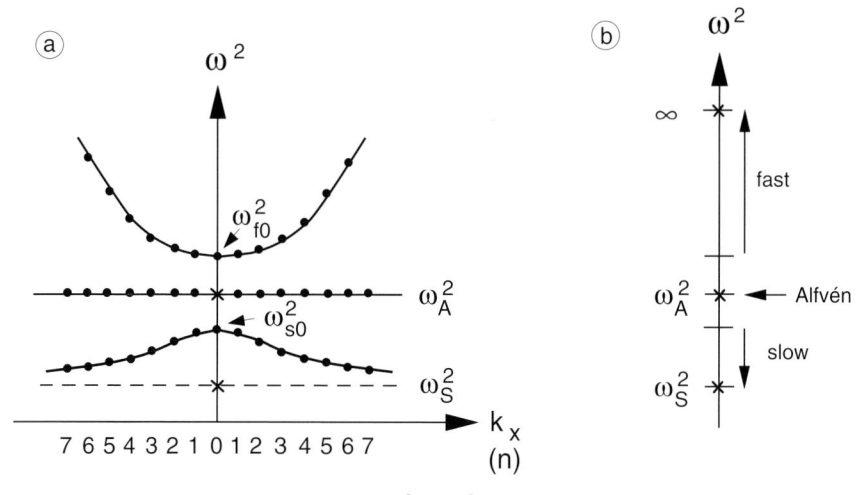

Fig. 7.8. (a) Dispersion diagram $\omega^2 = \omega^2(k_x)$ for k_y and k_z fixed; (b) corresponding structure of the spectrum.

However, the expressions (5.57) and (5.60) for the eigenvectors have to be modified to account for the changed direction of the plane spanned by **k** and **B** (previously, the x'–z' plane) since the Alfvén eigenvectors $\boldsymbol{\xi}_A$ are perpendicular to this plane and the magneto-sonic eigenvectors $\boldsymbol{\xi}_{s,f}$ are lying in it.

(b) Essential spectrum The spectrum of MHD waves is modified significantly by plasma inhomogeneity. If the inhomogeneity is in the x-direction, the assumption of uncoupled plane waves breaks down with respect to the wave number k_x so that we have to replace k_x by $-i\partial/\partial x$ in Eq. (7.64). Hence, the dispersion equation $\omega^2 = \omega^2(k_x, k_y, k_z)$ loses its meaning with respect to the dependence on k_x. The consequences of this will be worked out in the next section. However, one essential feature of the spectrum remains intact, that was already alluded to in the discussion of the asymptotic properties associated with the group velocity (Section 5.3.3). This relates to *localized wave motion in the direction of inhomogeneity*, which can be understood already from the homogeneous theory by considering the limit $k_x \to \infty$. To that end, we plot ω^2 as a function of k_x, keeping k_y and k_z fixed (Fig. 7.8(a)). Note that this picture is very similar to Fig. 5.7(b), but the degeneracy of Alfvén and fast modes at $k_\perp = 0$ is removed.

The effects of a confined inhomogeneous plasma can now be mimicked, within the context of homogeneous theory, by considering the plasma as a slab of finite extension enclosed by plates at $x = 0$ and $x = a$. The wave number k_x will then be quantized: $k_x = n\pi/a$, where n is the *number of nodes* of the eigenfunction ξ_x considered as a function of x. Hence, the eigenvalues will be labelled by n, rather than by k_x, and this labelling still makes sense for an inhomogeneous medium with equilibrium quantities that vary in the x-direction. This quantization produces a discrete spectrum, with three branches that behave distinctly differently (Fig. 7.8(b)). The essential features of these three sub-spectra of point eigenvalues, labelled by n, are the following ones.

(1) The discrete eigenvalues of the fast sub-spectrum monotonically increase, so that

$$\omega_F^2 \equiv \lim_{k_x \to \infty} \omega_f^2 \approx \lim_{k_x \to \infty} k_x^2(b^2 + c^2) = \infty \qquad (7.66)$$

is a formal *cluster point* of the fast wave point spectrum.

(2) The eigenvalues ω_a^2 of the Alfvén sub-spectrum are *infinitely degenerate*, because they do not depend on k_x, so that

$$\omega_A^2 \equiv \lim_{k_x \to \infty} \omega_a^2 = \omega_a^2 = k_\parallel^2 b^2 . \qquad (7.67)$$

Hence there is no need to distinguish between the point eigenvalues ω_a^2 and the limit ω_A^2. (That is why we did not even introduce the symbol ω_A^2 before.)

(3) The discrete eigenvalues of the slow wave sub-spectrum monotonically decrease with an *accumulation* or *cluster point* at

$$\omega_S^2 \equiv \lim_{k_x \to \infty} \omega_s^2 = k_\parallel^2 \frac{b^2 c^2}{b^2 + c^2} , \qquad (7.68)$$

where the subscript s denotes the slow discrete modes and S their cluster point.

Mathematically speaking, ω_F^2, ω_A^2, and ω_S^2 belong to what is called the *essential spectrum*, which is the manifestation of the continuous spectrum in this context. This turns out to be basic for the discussion of the inhomogeneous case, where the functions vary with x. Then, the infinite degeneracy of the Alfvén eigenvalues is lifted by the appearance of a *continuous spectrum of improper Alfvén modes* instead, and the cluster point of the slow point spectrum is spread out in a *continuous spectrum of improper slow modes*, whereas the fast cluster point $\omega_F^2 = \infty$ is unaffected.

The two values of ω^2 denoted by ω_{s0}^2 and ω_{f0}^2, where the slow and the fast modes cross the vertical axis in Fig. 7.8(a), also turn out to play a special role in the spectral theory of inhomogeneous plasmas. Their values are given by:

$$\omega_{s0, f0}^2 \equiv \omega_{s,f}^2(k_x = 0) = \tfrac{1}{2}k_0^2(b^2 + c^2)\left[1 \pm \sqrt{1 - \frac{4k_\parallel^2 b^2 c^2}{k_0^2(b^2 + c^2)^2}}\right], \qquad (7.69)$$

where $k_0 \equiv \sqrt{k_y^2 + k_z^2}$ is the horizontal wave number (see Fig. 7.7). The function of these *turning point frequencies* ω_{s0}^2 and ω_{f0}^2 appears to be to separate the three

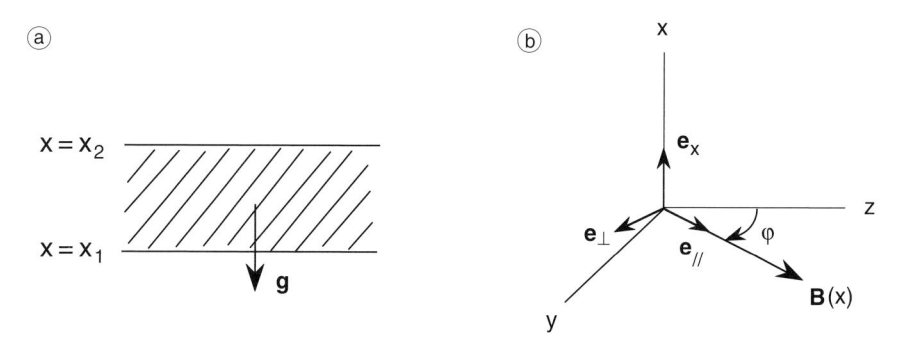

Fig. 7.9. (a) Gravitating plasma slab and (b) magnetic field line projection.

branches of the spectrum, as evidenced by the following sequence of inequalities:

$$0 \leq \omega_S^2 \leq \omega_s^2 \leq \omega_{s0}^2 \leq \omega_A^2 \leq \omega_{f0}^2 \leq \omega_f^2 \leq \omega_F^2 = \infty. \quad (7.70)$$

This clear separation of the three discrete sub-spectra for homogeneous media returns, in a modified form, when the plasma becomes inhomogeneous.

▷ **Exercise.** Check the above expressions to teach yourself the basic steps. All of this trivial algebra returns in operator form in later sections! ◁

7.3.2 Derivation of the MHD wave equation for a gravitating slab

We now introduce genuine inhomogeneity in the model. Consider a plasma slab, infinite and homogeneous in the y- and z-directions, and contained between two planes at $x = x_1$ and $x = x_2$ (Fig. 7.9(a)). The equilibrium is assumed to vary in the x-direction:

$$\mathbf{B} = B_y(x)\,\mathbf{e}_y + B_z(x)\,\mathbf{e}_z, \quad \rho = \rho(x), \quad p = p(x). \quad (7.71)$$

The magnetic field is confined to plane layers parallel to the y-z plane, but its direction is assumed to vary with height. This is caused by current layers that are confined to the same planes (to be considered as the *magnetic surfaces* for this configuration):

$$\mathbf{j} = \nabla \times \mathbf{B} = -B_z'(x)\,\mathbf{e}_y + B_y'(x)\,\mathbf{e}_z, \quad (7.72)$$

where primes denote differentiation with respect to x. In general, $\mathbf{j}$ has a different direction from $\mathbf{B}$ so that there is a Lorentz force $\mathbf{j} \times \mathbf{B} = (\nabla \times \mathbf{B}) \times \mathbf{B} = -\nabla(\frac{1}{2}B^2)$ in the x-direction. (The latter relation follows by application of Eq. (A.8) where the additional term $\mathbf{B} \cdot \nabla \mathbf{B}$, which is present for curved magnetic fields, vanishes here.) This force is to be balanced by the pressure gradient ∇p and the gravitational force $\rho(x)\mathbf{g}$, due to an external gravitational field Φ with constant acceleration

$$\mathbf{g} = -\nabla\Phi = -\hat{g}\mathbf{e}_x. \tag{7.73}$$

Here, a hat is put on the gravitational acceleration since the ordinary symbol g will be used for another purpose shortly. Consequently, equilibrium requires satisfaction of just one ODE,

$$(p + \tfrac{1}{2}B^2)' = -\rho\hat{g}, \tag{7.74}$$

which is the only restriction to be imposed on the possible choices of the functions $\rho(x)$, $p(x)$, $B_y(x)$, and $B_z(x)$.

In this manner, a generic 1D inhomogeneous model for confined plasmas is obtained. The presence of the two bounding plates may appear to be a rather artificial constituent of the model, in particular for astrophysical plasmas. Their function is to simply model plasma confinement, either by the presence of a conducting shell (as in tokamaks), or by the presence of an immobile (very dense) neighbouring plasma. One could always try to minimize their influence by moving them out to infinity, while keeping their essential role of eliminating the flow of energy in or out of the system. Of course, the ultimate remedy is to extend the analysis to an appropriate choice of one of the models (II)–(VI) of Section 4.6.

The wave and spectral equations for the gravitating plasma slab are obtained from Eqs. (6.23) and (6.28), which we repeat for convenience:

$$\mathbf{F}(\boldsymbol{\xi}) = \rho \frac{\partial^2 \boldsymbol{\xi}}{\partial t^2} = -\rho\omega^2\boldsymbol{\xi}. \tag{7.75}$$

Clearly, the most laborious part of the problem is the evaluation of the force operator

$$\mathbf{F}(\boldsymbol{\xi}) \equiv -\nabla\pi - \mathbf{B}\times(\nabla\times\mathbf{Q}) + (\nabla\times\mathbf{B})\times\mathbf{Q} + \nabla\Phi\nabla\cdot(\rho\boldsymbol{\xi}), \tag{7.76}$$

with

$$\pi = -\gamma p \nabla\cdot\boldsymbol{\xi} - \boldsymbol{\xi}\cdot\nabla p, \qquad \mathbf{Q} = \nabla\times(\boldsymbol{\xi}\times\mathbf{B}). \tag{7.77}$$

This will be done in the following sequence of steps.

(a) Field line projection First, normal modes are studied satisfying the spectral equation (7.75), but now the 3D Fourier representation (7.60) is replaced by a *2D Fourier representation* for the two homogeneous directions only:

$$\boldsymbol{\xi}(\mathbf{r}) = \frac{1}{2\pi}\iint \hat{\boldsymbol{\xi}}(x; k_y, k_z)\, e^{i(k_y y + k_z z)}\, dk_y dk_z. \tag{7.78}$$

7.3 MHD wave equation for a gravitating plasma slab

Our task is to find the x-dependence of the separate Fourier components, which simply will be denoted as $\boldsymbol{\xi}(x)\exp i(k_y y + k_z z)$, dropping the hat. The horizontal part of the differential operator ∇ then produces the algebraic substitutions $\partial_y \to ik_y$ and $\partial_z \to ik_z$, whereas the component ∂_x in the direction of inhomogeneity will produce (ordinary) differential equations.

Next, all vectors are projected onto the three unit vectors $\mathbf{e}_x$, $\mathbf{e}_\perp$, $\mathbf{e}_\parallel$ (*the field line triad*, Fig. 7.9(b)), which represent the physically relevant directions:

$$\begin{aligned}\mathbf{e}_x &\equiv \nabla x\,, \\ \mathbf{e}_\perp &\equiv (\mathbf{B}/B)\times\mathbf{e}_x = (0,\ B_z/B,\ -B_y/B)\,, \\ \mathbf{e}_\parallel &\equiv \mathbf{B}/B \qquad\qquad\ = (0,\ B_y/B,\ B_z/B)\,.\end{aligned} \qquad (7.79)$$

This *orthogonal projection*[3] is based on the physical significance of *magnetic field lines and magnetic surfaces* for the description of perturbations. It may be generalized to more complicated equilibria, like tokamaks with nested magnetic surfaces where $\mathbf{e}_x$ represents the normal, and $\mathbf{e}_\perp$ and $\mathbf{e}_\parallel$ the two tangential directions with respect to these surfaces. Hence, we will call the three directions defined by Eq. (7.79) by generic names, viz. *normal* (with respect to the magnetic surfaces), *perpendicular* (with respect to the field lines, but tangential to the magnetic surfaces so that the symbol $\perp$ is now more restrictive than in Section 7.3.1) and *parallel* (with respect to the field lines). In this projection, the result of the gradient operator on the perturbation $\boldsymbol{\xi}(x)\exp i(k_y y + k_z z)$ may be written as

$$\nabla = \mathbf{e}_x \partial_x + i\mathbf{e}_\perp(x)\, g + i\mathbf{e}_\parallel(x)\, f\,, \qquad (7.80)$$

where $\partial_x \equiv d/dx$ is the normal derivative, and f and g represent the perpendicular and parallel derivatives:

$$\begin{aligned}g = g(x) &\equiv -i\mathbf{e}_\perp\cdot\nabla = G/B\,, &\quad G &\equiv k_y B_z - k_z B_y\,, \\ f = f(x) &\equiv -i\mathbf{e}_\parallel\cdot\nabla = F/B\,, &\quad F &\equiv k_y B_y + k_z B_z\,.\end{aligned} \qquad (7.81)$$

(The lower case g and f will be replaced by the upper case G and F in later chapters where division by B becomes awkward.) The quantities g and f may be considered as the wave vectors in the perpendicular and parallel directions, although

---

[3] One frequently encounters the statement in the literature that a local field line *coordinate* representation is exploited when, at best, a *projection* like the present one is meant. At worst, the existence of coordinates $u(x,y,z)$, $v(x,y,z)$, $w(x,y,z)$ is really assumed and one of them, say w, is supposed to measure distance along the magnetic field. The existence of such coordinates would imply $\mathbf{B} = f(u,v,w)\nabla w$, where f is some scalar function. Hence $\mathbf{j} = \nabla\times\mathbf{B} = \nabla f\times\nabla w = (1/f)\nabla f\times\mathbf{B}$, so that $j_\parallel = 0$. Of course, such a severe restriction on the equilibrium is not justified in general. In plane slab geometry, that condition implies that $\mathbf{B}$ is uni-directional. Only for such trivial fields could one construct field line coordinates of that kind.

Field line coordinates of an entirely different kind, so-called *Clebsch coordinates* α, β, γ, do exist, though. Here, $\mathbf{B}$ is represented as the vector product of two gradients of them: $\mathbf{B} = \nabla\alpha\times\nabla\beta$. These coordinates are useful for the study of instabilities localized about magnetic field lines.

they are functions of x in general. However, the magnitude of the resulting horizontal wave vector is constant:

$$k_0 \equiv \sqrt{g^2 + f^2} = \sqrt{k_y^2 + k_z^2}. \qquad (7.82)$$

We also project $\boldsymbol{\xi}$ on the field line triad:

$$\boldsymbol{\xi} = \xi\, \mathbf{e}_x - i\eta\, \mathbf{e}_\perp - i\zeta\, \mathbf{e}_\|, \qquad (7.83)$$

where

$$\xi \equiv \mathbf{e}_x \cdot \boldsymbol{\xi} = \xi_x,$$
$$\eta \equiv i\mathbf{e}_\perp \cdot \boldsymbol{\xi} = i(B_z \xi_y - B_y \xi_z)/B,$$
$$\zeta \equiv i\mathbf{e}_\| \cdot \boldsymbol{\xi} = i(B_y \xi_y + B_z \xi_z)/B.$$

The factors i have been inserted here since this turns out to lead to a representation where ξ, η and ζ may be assumed to be real.

When manipulating with vector equations, it is important to remember that the directions of the unit vectors $\mathbf{e}_\perp$ and $\mathbf{e}_\|$ vary with x when $\mathbf{B}$ is not uni-directional:

$$\partial_x \mathbf{e}_\perp = -\varphi' \mathbf{e}_\|, \qquad \partial_x \mathbf{e}_\| = \varphi' \mathbf{e}_\perp, \qquad (7.84)$$

where $\varphi \equiv \arctan(B_y/B_z)$ is the angle between $\mathbf{B}$ and the z-axis (see Fig. 7.9(b)).

It remains to evaluate the components of the force operator $\mathbf{F}(\boldsymbol{\xi})$ explicitly, i.e. to exploit the consequences of the equilibrium relation (7.74), of the 2D Fourier representation (7.78), and of the orthogonal projection (7.79)–(7.83). This involves straightforward but rather tedious analysis, which is put in small print.

▷ **Explicit construction of the wave equation matrix.** In order to keep the analysis as mechanical as possible, we first derive the x, y, z components of the force operator $\mathbf{F}$ and only at the end compose the perpendicular and parallel components. Writing the different Fourier components of $\mathbf{F}$, we will drop the harmonic factor $\exp i(k_y y + k_z z)$, but, of course, only after the gradient, divergence and curl operations have been applied. Straightforward application of the operations (7.80)–(7.82) and (7.83) then yields, for example, the important relation

$$\nabla \cdot \boldsymbol{\xi} = \xi' + g\eta + f\zeta, \qquad (7.85)$$

so that the fluid terms of $\mathbf{F}$ may be written as

$$-\nabla \pi = [\, p'\xi + \gamma p(\xi' + g\eta + f\zeta)\,]'\, \mathbf{e}_x$$
$$+ ik_y[p'\xi + \gamma p(\xi' + g\eta + f\zeta)]\, \mathbf{e}_y + ik_z[p'\xi + \gamma p(\xi' + g\eta + f\zeta)]\, \mathbf{e}_z,$$
$$\nabla \Phi \nabla \cdot (\rho \boldsymbol{\xi}) = \hat{g}\,[(\rho\xi)' + \rho g\eta + \rho f\zeta]\, \mathbf{e}_x. \qquad (7.86)$$

The evaluation of the magnetic terms requires some more diligence because of the repeated

7.3 MHD wave equation for a gravitating plasma slab

curl operations:

$$\xi \times \mathbf{B} = -iB\eta\,\mathbf{e}_x - B_z\xi\,\mathbf{e}_y + B_y\xi\,\mathbf{e}_z\,,$$

$$\mathbf{Q} = iBf\xi\,\mathbf{e}_x - [(B_y\xi)' - k_z B\eta]\,\mathbf{e}_y - [(B_z\xi)' + k_y B\eta]\,\mathbf{e}_z\,,$$

$$\nabla \times \mathbf{Q} = -i[(Bg\xi)' + k_0^2 B\eta]\,\mathbf{e}_x$$
$$\quad + [(B_z\xi)'' - k_z Bf\xi + k_y(B\eta)']\,\mathbf{e}_y - [(B_y\xi)'' - k_y Bf\xi - k_z(B\eta)']\,\mathbf{e}_z\,,$$
(7.87)

$$-\mathbf{B}\times(\nabla\times\mathbf{Q}) = [B_y(B_y\xi)'' + B_z(B_z\xi)'' - B^2 f^2\xi + Bg(B\eta)']\,\mathbf{e}_x$$
$$\quad + iB_z[(Bg\xi)' + k_0^2 B\eta]\,\mathbf{e}_y - iB_y[(Bg\xi)' + k_0^2 B\eta]\,\mathbf{e}_z\,,$$

$$\mathbf{j}\times\mathbf{Q} = [B_y'(B_y\xi)' + B_z'(B_z\xi)' + B(Bg)'\eta]\,\mathbf{e}_x + iBB_y' f\xi\,\mathbf{e}_y + iBB_z' f\xi\,\mathbf{e}_z\,.$$

Inserting the above expressions in the spectral equation (7.75), using the equilibrium equation (7.74), produces the three vector components of $\mathbf{F}$ in the x, y, z directions:

$$-\rho\omega^2\xi_x = [(\gamma p + B^2)\xi']' - B^2 f^2\xi + [(\gamma p + B^2)g\eta]' + \rho\hat{g}g\eta + (\gamma pf\zeta)' + \rho\hat{g}f\zeta\,,$$
$$-\rho\omega^2\xi_y = iB_z Bg\,\xi' + ik_y(\gamma p\xi' - \rho\hat{g}\xi) + i(k_0^2 B_z B + k_y\gamma pg)\eta + ik_y\gamma pf\zeta\,,\quad (7.88)$$
$$-\rho\omega^2\xi_z = -iB_y Bg\,\xi' + ik_z(\gamma p\xi' - \rho\hat{g}\xi) - i(k_0^2 B_y B - k_z\gamma pg)\eta + ik_z\gamma pf\zeta\,.$$

Projecting the last two components onto $\mathbf{e}_\perp$ and $\mathbf{e}_\parallel$ finally yields the required representation given in Eq. (7.89) below. ◁

▷ **Exercise.** If you want to become conversant with spectral analysis of plasmas, you should not only reproduce the above steps, but also find alternative derivations. ◁

(b) Wave equation in field line projection The resulting wave or spectral equation for a plane gravitating slab, in orthogonal field line projection, assumes the following symmetric form:

$$\begin{pmatrix} \dfrac{d}{dx}(\gamma p + B^2)\dfrac{d}{dx} - f^2 B^2 & \dfrac{d}{dx}g(\gamma p + B^2) + g\rho\hat{g} & \dfrac{d}{dx}f\gamma p + f\rho\hat{g} \\ -g(\gamma p + B^2)\dfrac{d}{dx} + g\rho\hat{g} & -g^2(\gamma p + B^2) - f^2 B^2 & -gf\gamma p \\ -f\gamma p\dfrac{d}{dx} + f\rho\hat{g} & -fg\gamma p & -f^2\gamma p \end{pmatrix}\begin{pmatrix}\xi\\ \eta\\ \zeta\end{pmatrix}$$

$$= -\rho\omega^2\begin{pmatrix}\xi\\ \eta\\ \zeta\end{pmatrix}. \qquad (7.89)$$

The most important feature of this matrix representation of the force operator **F** is that it depends on x through $\rho(x)$, $p(x)$, $B^2(x)$, $f(x)$ and $g(x)$. If this dependence is neglected (which implies that the four gravitational terms $\rho\hat{g}$ must be neglected as well), the eigenvalue equation (7.64) for infinite homogeneous plasmas is recovered. The eigenvalue equation (7.32) for the gravito-acoustic modes of Section 7.2.1 is recovered in the limit $B \to 0$.

The matrix eigenvalue problem (7.89) for the three vector components ξ, η, ζ only contains normal differential operators d/dx in the first row and column. Hence, it can be reduced to a single second order differential equation in ξ by eliminating η and ζ by means of the second and third components, which are algebraic in η and ζ:

$$\eta = g \frac{[(b^2 + c^2)\omega^2 - f^2 b^2 c^2]\xi' - \hat{g}\,\omega^2 \xi}{D},$$

$$\zeta = f \frac{c^2(\omega^2 - f^2 b^2)\xi' - \hat{g}\,(\omega^2 - k_0^2 b^2)\xi}{D}, \qquad (7.90)$$

where $b \equiv B/\sqrt{\rho}$, $c \equiv \sqrt{\gamma p/\rho}$, and D is the determinant of the lower right 2×2 sub-matrix. Substituting these expressions into the first component gives the required *second order differential equation for ξ*, derived by Goedbloed [81](I):

$$\frac{d}{dx} \frac{N}{D} \frac{d\xi}{dx} + \left[\rho(\omega^2 - f^2 b^2) + \rho'\hat{g} - k_0^2 \rho \hat{g}^2 \frac{\omega^2 - f^2 b^2}{D} \right.$$

$$\left. - \left\{ \rho\hat{g}\,\frac{\omega^2(\omega^2 - f^2 b^2)}{D} \right\}' \right] \xi = 0, \qquad (7.91)$$

where

$$N = N(x;\omega^2) \equiv \rho(\omega^2 - f^2 b^2)\left[(b^2 + c^2)\omega^2 - f^2 b^2 c^2\right],$$

$$D = D(x;\omega^2) \equiv \omega^4 - k_0^2(b^2 + c^2)\omega^2 + k_0^2 f^2 b^2 c^2. \qquad (7.92)$$

This equation has to be solved subject to the model I boundary conditions

$$\xi(x_1) = \xi(x_2) = 0. \qquad (7.93)$$

This completes the formulation of this one-dimensional spectral problem.

Note that Eq. (7.91) contains the differential equation (7.34) for gravito-acoustic modes in the limit $b \to 0$. Also, the three MHD wave frequencies ω_s^2, ω_A^2, ω_f^2 for homogeneous plasmas, given in Eq. (7.65), are recovered in the limit $\rho\hat{g} \to 0$ and neglecting all x dependencies, so that $d/dx \to ik_x$, $g \to k_y$, and $f \to k_z$.

It is of some interest for later generalizations to observe the intriguing difference between the spectral problem posed by the '3D' vector eigenvalue equation (7.89) and the equivalent '1D' scalar eigenvalue equation (7.91), supplemented with the algebraic relations (7.90). Whereas the latter formulation is convenient for most of the analysis, and certainly for numerical integration, the eigenvalue character of the problem has been spoiled to a great extent because ω^2 is scattered all over the place in Eq. (7.91). (For that reason, equations of this type are sometimes called 'nonlinear' eigenvalue equations.) Hence, to prove or disprove general properties (like monotonicity of the discrete spectrum) it may be advisable to return to the original formulation (7.89).

(c) Singular frequencies The ordinary differential equation (7.91), with the boundary conditions (7.93), describes all gravito-magnetohydrodynamic modes of a gravitating magnetized plasma slab with completely arbitrary equilibrium profiles. Before trying to solve it by analytical or numerical methods, one first needs to pay attention to the physical meaning of the different terms and, in particular, to the different singularities that occur.

It is clear that the factor N/D in front of the highest derivative of the differential equation plays an important role. We may write this factor in terms of the four special frequencies (that were already introduced) in Eqs. (7.94)–(7.96) for the case of a homogeneous plasma:

$$\frac{N}{D} = \rho(b^2 + c^2) \frac{[\omega^2 - \omega_A^2(x)][\omega^2 - \omega_S^2(x)]}{[\omega^2 - \omega_{s0}^2(x)][\omega^2 - \omega_{f0}^2(x)]}. \quad (7.94)$$

The numerator N involves the *asymptotic Alfvén and slow magneto-sonic frequencies*,

$$\omega_A^2(x) \equiv f^2 b^2 \equiv F^2/\rho, \quad \omega_S^2(x) \equiv f^2 \frac{b^2 c^2}{b^2 + c^2} \equiv \frac{\gamma p}{\gamma p + B^2} \omega_A^2(x), \quad (7.95)$$

whereas the denominator D involves the *slow and fast turning point frequencies*,

$$\omega_{s0, f0}^2(x) \equiv \tfrac{1}{2} k_0^2 (b^2 + c^2) \left[1 \pm \sqrt{1 - \frac{4 f^2 b^2 c^2}{k_0^2 (b^2 + c^2)^2}} \right]. \quad (7.96)$$

For inhomogeneous plasmas, all of these four special frequencies depend on x, through $f^2(x)$, $b^2(x)$ and $c^2(x)$, so that their role is to be determined by a local analysis about the points where the different factors of N and D vanish. In particular, the ODE (7.91) becomes *singular* for $N \to 0$, when $\omega^2 \to \omega_A^2(x)$ or $\omega^2 \to \omega_S^2(x)$, i.e. when the local Alfvén or slow magneto-sonic frequencies are approached. These singularities deserve a separate treatment, given in Section 7.4, where it is shown that they give rise to non-square integrable solutions associated with *continuous spectra*. On the other hand, the ODE only develops *apparent* singularities for $D \to 0$, when $\omega^2 \to \omega_{s0}^2(x)$ or $\omega^2 \to \omega_{f0}^2(x)$, i.e. when the local magneto-sonic turning point frequencies are approached. An apparent singularity implies that there are cancellations in the series expansion of the solution, so that it remains finite in the end. This follows directly from the equivalent system of first order differential equations (next subsection), but can also be demonstrated explicitly from the second order differential equation (Section 7.4.1). The history of this topic is summarized at the end of Section 7.4.2.

The function $F(x)$, that occurs in the frequencies $\omega_A(x)$ and $\omega_S(x)$ of the Alfvén and slow continua, deserves special attention since it represents the projection of the gradient operator parallel to the magnetic field:

$$F \equiv -i\mathbf{B} \cdot \nabla = \mathbf{k}_0 \cdot \mathbf{B}. \tag{7.97}$$

The locations in the plasma where $F = 0$ play an important role in the stability of plasmas since the stabilizing magnetic field line bending energy vanishes there, favouring interchange instability (see Section 7.5.2). Accordingly, in stability theory, the continua $\{\omega_A^2(x)\}$ and $\{\omega_S^2(x)\}$ appear in the disguise of the 'interchange' values $\{F^2(x)\}$.

(d) Equivalent system of first order differential equations For numerical integration one usually converts a second order ODE into a pair of first order ODEs for the function and its first derivative. In our case, instead of ξ', it is much better to exploit a physical variable for that purpose, viz. the (Eulerian) *perturbation of the total pressure*:

$$\Pi \equiv \pi + \mathbf{B} \cdot \mathbf{Q} = -\gamma p \nabla \cdot \boldsymbol{\xi} - \boldsymbol{\xi} \cdot \nabla p + \mathbf{B} \cdot \mathbf{Q}. \tag{7.98}$$

Converting $\nabla \cdot \boldsymbol{\xi}$ by means of Eq. (7.85), $\mathbf{Q}$ with Eq. (7.87), and exploiting the equilibrium relation (7.74), this expression may be transformed into

$$\begin{aligned}\Pi &= -\rho(b^2 + c^2)\xi' + \rho\hat{g}\xi - \rho g(b^2 + c^2)\eta - \rho f c^2 \zeta \\ &= -\frac{N}{D}\xi' + \rho\hat{g}\frac{\omega^2(\omega^2 - f^2 b^2)}{D}\xi,\end{aligned} \tag{7.99}$$

where η and ζ have been eliminated in the second line by means of Eq. (7.90). Hence, the peculiar derivative term in curly brackets in the second order ODE (7.91) comes from this expression for the total pressure perturbation Π.

By means of the expression just derived, we may convert the second order differential equation (7.91) into the following pair of first order differential equations:

$$\begin{pmatrix}\xi \\ \Pi\end{pmatrix}' = -\frac{1}{N}\begin{pmatrix}C & D \\ E & -C\end{pmatrix}\begin{pmatrix}\xi \\ \Pi\end{pmatrix}, \tag{7.100}$$

where

$$C \equiv -\rho \hat{g} \omega^2 (\omega^2 - f^2 b^2),$$
$$E \equiv -\left[\rho(\omega^2 - f^2 b^2) + \rho' \hat{g} \right] N - \rho^2 \hat{g}^2 (\omega^2 - f^2 b^2)^2. \quad (7.101)$$

The determinant of the matrix on the RHS of Eq. (7.100) is proportional to N:

$$DE + C^2 = -\left\{ \left[\rho(\omega^2 - f^2 b^2) + \rho' \hat{g} \right] D - k_0^2 \rho \hat{g}^2 (\omega^2 - f^2 b^2) \right\} N. \quad (7.102)$$

This guarantees that no additional singularities have been introduced in this formulation. The vector eigenvalue equations (7.89) for ξ, η, ζ, the scalar eigenvalue problem (7.91) for ξ, and the two-variable eigenvalue problem (7.100) for ξ, Π are equivalent.

The formulation in terms of two first order ODEs was first exploited by Appert, Gruber and Vaclavik [9] for the analogous problem in cylindrical geometry. Its merit is that it shows right away that nothing blows up for $D = 0$ so that these singularities of Eq. (7.91) must be *apparent*, whereas the $N = 0$ singularities are genuine.

7.3.3 Gravito-MHD waves

Before we start the investigation of the general structure of the spectrum of MHD waves in a gravitating medium, where the mentioned singularities will be in the centre of our attention, it is useful to discuss one simple but non-trivial example of a complete spectrum where the singularities are still absent (or, rather, degenerate enough to permit a simple representation). This is provided by the generalization of the problem considered in Section 7.2.3 to an *exponentially stratified atmosphere with constant sound and Alfvén speeds*:

$$\rho = \rho_0 e^{-\alpha x}, \quad p = p_0 e^{-\alpha x}, \quad \mathbf{B} = B_0 e^{-\frac{1}{2}\alpha x} \mathbf{e}_z, \quad (7.103)$$

so that

$$c^2 = \frac{\gamma p}{\rho} = \frac{\gamma p_0}{\rho_0}, \quad b^2 = \frac{B^2}{\rho} = \frac{B_0^2}{\rho_0}. \quad (7.104)$$

It is useful to exploit the parameter β as a measure for the relative magnitude of the sound speed with respect to the Alfvén speed:

$$\beta \equiv \frac{2p}{B^2} = \frac{2p_0}{B_0^2}, \quad \text{so that} \quad \frac{c^2}{b^2} = \tfrac{1}{2}\gamma\beta. \quad (7.105)$$

The equilibrium relation (7.74) then fixes the relationship between the parameter α and the gravitational acceleration $\hat{g}$:

$$\alpha = \frac{\rho \hat{g}}{p + \frac{1}{2}B^2} = \frac{\rho_0 \hat{g}}{p_0 + \frac{1}{2}B_0^2} = \frac{\gamma \hat{g}}{c^2 + \frac{1}{2}\gamma b^2} = \frac{2}{1+\beta}\frac{\hat{g}}{b^2}. \tag{7.106}$$

Also, the wave numbers f ($= k_z$) and g ($= k_y$) are constants now so that the continuous spectra $\omega_S^2 \equiv f^2 b^2 c^2/(b^2 + c^2)$ and $\omega_A^2 \equiv f^2 b^2$, defined in Eq. (7.95), just degenerate into two points on the ω^2-axis.

Through these simplifications, the basic spectral equation (7.91) transforms into a differential equation *with constant coefficients* (after elimination of the factor $e^{-\alpha x}$):

$$\frac{N_0}{\rho_0 D_0}\frac{d}{dx}\left(e^{-\alpha x}\frac{d\xi}{dx}\right) + \left[\omega^2 - f^2 b^2 - k_0^2 \hat{g}^2 \frac{\omega^2 - f^2 b^2}{D_0} - \alpha \hat{g}\right.$$
$$\left. + \alpha \hat{g}\frac{\omega^2(\omega^2 - f^2 b^2)}{D_0}\right]e^{-\alpha x}\xi = 0, \tag{7.107}$$

where the numerator and denominator coefficients $N_0 \equiv N(0)$ and $D_0 \equiv D(0)$. This yields an exponential solution $\xi = Ce^{(\frac{1}{2}\alpha \pm iq)x}$ with a vertical 'wave number'

$$q \equiv \sqrt{-\frac{1}{4}\alpha^2 + \frac{\rho_0}{N_0}\left[(\omega^2 - f^2 b^2)(D_0 + k_0^2 c^2 N_B^2) + \alpha \hat{g}\, g^2 b^2 \omega^2\right]} = n\frac{\pi}{a} \tag{7.108}$$

that has to satisfy the quantization condition (7.47) because of the model I boundary conditions on ξ. Here, we have introduced the Brunt–Väisälä frequency N_B again (with an index B on it to avoid confusion with our other symbol N):

$$N_B^2 \equiv -\frac{1}{\rho}\left(\rho'\hat{g} + \frac{\rho^2 \hat{g}^2}{\gamma p}\right) = \alpha \hat{g} - \frac{\hat{g}^2}{c^2} = \frac{(\gamma - 1)\beta - 1}{1 + \beta}\frac{\hat{g}^2}{c^2}. \tag{7.109}$$

Note that in the HD limit ($\beta \to \infty$), the expression (7.44) of Section 7.2.3 is recovered: $N_B^2 \to (\gamma - 1)\hat{g}^2/c^2$. However, due to the finite β MHD modifications of the equilibrium, positivity of N_B^2 is no longer guaranteed now but depends on the value of β:

$$N_B^2 \begin{cases} \geq 0 & \text{for } \beta \geq (\gamma - 1)^{-1} \\ < 0 & \text{for } 0 \leq \beta < (\gamma - 1)^{-1}. \end{cases} \tag{7.110}$$

This permits some analysis of the MHD version of the convective instability, but one should be aware of the intrinsic limitation of the stability results obtained on the assumption (7.103) of exponential profiles. Obviously, this assumption is not dictated by physical necessity but is just made for analytical convenience.

7.3 MHD wave equation for a gravitating plasma slab

(a) Dispersion equation and special solutions Inversion of Eq. (7.108) for q yields the *dispersion equation for the gravito-MHD waves*:

$$(\omega^2 - f^2 b^2)\left[\omega^4 - k_{\text{eff}}^2(b^2 + c^2)\omega^2 + k_{\text{eff}}^2 f^2 b^2 c^2 + k_0^2 c^2 N_B^2\right] + \alpha \hat{g}\, g^2 b^2 \omega^2 = 0, \qquad (7.111)$$

where $f = k_0 \cos\vartheta$ and $g = k_0 \sin\vartheta$ (with ϑ indicating the angle between $\mathbf{k}_0$ and $\mathbf{B}$) and we recall the definition of the effective total 'wave number',

$$k_{\text{eff}}^2 \equiv k_0^2 + q^2 + \tfrac{1}{4}\alpha^2.$$

Since the dispersion equation (7.111) is a cubic equation in ω^2, with three independent solutions, we immediately conclude that gravity does not increase the number of degrees of freedom of the MHD system. We had three MHD waves without gravity and we keep three waves with gravity. These waves transform into the ordinary MHD waves of Chapter 5 when $\hat{g} \to 0$ and into the gravito-acoustic waves of Section 7.2.3 when $b \to 0$ ($\beta \to \infty$). However, the latter limit deserves further analysis to clarify how the twofold HD spectral structure of Fig. 7.3 (with clustering at $\omega = \infty$ and $\omega = 0$) relates to the threefold MHD spectral structure (with clustering at $\omega = \infty$, $\omega = \omega_A$ and $\omega = \omega_S$). This will be discussed after we have obtained the MHD counterpart in Fig. 7.10.

Two misleadingly simple expressions are obtained from the dispersion equation (7.111) for the limiting cases of purely parallel and purely perpendicular propagation. For *parallel propagation* ($\mathbf{k}_0 \parallel \mathbf{B} \Rightarrow \vartheta = 0$, $f = k_0$, $g = 0$), the following solutions are obtained:

$$\omega_1^2 = k_0^2 b^2, \qquad \omega_{2,3}^2 = \tfrac{1}{2} k_{\text{eff}}^2 (b^2 + c^2)\left[1 \pm \sqrt{1 - \frac{4k_0^2 c^2 (k_{\text{eff}}^2 b^2 + N_B^2)}{k_{\text{eff}}^4 (b^2 + c^2)^2}}\right]. \qquad (7.112)$$

These *appear* to be the unaffected Alfvén waves and two gravitational modifications of the fast and slow magneto-sonic waves (one of which will give rise to the Parker instability). For *perpendicular propagation* ($\mathbf{k}_0 \perp \mathbf{B} \Rightarrow \vartheta = \tfrac{1}{2}\pi$, $f = 0$, $g = k_0$) the dispersion equation yields the following solutions:

$$\omega_1^2 = 0, \qquad \omega_{2,3}^2 = \tfrac{1}{2} k_{\text{eff}}^2 (b^2 + c^2)\left[1 \pm \sqrt{1 - \frac{4k_0^2 N_m^2}{k_{\text{eff}}^4 (b^2 + c^2)}}\right], \qquad (7.113)$$

where another characteristic expression appears, viz.

$$N_m^2 \equiv -\frac{1}{\rho}\left(\rho' \hat{g} + \frac{\rho^2 \hat{g}^2}{\gamma p + B^2}\right) = \alpha \hat{g} - \frac{\hat{g}^2}{b^2 + c^2} = \frac{(\gamma - 1)\beta + 1}{1 + \beta}\frac{\hat{g}^2}{b^2 + c^2}, \qquad (7.114)$$

which is called the magnetically modified Brunt–Väisälä frequency; see Priest [190]. (Note the different sign in the numerator compared to the expression (7.109) for N_B^2, so that $N_m^2 > 0$ for this class of equilibria, independent of the value of β.) One of the solutions degenerates into a marginal mode and the two other solutions are magnetic compressional modifications of the p- and g-modes given by Eq. (7.49). The dispersion diagram of Fig. 7.3 remains valid for these solutions if we replace the sound speed by the magneto-sonic speed ($c^2 \to b^2 + c^2$), and adapt the definitions of the Lamb and Brunt–Väisälä frequencies accordingly.

We have termed the expressions (7.112) and (7.113) misleadingly simple since they represent limiting cases of intricate mode couplings where the labels A, s and f may no longer be appropriate (which is why we have used the neutral labels 1, 2 and 3 instead) and they create a false impression on stability (with only the parallel expression admitting unstable solutions, whereas MHD instabilities preferentially operate in the perpendicular direction). To obtain the full spectral picture of the different HD and MHD effects operating in gravitating plasmas, and their relation to stability, we should study the cubic dispersion equation (7.111) for *oblique propagation*. This is done, most effectively, by means of a numerical representation (exploiting Cardano's explicit solutions) for a typical choice of the parameters. Since the latter part of the problem is less trivial than it may appear, we first pay attention to the construction of dimensionless parameters.

(b) Dimensionless scaling of the dispersion equation and general solutions
When handling the final result of an MHD calculation, like the dispersion equation (7.111), it is always useful to convert it to a dimensionless form exploiting the scale independence introduced in Section 4.1.2 to get rid of the trivial dimensional factors. To that end, we exploit the thickness a ($\equiv x_2 - x_1$) of the slab as a measure for lengths, and the density ρ_0 and the magnetic field strength B_0 at $x = 0$ as measures for the density and magnetic field determining the (constant) Alfvén speed $b \equiv B_0/\sqrt{\rho_0}$. The relevant, dimensionless, parameters then become

$$\bar{\omega} \equiv (a/b)\omega, \quad \bar{k}_{\text{eff}} \equiv k_{\text{eff}} a, \quad \bar{k}_0 \equiv k_0 a, \quad \bar{q} \equiv qa, \quad \bar{\alpha} \equiv \alpha a, \quad (7.115)$$

in terms of which the implicit form of the dimensionless dispersion equation reads:

$$\bar{\omega}^6 - \left[(1 + \tfrac{1}{2}\gamma\beta)\bar{k}_{\text{eff}}^2 + \bar{k}_0^2 \cos^2\vartheta \right] \bar{\omega}^4$$
$$+ \left[(1 + \gamma\beta)\bar{k}_{\text{eff}}^2 \cos^2\vartheta + \tfrac{1}{4}(1 + \beta)(\gamma\beta - \beta + 1 - 2\cos^2\vartheta)\bar{\alpha}^2 \right] \bar{k}_0^2 \bar{\omega}^2$$
$$- \left[\tfrac{1}{2}\gamma\beta\bar{k}_{\text{eff}}^2 \cos^2\vartheta + \tfrac{1}{4}(1 + \beta)(\gamma\beta - \beta - 1)\bar{\alpha}^2 \right] \bar{k}_0^4 \cos^2\vartheta = 0. \quad (7.116)$$

7.3 MHD wave equation for a gravitating plasma slab

The solution of this cubic provides the explicit dimensionless dispersion equation:

$$\bar{\omega} = \bar{\omega}(\bar{k}_0, \vartheta, \bar{q}; \bar{\alpha}, \beta, \gamma). \quad (7.117)$$

Here, the parameters in front of the semicolon refer to the perturbation and the ones behind it refer to the equilibrium. For the numerical investigation of the stability of a particular equilibrium, one could fix the parameters $\bar{\alpha}, \beta, \gamma \, (= 5/3)$ to correspond to that equilibrium, but one should investigate the full range of the parameters $\bar{q}, \vartheta, \bar{k}_0$, since that is what nature does. Using the scale independence in this manner, we obtain certainty about the completeness of parameter space and avoid useless scanning of redundant parameters.

In terms of the dimensionless parameters, the 'singular' frequencies of Section 7.3.2 become

$$N_0 = 0 \Rightarrow \bar{\omega}_A^2 = \bar{k}_0^2 \cos^2\vartheta, \qquad \bar{\omega}_S^2 = \frac{\frac{1}{2}\gamma\beta}{1 + \frac{1}{2}\gamma\beta} \bar{k}_0^2 \cos^2\vartheta, \quad (7.118)$$

$$D_0 = 0 \Rightarrow \bar{\omega}_{s0, f0}^2 = \frac{1}{2}\bar{k}_0^2(1 + \frac{1}{2}\gamma\beta)\left[1 \pm \sqrt{1 - \frac{2\gamma\beta}{(1 + \frac{1}{2}\gamma\beta)^2}\cos^2\vartheta}\right]. \quad (7.119)$$

The dimensionless expressions of the two kinds of Brunt–Väisälä frequencies, involving the gravitational parameter $\bar{\alpha}$, read:

$$\bar{N}_B^2 = \frac{1}{4}\bar{\alpha}^2 \frac{(1+\beta)[(\gamma-1)\beta - 1]}{\frac{1}{2}\gamma\beta}, \qquad \bar{N}_m^2 = \frac{1}{4}\bar{\alpha}^2 \frac{(1+\beta)[(\gamma-1)\beta + 1]}{1 + \frac{1}{2}\gamma\beta}. \quad (7.120)$$

Whereas the 'singular' frequency expressions (7.118) and (7.119) do not involve $\bar{\alpha}$, they play a decisive role in the clarification of the MHD spectral structure, as we will see.

In Fig. 7.10 we show plots of $\bar{\omega}^2(\bar{k}_0^2)$ for oblique propagation ($\vartheta = \pi/4$) for thirty values of the vertical mode number n. The value of $\bar{\alpha} \, (= 20)$ has been chosen to be approximately the same as for the HD p- and g-modes of Fig. 7.3. To demonstrate the connection with the latter diagram, a high value of β is taken in Fig. 7.10(a), whereas β is chosen to be unity in Fig. 7.10(b) to exhibit the more usual MHD spectral structure. The singular frequencies $\bar{\omega}_A^2$ and $\bar{\omega}_S^2$ are indicated by long dashes and the magneto-sonic turning point frequencies $\bar{\omega}_{s0, f0}^2$ are indicated by short dashes. The three dotted lines represent the zeroth ($n = 0$) 'member' of each family of solutions. Those are actually not solutions, since they do not satisfy

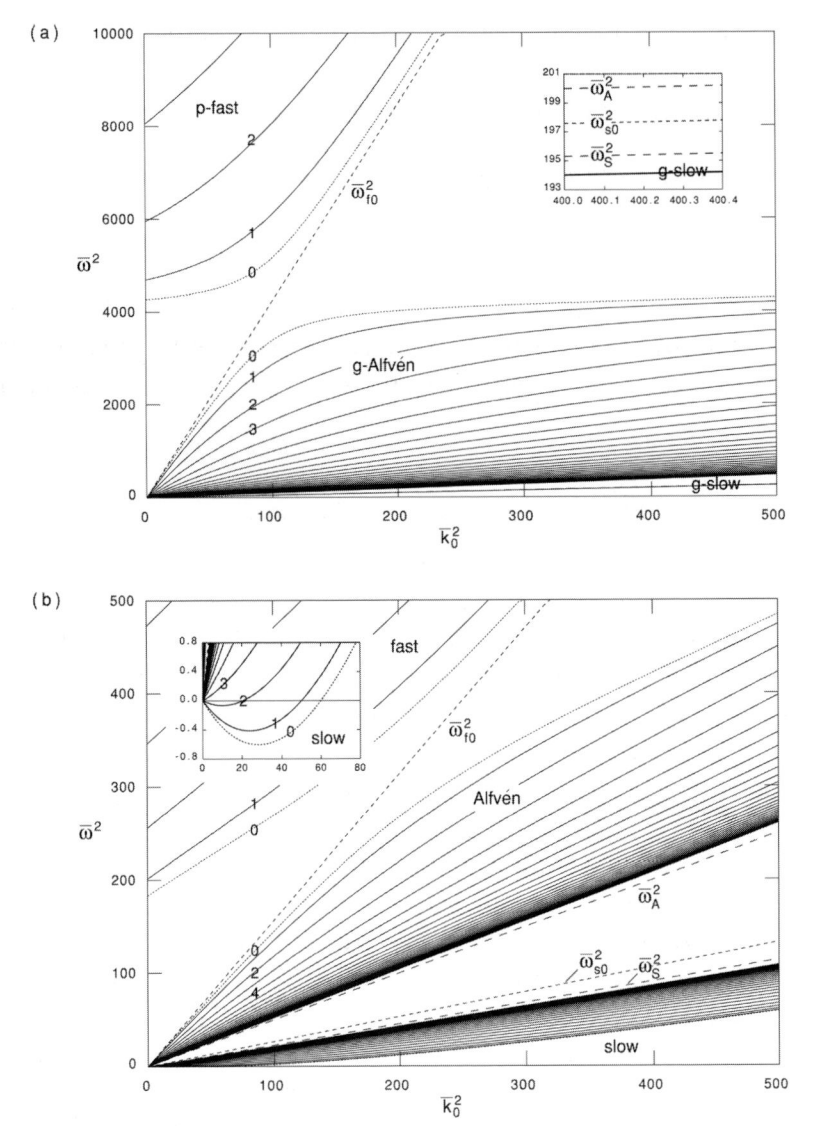

Fig. 7.10. Dispersion diagrams for the oblique gravito-MHD modes of an exponential plane plasma layer. The squared frequencies in normalized units, $\bar{\omega}^2 = \bar{\omega}^2(\bar{k}_0^2, \bar{q}^2, \vartheta)$, are plotted for thirty values of the vertical 'wave number' $\bar{q} = n\pi$ ($n = 1, 2, \ldots, 30$); $\bar{\alpha} = 20$, $\vartheta = \pi/4$. (a) Near HD regime ($\beta = 50$), inset: blow-up of low-frequencies (slow wave spectrum not yet resolved on this scale); (b) regular MHD regime ($\beta = 1$), inset: unstable slow modes.

the boundary conditions, but they conveniently mark the boundaries of the most global modes. The spectral structure is determined by how these three families of solutions for $\bar{\omega}^2$ are distributed from $n = 1$ to $n \to \infty$, where the latter limit yields the essential spectrum (see Section 7.3.1).

7.3 MHD wave equation for a gravitating plasma slab

By means of Fig. 7.10, the following matters may be elucidated:

(1) the relationship between the HD and the MHD spectral structures (Fig. 7.10(a));
(2) the influence of gravity and the essential spectrum in MHD (Fig. 7.10(b));
(3) the convective/Rayleigh–Taylor instability in the presence of a magnetic field.

We will discuss them, one by one.

(1) The HD dispersion equation (7.48) is obtained from the MHD dispersion equation (7.116) by scaling with β, $\widetilde{\omega}^2 \equiv \frac{1}{2}\gamma\beta\,\bar{\omega}^2$ (so that the vertical sizes in Fig. 7.10(a) are a factor $\frac{1}{2}\gamma\beta$ larger than those in Fig. 7.3), and taking the limit $\beta \to \infty$:

$$\widetilde{\omega}^2 \left[\widetilde{\omega}^4 - \bar{k}_{\mathrm{eff}}^2 \widetilde{\omega}^2 + \tfrac{1}{4}(\gamma-1)\bar{\alpha}^2 \bar{k}_0^2 \right] = 0. \tag{7.121}$$

This shows that, in addition to the p- and g-modes discussed in Section 7.2.3, the marginal horizontal shifts perpendicular to $\mathbf{k}_0$, that were discarded in Section 7.2.1, are needed to obtain a *threefold essential spectrum* ($\widetilde{\omega}_{1,2}^2 \equiv 0$, $\widetilde{\omega}_3^2 \equiv \infty$) that may be connected to the threefold essential spectrum of MHD. In fact, in the limit $\beta \to \infty$, the Alfvén and slow mode frequencies $\bar{\omega}_{A,S}^2 \to 0$, where the displacement $\boldsymbol{\xi}$ becomes dominantly horizontal, whereas $\bar{\omega}_F^2 \equiv \infty$, corresponding to dominantly vertical displacements. Fig. 7.10(a) shows that, for large but finite β, the p-modes transform into the fast MHD modes and the g-modes transform into the Alfvén modes. In addition, a new branch of gravitational slow MHD modes springs forth in the low-frequency domain. From the inset in Fig. 7.10(a) it is clear that for the value of the gravitational parameter chosen, $\bar{\alpha}^2 \sim \bar{k}_0^2$, the spectrum is strongly affected by gravity both for the Alfvén modes (infinitely degenerate in the absence of gravity) and for the slow modes (anti-Sturmian in the absence of gravity). Notice that there is still quite a lot of empty space left between the lowest Alfvén curve and $\bar{\omega}_A^2$ and between the highest slow curve and $\bar{\omega}_S^2$, which will be occupied by the curves for $n > 30$ (not shown) establishing the link with the essential spectrum.

(2) Decreasing the value of β to unity (Fig. 7.10(b)), the three MHD sub-spectra (separated by empty evanescent regions containing the turning point frequencies $\bar{\omega}_{s0}^2$ and $\bar{\omega}_{f0}^2$) become more clearly distinguished. The three sequences can be labelled as fast, Alfvén and slow by virtue of the monotonicity of $\bar{\omega}^2$ as a function of n (to be proved in general in Section 7.4.3), where $\bar{\omega}_F^2$, $\bar{\omega}_A^2$ and $\bar{\omega}_S^2$ are obtained in the limit $n \to \infty$. By expanding the solutions of the dispersion equation (7.116) around these frequencies, one obtains the following expressions for the *approach*

to the essential spectrum:

$$\bar{\omega}_f^2 \approx (1 + \tfrac{1}{2}\gamma\beta)\bar{q}^2 \qquad \rightarrow \qquad \bar{\omega}_F^2 \equiv \infty,$$

$$\bar{\omega}_a^2 \approx \bar{\omega}_A^2 + \tfrac{1}{2}(1 + \beta)\left(\frac{\bar{k}_0}{\bar{q}}\right)^2 \bar{\alpha}^2 \sin^2\vartheta \qquad \rightarrow \qquad \bar{\omega}_A^2, \qquad (7.122)$$

$$\bar{\omega}_s^2 \approx \bar{\omega}_S^2 + \frac{\bar{\omega}_S^4 - \tfrac{1}{4}(1+\beta)(1+\beta-\gamma\beta\cos^2\vartheta)\bar{k}_0^2\bar{\alpha}^2}{(1+\tfrac{1}{2}\gamma\beta)\bar{q}^2} \qquad \rightarrow \qquad \bar{\omega}_S^2.$$

(For the validity of the expressions for $\bar{\omega}_a^2$ and $\bar{\omega}_s^2$, the second term should be much smaller than the first one so that they are not valid in the limit $\vartheta \rightarrow \pi/2$, when $\bar{\omega}_A^2$ and $\bar{\omega}_S^2 \rightarrow 0$. This subject abounds with non-uniform limits; the reader is strongly advised to build up experience by studying this dispersion equation in the many different limits, virtually all of which turn out to be physically relevant.) Whereas the deviation from the limiting frequencies is determined by the gravitational parameter $\bar{\alpha}$, the essential spectrum itself is not affected by gravity; it is robust in that sense.

The expressions (7.122) clearly show how the infinite degeneracy of the Alfvén frequencies is lifted by gravity so that an anti-Sturmian sequence ($\bar{\omega}^2 \downarrow \bar{\omega}_A^2$ as $n \rightarrow \infty$) is obtained. The slow frequencies, which are anti-Sturmian ($\bar{\omega}^2 \downarrow \bar{\omega}_S^2$) in the absence of gravity, either become Sturmian ($\bar{\omega}^2 \uparrow \bar{\omega}_S^2$), as in the range of $\bar{k}_0^2$ plotted in Fig. 7.10(b), or remain anti-Sturmian when $\bar{k}_0^2$ is large enough (far outside the range shown). However, for monotonicity of $\bar{\omega}^2(n)$ to remain valid, the transition from Sturmian to anti-Sturmian behaviour should occur at a point of *infinite degeneracy of the slow modes*. This is exactly what happens when the numerator of the second term of the expression (7.122)(c) for $\bar{\omega}_s^2$ vanishes. Hence, for this class of equilibria, the sequence of inequalities (7.70) for homogeneous equilibria is modified to permit both Sturmian and anti-Sturmian slow frequencies ($\bar{\omega}_s^2 < \bar{\omega}_S^2$ and $\bar{\omega}_s^2 > \bar{\omega}_S^2$), and anti-Sturmian Alfvén frequencies ($\bar{\omega}_a^2 > \bar{\omega}_A^2$).

Having obtained a rather compelling way of labelling the MHD sub-spectra, we now return to our reservations when discussing the simple expressions (7.112) for parallel propagation. In particular, it would appear logical to associate ω_1^2 with the Alfvén modes and ω_2^2 (the solution with the minus sign) with the slow modes. However, as shown by Adam [3], when $N_B^2 c^2 > \tfrac{1}{4}\alpha^2 b^4$ (i.e. $\beta > 2$ for $\gamma = 5/3$), the curve for $\omega_2^2(q)$ crosses the value ω_1^2 so that some of the slow frequencies would exceed the Alfvén frequency, in conflict with the very name 'slow'. (There is no conflict with the direction of monotonicity determined by the sign of N_0/D_0 since $\omega_A^2 = \omega_{s0}^2$ in this case, so that two 'singular' frequencies are crossed simultaneously.) We eliminate this paradox by considering near parallel propagation ($|\vartheta| = \epsilon \ll 1$) in the limit $\epsilon \rightarrow 0$. For $\epsilon \neq 0$, the two solutions split apart into an anti-Sturmian Alfvén branch approaching ω_A^2 for $q \to \infty$ and an anti-Sturmian slow branch starting just below ω_A^2 ($= \omega_{s0}^2$) at $q = 0$ and approaching ω_S^2 for $q \to \infty$. For purely parallel propagation, in the limit $\epsilon \to 0$, these curves develop a kink at the cross-over point $q = q_c$ so that we should call the upper part (consisting of ω_2^2 for $0 \leq q < q_c$ and of ω_1^2 for $q_c \leq q < \infty$) 'Alfvén', and the lower part (consisting of ω_1^2 for $0 \leq q < q_c$ and of ω_2^2 for $q_c \leq q < \infty$) 'slow'.

(3) The transition to Sturmian monotonicity of the slow modes, exhibited by both Figs. 7.10(a) and (b), is crucial for the well-posedness of the ideal MHD stability problem. It guarantees that, for $\bar{\omega}^2 < 0$, *the largest growth rates are obtained for the most global* ($n = 1$) *instability*. (If this were not the case, making the plasma layer thinner by moving in the walls, would make the growth rates larger, in conflict with physical intuition.) For example, as shown by the inset of Fig. 7.10(b), the modes $n = 1$ and $n = 2$ are unstable but all the higher ones ($n \geq 3$) are stable for this particular choice of the parameters.

To systematically investigate the unstable range of parameters, one needs a general stability criterion, which is usually not a closed expression but a rather involved procedure like investigating the quadratic form of the energy (Section 6.4.4). Exceptionally, for the present case of the exponential equilibrium (7.103), an instability criterion is easily obtained from the dispersion equation (7.116). For unstable solutions ($\bar{\omega}^2 < 0$), the last coefficient (including the minus sign) should be positive:

$$\left(\tfrac{1}{2}\gamma\beta \bar{k}_{\text{eff}}^2 \cos^2\vartheta + \tfrac{1}{2}\gamma\beta \bar{N}_B^2 \right) \cos^2\vartheta \equiv$$

$$\left\{ \tfrac{1}{2}\gamma\beta (\bar{k}_0^2 + \bar{q}^2) \cos^2\vartheta + \tfrac{1}{4}\left[(1+\beta)(\gamma\beta - \beta - 1) + \tfrac{1}{2}\gamma\beta \cos^2\vartheta \right] \bar{\alpha}^2 \right\} \cos^2\vartheta < 0. \tag{7.123}$$

This expression shows that gravitational instability is obtained when the first term in square brackets is not only negative (i.e., $\bar{N}_B^2 < 0$) but also dominates over the remaining terms proportional to $\cos^2\vartheta$: the gravitational instability drive should dominate over the stabilizing field-line bending contribution of the slow mode perturbations. Hence, one expects the worst instability for perpendicular propagation ($\vartheta = \pi/2$). On the other hand, recalling from Section 7.2.3 that the exponential atmosphere is stable in the HD limit ($\beta \to \infty$), magnetic field perturbations are apparently essential to create gravitational instability in the MHD domain. Accordingly, the whole potentially negative expression in curly brackets is multiplied by $\cos^2\vartheta$ so that, at worst, marginal stability is obtained for $\vartheta = \pi/2$. Consequently, *for this class of equilibria, the MHD version of the Rayleigh–Taylor instability occurs in a range of near-perpendicular wave vectors* $\mathbf{k}_0$, *excluding the direction*

$\mathbf{k}_0 \cdot \mathbf{B} = 0$ itself. This range may even include parallel propagation ($\vartheta = 0$) when $\bar{N}_B^2$ is negative enough, giving rise to the Parker instability (see below). The stabilizing slow mode contributions also vanish in the extreme limit $\beta \to 0$, opposite to HD, when the plasma becomes unstable for all $\vartheta \neq \pi/2$ and any value of $\bar{\alpha}$. We leave it to the reader to derive the expression for the growth rate for that case.

(c) Parker instability According to Parker [175], "the interstellar magnetic field in the general neighbourhood of the Sun is, on average, parallel to the plane of the Galaxy ... This galactic magnetic field must be confined by the weight of the gas (plasma) threaded by the field. This interstellar gas-field system is subject to a universal Rayleigh–Taylor instability such that the interstellar gas tends to concentrate into pockets suspended in the field. The cause of the instability may be thought of as a hydromagnetic self-attraction in the interstellar gas, which may be ten times larger than the gravitational self-attraction of the gas. It is this hydromagnetic self-attraction which produces the observed tendency of the interstellar gas to be confined in discrete clouds."

The gas, which consists of electrons, ions and neutrals, is described by means of the ideal MHD equations, so that the equations of the present section apply. The instability is investigated for motions in the vertical x-z plane only, i.e. longitudinal perturbations ($\vartheta = 0$). Parker also includes the effects of a relativistic cosmic ray particle population on the pressure of the background equilibrium, as well as on the dynamics. Neglecting the latter, we obtain from Eq. (7.123) the criterion for the Parker instability in our notation:

$$\bar{k}_{\text{eff}}^2 + \bar{N}_B^2 < 0 \quad \Rightarrow \quad \bar{k}_0^2 + \bar{q}^2 - \tfrac{1}{4}\bar{\alpha}^2 \cdot \frac{2(1+\beta)(1+\beta-\gamma\beta) - \gamma\beta}{\gamma\beta} < 0. \tag{7.124}$$

The growth rate of the instabilities may be obtained from Eq. (7.112) (with the minus sign). This shows the connection with the slow magneto-sonic wave which is essential for the instability since it involves dominant perturbed flows parallel to the magnetic field. Hence, 'the gas tends to drain downward along the magnetic lines of force into the lowest region along each line'.

Obviously, at this point, numerical values for the wave numbers k_0, q and for the equilibrium parameters α, β and γ have to be chosen to demonstrate the viability of the mechanism for the description of cloud formation along galactic spiral arms. We leave the subject here and refer to the original literature [175] and the review paper by Mouschovias [160] for a discussion of all the astrophysical ramifications.

(d) Interchanges and quasi-interchanges We return to the discussion of the full dispersion equation (7.116), for $\vartheta \neq 0$, since perpendicular and near-perpendicular propagation properties are of generic interest for the stability of confined plasmas

in general. Expanding the dispersion equation for $\vartheta = \pi/2 \pm \epsilon$, with $\epsilon \ll 1$, we obtain two potentially unstable solutions:

$$\bar{\omega}^2 \approx \frac{\bar{k}_0^2}{\bar{k}_{\text{eff}}^2} \bar{N}_m^2 \qquad \text{(pure interchanges, } \vartheta = \pi/2\text{)},$$

$$\bar{\omega}^2 \approx \frac{\bar{N}_B^2}{\bar{N}_m^2} \bar{\omega}_S^2 = \frac{(\gamma-1)\beta - 1}{(\gamma-1)\beta + 1} \bar{k}_0^2 \cos^2\vartheta \quad \text{(quasi-interchanges, } \vartheta \neq \pi/2\text{)}.$$

(7.125)

This reveals one of the limitations of the equilibrium with exponential dependence of the background quantities. Pure interchanges, with a maximum growth rate at $\vartheta = \pi/2$, are trivially stable since they require $\bar{N}_m^2 < 0$, which is not possible for these equilibria. However, the second branch, with maximum growth rate at $\vartheta \neq \pi/2$, may be unstable since it requires $\bar{N}_B^2 < 0$. They have been called *gravitational quasi-interchanges* by Newcomb [165]. Both the regular and the modified Brunt–Väisälä frequencies, $\bar{N}_B$ and $\bar{N}_m$, enter the expression for their growth rate, indicating the possibility of intricate stability transitions.

In the stability analysis of general equilibria, the crossing of these two branches of the dispersion equation plays an important role in clarifying the apparent discrepancies that occur in the stability criteria for gravitational interchanges. We defer further discussion to Section 7.5.2, when we are no longer restricted to the choice of uni-directional magnetic field and exponential equilibrium profiles.

7.4 Continuous spectrum and spectral structure

'Singularity is almost invariably a clue.'

(Bender and Orszag [23], at the beginning of their exposition of approximate solutions of linear differential equations, quoting Sherlock Holmes in *The Boscombe Valley Mystery* by Sir Arthur Conan Doyle.)

7.4.1 Singular differential equations

We have obtained important insight in the structure of the MHD spectrum for equilibria with exponential profiles, where the 'singular' frequencies determined by $N_0 = 0$ and $D_0 = 0$ are constant. However, in this manner we have evaded the most important issues of inhomogeneity in MHD, which are, in general, that these frequencies depend on the coordinate x of inhomogeneity and that the magnetic field is not uni-directional: *continuous spectrum* and *magnetic shear* are essential constituents of the MHD description of plasmas. We face these issues in this section and the next.

When a problem has been reduced to a *non-singular* ordinary second order differential equation it may be considered to have been solved, because one can always obtain the explicit answers numerically to any relevant degree of accuracy. A specific example has just been given in Section 7.3.3. Since this example actually hinges on algebraic simplification through the exponential factor, let us briefly consider how the numerical solution might proceed with a more general inhomogeneity. To that end, we write the basic differential equation (7.91) symbolically as

$$\frac{d}{dx}\left[P(x;\omega^2)\frac{d\xi}{dx}\right] - Q(x;\omega^2)\xi = 0, \qquad (7.126)$$

where $P \equiv N/D$, and $Q \equiv -[\cdots]$ with the expression in square brackets as in Eq. (7.91). This equation is to be solved subject to the model I boundary conditions (7.93). Specifying the background equilibrium by a particular choice of the profiles $\rho(x)$, $p(x)$, $B_y(x)$ and $B_z(x)$, satisfying the equilibrium constraint (7.74) of Section 7.3.2, *and prescribing a particular value $\omega^2 = \omega^{2(0)}$* for the unknown eigenfrequency, the functions $P(x;\omega^2)$ and $Q(x;\omega^2)$ are completely determined. One may then solve Eq. (7.126) by means of a standard numerical library routine, starting from the left boundary condition $\xi(x_1) = 0$ (where the derivative $\xi'(x_1)$ may be prescribed arbitrarily since the differential equation is homogeneous and linear), and stepping towards the right end of the interval $x_1 \le x \le x_2$. Of course, arriving there, the right boundary condition $\xi(x_2) = 0$ will not be satisfied in general. This necessitates the construction of an algorithm for the choice of a new value $\omega^2 = \omega^{2(1)}$ that will bring the solution closer to satisfying that boundary condition in the next iteration. This part of the problem is important for the determination of the eigenvalue, but it does not present any difficulty. The point is that an extremely effective algorithm exists, based on shooting (Section 7.5.1) and the oscillation theorem (Section 7.4.3), which is valid *as long as the ODE is non-singular*. The essential problem left is, therefore, a proper treatment of the singularities of the basic differential equation. Hence, we now resume our discussion of the singularities that was started in Section 7.3.2(c).

In contrast to the non-singular integration procedure just sketched, where the boundary conditions determine the solution on the entire interval $[x_1, x_2]$, a singularity $\omega^2 = \omega_A^2(x_0)$ or $\omega^2 = \omega_S^2(x_0)$ at a point $x = x_0$ splits the interval into two parts that become independent, in a sense to be discussed. Hence, there is no way to produce the singular problem from the non-singular one by a small perturbation. There either is a singular point or there is not, and the presence of such a point changes the analysis completely. For example, in marginal stability theory (where $\omega^2 \equiv 0$, so that the mentioned singularities become zeros of the Alfvén frequency $\omega_A = F/\sqrt{\rho}$ with F defined in Eq. (7.97)) there is always a distinct difference

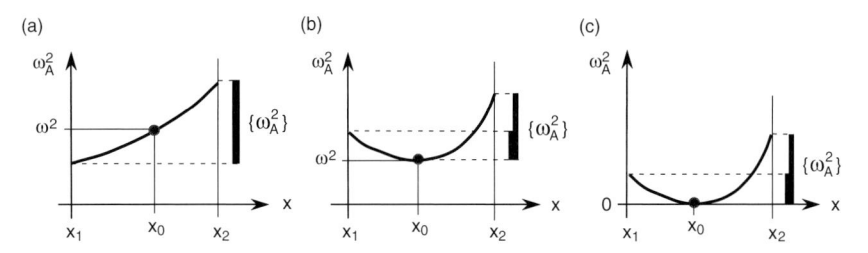

7.4 Continuous spectrum and spectral structure

Fig. 7.11. Normal dependence of $\omega_A^2(x)$ determines: (a) continuous spectrum singularity; (b) potential cluster point for stable waves; (c) potential cluster point at marginal stability.

between configurations with magnetic shear ($F' \neq 0$) and without ($F' = 0$), where the limit $F' \to 0$ produces complicated boundary layers. We will analyse this in Section 7.5.

Before we delve into the singularity theory of Eq. (7.126), an explicit example may illustrate the different kinds of singularities encountered in MHD of inhomogeneous plasmas. For that purpose, we concentrate on the Alfvén singularity. In Fig. 7.11 three generic cases are depicted for the distribution of $\omega_A^2(x)$, where the slow singularities $\omega_S^2(x)$ (not shown) may be assumed to be well separated from the Alfvén ones (cases (a) and (b)), but they necessarily coalesce at the marginal point (case (c)).

(a) If $\omega_A^2(x)$ is monotonic, any choice of ω^2 in the range $\omega_A^2(x_1) \leq \omega^2 \leq \omega_A^2(x_2)$ will lead to a singular point $x_1 \leq x_0 \leq x_2$ where the Alfvén factor of the function P may be expanded as follows:

$$\omega^2 - \omega_A^2 \approx -(\omega_A^2{}')_0 (x - x_0) \quad \Rightarrow \quad P \sim x - x_0. \tag{7.127}$$

This range will determine the *continuous spectrum* $\{\omega_A^2\}$ of Alfvén modes, and similarly for the frequency range $\{\omega_S^2\}$ of the slow modes, as will be shown in Section 7.4.2.

(b) If $\omega_A^2(x)$ is not monotonic, the continuous spectrum develops a maximum or minimum at some point x_0 where it starts to fold over onto itself. Choosing ω^2 to correspond to that extremum permits expansion of the Alfvén factor as

$$\omega^2 - \omega_A^2 \approx -\tfrac{1}{2}(\omega_A^2{}'')_0 (x - x_0)^2 \quad \Rightarrow \quad P \sim (x - x_0)^2. \tag{7.128}$$

The value $\omega^2 = \omega_A^2(x_0)$ corresponds to an internal edge of the continuous spectrum which may become a *cluster point* of the discrete spectrum, depending on further conditions, as will be discussed in Section 7.4.4.

(c) For marginal stability ($\omega^2 = 0$), such candidate cluster points always occur in sheared magnetic fields at an *interchange point* $F^2(x_0) = 0$, where both the

Alfvén and the slow continuum reach the origin: $\omega_S^2 \equiv \gamma p (\gamma p + B^2)^{-1} \omega_A^2 = 0$. Since, according to Eq. (7.96), also $\omega_{s0}^2 \approx \omega_S^2$ there, the factor ω_S^2 of the numerator $N(\omega^2 = 0)$ is cancelled by the leading order of the factor ω_{s0}^2 of the denominator $D(\omega^2 = 0)$ so that the Alfvén factor determines the behaviour at the singularity:

$$-\omega_A^2 \approx -\tfrac{1}{2}(\omega_A^2{}'')_0 (x - x_0)^2 = -(\omega_A')_0^2 (x - x_0)^2 \quad \Rightarrow \quad P \sim (x - x_0)^2 \,. \tag{7.129}$$

By means of this expansion, the conditions for *local interchange stability* may be derived, as will be demonstrated in Section 7.5.2.

We are now prepared to appreciate the physical significance of the different steps in the following mathematical analysis of Eq. (7.126).

Singularity of this differential equation is best discussed in terms of the standard form used in classical texts (i.e. dating from the pre-electronic computing era, e.g. Ince [115]) as well as in modern ones (e.g. Bender and Orszag [23]) on linear differential equations:

$$\xi'' + \frac{P'}{P}\xi' - \frac{Q}{P}\xi = 0\,. \tag{7.130}$$

This equation is non-singular if the two functions P'/P and Q/P are analytic on the whole domain of interest, i.e. in a finite region of the complex z-plane containing the interval $x_1 \le x \le x_2$ of the real axis (where the prime is redefined as differentiation with respect to z). Theoretically, the solution may be carried from (ordinary) point to (ordinary) point by Taylor series expansion with a radius of convergence determined by the distance to the nearest singularity of the two mentioned functions. Hence, such singularities interfere with the numerical procedure described earlier. Obviously, the zeros of

$$N \equiv \rho(b^2 + c^2)\left[\omega^2 - \omega_A^2(x)\right]\left[\omega^2 - \omega_S^2(x)\right] \sim (x - x_0)^l\,, \quad (l = 1 \text{ or } l = 2)\,, \tag{7.131}$$

produce singularities of the kind $P'/P \sim z^{-1}$ and $Q/P \sim z^{-l}$. Here, z is the complex continuation of the real variable $x - x_0$, which is the distance to a point x_0 where either ω_A^2 or ω_S^2 coincides with the eigenvalue parameter ω^2. We have seen above that it is sufficient for virtually all cases of interest to consider $l = 1$ and $l = 2$ only, i.e. *regular singularities*. In the neighbourhood of such singularities, the differential equation may be written as

$$\xi'' + \frac{1}{z} p(z) \xi' - \frac{1}{z^2} q(z) \xi = 0\,, \tag{7.132}$$

where $p(z)$ and $q(z)$ are analytic functions with the following Taylor expansion:

$$\begin{cases} p(z) \equiv z \dfrac{P'}{P} = p_0 + p_1 z + \cdots \\ q(z) \equiv z^2 \dfrac{Q}{P} = q_0 + q_1 z + \cdots \end{cases}, \quad \text{where} \quad \begin{cases} p_0 = 1, \; q_0 = 0 \; (l = 1) \\ p_0 = 2, \; q_0 \ne 0 \; (l = 2). \end{cases}$$

(7.133)

The behaviour of the solutions in the neighbourhood of such points may be obtained by means of the *Frobenius method of expansion*.

The Frobenius expansion about a regular singularity assumes

$$\xi = z^\nu \sum_{n=0}^{\infty} a_n z^n,$$

(7.134)

where, in general, the index ν is not an integer and may even be complex. The values of ν and the coefficients a_n are obtained by substituting the expansion into the differential equation (7.132):

$$\sum_{n=0}^{\infty} (\nu + n)(\nu + n - 1) a_n z^{\nu+n-2} + (p_0 + p_1 z + \cdots) \sum_{n=0}^{\infty} (\nu + n) a_n z^{\nu+n-2}$$

$$- (q_0 + q_1 z + \cdots) \sum_{n=0}^{\infty} a_n z^{\nu+n-2} = 0. \quad (7.135)$$

Balancing the different powers yields the following sequence of equalities:

$$\begin{aligned} z^{\nu-2} &: [\nu^2 + (p_0 - 1)\nu - q_0] a_0 = 0, \\ z^{\nu-1} &: [(\nu+1)\nu + p_0(\nu+1) - q_0] a_1 = (-\nu p_1 + q_1) a_0, \quad \text{etc.} \end{aligned}$$

(7.136)

Hence, in general, the two solutions ξ_1 and ξ_2 start off with a term $a_0 z^{\nu_{1,2}}$, where a_0 is arbitrary and the indices $\nu_{1,2}$ are the solutions of the *indicial equation*, obtained by putting the factor in square brackets of (7.136)(a) equal to zero. The coefficients a_n ($n \ge 1$) follow by recursion from the relations (7.136)(b), etc. The solutions of the indicial equation, and hence the solutions ξ_1 and ξ_2 themselves, are qualitatively different for the case (a) and the cases (b), (c) illustrated in Fig. 7.11:

$$l = 1 \text{ (continuous spectrum)}: \quad \nu_1 = \nu_2 = 0, \quad (7.137)$$

$$l = 2 \text{ (cluster points)}: \quad \nu_{1,2} = -\tfrac{1}{2} \pm \tfrac{1}{2}\sqrt{1 + 4q_0}. \quad (7.138)$$

For the cluster point analysis ($l = 2$), this is virtually sufficient information. It will be completed in Section 7.4.4 for the waves (case (b)) and in Section 7.5.2 for the marginal states (case (c)). For the analysis of the continuous spectrum ($l = 1$), however, we are not done at all since we have just obtained one value ($\nu = 0$) corresponding to an analytic function (called the 'small' solutions) which obviously does not exhibit singular behaviour. The 'large' solution, revealing the singularity, is to be obtained yet.

To complete the analysis for the first case, with equal indices $\nu_1 = \nu_2 = 0$, the quoted literature (Ince [115], p. 397, Bender and Orszag [23], p. 73) shows how the 'large' solution is obtained. This involves the special trick of writing the small solution obtained as a function of ν, not substituting the value $\nu_1 = 0$ yet, differentiating it with respect to ν, and then inserting the value $\nu = 0$. This produces a second, independent, solution containing a logarithmic contribution, since

$$\frac{\partial}{\partial \nu}(z^\nu) = \frac{\partial}{\partial \nu}\left(e^{\nu \ln z}\right) = z^\nu \ln z \;\rightarrow\; \ln z \;\; \text{for} \;\; \nu \rightarrow 0. \tag{7.139}$$

Hence, in the neighbourhood of the regular singular point $z = x - x_0 = 0$, two solutions are obtained:

$$\begin{aligned}
\xi_1 &= u(z), & u &= u_0 + u_1 z + \cdots, & &\text{('small' solution)}, \\
\xi_2 &= u(z) \ln z + v(z), & v &= v_0 + v_1 z + \cdots, & &\text{('large' solution)},
\end{aligned} \tag{7.140}$$

where the explicit form of the coefficients u_n and v_n is much less important than the appearance of the logarithmic factor. The latter introduces a branch point at $z = 0$ where a decision is to be taken on how to continue the solution past this point. This choice depends on the physical context in which the singularity is encountered.

In the solution of the initial value problem by means of the Laplace transform, as sketched in Section 6.3.2 and elaborated in Chapter 10, the parameter ω is assumed to be complex and the continuation is obtained by deforming the Laplace contour around the singularity. This procedure is completely analogous to that exploited by Landau in his solution of the initial value problem for the Vlasov equation (Section 2.3.3), where the velocity variable v of the Vlasov problem corresponds to the spatial coordinate x of the MHD problem.

In the determination of the spectrum of ideal MHD, since there is no dissipation, ω^2 remains real, so that we should consider the solutions (7.140) along the real axis, exploiting the principal value of $\ln z$, i.e. $\ln |x - x_0|$. The logarithmic singularity at $x = x_0$ then divides the real x-interval in two parts such that the solution ξ obtains just enough additional freedom (compared to that at an ordinary point) to satisfy the boundary conditions for any value of ω^2. This produces the continuous spectrum, as will be demonstrated in the next section.

7.4 Continuous spectrum and spectral structure

▷ **Apparent singularities** $D = 0$ singularities occur when either one of the factors $\omega^2 - \omega_{s0}^2$ or $\omega^2 - \omega_{f0}^2$ vanishes on the interval:

$$D \equiv \left[\omega^2 - \omega_{s0}^2(x)\right]\left[\omega^2 - \omega_{f0}^2(x)\right] \sim x - x_0 \Rightarrow P \sim (x-x_0)^{-1}, \quad Q \sim (x-x_0)^{-2}.$$
(7.141)

This also produces singularities of the kind $P'/P \sim z^{-1}$ and $Q/P \sim z^{-1}$. The coefficients of the analytic functions $p(z)$ and $q(z)$ then yield solutions of the indicial equation with an integer difference between the indices:

$$p_0 = -1, \ q_0 = 0 \Rightarrow \nu_1 = 2, \ \nu_2 = 0.$$
(7.142)

This implies that only one 'small' solution $\sim z^2$ is obtained, whereas the other one again contains a logarithmic contribution, unless the coefficients of the expansion of $p(z)$ and $q(z)$ possess the very special property

$$(p_1 - q_1)q_1 - (p_0 - q_0)q_2 = 0.$$
(7.143)

In that case, none of the solutions is singular; see Ince [115], p. 404. The MHD wave equation (7.91) turns out to have precisely that property! This follows from the way in which the expression D appears in the functions P and Q:

$$P \equiv \frac{N}{D}, \qquad Q \equiv -U - \frac{V}{D} - \left(\frac{W}{D}\right)',$$
(7.144)

where the explicit expressions for U, V, and W can be read off from Eq. (7.91). These coefficients are related by

$$W^2 + NV = \rho^2 \hat{g}^2 (\omega^2 - f^2 b^2)^2 D \to 0 \quad \text{as } x \to x_0.$$
(7.145)

Translated in terms of the functions p and q, this produces the equality (7.143). Consequently, *the $D = 0$ singularities are apparent*. ◁

7.4.2 Alfvén and slow continua

For simplicity, we choose the equilibrium quantities such that the singular frequency functions $\omega_A^2(x)$ and $\omega_S^2(x)$ are well separated and monotonically increasing, as in Fig. 7.11(a). We will prove that the collection of frequencies $\omega^2 \in \{\omega_A^2(x) | x_1 \leq x \leq x_2\}$ and $\omega^2 \in \{\omega_S^2(x) | x_1 \leq x \leq x_2\}$ constitutes the continuous spectrum, i.e. the set of *improper eigenvalues* of the MHD force operator $\rho^{-1} \mathbf{F}$, with associated non-square integrable 'eigenfunctions'.

We concentrate again on the Alfvén singularities $\omega^2 \in \{\omega_A^2(x)\}$, noting that the analysis for the slow singularities $\omega^2 \in \{\omega_S^2(x)\}$, lying in a narrow band below the Alfvén ones, is completely analogous. For a given value of $\omega^2 = \omega_0^2$ from this set, the ODE (7.91) is singular ($N = 0$) at the position $x = x_0$ where $\omega_A^2(x_0) = \omega_0^2$, as shown once more in Fig. 7.12(a). The monotonically increasing function $\omega_A^2 = \omega_A^2(x)$ may be inverted to give a monotonically increasing function $x_A = x_A(\omega^2)$, as illustrated in Fig. 7.12(b). In this manner, the position x

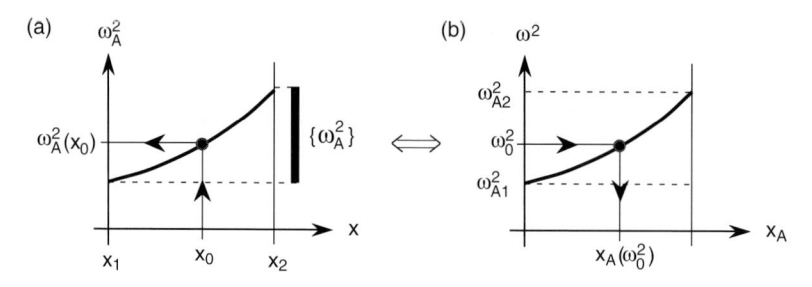

Fig. 7.12. Inversion of the Alfvén frequency function: (a) $\omega_A^2 = \omega_A^2(x)$; (b) $x_A = x_A(\omega^2)$.

of the singularity becomes a function of ω^2. As we have seen in Section 7.4.1, expansion about the singularity gives the dominant coefficient of the differential equation:

$$\omega^2 - \omega_A^2(x) \approx -(\omega_A^{2'})_0(x - x_0) = -(\omega_A^{2'})_0 s, \quad (7.146)$$

where, since everything is real now, the complex variable z is replaced by the real variable s, which is the distance to the singularity:

$$s \equiv x - x_0 = x - x_A(\omega^2). \quad (7.147)$$

Instead of the complex expressions (7.140), we now obtain real independent solutions of the form

$$\begin{cases} \xi_1 = u(s; \omega^2) & \text{('small' solution)}, \\ \xi_2 = u(s; \omega^2) \ln|s| + v(s; \omega^2) & \text{('large' solution)}, \end{cases} \quad (7.148)$$

to the right ($s > 0$) as well as to the left ($s < 0$) of the singularity.

The most important consequence of the singularities $N = 0$ of the differential equation (7.91) is that certain jumps of the solutions are permitted, so that satisfaction of the boundary conditions (7.93) is always possible for ω^2 in the singular domain. To prove this, we exploit the series expansions just obtained. Because of the assumed monotonicity of $\omega_A^2(x)$, the interval (x_1, x_2) contains only one singular point for a fixed value of ω^2 so that the general solution may be written as

$$\xi = \Big[A_1 u + C_1(u \ln|s| + v)\Big] H(-s) + \Big[A_2 u + C_2(u \ln|s| + v)\Big] H(s), \quad (7.149)$$

where $H(s)$ is the Heaviside step function (defined as $H \equiv 0$ for $s < 0$ and $H \equiv 1$ for $s > 0$), and we have to determine the values of A_1, C_1, A_2, and C_2. Of course, for a non-singular second order differential equation the solution should be continuous, so that $A_1 = A_2$ and $C_1 = C_2$. We will now prove that *the small*

7.4 Continuous spectrum and spectral structure

solution may jump whereas the large solution has to be continuous: $A_1 \neq A_2$, $C_1 = C_2$.

▷ **Proof** Write the original differential equation (7.91) as

$$(P\xi')' - Q\xi = 0, \qquad (7.150)$$

where

$$P \approx P_1 s + \cdots, \qquad Q \approx Q_0 + \cdots.$$

Substitution of a 'small' solution $\xi = uH(s)$ leads to the following expressions, successively:

$$\xi' = u'H(s) + u\delta(s),$$
$$P\xi' = Pu'H(s) + Pu\delta(s) = Pu'H(s),$$
$$(P\xi')' = (Pu')'H(s) + Pu'\delta(s) = (Pu')'H(s),$$
$$(P\xi')' - Q\xi = \big[(Pu')' - Qu\big]H(s) = 0,$$

by virtue of the fact that $u(s)$ is a solution of Eq. (7.150), and properties such as $H'(s) = \delta(s)$ and $s\delta(s) = 0$. Consequently, $A_1 u H(-s)$ is a solution of Eq. (7.150) but, likewise, $A_2 u H(s)$ is also a solution, so that the small solutions left and right are totally unrelated: $A_1 \neq A_2$. Performing a similar analysis for the large solution, one finds that the term $u \ln|s| H(s)$ produces a δ-function contribution that does not vanish, so that $C_1 = C_2$, QED. ◁

Consequently, the general solution to Eq. (7.150) may be written as

$$\xi = A_1 u\, H(-x + x_A) + A_2 u\, H(x - x_A) + C\left[u \ln|x - x_A| + v\right]. \qquad (7.151)$$

Due to the fact that we have now three (rather than the usual two) constants available, the two boundary conditions $\xi(x_1) = 0$ and $\xi(x_2) = 0$ may always be satisfied for $\omega^2 \in \omega_A^2(x)$, i.e. when there is a singular point on the interval (x_1, x_2). By imposing these two boundary conditions, the two constants A_1 and A_2 may be eliminated so that *the improper eigenfunctions for an Alfvén continuum mode become*:

$$\xi_A(x; \omega^2) = C(\omega^2) \left\{ \left[\ln \frac{x - x_A(\omega^2)}{x_1 - x_A(\omega^2)} - \frac{v_1(\omega^2)}{u_1(\omega^2)}\right] u(x; \omega^2) H(-x + x_A(\omega^2)) \right.$$
$$\left. + \left[\ln \frac{x - x_A(\omega^2)}{x_2 - x_A(\omega^2)} - \frac{v_2(\omega^2)}{u_2(\omega^2)}\right] u(x; \omega^2) H(x - x_A(\omega^2)) + v(x; \omega^2) \right\},$$
$$\qquad (7.152)$$

where

$$u(x; \omega^2) \equiv u(x - x_A(\omega^2)), \qquad u_1(\omega^2) \equiv u(x_1; \omega^2), \quad \text{etc.}$$

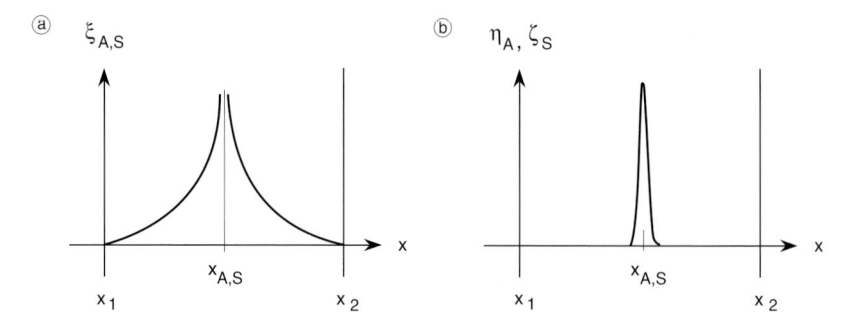

Fig. 7.13. Schematic representation of the (a) normal and (b) tangential components of the improper Alfvén and slow continuum 'eigenfunctions'.

The factor $C(\omega^2)$ may be fixed by 'normalizing' the eigenfunctions:

$$\langle \boldsymbol{\xi}_A(x; \omega^2), \boldsymbol{\xi}_A(x; \tilde{\omega}^2) \rangle = \delta(\omega^2 - \tilde{\omega}^2). \tag{7.153}$$

Similar expressions may be derived for the improper eigenfunctions of the slow continuum modes. Therefore, 'solutions' have been obtained which satisfy the boundary conditions (see Fig. 7.13(a)) for any $\omega^2 \in \{\omega_A^2(x)\}$ and $\omega^2 \in \{\omega_S^2(x)\}$, QED. The continuous ranges of frequencies $[\omega_{A\,\min}^2, \omega_{A\,\max}^2]$ and $[\omega_{S\,\min}^2, \omega_{S\,\max}^2]$ are called *the Alfvén continuum* and *the slow magneto-sonic continuum*, respectively.

This appears to establish the existence of two continuous spectra, but the most characteristic part of the eigenfunctions is not yet discussed. Actually, if we restrict the analysis to the normal component of that function, we cannot even prove that we have 'improper' eigenfunctions because the singularities $\ln|s|$ and $H(s)$ are square integrable. The dominant, non-square integrable, part of the 'improper' eigenfunction resides in the tangential components η and ζ, given by Eqs. (7.90), which involve the derivative $\xi' \gg \xi/a$. From Eq. (7.151), this derivative produces a $1/(x - x_A)$ singularity and a δ-function:

$$\xi_A' = C \left\{ \left[\mathcal{P} \frac{1}{x - x_A(\omega^2)} - \frac{A_1 - A_2}{C} \delta(x - x_A(\omega^2)) \right] u \right.$$

$$+ \left[\frac{A_1}{C} H(-x + x_A(\omega^2)) + \frac{A_2}{C} H(x - x_A(\omega^2)) + \ln|x - x_A(\omega^2)| \right] u' + v' \right\}$$

$$\approx C(\omega^2) \left\{ \mathcal{P} \frac{1}{x - x_A(\omega^2)} + \mu(\omega^2) \delta(x - x_A(\omega^2)) \right\}, \tag{7.154}$$

where $\mathcal{P}$ indicates that the principal part is to be taken when integrating this function, and $\mu(\omega^2)$ is a function involving the boundary data of u and v:

$$\mu \equiv \frac{-A_1 + A_2}{C} = \ln\left|\frac{x_2 - x_A(\omega^2)}{x_1 - x_A(\omega^2)}\right| - \frac{v_1(\omega^2)}{u_1(\omega^2)} + \frac{v_2(\omega^2)}{u_2(\omega^2)}. \quad (7.155)$$

For the Alfvén continuum modes, the tangential component η perpendicular to the magnetic field dominates (since ξ'_A is multiplied by $\omega^2 - \omega_A^2$ in the expression for ζ):

$$\eta_A \approx \frac{g(b^2 + c^2)(\omega_A^2 - \omega_S^2)}{D(\omega_A^2)} \xi'_A = -\frac{1}{g} \xi'_A \gg \xi_A \sim \zeta_A \quad \text{for} \quad \omega^2 \to \omega_A^2. \quad (7.156)$$

Similarly, for the slow continuum modes, the component ζ parallel to the magnetic field dominates (since ξ'_S is multiplied by $\omega^2 - \omega_S^2$ in the expression for η):

$$\zeta_S \approx \frac{f c^2(\omega_S^2 - \omega_A^2)}{D(\omega_S^2)} \xi'_S = -\frac{1}{f}\frac{b^2 + c^2}{c^2} \xi'_S \gg \xi_S \sim \eta_S \quad \text{for} \quad \omega^2 \to \omega_S^2, \quad (7.157)$$

where the formal expression for ξ'_S is obtained from Eq. (7.154) by replacing the subscripts A by S. The tangential components of these 'improper' eigenfunctions are represented schematically in Fig. 7.13(b).

In conclusion: the continuum modes are characterized by *a non-square integrable tangential component perpendicular to the magnetic field for the Alfvén modes* and *a non-square integrable parallel tangential component for the slow modes*. This shows the extreme anisotropy of ideal MHD waves as regards motion inside and across magnetic surfaces. This is a quite general property which remains true for other plasma geometries, like cylindrical and toroidal ones.

We appear to have obtained the generalization of the essential spectrum, represented by the three cluster point singularities for the simpler inhomogeneous plasmas discussed in Section 7.3.3. What about the *fast magneto-sonic singularities* in the high frequency limit? In that limit, as we will show below, ξ' also blows up but the normal component itself still dominates over the tangential components:

$$\xi'_F \gg \xi_F \gg \eta_F \sim \zeta_F \quad \text{for} \quad \omega^2 \to \omega_F^2 \equiv \infty. \quad (7.158)$$

Hence, the three expressions (7.156)–(7.158) for the slow and Alfvén continuum modes and the fast cluster modes together establish the generalization for inhomogeneous plasmas of the orthogonality property (5.63) and of the asymptotic properties (5.88) for the three MHD waves that were introduced in Chapter 5.

▷ **Fast magneto-sonic cluster point singularities** For large ω^2, the derivatives with respect to x also become large, so that the differential equation for ξ reduces to

$$\xi'' + \frac{\omega^2}{b^2 + c^2}\xi = 0. \tag{7.159}$$

The solution of this equation may be obtained by means of WKB analysis; see e.g. Bender and Orszag [23], p. 484. The leading order solution reads:

$$\xi \approx C(b^2+c^2)^{1/4} \sin\left[\omega_n \int_{x_1}^{x} (b^2+c^2)^{-1/2}\,dx\right], \qquad \omega_n \approx \frac{n\pi}{\int_{x_1}^{x_2}(b^2+c^2)^{-1/2}\,dx}. \tag{7.160}$$

Although the sequence (7.160) contains functions of finite norm, it does not converge to an element of Hilbert space for $n \to \infty$ (see Section 6.3.1). Hence, the fast cluster point $\omega_F^2 \equiv \lim_{n\to\infty} \omega_n^2 = \infty$ belongs to the continuous or essential spectrum. From these expressions, the singular behaviour (7.158) of the fast magnetosonic modes follows directly. ◁

▷ **Historical note** It is clear now that the proper description of the singularities of the MHD wave equation is crucial for the theory of inhomogeneous plasmas. The correct picture gradually evolved, mostly from the analysis of the analogous problem in cylinder geometry with various simplifying assumptions. The presence of a continuous spectrum of Alfvén waves in MHD was pointed out by Uberoi [233], who also drew attention to the analogy with the problem of electrostatic oscillations in inhomogeneous cold plasmas (another fluid model). The latter was analysed by Barston [16] by means of singular ('improper') normal modes, like the Van Kampen modes [237] in the Vlasov description of plasmas, and by Sedláček [206] by means of the Laplace transform of the initial value problem, like in the celebrated Landau solution of the damping [136]. All this made it clear that these kinds of singularities are not restricted to the microscopic velocity space details of a kinetic description of plasmas, but also occur in fluid descriptions when the plasma is inhomogeneous in ordinary space. (Of course, always with the caveat, also made in Section 7.2.4 on the analogy with quantum mechanical spectral theory, that the analogy is through the mathematics of linear operators, not necessarily through the physics.) Accordingly, the possibility of dissipationless damping of Alfvén waves was pointed out by Tataronis and Grossmann [225], and the complementary theory of heating by Alfvén waves was developed by Chen and Hasegawa [54]. From there, applications to laboratory and astrophysical plasmas multiplied. These topics will be discussed in Chapters 10 and 11. In the meanwhile, Grad [99] had put the subject, including stability, in the context of MHD spectral theory, pointing out the presence of four types of singularities (for $N = 0$ and $D = 0$) and, unfortunately, associating a continuous spectrum with each of them. The mistake concerning the $D = 0$ singularities was pointed out immediately by Appert, Gruber and Vaclavik [9] (by means of the system of first order differential equations discussed in Section 7.3.2), but Grad maintained that this was insufficient proof since a full construction of the resolvent operator (appearing in the solution of the initial value problem; see Section 6.3) would be required. Such a proof was given by Goedbloed in a memorandum, that was only published 24 years later [85] when the mistake was repeated by other

7.4 Continuous spectrum and spectral structure

authors. Since the form of the resolvent operator is another manifestation of the consistency of MHD spectral theory, it is reproduced in Section 10.2. ◁

7.4.3 Oscillation theorems

In Section 7.3.1, we have shown that the sequence of inequalities (7.70) already exhibits the central position of the essential spectrum for homogeneous plasmas, where it just consists of the single frequencies ω_S^2, ω_A^2, and ω_F^2. In the MHD wave equation (7.91) for inhomogeneous plasmas, these frequencies returned in the form of the *genuine singularities* $\omega_S^2(x)$ and $\omega_A^2(x)$, which give rise to continuous spectra consisting of their values on the entire plasma interval $x_1 \leq x \leq x_2$. In addition, certain *apparent singularities* $\omega_{s0}^2(x)$ and $\omega_{f0}^2(x)$ turned up, which could be proved not to give rise to logarithmic contributions in the normal component, nor to any non-square integrable contributions in the tangential components, so that they do not represent continuous spectra. In this section, we will get to appreciate the positive role they play in the determination of the spectral structure. For inhomogeneous plasmas, similar to the inequalities (7.70) (of course, omitting the discrete eigenvalues ω_s^2, ω_a^2, ω_f^2 since they are unknown now), one easily demonstrates that the mentioned frequencies are well ordered at each position x of the plasma slab:

$$0 \leq \omega_S^2(x) \leq \omega_{s0}^2(x) \leq \omega_A^2(x) \leq \omega_{f0}^2(x) \leq \omega_F^2 = \infty. \quad (7.161)$$

Hence, for a weakly inhomogeneous plasma, or for a thin slice of the plasma not containing points $F = 0$ (since ω_S, ω_{s0}, and ω_A coalesce there), the values of the frequencies (7.161) on the interval $x_1 \leq x \leq x_2$ might be distributed along the real ω^2-axis as schematically illustrated in Fig. 7.14. In the present section, we wish to determine how the empty spaces of this diagram will be filled up (or not) by the discrete spectrum.

(a) Sturm's theorems To do that, we need to study the qualitative behaviour of the eigenfunctions ξ of the differential equation (7.91), subject to the boundary conditions (7.93), as a function of the eigenvalue parameter ω^2. The kind of qualitative behaviour we envision is exemplified by the classical *Sturm–Liouville system* (see, e.g. Morse and Feshbach [159], p. 719) which is described by the non-singular

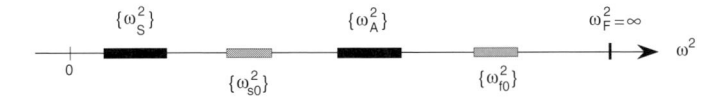

Fig. 7.14. Schematic representation of the ranges of the genuine (black) and apparent (grey) singularities of the MHD spectral equation for a weakly inhomogeneous plasma.

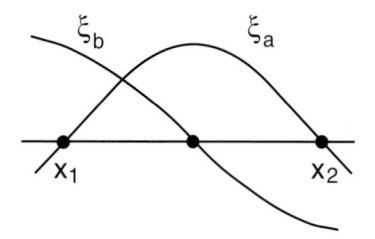

Fig. 7.15. Sturm's separation theorem.

second order differential equation

$$(P\xi')' - (Q - \lambda R)\xi = 0, \tag{7.162}$$

where λ is the eigenvalue parameter, P, Q and R are functions of x, with $P > 0$ and $R > 0$.

Let $\xi^{(1)}$ and $\xi^{(2)}$ be two linearly independent solutions of Eq. (7.162) for a fixed value of λ. Denote the two linear combinations of these two solutions by

$$\xi_a = a_1 \xi^{(1)} + a_2 \xi^{(2)}, \qquad \xi_b = b_1 \xi^{(1)} + b_2 \xi^{(2)}. \tag{7.163}$$

If $b_2/b_1 \neq a_2/a_1$, these solutions are linearly independent, i.e. the Wronski determinant $\xi_a \xi_b' - \xi_a' \xi_b \neq 0$ on the interval considered. *Sturm's separation theorem* states that the zeros of these solutions separate each other: *if x_1 and x_2 are consecutive zeros of ξ_a, then ξ_b vanishes once in the open interval (x_1, x_2)* (Ince [115], p. 223; see Fig. 7.15). (Temporarily, we use x_1 and x_2 to indicate a sub-interval of the complete interval.)

Proof Suppose ξ_b does not vanish on (x_1, x_2). Then, x_1 and x_2 are consecutive zeros of the finite function ξ_a/ξ_b. Hence, $d(\xi_a/\xi_b)/dx$ must vanish at least once on the interval. However,

$$\frac{d}{dx}\left(\frac{\xi_a}{\xi_b}\right) = \frac{\xi_a \xi_b' - \xi_a' \xi_b}{\xi_b^2} \tag{7.164}$$

cannot vanish because that would imply that the Wronski determinant vanishes somewhere. This contradiction proves that ξ_b must vanish at least once. It cannot vanish more than once since then we could interchange the roles of ξ_a and ξ_b and again get a contradiction. Hence, ξ_b vanishes once, and only once; QED.

As far as the oscillatory properties of Eq. (7.162) are concerned, one could say that all solutions oscillate equally fast if λ is kept fixed. Considering solutions of Eq. (7.162) for different values of λ, one may compare their oscillatory behaviour by means of *Sturm's fundamental oscillation theorem* stating the following. *If x_1 and x_2 are two consecutive zeros of the function ξ_1 satisfying*

$$(P\xi_1')' - (Q - \lambda_1 R)\xi_1 = 0, \tag{7.165}$$

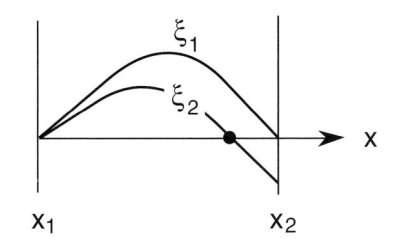

7.4 Continuous spectrum and spectral structure

Fig. 7.16. Sturm's oscillation theorem.

then the solutions ξ_2 of the equation

$$(P\xi_2')' - (Q - \lambda_2 R)\xi_2 = 0 \tag{7.166}$$

oscillate faster than ξ_1 if $\lambda_2 > \lambda_1$ (Ince [115], p. 224; see Fig. 7.16). Here, 'faster oscillating' means that the solution ξ_2 that vanishes at the left endpoint $x = x_1$ vanishes at least once on the interval (x_1, x_2). (Note that λ_1 is an eigenvalue if the interval (x_1, x_2) corresponds to the complete interval of the problem.)

Proof Multiply Eq. (7.165) by ξ_2 and Eq. (7.166) by ξ_1, integrate over (x_1, x_2) and subtract:

$$\int_{x_1}^{x_2} \left[\xi_2 (P\xi_1')' - \xi_1 (P\xi_2')' \right] dx = \left[\xi_2 P\xi_1' - \xi_1 P\xi_2' \right]_{x_1}^{x_2}$$

$$= \left(\xi_2 P\xi_1' \right)\Big|_{x=x_2} = (\lambda_2 - \lambda_1) \int_{x_1}^{x_2} R \xi_1 \xi_2 \, dx. \tag{7.167}$$

Suppose that the solution ξ_2, which vanishes at $x = x_1$, does not vanish in the open interval (x_1, x_2). Then, the LHS of Eq. (7.167) is negative, whereas the RHS is positive. This contradiction proves that ξ_2 has to vanish at least once on the open interval (x_1, x_2); QED.

Sturm's oscillation theorem gives the behaviour of the solutions of the differential equation (7.162) on any sub-interval of the interval (x_1, x_2). Such an equation, with the property that the solutions oscillate faster upon increasing the eigenvalue parameter λ, is called *Sturmian*. We will call differential equations that have the opposite property (e.g. Eq. (7.162) when the sign of λ is reversed) *anti-Sturmian*. An immediate consequence of these properties is that one can label different discrete modes by just counting the number of nodes on the interval (x_1, x_2). If the system is Sturmian the eigenvalue λ is an increasing function of the number of nodes n, whereas for anti-Sturmian systems λ decreases as a function of n (Fig. 7.17). The classical example of the first kind of behaviour is the vibrating string with characteristic wave speed c and length L, described by the equation $c^2 \partial^2 \xi / \partial x^2 = \partial^2 \xi / \partial t^2 = -\omega^2 \xi$, having the eigenvalues $\omega^2 = n^2 \pi^2 c^2 / L^2$.

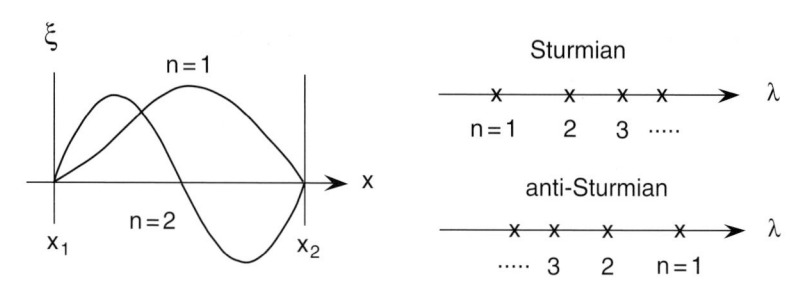

Fig. 7.17. Sturmian and anti-Sturmian dependence of the eigenvalue λ on the number of nodes of the eigenfunctions.

Examples of the second kind of behaviour are much less familiar. However, in Sections 7.2.3 and 7.3.3, we have encountered some significant ones for the HD and MHD waves in a gravitating exponential atmosphere. We will now generalize these properties to arbitrary inhomogeneity.

(b) Oscillation theorem for the MHD wave equation Turning to the MHD wave equation (7.91), it is immediately clear that it is not an equation of the simple Sturm–Liouville kind (7.162). Nevertheless, we may ask whether it still has the Sturmian, or anti-Sturmian, property. It is clear that in order to prove such a property we certainly have to exclude regions of ω^2 corresponding to the continuous spectrum, where the differential equation becomes singular, i.e. where the coefficient N/D in front of the highest derivative develops zeros ($N = 0$). Moreover, it turns out that we also will have to exclude the regions of ω^2 corresponding to the apparent singularities, where N/D becomes infinite ($D = 0$). Let us then study the monotonicity properties, if any, of the discrete spectrum of Eq. (7.91) for values of ω^2 outside the continua $\{\omega_A^2\}$ and $\{\omega_S^2\}$ and also outside the ranges $\{\omega_{s0}^2\}$ and $\{\omega_{f0}^2\}$ shown in Fig. 7.14. For those values of ω^2 the wave equation is free of both genuine and apparent singularities, but the way in which ω^2 appears in the equation makes it virtually impossible to prove anything directly from the equation itself. The reason is, of course, that the wave equation (7.91) has been derived by reducing the original vector eigenvalue problem (with linear dependence on ω^2) to a scalar one. In such a case, the only hope to prove general properties about the spectrum is to go back to first principles and, in particular, to exploit the only general property of the original operator $\rho^{-1}\mathbf{F}$ that we have, viz. that it is self-adjoint.

We prove the following theorem (Goedbloed and Sakanaka [89]). *If x_1 and x_2 are two consecutive zeros of the function ξ_1 satisfying the MHD wave equation (7.91) for $\omega^2 = \omega_1^2$, then the solutions ξ_2 of the MHD wave equation for $\omega^2 = \omega_2^2$*

oscillate faster than $\boldsymbol{\xi}_1$ if $\omega_2^2 > \omega_1^2$ and $N/D > 0$ *(Sturmian)*, and slower if $N/D < 0$ *(anti-Sturmian)*.

Proof Recall the proof of self-adjointness of the operator $\rho^{-1}\mathbf{F}$ in Section 6.2.3. In particular, let us start from the expression (6.57) for the inner product involving the two displacement vectors $\boldsymbol{\xi}$ and $\boldsymbol{\eta}$, which we now replace by $\boldsymbol{\xi}_1$ and $\boldsymbol{\xi}_2$:

$$\langle \boldsymbol{\xi}_2, \rho^{-1}\mathbf{F}(\boldsymbol{\xi}_1) \rangle - \langle \boldsymbol{\xi}_1, \rho^{-1}\mathbf{F}(\boldsymbol{\xi}_2) \rangle = \tfrac{1}{2} \int \mathbf{n} \cdot \boldsymbol{\xi}_2 \left(\gamma p \nabla \cdot \boldsymbol{\xi}_1 + \boldsymbol{\xi}_1 \cdot \nabla p - \mathbf{B} \cdot \mathbf{Q}_1 \right) dS$$

$$- \tfrac{1}{2} \int \mathbf{n} \cdot \boldsymbol{\xi}_1 \left(\gamma p \nabla \cdot \boldsymbol{\xi}_2 + \boldsymbol{\xi}_2 \cdot \nabla p - \mathbf{B} \cdot \mathbf{Q}_2 \right) dS. \quad (7.168)$$

For simplicity, we here exploit the inner product definition (6.35) but assume the vectors to be real. The surface S bounding the plasma actually consists of two surfaces, viz. the planes $x = x_1$ and $x = x_2$. Note that the expressions in the brackets on the RHS correspond to the Eulerian perturbation Π of the total pressure defined in Eq. (7.98), and reduced in Eq. (7.99). Therefore, we may write

$$\langle \boldsymbol{\xi}_2, \rho^{-1}\mathbf{F}(\boldsymbol{\xi}_1) \rangle - \langle \boldsymbol{\xi}_1, \rho^{-1}\mathbf{F}(\boldsymbol{\xi}_2) \rangle = -\tfrac{1}{2} \int \mathbf{n} \cdot \boldsymbol{\xi}_2 \Pi(\boldsymbol{\xi}_1) \, dS + \tfrac{1}{2} \int \mathbf{n} \cdot \boldsymbol{\xi}_1 \Pi(\boldsymbol{\xi}_2) \, dS$$

$$= -\tfrac{1}{2} \Big[\xi_2 \, \Pi(\boldsymbol{\xi}_1) \Big]_{x_1}^{x_2} + \tfrac{1}{2} \Big[\xi_1 \, \Pi(\boldsymbol{\xi}_2) \Big]_{x_1}^{x_2}$$

$$= \tfrac{1}{2} \Big[\xi_2 \, \tfrac{N}{D} \Big|_{\omega_1^2} \xi_1' \Big]_{x_1}^{x_2} - \tfrac{1}{2} \Big[\xi_1 \, \tfrac{N}{D} \Big|_{\omega_2^2} \xi_2' \Big]_{x_1}^{x_2}, \quad (7.169)$$

where we have suppressed the (infinite) area $A \equiv \int dS$ of the y-z plane by tacitly renormalizing the quadratic forms.

Let us now consider two solutions $\boldsymbol{\xi}_1$ and $\boldsymbol{\xi}_2$ of the MHD normal mode equation corresponding to different values ω_1^2 and ω_2^2 of the eigenvalue parameter:

$$\mathbf{F}(\boldsymbol{\xi}_1) = -\rho \omega_1^2 \boldsymbol{\xi}_1,$$
$$\mathbf{F}(\boldsymbol{\xi}_2) = -\rho \omega_2^2 \boldsymbol{\xi}_2, \quad (7.170)$$

but not necessarily satisfying the boundary conditions (7.93). Then, the LHS of Eq. (7.169) transforms to

$$(\omega_2^2 - \omega_1^2) \langle \boldsymbol{\xi}_1, \boldsymbol{\xi}_2 \rangle.$$

Consider a sub-interval (x_1, x_2) of the complete physical interval, temporarily exploiting x_1 and x_2 again to indicate two consecutive zeros of the normal component ξ_1 of $\boldsymbol{\xi}_1$. Let ω_2^2 be close to ω_1^2, so that $\boldsymbol{\xi}_2$ is close to $\boldsymbol{\xi}_1$ and, hence, $\langle \boldsymbol{\xi}_1, \boldsymbol{\xi}_2 \rangle > 0$. We will also choose $\boldsymbol{\xi}_2$ such that ξ_2 vanishes at $x = x_1$. We now wish to find out whether or not ξ_2 has another zero on (x_1, x_2), i.e., we want to investigate whether ξ_2 oscillates faster or slower than ξ_1 for a given difference of ω_1^2 and ω_2^2. Under the mentioned conditions all that remains of Eq. (7.169) is

$$(\omega_2^2 - \omega_1^2) \langle \xi_1, \xi_2 \rangle = \tfrac{1}{2} \left(\xi_2 \frac{N}{D} \bigg|_{\omega_1^2} \xi_1' \right) \bigg|_{x=x_2}. \tag{7.171}$$

Let $\xi_1 > 0$ on the open interval (x_1, x_2) so that $\xi_1'(x=x_2) < 0$. If $N/D > 0$ and $\omega_2^2 - \omega_1^2 > 0$, this implies that $\xi_2(x=x_2) < 0$ so that ξ_2 oscillates faster than ξ_1 (Sturmian behaviour). If, on the other hand, $N/D < 0$, ξ_2 oscillates slower than ξ_1 (anti-Sturmian behaviour); QED.

An important property, directly following from Eq. (7.171), concerns the *orthogonality* of the eigenfunctions of the discrete spectrum. Returning again to using x_1 and x_2 for the complete physical interval, if ξ_1 and ξ_2 both satisfy the left and the right hand boundary conditions (7.93), the RHS of Eq. (7.171) vanishes, so that

$$\langle \xi_1, \xi_2 \rangle = 0 \quad \text{for} \quad \omega_1^2 \neq \omega_2^2. \tag{7.172}$$

Hence, the discrete eigenfunctions form an orthogonal set, which may also be normalized to obtain an orthonormal set.

(c) Spectral structure From the oscillation theorem it follows that the discrete spectrum outside the ranges $\{\omega_A^2\}$, $\{\omega_S^2\}$, $\{\omega_{s0}^2\}$, $\{\omega_{f0}^2\}$ shown in Fig. 7.14 is Sturmian for $N/D > 0$, so that the eigenvalue ω^2 increases with the number of nodes of the eigenfunction, and anti-Sturmian for $N/D < 0$, with opposite behaviour of the eigenfunctions. Consequently, the discrete spectrum changes from Sturmian to anti-Sturmian, and vice versa, every time ω^2 crosses one of those four special frequency regions. Thus, the frequency regions $\{\omega_{s0}^2\}$ and $\{\omega_{f0}^2\}$ act as separators of the discrete spectra where non-monotonicity may occur. Consequently, the spectrum of an inhomogeneous plasma slab schematically may look like that shown in Fig. 7.18; see Goedbloed [82].

In conclusion, compared to the spectrum of eigenoscillations of a uniform plasma, the spectrum of an inhomogeneous plasma slab exhibits the following features.

- The infinite degeneracy of the Alfvén point spectrum is lifted and replaced by a continuum of improper modes; in addition, a finite or infinite number of Sturmian, as well as anti-Sturmian, discrete Alfvén modes may occur.
- The accumulation point of the slow magneto-sonic point eigenvalues is spread out into a continuum of improper slow magneto-sonic modes; in addition, a finite or infinite number of Sturmian discrete slow modes may occur.
- The fast magneto-sonic point spectrum still accumulates at infinity.

7.4 Continuous spectrum and spectral structure

▬▬ continuum
▬▬ non-monotonic
→ Sturmian
← anti-Sturmian

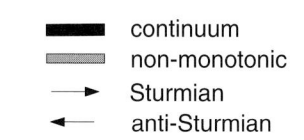

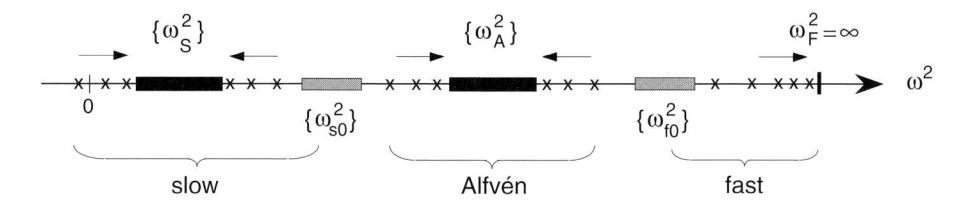

Fig. 7.18. Schematic structure and monotonicity properties of the spectrum of a slightly inhomogeneous plasma layer with gravity. (Without gravity the Sturmian slow modes and all discrete Alfvén modes are missing.)

– The cutoff frequencies ω_{s0}^2 and ω_{f0}^2 are spread out into ranges, which are *not* part of the spectrum, where the discrete spectrum may be non-monotonic.

The clear separation of the sub-spectra shown in Fig. 7.18 is only obtained if the inhomogeneity is not too strong. For strong inhomogeneity the different parts of the spectrum, and their separators, in general fold over each other so that very complicated structures may result. To analyse those, one could consider the inhomogeneous plasma as consisting of a sequence of thin layers, each of which is only weakly inhomogeneous. For each of those layers the local modes (continua and cluster spectra) are well ordered according to the inequalities (7.161), but the non-local discrete modes may appear at apparently arbitrary positions in the spectrum since they may be due to composite structures displaying Alfvénic properties in one part of the plasma and magneto-sonic properties in another part.

To study the problem of stability, in particular to compute the growth rates of instabilities ($\omega^2 < 0$), the relevant part of the spectral structure is extremely simple. There are no continuous spectra for $\omega^2 < 0$ and the discrete modes are Sturmian there. This is exploited in the shooting method, discussed in Section 7.5.1(b).

7.4.4 Cluster spectra*

To show that all four branches of the Alfvén and slow discrete sub-spectra actually may occur, we now derive the conditions for clustering at the edges of the Alfvén and slow continua. To that end, we continue with the analysis of case (b) of Section 7.4.1, where ω^2 is chosen to correspond with the maximum or minimum of the Alfvén or slow continuum; see Fig. 7.11(b). According to Eq. (7.128), at such an extremum the expression for the singular Alfvén factor

$$\omega^2 - \omega_A^2 \approx -\tfrac{1}{2}(\omega_A^2)''_0 s^2 , \quad (7.173)$$

and there is a similar expression for the singular slow factor, which yields a Frobenius expansion with indices $\nu_{1,2}$ given by Eq. (7.138). These indices become complex when the factor under the square root sign is negative, i.e. for

$$q_0 \equiv \left(\frac{s^2 Q}{P}\right)_0 < -\frac{1}{4} . \quad (7.174)$$

We will show in Section 7.5.2, for the similar case (c) of a cluster point at marginal stability, that complex indices imply that ξ oscillates infinitely rapidly when the singular point is approached ($s \to \infty$). This implies, through the application of the oscillation theorem, that an infinity of discrete modes is found accumulating at the chosen values of ω^2. We will work out the explicit forms of this inequality for the extrema of the Alfvén and slow continua.

Recall that $P \equiv N/D$. For the Alfvén continuum, one then finds from the expressions (7.91) and (7.92) that

$$\left(\frac{N}{s^2}\right)_0 = f^2 b^4 \cdot -\tfrac{1}{2}\rho(\omega_A^2)'' , \quad D_0 = -f^2 g^2 b^4 , \quad Q_0 = -\rho' \hat{g} , \quad (7.175)$$

so that

$$q_0 = -\frac{g^2 \cdot \rho' \hat{g}}{\tfrac{1}{2}\rho(\omega_A^2)''} . \quad (7.176)$$

Hence, in order to have a *Sturmian sequence of discrete Alfvén modes clustering at the lower edge of the Alfvén continuum*, i.e. at the minimum of ω_A^2 (where $\omega_A^{2\,\prime\prime} > 0$), the following condition should be satisfied:

$$g^2 \cdot \rho' \hat{g} > \tfrac{1}{8} \rho(\omega_A^2)'' > 0 ; \quad (7.177)$$

vice versa, for an *anti-Sturmian sequence at the upper edge of the Alfvén continuum*, i.e. at the maximum of ω_A^2 (where $\omega_A^{2\,\prime\prime} < 0$):

$$g^2 \cdot \rho' \hat{g} < \tfrac{1}{8} \rho(\omega_A^2)'' < 0 . \quad (7.178)$$

We have encountered an example of the latter in Fig. 7.10, with the cluster spectrum given by Eq. (7.122)(b), for the constant α exponential atmosphere (where $\omega_A^{2\,\prime\prime} = 0$). The counterpart of a Sturmian Alfvén sequence was missing there because that requires an inverted density profile (which is very well possible for more general equilibria).

Similarly, for extrema of the slow continua, one finds the following expressions:

$$\left(\frac{N}{s^2}\right)_0 = f^2 b^4 \cdot \tfrac{1}{2}\rho(\omega_S^2)'', \quad D_0 = -\omega_S^4,$$

$$Q_0 = \frac{\rho b^2}{c^2 \omega_S^2}(\omega_S^4 - k_0^2 \hat{g}^2) - \left(\frac{b^2 + c^2}{c^2}\rho \hat{g}\right)'. \quad (7.179)$$

Hence, in order to have a *Sturmian sequence of discrete slow modes clustering at the lower edge of the slow continuum*, the following condition should be satisfied:

$$\frac{1}{b^2 + c^2}\left[-\rho(\omega_S^4 - k_0^2\hat{g}^2) + \frac{c^2}{b^2}\omega_S^2\left(\frac{b^2+c^2}{c^2}\rho\hat{g}\right)'\right] > \tfrac{1}{8}\rho(\omega_S^2)'' > 0; \quad (7.180)$$

vice versa, for an *anti-Sturmian sequence at the upper edge of the slow continuum*:

$$\frac{1}{b^2 + c^2}\left[-\rho(\omega_S^4 - k_0^2\hat{g}^2) + \frac{c^2}{b^2}\omega_S^2\left(\frac{b^2+c^2}{c^2}\rho\hat{g}\right)'\right] < \tfrac{1}{8}\rho(\omega_S^2)'' < 0. \quad (7.181)$$

We have encountered an example of the latter behaviour for homogeneous plasmas (where the gravity terms are missing) and of the forms in Fig. 7.10(b), with the cluster spectrum given by Eq. (7.122)(c), for the exponential atmosphere (where $\omega_S^{2\prime\prime} = 0$). Also recall from that example that transition from Sturmian to anti-Sturmian behaviour occurs for a value of $\bar{k}_0^2$ far to the right of the plotted region in Fig. 7.10(b).

Hence, for general equilibria, the five discrete cluster spectra indicated in Fig. 7.18 do actually occur. In the absence of complex indices, the Alfvén and slow cluster points plus the high n modes will be missing, but global (lower n) modes may still be present. To compute those requires numerical solution of the MHD wave equation (7.91).

7.5 Gravitational instabilities of plasmas with magnetic shear

In this section, we apply the stability theory by means of the *energy principle*, as presented in Chapter 6, to the plane inhomogeneous gravitating plasma slab. This also provides another opportunity to illustrate the equivalence of the variational approach with that of the differential equations, as developed in Sections 7.2–7.4.

We have seen that gravity does not influence the continua (i.e. the essential spectrum), but it does influence the appearance or disappearance of discrete cluster spectra (Section 7.4.4). This is clear from the fact that gravity does not appear in the expressions (7.95) and (7.96) for the genuine and apparent singularities ω_A^2, ω_S^2, ω_{s0}^2, ω_{f0}^2 of the basic differential equation (7.91). However, gravity does appear in the other coefficients of the ODE determining whether the edge of a continuum is a cluster point or not. Hence, in the logical exposition of the theory, we first had to pay attention to the more difficult issue of the singularities (Section 7.4) whereas the simpler problems of local (related to cluster points) and global stability could be delayed until the present section.

7.5.1 Energy principle for a gravitating plasma slab

Recall that stability can be determined by just studying the sign of the quadratic form $W[\boldsymbol{\xi}]$ for the potential energy for all possible perturbations $\boldsymbol{\xi}$. For convenience, we here repeat the general expression (6.85):

$$W = \tfrac{1}{2} \int \Big[\gamma p \, |\nabla \cdot \boldsymbol{\xi}|^2 + |\mathbf{Q}|^2 + (\boldsymbol{\xi}^* \cdot \nabla p) \nabla \cdot \boldsymbol{\xi} \\ + \mathbf{j} \cdot \boldsymbol{\xi}^* \, \mathbf{Q} - (\boldsymbol{\xi}^* \cdot \nabla \Phi) \nabla \cdot (\rho \boldsymbol{\xi}) \Big] dV \,.$$

We now reduce this expression for the special case of a gravitating plasma slab. This will be the quadratic forms counterpart of the differential equation analysis of Sections 7.3 and 7.4. Of course, the starting point is the same, so that the equilibrium equations (7.71)–(7.74), the 2D Fourier expressions (7.78) for the perturbations, and the projections (7.79)–(7.83) can be maintained. Again, we study separate harmonics, denoted by $\boldsymbol{\xi}(x) \exp i(k_y y + k_z z)$.

Note that the integrand of the expression (6.85) for W consists of products of the unknowns with their complex conjugates. Of course, the energy itself has to be real so that it should be equal to its complex conjugate. This follows from the self-adjointness of the force operator. When working out quadratic forms, it is important to keep track of this reality of the final expression.

(a) Reduction to a one-dimensional variational problem Consider the different contributions to the integrand of W. Recall that we defined the three components ξ, η and ζ such that they can be assumed to be real. Hence, the expression for the compressibility,

$$\nabla \cdot \boldsymbol{\xi} = \xi' + g\eta + f\zeta \,, \tag{7.182}$$

is real. The Cartesian components of the magnetic field perturbation $\mathbf{Q}$ were worked out in Eq. (7.87):

$$Q_x = \mathrm{i} f B \xi \,, \qquad Q_y = -(B_y \xi)' + k_z B \eta \,, \qquad Q_z = -(B_z \xi)' - k_y B \eta \,,$$

$$\tag{7.183}$$

7.5 Gravitational instabilities of plasmas with shear

so that Q_x is imaginary, whereas Q_y and Q_z are real. A typical term of the integrand,

$$\mathbf{j} \cdot \boldsymbol{\xi}^* \times \mathbf{Q} = j_y(\xi_z^* Q_x - \xi_x^* Q_z) + j_z(\xi_x^* Q_y - \xi_y^* Q_x)$$

$$= -B_z'\left[-\frac{\mathrm{i}}{B}(B_y\eta - B_z\zeta)\,Q_x - \xi\,Q_z\right] + B_y'\left[\xi\,Q_y - \frac{\mathrm{i}}{B}(B_z\eta + B_y\zeta)\,Q_x\right],$$

is now easily seen to be real when the expressions (7.183) are inserted. The remaining contributions to the integrand are also real:

$$(\boldsymbol{\xi}^* \cdot \nabla p)\,\nabla \cdot \boldsymbol{\xi} = p'\xi\,\nabla \cdot \boldsymbol{\xi},$$

$$(\boldsymbol{\xi}^* \cdot \nabla \Phi)\,\nabla \cdot (\rho\boldsymbol{\xi}) = -\hat{g}\xi\,(\rho\nabla \cdot \boldsymbol{\xi} + \rho'\xi).$$

Putting everything together, and cancelling terms by exploiting the equilibrium relation (7.74), the following expression for W is obtained:

$$W = \tfrac{1}{2}\int\left[\frac{f^2 B^2}{k_0^2}\xi'^2 + \left(f^2 B^2 - \rho'\hat{g} - \frac{\rho^2\hat{g}^2}{\gamma p}\right)\xi^2\right.$$

$$\left. + B^2\left(k_0\eta + \frac{g}{k_0}\xi'\right)^2 + \gamma p\left(\nabla \cdot \boldsymbol{\xi} - \frac{\rho\hat{g}}{\gamma p}\xi\right)^2\right]dV. \qquad (7.184)$$

This quadratic form only depends on ξ, ξ', η and ζ so that, upon minimization, one may expect a differential equation for ξ, and algebraic relations for η and ζ.

▷ **Exercise.** Check all of the above steps for yourself! ◁

In fact, minimization with respect to the transverse variables η and ζ is algebraic and, hence, trivial:

$$\eta = -\frac{g}{k_0^2}\xi', \quad \text{and} \quad \nabla \cdot \boldsymbol{\xi} = \frac{\rho\hat{g}}{\gamma p}\xi \;\Rightarrow\; \zeta = -\frac{f}{k_0^2}\xi' + \frac{\rho\hat{g}}{\gamma p f}\xi. \qquad (7.185)$$

Consequently, only the first two terms remain in the expression (7.184) for the energy:

$$W = \tfrac{1}{2}\int_{x_1}^{x_2}\left[\frac{f^2 B^2}{k_0^2}\xi'^2 + \left(f^2 B^2 - \rho'\hat{g} - \frac{\rho^2\hat{g}^2}{\gamma p}\right)\xi^2\right]dx, \qquad (7.186)$$

where we have suppressed again the infinite area A of the horizontal plane by renormalizing ($\overline{W} \equiv W/A$) and dropping the bar. One immediately notices that the plasma slab is trivially stable in the absence of gravity ($\hat{g} = 0$) since $W \geq 0$ then. An inequality like this illustrates a typical use of the energy principle since it has no simple counterpart in the differential equation approach.

Systematic minimization of the remaining 1D problem (7.186) is a standard variational problem, as explained in Section 6.6.4, Eqs. (6.198)–(6.201). For a general quadratic form

$$W = \tfrac{1}{2} \int_{x_1}^{x_2} \left(P_0 \xi'^2 + Q_0 \xi^2 \right) dx, \qquad (7.187)$$

the minimizing solution is a solution of the *Euler–Lagrange equation*

$$\frac{d}{dx}\left(P_0 \frac{d\xi}{dx} \right) - Q_0 \xi = 0, \qquad (7.188)$$

subject to the boundary conditions

$$\xi(x_1) = \xi(x_2) = 0. \qquad (7.189)$$

For the present problem, the Euler–Lagrange equation corresponding to the quadratic form (7.186) reads

$$\frac{d}{dx}\left(\frac{F^2}{k_0^2} \frac{d\xi}{dx} \right) - \left(F^2 - \rho' \hat{g} - \frac{\rho^2 \hat{g}^2}{\gamma p} \right) \xi = 0, \qquad (7.190)$$

where we now start to exploit the notation $F \equiv -i\mathbf{B}\cdot\nabla \equiv k_y B_y + k_z B_z$, already introduced in Eq. (7.97), to stress the central importance of the parallel gradient operator in stability theory.

Note that the Euler–Lagrange equation (7.190) is just the *marginal equation of motion*, obtained from the general MHD wave equation (7.91) by substituting $\omega^2 = 0$:

$$P_0 \equiv P(\omega^2{=}0), \qquad Q_0 \equiv Q(\omega^2{=}0). \qquad (7.191)$$

(In contrast to Section 7.4, the index 0 on P_0 and Q_0 now denotes this substitution.) This connection nearly closes the circle of our presentation. We just need to interpret the physical meaning of the 'solutions' of Eq. (7.190) and of the singularities that occur when $F = \mathbf{k}_0 \cdot \mathbf{B} = 0$ somewhere. The latter problem will be addressed in Section 7.5.2.

(b) Variational procedure (absence of singularities) We put quotation marks on 'solutions' since, in general, it is impossible to solve Eq. (7.190) if the boundary conditions (7.189) are imposed. This follows from the simple fact that *there is no eigenvalue parameter* in that differential equation. Hence, all one can do is, for example, start from the left by satisfying the left boundary condition and then integrate to the right and check whether or not more zeros are encountered on the interval (x_1, x_2). Of course, in general, the right boundary condition will not be satisfied for that 'solution'. By means of the calculus of variations, as carried out

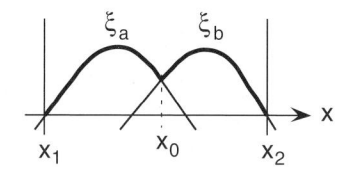

7.5 Gravitational instabilities of plasmas with shear

Fig. 7.19. Composite trial function.

systematically by Newcomb [164] for the analogous case of a cylindrical plasma, one may relate the oscillatory behaviour of ξ to stability as follows.

Insert the 'solution' ξ of the Euler–Lagrange equation (7.188) in Eq. (7.187) for the energy W, and integrate by parts:

$$W = \tfrac{1}{2} \int_{x_1}^{x_2} \left(P_0 \xi'^2 + Q_0 \xi^2 \right) dx$$

$$= \tfrac{1}{2} \int \left[P_0 \xi'^2 + \xi \left(P_0 \xi' \right)' \right] dx = \tfrac{1}{2} \left[P_0 \xi \xi' \right]_{x_1}^{x_2}. \qquad (7.192)$$

If the interval (x_1, x_2) is larger than the distance between two consecutive zeros of a solution to the Euler–Lagrange equation, one may split it into two sub-intervals (x_1, x_0) and (x_0, x_2) such that a solution ξ_a which vanishes at $x = x_1$ does not vanish again on (x_1, x_0) and a solution ξ_b which vanishes at $x = x_2$ does not vanish a second time on (x_0, x_2). At $x = x_0$ the amplitudes of the two solutions may be chosen equal (Fig. 7.19). By applying Eq. (7.192) to a solution composed of ξ_a on (x_1, x_0) and ξ_b on (x_0, x_2), one then obtains

$$W = \tfrac{1}{2} \left(P_0 \xi_a \xi_a' \right) \Big|_{x=x_0} - \tfrac{1}{2} \left(P_0 \xi_b \xi_b' \right) \Big|_{x=x_0} = \tfrac{1}{2} \left(P_0 \xi (\xi_a' - \xi_b') \right) \Big|_{x=x_0} < 0, \qquad (7.193)$$

so that the contribution to the energy is negative. In conclusion:

(1) if the 'solution' ξ_0 of the Euler–Lagrange equation (7.190) (i.e. the marginal equation of motion) that satisfies the left boundary condition $\xi_0(x_1) = 0$ has another zero on the interval (x_1, x_2), then a trial function ξ_1 can be constructed (the composite function of Fig. 7.19) that satisfies both boundary conditions and for which the energy $W(\xi_1) < 0$: the system is unstable;

(2) if ξ_0 has no other zeros on the interval, that construction fails and $W(\xi_1) \geq 0$ for all trial functions: the system is stable.

This is the simple part of Newcomb's variational procedure, valid in the absence of singularities ($F \neq 0$ on the entire interval). The complicated part is the proper handling of those singularities.

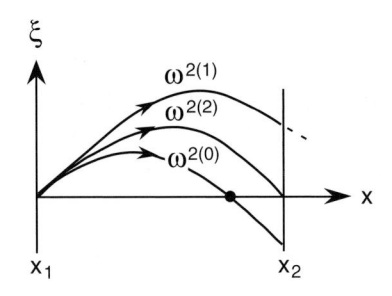

Fig. 7.20. 'Shooting' for discrete eigenvalues.

(c) 'Shooting' method By means of the oscillation theorem of Section 7.4.3, the variational procedure can be replaced by a more transparent one related to the calculation of the *growth rates of the instabilities*. This approach is just a special case of solving for the discrete modes for values of ω^2 outside the ranges of $\omega_A^2(x)$, $\omega_S^2(x)$, $\omega_{s0}^2(x)$, $\omega_{f0}^2(x)$ on the interval $x_1 \leq x \leq x_2$ depicted in Fig. 7.14. In that case, as already alluded to in the introduction of Section 7.4.1, the full MHD wave equation (7.91) is non-singular and a straightforward method to find discrete eigenvalues is to apply the so-called *shooting method*. This amounts to a procedure in which the second order differential equation (7.91) is first solved as an initial value problem where a value of ω^2 is guessed, say $\omega^2 = \omega^{2(i)}$, so that the functions appearing in Eq. (7.91) are determined, and the differential equation is solved numerically starting from the left 'initial' data $\xi(x_1) = 0$, $\xi'(x_1) = \text{const.}$ By some numerical integration scheme the solution $\xi(x; \omega^{2(i)})$ is found. Since the RHS boundary condition (7.93) will not be satisfied in general, $\xi(x_2; \omega^{2(i)}) \neq 0$, one needs to shoot again until this boundary condition is satisfied as well (see Fig. 7.20). To stay with the metaphor: this is not done by changing the inclination of the gun (i.e., by changing the value of $\xi'(x_1)$, which would just change the amplitude of the solution and not the location of the zeros) but by changing the amount of gun powder (i.e., by changing the value of ω^2). Since, according to the oscillation theorem, the eigenvalue ω^2 is monotonic in the distance between the zeros of ξ, one knows precisely in which direction this change has to go: for Sturmian frequency ranges, one should decrease ω^2 to move out the next zero to the position $x = x_2$ to get a genuine eigenfunction, for anti-Sturmian frequency ranges one should increase ω^2.

Unstable discrete modes are always Sturmian because $N/D < 0$ for $\omega^2 < 0$. Since the Euler–Lagrange equation (7.190) is just the marginal form ($\omega^2 = 0$) of the general MHD wave equation (7.91), the connection with the variational procedure for stability is obvious. Referring to Fig. 7.20: if one shoots with $\omega^2 = 0$ and one finds a zero (the curve labelled $\omega^{2(0)}$), there will be a genuine eigenvalue $\omega^{2(2)} < 0$ so that the system is unstable. On the other hand, if one shoots with

$\omega^2 = 0$ and one finds no zero (the curve labelled $\omega^{2(1)}$), the eigenvalue $\omega^{2(2)} > 0$ so that the system is stable. Hence, the shooting method not only supplies an answer to the stability problem, but also an answer to the question about the 'danger' of the instabilities, i.e. a value for their growth rate. In this manner, the global stability problem appears to have been solved completely for one-dimensional inhomogeneities.

The singularities $F = 0$ of the marginal stability equation are absent in the shooting method as long as $\omega^2 < 0$. From a practical point of view, this is sufficient to solve any stability problem because one can always exploit the σ-stability concept (Section 6.5.3) to stay away from these singularities (any small σ^2 suffices). However, singularities also offer the advantage of increased analytical insight in stability problems, as will be shown in the next section.

7.5.2 Interchange instabilities in sheared magnetic fields

The remaining problem in the stability theory of gravitating plasmas is the proper interpretation of the *singularities* $F \equiv \mathbf{k}_0 \cdot \mathbf{B} = 0$ of the marginal equation (7.190). This is again related to the continuous spectrum since the expressions (7.95) for the Alfvén and slow continuum frequencies both degenerate into the marginal frequency $\omega^2 = 0$ when $F = 0$. Hence, if there is such a point on the interval $[x_1, x_2]$, the Alfvén and slow continua both extend to the origin $\omega^2 = 0$. The physical significance of these points is that the horizontal wave vector $\mathbf{k}_0$ is perpendicular to $\mathbf{B}$, so that the perturbations do not disturb the magnetic field. The magnetic part of the potential energy of the Alfvén wave perturbations vanishes there because the field lines are not bent. Consequently, these are the positions where the driving forces of instability are minimally counterbalanced by magnetic tensions so that instabilities are predominantly localized there.

On the other hand, by means of *magnetic shear* ($F' \neq 0$), the region of minimal field line bending can be minimized so that stability can be restored, as we will see. We have already encountered this effect in Section 6.6.4 when discussing the stabilization of the Rayleigh–Taylor instability of interface plasmas by means of a magnetic field with different directions in the plasma and the vacuum (Fig. 6.21). In that particular case, the magnetic shear was entirely localized in the infinitely narrow surface layer separating plasma and magnetic field. In the present section, we deal with diffuse plasmas where shear is present everywhere.

To establish the significance of the singularities $F(x_0) = 0$, we expand all functions in the neighbourhood of that singularity. Hence, we continue the analysis of case (c) of Section 7.4.1, with the expansion (7.129) of the Alfvén factor, illustrated in Fig. 7.11(c). Introducing the variable angle $\varphi(x) \equiv \arccos(B_z/B)$

between the magnetic field and the z-axis (Fig. 7.9(b)), and the constant angle $\theta \equiv \arccos(k_z/k_0)$ between the horizontal wave vector and the z-axis, we have

$$F \equiv \mathbf{k}_0 \cdot \mathbf{B} = k_0 B(x) \cos[\varphi(x) - \theta]. \tag{7.194}$$

At the singularity

$$\varphi(x_0) = \theta \pm \pi/2 \quad \Rightarrow \quad F'(x_0) = \mp k_0 (B\varphi')_0, \tag{7.195}$$

so that

$$\rho \omega_A^2 \equiv F^2(x) \approx (F'^2)_0 s^2 = k_0^2 (B^2 \varphi'^2)_0 s^2, \tag{7.196}$$

where $s \equiv x - x_0$. The quantity φ' represents *the shear of the magnetic field*, which is the change of the angle φ when x is increased. This is caused by the parallel component of the current: $\varphi' = (B_y B'_z - B'_y B_z)/B^2 = -\mathbf{j} \cdot \mathbf{B}/B^2$, according to Eq. (7.72).

In the neighbourhood of the singularity, the marginal stability equation (7.190) is approximated by

$$\frac{d}{ds}\left[s^2(1+\cdots)\frac{d\xi}{ds}\right] - q_0(1+\cdots)\xi = 0, \qquad q_0 \equiv -\left(\frac{\rho'\hat{g} + \dfrac{\rho^2 \hat{g}^2}{\gamma p}}{B^2 \varphi'^2}\right)_0, \tag{7.197}$$

which has the form of the standard differential equation (7.132) of Section 7.4.1. Consequently, the Frobenius expansion (7.134) may be exploited. To leading order, the solutions of equation (7.197) behave as s^{ν_1} and s^{ν_2}, where ν_1 and ν_2 are the roots of the indicial equation $\nu(\nu+1) - q_0 = 0$, as given by Eq. (7.138). Depending on whether $1 + 4q_0$ is positive or negative, the indices are real or complex.

(a) Local stability (complex indices) The most interesting case is obtained for $1 + 4q_0 < 0$, *when the indices are complex*, because the real solutions of the marginal stability equation are oscillatory then:

$$\begin{aligned}\xi_1 &= s^{-1/2+iw} + s^{-1/2-iw} = 2s^{-1/2}\cos(w \ln s), \\ \xi_2 &= i(s^{-1/2+iw} - s^{-1/2-iw}) = -2s^{-1/2}\sin(w \ln s),\end{aligned} \tag{7.198}$$

where $w \equiv \frac{1}{2}\sqrt{-(1+4q_0)}$. The kind of oscillatory behaviour obtained is quite extreme, since the solutions not only oscillate infinitely rapidly but their amplitude also blows up when $s \to 0$, as schematically illustrated in Fig. 7.21(a). According to the oscillation theorem of Section 7.4.3(b), such marginal 'solutions' signal instability, since the zeros of the solutions of the full MHD spectral equation are peeled off one by one as the value of ω^2 is decreased from 0 to the actual growth

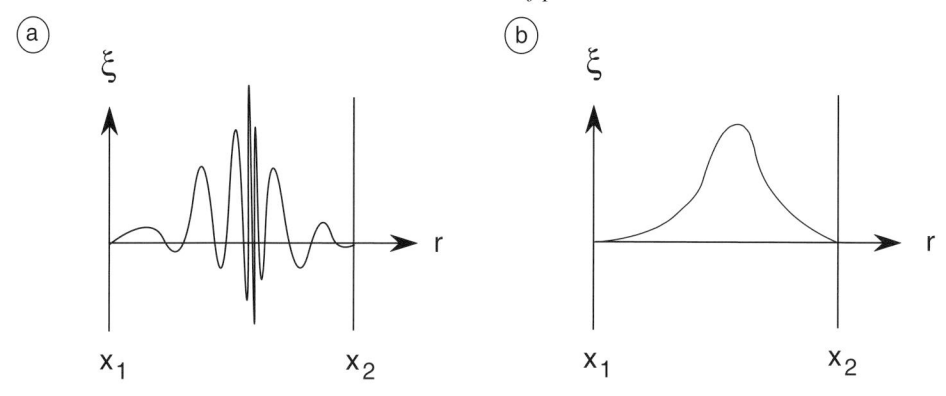

7.5 Gravitational instabilities of plasmas with shear

Fig. 7.21. Violation of the interchange criterion: (a) marginal mode ($\omega^2 = 0$, $n = \infty$), (b) associated most global mode ($\omega^2 < 0$, $n = 1$).

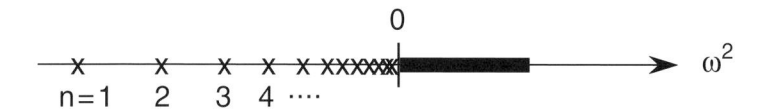

Fig. 7.22. Violation of the interchange criterion: appearance of a cluster point at $\omega^2 = 0$.

rate(s). Consequently, one obtains an infinity of unstable point eigenvalues, accumulating at $\omega^2 = 0$ for $n \to \infty$, where n is the number of nodes of the corresponding eigenfunctions: *when the indices are complex, the marginal point is a cluster point of the unstable modes of the discrete spectrum* (Fig. 7.22). This should not distract the attention from the most important physical fact, viz. that the fastest growing instability is obtained for $n = 1$. Although the condition for instability is local, that $n = 1$ instability need not be local at all, as illustrated by Fig. 7.21(b). Generally, the width of this mode is determined by the width of the region over which the local stability criterion is violated.

To avoid these instabilities, one should demand that $1 + 4q_0 > 0$, so that the indices are real. This leads to the following *stability criterion for interchange modes*:[4]

$$\rho'\hat{g} + \frac{\rho^2 \hat{g}^2}{\gamma p} \left(\equiv -\rho N_B^2 \right) \leq \tfrac{1}{4} B^2 \varphi'^2 . \tag{7.199}$$

The three respective terms represent the driving force of the *gravitational* or *Rayleigh–Taylor instability* (heavy fluid on top of a lighter one), *modified by adiabatic effects* (term with γ), and *stabilized by magnetic shear* (term on the right

---

[4] A rather imprecise term dating from the early days of fusion research when there were hopes of confining extremely high-β plasmas, where the plasma and the magnetic field are nearly completely separated in space. Such plasmas are virtually always unstable with respect to 'interchange' of plasma and magnetic field.

hand side). To return to our introductory discussion of Section 6.1.1: a glass of magnetized plasma may be turned upside down without the contents dropping out, if the magnetic shear is large enough! Note that the criterion reduces to the HD criterion (7.40) of Section 7.2.2 for convective stability, $N_B^2 > 0$, when $\varphi' = 0$. Here, N_B^2 is the square of the Brunt–Väisälä frequency, which also turned up in the condition (7.110) of Section 7.3.3 for the stability of gravito-MHD modes.

The analysis has been a *local* one, so that it may be repeated for every point on the interval $[x_1, x_2]$. Hence, for overall stability, one should at least demand that the inequality (7.199) is satisfied everywhere: the criterion is a necessary one for stability. A local criterion of this kind, known as *Suydam's criterion*, was first derived by Suydam (1958) [222] for a diffuse cylindrical plasma column (pinch) where the driving force of the instability is the pressure gradient in combination with curvature of the magnetic field lines. The derivation is completely analogous to the one given here. We will discuss it more extensively in Chapter 9.

(b) Global stability (real indices) For $1 + 4q_0 > 0$, when *the indices are real*, the two solutions at the singularity behave as

$$\xi_s \sim s^{\nu_s}, \qquad \nu_s = -\tfrac{1}{2} + \tfrac{1}{2}\sqrt{1 + 4q_0} > -\tfrac{1}{2} \qquad \text{('small' solution)},$$
$$\xi_\ell \sim s^{\nu_\ell}, \qquad \nu_\ell = -\tfrac{1}{2} - \tfrac{1}{2}\sqrt{1 + 4q_0} < -\tfrac{1}{2} \qquad \text{(large solution)}. \tag{7.200}$$

Hence, the large solution ξ_ℓ always blows up at $s = 0$, whereas the 'small' solution may or may not blow up depending on whether the square root is smaller or larger than 1. Similar to the singularities for $\omega^2 > 0$, discussed in Section 7.4.2, the present ones (at $\omega^2 = 0$) also split the interval (x_1, x_2) into independent sub-intervals (x_1, x_0) and (x_0, x_2), in the following sense. Consider the sub-interval to the left of the singularity. According to Eq. (7.192), the contribution of that sub-interval to the energy is:

$$W(x_1, x_0) = \tfrac{1}{2}\Big[P_0 \xi \xi' \Big]_{x_1}^{x_0}. \tag{7.201}$$

At the singularity, this expression behaves as

$$W(x_0) = \tfrac{1}{2}\Big(P_0 \xi \xi' \Big)(x_0) \sim s^{2\nu+1} \rightarrow \begin{cases} 0 & \text{for } \nu = \nu_s, \\ \infty & \text{for } \nu = \nu_\ell, \end{cases} \tag{7.202}$$

so that the energy contribution of the large solution diverges, but that of the 'small' one vanishes (hence: 'small', also when $-\frac{1}{2} < \nu_s < 0$). Consequently, testing for stability while keeping the energy of the perturbations finite implies that we have to exclude the large solution. This means that a kind of internal boundary condition is to be imposed at the singularity, viz. that ξ should be 'small' there. This can be done since jumps in the 'small' solution are allowed by an argument similar to that of Section 7.4.2. Such jumps do not contribute to the energy:

$$P_0 \xi_s H(s) \left[\xi_s H(s) \right]' \sim s^{\nu+2} H(s) \left[\nu s^{\nu-1} H(s) + s^\nu \delta(s) \right]$$
$$= s^{2\nu+1} \left[\nu H(s) + s \delta(s) \right] H(s) \to 0. \quad (7.203)$$

Therefore, the intervals (x_1, x_0) and (x_0, x_2) may be tested separately with respect to stability (i.e. the sign of W) by means of trial functions that are 'small' at $x = x_0$ and vanish identically either to the right or to the left of $x = x_0$.

The stability test is carried out with the following modification of the non-singular case (described in Section 7.5.1(b), Fig. 7.19), where we now exploit the symbol x_s for the singularity and x_0 for an ordinary interior point. Consider a solution ξ_a of the Euler–Lagrange equation (7.190) which vanishes on the left interval (x_1, x_s), is 'small' to the right of the singularity $x = x_s$, and vanishes once in the interval (x_s, x_2). Such a solution may be joined at a point x_0 in between the singularity x_s and the zero point of ξ_a to another solution ξ_b which vanishes at the right endpoint $x = x_2$, but does not vanish in the open interval (x_0, x_2) (Fig. 7.23). The energy of the Euler–Lagrange solution consisting of $\xi = 0$ on $(0, x_s)$, $\xi = \xi_a$ on (x_s, x_0), and $\xi = \xi_b$ on (x_0, x_2) may be shown to be negative by a completely analogous argument to that used in the derivation of Eq. (7.193). Hence, on independent sub-intervals the 'smallness' of a solution should be counted as a zero,

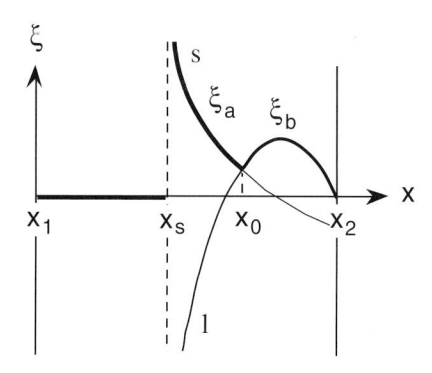

Fig. 7.23. Composite trial function in the presence of a singularity.

so that for stability a solution that is 'small' at the singularity should not vanish somewhere in the interval. Thus, using Newcomb's wording, we obtain the following stability theorem for the case that the interval (x_1, x_2) contains one singularity $F = 0$ at $x = x_s$.

Theorem. *For specified values of k_y and k_z such that $F \equiv k_y B_y + k_z B_z = 0$ at some point $x = x_s$ of the interval (x_1, x_2), the gravitating plasma slab is stable if, and only if, (1) the interchange criterion (7.199) is satisfied at $x = x_s$; (2) the non-trivial solution ξ_L of the Euler–Lagrange equation (7.190) that is 'small' to the left of $x = x_s$ does not vanish in the open interval (x_1, x_s); (3) the non-trivial solution ξ_R that is 'small' to the right of $x = x_s$ does not vanish in the open interval (x_s, x_2).*

Of course, if there is more than one singularity, there will be more than two independent sub-intervals that have to be tested for oscillatory behaviour in the extended sense.

7.5.3 Interchanges in the absence of magnetic shear

A special case occurs when the magnetic field has *no shear*, i.e. it is uni-directional (like the exponential atmosphere considered in Section 7.3.3). Then, perturbations may be found with $F \equiv 0$ over the whole domain, so that the horizontal wave vector $\mathbf{k}_0$ is perpendicular to the magnetic field everywhere. Naively, one would expect this to be the worst case scenario with stability determined by the condition (7.199) in the no-shear limit ($\varphi' = 0$), i.e. the familiar condition $N_B^2 \geq 0$. However, if $F \equiv fB = 0$, the variable ζ disappears from the expression (7.182) for $\nabla \cdot \boldsymbol{\xi}$ and, thereby, also from the expression (7.184) for the energy. Hence, the two last terms of Eq. (7.184) cannot be considered as independent, so that the minimization proceeds differently than for the case with shear. For $f = 0$, $g = k_0$, the expression for the energy reduces to

$$W = \tfrac{1}{2} \int \left[-\left(\rho' \hat{g} + \frac{\rho^2 \hat{g}^2}{\gamma p} \right) \xi^2 + B^2 \left(k_0 \eta + \xi' \right)^2 + \gamma p \left(\xi' + k_0 \eta - \frac{\rho \hat{g}}{\gamma p} \xi \right)^2 \right] dV$$

$$= \tfrac{1}{2} \int \left\{ -\left(\rho' \hat{g} + \frac{\rho^2 \hat{g}^2}{B^2 + \gamma p} \right) \xi^2 \right.$$

$$\left. + k_0^2 (B^2 + \gamma p) \left[\eta + \frac{(B^2 + \gamma p) \xi' - \rho \hat{g} \xi}{k_0 (B^2 + \gamma p)} \right]^2 \right\} dV. \qquad (7.204)$$

7.5 Gravitational instabilities of plasmas with shear

By minimization with respect to η, the last term disappears and we get

$$W = -\tfrac{1}{2} \int \left(\rho' \hat{g} + \frac{\rho^2 \hat{g}^2}{B^2 + \gamma p} \right) \xi^2 \, dx , \qquad (7.205)$$

so that the stability criterion becomes

$$\rho' \hat{g} + \frac{\rho^2 \hat{g}^2}{B^2 + \gamma p} \quad \left(\equiv -\rho N_m^2 \right) \leq 0 . \qquad (7.206)$$

Consequently, the stability criterion obtained, for stability with respect to *pure interchanges*, is less severe than the condition (7.199) in the no-shear limit. The expression is again familiar: $N_m^2 \geq 0$, where N_m^2 is the square of the magnetically modified Brunt–Väisälä frequency defined in Eq. (7.114) of Section 7.3.3.

The criterion $N_m^2 \geq 0$ is automatically satisfied for the exponential atmosphere analysed in that section, consistent with the stability of the solutions (7.113) for perpendicular propagation. However, we also found out there, in the intriguing equation (7.125) where both N_B^2 and N_m^2 appeared, that *the worst instabilities sometimes occur for near-perpendicular*, not for purely perpendicular, *propagation*. This subtlety in the stability of general (not restricted to exponential profiles) shearless magnetic fields was already pointed out in 1961 by Newcomb [165]. We summarize the results of that paper since it throws light on the relationship between stability and spectral analysis.

As is clear from Section 7.3.3, the relevant modes for instability are the low-frequency Alfvén and slow modes. Hence, it is expedient to get rid of the high-frequency fast modes by means of an ordering. This we do by introducing a scale length L of the equilibrium variations,

$$L^{-1} \equiv \frac{\rho \hat{g}}{p + \tfrac{1}{2} B^2} \qquad (7.207)$$

(the parameter α of Section 7.3.3), and imposing the following ordering on the frequencies and wave numbers:

$$\rho \omega^2 \sim L^{-1} \rho \hat{g} \sim f^2 b^2 \ll k_0^2 b^2 . \qquad (7.208)$$

This implies that we assume near perpendicular propagation of the modes ($|f| \ll |g|$). The MHD wave equation (7.91) then reduces to

$$\frac{d}{dx} \rho(\omega^2 - f^2 b^2) \frac{d\xi}{dx}$$
$$- k_0^2 \left[\rho(\omega^2 - f^2 b^2) + \rho' \hat{g} + \rho \hat{g}^2 \frac{\omega^2 - f^2 b^2}{(b^2 + c^2)\omega^2 - f^2 b^2 c^2} \right] \xi = 0. \qquad (7.209)$$

This equation still contains the exact marginal stability equation (7.190) for $\omega^2 \to 0$. We solve it for local modes, varying rapidly over the length scale L:

$$\xi \sim e^{iqx}, \qquad qL \gg 1. \tag{7.210}$$

This yields the following dispersion equation:

$$\omega^4 - \left(\frac{b^2 + 2c^2}{b^2 + c^2} f^2 b^2 + \frac{k_0^2}{k_0^2 + q^2} N_m^2 \right) \omega^2$$
$$+ \frac{c^2}{b^2 + c^2} f^2 b^2 \left(f^2 b^2 + \frac{k_0^2}{k_0^2 + q^2} N_B^2 \right) = 0, \tag{7.211}$$

having two qualitatively different solutions. For $F^2/\rho \equiv f^2 b^2 \ll |N_B^2| \sim |N_m^2|$, we recover the solutions (7.125) found in Section 7.3.3:

$$\omega_1^2 \approx \frac{k_0^2}{k_0^2 + q^2} N_m^2, \qquad \omega_2^2 \approx \frac{\gamma p}{\gamma p + B^2} \frac{N_B^2}{N_m^2} F^2/\rho. \tag{7.212}$$

The first mode is a *pure interchange* ($F = 0$) which becomes unstable when $N_m^2 < 0$. The second mode is called a *quasi-interchange* ($F \neq 0$) since only finite segments of the field lines can be involved in interchanging plasma and magnetic field. They become unstable when $N_B^2 < 0$. Since this is the more severe criterion, it should be considered as the overall boundary for stability of gravitational instabilities in shearless magnetic fields. With the consideration of the quasi-interchanges, the discrepancy in the stability criteria between sheared magnetic fields in the limit $\varphi \to 0$ and shearless magnetic fields disappears.

To establish the danger of instabilities, it is not sufficient to derive stability criteria, one should also calculate the maximum growth rate to find out which mode dominates. This involves maximizing the growth rate obtained from Eq. (7.211) with respect to the mode numbers f, g and q. This is left as an exercise for the reader. It is expedient to introduce the following notation:

$$\Gamma \equiv -\frac{\rho'}{\rho} \hat{g}, \qquad \Gamma_B \equiv \frac{\rho \hat{g}^2}{\gamma p}, \qquad \Gamma_m \equiv \frac{\rho \hat{g}^2}{\gamma p + B^2}, \qquad \Gamma_0 \equiv \frac{\gamma p \rho \hat{g}^2}{(\gamma p + B^2)^2}, \tag{7.213}$$

where it is to be noted that $\Gamma_B \geq \Gamma_m \geq \Gamma_0$. The result of the optimization can then be summarized as follows:

(1) if $\Gamma \geq \Gamma_B$, the plasma is stable;

(2) if $\Gamma_0 \leq \Gamma < \Gamma_B$, the most unstable mode is a quasi-interchange ($k_\parallel \neq 0$) with

growth rate
$$\omega^2 = -\frac{\rho \hat{g}^2}{B^2}\left(1 - \sqrt{\Gamma/\Gamma_B}\right)^2; \qquad (7.214)$$

(3) if $\Gamma \leq \Gamma_0$, the most unstable mode is a pure interchange ($k_\parallel = 0$) with growth rate
$$\omega^2 = \Gamma - \Gamma_m \equiv N_m^2. \qquad (7.215)$$

Hence, calculating the eigenfrequencies of the modes not only removes apparent discrepancies from stability theory, but also introduces new transitions ($\Gamma = \Gamma_0$) that are relevant for the description of the dynamics. Spectral theory is not only a mathematical beauty but also a physical necessity for understanding the dynamics of inhomogeneous plasmas.

7.6 Literature and exercises

Notes on literature

Hydrodynamics of the solar interior and corona:

- Priest, *Solar Magnetohydrodynamics* [190], Chapter 4 on waves.
- Stix, *The Sun* [217], Chapter 2 on internal structure of the Sun, Chapter 5 on oscillations.
- Christensen-Dalsgaard, *Stellar Oscillations* [56], Chapter 2 on the analysis of oscillation data, Chapter 5 on properties of solar and stellar oscillations.

Spectral theory of gravitating plasma slab:

- Goedbloed [81], in a series of papers on 'Stabilization of magnetohydrodynamic instabilities by force-free magnetic fields', derives the MHD wave equation for a gravitating plasma slab (I), and discusses quasi-interchanges in shearless magnetic fields (III).
- Goedbloed & Sakanaka [89], in a 'New approach to magnetohydrodynamic stability', prove the oscillation theorem for the MHD wave equation and introduce the concept of σ-stability.
- Lifschitz, *Magnetohydrodynamics and Spectral Theory* [146], Chapter 7 on MHD oscillations of a gravitating plasma slab.

Singular differential equations:

- Ince, *Ordinary Differential Equations* [115], Chapter 18 on the solution of linear differential equations in series.
- Bender & Orszag, *Advanced Mathematical Methods for Scientists and Engineers* [23], Chapter 3 on approximate solutions of linear differential equations.

Gravitational instabilities:

- Newcomb [165], on quasi-interchanges in 'Convective instability induced by gravity in a plasma with a frozen-in magnetic field'.

380   *Waves and instabilities of inhomogeneous plasmas*

- Parker [175], on the Parker instability in 'The dynamical state of the interstellar gas and field'.
- Mouschovias [160], on 'The Parker instability in the interstellar medium'.

Exercises

[7.1] *The Sun – energy source*

The Sun appears to be a regularly shaped sphere subject to the laws of hydrodynamics.

- Describe the mechanism which produces this shape. Looking at the Sun more carefully, it appears flattened at the poles. Why is that?

Before nuclear physics was understood, it was thought that the energy of the Sun comes from slow contraction due to gravity.

- Using the Sun's luminosity and Newton's law of gravity, estimate the speed at which the Sun would have to shrink. How long would the Sun be able to burn like this?

The light emanating from the Sun exerts pressure.

- Using physical arguments, find the maximum luminosity (called the Eddington luminosity) the Sun could have before it would blow away its outer layers.
- Using the luminosity of the Sun, calculate the heat flux at the Earth. Estimate the surface of exposed skin on your body. When sunbathing for 15 minutes, how much energy would have been transported to your body? How long could a light bulb of 60 watts burn on that amount of energy?

The source of all this radiation is not located near the surface of the Sun.

- Draw a cross-section of the Sun, indicate how and where this energy is produced, and name the different regions resulting from that.

[7.2] *The Sun – radiative transport*

The transport of radiation through a medium can be regarded as a random walk process. On average, a photon will travel one mean free path λ_{mfp} and collide with a particle, which redistributes it into a random direction. Due to the gradient in particle density, the net flux of radiation will be pointing outward. (This process is often compared to a drunkard trying to make his way through a crowded room.)

- The mean free path is defined by $\lambda_{\text{mfp}} \equiv (\kappa \rho)^{-1}$. Choosing the value $\kappa = 0.12 \, \text{m}^2 \, \text{kg}^{-1}$ for the opacity (also called the Rosseland absorption coefficient), calculate the mean free path of a photon, using the solar radius and mass.
- In n steps, the photon travels a distance $d = \sqrt{n} \, \lambda_{\text{mfp}}$. Estimate the number of steps needed for a photon to travel one solar radius and the time needed to cover this distance.

[7.3] *The Sun – convection*

At some radius, the energy produced no longer escapes as radiation, but is transported by convection. We are going to derive the Schwarzschild criterion for the onset of convection.

- Write down the equation of state for an ideal gas in terms of ρ and T, and derive the temperature gradient from this.
- The bulk motion of gas is only stable if the specific entropy, $S \equiv p \rho^{-\gamma}$, is constant in space. Derive the density gradient and the isentropic temperature gradient from this.

7.6 Literature and exercises

– By comparing these temperature gradients, the criterion for instability is obtained. Convert this criterion by means of the gravitational equilibrium equation to another form, relating to an apparently different instability. Which one is it?

[7.4] *Waves in a gravitating hydrodynamic slab*
Derive the HD wave equation for a plane parallel atmosphere from the nonlinear equations for the density, pressure, and velocity, ignoring non-adiabatic effects.

– What does the latter assumption imply for the description of the Sun?
– Using the Cowling approximation (what does that mean?), derive the differential equations in terms of the displacement field $\boldsymbol{\xi}$ for the appropriate horizontal Fourier modes.
– Rotate the coordinate system such that one of the axes is in the direction of the horizontal wave vector (why is that expedient?), and then derive the second order differential equation for the vertical displacement.

[7.5] *HD waves in an exponentially stratified atmosphere*
We continue with the HD wave equation obtained in the previous exercise.

– What is the physical significance of choosing exponential dependence on height, $e^{-\alpha x}$, for both the pressure and the density? What does it imply for the sound speed? Using the equilibrium condition, express α in the gravitational parameter $\hat{g}$.
– Derive the second order differential equation for ξ_x in terms of the Brunt–Väisälä frequency, $N_B^2 = \alpha \hat{g} - \hat{g}^2/c^2$. Obtain the dispersion equation from it and solve it.

[7.6] *MHD waves in an exponentially stratified atmosphere*
Look up the MHD wave equation for general one-dimensional inhomogeneity.

– In addition to the exponential pressure and density profiles of the previous exercise, introduce the appropriate form of the magnetic field as a function of the vertical coordinate to have an atmosphere with constant Alfvén speed. Calculate the new expression for α and use it to simplify the MHD wave equation and resulting dispersion equation.
– For these magnetic field, pressure, and density profiles, calculate the different MHD singularities. What do they represent?
– Give the solutions for purely parallel and purely perpendicular propagation of the waves, exploiting the magnetically modified Brunt–Väisälä frequency, $N_m^2 = \alpha \hat{g} - \hat{g}^2/(b^2 + c^2)$, as well as N_B^2. Which wave may become unstable? Why not the other one as well?

[7.7]★ *Dispersion equation for gravito-MHD waves*
Write a numerical program, exploiting the explicit solutions given in the text, to solve for the three gravito-MHD waves in an exponential atmosphere of magnetized plasma.

– Plot the three branches $\bar{\omega}^2(\bar{k}_0^2)$, similar to Fig. 7.10, for $\bar{\alpha} = 20$, $\beta = 1$, $\vartheta = \pi/2 - 0.1$ with $\bar{q} = \pi, 2\pi, \ldots$. How wide is the unstable region? What kind of modes are they?
– Estimate the value of $\bar{k}_0^2$ where the slow modes change from Sturmian to anti-Sturmian. Continue the numerical scan to that value to check the answer. What do you learn?

[7.8]* *Continuous spectra and spectral structure*

Assuming weak inhomogeneity of the equilibrium, make a sketch of a possible normal dependence of the singular frequencies ω_A^2 and ω_S^2. Also make a sketch of the associated MHD spectrum, indicating continua, discrete spectra, and the separating non-monotonicity regions.

– Starting with the first picture: the second order differential equation $(P\xi')' - Q\xi = 0$ for the MHD spectrum becomes singular at a point $x = x_0$ where $\omega = \omega_A(x_0)$, giving rise to the Alfvén continuum. Expand $\omega^2 - \omega_A^2$ around that singularity, use it to express $P \equiv N/D \sim s \equiv x - x_0$, and expand the other function, Q, as well (only the constant is needed). Solve the differential equation by means of Frobenius' method and discuss the meaning of the two solutions obtained.

– Concerning the second picture: what is the difference between the solutions of the Alfvén and slow continua and those of the turning point frequencies?

– Indicate Sturmian and anti-Sturmian regions in the diagram. (What do those terms mean?) How does the picture change when gravity is introduced? Show that gravity does not influence the continua, but it may effect the appearance of discrete spectra.

[7.9] *Spectrum of an incompressible gravitating slab*

Derive the MHD wave equation for an incompressible ($\gamma \to \infty$) gravitating plasma slab from the general wave equation. (This is a significant simplification, but watch out with the confluences!)

– Show that the Alfvén and slow singularities coincide so that the two continua are degenerate. How do the associated 'improper' modes differ?

– Neglecting gravity, assuming $\rho = \text{const}$, and a linear dependence of $\omega_A^2(x)$ about the singularity at $x = x_0$, show that the wave equation reduces to the zero order modified Bessel equation,

$$\xi'' + z^{-1}\xi' - \xi = 0, \quad z \equiv k_0(x - x_0),$$

with solutions

$$I_0 = 1 + \tfrac{1}{4}z^2 + \tfrac{1}{32}z^4 + \cdots,$$
$$K_0 = -[\ln(\tfrac{1}{2}z) + \gamma_0] I_0(z) + \tfrac{1}{4}z^2 + \tfrac{3}{128}z^4 + \cdots \quad (\gamma_0 \approx 0.5772 \text{ is Euler's constant}).$$

Are there any 'solutions' other than the ones associated with the continua?

– What changes when you allow for gravity?

[7.10]* *Shooting method*

One can investigate the discrete spectrum without worrying about the singular solutions by exploiting the 'shooting method', outside the ranges of the continua.

– Explain how this method works.

– In a numerical shooting procedure, what complications do you foresee when solving in the ranges of the turning point frequencies? How would you handle that?

[7.11] *Interchange instabilities*

Unstable modes have frequencies that are automatically outside the ranges of the continua, at least when their growth rate is finite. At $\omega^2 = 0$, one may exploit the potential energy of the perturbations to investigate stability, but one has to worry about a particular singularity.

- Assume we have a plane gravitating plasma with a sheared magnetic field. Write down the expression for the potential energy, W, and reduce it to a one-dimensional form.
- Explain why, usually, instabilities are localized at positions $x = x_0$ where $F(x_0) = 0$, when $F \equiv \mathbf{k}_0 \cdot \mathbf{B}$. Expand F around $x = x_0$ exploiting the angle $\varphi(x) \equiv \arccos(B_z/B)$.
- Insert this expansion in the Euler–Lagrange equation, and solve by means of a Frobenius expansion. Show that this yields a leading order equation of the form $(s^2 \xi')' - q_0 \xi = 0$ with indicial equation $\nu(\nu + 1) - q_0 = 0$, where $q_0 \equiv -(\rho' \hat{g} + \rho^2 \hat{g}^2/\gamma p)_0/(B^2 \varphi'^2)_0$.
- How do you obtain the stability criterion for interchange modes from that equation? Explain the meaning of the three terms.

8
Magnetic structures and dynamics

8.1 Plasma dynamics in laboratory and nature

In this chapter we will make an excursion to the vast territory of magnetic structures and dynamics of the different plasmas encountered in the solar system, in particular the Sun and the planetary magnetospheres. While laboratory plasma confinement for the eventual goal of energy production also provides a rich diversity of magnetic structures, their topology and dynamics is always constrained by the presence of a fixed set of coils with programmed currents that should control the spatial and temporal behaviour of the magnetic fields. The reason is clear: for the success of thermonuclear energy production, plasma dynamics and complexity are not really desired. The best thing would be to extract energy from a plasma that just sits quietly inside a toroidal vessel and the engineering approach to plasma confinement is to try to approach this ideal as closely as possible. The history of thermonuclear fusion research demonstrates impressive progress along this line but also the immense obstacles, due to complex plasma dynamics, that have to be overcome. In astrophysical plasmas, on the other hand, no such human engineering constraints exist: plasmas and their associated magnetic structures appear to be almost free to exhibit the bewildering variety of different dynamics that are observed on virtually all length and time scales.

Space missions in the second part of the twentieth century have played an important role in demonstrating the different magnetic structures and dynamics of plasmas in the solar system. The Skylab missions of 1973 revealed new solar structures in X-rays (like coronal holes) due to magnetic fields, the Voyager missions of the 1980s and 1990s provided completely new facts on the magnetic fields and magnetospheres of the planets, while the SOHO satellite, launched in 1995, provided visualizations of the dynamics of the solar corona that have become box office and website hits.

With the tremendous progress in satellite and ground-based observations of solar system plasma dynamics, the need for a theoretical framework to describe it all has become pressing. At this point, it is probably wise not to hide the embarrassment about the absence of such an all-encompassing vision. Theory is lagging far behind observations at the present moment. Frequently, cartoons representing a particular plasma phenomenon take the place of genuine theoretical analysis. Whereas cartoons may be a useful way of communicating ideas, they cannot replace genuine physical understanding based on mathematical analysis. (Just recall how the Bohr picture of the electron orbits in atomic theory had to be replaced by the quantum mechanical picture '$H\psi = E\psi$', standing for an extensive body of theoretical analysis which is much closer to physical reality than the cartoon it replaced.) Computational MHD, stimulated by recent programs on *space weather*, is presently filling in some gaps. The positive way of looking at it is that the field is open for lots of new ideas.

Hence, the purpose of this chapter is not to provide a detailed description of the observational facts (excellent textbooks exist: Parker [176], Priest [190], Friedman [75], Zirin [250], Stix [217], Foukal [69], Mestel [154], Schrijver and Zwaan [204], from which we have freely borrowed in this chapter), but just to remind the reader that all theory eventually has to be confronted with empirical reality. This should lead to an attempt to answer the following questions: 'Is the MHD model developed in Chapters 4–7 an adequate starting point for the description of observed plasma dynamics?', 'Are important theoretical pieces missing in this approach?', and 'What should be the main goals to be pursued in the following chapters?' The phenomenology of magnetic structures and associated dynamics presented in the present chapter is used in later chapters, on the dynamics of inhomogeneous plasmas, to provide some flesh and blood to the model problems presented in Section 4.6.

8.2 Solar magnetism

Let us start with the central object of the solar system, the Sun, and ask the central question: *where does its magnetism come from?* To answer that question, we first recall some basic facts about the solar structure and then discuss some of the observational evidence leading to a model of solar magnetism.

Recall the standard solar model depicted in Fig. 8.1, which we already encountered in Chapter 7. The *interior* of the Sun cannot be observed directly and our knowledge of its structure is based on theoretical models and helioseismology. On the basis of differences in physical properties, the dense interior of the Sun can be divided into three layers, viz. the core, the radiative zone and the convection zone.

386   *Magnetic structures and dynamics*

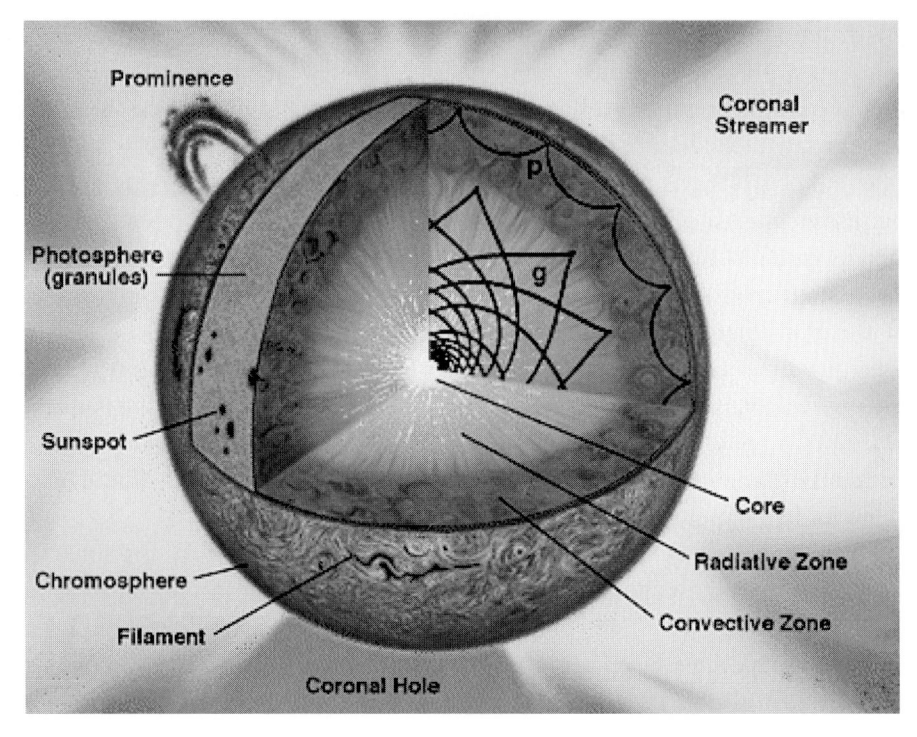

Fig. 8.1. Global model of the Sun showing gravity waves (g-modes) and sound waves (p-modes) propagating in the interior, and photospheric, chromospheric, and coronal magnetic features in the exterior. (Courtesy of SOHO (ESA-NASA).)

The *core* of the Sun ($r \leq 0.25\,R_\odot$) is the region where the solar energy is generated by means of thermonuclear conversion of hydrogen into helium (see Section 1.2.1). This region is surrounded by the *radiative zone* ($0.25\,R_\odot \leq r \leq 0.71\,R_\odot$) where the produced energy is radiatively transported outward. Finally, in the relatively shallow *convection zone* ($0.71\,R_\odot \leq r \leq R_\odot$) the temperature gradient is so steep that the plasma becomes convectively unstable. This region is considered to be the seat of the solar dynamo (see Section 8.2.1).

Recent results of helioseismology have led to the idea that the shear in the differential rotation, which is the main driving force of the solar dynamo, is concentrated at the bottom of the convection zone in a relatively thin (just 13 000 km thick) and narrow (25–30° latitude) layer. Estimates of the azimuthal field strength in that shear layer yield fields up to 5–10 T. However, there is no explanation yet for such high fields at the bottom of the convection zone.

The visible *atmosphere* of the Sun also consists of three layers. The *photosphere* is the region where the visible light of the Sun escapes. It is only 500 km thick.

8.2 Solar magnetism

At the bottom of the photosphere the temperature is about 6600 K and it decreases to about 4300 K at the top (see Fig. 8.14). The next layer is the *chromosphere*, with a thickness of about 2500 km, where the temperature starts to increase again (from 4300 K to about 10^6 K) connecting smoothly onto the very hot (millions of degrees) and very tenuous *corona*, which stretches out into the whole *heliosphere*. The mechanism for this dramatic temperature rise is one of the major, as yet not really resolved, issues of solar MHD but it is generally agreed that the magnetic field in the corona is the essential carrier of the energy transport.

According to our present knowledge all of the mentioned regions, except for the core and the radiative zone, form the scene of spectacular magnetic activity of different kinds which are characteristic for each of them. Since the magnetism of the solar atmosphere is a result of that of the solar interior it is appropriate to consider the latter first, even though it presents the hardest of all theoretical questions, viz. the mechanism of the solar dynamo.

8.2.1 The solar cycle

(a) Sunspots Throughout the centuries people have been observing dark spots on the Sun: the *sunspots* (in fact, they were already mentioned in Chinese chronicles of 800 BC). Since they have typical sizes in the order of the diameter of the Earth and sometimes as large as 40 000 km, they can be seen with the naked eye, e.g. at sunset or sunrise during hazy weather. When the telescope was invented in the seventeenth century it was immediately put to use (by Fabricius, Galilei and Scheiner) to observe these structures. It was found that they can last from 2 to 30 days, i.e. long enough to observe their systematic motion from west to east across the disc, demonstrating that the Sun is actually rotating about a fixed axis. Moreover, these early astronomers observed that the sunspot motion depends on the solar latitude: they occur in two bands around the equator and they move faster at the solar equator (with a rotation period of 25 days) than at higher latitudes (with periods of 27 days at 40° and 30 days at 70°). Hence, already the very first observations – with very primitive telescopes, according to present day standards – revealed the basic fact behind solar magnetism, viz. that the *Sun rotates differentially*.

The *association of sunspots with magnetic fields* only came in the twentieth century, notably through the spectroscopic work of George Ellery Hale at the Mt Wilson Observatory (1908) and culminating in the development of the magnetograph by Harold Babcock (1953). Whereas the magnetic fields inside sunspots are relatively easy to measure because they are so intense (1000–4000 gauss), the magnetograph enables one to measure the Zeeman splitting by the much weaker magnetic fields outside the sunspots. Thus, the spatial variation of the magnetic field across sunspots may be determined. It has been firmly established that sunspots represent the most intense magnetic flux concentrations on the surface of the Sun. The average field strength increases with the area of the spot.

From the numbers quoted above one easily calculates that the speed of sunspots (i.e., the solar rotation rate) at the equator is about $2\,\mathrm{km\,s^{-1}}$, which corresponds to $10''\,\mathrm{h^{-1}}$. However, this simple dynamics is embedded in the much more complicated one of the surrounding photosphere which exhibits *a cellular convection pattern* due to the thermal instability of the underlying convection zone. High-resolution photographs of the photosphere by means of telescopes on board balloons (early 1960s) and dedicated solar telescopes (like the Dutch Open Telescope at present, see Fig. 8.8) reveal the different scales of the *granules* with a typical size of 1000 km, horizontal flow velocities of $1.5\,\mathrm{km\,s^{-1}}$, and a turnover time ($\approx$ a life time) of the eddies of the order of 5 min and of the *supergranules* with characteristic sizes of 30 000 km, flow velocities of $0.5\,\mathrm{km\,s^{-1}}$, and turnover times of the order of 20 hours. The latter structures are the result of an outflowing velocity field pushing the upwelling magnetic field to the boundaries of the eddies, which become visible as the *photospheric network*. Note that these convective motions take place at widely separated space and time scales.

Returning to the sunspots, the most important aspect of solar magnetism is yet to be mentioned, viz. the long time scale periodicity of the large scale solar magnetic field and the associated magnetic activity: the *solar cycle*. This periodicity was discovered by Heinrich Schwabe, who systematically recorded the occurrence of sunspots during the period 1826–1851 in the hope of detecting a planet inside the orbit of Mercury. Instead, he found the more lasting result that the *number of sunspots varies periodically in time* with a periodicity of about 11 years. Important for our subject is the observation that maxima and minima of sunspot numbers coincide with increased and decreased magnetic activity of the Sun as a whole so that they are appropriately called *solar maxima* and *solar minima*. By means of historical records one has been able to reconstruct the solar cycle back to the time of Galilei (Fig. 8.2). The periodicity is not precise, shorter and longer periods do occur (from 7 to 17 years). Also, the amplitudes vary considerably. A particularly quiet time occurred in the second part of the seventeenth century, the Maunder minimum, which coincided with the cold period of the little ice age on Earth.

Most sunspots appear in two belts between the equator and the latitudes $\pm 35°$. However, in 1859 Christopher Carrington discovered that the average latitude of occurrence of the sunspots depends on the phase of the solar cycle. During the 11 year period the sunspots gradually drift from latitudes between 25–30°, where they first appear, to the equator, where they disappear again at the end of the cycle.

8.2 Solar magnetism

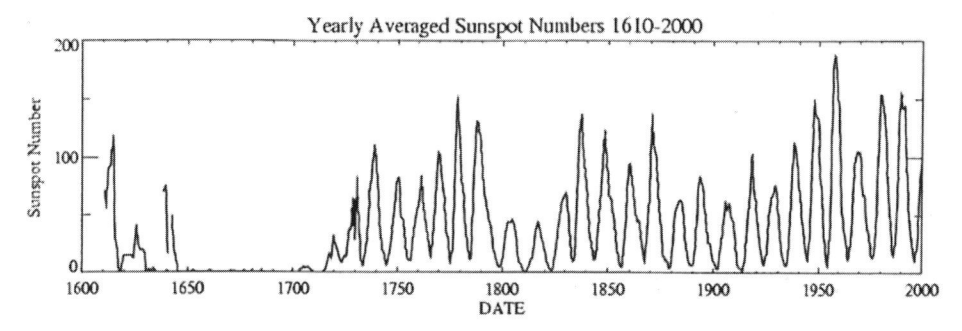

Fig. 8.2. The solar cycle: number of sunspots versus time. (Courtesy of David H. Hathaway (NASA, Huntsville).)

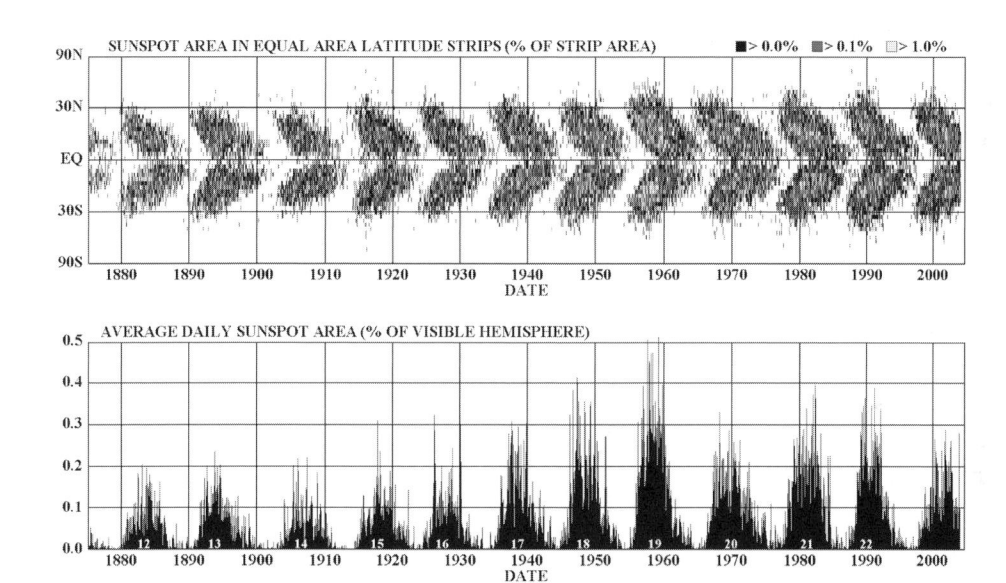

Fig. 8.3. Butterfly diagram: daily sunspot area averaged over individual solar rotations. Detailed observations of sunspots have been obtained by the Royal Greenwich Observatory since 1874. These data show that sunspots are concentrated in two latitude bands which first form at mid-latitudes, widen, and then move toward the equator as each cycle progresses. (Courtesy of David H. Hathaway (NASA, Huntsville).)

This drift towards the equator (called Spörer's law) yields the *butterfly diagram* (see Fig. 8.3).

Again, these phenomena would only be associated with magnetic fields in the twentieth century. In particular, only when the polarity of the fields was taken into account, it was realized that the physically relevant period of the solar cycle is actually 22 years with a reversal of the overall magnetic field direction every

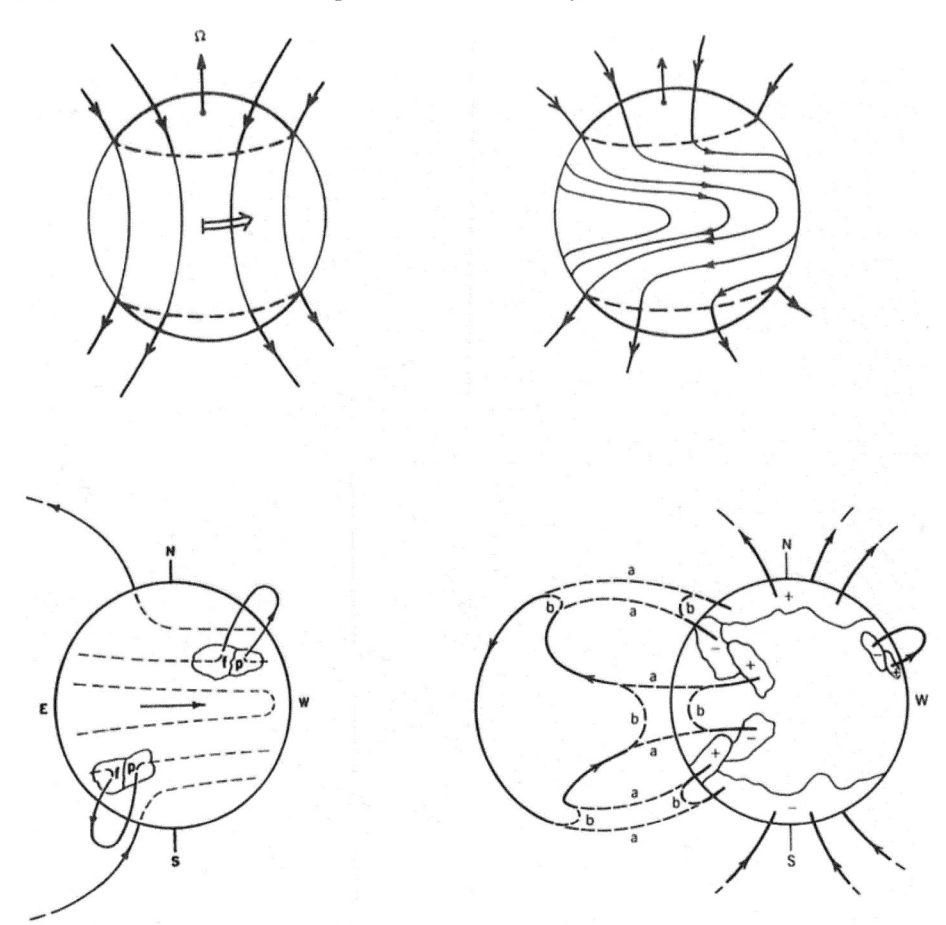

Fig. 8.4. Babcock model for solar magnetism. (From P. Foukal [69], after H. Babcock, *Astrophys. J.* **133**, 572 (1961).)

11 years. In general, sunspots occur in pairs with a leading spot of a certain polarity and a following one at a slightly higher latitude of opposite polarity. Moreover, it was found by Hale and Nicholson (1925) that the leading sunspots on the northern hemisphere all have the same polarity whereas those on the southern hemisphere all have the opposite polarity during the first 11 years of the cycle, and the roles are reversed during the second period of 11 years. Hence, adding the sign of the sunspot magnetic field to the picture given in Fig. 8.3 reveals that the *solar cycle is really a magnetic oscillation*.

(b) The solar dynamo By way of introduction, consider a particular example of the cartoon approach to dynamo action provided by the Babcock model of the solar cycle (Fig. 8.4). Due to differential rotation and the 'frozen in' condition of magnetic

fields in highly conductive plasmas, an initially poloidal magnetic field (like the usual dipole field created by toroidal currents in the convection zone) is stretched and wound up at the equator. After many periods of the solar rotation, the toroidal magnetic field component, which was negligible at first, now has become the dominant one. This field is expelled from the convection zone and breaks through the photosphere at isolated places where it forms sunspots with preceding (p) and following (f) polarities of different but fixed signs on the northern hemisphere, and of the opposite signs on the southern hemisphere. The outer portions of the flux loops expand into the corona and reconnect with the original poloidal field. This results in reversal of the direction of the latter field: the second half of the solar cycle has started. Note that reconnection, i.e. breaking and rejoining of field lines due to some anomalous resistive process in the corona, is necessary here.

We have seen that the Sun rotates differentially, that the photospheric surface exhibits cellular convection patterns characteristic of thermal convection, and that the solar magnetic field is periodic in time. These are the main facts behind the *solar dynamo*, i.e., the conversion of mechanical energy into magnetic energy. We now come to one of the main questions, viz. can all this be explained by the MHD equations? Recall the (near) conservation laws of Sections 4.1 and 4.4:

$$\frac{\partial \rho}{\partial t} = -\nabla \cdot (\rho \mathbf{v}) \qquad \text{(mass)}, \qquad (8.1)$$

$$\rho \frac{D\mathbf{v}}{Dt} = -\nabla p - \frac{1}{\mu_0} \mathbf{B} \times (\nabla \times \mathbf{B}) + \rho \mathbf{g} \qquad \text{(momentum)}, \qquad (8.2)$$

$$\frac{Dp}{Dt} = -\gamma p \nabla \cdot \mathbf{v} + (\gamma - 1)\left[H - \nabla \cdot \mathbf{h} \right] \qquad \text{(entropy)}, \qquad (8.3)$$

$$\frac{\partial \mathbf{B}}{\partial t} = \nabla \times (\mathbf{v} \times \mathbf{B}) - \nabla \times \left[\frac{\eta}{\mu_0}(\nabla \times \mathbf{B}) \right], \quad \nabla \cdot \mathbf{B} = 0 \quad \text{(flux)}. \qquad (8.4)$$

Here, the terms in square brackets are the ones which spoil the conservation of entropy and magnetic flux. In particular, the pressure evolution equation (8.3), which expresses (near) entropy conservation, and which also can be expressed in terms of the internal energy $e \equiv p/[(\gamma - 1)\rho]$ or the temperature $T \equiv e/C_V$, contains the generated heat per unit volume, H, and the heat flow $\mathbf{h}$. For Ohmic dissipation, $H = \eta j^2$, with $\mathbf{j} = \mu_0^{-1} \nabla \times \mathbf{B}$, and $\mathbf{h} = -\lambda \nabla T$, where λ is the coefficient of thermal conductivity[1] ($[\lambda] = \text{W K}^{-1}\text{m}^{-1}$). The resistive term in the (near) flux conservation equation (8.4) results in magnetic field diffusion. Clearly, our use of

---

[1] Recall that we replaced the symbol κ, introduced in Section 4.4.2 for the thermal conductivity, by λ in Section 7.1.1.

the words 'near conservation' implies that the transport coefficients λ and η are considered to be small.

One part of the dynamo problem involves the combined effects of differential rotation, gravitational contraction, and the presence of a highly conducting medium. This part is well described by the combined equations of gas dynamics and electrodynamics, i.e. by the *partial differential equations* (8.1)–(8.4) of magnetohydrodynamics. The other part, which is just as important, involves the coupling to the external world through the *boundary conditions*. Together this constitutes a complicated nonlinear problem in space and time which is essentially four-dimensional and, hence, necessarily numerical.

Let us approach this problem in three steps. First, the crudest approximation would be to neglect the velocity $\mathbf{v}$. Eq. (8.2) then results in the magneto-static equilibrium equation where gravitational forces are to balance pressure gradients and magnetic expansion forces (so-called magnetic buoyancy by which flux tubes are expelled from the solar interior). The equations (8.3) and (8.4), with $\mathbf{v} = 0$, are diffusion equations for the pressure (or temperature) and the magnetic field. In particular, assuming constant resistivity for simplicity, Eq. (8.4) becomes

$$\frac{\partial \mathbf{B}}{\partial t} = -\nabla \times \left[\frac{\eta}{\mu_0} (\nabla \times \mathbf{B}) \right] = -\frac{\eta}{\mu_0} \nabla \times \nabla \times \mathbf{B} \stackrel{(A.5)}{=} \frac{\eta}{\mu_0} \nabla^2 \mathbf{B} \equiv \tilde{\eta} \nabla^2 \mathbf{B}, \tag{8.5}$$

where the reference to Eq. (A.5) above the last equal sign refers to a vector identity of Appendix A. Hence, the *inhomogeneity of the magnetic field* (created by currents) *will decay* on a time scale τ_D determined by the resistivity η and the length scale $l_0 \sim \nabla^{-1}$ of the inhomogeneity:

$$\tau_D = \mu_0 l_0^2 / \eta = l_0^2 / \tilde{\eta}. \tag{8.6}$$

We have introduced a new quantity $\tilde{\eta} \equiv \eta/\mu_0 = \eta/(4\pi \times 10^{-7})$ which just absorbs the awkward factor μ_0. It is called the *magnetic diffusivity* and it has the convenient dimension of $m^2\,s^{-1}$ (since $[\eta] = \Omega\,m$ and $[\mu_0] = H\,m^{-1} = \Omega\,s\,m^{-1}$). In the astrophysical literature the tilde on $\tilde{\eta}$ is usually omitted and the conductivity $\sigma \equiv 1/\eta$ rather than the resistivity η is exploited. Here, we stick to the plasma physics convention by considering the resistivity η as the basic parameter from which the magnetic diffusivity $\tilde{\eta}$ is derived.

The very first question to be answered is: what value to take for the resistivity? Spitzer and Härm (1953) [214] derived an expression for the resistivity of plasmas due to collisions between electrons and ions which has become known as the

8.2 Solar magnetism

Spitzer, or *classical*, *resistivity* for reasons that soon will become apparent:

$$\eta_\| = \frac{e^2 \sqrt{m_e}}{6\epsilon_0^2 (2\pi k)^{3/2}} Z \ln \Lambda \, T_e^{-3/2} \approx 65 \, Z \ln \Lambda \, T_e^{-3/2}. \tag{8.7}$$

Here, $[\eta] = \Omega\,\text{m}$, $[T_e] = \text{K}$, Z is the charge number of the ions, and $\ln \Lambda$ is the Coulomb logarithm, which only weakly depends on the electron temperature T_e and the particle density n. It has a value between 10 and 20 for plasmas of interest.

For the solar corona, with $T_e \approx 10^6$ K, $n_e \approx 10^{16}$ m^{-3} (see Table B.5), $\ln \Lambda \approx 18$, and the resistivity $\eta \approx 10^{-6}\,\Omega\,\text{m}$ (i.e. $\tilde{\eta} \approx 0.8$ m^2 s^{-1}), so that $\tau_D \approx 10^{14}$ s $\approx 3 \times 10^6$ y for a loop with a transverse length scale $l_0 \approx 10\,000$ km.[2]

For the solar interior, in particular the bottom of the convection zone (at present considered to be the origin of the solar dynamo), where $T_e = 1.9 \times 10^6$ K and $n = 1.9 \times 10^{29}$ m^{-3}, the value of the resistivity $\eta \approx 1.7 \times 10^{-7}\,\Omega\,\text{m}$ (i.e. $\tilde{\eta} \approx 0.14$ m^2 s^{-1}) and the length scale of the inhomogeneity $l \approx R_\odot = 700\,000$ km, so that $\tau_D \approx 3.6 \times 10^{18}$ s $\approx 2 \times 10^{11}$ y ! It is clear that the factor l_0^2 in the expression (8.6) for τ_D beats everything. In astrophysical plasmas, effects of classical resistivity are extremely small, so that ideal MHD (i.e. $\eta = 0$) is an excellent approximation for many purposes. However, for the present purpose of explaining why the solar dynamo has a 22 year period, it is clear that classical resistive diffusion cannot be the controlling factor since it is completely negligible. We would be left with a static equilibrium at this point: certainly not a dynamo.

The second step is to add flow. One would have hoped that the apparent rotational symmetry of the differential rotation of the Sun would get us rid of at least one coordinate so that we could assume this flow to be axi-symmetric. However, it is clear from the behaviour of the polarity of the magnetic field that this symmetry is not respected by the solar cycle. In fact, it cannot be respected since *axi-symmetric motion does not lead to a dynamo*. The latter statement is the content of *Cowling's theorem*, which we will not prove here. Thus, *non-axi-symmetry of the flow* is the necessary third step to get dynamo action. Such a flow is created automatically by the convective motions of the convection zone. Here, the main complication is the fact that these motions are *turbulent*, involving the interaction of many small scale vortices. It is generally assumed that these turbulent processes are also responsible for an anomalous increase of the magnetic diffusivity and associated decay of the magnetic field so as to correspond to the time scales of the solar cycle. Note that this implies an increase from the classical value computed above, $\tilde{\eta}_{\text{class}} \sim 0.1$ m^2 s^{-1}, to $\tilde{\eta}_{\text{turb}} \sim 10^9$ m^2 s^{-1}, i.e. an increase by a factor of 10^{10}!

---

[2] Yet, coronal loops may disrupt in a solar flare on a time scale of minutes: this certainly cannot be described by classical resistive diffusion.

Although the latter fact leads us far away from our present theme (which is to present a simple survey of solar phenomena which are, in principle, tractable by means of the equations of magnetohydrodynamics), we will just indicate the kind of arguments involved in the analysis of a dynamo based on *turbulent eddy magnetic diffusivity*. This theory is based on the idea that in a turbulent fluid the mean values of the variables can be distinguished from the fluctuating, turbulent, ones. Thus, one writes

$$\mathbf{v} = \langle \mathbf{v} \rangle + \mathbf{v}', \qquad \mathbf{B} = \langle \mathbf{B} \rangle + \mathbf{B}', \qquad (8.8)$$

where angular brackets denote average values and primes denote fluctuating parts. Averaging is to be understood in a statistical sense but, in the solar case, it could be interpreted as averaging over a solar rotation period (~ 27 days). Inserting the expressions (8.8) into Eq. (8.4), one obtains the following averaged form of the induction equation:

$$\frac{\partial \langle \mathbf{B} \rangle}{\partial t} = \nabla \times \left(\langle \mathbf{v} \rangle \times \langle \mathbf{B} \rangle \right) + \nabla \times \langle \mathbf{v}' \times \mathbf{B}' \rangle - \nabla \times \left(\tilde{\eta} \nabla \times \langle \mathbf{B} \rangle \right). \qquad (8.9)$$

Obviously, all complications of the theory are hidden in the second term on the right hand side which involves the average cross-product of the fluctuating parts of the velocity and the magnetic field. These quantities have to come from the counterparts of Eq. (8.9) describing the evolution of the turbulent variables $\mathbf{v}'$ and $\mathbf{B}'$. In particular, by means of a number of drastic assumptions, this cross-product is written as

$$\langle \mathbf{v}' \times \mathbf{B}' \rangle \approx \alpha \langle \mathbf{B} \rangle - \beta \nabla \times \langle \mathbf{B} \rangle + \cdots, \qquad (8.10)$$

where the coefficients α and β are correlation functions of the turbulent velocity $\mathbf{v}'$. This transforms Eq. (8.9) into the following form:

$$\frac{\partial \langle \mathbf{B} \rangle}{\partial t} = \nabla \times \left(\langle \mathbf{v} \rangle \times \langle \mathbf{B} \rangle \right) + \nabla \times \left(\alpha \langle \mathbf{B} \rangle \right) - \nabla \times \left[(\tilde{\eta} + \beta) \nabla \times \langle \mathbf{B} \rangle \right]. \qquad (8.11)$$

This equation describes both magnification of the magnetic field, i.e. possible dynamo action, through the term with α, and decay due to turbulent magnetic diffusivity $\tilde{\eta}_{\mathrm{turb}} = \tilde{\eta} + \beta \approx \beta$, through the term with β. Needless to say, everything is hidden now in the derivation of expressions for the coefficients α and β. Order of magnitude estimates based on the length scale $l \sim 1000$ km and the time scale $\tau \sim 1000$ s of the turbulent velocity fluctuations $v' \approx l/\tau \sim 1 \,\mathrm{km\, s^{-1}}$ are $\alpha \sim l\Omega \sim 3 \,\mathrm{m\, s^{-1}}$ (where $\Omega \sim 2.6 \times 10^{-6} \,\mathrm{rad\, s^{-1}}$ is the angular velocity corresponding to a solar rotation period of 27 days) and $\tilde{\eta}_{\text{turb}} \approx \beta \sim v'l \sim 10^9 \, \text{m}^2 \, \text{s}^{-1}$. The miracle has been performed: the required enhancement factor of 10^{10} for $\tilde{\eta}$ has been obtained!

For the demonstration of an α-effect and a turbulent diffusivity, the back-reaction of the magnetic field on the plasma flow has been ignored. This reduced problem is called the *kinematic dynamo problem*. Even if this part is taken for granted, one still has to solve the full nonlinear *MHD dynamo problem*, i.e. one needs to show that plasma motion and a quasi-oscillatory magnetic field can maintain each other. This is presently a 'hot' item in computational plasma-astrophysics, involving large scale numerical simulations. Nevertheless, the subject of solar and stellar dynamo theory is far from nearing completion.

8.2.2 Magnetic structures in the solar atmosphere

The resulting magnetic structures and their dynamics in the solar atmosphere will now be described.

(a) Photosphere and chromosphere The energy radiated by the Sun comes from the photosphere. The Sun emits a continuous spectrum but in the spectrum received on Earth dark absorption lines occur, the Fraunhofer lines. They are dark because the lower gas transparency, due to the absorption at specific wavelengths by specific particles such as iron atoms, causes radiation escape at greater height where the temperature is lower. Most of the absorption lines are formed in the photosphere. In the optical solar spectrum there are about 20 000 of these Fraunhofer spectral lines. Many of them are overlapping but sufficient of them are single. From the comparison with laboratory experiments these absorption lines allow us not only to study the abundance of the elements of the Sun (all 92 elements are present although most of them are very rare), but they also allow us to measure density, temperature, velocity and field vectors.

The *intensity* of the absorption lines in the solar spectrum provides us with information on the *temperature*, their *Zeeman splitting* provides information on the *strength of the magnetic field*, and their *Doppler shifts* provide information on the *velocity* along the line of sight. Different absorption lines are formed at different heights so that one can look at different levels in the solar atmosphere by using different filters. For instance, selecting the Hα Balmer line (due to a transition in hydrogen atoms from the second to the third quantum level) one can look at the middle chromosphere. By observing in the Ca$^+$ K line one sees the low chromosphere, and in soft X-rays one gets a direct view of the corona.

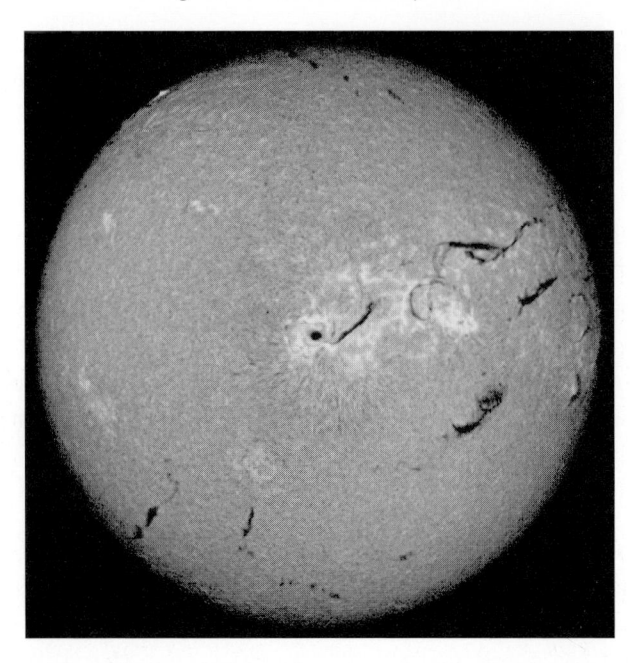

Fig. 8.5. Image of the Sun taken through a filter centred on a spectral line of hydrogen (Hα, wavelength = 6563 Å). This line forms above the surface of the Sun, but large *sunspots* are still visible and *active regions* and *plages* show up brighter than their surroundings. Also visible are condensations of cooler gas high up in the solar atmosphere which show up as *filaments*, dark string-like structures visible on the disc, and *prominences*, bright structures extending outward over the limb. (Courtesy of National Solar Observatory/Sacrament. Peak.)

Fig. 8.5 is an Hα picture of the lower chromosphere. The bright areas are active regions (above sunspots). The elongated dark filaments are prominences which, in the line of sight, appear to be vertically flat structures. Their protruding shape is visible though in observations of the limb (see below).

In Section 8.2.1 sunspots were just considered as 'tracers' of the magnetic processes going on in the interior of the Sun. However, sunspots can also be studied as entities by themselves. They are also of interest as emerging magnetic flux. Most sunspots disappear again after a couple of days but the larger ones can last much longer, even up to several months. These large sunspots have diameters between 40 000 km and 60 000 km. Such a large sunspot is displayed in Fig. 8.6. The central dark area is called the *umbra*. It has a diameter of 10 000–20 000 km (about 40% of the total diameter of the spot) and the magnetic field and the temperature are almost uniform in the umbra. The magnetic field is typically 2000–3000 gauss but can be as high as 4000 gauss. The temperature is typically 3700 K, which is lower than the temperature of the surrounding photosphere. This is the reason why the

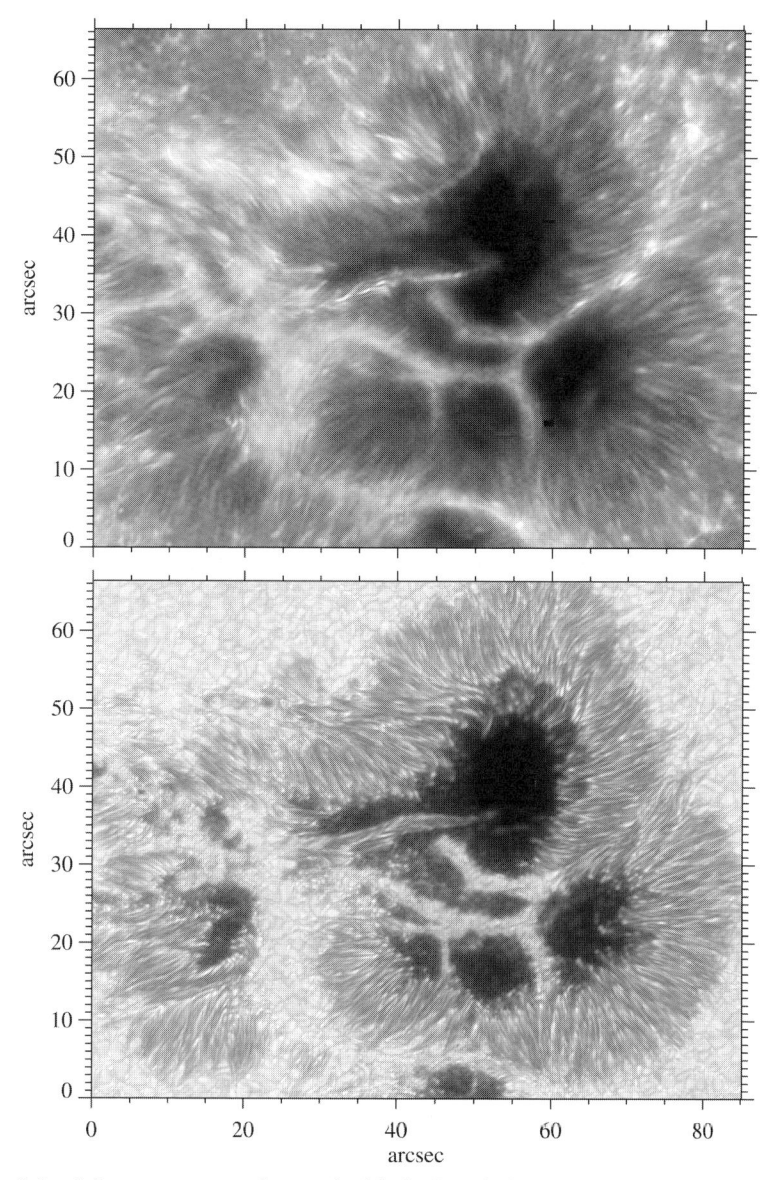

Fig. 8.6. Mature sunspot, observed with the Dutch Open Telescope on La Palma, 1 May 2003. Axes: arcsec (1 arcsec = 725 km on the Sun). Upper: Ca IIH (3968 Å) sampling the low chromosphere. Lower: G band around 4305 Å sampling the deep photosphere. At the photospheric level the solar magnetic fields are very finely structured; they spread and become more diffuse in the chromosphere. (Courtesy of R. Rutten (Utrecht University).)

umbra appears dark: the brighter surroundings dictate short exposure times. By itself, against a non-emitting background, the umbra would be bright. Longer exposure times reveal that the umbra is not uniformly dark. Bright dots with diameters

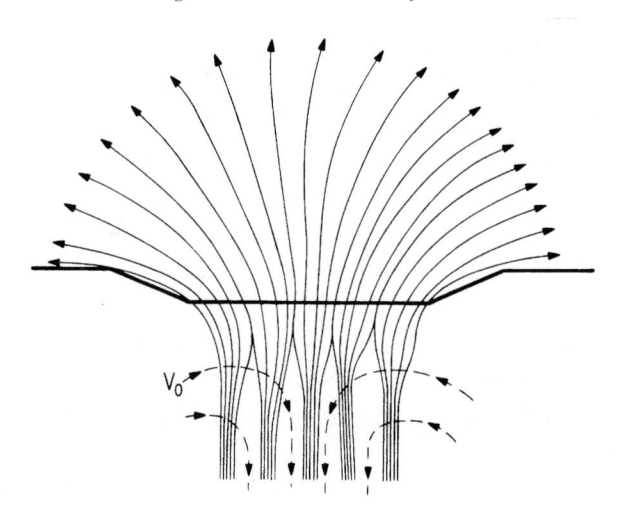

Fig. 8.7. Parker's sunspot model. (Courtesy of E. N. Parker, *Astrophys. J.* **230**, 905 (1979).)

of 150–200 km and life times of typically 20 minutes occur in the umbra and may be indications of the 'spaghetti model' (see below). The umbra is surrounded by the *penumbra* which consists of light and dark radial filaments. These filaments are 5000–7000 km long and 300–400 km wide and live from half an hour up to six hours. The magnetic field strength decreases in the penumbra and at the interface between the penumbra and the photosphere it is about 1000–1500 gauss.

Standard magnetohydrostatic models of the equilibrium configuration of sunspots picture them as a single 'monolithic' flux tube with almost vertical magnetic field lines under the photosphere. Above the photosphere the external pressure decreases exponentially with height and, as a result, the magnetic field lines fan out in the chromosphere. However, even the most sophisticated sunspot models based on such monolithic flux tube configurations are not compatible with the observed external pressure stratification. Moreover, these standard models yield no explanation for the non-axi-symmetric fine structure of the penumbra. In 1979 Parker suggested an alternative model for sunspots, depicted in Fig. 8.7. In this alternative model, the single flux tube splits up into a large number of narrow flux tubes at a depth of about 1000 km under the photosphere. This bundle of magnetic flux tubes is held together by the convective motions at the granulation boundaries, indicated by the dashed arrows in the figure. The mentioned bright dots in the umbra are interpreted as observational evidence for this 'spaghetti model' for the sunspot. This structure is at present investigated by means of a new branch of solar physics called *sunspot seismology* (to be discussed in Section 11.3.3), which exploits the fact that p-modes (the pressure driven acoustic oscillations of the Sun as a whole) sample the different layers of the Sun. These modes are scattered and absorbed by sunspots so that one can learn something about the internal structure of the sunspots. The radial structure can be simulated by 1D models, but the difficult (2D) part is the dependence on the vertical inhomogeneity (along the magnetic field).

The solar photosphere is covered by 3 to 4 million granules at any time. A close-up of the granulation pattern is shown in Fig. 8.8: clear evidence of the nonlinear dynamics of the underlying convection zone. The granules are the tops of convection cells overshooting the convection zone. Their centres appear brighter because of the hot, rising plasma there. The diameter of these cells is typically for 1 to 2″, corresponding to 700–1500 km. They exist typically for 5 minutes, i.e. shorter than the turnover time.

In the quiet photosphere, the magnetic field is concentrated in intense flux tubes in the intergranular lanes, i.e. the downdrafts of the granulation. They are driven to the boundaries of the supergranulation cells by horizontal outflows, as indicated in Fig. 8.9. Supergranulation cells were discovered by studying the vertical motions of the photosphere by means of the Doppler shifts of the absorption lines in the spectrum. Horizontal outflows and large-scale velocity patterns were identified as the tops of large convection cells. The diameter of these supergranules varies from 20 000 to 54 000 km and is about 30 000 km on average. In the centre, hot plasma rises and then flows out horizontally at 1.3–$1.5\,\mathrm{km\,s^{-1}}$. The typical life time of the supergranules is 1 to 2 days and the turnover time is of the same order of magnitude. Their boundaries are very prominent in the chromosphere in which the magnetic field lines fan out due to the decrease of the external plasma pressure. As a result, the magnetic field becomes more uniform in the upper chromosphere and corona, as illustrated in Fig. 8.9, which also displays the temperature contours in the atmosphere above a supergranule cell.

(b) Corona Spectacular evidence for magnetic fields in the corona comes from the shapes of *prominences* and *loops* protruding into the corona. An old example (from 1871!) is shown in Fig. 8.10(a), where coronal loops and chromospheric spicules (shooting jets) are *drawn* from spectrohelioscope observations. A modern counterpart is shown in Fig. 8.10(b). The loops extend to heights of 500 000 km and contain hot plasma that is heated and rises along the magnetic field and then cools again and falls back with speeds of about 100 km per second. We now know that we are looking at *magnetic* structures with properties that are extremely well known. However, even though we now have these high-resolution observations, a satisfactory theory describing the dynamics and the heating of these loops, and predicting when they develop into flares, does not yet exist.

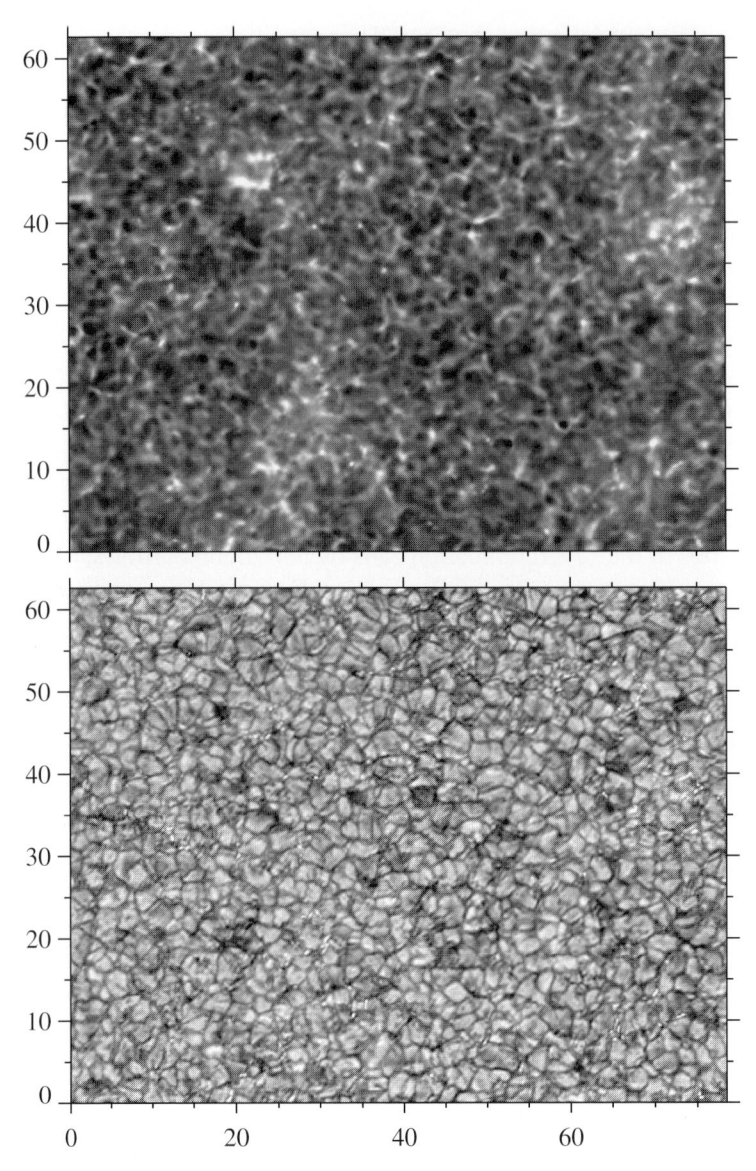

Fig. 8.8. Granulation with network of the quiet Sun, observed with the Dutch Open Telescope on La Palma, 3 July 2002. Axes: arcsec (1 arcsec = 725 km on the Sun). Upper: Ca II H (3968 Å) sampling the low chromosphere. The magnetic flux tubes become relatively bright at this height due to a not understood heating mechanism. The granulation is replaced by a brightness pattern due to convective overshoot, acoustic waves and gravity waves. Lower: G band around 4305 Å sampling the deep photosphere. The solar surface granulation is caused by the abrupt transition from convective to radiative energy transport. The tiny bright points in some dark intergranular lanes mark magnetic flux tubes with field strengths around 1400 gauss. (Courtesy of R. Rutten (Utrecht University).)

8.2 Solar magnetism

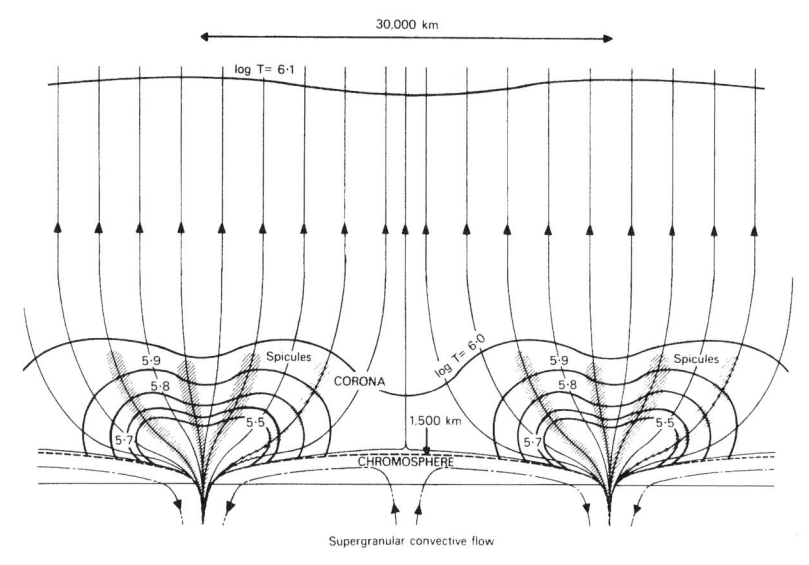

Fig. 8.9. Canopy fields. (From A. H. Gabriel, *Phil. Trans. Roy. Soc. London* **A281**, 339 (1976).)

In Fig. 8.5, prominences showed up as thin, dark, filaments in Hα pictures of the photosphere but we now consider their extension in the corona. In eclipse or coronagraph pictures they appear bright at the limb. Prominences are cool and dense structures. Their temperature is about 100 times lower than that of the surrounding corona and their density is 100 to 1000 times higher than coronal values. There are two types of prominences: quiescent and active. The active prominences are located in active regions and exhibit violent motions that may give rise to *solar flares*. An example of an erupting prominence is shown in Fig. 8.11. Active prominences exist for minutes or hours. Quiescent prominences, on the other hand, can last much longer, up to 200 days. These huge structures are typically 200 000 km long, 50 000 km high, and 6000 km wide.

Solar flares are amongst the most impressive explosive phenomena generally believed to result from the *release of huge amounts of magnetic energy*. Frequently, the irrelevant, but anyway impressive, comparison is made with nuclear explosions (with energy releases of 10^{13}–10^{15} J). Large solar flares of 10^{24} J release the energy equivalent to a billion hydrogen bombs. Of course, the relevant comparison is with the solar luminosity $L_\odot = 4 \times 10^{26}$ J s^{-1}, which is hardly affected by a large flare on a time scale of minutes. However, recently discovered superflares of 10^{26}–10^{31} J on ordinary main-sequence stars, including those that were discovered to have planets, have evoked some discussion on the consequences for possible life on

Fig. 8.10. (a) Coronal loops drawn by A. Secchi in 1871 (from C. Young, *The Sun* (London, 1882)). (b) Coronal loops observed in unprecedented detail with NASA's Transition Region and Coronal Explorer (TRACE) spacecraft. Coronal heating is deduced to be located at the bases of these loops. (From Vestige.lmsal.com/TRACE.)

those planets (Schaeffer, King and Deliyannis [202]). For our subject, the effects on our own magnetosphere of *coronal mass ejections* (CMEs), often concomitant with solar flares, are more relevant for human enterprises like radio communication and power transmission.

8.2 Solar magnetism

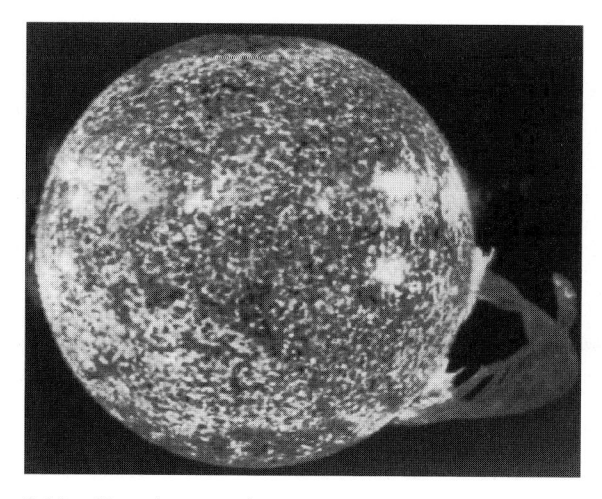

Fig. 8.11. Erupting prominence. (From www.solarviews.com.)

The solar corona emits thermally in soft X-rays. This means that it can be observed directly in this frequency range since the contribution of the much colder lower atmosphere is negligible. Soft X-ray pictures of the solar corona, such as routinely taken since the Skylab missions in 1973, show that the corona is highly inhomogeneous and structured. Based on the topology of the magnetic field, the corona can be divided into two types of regions, viz. open and closed regions. These regions are associated with magnetic field lines that either are fanning out into interplanetary space or return to the photosphere, forming closed magnetic loops. The 'open' regions are colder and, hence, appear dark on soft X-ray pictures of the whole Sun. They are called *coronal holes* since the solar wind escapes here along the 'open' field lines and is accelerated to enormous speeds (e.g. at 1 AU the wind speed is about $400 \,\text{km}\,\text{s}^{-1}$). The *closed* or *active* regions appear brighter because they are hotter. They consist of bundles of hot magnetic loops with temperatures of 2 to 3 million Kelvin. They are typically 200 000 km long and have a life time of the order of 1 day, although a system of loops may last for many months.

In white light the corona can only be seen during an eclipse or by means of a coronagraph (a telescope with a small disc in it which creates an artificial eclipse) because the much higher photospheric emission in white light completely overwhelms the coronal emission. Fig. 8.12 displays a white-light eclipse photograph of the corona. In general, such photographs show radial structures, stretching out to 1–10 $R_\odot$, which are called coronal streamers. Special ones are the 'helmet streamers', that exhibit a cusp-like structure and usually appear above prominences. Near the poles one sees 'polar plumes', but otherwise these regions are dark and associated with 'coronal holes'. In Fig. 8.13 a predicted polarization brightness is

Fig. 8.12. Total solar eclipse of 11 July 1991 as seen from Baja California. This digital mosaic is processed by Steve Albers (Boulder, CO) and derived from five individual photographs, each exposed correctly for a different radius in the solar corona. It shows helmet streamers, prominences and coronal holes. (From www.solarviews.com.)

shown together with magnetic field line traces for the total solar eclipse in 1999. This illustrates the correspondence between the open or closed magnetic field configurations and the coronal structures, and convincingly shows that these structures are due to these two types of magnetic fields. The 3D MHD simulation was performed before the eclipse to predict what the solar corona would look like during the eclipse.

The closed and open magnetic field configurations have been the subject of many investigations. It has been remarked by B. C. Low [148] that the helmet streamer structure plays an important role in the dynamics of coronal plasmas. It both represents the large scale relaxed state of the quiescent corona and acts as an agent for coronal reconfiguration by flows and CMEs. Concerning the underlying filament-prominence configuration, it has been noted by Martens and Zwaan [152] that they are usually found at the location of inversion of the magnetic polarity in the active regions of the photosphere. At those locations, emergent magnetic flux produces loop-like filaments with helical magnetic field lines that are stable through line-tying to the photosphere. Through some form of reconnection, several such filaments with the same sense of helicity of the magnetic field join to form long structures that cannot be stabilized any more by line-tying so that they erupt. Clearly, the subject of erupting magnetic structures is a treasury for solar MHD.

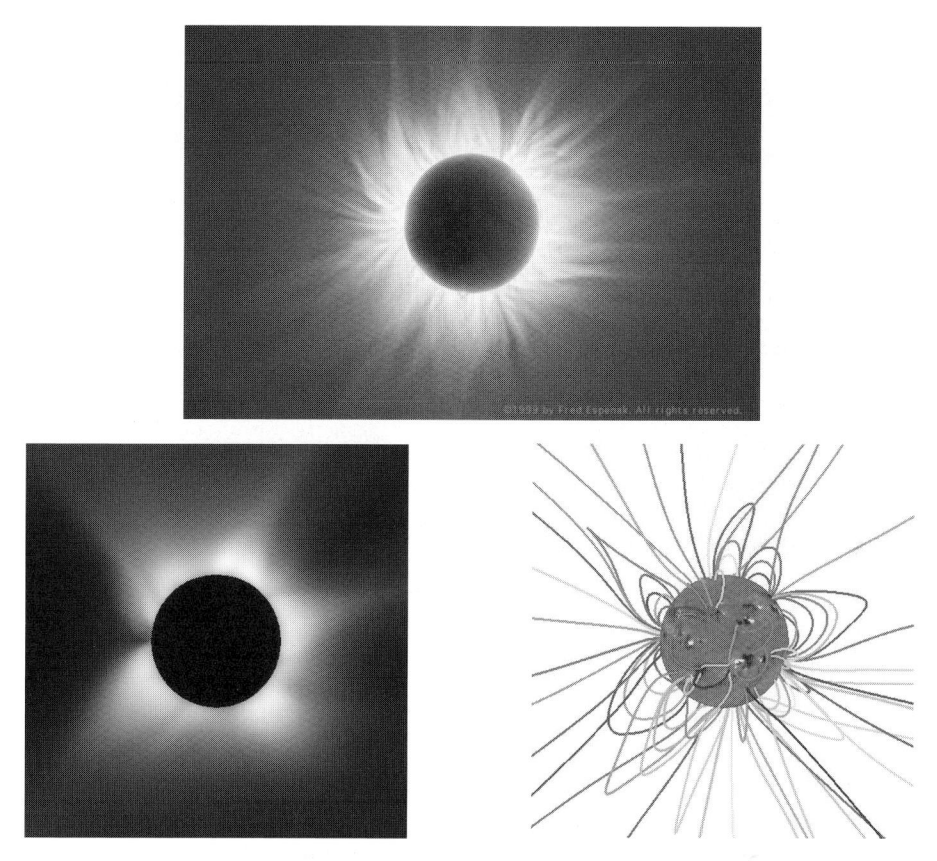

Fig. 8.13. Comparison of an MHD model prediction with a photograph for the total solar eclipse on 11 August 1999 in central Europe, the Middle East and India. The 3D MHD simulation was performed on 28 July 1999. The prediction model used photospheric magnetic field data from Carrington rotation 1951 corresponding to 24 June–21 July 1999. Top: eclipse image constructed by adding 22 separate photographs taken in Turkey at different exposures to compensate for the rapid radial fall-off of the brightness in the corona and digitally processed to enhance the fine details of the corona (from F. Espenak). Bottom left: predicted polarization brightness for 11 August 1999 at 11:38 UT, corresponding to totality in Eastern Turkey. Bottom right: magnetic field line traces in the 3D MHD model prediction (courtesy of Z. Mikic, haven.saic.com/corona).

At the surface of the Sun, the temperature is about 6600 K and it decreases further in the photosphere to about 4300 K at the top of the photosphere. Above the photosphere, however, the temperature starts to increase again, as shown in Fig. 8.14 computed by Athay. (Note the logarithmic scaling of the vertical axis.) In the lower and middle chromosphere the temperature first increases relatively slowly, but in the higher chromosphere and the transition region to the corona the final increase to $T \sim 2\text{–}3 \times 10^6$ K is so counter-intuitive from a thermodynamic

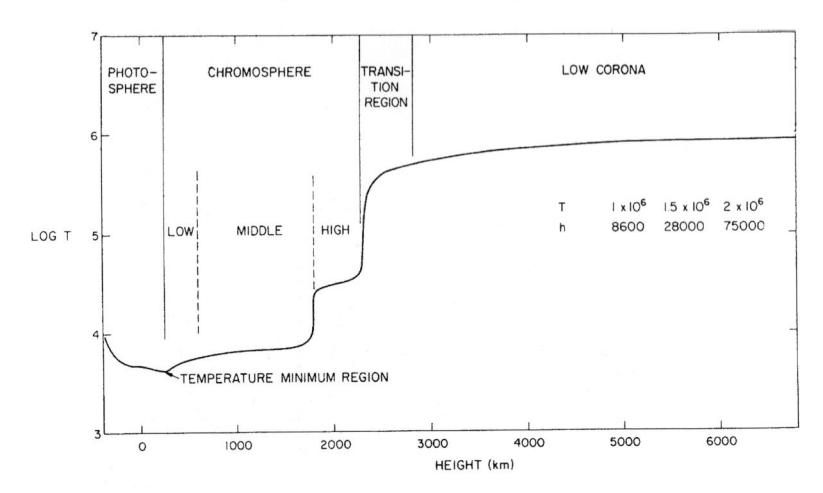

Fig. 8.14. Coronal temperature variation with height. (From R. G. Athay, *The Solar Chromosphere and Corona: Quiet Sun* (Reidel, Dordrecht, 1976).)

point of view that it has become known as the *coronal heating problem*. We have already mentioned that the mechanism behind this increasing temperature with increasing distance from the Sun is still one of the major problems in solar MHD research. It will be more extensively discussed in later chapters (Chapters 10 and 11).

(c) Heliosphere We mentioned that the solar wind escapes from the solar coronal holes along the open field lines. These field lines again form a giant magnetic structure. Looking down to the Sun from a position above one of the poles, one would see that the wind rotates with the Sun and that the magnetic field lines form so-called Archimedes spirals. Since the magnetic field has a different polarity in the two hemispheres, a thin magnetically neutral current sheet separates these opposite polarities. Because the flux in the solar wind is not nicely up–down symmetric, this *neutral current sheet* is warped. Magnetometers on board satellites in the 1960s have shown that the rotating solar wind had four magnetic sectors at that time. During the Skylab mission in the summer of 1973, this warped neutral current sheet looked schematically as shown in Fig. 8.15. Clearly, this large scale rotating magnetic structure already provides a very complex system by itself, but at the planetary magnetospheres it produces solar wind magnetic field orientations that systematically change direction with respect to the dipolar fields of the planets. We will return to this large scale structure of the solar wind and the interaction with the magnetospheres in Section 8.4, after the exposition of the planetary magnetic fields.

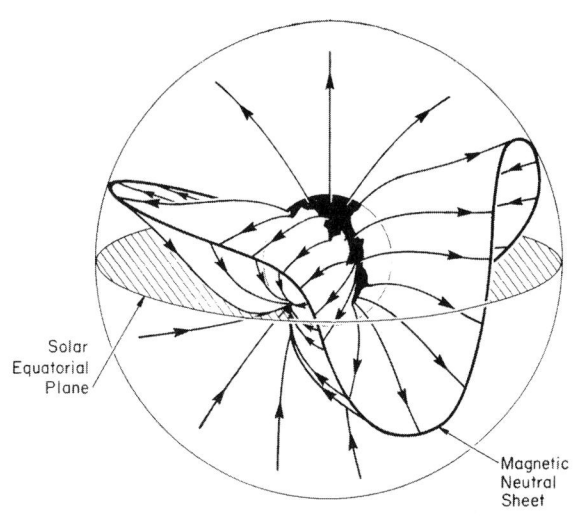

Fig. 8.15. Neutral current sheet. (From A. J. Hundhausen, in *Encyclopedia of Science and Technology* (McGraw-Hill, 1981).)

8.3 Planetary magnetic fields

We now turn to the magnetic fields of the planets. From the numbers presented in Table B.8 it is clear that the largest magnetic fields are found for the *Earth* and the *Jovian or giant planets*. Of the latter, Jupiter has by far the largest magnetic dipole moment and, consequently, the most extended magnetospheric system. The question of why these planets have large magnetic fields, whereas the fields of the remaining terrestrial planets are very weak, is a difficult one that has not been satisfactorily answered yet. The answer would require a detailed knowledge of both the *interior structure of the planets* and of the theoretical solutions of the *nonlinear dynamo equations*. In a sense, these problems are *mathematically ill-posed*: solutions depend extremely sensitively on the boundary conditions imposed on the outside, which is the only place where magnetic field or seismological measurements can be made. It is beyond the scope of the present book to analyse these problems in detail. We will be content with a phenomenological approach.

For many purposes we do not need to go into the details of the dynamo mechanism of the planetary magnetic fields. It is sufficient to have an equivalent representation of the magnetic field which is valid on time scales shorter than the dynamo period (which is much longer for the planets than for the Sun).

First, consider the magnetic field produced by a current flowing in a straight infinitely long wire (Fig. 8.16(a)). According to the law of Biot and Savart,

$$B = \frac{\mu_0 I}{2\pi r}. \tag{8.12}$$

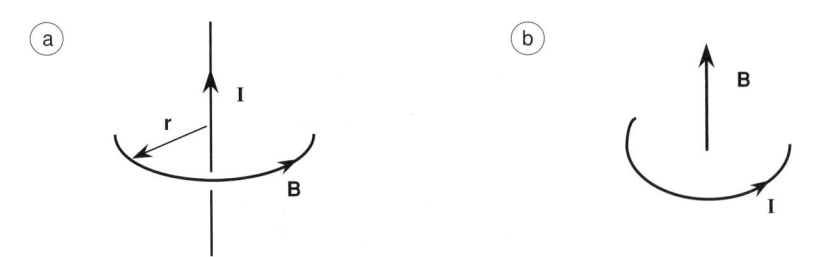

Fig. 8.16. Currents and associated magnetic fields.

Insert for the distance r the radius of the Earth, $R_E = 6.4 \times 10^6$ m, and for the current $I = 1000$ MA. This gives a magnetic field strength $B = 3 \times 10^{-5}$ T ($= 0.3$ gauss), which is the magnitude of the *geomagnetic field* at the equator (see Table B.8). Of course, we do not suggest that the magnetic field of the Earth is produced in this manner, but equation (8.12) does give an impression of the amount of current that is needed to produce magnetic fields of the right order of magnitude.

We come a little closer to reality by considering a *magnetic dipole* produced by a current loop (Fig. 8.16(b)):

$$B = \frac{\mu_0 I}{2r}. \tag{8.13}$$

Again inserting the radius of the Earth for r and a current of $I = 630$ MA this time, we find $B = 6.2 \times 10^{-5}$ T ($= 0.62$ gauss), which is the magnitude of the geomagnetic field at the magnetic poles. For such a loop we may define the *magnetic dipole moment*, which is the current in the loop multiplied by the area spanned by the loop:

$$m \equiv \pi r^2 I. \tag{8.14}$$

(In Section 2.2.3, Eq. (2.33), we have exploited the symbol μ for this quantity.) Inserting the same values for r and I as above, the magnetic moment of the loop is $m = 8.1 \times 10^{22}$ A m^2 ($= 8.1 \times 10^{25}$ gauss cm^3), which is the magnetic dipole moment of the Earth (see Table B.8). The idea is to consider the magnetic field external to the Earth to be produced by an equivalent magnetic dipole situated at the centre of the Earth. This imaginary dipole is the limit of a current loop in which the current $I \to \infty$ and the radius $r \to 0$ in such a manner that the value of the magnetic dipole moment given by Eq. (8.14) is kept fixed. The magnetic field produced by such a dipole is given by the expression

$$\mathbf{B(r)} = \frac{\mu_0}{4\pi r^3} (3\mathbf{m} \cdot \mathbf{e}_r \, \mathbf{e}_r - \mathbf{m}), \tag{8.15}$$

8.3 Planetary magnetic fields

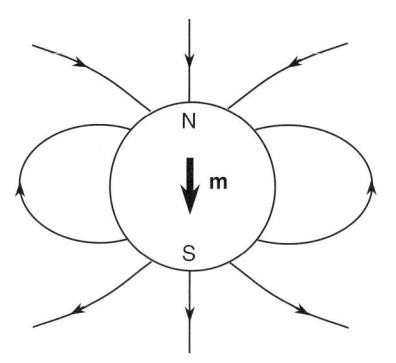

Fig. 8.17. Magnetic field of a dipole.

where **m** is the magnetic dipole moment vector,[3] $\mathbf{e}_r$ is the unit vector in the direction of the position vector **r**, and $r \equiv |\mathbf{r}|$. Whatever the actual mechanism by which the geomagnetic field is generated, it turns out that Eq. (8.15) with the given value of m reproduces the magnetic field external to the Earth quite well. Note that the magnetic field lines point to the magnetic north pole so that the equivalent dipole at the centre of the Earth should be pointing south, i.e. $\mathbf{m}_E = -m_E \mathbf{e}_z$, where $\mathbf{e}_z$ is the unit vector pointing north (Fig. 8.17). For Jupiter, which has a magnetic dipole moment of 1.5×10^{27} A m^2 pointing approximately north (with respect to the ecliptic) we find: $B_{\text{pole}} = 8.4 \times 10^{-4}$ T and $B_{\text{equator}} = 4.2 \times 10^{-4}$ T.

8.3.1 The geomagnetic dynamo

The Earth (Fig. 8.18) probably consists of *a solid inner core* of iron and nickel ($0 < R < 1300$ km), *a fluid outer core* of liquid iron ($1300 < R < 3400$ km), *a solid mantle* of silicates ($3400 < R < 6400$ km), and, outermost, *a rocky crust* of some tens of kilometres thick. The fluid outer core is considered to be the seat of the *geodynamo*, which maintains the magnetic field against resistive decay. This dynamo operates when an electrically conducting fluid is kept in motion by some source (in this case, heat from radioactive decay and rotation of the Earth). In the presence of a magnetic field, this fluid flow generates electric fields and currents which, in turn, may amplify the existing magnetic field. Magnetohydrodynamics (MHD) is the theoretical tool to describe the motion of a conducting fluid in a magnetic field. However, one needs substantial computer programs for the solution

---

[3] Note that it is convenient to define another quantity $\mathbf{M} \equiv (\mu_0/(4\pi))\mathbf{m}$ to get rid of the awkward factor $\mu_0/(4\pi) = 10^{-7}$ H m^{-1}. This quantity is also called the magnetic dipole moment by some authors. In the Gaussian system of units there is no need for this distinction since the connection between **B** and **m** is then given by an equation like (8.15), however without the factor $\mu_0/(4\pi)$. Consequently, the values of the planetary dipole moments given in Table B.8 should be multiplied by 10^{-7} to get the values of M in T m^3 and by 10^3 to get the values of m in gauss cm^3.

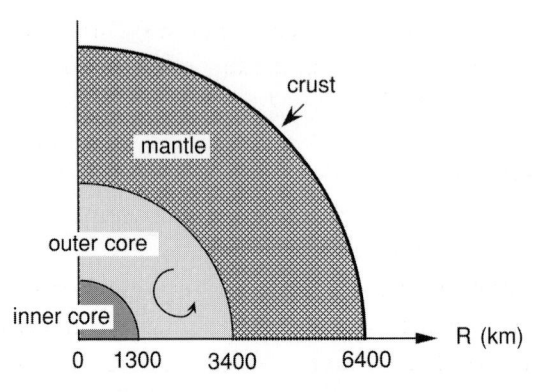

Fig. 8.18. Interior of the Earth.

because the dynamo is essentially nonlinear and the required solutions are truly three-dimensional, i.e. they do not exhibit a symmetry.

Paleomagnetic research, i.e. the study of the magnetization of ancient rocks, has revealed that the geomagnetic field reverses sign about every 400 000 years, whereas the reversal itself requires some 10 000 years to take place. Investigation of the magnetization of historical artefacts (e.g. magnetized iron particles in Roman pottery) has shown that the magnetic field of the Earth is at present decaying at a fast rate, implying that it would reverse sign in the next 1500 years.[4] Moreover, about 10% of the field at the surface of the Earth is non-dipolar, i.e. deviates from the form given by Eq. (8.15). Retracing the geomagnetic field from the surface of the Earth back to the outer core (a procedure where the caveat of ill-posedness should be heeded!) one finds so-called magnetic core spots where the polarity of the field is reversed. These spots might grow and eventually cause a global field reversal.

The described time-dependence and inhomogeneity of the geomagnetic field clearly cannot be explained by a stationary electrodynamic model without fluid flow ($\mathbf{v} = 0$) and only having electric currents flowing in the molten iron outer core of the Earth. Such a model would again lead to Eq. (8.5), which implies that an inhomogeneous magnetic field will decay with a time scale determined by the resistivity η and the length scale $l_0 \sim |\nabla|^{-1}$ of the inhomogeneity. The latter is related to the amount of current flowing in the medium: $|j| \sim |B|/(\mu_0 l_0)$. Equation (8.5) then provides the simple estimate of the decay time of the current and, hence, of the

---

[4] Subsequent issues of *Scientific American* provide an interesting account of present knowledge on this subject: G. N. Parker, *Scientific American* (August 1983) 44, 'Magnetic fields in the cosmos'; R. Jeanloz, *Scientific American* (September 1983) 56, 'The earth's core'; K. A. Hoffman, *Scientific American* (May 1988) 50, 'Ancient magnetic reversals: clues to the geodynamo'; J. Bloxham and D. Gubbins, *Scientific American* (December 1989) 30, 'The evolution of the earth's magnetic field'.

associated magnetic field[5] already given in Eq. (8.6). Taking values characteristic for the outer core, viz. the resistivity of molten iron, $\eta \sim 10^{-5}\,\Omega\,\text{m}$, and the size of the outer core, $l_0 \sim 2000\,\text{km}$, this gives $\tau_D \sim 5 \times 10^{11}\,\text{s} = 16\,000$ years. Now, we have an entirely different situation than in the solar case. The time scale of resistive decay is much shorter than the time scale of the periodicity of the geomagnetic field. One needs to explain why the field would recover its initial strength and be maintained during times of the order of 500 000 years (until the next reversal).

To explain that one needs dynamo theory, i.e. equations describing fluid motion, in much the same way as the treatment of the solar dynamo with the Eqs. (8.1)–(8.4). In particular, one should exploit the appropriate form of Ohm's law for a moving conducting fluid:

$$\mathbf{E} + \mathbf{v} \times \mathbf{B} = \eta \mathbf{j}. \tag{8.16}$$

(Note that, since molten iron is not a plasma, we do not expect, nor do we need, a huge anomalous increase of the resistivity η as in the solar case.) Now, fluid flow introduces a magnetic field evolution that is entirely different from simple resistive decay due to the first term on the RHS of Eq. (8.4). Resistive decay may be counteracted by convection of magnetic field through fluid motion. As for the solar dynamo problem, Eq. (8.4) for the evolution of $\mathbf{B}$ is to be considered together with the equations (8.1)–(8.3) which describe the evolution of the density ρ, the velocity $\mathbf{v}$, and the pressure p. Most important, one has to show that this system confined within the boundaries of the solid inner core and the mantle, which both may be considered as electric insulators but which do exchange heat with the fluid, gives rise to solutions exhibiting the highly non-stationary behaviour of the geomagnetic field described above.

The fluid flow may be driven either by heat from radioactive decay of potassium nuclei in the inner core or, according to an alternative theory, by the unmixing of the dense iron crystals forming at the interface of the inner and outer core from the remaining lighter fluid. In both cases there is enough energy available to regenerate the magnetic field. This happens through a highly turbulent process in which the upwelling fluid is put into cyclonic motion by the non-uniform rotation of the outer core, as schematically shown in Fig. 8.19. As a result, the field lines, carried with the fluid, are first stretched out in the equatorial direction so that an azimuthal

---

[5] A more explicit example of such decay may be obtained for the special model of *a force-free magnetic field*, obeying $\mathbf{j} \times \mathbf{B} = 0$ so that $\nabla \times \mathbf{B} = \alpha \mathbf{B}$. (Do not confuse the force-free field parameter α with the dynamo parameter α of Section 8.2.1. Unfortunately, both notations are standard in their respective contexts.) If, by some magic, $\alpha = $ const in space and time, Eq. (8.5) would lead to

$$\frac{\partial \mathbf{B}}{\partial t} = -\frac{\alpha^2 \eta}{\mu_0} \mathbf{B},$$

so that $\mathbf{B} \sim \exp(-t/\tau)$ with $\tau_D = \mu_0/(\alpha^2 \eta)$. Since α measures the current and, hence, the inhomogeneity of the field: $\alpha \sim 1/l_0$, and we recover Eq. (8.6).

412   *Magnetic structures and dynamics*

Fig. 8.19. A snapshot of the 3D magnetic field structure simulated with the Glatzmaier–Roberts geodynamo model. The rotation axis of the model Earth is vertical and through the centre. A transition occurs at the core–mantle boundary from the intense, complicated field structure in the fluid core, where the field is generated, to the smooth, potential field structure outside the core. The field lines are drawn out to two Earth radii. (Courtesy of Gary A. Glatzmeier (UCSC) and Paul H. Roberts (UCLA).)

magnetic field component develops (directed $W \to E$ on the northern and $W \leftarrow E$ on the southern hemisphere). Next, this azimuthal field is deformed into helices having a meridional component (in the same direction on both hemispheres) which would amplify the original dipolar field.

Of course, this qualitative argument cannot replace genuine analysis. In particular, it does not explain why and when the field reverses sign. Some kind of a nonlinear triggering mechanism is needed here. Recently, such reversal has been found in computer simulations of the geodynamo (G. A. Glatzmaier and P. H. Roberts [79]).

For the purpose of the present chapter we do not need to go into more detail about the geodynamo problem because we will be concerned mainly with magnetospheric physics where the relevant phenomena occur on time scales of the order of hours (magnetospheric substorms), days (solar flares and geomagnetic storms), or, at most, decades (the 11 year cycle of the solar magnetic field). It is clear that on those time scales the magnetic field of the Earth may be considered as a static dipole field of the form given in Eq. (8.15) with a given strength. Spatial deviations from the dipolar form could be described by adding higher order multipole terms.

8.3.2 *Magnetic fields of the other planets*

One of the interesting features of the solar system is the variety of the physical properties of the planets. It is certainly not so that 'if you have seen one you have seen them all'. From the discussion of the geodynamo it is clear that the existence and the strength of a planetary field will depend on the *size*, the *interior structure and composition*, and the *rotation* of the planet. The relative importance of these three factors is still far from obvious at present. From this point of view it is nice that there is more than one planet so that we are prevented from drawing oversimplified and erroneous conclusions. For example, the fast rotation of the *giant planets* correlates well with their sizeable magnetic fields, whereas the slow rotation of *Venus* also correlates with the virtual absence of an internal magnetic field. (See Tables B.7 and B.8.) However, the similarity in the rotation periods of *Mars* and *Earth* does not give a clue as to why the magnetic field of Mars is three orders of magnitude smaller. Also notice that the smaller planet *Mercury*, which also rotates much slower than Mars, still has a magnetic field larger than that of Mars.

With respect to size and interior structure, Eq. (8.5) shows that we need a relatively large cross-sectional area of a well-conducting material where currents can flow without too much resistive decay so that the generated magnetic field does not disappear too fast. On the other hand, this material should preferably be in a liquid phase in order for the magnetic field to be regenerated by means of the dynamo process through convective fluid flow. Broadly speaking, these conditions are provided for the *terrestrial planets* by the presence of an outer core of molten iron, whereas the giant planets *Jupiter and Saturn* have extended outer cores of liquid metallic hydrogen. The relatively smaller fraction of metallic hydrogen as compared to molecular hydrogen in Saturn could explain its smaller magnetic field. However, the recently discovered magnetic fields of *Uranus and Neptune*, which are of the same order of magnitude as the magnetic field of Saturn, are apparently generated by a dynamo operating in a quite different environment since these planets are not supposed to have a metallic outer core but, instead, a thick shell consisting of water, ammonia and methane.

The magnetic field of *Venus*, although very small, deserves separate treatment because it does not originate from an internal dynamo (the rotation period of this planet is too small as we have noted above) but from the interaction between the solar wind and the ionosphere of this planet. This interaction induces currents in the upper atmosphere which are responsible for the observed magnetic field.

414   *Magnetic structures and dynamics*

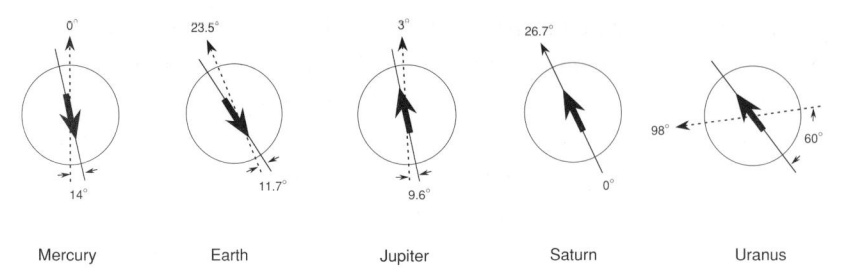

Fig. 8.20. Orientation of the main planetary magnetic moments.

With respect to these magnetospheric properties, Venus resembles more comets like Giacobini–Zinner and Halley, which also do not have internal magnetic fields but magnetospheric fields induced by the interaction with the solar wind.

Another interesting feature of the planetary magnetic fields is the orientation of the magnetic axis with respect to the rotation axis of the planet (see Table B.8 and Fig. 8.20). Since dynamo action requires rotation one would expect that the two axes would be more or less aligned. This turns out to be the case for most of the planets, where it should be noted that Mercury and Earth have their magnetic moment oriented anti-parallel to the angular momentum vector whereas the giant planets have an approximately parallel orientation. (For historical reasons, the latter orientation is indicated with a minus sign in front of the value of the tilt angle given in Table B.8.) Since dynamo action allows for field reversal one probably should not make too much of this difference: apparently, two planets are accidentally in a reversed magnetic state with respect to the others at this particular moment in time.

An exception to the rule of aligned magnetic and rotation axes is provided by Uranus, where the orientation of the two axes appears to be totally unrelated. In a sense, this is too bad since, with alignment, the special orientation of the rotation axis of Uranus (approximately lying in the ecliptic plane) would have given rise to the peculiar phenomenon of the solar wind impinging on one of the magnetic poles of a planet.[6] On the other hand, it is good to have this counterexample since it shows us that the dynamo mechanism operates under a wide variety of circumstances. Apparently, rotation plus an internal energy source of some kind is already enough to set it off. Another possibility, which cannot be excluded at present, is that we are observing Uranus at a period of magnetic reversal so that the magnetic axis is somewhere in the middle of its wandering motion. Uranus is also exceptional with respect to the deviation from the dipolar form given by Eq. (8.15).

---

[6] Also, too bad for the theoretician who would have been excused for studying the problem of the interaction of the solar wind with a planetary magnetic field for a two-dimensional system, conserving the axi-symmetry of the dipole field, rather than facing the real three-dimensional problem.

Large quadrupole and octupole components have been observed which can be represented by an effective magnetic dipole displaced off centre by $0.3 \times R_U$. This could also be considered as evidence for a field reversal period.

In conclusion, the preliminary material collected in Tables B.7 and B.8 gives a glimpse of the vast territory of magnetic properties of the planets, which is largely uncharted at present. The results of Voyager 2 have already changed our picture substantially.[7] It is likely that this picture will be substantially modified in the near future, in particular by further space missions (like Cluster).

8.4 Magnetospheric plasmas

So far we have discussed the planetary magnetic fields outside the planet from the point of view of a simple dipolar representation, as given in Eq. (8.15). This representation is an idealization. It breaks down in the interior of the planet close to the source of the field, i.e. in the neighbourhood of the distributed dynamo currents in the outer core of the planet. More important, and much better accessible to observation, is the huge deviation from a dipole field outside the planet. This deviation is due to the interaction with the solar wind. This is a plasma consisting of protons and electrons, of different temperatures, emitted by the corona of the Sun. We will first discuss the mechanism of the solar wind generation (and continue the discussion of Section 8.2.2 on the large scale structure of the heliosphere) in Section 8.4.1, and then turn to the interaction of the solar wind with the planetary magnetospheres in Section 8.4.2.

8.4.1 The solar wind and the heliosphere

We have seen that the density of the corona is many orders of magnitude smaller than that of the photosphere and that the temperature increases by a factor of 100 to 1000. Typical numbers are:

$$n_{photosphere} \sim 10^{23} \text{ m}^{-3}, \quad n_{corona} \sim 10^{14} \text{ m}^{-3} \text{ (at 3000 km)},$$
$$T_{photosphere} \sim 6000 \text{ K}, \quad T_{corona} \sim 10^6 \text{ K}.$$

Whatever the mechanism behind this remarkable phenomenon, these numbers imply that the solar corona cannot be in static equilibrium but suffers a continuous outflow of mass, the *solar wind*. It was predicted by E. N. Parker in 1958 [174] and observed by satellites in 1959. This plasma escapes along the open magnetic field lines, mainly originating in the coronal holes (the dark regions of reduced X-ray emission). The question then arises as to what kind of stationary equilibria

---

[7] See E. D. Miner, *Physics Today* (July 1990) 'Voyager 2's encounter with the gas giants'.

(i.e. equilibria with a time-independent outflow velocity) such a corona permits. To sketch the answer to that question, we will consider the simplest possible gas dynamic model of the corona.

The principal arguments (A. J. Hundhausen [113]) can be obtained from a model in which the energy equation is ignored, i.e. an *isothermal* ($T \approx T_e \approx T_i$) *model of the corona* with a hydrogen plasma ($n \equiv n_e \approx n_i$):

$$\frac{\partial \rho}{\partial t} + \nabla \cdot (\rho \mathbf{v}) = 0 \qquad \text{(mass conservation)}, \tag{8.17}$$

$$\rho \frac{\partial \mathbf{v}}{\partial t} + \rho \mathbf{v} \cdot \nabla \mathbf{v} + \nabla p - \rho \mathbf{g} = 0 \quad \text{(momentum conservation)}, \tag{8.18}$$

$$p = n_e k T_e + n_i k T_i \approx 2nkT \qquad \text{(equation of state with given } T\text{)}, \tag{8.19}$$

$$\rho \equiv n_e m_e + n_i m_i \approx nm \qquad \text{(definition of the density)}, \tag{8.20}$$

where m is the sum of the proton and electron mass. Hence, p and ρ are related through the isothermal sound speed,

$$p/\rho \approx 2kT/m \equiv v_{\text{th}}^2, \tag{8.21}$$

which is assumed to be constant in this model.

Spherically symmetric *static* ($\mathbf{v} = 0$) equilibrium would imply

$$\nabla p = \rho \mathbf{g} \quad \Rightarrow \quad \frac{dp}{dr} = -\rho \frac{GM_\odot}{r^2} = -\alpha R_\odot \frac{p}{r^2}, \quad \alpha \equiv \frac{GM_\odot}{R_\odot v_{\text{th}}^2}, \tag{8.22}$$

so that

$$p = p_0 e^{-\alpha(1 - R_\odot/r)}. \tag{8.23}$$

With typical parameters for the base of the corona, $T = 10^6$ K and $n = 1.5 \times 10^{15}$ m^{-3}, we have $p_0 = 4 \times 10^{-3}$ N m^{-2} and $\alpha = 11.53$. Hence, far away from the Sun, in interstellar space, we would have $p_\infty = e^{-\alpha} p_0 = e^{-11.53} \times 4 \times 10^{-3} = 4 \times 10^{-8}$ N m^{-2}, i.e. many orders of magnitude too big as compared to the actual values there, viz. $p \approx 3 \times 10^{-14}$ N m^{-2}. Consequently, a hot corona (10^6 K) cannot be in static equilibrium with the interstellar pressure. (For a cold atmosphere at photospheric temperature, $T = 6000$ K, such a problem does not arise: $\alpha = 1920$, and p_∞ is completely negligible.) This is the reason for the solar wind solutions, i.e. stationary equilibria with $\mathbf{v} \neq 0$.

Hence, consider a spherically symmetric *stationary state* with $v \equiv v_r$. The mass conservation equation (8.17) then gives

$$\frac{d}{dr}(r^2 \rho v) = 0 \quad \Rightarrow \quad r^2 \rho v = \text{const}, \tag{8.24}$$

8.4 Magnetospheric plasmas

where the constant is related to the total mass loss of the Sun:

$$\dot{M}_\odot \equiv \frac{d}{dt} \int \rho \, dV = -\oint \rho \mathbf{v} \cdot \mathbf{n} \, dS = -4\pi r^2 \rho v. \tag{8.25}$$

Note that, in contrast to Eq. (4.84) of Section 4.3.2, we now have an open system where mass is lost. Obviously, the representation by a stationary state is restricted to time scales $\tau \ll M_\odot / \dot{M}_\odot$. The momentum equation (8.18) with the relation (8.21) then yields

$$\rho v \frac{dv}{dr} + v_{\text{th}}^2 \frac{d\rho}{dr} + GM_\odot \frac{\rho}{r^2} = 0. \tag{8.26}$$

Introducing dimensionless variables $\bar{v} \equiv v/v_{\text{th}}$ (which is actually the Mach number of the flow) and $\bar{r} \equiv r/R_\odot$, the latter equation becomes

$$\bar{v} \frac{d\bar{v}}{d\bar{r}} + \frac{1}{\rho} \frac{d\rho}{d\bar{r}} + \frac{\alpha}{\bar{r}^2} = 0, \tag{8.27}$$

which gives a one-parameter ordinary differential equation after elimination of ρ by means of Eq. (8.24):

$$\left(\bar{v} - \frac{1}{\bar{v}} \right) \frac{d\bar{v}}{d\bar{r}} - \frac{2}{\bar{r}} + \frac{\alpha}{\bar{r}^2} = 0. \tag{8.28}$$

The solutions are implicitly obtained from the first integral:

$$F(\bar{v}, \bar{r}) \equiv \tfrac{1}{2} \bar{v}^2 - \ln \bar{v} - 2 \ln \left(\frac{\bar{r}}{\bar{r}_c} \right) - 2 \frac{\bar{r}_c}{\bar{r}} + \frac{3}{2} = C, \qquad \bar{r}_c \equiv \tfrac{1}{2} \alpha, \tag{8.29}$$

where the constant C labels the solutions corresponding to the different boundary data at $\bar{r} = 1$ (the solar surface or, rather, the base of the corona). The normalization has been chosen such that solutions *going through the critical point*, where $\partial F / \partial \bar{v} = 0$, $\partial F / \partial \bar{r} = 0$, and located at $\bar{v} = \bar{v}_c = 1$, $\bar{r} = \bar{r}_c = \tfrac{1}{2} \alpha$, are labelled by $C = 0$. At this point the flow speed crosses the sound speed (hence, it is also called the *sonic point*) and the position is determined by the parameter α, i.e. by the temperature of the corona (assumed to be constant for the present purpose). The solutions are schematically represented in Fig. 8.21.

Solutions of interest are those which connect the solar surface with interplanetary space. They are of two types, viz.

(a) *Solar breeze*, i.e. subsonic solutions which decelerate for $\bar{r} \to \infty$:

$$-\ln \bar{v} \approx 2 \ln \bar{r} + \text{const} \quad \Rightarrow \quad \bar{v} \sim \bar{r}^{-2}.$$

For these solutions, $\rho \sim \text{const}$, $p \sim \text{const}$ as $\bar{r} \to \infty$. Obviously, the static solution (8.23) is one of them so that the pressure is again too high to be balanced

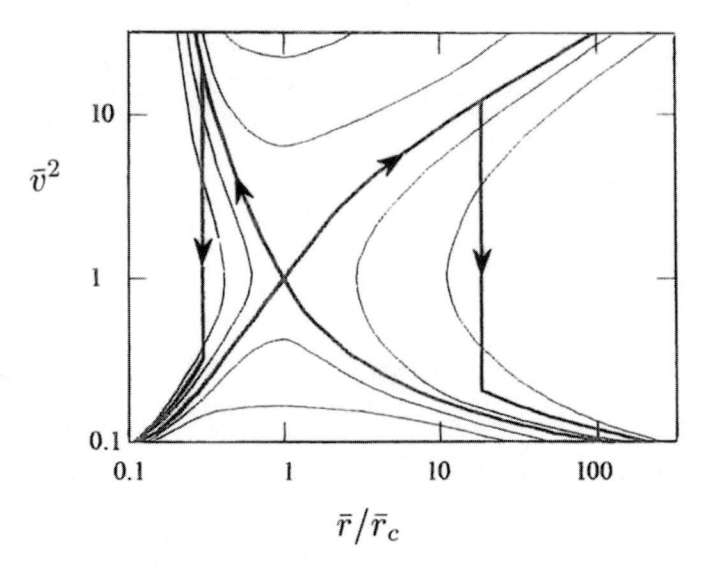

Fig. 8.21. Continuous solar wind solutions (thick lines). The lines with arrows show a modification to indicate the possibility of shocked wind outflow and accretion flow. (Adapted from Holzer and Leer [112].)

by the interstellar pressure. However, this behaviour at ∞ could be cured by the consideration of a decreasing temperature profile $v_{\text{th}}^2(r)$.

(b) *Solar wind*, i.e. transonic solutions going through the critical point and accelerating for $\bar{r} \to \infty$:
$$\bar{v}^2 \sim 4 \ln \bar{r} \quad \Rightarrow \quad \bar{v} \sim (\ln \bar{r})^{1/2}.$$
For these solutions, $p \sim \bar{r}^{-2}(\ln \bar{r})^{-1/2}$ as $\bar{r} \to \infty$, which is the acceptable behaviour at ∞. The flow is *subsonic* at the Sun, and *supersonic* at 1 AU.

As an aside, the two critical solutions shown in Fig. 8.21 may be modified to illustrate an important aspect of transonic flow, viz. the possibility of the formation of shocks (lines with arrows). Since the direction of the flow may be reversed in Fig. 8.21, the accelerating solar wind solution represents a typical example of stellar outflow, whereas the reversed solar breeze critical solution could represent a transonic accretion flow. In the first case, the supersonic flow with ever decreasing pressure eventually meets the small but finite pressure of the interstellar medium and ends there with a shock (the termination shock). Similarly, supersonic accretion (e.g. onto a compact object) is stopped by a shock situated at some location inside the sonic point.

The solutions shown are just the simplest examples of transonic flows, considered from a gas dynamic point of view. However, the solar wind is a tenuous plasma, that will convect the *interplanetary magnetic field*. Hence, a highly complex magnetic structure arises which engulfs all of the planets: the *heliosphere*

8.4 Magnetospheric plasmas

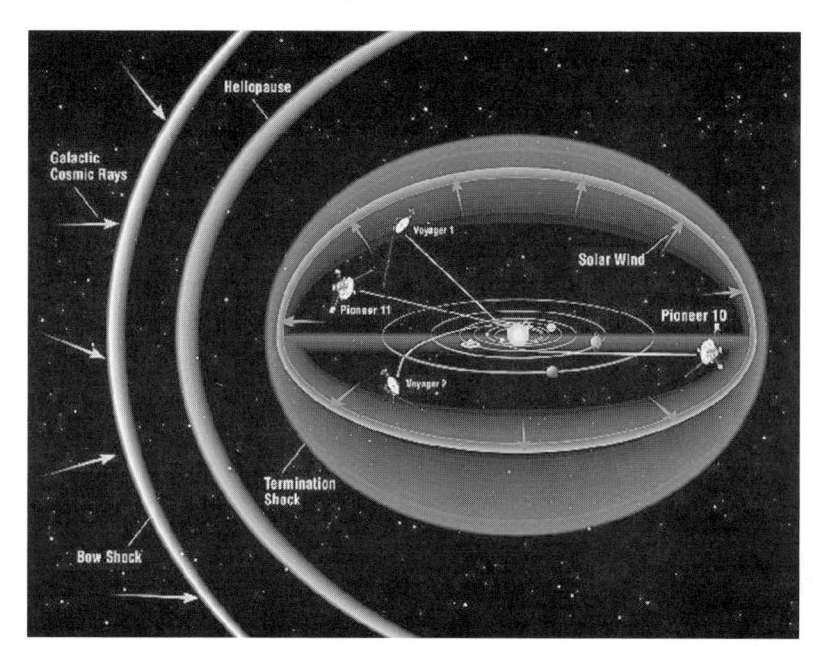

Fig. 8.22. Model of the heliosphere with spacecraft. (From www.science-at-home.de.).

(Fig. 8.22). Obviously, for the description of the heliosphere and the interaction of the solar wind with the magnetospheres of the planets, the magnetic field cannot be neglected and the much richer arsenal of MHD shocks is needed. This will be the subject of one of the later chapters. Here, we just present some order of magnitude estimates of the shape of the magnetospheric structures.

8.4.2 Solar wind and planetary magnetospheres

In a sense, the solar wind and the heliosphere are just extensions of the corona over the whole solar system. Of course, the density of this plasma rapidly decreases from the value quoted above. Nevertheless, the influence of this plasma on the global structure of the magnetospheres of the planets is considerable. To study this effect, in particular on the magnetosphere of the Earth, consider the numbers of Table 8.1.

It is to be noted that large variations of these values occur since the solar wind is a highly non-stationary phenomenon. For example, after solar flares the intensity of the solar wind may increase by an order of magnitude giving rise to geomagnetic storms since the magnetic field intensity suddenly increases.

To understand why the solar wind could influence the value of the geomagnetic field let us make a small excursion to a standard problem of plasma confinement

Table 8.1.

Typical values for solar wind parameters at 1 AU
[adapted from E.R. Priest, *Solar Magnetohydrodynamics* (1984).]

velocity	v	$= 3 \times 10^5 \text{ m s}^{-1}$	$(= 300 \text{ km s}^{-1})$
number of particles	n	$= 5 \times 10^6 \text{ m}^{-3}$	$(= 5 \text{ cm}^{-3})$
electron temperature	T_e	$= 2 \times 10^5 \text{ K}$	$(= 20 \text{ eV})$
proton temperature	T_p	$= 5 \times 10^4 \text{ K}$	$(= 5 \text{ eV})$
magnetic field	B	$= 6 \times 10^{-9} \text{ T}$	$(= 6 \times 10^{-5} \text{ gauss})$
Alfvén speed	v_A	$= 6 \times 10^4 \text{ m s}^{-1}$	$(= 60 \text{ km s}^{-1})$
ion sound speed $[\equiv (kT_e/m_p)^{1/2}]$	v_s	$= 4 \times 10^4 \text{ m s}^{-1}$	$(= 40 \text{ km s}^{-1})$

in the laboratory. In Section 2.4.3 we considered one of the oldest ideas of plasma confinement for the purpose of producing thermonuclear energy, viz. the z-pinch. An alternative, also old, idea is the θ-pinch, where plasma currents are induced in the θ-direction. The static equilibrium equations (2.152)–(2.154) for that case yield the following relations:

$$\frac{dp}{dr} = j_\theta B_z, \qquad j_\theta = -\frac{1}{\mu_0}\frac{dB_z}{dr}. \tag{8.30}$$

Hence,

$$\frac{dp}{dr} = -\frac{d}{dr}\left(\frac{B_z^2}{2\mu_0}\right), \quad \text{so that} \quad p + \frac{B_z^2}{2\mu_0} = \text{const}. \tag{8.31}$$

In other words: a θ-pinch is characterized by radial profiles of the kinetic pressure $p(r)$ and of the *magnetic pressure* $B_z^2/2\mu_0$ which are each other's opposites (Fig. 8.23). An extreme case, which can be rather closely approached in the laboratory, is the skin-current θ-pinch where the current is induced so fast that there is no time for it to diffuse into the plasma. Then, a 'magnetic piston' drives the plasma inward, and a final equilibrium state is obtained which satisfies the relation

$$p_1 + \left(\frac{B_z^2}{2\mu_0}\right)_1 = \left(\frac{B_z^2}{2\mu_0}\right)_2. \tag{8.32}$$

8.4 Magnetospheric plasmas

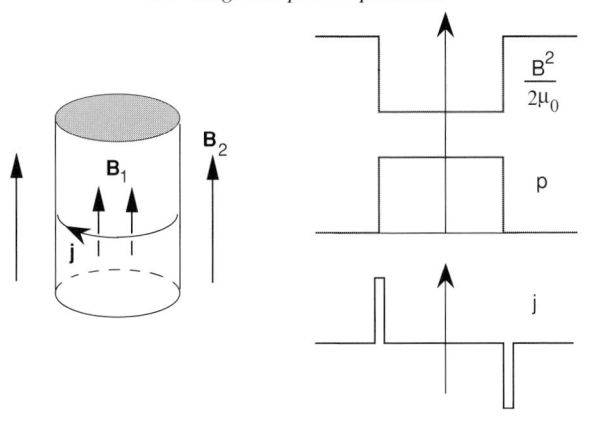

Fig. 8.23. Magnetic pressure in a θ-pinch.

Since $p = nkT$, it is just a matter of inducing a large enough current to obtain thermonuclear values of nT. This has been well within reach since the 1950s. Unfortunately, the third parameter needed for fusion, viz. the plasma confinement time τ, falls short of many orders of magnitude for this particular configuration (the plasma is squeezed out of the ends like toothpaste). This is the reason that this line of research has been abandoned in thermonuclear research.

Returning to the problem of the interaction of the solar wind with the geomagnetic field, we now appreciate that *a magnetic field exerts a pressure on a plasma* (as already discussed in Section 4.3.2). Vice versa, a plasma which is propelled by some mechanism will squeeze a magnetic field that is anchored in a fixed body. In the case of the solar wind, several other ingredients contribute to the effective pressure that is exerted on the day-side of the planetary magnetic field, viz., except for the magnetic pressure $B^2/(2\mu_0)$ perpendicular to the magnetic field that is carried with the flow, a pressure $\frac{1}{2}\rho v^2$ due to the flow and exerted in the direction of the flow, and an isotropic kinetic pressure nkT associated with the hottest particle population (the electrons in this case). From the numbers given in Table 8.1 we obtain the following estimates for these pressures:

$$(B^2/(2\mu_0))_{sw} = 1.4 \times 10^{-11}\,\mathrm{N\,m^{-2}},$$
$$\left(\tfrac{1}{2}\rho v^2\right)_{sw} \approx \tfrac{1}{2}nm_p v^2 = 3.8 \times 10^{-10}\,\mathrm{N\,m^{-2}}, \quad (8.33)$$
$$nk(T_i + T_e)_{sw} \approx nkT_e = 1.4 \times 10^{-11}\,\mathrm{N\,m^{-2}}.$$

Hence, $\frac{1}{2}\rho v^2 \gg B^2/(2\mu_0) \sim nkT_e$ so that the dominant effect is the pressure $\frac{1}{2}\rho v_{sw}^2$ directed away from the Sun.

At this point, we would need to interrupt our presentation based on magnetic pressures alone by the consideration of shocks associated with a sudden change of

Fig. 8.24. Schematics of the interaction of the solar wind with the magnetosphere. (From science.nasa.gov.)

the velocity and the magnetic field at the magnetospheric *bow shock*. We would need to continue the analysis of Section 4.5.1 on MHD discontinuities but this will only be done in conjunction with the intricate problem of transonic MHD flows, to be analysed in the companion Volume 2. For the moment we will have to be content with a cartoon argument, viz. Fig. 8.24, which schematically shows the complete magnetospheric system with the solar wind impinging on the planetary magnetic field, compressing it on the day-side and stretching it out on the night-side. (We remind the reader that all talk of magnetic field being compressed or stretched out really refers to the composite system of magnetic field plus plasma.) Fig. 8.24 shows the bow shock where the solar wind is blocked and a tangential discontinuity, closer to the planet.

In general terms, the solar wind pressure is balanced at some distance from the planet by the magnetic pressure $B^2/(2\mu_0)$ exerted by the geomagnetic field. This balance is a delicate process which gives rise to the *magnetopause*, which constitutes the boundary between the region dominated by the solar wind flow and the geomagnetic field region (the magnetosphere). Here, we can only give an order of magnitude estimate for the position of the magnetopause on the day-side of the planet which, roughly speaking, is the position where the solar wind develops a stagnation point due to the obstacle of the planetary magnetic field which is fixed

8.4 Magnetospheric plasmas

in the outer core of the planet. This mechanism is the same for all the planets, but we will derive the numbers for the Earth.

We express the components of the dipole field, given in Eq. (8.15), in terms of the spherical coordinates r, θ, ϕ (see Eq. (A.51) of Appendix A.2), related to the Cartesian coordinates by

$$x = r\sin\theta\cos\phi, \quad y = r\sin\theta\sin\phi, \quad z = r\cos\theta. \tag{8.34}$$

Exploiting the definition for the dipole moment M given in the footnote on Eq. (8.15) and the numbers of Table B.8, in particular for the geomagnetic field, i.e.

$$\mathbf{M}_E = -M_E \mathbf{e}_z, \quad M_E \equiv (\mu_0/4\pi)\, m_E = 8.1 \times 10^{15}\, \text{T m}^3, \tag{8.35}$$

we obtain:

$$B_r = -2M_E \cos\theta / r^3, \quad B_\theta = -M_E \sin\theta / r^3, \quad B_\phi = 0. \tag{8.36}$$

The pertinent numerical value of the geomagnetic field on the day-side of the magnetosphere is given by

$$B_\theta(r, \theta = \pi/2) = -M_E/r^3 = -(M_E/R^3)(R/r)^3 = -3.1 \times 10^{-5}\,(R/r)^3\,\text{T}, \tag{8.37}$$

corresponding to a magnetic pressure of

$$\left(\frac{B^2}{2\mu_0}\right)_{\text{geo}} = \frac{1}{2\mu_0}\left(\frac{M_E}{R^3}\right)^2\left(\frac{R}{r}\right)^6 = 3.8 \times 10^{-4}\left(\frac{R}{r}\right)^6 \text{N m}^{-2}. \tag{8.38}$$

Clearly, this exceeds the pressure $(\tfrac{1}{2}\rho v^2)_{\text{sw}} = 3.8 \times 10^{-10}\,\text{N m}^{-2}$ given in Eq. (8.20) by many orders of magnitude at the surface of the Earth. However, since the magnetic pressure rapidly decays, with a power of 6 according to Eq. (8.38), balance is already reached at a distance of some planetary radii, so that

$$r_{\text{magnetopause}} \approx 10R. \tag{8.39}$$

This agrees well with the number given in the last column of Table B.8.

The positions of the day-side magnetopause for the other planets, given in B.8, can be estimated in the same way, provided one takes the appropriate values of the density and the velocity of the solar wind into account. Note that the magnitude of the planetary magnetic field is reflected in the size of the magnetospheres on the day-side. For Mercury and Venus the day-side magnetopause is situated very close to the planet, whereas for the giant planets the distance is some tens of planetary radii.

It should be stressed once more that we have sketched here a very crude model of the structure of the magnetosphere, neglecting all the details of the shocks which develop in front of the magnetopause. The picture is even less complete with respect to the night-side of the magnetosphere. Here, the deviation from a dipole field is much more pronounced, with field lines stretching out over hundreds of planetary radii to form what is called the *magnetotail*. We have seen that the discontinuity of the magnetic field at the magnetopause is due to surface currents flowing there. The magnetotail exhibits a similar discontinuity at the equatorial plane, viz. a discontinuity of the direction of the field (see Fig. 8.24). Such a discontinuity requires surface currents (see Section 4.5) flowing perpendicular to the field:

$$j_x^\star \equiv \lim_{\delta \to 0, \, j_x \to \infty} (\delta \cdot j_x) = B_y \big|_{z<0} - B_y \big|_{z>0}. \tag{8.40}$$

Because of the associated null in the magnetic field at the equatorial plane, these currents are called *neutral sheet currents*. It is beyond the scope of the present section to discuss the origin of these currents, but their existence is a simple requirement for the extended structure of the magnetotail. In conclusion: the global structure of the magnetosphere is mainly due to the surface currents flowing on the magnetopause and on the neutral sheet. These currents constitute the response to the huge MHD generator driven by the solar wind and the geomagnetic obstacle.

The magnetospheric structure may be disturbed by sudden changes of the solar wind variables (e.g. by CMEs, see below), giving rise to *MHD waves* like ultra low frequency ($< 10\,\text{Hz}$) waves, or higher frequency plasma waves, or may be disrupted by *reconnection* of field lines causing so-called flux transfer events on the day-side as well as on the night-side of the magnetosphere, *Kelvin–Helmholtz instabilities* driven by shear flow at the magnetopause, etc. All of these topics will return as genuine plasma dynamics in later chapters.

Coronal mass ejections (CMEs) are huge plasma bubbles threaded with magnetic field lines that are ejected from the Sun over the course of several hours. CMEs are spectacular manifestations of solar activity and the most energetic phenomena observed in the solar system. In large CMEs, such as the one depicted in Fig. 8.25, up to 10^{13} kg of coronal material is ejected, although the average value is closer to 10^{12} kg. The average speed of a CME is about 300–400 km s^{-1}, but it can be as low as 50 km s^{-1}, while fast CMEs have speeds up to 2000 km s^{-1}. The energy associated with CMEs amounts to 10^{24}–10^{25} Joule. In spite of this, the existence of CMEs was not realized until the space age and the earliest evidence for these dynamical events came from coronagraph observations made by the 7th Orbiting Solar Observatory (OSO 7) from 1971 to 1973, see Brueckne [45]. (Ground based coronagraphs only make the innermost corona visible.) The term CME comes from Burlaga *et al.* [47].

8.4 Magnetospheric plasmas

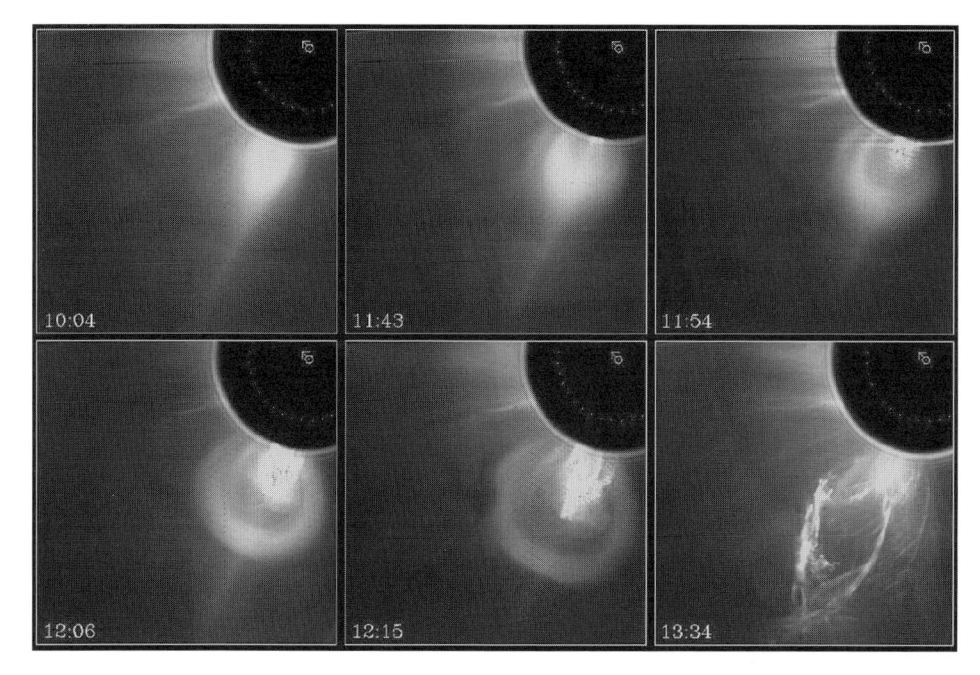

Fig. 8.25. Six snapshots showing the evolution of a *coronal mass ejection*. The dark disc in the upper right corner is the occulting disc (radius 60% larger than the solar disc) of the Solar Maximum Mission (SMM) coronagraph, used to take these images. (From www.hao.ucar.edu.)

Close to the Sun, many CMEs have a three-part structure consisting of a bright core (the eruptive prominence), a dark cavity, and a bright loop. However, this structure is lost in interplanetary space. CMEs are often associated with solar flares or prominence eruptions but they can also occur independently. The frequency of CMEs varies with the sunspot cycle from about one CME a week at solar minimum to an average of 2 to 3 CMEs per day near solar maximum.

CMEs play an important role in *space weather*, which is defined as follows: 'Space weather refers to conditions on the Sun and in the solar wind, magnetosphere, ionosphere and thermosphere that can influence the performance and reliability of space-borne and ground-based technological systems and can endanger human life or health' (US NSWP Strategic Plan). Coronal mass ejections disrupt the flow of the solar wind and produce disturbances and MHD shocks. About 10% of these shocks strike the Earth with sometimes catastrophic results for navigation systems, power lines, radio traffic, functioning of oil pipelines, technology aboard space vehicles, etc.

426   *Magnetic structures and dynamics*

8.5 Perspective

At present, quite a number of important space missions of the Solar Terrestrial Physics Program of NASA (USA), ESA (Europe) and ISAS (Japan) are being carried out and planned. The satellite SOHO, launched in December 1995 to study solar phenomena from the core to beyond the Earth's orbit, turned out to be a great success. The original launch of the Cluster satellites in 1996 failed, but was successfully repeated in 2000. Its purpose is to study the 3D spatial structure of the magnetosphere of the Earth (Fig. 8.26). Ulysses (already launched in 1990) provided in situ investigations of the inner heliosphere from the solar equator to the poles. The Solar Orbiter (planned for 2012–2017), will provide the highest resolution solar observations and first images of the Sun's polar regions.

From a theoretical point of view, it is quite satisfactory that an important part of the observed dynamics clearly demonstrates the validity of *magnetic flux conservation* and of dynamics controlled by the motion of *magnetic flux tubes*. In essence, the magnetic structures observed are magnetic flux tubes, but they usually do not appear singly (as the theoreticians prefer) but in large numbers. Hence, many problems remain unresolved: a quantitative *solar dynamo* theory, a basic theory of *coronal heating*, prediction of *solar flares*, a detailed theory of *solar wind generation and heating*, the dynamics of the *interaction of the solar wind with the*

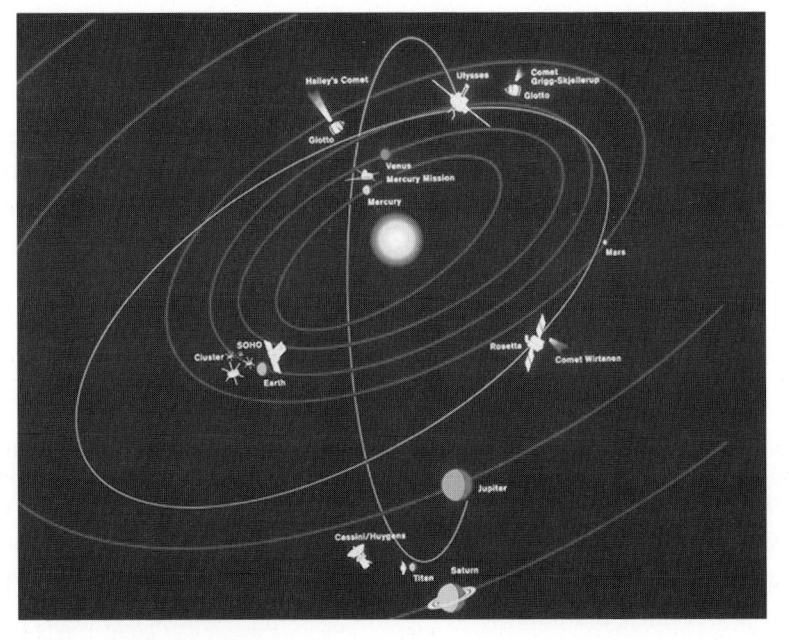

Fig. 8.26.   Schematic showing ESA's planetary and solar missions. (From ESA Photo Archive.)

planetary magnetospheres, and, ultimately, the prediction of *space weather*. All of these problems can be translated to other stars and to magnetic phenomena of distant objects in the Universe. The solar system is just special in that it provides the necessary spatial resolution to investigate plasma dynamics. (So much is clear from laboratory plasma research: all dynamics is determined by fine details of the distribution of the magnetic field. Without knowledge of that, theory just amounts to fantasy.) MHD does apply as a leading order description, some extensions are needed to account for small scales with different electron and ion dynamics, other extensions concern particle acceleration. However, the bottom line is: with present high-resolution observations there is no excuse any more for cartoon theories. A lot remains to be done and one can be sure that a very different view will emerge from the present one.

8.6 Literature and exercises

Notes on literature

Solar MHD:

– Priest [190], *Solar Magnetohydrodynamics*, is still the basic text on this subject.

Solar physics:

– Stix [217], *The Sun* (second edition), Chapter 8 on solar magnetism treats the subjects of flux tubes, sunspots, and the solar cycle, and has a discussion of mean field electrodynamics and the solar dynamo.
– Foukal [69], *Solar Astrophysics*, Chapter 11 on dynamics of the solar magnetic field reviews the concepts of solar flux tubes, sunspots, butterfly diagram, and the Babcock model for the solar dynamo.
– Schrijver & Zwaan [204], *Solar and Stellar Magnetic Activity*, is a comprehensive review putting solar magnetism in the wider context of magnetic fields in stars that have a convective envelope immediately below their photosphere.

Magnetospheric physics:

– Busse [48], on 'Problems of planetary dynamo theory', is a theoretical review of magnetic field generation by buoyancy-driven convection flows in rotating spherical shells subject to a spherically symmetric gravity force, with discussion of the problem of core–mantle coupling.
– Saunders [201], in 'The earth's magnetosphere', contains an MHD description of the magnetosphere, with the magnetic field and electric current structures coupled to the solar wind, and the dynamics of MHD waves and shocks, reconnection, FTEs, and substorms.

Solar and stellar winds:

– Hundhausen [113], in 'The solar wind', reviews the properties of the solar wind, fluid theories on formation in the corona, the large-scale magnetic structure of the heliosphere, and the termination shock.

Exercises

[8.1] *The solar cycle*

The solar cycle is a periodic phenomenon which can be related to the number of sunspots on the surface of the Sun. They are present in a narrow strip around the equator only. The mechanism of the solar cycle is probably an oscillation of the vector of the solar magnetic dipole moment.

- Sketch the so-called 'butterfly diagram' and explain what it shows.
- Discuss the cartoons illustrating the Babcock model of the solar cycle. Does it explain the narrow strip around the equator?
- Write down the evolution equation for the magnetic field, including resistivity, but assuming a static configuration. Express the typical time scale for the decay of the magnetic field inhomogeneity using the Spitzer resistivity $\eta \approx 64 Z \ln \Lambda \, T_e^{-3/2}$, where $\ln \Lambda \approx 15$. Find the required quantities at the bottom of the convection zone, where the source of the solar dynamo is believed to be, and estimate the time scale. Does it approach the required time scale of the solar cycle? Comment on the implications.

[8.2] *Turbulent magnetic diffusivity*

Next, consider the evolution of the solar magnetic field including turbulent flow. Split the magnetic field and the velocity field into an average part, indicated by $\langle \ldots \rangle$, and a perturbation, indicated by a prime. Assume $\langle \mathbf{v}' \times \mathbf{B}' \rangle \approx \alpha \langle \mathbf{B} \rangle - \beta \nabla \times \langle \mathbf{B} \rangle$, where β functions as the turbulent magnetic diffusivity.

- Comment on the interpretation of the coefficients α and β, and give estimates of their order of magnitude from the solar parameters you can find. How did the situation improve with respect to the previous exercise? What would be your next step validating this approach?

[8.3] *Alfvén waves in flux tubes in the solar corona and in the magnetosphere (analytical)*

Consider an MHD wave propagating in a thin flux tube in the dipolar magnetic field of the Sun or the magnetosphere of the Earth. The magnitude of the magnetic field varies both along and across the flux tube. However, in this problem, we ignore the latter dependence. (The intricacies of wave dynamics connected with inhomogeneity in the perpendicular direction are treated in Chapter 11.) Effectively, the problem is then one-dimensional, with variations of the background quantities in the z-direction only. We assume that the flux tube is pressureless and carries no current.

- Starting from the general equation $\rho \partial^2 \boldsymbol{\xi}/\partial t^2 = \mathbf{F}(\boldsymbol{\xi})$, derive the wave equation for the MHD waves with magnetic field $\mathbf{B} = B(z) \mathbf{e}_z$ and density $\rho = \rho(z)$. Show that only two of the three MHD waves survive, viz. the Alfvén wave and one of the magneto-sonic waves: which one? Choose the perpendicular coordinates x and y such that the Alfvén wave is represented by ξ_x and the other one by ξ_y. What happened to ξ_z?
- Give reasons to ignore the magneto-sonic wave in favour of the Alfvén wave. Show that the latter obeys the wave equation

$$\frac{\partial^2 \xi_x}{\partial t^2} = v_A^2 \frac{\partial^2 \xi_x}{\partial z^2}.$$

Discuss the qualitative properties of this equation. For short wavelength variations of ξ_x, you may exploit a WKB solution to find out about the eigenfrequency (not very exciting). Try to improve on this by means of a perturbation analysis for the changes of the frequencies and eigenfunctions for long wavelengths with small variations of $B(z)$ and $\rho(z)$.

[8.4] *Alfvén waves in flux tubes in the solar corona and in the magnetosphere (numerical)*

We continue with the previous exercise (possibly skipping the last part since the numerical solution will be more satisfactory) and now insert the actual dependence of the magnitude of **B** of the dipolar field given by Eq. (8.15).

– Find an expression for the effective coordinate z in terms of the spherical coordinates $r(\theta)$, θ along a field line, where θ runs from $\theta = \theta_0$ to $\pi - \theta_0$. This will give you an expression for $B(z)$, or $B(\theta)$.

– Numerically solve the wave equation for the Alfvén waves with the magnetic field you just found. Assume constant density and periodic boundary conditions on ξ_x. This may represent standing Alfvén waves in a solar or magnetospheric loop. What is the influence of the variation of $B(z)$?

– Now specify to a solar coronal loop, where the photospheric boundary conditions are simulated by a large density increase at the ends of the loop. Use, for example, the density profile exploited by Beliën et al. [21]:

$$\rho(z) = 1 + (\rho^p - 1)\exp\left[-\frac{\sin^2(\pi z/L)}{2\sigma^2}\right],$$

where ρ^p is the density at the photospheric boundaries $z = 0$ and $z = L$ and σ is the density scale length. For large σ, how are the waves modified compared to the case of periodic boundary conditions?

[8.5]* *Solar wind interaction with the magnetosphere*

The interaction of the solar wind with the magnetosphere of the Earth is a complicated three-dimensional MHD problem giving rise to many kinds of time-dependent disturbances that may influence the magnetic structure of the magnetosphere as a whole. For example, when the solar wind carries a magnetic field pointing in the same direction as the dipole field of the Earth, the magnetosphere is compressed and the structure remains closed. However, when the solar wind magnetic field has opposite direction, a flux transfer event may occur where the day-side magnetosphere is suddenly opened up and an entirely different magnetic configuration is formed. In this problem, we will model the two types of magnetospheric structures (closed and open) by means of a very crude (static and two-dimensional) model where the one-sided solar wind with embedded magnetic field is replaced by a rotationally symmetric vertical magnetic field that exerts pressure on the dipole field of the Earth from all sides. We will exploit spherical coordinates (see Appendix A.2), where ϕ is the ignorable coordinate.

– Show that $\nabla \cdot \mathbf{B} = 0$ is solved in these coordinates by

$$B_r = -\frac{1}{Rr}\frac{\partial \psi}{\partial \theta}, \quad B_\theta = \frac{1}{R}\frac{\partial \psi}{\partial r}, \quad \text{i.e. } \mathbf{B} = -\frac{1}{R}\mathbf{e}_\phi \times \nabla \psi,$$

where ψ is the poloidal flux (evaluated through a circle R in the ecliptic plane).

– Show that $(\nabla \times \mathbf{B})_\phi = 0$ yields a second order partial differential equation for ψ,

$$\frac{\partial^2 \psi}{\partial r^2} + \frac{\sin\theta}{r^2}\frac{\partial}{\partial \theta}\frac{1}{\sin\theta}\frac{\partial \psi}{\partial \theta} = 0,$$

which is the Grad–Shafranov equation for this problem. (In plasmas with pressure and current, like in tokamaks, this equation has a non-vanishing RHS.)

- Show that the dipole field $\mathbf{B}_d = r^{-3}(3\mathbf{M} \cdot \mathbf{e}_r \, \mathbf{e}_r - \mathbf{M})$ is obtained from the particular solution

$$\psi_d = C + M \frac{\sin^2 \theta}{r},$$

where we exploit the value $M = -8.1 \times 10^{15} \, \text{T m}^3$ for the dipole moment of the Earth.

- Now add a vertical field $\mathbf{B}_v = B_0 \mathbf{e}_z = B_0(\cos\theta \, \mathbf{e}_r - \sin\theta \, \mathbf{e}_\theta)$, derivable from the flux function

$$\psi_v = \tfrac{1}{2} B_0 R^2,$$

representing the magnetic pressure of the solar wind.

- Find out what the relevant dimensionless parameters are for this problem. Make contour plots of $\psi \equiv \psi_d + \psi_v$ and study the magnetic structures you obtain for $B_0 > 0$ and $B_0 < 0$.

- Having obtained these qualitatively different solutions, determine the special points (x-points and stagnation points) analytically. Insert numbers and make estimates for solar wind parameters. Comment on the results obtained.

9
Cylindrical plasmas

9.1 Equilibrium of cylindrical plasmas

We have considered the effects of plasma inhomogeneity on MHD waves and instabilities in Chapter 7 for the model of a plane gravitating plasma slab where inhomogeneity is restricted to the vertical direction. For the description of laboratory and astrophysical plasma dynamics, the concept of *magnetic flux tubes* is quite central, as we have seen in Chapter 8. This automatically leads to the consideration of cylindrical plasmas where the inhomogeneities are operating in the radial direction. Whereas the model remains one-dimensional, so that most of the analytical techniques developed in Chapter 7 remain valid, the introduction of *curvature of the magnetic field* brings in qualitatively different physical effects that significantly influence the dynamics of flux tubes. We will now neglect gravity since it plays no role in laboratory plasmas and, for astrophysical plasmas, it is more adequately incorporated in an axi-symmetric model with a central gravitating object. The latter requires a two-dimensional model, which has to be relegated to the more advanced chapters. We will see that curvature of the magnetic field enters the equations in a very similar way to gravity in the plasma slab of Chapter 7.

9.1.1 Diffuse plasmas

For the study of confined plasmas, the diffuse cylindrical plasma column (called 'diffuse linear pinch' in the older plasma literature) is one of the most useful models. It is probably the most widely studied model in plasma stability theory. Since we have obtained a basic understanding of the spectrum of inhomogeneous one-dimensional systems, the analysis of the diffuse linear pinch can now be undertaken with more fruit than was possible in the early days of fusion research when this configuration was first investigated. Also, we will consider this configuration as a first approximation to toroidal systems, where the addition of a second

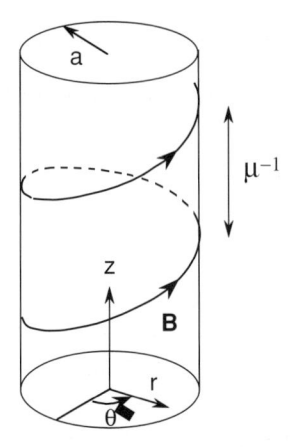

Fig. 9.1. Diffuse cylindrical plasma column with a helical magnetic field **B** with inverse pitch $\mu = \mu(r)$, drawn here at the wall radius $r = a$.

direction of inhomogeneity leads to partial differential equations and, therefore, to substantial complications of the analysis. For those systems, the construction of a coherent picture of the spectrum of waves and instabilities is a very demanding task, which is still far from completion.

Except for historical reasons in laboratory fusion research, where diffuse linear pinches were originally considered to be quite promising because of their high temperatures and high β-values (we have already indicated in Chapters 1 and 2 that the crucial issue of long duration stable confinement was somewhat underestimated in those days), these configurations are of intrinsic interest for the study of solar and astrophysical flux tubes and, recently, in the use of capillary discharges for laser wake-field acceleration. For those applications, toroidicity is not a prime factor, but end effects may not be negligible. Again, those effects make the model two-dimensional so that they have to await the advanced theory of Volume 2. However, it makes no sense to dwell on that before we have a firm grasp on the one-dimensional theory.

Consider a diffuse plasma in an infinite cylinder of radius a (Fig. 9.1). In cylindrical r, θ, z-coordinates, with rotational symmetry in θ and translational symmetry in z, the equilibrium equations,

$$\mathbf{j} \times \mathbf{B} = \nabla p, \qquad \mathbf{j} = \nabla \times \mathbf{B}, \qquad \nabla \cdot \mathbf{B} = 0, \tag{9.1}$$

reduce to

$$p' = j_\theta B_z - j_z B_\theta, \qquad j_\theta = -B_z', \qquad j_z = \frac{1}{r}(r B_\theta)', \tag{9.2}$$

where the prime denotes derivatives with respect to r. Eliminating j_θ and j_z, the equilibrium turns out to be characterized by the pressure profile $p(r)$ and the

9.1 Equilibrium of cylindrical plasmas

magnetic field profiles $B_\theta(r)$ and $B_z(r)$, subject to just one differential equation:

$$\left[p(r) + \tfrac{1}{2}B^2(r) \right]' + \frac{B_\theta^2(r)}{r} = 0. \tag{9.3}$$

Hence, we may choose two of these three profiles arbitrarily, whereas the density profile $\rho(r)$ may be chosen arbitrarily as well since it does not appear in the equilibrium equations when gravity is neglected. Special cases of such diffuse cylindrical equilibria are the z-pinch and θ-pinch configurations introduced in Section 1.2.3 (Fig. 1.4) and Sections 2.4.3 (Fig. 2.8) and 8.4.2 (Fig. 8.23).

We here assume a model I plasma (Section 4.6.1), where the magnetic field is not considered beyond the radius $r = a$. For the equilibrium, this implies the presence of a rigid wall absorbing the mechanical forces (laboratory plasmas) or anything else that justifies the assumption of a radially confined plasma. For the perturbations, this implies that the flows have to be tangential at that radius.

Instead of the magnetic field profiles $B_\theta(r)$ and $B_z(r)$, it is expedient to introduce a function that describes the radial variation of the helicity of the magnetic field. For an infinite cylinder, the most appropriate choice is the *inverse pitch of the magnetic field lines*:

$$\mu(r) \equiv \frac{B_\theta(r)}{r B_z(r)}. \tag{9.4}$$

Another useful variable is the *kinetic pressure contained versus the magnetic pressure on axis*:

$$\beta(r) \equiv \frac{2p(r)}{B_0^2}. \tag{9.5}$$

The diffuse cylindrical equilibrium is now completely determined by, for example, prescribing the functions $p(r)$ (or $\beta(r)$) and $B_\theta(r)$ (or $\mu(r)$), whereas $B_z(r)$ follows from the solution of the equilibrium equation (9.3) and $\rho(r)$ is arbitrary.

▷ **Dimensionless scaling of the equilibrium.** It is important (e.g. for numerical applications) to construct the *smallest set of parameters* that characterize these equilibria. To that end, as before, we exploit scale independence (Section 4.1.2) to eliminate the three *trivial* parameters

$$a, \quad B_0 \, (\equiv B_{z0}), \quad \rho_0 \tag{9.6}$$

by normalizing everything with respect to those scales of lengths, field strengths and densities (i.e. Alfvén speeds). Here, we exploit the subscript 0 to indicate function values on axis ($r = 0$). On the other hand, the two *essential* parameters

$$\beta_0 \equiv \beta(r=0) = \frac{2p_0}{B_0^2}, \quad \bar\mu_0 \equiv a\mu(r=0) = \frac{aj_{z0}}{2B_0}, \tag{9.7}$$

fix the magnitude of the plasma pressure contained and the inverse pitch of the magnetic field lines ($\sim$ the current density) on axis. Having thus fixed the amplitudes, all physical functions are distributed on the unit plasma interval $0 \leq \bar{r} \equiv r/a \leq 1$ according to *shape functions*:

$$\bar{p}(\bar{r}) \equiv \frac{1}{p_0} p(r), \qquad \bar{\mu}(\bar{r}) \equiv a\,\mu(r), \qquad \bar{\rho}(\bar{r}) \equiv \frac{1}{\rho_0} \rho(r), \tag{9.8}$$

$$\stackrel{(9.3)}{\Longrightarrow} \bar{B}_z(\bar{r}) \equiv \frac{1}{B_0} B_z(r), \qquad \stackrel{(9.4)}{\Longrightarrow} \bar{B}_\theta(\bar{r}) \equiv \bar{\mu}(\bar{r})\,\bar{r}\,\bar{B}_z(\bar{r}).$$

In conclusion: the equilibrium is determined by choosing the two parameter values β_0 and $\bar{\mu}_0$, and the (infinitely many parameters of the) shape functions $\bar{p}(\bar{r})$ and $\bar{\mu}(\bar{r})$. In the dynamics of the perturbations, the additional shape function $\bar{\rho}(\bar{r})$ enters. ◁

(a) Force-free magnetic fields An interesting class of cylindrical equilibria is obtained for very low β when pressure gradients can be completely neglected so that the magnetic field becomes 'force-free' (i.e. in the interior; there will be significant forces exerted on the wall). The condition for force-free magnetic fields is

$$\mathbf{j} = \alpha\,\mathbf{B}, \tag{9.9}$$

where the function $\alpha(r)$ is completely free. From this expression and the components of $\mathbf{j} = \nabla \times \mathbf{B}$, the relationship between α and μ is obtained:

$$\left. \begin{array}{l} j_\theta = -B'_z = \alpha B_\theta \\[4pt] j_z = \frac{1}{r}(r B_\theta)' = \alpha B_z \end{array} \right\} \quad \Rightarrow \quad \alpha = \frac{2\mu + \mu' r}{1 + \mu^2 r^2}. \tag{9.10}$$

For the choice of constant pitch, $\mu' = 0$, one obtains from these first order differential equations the following explicit solutions:

$$B_z(r) = \frac{B_0}{1 + \mu^2 r^2}, \qquad B_\theta(r) = \frac{B_0 \mu r}{1 + \mu^2 r^2}. \tag{9.11}$$

For the special choice of constant α, another one-parameter family of equilibrium solutions is obtained:

$$B_z(r) = B_0 J_0(\alpha r), \qquad B_\theta(r) = B_0 J_1(\alpha r), \tag{9.12}$$

where J_0 and J_1 are the zeroth and first order Bessel functions. These force-free magnetic field solutions were already discussed in Section 4.3.4, in the context of magnetic helicity (Fig. 4.9). Because of their apparent simplicity, they have been the subject of numerous investigations with respect to stability and slow dissipative dynamics. On axis, the inverse pitch μ is simply related to α through $\mu_0 = \frac{1}{2}\alpha$, but away from the axis, μ varies from $+\infty$ to $-\infty$ when αr progresses through the various zeros of the Bessel function J_0: an entirely non-trivial class of current-carrying equilibria.

9.1 Equilibrium of cylindrical plasmas

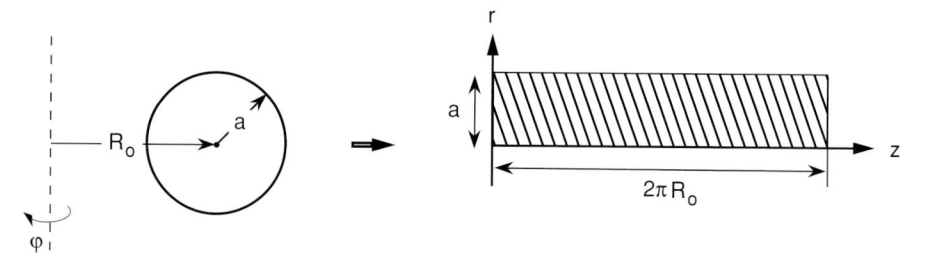

Fig. 9.2. Slender torus with inverse aspect ratio $\epsilon \equiv a/R_0 \ll 1$ represented as a periodic cylinder with length $2\pi R_0$.

(b) 'Straight tokamak' limit A slender torus, with small *inverse aspect ratio* $\epsilon \equiv a/R_0 \ll 1$ (Fig. 9.2, left frame), may be approximated by a straight cylinder of finite length $L = 2\pi R_0$ (Fig. 9.2, right frame). This becomes a mathematical torus when the ends are identified. Thus, a first approximation of toroidal equilibrium is obtained. Some of the toroidal dynamical effects are well represented in this manner (e.g. the fact that the wave number has to be quantized in the toroidal direction). By consistently developing all physical quantities to the relevant order in ϵ, the dynamics of these equilibria may be computed to leading order. A meaningful choice of the order of magnitude of the two essential parameters β_0 and μ_0 then becomes crucial.

In the periodic cylinder representation of tokamaks, called the 'straight tokamak' limit, the variable μ is replaced by a variable that measures the pitch of the field lines relative to the circumference of the torus (Fig. 9.3), i.e. the *'safety factor'*

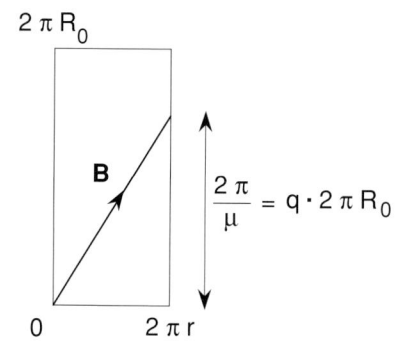

Fig. 9.3. Inverse pitch μ and safety factor q of the magnetic field lines in a 'straight tokamak' periodic cylinder model of a toroidal plasma.

(already introduced in Section 2.4.3, Fig. 2.12),

$$q(r) \equiv \frac{rB_z(r)}{R_0 B_\theta(r)} \equiv \frac{1}{\mu(r)R_0} \left[\equiv \frac{\epsilon}{\mu(r)a} \right]. \qquad (9.13)$$

In the so-called *low-beta tokamak regime*, the order of magnitude of the two essential parameters is chosen as

$$\beta_0 \sim \epsilon^2 \ll 1, \qquad q_0 \sim 1. \qquad (9.14)$$

In the cylindrical 'straight tokamak' approximation, the normalized q-profile,

$$\bar{q}(\bar{r}) \equiv q(r)/q_0, \qquad (9.15)$$

then becomes the only shape function entering the leading order expressions of the equilibrium and the perturbations. The reason is that, because β is small, pressure effects, leading to an outward shift of the magnetic axis in a torus (the Shafranov shift), only enter in a higher order where poloidal θ-variations of the equilibrium through toroidal curvature are permitted. The proper treatment of the latter toroidal effects requires the two-dimensional theory developed in Volume 2.

9.1.2 Interface plasmas

In laboratory fusion research, the plasma is usually isolated from the wall by means of a region of rather cold plasma with low pressure and small current density. Such a plasma configuration can be idealized by means of either one of the two interface models II or II* introduced in Section 4.6.1. Here, we will assume that the outer plasma is pressureless and carries no current, so that there is no difference with respect to the equilibrium between model II and model II*: the outer region is characterized by a *vacuum magnetic field* configuration. With respect to the perturbations, model II and model II* interface plasmas may behave very differently, though, even with identical equilibrium fields, since a plasma (model II*) allows for induction of perturbed currents whereas a vacuum (model II) does not. We will see in Section 9.2.2 that this leads to major differences in the stability properties.

We consider cylindrical model II and model II* configurations as sketched in Fig. 9.4: a diffuse plasma in an infinite cylinder of radius a is surrounded by a vacuum magnetic field $\hat{\mathbf{B}}$, enclosed by a perfectly conducting wall at $r = b$. In the plasma region $0 \leq r \leq a$ the equilibrium is described by the functions $p(r)$, $B_\theta(r)$, $B_z(r)$ (and $\rho(r)$), that satisfy the equilibrium equation (9.3). At the plasma surface $r = a$, surface currents produce jumps in the variables p, B_θ, and B_z which are restricted to satisfy pressure balance:

$$p_1 + \tfrac{1}{2}(B_{\theta 1}^2 + B_{z1}^2) = \tfrac{1}{2}(\hat{B}_{\theta 1}^2 + \hat{B}_{z1}^2), \qquad (9.16)$$

9.1 Equilibrium of cylindrical plasmas

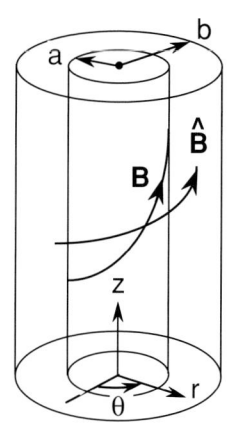

Fig. 9.4. Cylindrical interface model with diffuse inner plasma with a helical magnetic field **B**, surface currents at $r = a$, and surrounded by a vacuum magnetic field $\hat{\mathbf{B}}$.

where the subscript 1 indicates equilibrium values at the plasma surface. The surface currents are given by

$$j_\theta^\star = -[\![B_z]\!] \equiv -\hat{B}_{z1} + B_{z1}, \qquad j_z^\star = [\![B_\theta]\!] \equiv \hat{B}_{\theta 1} - B_{\theta 1}. \qquad (9.17)$$

The outer vacuum magnetic field in the region $a < r \leq b$ is given by

$$\hat{B}_z(r) = \hat{B}_{z1}, \qquad \hat{B}_\theta(r) = \frac{\hat{\mu}_1 a^2}{r} \hat{B}_{z1}. \qquad (9.18)$$

Assuming the interior plasma to be prescribed as in the previous section, the pressure jump condition (9.16) fixes the magnitude of the vacuum magnetic field $\hat{\mathbf{B}}$ but leaves its direction free. Hence, the interface model with an outer vacuum magnetic field adds two parameters to the problem: the pitch of the outer magnetic field, $\hat{\mu}_1 \neq \mu_1$, and the relative wall position b/a.

A special case is the early sharp-boundary model of pinch discharges, investigated by Kruskal and Schwarzschild [130], Kruskal and Tuck [131], Rosenbluth [197] and Tayler [226]. The inner plasma was taken to be homogeneous, with $B_\theta = 0$, $B_z = B_0$, $p = p_0 = p_1$, so that the only relevant parameter for the inner plasma is $\beta \equiv 2p/B_0^2$, which was taken to be high since dreams of confining plasma by separating it completely from the magnetic field were still alive. The current is then exclusively confined to the surface $r = a$, and the vacuum magnetic field there is determined from Eq. (9.16) by the two parameters β

and $\hat{\mu}_1$:

$$\hat{B}_{\theta 1} = \hat{\mu}_1 a \cdot \hat{B}_{z1}, \qquad 1 + \beta = (1 + \hat{\mu}_1^2 a^2) \frac{\hat{B}_{z1}^2}{B_0^2}. \qquad (9.19)$$

This model successfully described the threat posed by external kink modes to high-β confinement (see Section 9.3.2).

9.2 MHD wave equation for cylindrical plasmas

9.2.1 Derivation of the MHD wave equation for a cylinder

We derive the equation of motion for cylindrical equilibria along the same lines as for the gravitating slab, given in Section 7.3.2. Our starting point is the equation of motion

$$\mathbf{F}(\boldsymbol{\xi}) \equiv -\nabla \pi - \mathbf{B} \times (\nabla \times \mathbf{Q}) + (\nabla \times \mathbf{B}) \times \mathbf{Q} = \rho \frac{\partial^2 \boldsymbol{\xi}}{\partial t^2}, \qquad (9.20)$$

with the usual abbreviations

$$\pi \equiv -\gamma p \nabla \cdot \boldsymbol{\xi} - \boldsymbol{\xi} \cdot \nabla p, \qquad \mathbf{Q} \equiv \nabla \times (\boldsymbol{\xi} \times \mathbf{B}). \qquad (9.21)$$

Because of the symmetry, we may study normal mode solutions of the form

$$\boldsymbol{\xi}(r, \theta, z, t) = \left(\xi_{r,mk}(r), \xi_{\theta,mk}(r), \xi_{z,mk}(r) \right) e^{i(m\theta + kz - \omega t)}, \qquad (9.22)$$

where the subscripts m and k will again be dropped in the following. For these separate modes the equation of motion may be reduced to an ordinary second order differential equation in terms of the component $\xi_r(r)$.

As in the analysis of the plasma slab, we exploit a projection based on the magnetic field lines with unit vectors

$$\mathbf{e}_r, \qquad \mathbf{e}_\perp \equiv (0, B_z, -B_\theta)/B, \qquad \mathbf{e}_\parallel \equiv (0, B_\theta, B_z)/B. \qquad (9.23)$$

In this projection, the result of the gradient operator applied to a perturbed quantity as given in Eq. (9.22) may be written as

$$\nabla = \mathbf{e}_r \, \partial_r + i\mathbf{e}_\perp g + i\mathbf{e}_\parallel f, \qquad (9.24)$$

where the perpendicular and parallel gradient operators become algebraic multipliers:

$$g \equiv \frac{1}{B}(mB_z/r - kB_\theta) = \frac{G}{B}, \qquad f \equiv \frac{1}{B}(mB_\theta/r + kB_z) = \frac{F}{B}. \qquad (9.25)$$

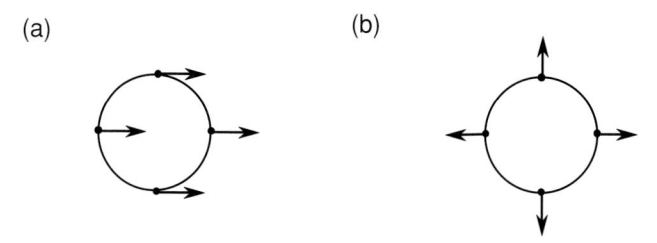

Fig. 9.5. Different vectors: (a) $\boldsymbol{\xi}_{mk}(r)$ and (b) $\boldsymbol{\xi}_{mk}(r,\theta)$.

The use of the symbols G and F instead of g and f will prove more convenient later on in the analysis (starting with Eq. (9.29)).

▷ **Pitfalls.** (1) One should not denote the vector in large brackets on the RHS of Eq. (9.22) as $\boldsymbol{\xi}_{mk}(r)$. On a circle $r = \text{const}$, this would indicate a vector of constant amplitude and direction (as, e.g., shown in Fig. 9.5(a)), which is not meant here. The correct notation is $\boldsymbol{\xi}_{mk}(r,\theta)$ (Fig. 9.5(b)). This incorporates the θ-dependence of the unit vectors: $\partial \boldsymbol{\xi}_{mk}(r,\theta)/\partial \theta \neq 0$.
(2) The representation (9.24) for the gradient operator should not be considered as a recipe to be applied blindly (if at all), but just as a kind of short-hand notation for the expressions obtained after the conversion to cylindrical coordinates has been carried out by means of Appendix A.2.2. Recall that in the analogous projection (7.80) for the plane slab with shear, this representation of the gradient operator could be used also for computing divergences and curls if one properly accounted for the dependence of the unit vectors $\mathbf{e}_\perp$ and $\mathbf{e}_\parallel$ on the normal coordinate x. Here, the situation is basically different since the cylindrical coordinate system has a scale factor h_2 $(= r)$, and the unit vectors $\mathbf{e}_r$ and $\mathbf{e}_\theta$ depend also on the ignorable coordinate θ: $\partial \mathbf{e}_r / \partial \theta = \mathbf{e}_\theta$, $\partial \mathbf{e}_\theta / \partial \theta = -\mathbf{e}_r$. ◁

The projection of the displacement vector on the field line triad is denoted by

$$\xi \equiv \mathbf{e}_r \cdot \boldsymbol{\xi} = \xi_r\,,$$
$$\eta \equiv i\mathbf{e}_\perp \cdot \boldsymbol{\xi} = i(B_z \xi_\theta - B_\theta \xi_z)/B\,, \qquad (9.26)$$
$$\zeta \equiv i\mathbf{e}_\parallel \cdot \boldsymbol{\xi} = i(B_\theta \xi_\theta + B_z \xi_z)/B\,.$$

In terms of these variables we obtain

$$\mathbf{Q} = ifB\xi\,\mathbf{e}_r - [(B_\theta \xi)' - kB\eta]\,\mathbf{e}_\theta - \frac{1}{r}[(rB_z\xi)' + mB\eta]\,\mathbf{e}_z\,,$$

$$\pi = -p'\xi - \gamma p \nabla \cdot \boldsymbol{\xi}\,, \qquad \nabla \cdot \boldsymbol{\xi} = \frac{1}{r}(r\xi)' + g\eta + f\zeta\,, \qquad (9.27)$$

where the factor $\exp[i(m\theta + kz - \omega t)]$ is dropped for notational simplicity. By means of these expressions, and the equilibrium relation (9.3), the equation of motion (9.20) can be evaluated in the same manner as for the plane gravitating slab in Section 7.3.2. This yields the following matrix formulation of the spectral problem:

$$\left(\begin{array}{ccc} \dfrac{d}{dr}\dfrac{\gamma p + B^2}{r}\dfrac{d}{dr}r - f^2 B^2 - r\left(\dfrac{B_\theta^2}{r^2}\right)' & \dfrac{d}{dr}g(\gamma p + B^2) - \dfrac{2k B_\theta B}{r} & \dfrac{d}{dr}f\gamma p \\[6pt] -\dfrac{g(\gamma p + B^2)}{r}\dfrac{d}{dr}r - \dfrac{2k B_\theta B}{r} & -g^2(\gamma p + B^2) - f^2 B^2 & -fg\gamma p \\[6pt] -\dfrac{f\gamma p}{r}\dfrac{d}{dr}r & -fg\gamma p & -f^2\gamma p \end{array} \right) \left(\begin{array}{c} \xi \\ \eta \\ \zeta \end{array} \right)$$

$$= -\rho\omega^2 \left(\begin{array}{c} \xi \\ \eta \\ \zeta \end{array} \right). \qquad (9.28)$$

This formulation is symmetric, apart from the occurrence of some factors r. These could be absorbed by exploiting the normal variable $r\xi$ instead of ξ, which would make the dimensions of the different matrix elements unequal.

Note the similarity of the matrices (9.28), for the cylindrical plasma, and (7.89), for the gravitating slab, where now off-diagonal curvature terms (with B_θ) appear instead of gravitational ones. Since these terms do not involve derivatives of ξ, they do not affect the essential spectrum associated with the singularities. Of course, the discrete spectrum and, hence, stability is quite significantly affected by these terms.

(a) Generalized Hain–Lüst equation The typical structure of Eq. (9.28), with lower order differential equations for the tangential components η and ζ, allows us again to reduce the system to a single second order differential equation by expressing the tangential components in terms of the radial variable $\chi \equiv r\xi$:

$$\eta = \dfrac{G\left[(\gamma p + B^2)\rho\omega^2 - \gamma p F^2\right] r\chi' + 2k B_\theta(B^2\rho\omega^2 - \gamma p F^2)\chi}{r^2 B D},$$

$$\zeta = \dfrac{\gamma p F\left[(\rho\omega^2 - F^2) r\chi' + 2k B_\theta G \chi\right]}{r^2 B D}, \qquad (9.29)$$

where

$$D \equiv \rho^2\omega^4 - (m^2/r^2 + k^2)(\gamma p + B^2)\rho\omega^2 + (m^2/r^2 + k^2)\gamma p F^2. \qquad (9.30)$$

9.2 MHD wave equation for cylindrical plasmas

Substituting these expressions into the first component of Eq. (9.28) yields the *generalized Hain–Lüst equation*:[1]

$$\frac{d}{dr}\left[\frac{N}{rD}\frac{d\chi}{dr}\right] + \left[\frac{1}{r}(\rho\omega^2 - F^2) - \left(\frac{B_\theta^2}{r^2}\right)' - \frac{4k^2 B_\theta^2}{r^3 D}(B^2\rho\omega^2 - \gamma p F^2)\right.$$

$$\left. + \left\{\frac{2k B_\theta G}{r^2 D}\left((\gamma p + B^2)\rho\omega^2 - \gamma p F^2\right)\right\}'\right]\chi = 0, \qquad (9.31)$$

where

$$N \equiv (\rho\omega^2 - F^2)\left((\gamma p + B^2)\rho\omega^2 - \gamma p F^2\right). \qquad (9.32)$$

Comparing this equation with the corresponding equation (7.91) for the plane slab, it is clear that the terms caused by the curvature of the poloidal field B_θ play a similar role as the gravitational terms $\rho \hat{g}$ (although one cannot simply translate one formulation into the other). These terms disappear when $B_\theta = 0$ (θ-pinch) and we obtain a problem of (almost) equal complication to a plane slab in the absence of gravity. Since the latter is stable, it follows that the linear θ-pinch is also stable.

When the wall is at the plasma (model I), the appropriate boundary conditions are:

$$\chi(0) = \chi(a) = 0. \qquad (9.33)$$

In the presence of an external vacuum (model II) or pressureless plasma (model II*), the boundary condition at $r = a$ becomes a rather complicated expression. It is derived in Section 9.2.2.

The boundary condition at $r = 0$ deserves extra stress since the wave equation has a *singularity* there which is *due to the cylindrical geometry*. For $m \neq 0$, the Frobenius expansion (see Section 7.4.1) around the origin yields an indicial equation $\nu^2 - m^2 = 0$, so that there is a small solution $\chi_1 \sim r^{|m|}$ and a singular solution with a logarithm that is excluded by the boundary condition (9.33). For $m = 0$, the indicial equation is $\nu(\nu - 2) = 0$, so that the small solution behaves like $\chi_1 \sim r^2$ and the solution with the logarithm is again excluded by the boundary condition. Although the boundary condition (9.33) on χ conveniently combines the different possibilities, it is useful to realize the different meaning of it for the different modes when expressed in terms of the physical variable ξ. (In the limit $r \to 0$, Fig. 9.5(a) illustrates the behaviour of ξ for an $m = 1$ mode, and Fig. 9.5(b) for an $m = 0$ mode.) Using ξ, one has to distinguish between the $|m| = 1$ modes, which have a finite value of ξ on axis (corresponding to the important feature of a finite displacement of the plasma column on axis by these modes) and all other modes

---

[1] The derivation by Hain and Lüst [102] was unnecessarily restricted to isothermal plasmas ($\gamma = 1$). This restriction was lifted in the derivation by Goedbloed [81](II), which is followed here.

which have $\xi(0) = 0$. Consequently, when using ξ, the appropriate boundary conditions become:

$$\xi'(0) = 0 \quad \text{for } |m| = 1, \qquad \xi(0) = 0 \quad \text{for } |m| \neq 1. \tag{9.34}$$

Clearly, the geometrical singularity at $r = 0$ is of a completely different nature from the physical singularities associated with the continuous spectra.

(b) Singularities For the purpose of the analysis, we again abbreviate the differential equation (9.31) as

$$\left[P(r; \omega^2) \chi' \right]' - Q(r; \omega^2) \chi = 0, \tag{9.35}$$

where

$$P \equiv \frac{N}{rD}, \quad N(r; \omega^2) \equiv \rho^2 (\gamma p + B^2) [\omega^2 - \omega_A^2(r)][\omega^2 - \omega_S^2(r)],$$
$$D(r; \omega^2) \equiv \rho^2 [\omega^2 - \omega_{s0}^2(r)][\omega^2 - \omega_{f0}^2(r)], \tag{9.36}$$

and $-Q$ denotes the second term in square brackets of Eq. (9.31). The expressions for the singular frequencies ω_A^2, ω_S^2, ω_{s0}^2 and ω_{f0}^2 are given by

$$\omega_A^2(r) \equiv F^2/\rho, \qquad \omega_S^2(r) \equiv \frac{\gamma p}{\gamma p + B^2} F^2/\rho,$$

$$\omega_{s0, f0}^2(r) \equiv \tfrac{1}{2}(m^2/r^2 + k^2) \frac{\gamma p + B^2}{\rho} \left[1 \pm \sqrt{1 - \frac{4 \gamma p F^2}{(m^2/r^2 + k^2)(\gamma p + B^2)^2}} \right]. \tag{9.37}$$

They are completely analogous, almost identical, to the expressions Eqs. (7.94)–(7.96) for the plane gravitating plasma slab (except that the square of the horizontal wave number, $k_0^2 = k_y^2 + k_z^2$, is replaced by the expression $m^2/r^2 + k^2$, which is not constant and suffers from the geometrical singularity at $r = 0$ discussed above). Consequently, we may refer to the analysis of Section 7.4 and conclude that the diffuse cylindrical plasma also has *two continuous spectra*, the Alfvén continuum $\{\omega_A^2(r)\}$ and the slow continuum $\{\omega_S^2(r)\}$. Furthermore, the sets $\{\omega_{s0}^2(r)\}$ and $\{\omega_{f0}^2(r)\}$ consist of apparent singularities, which are not continuous spectra but ranges of *turning point frequencies*.

For every radius r, these genuinely and apparently singular frequencies are well ordered according to the scheme

$$0 \leq \omega_S^2 \leq \omega_{s0}^2 \leq \omega_A^2 \leq \omega_{f0}^2 \leq \omega_F^2 \equiv \infty; \tag{9.38}$$

see Fig. 9.6. The collections of these frequencies for the whole interval (0, 1) may be represented by a diagram similar to Fig. 7.14, with one important difference: no

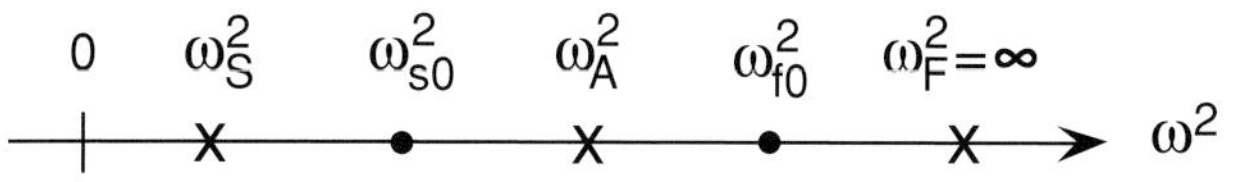

9.2 MHD wave equation for cylindrical plasmas

$$0 \quad \omega_S^2 \quad \omega_{s0}^2 \quad \omega_A^2 \quad \omega_{f0}^2 \quad \omega_F^2 = \infty$$

Fig. 9.6. Ordering of the genuine and apparent frequencies for fixed radius.

matter how small the inhomogeneity, the geometrical singularity at $r = 0$ causes overlap between the slow turning point frequencies and the slow continua and between the fast turning point frequencies and the (formal) fast continuum because $\omega_{s0}^2(r \to 0) \to \omega_S^2(0)$ and $\omega_{f0}^2(r \to 0) \to \omega_F^2 \equiv \infty$.

(c) Equivalent system of first order differential equations As in Section 7.3.2(d), we transform the second order differential equation (9.31) into a system of two first order equations. This turns out to be quite illuminating. Rather than just rewriting the equation in terms of the variables χ and χ', we use a variable with physical significance, viz. the perturbation Π of the total pressure $p + \tfrac{1}{2}B^2$:

$$\Pi = \pi + \mathbf{B} \cdot \mathbf{Q}. \tag{9.39}$$

(This is the Eulerian pressure perturbation, Π_E, related to the Lagrangian pressure perturbation, Π_L, by $\Pi_E = \Pi_L + B_\theta^2 \chi / r^2$.) Inserting the expressions (9.27) and (9.29) for $\mathbf{Q}$ and π gives

$$\Pi = -\frac{N}{rD}\chi' + \left\{ \frac{2B_\theta^2}{r^2} - \frac{2kB_\theta G}{r^2 D}\left[(\gamma p + B^2)\rho\omega^2 - \gamma p F^2\right] \right\}\chi. \tag{9.40}$$

Notice that all terms with radial derivatives that occur in the Hain–Lüst equation (9.31) also appear in the expression for Π, apart from the factor 2 in front of B_θ^2/r^2 which is due to the fact that we exploit the Eulerian rather than the Lagrangian pressure.

By straightforward algebra, the Hain–Lüst equation is then transformed into the following pair of first order differential equations:

$$\frac{N}{r}\begin{pmatrix}\chi\\ \Pi\end{pmatrix}' + \begin{pmatrix}C & D\\ E & -C\end{pmatrix}\begin{pmatrix}\chi\\ \Pi\end{pmatrix} = 0, \tag{9.41}$$

where

$$C \equiv -\frac{2B_\theta^2}{r^2}\rho^2\omega^4 + \frac{2mB_\theta F}{r^3}\left[(\gamma p + B^2)\rho\omega^2 - \gamma p F^2\right],$$

$$E \equiv -\frac{N}{r}\left[\frac{\rho\omega^2 - F^2}{r} + \left(\frac{B_\theta^2}{r^2}\right)'\right] - \frac{4B_\theta^4}{r^4}\rho^2\omega^4$$

$$+ \frac{4B_\theta^2 F^2}{r^4}\left[(\gamma p + B^2)\rho\omega^2 - \gamma p F^2\right], \tag{9.42}$$

and N and D were defined in Eqs. (9.32) and (9.30). The determinant of the matrix,

$$DE + C^2 = -\frac{N}{r}\left[D\left\{ U + 2\left(\frac{B_\theta^2}{r^2}\right)' \right\} + V \right] \to 0 \quad \text{when } N \to 0, \quad (9.43)$$

exhibits the proportionality with N required to cancel one of the two factors introduced by multiplying both derivatives by N.

This formulation, which is due to Appert, Gruber and Vaclavik [9], again shows that the slow and Alfvén continua $\{\omega_S^2\}$ and $\{\omega_A^2\}$ originate from the zeros of the factor N in front of the derivatives. The real virtue of this formulation, over that in terms of the second order differential equation, is that the singularities $D = 0$ are immediately seen to be apparent ones since nothing singular shows up there. In the numerical problem of solving Eq. (9.41) by means of a shooting method, one multiplies the equation by r/N and proceeds to calculate the derivatives. Giving initial data χ_0 and Π_0 at a certain point, one then calculates χ_0' and Π_0', from which one obtains new initial data χ_1 and Π_1, and so forth. Clearly, the only difficulty which may arise is the occurrence of $N = 0$ singularities. For $D = 0$ no problem turns up. This is much less evident in the Hain–Lüst formulation.

▷ **Apparent singularities.** In the formulation in terms of the second order differential equation, one has to prove that the expansion about the $D = 0$ locations obeys the special condition (7.143) of Section 7.4.1. To that end, the function $Q(r; \omega^2)$ of Eq. (9.35) is again expressed as

$$Q(r; \omega^2) = -U - \frac{V}{D} - \left(\frac{W}{D}\right)', \quad (9.44)$$

where the explicit expressions for U, V and W can be read off from Eq. (9.31). After straightforward algebra, these coefficients turn out to be related by

$$W^2 + \frac{1}{r}NV = -\frac{4k^2 B_\theta^2 B^2}{r^4}\left[(\gamma p + B^2)\rho\omega^2 - \gamma p F^2 \right] D \to 0 \quad \text{when } D \to 0, \quad (9.45)$$

which is the required condition. ◁

(d) Limiting forms of the generalized Hain–Lüst equation For reference purposes we list three significant limits of the Hain–Lüst equation.

(1) *Low frequency limit* $[\,|\rho\omega^2| \ll (m^2/r^2 + k^2)(\gamma p + B^2)\,]$:

$$\frac{d}{dr}\left[\frac{\rho\omega^2 - F^2}{m^2 + k^2 r^2} r \frac{d\chi}{dr} \right] - \frac{1}{r}\left[\rho\omega^2 - F^2 \right.$$
$$- \frac{4k^2 B_\theta^2}{m^2 + k^2 r^2} \frac{\gamma p \rho\omega^2}{(\gamma p + B^2)\rho\omega^2 - \gamma p F^2} - \frac{2B_\theta^2}{rB^2} p'$$
$$+ \frac{4B_\theta F}{m^2 + k^2 r^2}\left(\frac{k^2 m r}{m^2 + k^2 r^2} + \frac{m}{B^2} p' + \frac{B_\theta}{B^2} F - \frac{rB_z G}{2B^2}\frac{\mu'}{\mu} \right) \right] \chi = 0. \quad (9.46)$$

This equation was obtained by Goedbloed and Hagebeuk [86] by assuming small frequencies or growth rates, and systematically expressing all derivatives B'_θ and B'_z in terms of p' and μ' by means of the equilibrium relation (9.3) and the definition $\mu \equiv B_\theta/(rB_z)$. Although the number of terms is increased this way, this form has the advantage that the important terms involved in instabilities (p', F and μ') are clearly distinguished so that different orderings may be designed to optimize their effect. This equation still contains the exact limits of the incompressible and marginal equations (see below).

(2) *Incompressible limit* ($\gamma \to \infty$):

$$\frac{d}{dr}\left[\frac{\rho\omega^2 - F^2}{m^2 + k^2 r^2} r \frac{d\chi}{dr}\right] - \left[\frac{1}{r}(\rho\omega^2 - F^2) - \left(\frac{B_\theta^2}{r^2}\right)' \right.$$

$$\left. - \frac{4k^2 B_\theta^2 F^2}{r(m^2 + k^2 r^2)(\rho\omega^2 - F^2)} - \left(\frac{2k B_\theta G}{m^2 + k^2 r^2}\right)'\right]\chi = 0.$$

(9.47)

This equation was derived by Freidberg [71] (with a minor rearrangement of the derivative terms). It contains the exact marginal stability equation, listed below, in the limit $\omega^2 \to 0$. This shows that stability is not affected by compressibility, although growth rates are. (Note that the singularity $(\rho\omega^2 - F^2)^{-1}$ in the second term in square brackets is an apparent one since it originates from the factor $\omega^2 - \omega_{s0}^2$ of D.)

(3) *Marginal equation of motion* ($\rho\omega^2 = 0$):

$$\frac{d}{dr}\left[\frac{rF^2}{m^2 + k^2 r^2}\frac{d\chi}{dr}\right] - \left[\frac{1}{r}F^2 + \left(\frac{B_\theta^2}{r^2}\right)' \right.$$

$$\left. - \frac{4k^2 B_\theta^2}{r(m^2 + k^2 r^2)} + \left(\frac{2k B_\theta G}{m^2 + k^2 r^2}\right)'\right]\chi = 0.$$

(9.48)

This is one form of Newcomb's [164] Euler–Lagrange equation describing the stability of cylindrical plasmas (see Section 9.4.1). With respect to the singularity $F = 0$, the equation is analogous to the marginal equation of motion (7.190) for the gravitating slab, so that singular behaviour can be discussed in complete analogy with Section 7.5.

9.2.2 Boundary conditions for cylindrical interfaces

If there is an external vacuum (model II), or pressureless plasma (model II*), surrounding the central plasma column, the right (model I) boundary condition,

$\chi(a) = 0$, should be replaced by conditions determining the amplitude and the normal derivative of $\chi(a)$ describing the freely moving plasma surface. This problem turns up in the investigation of free-boundary modes (e.g. external kinks). For the cylindrical interface equilibrium of Section 9.1.2 (Fig. 9.4), the calculation of ξ (or χ) in the interior region should then be complemented with the appropriate 'extension' of the perturbation in the outer region. For model II, this involves the calculation of the perturbation $\hat{\mathbf{Q}}$ of the vacuum magnetic field plus the boundary conditions connecting $\hat{\mathbf{Q}}$ to ξ at the plasma–vacuum interface at $r = a$.

(a) Plasma–vacuum interface (model II) The appropriate boundary conditions were derived in Section 6.6.1, Eqs. (6.140) and (6.143), which we repeat for convenience:

$$\mathbf{n} \cdot \nabla \times (\xi \times \hat{\mathbf{B}}) = \mathbf{n} \cdot \hat{\mathbf{Q}} \qquad \text{(1st int. cond.)}, \qquad (9.49)$$

$$-\gamma p \nabla \cdot \xi + \mathbf{B} \cdot \mathbf{Q} + \xi \cdot \nabla(\tfrac{1}{2}B^2) = \hat{\mathbf{B}} \cdot \hat{\mathbf{Q}} + \xi \cdot \nabla(\tfrac{1}{2}\hat{B}^2) \quad \text{(2nd int. cond.)}. \quad (9.50)$$

The first boundary condition is easily transformed to

$$i\hat{F}\chi = r\hat{Q}_r \quad \text{(at } r = a\text{)}, \qquad \text{where} \quad \hat{F} = m\hat{B}_{\theta 1}/a + k\hat{B}_{z1}. \qquad (9.51)$$

The LHS of the second boundary condition is the Lagrangian perturbation of the total pressure, so that this condition may be transformed by means of Eq. (9.39) to:

$$\Pi - (B_\theta^2/r^2)\chi = \hat{B}_\theta \hat{Q}_\theta + \hat{B}_z \hat{Q}_z - (\hat{B}_\theta^2/r^2)\chi \qquad \text{(at } r = a\text{)}, \qquad (9.52)$$

with Π given by Eq. (9.40).

The equations (9.51) and (9.52) determine the plasma variables Π (or χ') and χ at the plasma surface completely if the vacuum solutions are known. This part of the problem can be carried out explicitly since the solutions in the vacuum are Bessel functions, as we will see. From the vacuum equations

$$\nabla \times \hat{\mathbf{Q}} = 0, \qquad \nabla \cdot \hat{\mathbf{Q}} = 0, \qquad (9.53)$$

we obtain the tangential components of $\hat{\mathbf{Q}}$ in terms of the radial component:

$$\hat{Q}_\theta = i\frac{m}{m^2 + k^2 r^2}(r\hat{Q}_r)', \qquad \hat{Q}_z = i\frac{kr}{m^2 + k^2 r^2}(r\hat{Q}_r)', \qquad (9.54)$$

so that

$$\hat{\mathbf{B}} \cdot \hat{\mathbf{Q}} = i\frac{r\hat{F}}{m^2 + k^2 r^2}(r\hat{Q}_r)'. \qquad (9.55)$$

9.2 MHD wave equation for cylindrical plasmas

The radial component satisfies the second order differential equation

$$\left[\frac{r}{m^2 + k^2 r^2}(r\hat{Q}_r)'\right]' - \hat{Q}_r = 0, \qquad (9.56)$$

which has derivatives of the modified Bessel functions as solutions:

$$\hat{Q}_r = C_1 I'_m(kr) + C_2 K'_m(kr). \qquad (9.57)$$

One of the constants is determined by the boundary condition (6.130) at the conducting wall,

$$\hat{Q}_r(b) = 0, \qquad (9.58)$$

so that the final solution for $\hat{Q}_r$ on the interval (a, b) becomes

$$\hat{Q}_r = C[\, I'_m(kb) K'_m(kr) - K'_m(kb) I'_m(kr) \,]. \qquad (9.59)$$

The constant C is eliminated by inserting this solution in Eq. (9.52) and dividing that equation by Eq. (9.51), which leads to a *single boundary condition*:

$$\left(\frac{\Pi}{\chi}\right)_{r=a} = \frac{B_\theta^2(a) - \hat{B}_\theta^2(a)}{a^2} - \frac{\hat{F}^2(a)}{ka} \frac{I_m(ka) K'_m(kb) - K_m(ka) I'_m(kb)}{I'_m(ka) K'_m(kb) - K'_m(ka) I'_m(kb)}. \qquad (9.60)$$

In this boundary condition, Π is to be expressed in terms of χ' and χ by means of Eq. (9.40), and χ' and χ, in turn, are found by solving the Hain–Lüst equation (9.31). This determines the free-boundary modes of model II.

The replacement of the two boundary conditions (9.51) and (9.52) by the single boundary condition (9.60) is possible because, for a homogeneous second order differential equation, the choice of the amplitude of the eigenfunction does not influence the eigenvalue. Equation (9.60) corresponds to normalizing the eigenfunctions with $\chi(a) = 1$. Obviously, if $\chi(a)$ happens to vanish one should not divide by it, but one should exploit a different normalization. This case corresponds to a situation where there is already an eigensolution in the absence of the vacuum, with the wall at the plasma ($b = a$).

For the numerical solution of the Hain–Lüst equation (9.31), or the equivalent system of first order differential equations (9.41), one may exploit a shooting method, as described in Section 7.5.1(c). One chooses a value of ω^2 and integrates into the outward direction, starting from $\chi = 0$ at $r = 0$. One then keeps changing ω^2 until $(\Pi/\chi)_{r=a}$ reaches the value prescribed by the RHS of Eq. (9.60). Then, ω^2 has become an eigenvalue. (In the absence of a vacuum (model I), one iterates until $\chi(r)$ goes through zero at $r = a$.) For this procedure to be useful, a guiding principle should exist on how to change the parameter ω^2 in such a way that the solution for the next try is closer to satisfying the boundary condition at $r = a$ than it was in the previous run. Such a principle is provided by the oscillation theorem, proved in Section 7.4.3 for the gravitating plasma slab, but equally valid for the cylindrical plasma (Section 9.3.1).

(b) Plasma–'ghost plasma' interface (model II*) We replace the vacuum by a pressureless plasma ($p_0 = 0$), carrying no current ($\hat{\mathbf{j}}_0 = 0$), with negligible density ($\hat{\rho}_0 = 0$), and having the same distribution of the magnetic field $\hat{\mathbf{B}}_0$ as the vacuum, given by Eq. (9.18). Let us call such a plasma a 'ghost plasma'. With respect to the equilibrium, there is evidently no difference with a vacuum. With respect to the perturbations, from Eqs. (6.19) and (6.21), the perturbed pressure and density also vanish: $\hat{p}_1 \equiv \hat{\pi} = 0$, $\hat{\rho}_1 = 0$. However, since the medium is assumed to be perfectly conducting, the magnetic field perturbation will be associated with the displacement $\hat{\boldsymbol{\xi}}$ of the 'ghost plasma' by $\hat{\mathbf{Q}} = \nabla \times (\hat{\boldsymbol{\xi}} \times \hat{\mathbf{B}})$, and there is nothing to prevent the development of a perturbed current density $\hat{\mathbf{j}}_1 = \nabla \times \hat{\mathbf{Q}}$. The question is: do those currents actually develop, and is a 'ghost plasma' any different from a vacuum with respect to stability?

Let us solve the pertinent limiting form of the Hain–Lüst equation for the 'ghost plasma'. Since we have assumed $\hat{\rho}_0 = 0$, that equation turns out to be identical with Newcomb's marginal equation of motion (9.48) (of course, with hats on all the physical variables). To facilitate the solution, we transform this equation to $\hat{Q}_r$. From Eq. (9.27),

$$r\hat{Q}_r = \mathrm{i}\hat{F}(r\hat{\xi}_r) \quad \Rightarrow \quad \hat{\chi} = -\mathrm{i}\hat{F}^{-1}(r\hat{Q}_r), \tag{9.61}$$

where $\hat{F}$ is the parallel gradient operator for a vacuum magnetic field distribution:

$$\hat{F} = (k + \hat{\mu}m)\hat{B}_{z1}, \qquad \hat{\mu} = \hat{\mu}_1(a^2/r^2). \tag{9.62}$$

Now, it is just a matter of diligent algebra to show that the resulting equation for $\hat{Q}_r$ is just Eq. (9.56) and, hence, that the expression (9.55) for $\hat{\mathbf{B}} \cdot \hat{\mathbf{Q}}$ entering the second interface condition (9.50) is unchanged! Apparently, no current perturbation is induced and the boundary condition (9.60), derived for the vacuum, also applies for the 'ghost plasma'. With respect to the perturbations, there appears to be no difference between a 'ghost' plasma and a vacuum.

That answer is WRONG. It is true that the vacuum field equation (9.56) for $\hat{Q}_r$ correctly describes the dynamics of the 'ghost plasma', so that no perturbed currents develop, but only *when the inversion (9.61) can be carried out*, i.e. for all points where $k + \hat{\mu}m \neq 0$. For mode number k and m such that there is a point

$r = r_s$ in the interval (a, b) where

$$k + \hat{\mu}(r_s)m = 0, \tag{9.63}$$

the inversion has to be reconsidered. Such points are not 'seen' by the magnetic field equation (9.56), so that $\hat{Q}_r$ is finite, in general, which implies that $\hat{\chi}$ blows up. The Frobenius expansion of the marginal equation of motion (9.48) for $\hat{\chi}$ around such points is a special case of Newcomb's expansion around interchange points, described in Section 7.5.2 for the analogous case of a plane gravitating plasma. For the cylinder, the parameter q_0 of the indicial equation becomes proportional to p' (see Section 9.4.1) which vanishes for the 'ghost plasma' so that the indices, given by Eq. (7.200), become $\nu_1 = 0$, and $\nu_2 = -1$. Hence, there is a small solution with a finite amplitude at the singularity and a large solution with a logarithmic term. To have physically acceptable solutions, the large solution has to be excluded because its contribution to the energy W blows up, according to Eq. (7.202) of Section 7.5.2, so that the physically acceptable small solution (with finite energy) becomes

$$\hat{\chi}_s = c_1 + c_2(r - r_s) + \cdots. \tag{9.64}$$

From Eq. (9.61), this implies that $\hat{Q}_r$ vanishes at $r = r_s$. In other words: the differential equation (9.56) for the magnetic field perturbation $\hat{Q}_r$ can be exploited, but it should be subjected to the boundary condition

$$\hat{Q}_r(r_s) = 0, \quad \text{when} \quad k + \hat{\mu}(r_s)m = 0. \tag{9.65}$$

However, since the solution (9.59) for $\hat{Q}_r$ is not oscillatory, this would be in conflict with also satisfying the boundary condition (9.58) at the wall. Similarly, when describing the problem in terms of $\hat{\chi}$, there would be a conflict with the model II* boundary condition (6.131), viz. $\hat{\chi}(b) = 0$. How can this be reconciled? Here, the other property of the singular solutions comes to the rescue. It has been proved in Section 7.5.2 that the small solution may jump at the singularity. Hence, the physically acceptable solution looks like that depicted in Fig. 9.7. On (a, r_s) the perturbation is finite, it jumps to zero at the singularity, and on (r_s, b) it vanishes identically. Hence, *the boundary condition for a plasma–'ghost' plasma interface is given by the boundary condition (9.60) for a plasma–vacuum interface with the following modification*:

$$b \to b^*, \text{ where } b^* \equiv \begin{cases} b & \text{if } k + \hat{\mu}(r)m \neq 0 \text{ for all } r \text{ on } (a, b), \\ r_s = a\sqrt{-\dfrac{\hat{\mu}_1 m}{k}} & \text{if } k + \hat{\mu}(r_s)m = 0 \text{ for } r_s \text{ on } (a, b). \end{cases}$$

$$\tag{9.66}$$

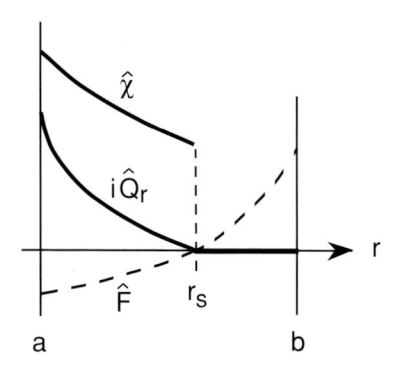

Fig. 9.7. Perturbations at an interchange singularity for a 'ghost' plasma.

Clearly, the 'ghost' plasma has a huge effect on the stability in that *the wall is effectively placed at the singularity*. This effect is due to the perturbed current density $\hat{\mathbf{j}}_1 = \nabla \times \hat{\mathbf{Q}}$, which vanishes nearly everywhere in the exterior region, but exhibits a surface current concentration (see Section 4.5.2) at $r = r_s$ due to the jump of $\hat{\mathbf{Q}}$:

$$\hat{\mathbf{j}}_1^\star = \mathbf{n} \times [\![\hat{\mathbf{Q}}]\!]. \tag{9.67}$$

This induced surface current has the same effect as a solid wall put at the position of the singularity.

We now have the complete machinery available to describe the waves and instabilities of cylindrical plasmas, to be applied in the following sections. The present subsection should have illustrated, once more, that singularities are nearly always present. They are not a mathematical frivolity, but they determine the dominant dynamics of the plasma. Bender and Orszag's motto on Section 7.4 stays with us.

9.3 Spectral structure

9.3.1 One-dimensional inhomogeneity

At this point in the exposition, the reader may well wonder why the algebra of the spectral analysis is so complicated and how one can be sure of results when so many factors contribute and subtle cancellations are a rule rather than an exception. We first point out, under (a), the intrinsic reason for these complications, but also the challenges this represents. Next, under (b), we point out several alleviating factors which help to restore confidence in the spectral enterprise. We are then ready to apply the developed formalism to problems of practical interest.

(a) Corresponding problems in quantum mechanics In Chapters 6 and 7, we have frequently stressed the analogy between MHD and quantum mechanical spectral theory. Since calculations with the Hain–Lüst equation have direct relevance for plasma confinement in realistic geometries, it is instructive to compare it with a corresponding concrete problem in quantum mechanics. To that end, we contrast the normal mode equation $\mathbf{F}(\boldsymbol{\xi}) = -\rho\omega^2\boldsymbol{\xi}$ with the Schrödinger equation $H\Psi = E\Psi$ which, for a particle in a potential field $V(\mathbf{r})$, becomes

$$\left[-\frac{\hbar^2}{2M}\Delta + V(\mathbf{r})\right]\Psi(\mathbf{r}) = E\Psi(\mathbf{r}). \tag{9.68}$$

One-dimensional problems are obtained for a potential that is spherically symmetric, like the H-atom where $V = V(r)$. In that case, one writes the wave function as a superposition of spherical harmonics which may be studied separately,

$$\Psi(r, \theta, \phi) = R(r)\, Y_\ell^m(\theta, \phi), \tag{9.69}$$

in much the same way as the separate Fourier components (9.22) for a cylindrical plasma (or the spherical harmonics (7.55) themselves exploited in helioseismology). Inserting the expression (9.69) in Eq. (9.68) leads to a second order differential equation for the radial wave function:

$$\frac{1}{r}\frac{d^2}{dr^2}(rR) - \left[\frac{\ell(\ell+1)}{r^2} + \frac{2M}{\hbar^2}\bigl(V(r) - E\bigr)\right]R = 0. \tag{9.70}$$

This is the equation that should be compared with the generalized Hain–Lüst equation.

It is clear that the spectral problem of calculating the waves and instabilities of a cylindrical plasma is a much more complicated one than the determination of the energy levels of the hydrogen atom, or even the general quantum mechanical problem of scattering of particles in an *arbitrary* one-dimensional potential field. In the latter case, the only controlling function is $V(r)$, whereas four such functions, $\rho(r)$, $p(r)$, $B_\theta(r)$ and $B_z(r)$, enter the MHD equation. More important, the reduction to the generalized Hain–Lüst equation from a vector equation (with three components ξ, η, ζ) implies that the eigenvalue ω^2 is scattered through the coefficients P and Q of Eq. (9.35) in a complicated manner. Consequently, whereas the radial wave equation (9.70) is a classical differential equation of the Sturm–Liouville type, where the linear occurrence of the eigenvalue E guarantees monotonicity with the number of nodes of the radial eigenfunction $R(r)$, the Hain–Lüst equation (9.31) is not of such a classical type, so that the dependence of ω^2 on the number of nodes of $\chi(r)$ is more complicated.

The vector character of ideal MHD is reflected in the occurrence of three sub-spectra. However, the general structure of each of these sub-spectra is very similar

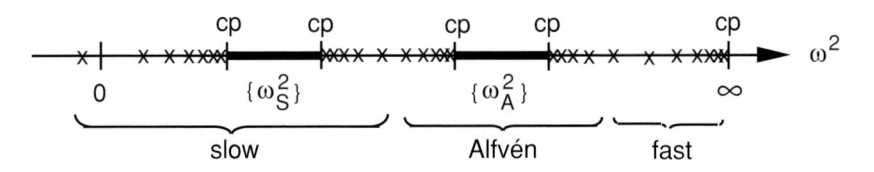

Fig. 9.8. Schematic spectra in quantum mechanics (fixed m and ℓ) and MHD (fixed m and k).

to the complete spectrum of quantum mechanical systems (Fig. 9.8). If one fixes the quantum numbers m and ℓ for the H-atom one finds a discrete spectrum of bound states for $E < 0$ clustering at $E = 0$, which is the edge of a continuum of free states for $E > 0$. Likewise, for the diffuse cylindrical plasma the Alfvén and slow sub-spectra consist of discrete modes that may cluster at the edge of the continua $\{\omega_A^2\}$ and $\{\omega_S^2\}$, whereas the fast sub-spectrum accumulates at $\omega^2 = \infty$.

Hence, there is much more in common to the two problems than suggested by the evident differences:

(1) both concern the determination of the spectrum of a *self-adjoint linear operator in Hilbert space*;
(2) the operators have a *discrete spectrum* as well as a *continuous spectrum*, with different but intrinsically physical reasons for the distinction between them;
(3) for one-dimensional inhomogeneity, *the discrete spectrum (or sub-spectra for vector problems) is asymptotically monotonic in the number of nodes of the eigenfunctions.*

The latter property connects the discrete spectrum, or sub-spectra, to the essential spectrum, consisting of the continuous spectrum of free states in quantum mechanics, and of the Alfvén and slow continua and the fast cluster point in the MHD case.

This structural unity of spectral theory is much more important than the algebraic complications due to the vector character of MHD. With the splendid example of the unravelling of the atomic structure in the twentieth century, there can be no doubt that nature still has a lot in store for us when MHD spectroscopy (Section 7.2.4) has outgrown its present state of infancy.

(b) Common properties of inhomogeneous plasmas With the extensive preparation of Chapter 7 on the plane gravitating plasma slab, for the analogous problem of the cylindrical plasma there is no need to repeat:

- the proof of the existence of continuous Alfvén and slow continua (Section 7.4.2);
- the demonstration of apparent singularities (Sections 7.3.2 and 7.4.1);
- the proof of the oscillation theorem (Section 7.4.3);
- the variational procedures for stability (Sections 7.5.1 and 7.5.2).

All this immediately carries over to the cylindrical case so that we can concentrate on the surprisingly many different effects of the curvature term associated with the transverse magnetic field component B_θ. The fact that all these features carry over from one inhomogeneous problem to another illustrates that there is good reason for trust in the final outcome of most spectral problems in MHD.

One should be aware, though, that these chapters focus on *one-dimensional* inhomogeneous plasmas described by *ideal* MHD. When toroidal curvature is introduced, separability usually fails and ODEs are replaced by PDEs. Naive expectations about similar monotonicity properties of the eigenvalues in the toroidal case, with 2D nodal lines taking the place of 1D nodal points, are quickly shattered when one realizes what could happen (see the example given by Courant and Hilbert [60], Vol. I, p. 455, on the peculiar behaviour of nodal lines of the Helmholtz equation on a square). When dissipation is admitted, self-adjointness is lost and spectral problems become much more complicated. However, this does not take away from the importance of the ideal, one-dimensional, problem because it usually returns in the form of a leading order contribution with toroidal or resistive corrections. For example, the large stabilizing effect of the singularity in the 'ghost' plasma of the previous section is lost when the resistivity of that plasma is taken into account. In this respect, it is similar to ordinary wall stabilization when the resistivity of the wall is taken into account. However, the resistive instabilities that develop in those cases exponentiate on a much slower time scale than the ideal MHD time scale so that they are much easier to control by means of feedback magnetic fields. Toroidal and dissipative spectral problems will be discussed more extensively in Volume 2.

9.3.2 Cylindrical model problems

In cylindrical geometry, analytically solvable models inevitably involve Bessel functions. We discuss three examples that are frequently used in cylindrical stability problems.

(a) Waves in a homogeneous θ-pinch We start with the simplest model, a linear θ-pinch with a homogeneous magnetic field, pressure and density,

$$B_\theta = 0, \qquad B_z = B_0, \qquad \rho = \rho_0, \qquad p = p_0, \qquad (9.71)$$

so that the Alfvén speed and sound speed are constant, related to each other by the parameter β:

$$b \equiv \sqrt{B_0^2/\rho_0}, \qquad c \equiv \sqrt{\gamma p_0/\rho_0}, \qquad \tfrac{1}{2}\gamma\beta \equiv c^2/b^2. \qquad (9.72)$$

The Hain–Lüst equation (9.31) then simplifies to

$$(\omega^2 - k^2 b^2)\left[\left(\frac{r}{m^2 + k^{*2} r^2}\chi'\right)' - \frac{1}{r}\chi\right] = 0, \qquad (9.73)$$

where a kind of modified longitudinal wave number appears:

$$k^* \equiv \left[\frac{(k^2 - \omega^2/b^2)(k^2 - \omega^2/c^2)}{k^2 - \omega^2/b^2 - \omega^2/c^2}\right]^{1/2}. \qquad (9.74)$$

For internal modes (model I), the differential equation (9.73) is to be solved subject to the boundary conditions

$$\chi(0) = \chi(a) = 0. \qquad (9.75)$$

Eq. (9.73) yields, first of all, *an infinitely degenerate spectrum of Alfvén waves* with frequency $\omega^2 = \omega_A^2 \equiv k^2 b^2$, and with a completely arbitrary radial dependence of the eigenfunction χ. They propagate along the axis of the cylinder with the Alfvén speed b.

For $\omega^2 \neq \omega_A^2$, the solutions of Eq. (9.73) are Bessel functions, where it depends on the sign of k^{*2} whether they are of the modified, exponential, kind (giving evanescence) or of the ordinary, oscillatory, kind (giving propagation):

$$\chi = Cr I_m'(k^* r) \quad \text{if } k^{*2} > 0, \quad \text{i.e. for } \begin{cases} \omega^2 < \omega_S^2 \\ k^2 c^2 < \omega^2 < \omega_A^2 \end{cases} \text{(evanescent)},$$

$$\chi = Cr J_m'(ik^* r) \quad \text{if } k^{*2} < 0, \quad \text{i.e. for } \begin{cases} \omega_S^2 < \omega^2 < k^2 c^2 \quad \text{(slow)}, \\ \omega_A^2 < \omega^2 \quad\quad\quad\quad \text{(fast)}, \end{cases} \qquad (9.76)$$

where $c^2 < b^2$ has been assumed in the inequalities. The first expression does not permit satisfaction of the boundary condition for internal modes. It will be used below in the expression for the external kink modes. From the second expression,

9.3 Spectral structure

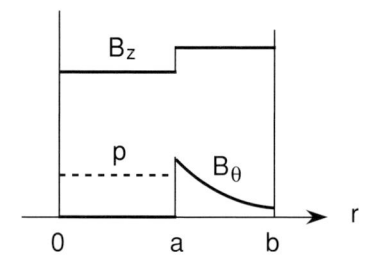

Fig. 9.9. Sharp-boundary skin-current model.

the boundary condition $\chi(a) = 0$ implicitly fixes the eigenfrequencies through

$$ik^* a = j'_{mn}, \qquad (9.77)$$

where j'_{mn} is the nth zero of the Bessel function $J'_m(x)$. This yields the dispersion equation for *the slow and fast magneto-acoustic waves in a homogeneous θ-pinch*:

$$\omega^4 - (k^2 + j'^2_{mn}/a^2)(b^2 + c^2)\omega^2 + k^2(k^2 + j'^2_{mn}/a^2)b^2 c^2 = 0. \qquad (9.78)$$

This equation is fully analogous to the magneto-sonic factor of the dispersion equation (5.53) for homogeneous plasmas, where the parallel wave number is now indicated by k and the effective total wavenumber by $\sqrt{k^2 + j'^2_{mn}/a}$.

(b) Free-boundary modes of interface plasmas For the modes of a plasma–vacuum (model II) or plasma–'ghost' plasma (model II*) system, we need to solve the Hain–Lüst equation (9.31) subject to the boundary conditions discussed in Section 9.2.2. As in model I, the perturbation on axis is restricted by the regularity condition $\chi(0) = 0$, whereas the boundary condition at the interface involves the perturbation of the total pressure, $(\Pi/\chi)_{r=a}$ as given by Eq. (9.60). For a 'ghost' plasma, the substitution (9.66) should be made in the latter expression.

We wish to study this problem for a sharp-boundary plasma where the current is confined to the plasma surface $r = a$ *(skin-current model, Fig. 9.9)*. Then, the equilibrium quantities for the interior of the plasma column are those of a homogeneous θ-pinch given by Eq. (9.71), whereas the external magnetic field is given by Eqs. (9.18) and (9.19). This model provides a very useful first approximation to the study of *external kink modes*, which are the most dangerous instabilities occurring in a cylindrical plasma column. Here, most dangerous is meant in the sense of affecting the bulk of the plasma and having large growth rates. For typical densities of high-β pinches they exponentiate on the μs time scale.

With a free boundary, exploiting the oscillatory Bessel function solutions (9.76)(b), the frequencies of the modes will be slightly shifted from those discussed under (a). However, the really interesting new feature of free-boundary plasmas is the appearance of a *surface mode*, like the f-mode in the gravito-acoustic spectrum (Section 7.2.3). This mode is just the lowest one in the evanescent region, which may even become unstable ($\omega^2 < 0$), so that one should exploit the exponential Bessel functions (9.76)(a) now.

To obtain the dispersion equation of the modes, we first compute the perturbation Π of the total pressure for the first solution (9.76) for χ,

$$\Pi = -\frac{N}{rD}\chi' = -\frac{\omega^2 - k^2 b^2}{m^2 + k^{*2}r^2} r\chi' = -\frac{\omega^2 - k^2 b^2}{k^*} \cdot C I_m(k^*r), \tag{9.79}$$

and then insert this expression and χ in the boundary condition (9.60). This yields *the dispersion equation for free-boundary modes*:

$$\omega^2 = \frac{k^2 B_0^2}{\rho_0} - \frac{k^* a\, I'_m(k^*a)}{\rho_0 I_m(k^*a)} \left[\frac{\hat{B}_\theta^2}{a^2} \right.$$

$$\left. + \frac{(m\hat{B}_\theta/a + k\hat{B}_z)^2}{ka} \frac{I_m(ka)K'_m(kb) - K_m(ka)I'_m(kb)}{I'_m(ka)K'_m(kb) - K'_m(ka)I'_m(kb)} \right]. \tag{9.80}$$

At this point, the dispersion equation is still a highly transcendental equation in the eigenvalue because of the dependence of k^* on ω^2. (Also note that the symbol b now indicates the wall position.)

Many different limits may be studied for this equation, but the most interesting one is obtained for *the tokamak approximation* where we again consider a cylindrical plasma of length $2\pi R_0$ as a first approximation to a torus of major radius R_0 (Fig. 9.2). In that case, *the wave number k is quantized with integer toroidal mode number n* (not to be confused with the radial node number of the Bessel functions):

$$k = n/R_0, \quad \text{so that} \quad ka = \epsilon n \quad \left[\ll 1 \text{ for } n \sim 1 \right]. \tag{9.81}$$

The approximation in square brackets is the long wavelength approximation for the longitudinal mode number, which is quite relevant here. Furthermore, the magnetic field components are ordered as

$$\hat{B}_\theta \sim \epsilon \hat{B}_z, \quad \text{so that} \quad \hat{q} = \epsilon \hat{B}_z/\hat{B}_\theta \sim 1. \tag{9.82}$$

We assume (and easily justify this from the result obtained below) that the eigenvalues of the modes (or the growth rates of the modes) are much smaller than the Alfvén frequency:

$$|\omega^2| \ll \omega_A^2 \equiv k^2 b^2 \quad \Rightarrow \quad k^* \approx k. \tag{9.83}$$

In view of Eq. (9.81), the arguments of all the occurring Bessel functions are small, so that we may use the following approximations for $m \neq 0$:

$$k^* a \, I'_m(k^*a)/I_m(k^*a) \approx |m|,$$

$$\frac{I_m(ka)K'_m(kb) - K_m(ka)I'_m(kb)}{I'_m(ka)K'_m(kb) - K'_m(ka)I'_m(kb)} \approx -\frac{\epsilon n}{|m|} \frac{(b/a)^{|m|} + (b/a)^{-|m|}}{(b/a)^{|m|} - (b/a)^{-|m|}}. \tag{9.84}$$

Inserting these approximations in Eq. (9.80) leads to the following approximate form of the dispersion equation:

$$\omega^2 \approx \frac{\epsilon^2 B_0^2}{a^2 \rho_0} \left\{ n^2 - \frac{\hat{B}_\theta^2}{\epsilon^2 B_0^2} \left[|m| - (m+n\hat{q})^2 \frac{(b/a)^{|m|} + (b/a)^{-|m|}}{(b/a)^{|m|} - (b/a)^{-|m|}} \right] \right\}. \tag{9.85}$$

From the equilibrium expressions (9.19), and Eq. (9.13) relating $\hat{\mu}$ to $\hat{q}$, one may convert the factor involving $\hat{B}_\theta$ in terms of β and $\hat{q}_1$:

$$\frac{\hat{B}_\theta^2}{\epsilon^2 B_0^2} = \frac{1+\beta}{\hat{q}_1^2 + \epsilon^2} \approx \frac{1}{\hat{q}_1^2}. \tag{9.86}$$

In the rightmost approximation, we have neglected small terms β and ϵ^2 in agreement with *the low-β tokamak ordering* (9.14). The dispersion equation for the 'straight tokamak' then becomes:

$$\omega^2 \approx \frac{\epsilon^2 B_0^2}{a^2 \rho_0 \hat{q}^2} \left[n^2 \hat{q}^2 - |m| + (m+n\hat{q})^2 \frac{(b/a)^{|m|} + (b/a)^{-|m|}}{(b/a)^{|m|} - (b/a)^{-|m|}} \right]. \tag{9.87}$$

In view of scale independence (Section 4.1.2), this expression is to be considered as an end product, since it has the trivial dimensional factors B_0, a, ρ_0 in the appropriate way to give the dimension of a growth rate squared, whereas the essential parameters $\hat{q}$, b/a, and ϵ, describing the equilibrium features, and the mode number m and n, describing the perturbations, appear in a physically significant way. (Note that there is no radial node number, in agreement with our choice for the evanescent solutions.)

Rearranging terms, Eq. (9.87) may be written as

$$\omega^2 \approx \frac{\epsilon^2 B_0^2}{a^2 \rho_0 \hat{q}^2} \left[\tfrac{1}{2}|m|(|m|-2) + \tfrac{1}{2}(2n\hat{q}+m)^2 + \frac{2(n\hat{q}+m)^2}{(b/a)^{2|m|} - 1} \right]. \tag{9.88}$$

This rearrangement reveals some of the physical mechanisms at work in this model. First, there is *the kink* term which is only negative when $|m| = 1$. Then, there is a stabilizing term representing *the average field line-bending* across the plasma boundary which disappears for modes that propagate perpendicular to the average direction of the field across the surface layer at $r = a$ (recall that $q = \infty$ for $r = a^-$ and $q = \hat{q}$ for $r = a^+$). The last term represents *the stabilizing*

458   Cylindrical plasmas

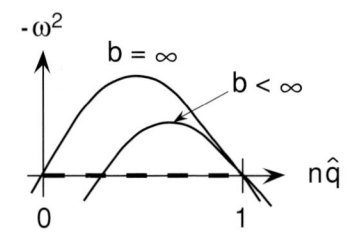

Fig. 9.10. Growth rate of the external kink mode for a skin-current plasma–vacuum model. Replacing the vacuum by a 'ghost plasma' yields complete stabilization (thick dashed line).

influence of the wall, ranging from infinitely stabilizing when $b/a = 1$ to no effect when $b/a \to \infty$.

Since only $|m| = 1$ is unstable, we may restrict the analysis to that mode:

$$\omega^2(m=-1) = \frac{2\epsilon^2 B_0^2}{a^2 \rho_0 \hat{q}^2} \frac{(n\hat{q} - 1)(n\hat{q} - a^2/b^2)}{1 - a^2/b^2}. \tag{9.89}$$

This growth rate is plotted in Fig. 9.10. Clearly, the external kink mode is always unstable for this model in the region

$$a^2/b^2 < n\hat{q} < 1. \tag{9.90}$$

This suggests a simple way of eliminating unstable external kink modes by prescribing the geometry of the torus and the total plasma current I_z such that

$$\hat{q} = 2\pi a^2 B_0/(R_0 I_z) > 1, \tag{9.91}$$

so that the unstable $n = 1$ modes (and, hence, all the $n > 1$ modes as well) simply do not fit into the torus. This condition is called *the Kruskal–Shafranov limit*. The limit imposed on the plasma currents by Eq. (9.91) is a quite important consideration in the operation of tokamaks. It is appropriate to repeat here the remark made in Section 2.4.3 that the fact that $\hat{q} = 1$ corresponds to a topology with closed magnetic field lines has nothing to do with the stability mechanism of the external kink mode. This is a purely accidental coincidence which disappears as soon as one introduces genuine toroidal effects in the theory (see the companion Volume 2).

Let us now consider the model II* version of the free-boundary external kink mode. According to Eq. (9.66), one should replace the actual wall position by a virtual wall position b^* because of stabilization by induced skin currents at the singular position given by $m + n\hat{q}(r_s) = 0$. The unstable region given by Eq. (9.90) indicates that $|m| = 1$ instability precisely occurs when there is such a singularity.

Hence, in that regime, we should exploit the virtual wall position given by

$$b^*/a = \sqrt{-\frac{\hat{\mu}_1 m}{k}} = \sqrt{-\frac{m}{n\hat{q}_1}}. \qquad (9.92)$$

Hence, in the expression (9.89) for the growth rate, the factor

$$n\hat{q} - a^2/b^2 \rightarrow n\hat{q} - (a/b^*)^2 = 0 \quad \text{for } |m| = 1. \qquad (9.93)$$

The external kink mode is completely stabilized by the singular currents in the 'ghost plasma'! This agrees with the theory of internal kink modes (Section 9.4.4), which are stable to leading order in the inverse aspect ratio ϵ. The external kink mode of the plasma–vacuum model II actually becomes an internal kink mode in the plasma–'ghost plasma' model II*.

Of course, for the operation of an actual fusion experiment, one would not rely on stabilization by an external 'ghost plasma' to push the current beyond the Kruskal–Shafranov limit: stabilization depends on perfect conductivity, whereas the outermost plasma is most likely to be subject to resistive instabilities. Moreover, stability with respect to internal kink modes is lost when the next order of the toroidal effects is taken into account. Nevertheless, it is true that an external 'ghost plasma' does slow down the growth rate of the kink mode compared to an external vacuum.

(c) Modes of an incompressible plasma with constant-pitch magnetic field In our next model, we move a bit closer to physical reality on the longer time scale by admitting a distributed current in the plasma. To facilitate the analysis, we exploit Freidberg's simplified form (9.47) of the Hain–Lüst equation for an incompressible plasma. Moreover, we assume a constant longitudinal magnetic field component B_z and a linearly increasing transverse magnetic field component B_θ, so that the pitch μ of the magnetic field lines is constant and the current is evenly distributed over the plasma:

$$B_\theta = Ar \quad \Rightarrow \quad j_z = \frac{1}{r}(rB_\theta)' = 2A. \qquad (9.94)$$

According to the equilibrium equation (9.3), the pressure distribution then becomes parabolic and may be chosen to vanish at the plasma–vacuum boundary:

$$p = p_0 - A^2 r^2 = a^2 A^2 (1 - r^2/a^2). \qquad (9.95)$$

The parallel gradient operator $F \equiv mB_\theta/r + kB_z = B_z(k + \mu m) = \text{const}$, so that the Alfvén factor $\rho\omega^2 - F^2$ also becomes constant if we assume constant density.

Then, the incompressible equation of motion (9.47) simplifies to

$$\frac{d}{dr}\left[\frac{r}{m^2+k^2r^2}\frac{d\chi}{dr}\right] - \frac{1}{r}\left[1 + \left(\frac{2\alpha}{m^2+k^2r^2} - \frac{\alpha^2}{m^2}\right)\frac{k^2r^2}{m^2+k^2r^2}\right]\chi = 0, \tag{9.96}$$

where

$$\alpha \equiv \frac{2mAF}{\rho\omega^2 - F^2}. \tag{9.97}$$

The solution of this equation may again be expressed in terms of Bessel functions, in the following combination:

$$\chi = C\left[k^*rI'_m(k^*r) - \alpha I_m(k^*r)\right], \qquad k^* \equiv k\sqrt{1-\alpha^2/m^2}. \tag{9.98}$$

(How does one produce such a miraculous answer? By transforming to another variable! The reader may wish to check that, in this case, one obtains a simpler ODE (Bessel's equation itself) for the total pressure perturbation, with the solution $\Pi \sim I_m(k^*r)$.) Note that, in the absence of a transverse magnetic field, $A = 0$ and $k^* = k$, so that the solution (9.76) for the θ-pinch in the incompressible limit is recovered.

This model was probably considered first by Alfvén to study unstable loops as a mechanism for the generation of cosmic magnetic fields; see the second (1963) edition of Ref. [6] co-authored with C. Fälthammar. For internal modes, it is clear that the boundary condition $\chi(a) = 0$ can only be satisfied if the Bessel functions are oscillatory. This requires $\alpha^2 > m^2$, or $4A^2F^2 > (\rho\omega^2 - F^2)^2$. For $F \to 0$, this implies $-\rho\omega^2 < 2|AF|$ and it follows that the growth rate tends to zero. On the other hand, for $\omega^2 = 0$, one always finds oscillatory solutions in the limit $F \to 0$, so that there will be infinitely many unstable branches according to the oscillation theorem. We have encountered this behaviour before, in Chapter 7 for the gravitational quasi-interchanges (Section 7.3.3).

Alfvén's model for the incompressible plasma cylinder with a constant pitch magnetic field is unstable with respect to *quasi-interchanges* (see Section 9.4.2), having maximum growth rate for $F \neq 0$. If one relaxes the incompressibility constraint, the plasma becomes unstable with respect to *pure interchanges*, having maximum growth rate at $F = 0$. This model has been used for the benchmarking of large scale eigenvalue solvers exploiting finite elements. One particularly nice example, due to Chance et al. [50], is reproduced in Fig. 9.11 in the representation by Kerner [125]. It shows the full structure of the ideal MHD spectrum in the presence of an instability. The spectrum is resolved in a practical sense (where continua are represented by as many densely spaced eigenvalues as there are grid points in the calculation) over the many orders of magnitude from high-frequency

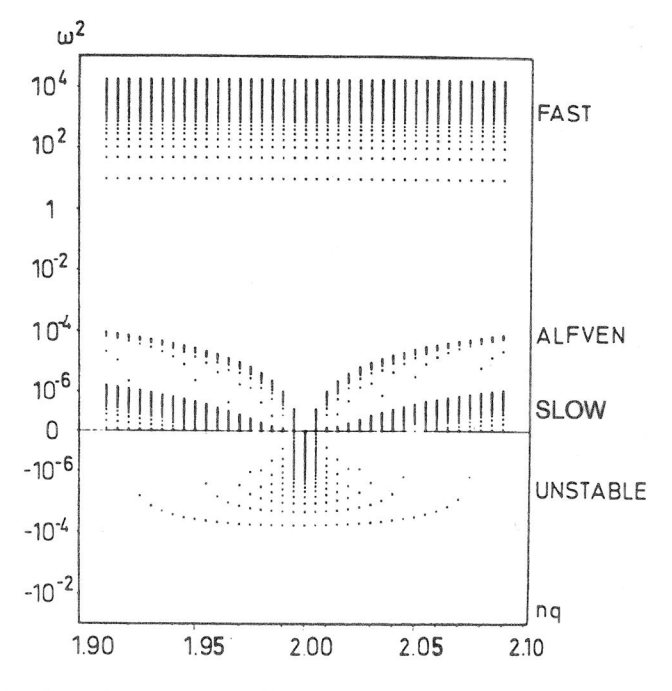

Fig. 9.11. Complete spectrum of modes for a compressible plasma with a shearless magnetic field, in the presence of interchange instabilities; $m = -2$. (From Kerner [125].)

fast modes to unstable interchanges. Note that the interchange instabilities occur in the region where $F = 0$, so that the Alfvén and slow continua extend to the origin. In a sense, these instabilities belong to both the Alfvén and the slow sub-spectra.

For external modes, exploiting again the low-β tokamak ordering to simplify the Bessel function expressions of the incompressible model, we obtain the following dispersion equation for the distributed current model:

$$\omega^2 = \frac{2\epsilon^2 B_0^2}{a^2 \rho_0 q^2} \frac{m+nq}{1-(a/b)^{2|m|}} \left[m + nq - sg(m)\left(1 - (a/b)^{2|m|}\right) \right]. \quad (9.99)$$

Here, the constant value of q in the interior plasma has been chosen equal to the value $\hat{q}_1$ of the vacuum magnetic field at $r = a$. Note that, for $m = -1$, the expression reduces to that of Eq. (9.89) so that this mode is not sensitive for the current distribution. However, we now obtain external instabilities for all values of m, in regions to the left of the integer values of nq, as for the $m = -1$ mode of the skin-current model shown in Fig. 9.10. The constant current distribution is violently unstable, with no easy stabilization in the manner of the Kruskal–Shafranov limit for the $|m| = 1$ external kink modes.

9.3.3 Cluster spectra*

A cluster point analysis fully analogous to that of Section 7.4.4 for the plane gravitating plasma slab may also be carried out for cylindrical plasmas. The result is as follows, see Goedbloed [84].

In order to have a *Sturmian sequence of discrete Alfvén modes clustering at the lower edge of the Alfvén continuum*, i.e. at the minimum of ω_A^2 (where $\omega_A^{2''} > 0$), the following condition should be satisfied:

$$\frac{4k^2 B_\theta^2 (B^2 - \gamma p)}{r^2 B^2} - \frac{rG^2}{B^2}\left(\frac{B_\theta^2}{r^2} + \frac{2k B_\theta B^2}{r^2 G}\right)' > \frac{1}{8}\rho(\omega_A^2)'' > 0; \quad (9.100)$$

vice versa, for an *anti-Sturmian sequence at the upper edge of the Alfvén continuum*, i.e. at the maximum of ω_A^2 (where $\omega_A^{2''} < 0$), the inequality signs should be reversed.

In order to have a *Sturmian sequence of discrete slow modes clustering at the lower edge of the slow continuum*, the following condition should be satisfied:

$$-\left(\frac{\gamma p}{\gamma p + B^2}\right)^2 \left[\frac{F^4}{\gamma p + B^2} + \frac{rF^2}{B^2}\left(\frac{B_\theta^2}{r^2}\right)' - \frac{4k^2 B_\theta^2 (\gamma p + B^2)}{r^2 B^2}\right]$$

$$> \frac{1}{8}\rho(\omega_S^2)'' > 0; \quad (9.101)$$

vice versa, for an *anti-Sturmian sequence at the upper edge of the slow continuum*, the inequality signs should be reversed.

These conditions show that, in principle, all branches of the discrete spectrum may occur in a cylindrical plasma with $B_\theta \neq 0$. The Alfvén cluster spectra are associated with the occurrence of global Alfvén eigenmodes (GAEs); see Appert et al. [10].

9.4 Stability of cylindrical plasmas

9.4.1 Oscillation theorems for stability

We have seen in Chapter 7 that most of the classical stability theory for plasmas with one-dimensional inhomogeneity can be derived from the MHD oscillation theorem (Section 7.4.3), where the occurrence of the interchange singularities $F = 0$ is the main complicating factor. We will not repeat the analysis for cylindrical plasmas, but only present the main steps in the derivations in so far as they lead to different equations, in particular with respect to magnetic curvature terms (B_θ) instead of the gravitational terms ($\rho' \hat{g}$) of Chapter 7.

9.4 Stability of cylindrical plasmas

Recall that the eigenfrequencies of discrete modes are monotonic in the number of nodes of the radial component ξ of the eigenfunction for frequencies outside the ranges of the continua and the turning point frequencies. Hence, the discrete sub-spectra are asymptotically either Sturmian or anti-Sturmian. Since the unstable range is always Sturmian, there is an immediate connection between the MHD oscillation theorem and Newcomb's stability theory.

(a) Newcomb's variational procedure and Suydam's criterion For the study of stability, we start from the energy principle (6.96). For the diffuse cylindrical plasma, the reduction of W proceeds along the same lines as for the plane gravitating slab (Section 7.5.1). Exploiting the expressions (9.27) for $\mathbf{Q}$ and $\nabla \cdot \boldsymbol{\xi}$, consistently expressing all variables in ξ, η and ζ, and integrating by parts to get rid of a term $\xi'\xi$, one obtains the following form:

$$W = \pi L \int_0^a \Bigg\{ P_0[(r\xi)']^2 + Q_0(r\xi)^2$$
$$+ (m^2 + k^2 r^2)\left[\frac{B}{r}\eta + \frac{G(r\xi)' + 2k B_\theta \xi}{m^2 + k^2 r^2}\right]^2$$
$$+ \gamma p \left[\frac{1}{r}(r\xi)' + \frac{G\eta + F\zeta}{B}\right]^2 \Bigg\} r\, dr, \qquad (9.102)$$

where the length $L \to \infty$ for the full cylinder. Minimization with respect to η and ζ is again trivial (it leads to the expressions (9.29) in the limit $\omega^2 \to 0$), so that W reduces to

$$W = \pi L \int_0^a \Big\{ P_0[(r\xi)']^2 + Q_0(r\xi)^2 \Big\} r\, dr. \qquad (9.103)$$

This expression is minimized by solutions to the Euler–Lagrange equation (9.48), which is the generalized Hain–Lüst equation in the limit $\omega^2 \to 0$. The explicit form of P_0 and Q_0 may be read off from the marginal equation of motion (9.48) given in Section 9.2.1(d).

For many stability applications it is convenient to transform to the variable ξ again, where we now exploit the notation f_0 and g_0 of Newcomb:

$$W = \pi L \int_0^a \left(f_0 \xi'^2 + g_0 \xi^2 \right) dr, \qquad (9.104)$$

leading to the Euler–Lagrange equation

$$(f_0\, \xi')' - g_0\, \xi = 0, \qquad (9.105)$$

where f_0 and g_0 are obtained from P_0 and Q_0 of Eq. (9.48) by a straightforward transformation involving the equilibrium relation (9.3):

$$f_0 \equiv r^2 P_0 = \frac{r^3 F^2}{m^2 + k^2 r^2}, \tag{9.106}$$

$$g_0 \equiv r^2 Q_0 - r P_0'$$
$$= \frac{2k^2 r^2}{m^2 + k^2 r^2} p' + \frac{m^2 + k^2 r^2 - 1}{m^2 + k^2 r^2} r F^2 - \frac{2k^2 r^3 (m B_\theta / r - k B_z)}{(m^2 + k^2 r^2)^2} F. \tag{9.107}$$

Since Eq. (9.105) is equivalent to the Hain–Lüst equation for $\omega^2 = 0$, we obtain Newcomb's stability theorem [164] directly from the MHD oscillation theorem for the case that the interval $(0, a)$ contains no singularity $F = 0$.

Theorem. *For specified values of m and k such that $F \equiv m B_\theta / r + k B_z \neq 0$ on the interval $(0, a)$, the diffuse cylindrical plasma is stable if, and only if, the non-trivial solution $\chi = r\xi$ of the marginal equation of motion (9.48), corresponding to the Euler–Lagrange equation (9.105) that vanishes at $r = 0$ does not have a zero in the open interval $(0, a)$.*

Not all values of m need to be investigated since W has two convenient monotonicity properties.

(1) For $m = 0$, the energy integral becomes

$$W_{m=0} = W_0 + \pi L k^2 \int r B_z^2 \, dr, \quad W_0 \equiv \pi L \int \left[r B_z^2 \xi'^2 + \left(2p' + \frac{B_z^2}{r} \right) \xi^2 \right] dr. \tag{9.108}$$

Hence, if W is positive for $m = 0$, $k \to 0$, it is positive for $m = 0$, all $k \neq 0$. The case $m = 0$, $k = 0$ is not contained since the derivation of Eq. (9.102) is invalid then (there is a division by 0). Starting anew from the original expression for W, one finds:

$$W_{m=0,k=0} = W_0 + \pi L \int \left\{ B_\theta^2 \left[r \left(\frac{\xi}{r} \right)' \right]^2 + \gamma p \left[\frac{1}{r} (r\xi)' \right]^2 \right\} r \, dr. \tag{9.109}$$

This shows that, if W is positive for $m = 0$, $k \to 0$, it is also positive for $m = 0$, $k = 0$.

(2) For $m \neq 0$, a different trick is performed. Keeping the mode number m, but replacing k by the parameter $\lambda \equiv k/m$, one finds:

$$f_0 = f_0(r; \lambda), \quad g_0 = h_0(r; \lambda) + r m^2 (B_\theta + \lambda B_z)^2. \tag{9.110}$$

9.4 Stability of cylindrical plasmas

Since the only term in m is positive definite, the worst case is $m = 1$, so that one may restrict the analysis to that mode.

In conclusion, *the cylindrical plasma is stable for all m and k, if it is stable for the cases* $m = 0$, $k \to 0$ *and* $m = 1$, *all k*.

As in Section 7.5, the main complication is the proper analysis of the singularities $F = 0$. These singularities are just the lower edges of the Alfvén and slow continua $\{\omega_A^2\}$ and $\{\omega_S^2\}$, which extend to $\omega^2 = 0$ if the interval $(0, a)$ contains a point where $F = 0$, i.e.

$$k + \mu m = 0. \tag{9.111}$$

For these values of the wave numbers m and k, the tangential wave vector is perpendicular to $\mathbf{B}$. In that case the phase of the perturbation is constant along the field lines at the position $r = r_s$ of the singularity. Expanding all quantities in terms of the variable

$$s \equiv r - r_s, \tag{9.112}$$

so that

$$F \approx m B_z \mu' s, \qquad m^2 + k^2 r^2 \approx m^2(1 + \mu^2 r^2), \tag{9.113}$$

we obtain

$$f_0 \approx \frac{r^3 B_z^2 \mu'^2}{1 + \mu^2 r^2} s^2, \qquad g_0 \approx \frac{2\mu^2 r^2}{1 + \mu^2 r^2} p'. \tag{9.114}$$

Consequently, close to the singularity, the Euler–Lagrange equation (9.105) reduces to

$$(s^2 \xi')' - \alpha \xi = 0, \tag{9.115}$$

where

$$\alpha \equiv 2\mu^2 p' / (r B_z^2 \mu'^2). \tag{9.116}$$

The solutions of the equation (9.115) are s^{ν_1} and s^{ν_2}, where ν_1 and ν_2 are the roots of the indicial equation $\nu(\nu + 1) - \alpha = 0$:

$$\nu_{1,2} = -\tfrac{1}{2} \pm \tfrac{1}{2}\sqrt{1 + 4\alpha}. \tag{9.117}$$

The discussion of the implications of real or complex indices is again identical to that given in Section 7.5.2.

The condition $1 + 4\alpha > 0$, which is necessary for the absence of the oscillatory solutions, was derived first by Suydam [222] and is, therefore, known as *Suydam's*

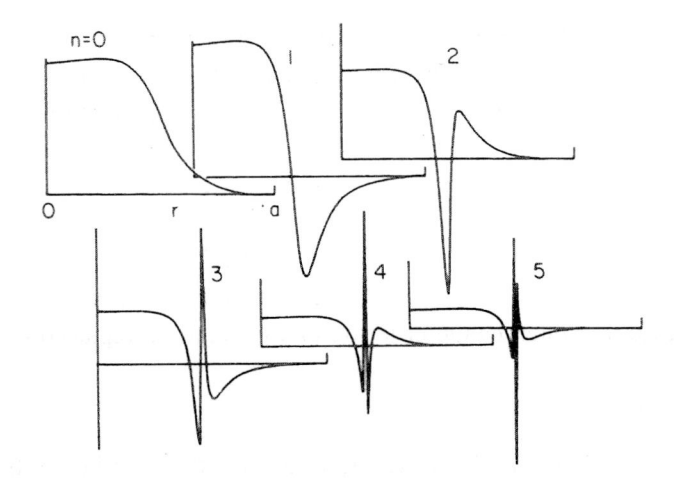

Fig. 9.12. Suydam unstable $m = 1$ modes with an increasing number of radial nodes of the eigenfunctions. (From Goedbloed and Sakanaka [89].)

criterion:

$$p' + \tfrac{1}{8} r B_z^2 \left(\frac{\mu'}{\mu} \right)^2 > 0 . \qquad (9.118)$$

Its violation implies the existence of highly localized instabilities close to a singular surface where $k + \mu m = 0$. These instabilities are so-called flute modes which interchange the magnetic field lines without appreciable bending. They are driven by the pressure gradient p' and stabilized by the magnetic shear, if the second term is large enough. One of the merits of Suydam's criterion is that it provides a simple explicit condition that may be tested easily and that, at least for laboratory fusion research, suggests measures (like increasing the shear or lowering the value of the pressure gradient) to be taken to ensure its satisfaction. A considerably more complicated toroidal version of this condition is known as *the Mercier criterion* (1960) [153].

The real importance of the localized Suydam solutions, however, resides in the implications obtained from the MHD oscillation theorem. If Suydam's criterion is violated, so that the marginal equation of motion has solutions that oscillate infinitely rapidly, the oscillation theorem asserts that a global $n = 0$ solution to the full equation of motion exists for which the growth rate $-\omega^2$ is larger than that of all the higher node solutions. In other words: *violation of Suydam's criterion implies the existence of a global $n = 0$ instability* (Fig. 9.12). This instability may also be global in the azimuthal direction (e.g. $m = 1$) if the mode number k may be chosen such that $k + \mu m = 0$ somewhere on the interval $(0, a)$.

Hence, Suydam's criterion provides a first test of stability which is quite significant. Clearly, violation of Suydam's criterion (9.102) is the condition that the marginal point $\omega^2 = 0$ is *an accumulation (or cluster) point* of the unstable side of the discrete spectrum.

Finally, Newcomb's stability test in the presence of a singular point $F = 0$ involves the consideration of both complex and real indices. This leads to the general stability theorem stated at the end of Section 7.5.2(b) for the analogous case of a gravitating slab. We will not repeat that theorem here but just note that, for the cylindrical case, the interchange criterion (7.199) is to be replaced by Suydam's criterion (9.118).

(b) σ-stability In Section 7.4.3 we have proved the oscillation theorem for the plane inhomogeneous plasma slab. This proof carries over to inhomogeneous cylindrical plasmas with appropriate modifications, exploiting the expressions (9.39) and (9.40) for the perturbation Π of the total pressure. Sturmian branches of the slow and Alfvén sub-spectra were foreseen in the proof of Chapter 7, when it still had to be shown that such branches actually exist. In the meantime, we have encountered plenty of examples demonstrating this. The most important one has just been discussed, viz. *instabilities* for values of m and k such that $F = 0$ at some point in the interval $(0, a)$. Since the continua $\{\omega_S^2\}$ and $\{\omega_A^2\}$ then stretch out to $\omega^2 = 0$, the mere existence of instabilities indicates that at least one of the Alfvén or slow branches of the discrete spectrum has become Sturmian. It is convenient that the function N/D never changes sign on the unstable side of the spectrum, so that *unstable modes are always Sturmian*. This is also in agreement with our intuition that moving the wall inward does not increase the growth rate of an unstable mode, which would be the case if the unstable side of the spectrum were anti-Sturmian.

Since the unstable side of the spectrum is non-singular, we immediately obtain a theorem for σ-stability of the diffuse cylindrical plasma. To that end, we notice that the σ-*marginal equation of motion* (6.116) for the diffuse cylindrical plasma is obtained from the Hain–Lüst equation (9.31) by just replacing ω^2 by $-\sigma^2$:

$$\left[P(r; -\sigma^2) \chi' \right]' - Q(r; -\sigma^2) \chi = 0, \tag{9.119}$$

where

$$\chi(0) = \chi(a) = 0. \tag{9.120}$$

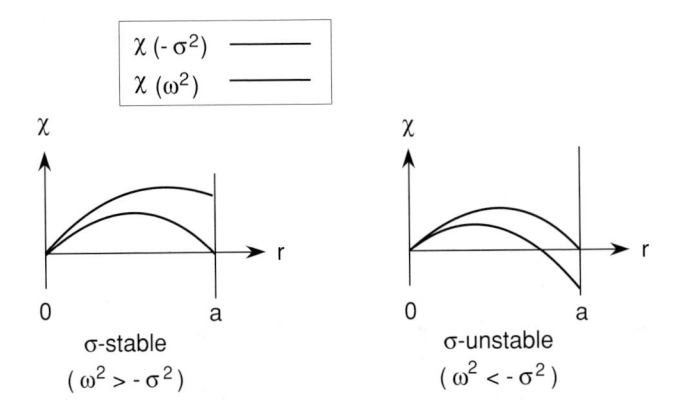

Fig. 9.13. Relationship between σ-marginal solutions and eigenfunctions.

The one-dimensional *modified energy principle* corresponding to this equation reads:

$$W^\sigma[\chi] = \pi L \int_0^a \left[P(r; -\sigma^2) \chi'^2 + Q(r; -\sigma^2) \chi^2 \right] r \, dr . \tag{9.121}$$

It could have been derived from Eq. (6.117) by a similar analysis to the one leading to the generalized Hain–Lüst equation. Here, we have simply posed it directly as that functional which produces Eq. (9.119) as the σ-Euler equation.

In general, analogous to the Euler–Lagrange equation (9.105), Eq. (9.119) does not have solutions satisfying both boundary conditions (9.120). This problem is solved in the same way as in the ordinary stability theory (see Section 7.5.1). Suppose that we integrate Eq. (9.119) starting from the left endpoint $r = 0$ where we satisfy the boundary condition $\chi = 0$. If the solution $\chi(r)$ thus obtained does not develop a zero in the open interval $0 < r < a$, our oscillation theorem asserts that a discrete eigenvalue $\omega^2 < -\sigma^2$ does not exist, so that the system is σ-stable. On the other hand, if the solution $\chi(r)$ vanishes somewhere on the open interval $0 < r < a$, a discrete eigenvalue $\omega^2 < -\sigma^2$ does exist for which both boundary conditions (9.120) are satisfied (Fig. 9.13). This result could also have been obtained from Eq. (9.121) where it just coincides with Jacobi's minimization condition from the calculus of variations (see, e.g., Smirnov [208]). We then have the following theorem for σ-stability of the diffuse cylindrical plasma.

Theorem. *For specified values of m and k, the diffuse cylindrical plasma is σ-stable if, and only if, the non-trivial solution χ of the σ-marginal equation of motion (9.119) that vanishes at $r = 0$ does not have a zero in the open interval $(0, a)$.*

The wording of this theorem is the same as that of the parallel theorem of Newcomb, discussed above, for the theory of marginal stability in the usual sense.

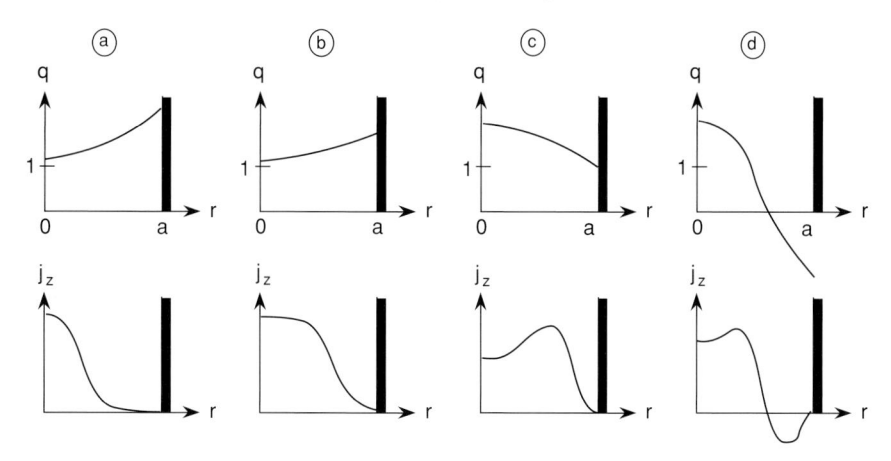

Fig. 9.14. Schematic overview of the different σ-stable cylindrical configurations: (a) tokamak; (b) flux-conserving tokamak; (c) screw pinch; (d) reversed field pinch. (After Sakanaka and Goedbloed [199].)

However, since in the latter theory the singularities associated with the continua at $\omega^2 = 0$ have to be accounted for, the marginal theory in the usual sense is much more complicated than the corresponding theory for σ-stability.

On the basis of the σ-stability theorem it is possible to systematically search for σ-stable configurations while taking a reasonable choice for σ, e.g. one which corresponds to the msec time scale. From a large number of numerical runs the following qualitative picture emerged. There are, broadly speaking, four categories of diffuse cylindrical configurations that are σ-stable with respect to internal modes. All four of them are characterized by a monotonically increasing or decreasing q-profile, representing shear of the field lines, which turns out to facilitate stability. (Conforming to present conventions, the parameter q is used instead of μ, although the results presented here strictly refer to infinite cylinder theory.) The q and j_z profiles for these configurations are the most characteristic ones to distinguish the different configurations, as illustrated in Fig. 9.14. As the current profile is broadened the maximum allowable β for stability in general increases from a few per cent for tokamaks to some 40% for the reversed field pinch. Except for the latter configuration all other configurations require $q > 1$, either on axis when the q-profile is increasing as in a tokamak, or at the wall when the q-profile is decreasing as in a screw pinch.

9.4.2 Stability of plasmas with shearless magnetic fields

(a) Instabilities of a z-pinch The most well-known MHD instabilities are the sausage ($m = 0$) and kink ($m = 1$) instabilities of a z-pinch ($B_z = 0$, $B = B_\theta$).

Since the magnetic field is exclusively azimuthal, it has no shear so that Newcomb's theory does not apply to the $m = 0$ modes, which are pure interchanges (similar to the gravitational interchanges in the absence of shear in Section 7.5.3). For those modes, $F = 0$, so that the variable ζ disappears from the expression (9.102) for the energy and the last two terms have to be combined:

$$W = \pi L \int_0^a \left\{ \frac{2p'}{r} \xi^2 + k^2 B^2 \left[\eta - \frac{1}{kr}\left((r\xi)' - 2\xi\right) \right]^2 + \gamma p \left[\frac{1}{r}(r\xi)' - k\eta\right]^2 \right\} r\, dr$$

$$= \pi L \int_0^a \left\{ \left[\frac{2p'}{r} + \frac{4\gamma p B^2}{r^2(\gamma p + B^2)}\right] \xi^2 \right.$$

$$\left. + k^2(\gamma p + B^2)\left[\eta - \frac{1}{kr}\left((r\xi)' - \frac{2B^2}{\gamma p + B^2}\xi\right)\right]^2 \right\} r\, dr. \quad (9.122)$$

Upon minimization, the last term disappears and we obtain the following stability criterion for the $m = 0$ sausage modes:

$$-rp' < \frac{2\gamma p B^2}{\gamma p + B^2} \quad \text{(everywhere)}. \quad (9.123)$$

This implies that there is a limit on each point of the radial pressure profile of a confined plasma (which requires a negative pressure gradient) in a z-pinch. Again, as in Section 7.5.3, the stability criterion for pure interchanges is less severe than the local interchange condition (Suydam's criterion in this case) in the limit of no shear.

For the $m \neq 0$ modes, Newcomb's analysis applies. Here, it is interesting to consider the limit $k \to \infty$ since that approaches the interchange condition most closely. In that case,

$$f_0 \to 0, \qquad g_0 \approx 2p' + \frac{m^2 B^2}{r}, \quad (9.124)$$

so that the stability criterion for $m \neq 0$ modes becomes:

$$-rp' < \tfrac{1}{2} m^2 B^2 \quad \text{(everywhere)}. \quad (9.125)$$

Notice that the limit of Suydam's criterion, $p' > 0$, would be obtained for $m \to 0$, just like the quasi-interchange stability criterion of Section 7.5.3. However, in a cylinder, the smooth approach of this limit is excluded because of azimuthal periodicity, and the limit $m = 0$ itself is governed by the stability criterion (9.123). Hence, $m = 1$ becomes the worst case.

Comparing the two stability criteria: for local values of $\gamma\beta \equiv c^2/b^2 < 1/3$, the $m = 0$ criterion (9.123) is the more restrictive condition. For $\gamma\beta > 1/3$, the criterion (9.125) with $m = 1$ is the more restrictive one; see Kadomtsev [119].

9.4 Stability of cylindrical plasmas

Hence, it is possible to construct stable pressure profiles for a z-pinch, but they 'live' in a sea of violently unstable profiles, so that experimental control would be extremely complicated and risky.

It is instructive to consider the special case of an incompressible z-pinch with the current and pressure distributions (9.94) and (9.95) of Section 9.3.2(c), but taking $B_z = 0$. For that equilibrium, $-rp' = 2A^2r^2 = 2B^2$, so that the $m = 0$ criterion is marginally satisfied, but the $m = 1$ criterion is violated. To compute the instability threshold $k = k_0$ for the $m = 1$ modes, we exploit the eigenfunction (9.98) with

$$B_z = 0 \quad \Rightarrow \quad \alpha = \frac{2A^2}{\rho\omega^2 - A^2}, \quad k^{*2} = k^2(1 - \alpha^2). \tag{9.126}$$

For marginal stability, $\alpha = -2$ so that $k^* = ik\sqrt{3}$ and the solution becomes:

$$\chi_0 = Cx^{-1}[x^2 J_1(x)]', \quad \text{where } x \equiv kr\sqrt{3}. \tag{9.127}$$

Increasing the value of $|k|$, instability sets in when $\chi_0(a) = 0$ for the first time, i.e. for $|k_0|a = 1.58$. Hence, this particular z-pinch is $m = 1$ unstable for $|k|a > 1.58$.

An estimate of the growth rate of the $m = 1$ modes may be obtained as well from the eigenfunction (9.98). Since $1 - \alpha^2 < 0$ for $-A^2 < \rho\omega^2 < 0$, the Bessel function oscillates infinitely rapidly in that range when $k \to \infty$, unless $\alpha^2 \approx 1$. Therefore, the growth rate of the $m = 1$ kink mode in the limit $k \to \infty$ is given by:

$$\rho\omega^2 \approx -A^2 = -\frac{B_\theta^2}{\mu_0 r^2} \quad \Rightarrow \quad \omega \approx i\frac{B_\theta}{r\sqrt{\mu_0\rho}} = \tfrac{1}{2}i\sqrt{\frac{\mu_0}{\rho}}\, j_z. \tag{9.128}$$

This expression shows that the internal kink modes of a z-pinch exponentiate on the same time scale as the external kink modes, for which the growth rate was given by Eq. (2.159). Since growth rates of this order of magnitude cannot be tolerated in magnetic fusion devices, the z-pinch has been abandoned there long ago. However, in inertial confinement fusion and in discharges for laser wake-field acceleration, where time scales are very much shorter, the z-pinch is a valuable plasma confinement scheme.

(b) Interchanges and quasi-interchanges in a constant-pitch magnetic field The peculiar crossing of the stability criteria for the z-pinch, at the interchange point $F = 0$, is a general property of shearless magnetic fields. As we have seen, for sheared magnetic fields, the singular points $F = 0$ significantly influence the stability properties of the plasma. In shearless magnetic fields, either such singularities are absent or the whole interval is singular. This causes the discontinuous

behaviour of the stability criteria. These discontinuities disappear in the expressions for the growth rates of the instabilities. We have already encountered this effect in Section 7.5.3 for the gravitational interchanges. Here, we present the cylindrical counterpart, developed in the papers by Ware [242], Goedbloed [81](III), and Goedbloed and Hagebeuk [86].

For analytical calculations, it is expedient to exploit the simplified form (9.46) of the generalized Hain–Lüst for low-frequency waves or instabilities. We here exploit the special case of a shearless magnetic field $\mu' = 0$:

$$\frac{d}{dr}\left[\frac{\rho\omega^2 - F^2}{m^2 + k^2 r^2} r \frac{d\chi}{dr}\right]$$

$$-\frac{1}{r}\left[\rho\omega^2 - F^2 - \frac{4k^2 B_\theta^2}{m^2 + k^2 r^2} \frac{\gamma p \rho \omega^2}{(\gamma p + B^2)\rho\omega^2 - \gamma p F^2} - \frac{2B_\theta^2}{rB^2} p'\right.$$

$$\left. + \frac{4B_\theta F}{m^2 + k^2 r^2}\left(\frac{k^2 mr}{m^2 + k^2 r^2} + \frac{m}{B^2} p' + \frac{B_\theta}{B^2} F\right)\right]\chi = 0. \qquad (9.129)$$

This form of the equation of motion clearly exhibits the terms driving the interchange instabilities ($\sim p'$) and the terms stabilizing them ($\sim F$). To further simplify the equation, we exploit the *low-β tokamak ordering* (introduced in Section 9.1.1) for the equilibrium quantities, and an ordering of the wave numbers of the perturbations to focus on *the range of the interchanges ($F = 0$) and quasi-interchanges ($F \neq 0$)*.

We demonstrate these orderings with a particular equilibrium obtained by generalizing the constant-pitch force-free field of Eq. (9.11):

$$B_z = \frac{B_0}{1 + \delta^2 r^2} \approx B_0(1 - \delta^2 r^2), \qquad B_\theta = \frac{B_0 \mu r}{1 + \delta^2 r^2} \approx B_0 \mu r,$$

$$p = \text{const} + \frac{B_0^2(\mu^2 - \delta^2)}{2\delta^2(1 + \delta^2 r^2)^2} \approx B_0^2 [\tfrac{1}{2}\beta - (\mu^2 - \delta^2) r^2], \qquad (9.130)$$

$$p' = -\frac{2B_0^2(\mu^2 - \delta^2) r}{(1 + \delta^2 r^2)^3} \approx -2B_0^2(\mu^2 - \delta^2) r.$$

For $\delta^2 = \mu^2$, the pressure gradient vanishes and the force-free field (9.11) is recovered. The low-β tokamak ordering implies

$$\beta \sim \mu^2 a^2 \sim \delta^2 a^2 \sim \epsilon^2, \qquad (9.131)$$

giving the approximations indicated in Eqs. (9.130). Since the pitch of the field lines is constant, we may introduce a parallel wave number that is approximately constant:

$$F \equiv k_\| B, \qquad k_\| = \frac{k + \mu m}{\sqrt{1 + \mu^2 r^2}} \approx k + \mu m. \qquad (9.132)$$

Next, we order the wave numbers as follows:

$$k_\| a \ll ka \ll m^2 \sim 1 \quad \Rightarrow \quad m^2 + k^2 r^2 \approx m^2, \quad k^2/m^2 \approx \mu^2. \qquad (9.133)$$

With the orderings (9.131) and (9.133), the last three terms of Eq. (9.129) are negligible. Further, with $\rho = $ const and $F = $ const, the Alfvén factor $\rho \omega^2 - F^2$ may be extracted and we obtain the following form of the eigenvalue problem:

$$\frac{d}{dr}\left(r \frac{d\chi}{dr}\right) - \frac{1}{r}(m^2 - \lambda^2 r^2)\chi = 0, \qquad \chi(0) = \chi(a) = 0, \qquad (9.134)$$

where

$$\lambda \equiv \frac{m^2 B_\theta^2}{r^2(\rho\omega^2 - F^2)}\left[4\mu^2 \frac{\gamma p \rho \omega^2}{(\gamma p + B^2)\rho\omega^2 - \gamma p F^2} + \frac{2p'}{rB^2}\right] \approx \text{const}. \qquad (9.135)$$

The assumption $\lambda \approx $ const is justified for the particular equilibrium chosen, but it may also be assumed for radially localized modes in more general equilibria.

The solutions of Eq. (9.134) are Bessel functions:

$$\chi = J_m(\sqrt{\lambda} r), \qquad \lambda = j_{mn}^2/a^2, \qquad (9.136)$$

where j_{mn} are the consecutive zeros of J_m. This yields the following dispersion equation:

$$\rho^2 \omega^4 - \left[\frac{2\gamma p + B^2}{\gamma p + B^2} F^2 + \frac{m^2}{j_{mn}^2}\frac{2B_\theta^2}{r^2}\left(\frac{2\mu^2 a^2 \gamma p}{\gamma p + B^2} + \frac{a^2 p'}{rB^2}\right)\right]\rho\omega^2$$

$$+ \frac{\gamma p}{\gamma p + B^2} F^2 \left(F^2 + \frac{m^2}{j_{mn}^2}\frac{2B_\theta^2}{r^2}\frac{a^2 p'}{rB^2}\right) = 0. \qquad (9.137)$$

Comparing this dispersion equation with the analogous dispersion equation (7.211) for the gravitational interchanges (Section 7.5.3) shows that the first term with p' plays the role of the magnetically modified Brunt–Väisälä frequency N_m^2, whereas the second term with p' plays the role of the Brunt–Väisälä frequency N_B^2 itself. Here, we may exploit a similar ordering, $F^2 \ll a^2 B_\theta^2 p'/(r^3 B^2)$, to obtain the growth rates of the pure interchanges ($F = 0$) and quasi-interchanges

($F \neq 0$):

$$F = 0: \quad \rho\omega_1^2 \approx \frac{m^2}{j_{mn}^2} \frac{4B_\theta^2}{r^2}(\Pi - \Pi_1), \quad \text{where} \quad \Pi \equiv \frac{a^2 p'}{2rB^2}, \quad \Pi_1 \equiv -\frac{\mu^2 a^2 \gamma p}{\gamma p + B^2},$$

$$F \neq 0: \quad \rho\omega_2^2 \approx \frac{\gamma p}{\gamma p + B^2} \frac{\Pi}{\Pi - \Pi_1} F^2. \tag{9.138}$$

The pure interchanges are stable when $\Pi - \Pi_1 > 0$, which is the generalization of Kadomtsev's criterion (9.123) for the z-pinch. The quasi-interchanges are stable when $\Pi > 0$, which is the constant-pitch limit of Suydam's criterion. In contrast to the z-pinch, taking the latter limit makes sense now.

One can also compute the maximum growth rate for arbitrary values of $k_\parallel$, like the expressions (7.214) and (7.215) of Section 7.5.3. This yields the following results:

(1) for $\Pi \geq 0$, the plasma is stable;
(2) for $\Pi_2 \equiv -\mu^2 a^2 \gamma p (2\gamma p + B^2)/(\gamma p + B^2)^2 \leq \Pi < 0$, the most unstable mode is a quasi-interchange ($k_\parallel \neq 0$) with growth rate

$$\rho\omega_{1,\max}^2 = -\frac{m^2}{j_{m1}^2} \frac{B^2}{a^2} \left[\frac{2\gamma p \mu^2 a^2}{B^2} \left(1 - \sqrt{1 + \frac{B^2}{\gamma p \mu^2 a^2} \Pi} \right) \right]^2; \tag{9.139}$$

(3) for $\Pi \leq \Pi_2$, the most unstable mode is a pure interchange ($k_\parallel = 0$) with a growth rate given by Eq. (9.138)(a), substituting j_{m1}.

The degenerate example of an equilibrium with constant-pitch magnetic field demonstrates that stability conditions, considered for fixed wave number, may exhibit discontinuities that disappear when the growth rates are considered. Fig. 9.15 clearly shows the gradual change of the growth rate $-\omega^2(k_\parallel)$ when the negative pressure gradient is increased. The maxima of these growth rates are plotted in Fig. 9.16. This picture highlights another feature: if one wishes to establish overall stability of a particular magnetic configuration, the value chosen for the threshold $\sigma^2 \equiv -\omega^2$, of growth rates considered to be intolerable, decides the outcome. For a constant-pitch magnetic field, Suydam's stability criterion (9.118) degenerates into the quasi-interchange stability condition $\Pi > 0$ (i.e. $p' > 0$) and not into the pure interchange condition $\Pi > \Pi_1$, as one might have expected. However, the growth rates of the quasi-interchanges are rather small, as is clear from Fig. 9.16, so that one could imagine that the instabilities of the range $\Pi_1 < \Pi < 0$ would be acceptable for certain experimental purposes. On the other hand, if one insisted on absolute stability, the quasi-interchange condition $\Pi > 0$ would not be enough either, because other instabilities (the quasi-kinks depicted in the inset of Fig. 9.16) are still unstable in the absence of pressure gradients [86].

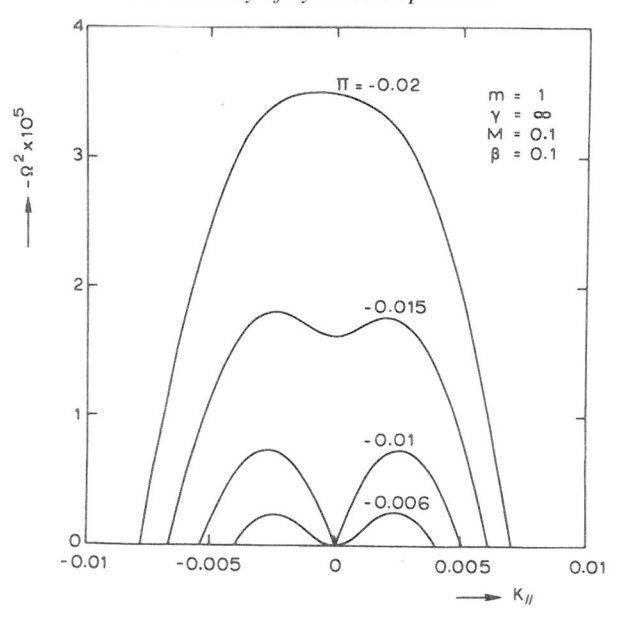

Fig. 9.15. Normalized growth rate, $-\Omega^2 \equiv -(\rho a^2/B^2)\omega^2$, of the interchange and quasi-interchange instabilities in a constant-pitch magnetic field as a function of $K_\| \equiv k_\| a$ for different values of the normalized pressure gradient $\Pi \equiv a^2 p'/(2rB^2)$; $M \equiv \mu a$ is the normalized pitch of the field lines. (From Goedbloed and Hagebeuk [86].)

9.4.3 Stability of force-free magnetic fields

Stability analysis becomes more complicated when the restriction of a constant-pitch magnetic field is dropped, which usually means that solutions can only be obtained numerically (as in the final analysis of Section 9.4.1). In this section and the next, we discuss two important configurations, one relevant for astrophysical plasmas and one for laboratory fusion plasmas, where semi-analytical solutions can be constructed which, again, centre about the $F = 0$ singularities.

For many astrophysical plasmas, like magnetic flux loops in the solar corona, gravity and the kinetic pressure of the plasma are negligible compared to the magnetic pressure, so that $\nabla p \approx 0$ and the magnetic field is force-free:

$$\mathbf{j} \times \mathbf{B} = 0 \quad \Rightarrow \quad \mathbf{j} = \alpha(\mathbf{r})\mathbf{B}. \tag{9.140}$$

One of the simplest, non-trivial, examples is the Lundquist field, a cylindrical force-free magnetic field with a constant value of α. This field has been discussed extensively in Section 4.3.1, in the context of magnetic helicity, and in Section 9.1.1(a). The explicit Bessel function expressions for B_θ and B_z are given in Eq. (4.115), or Eq. (9.12), and illustrated in Fig. 4.9. Since the instabilities of

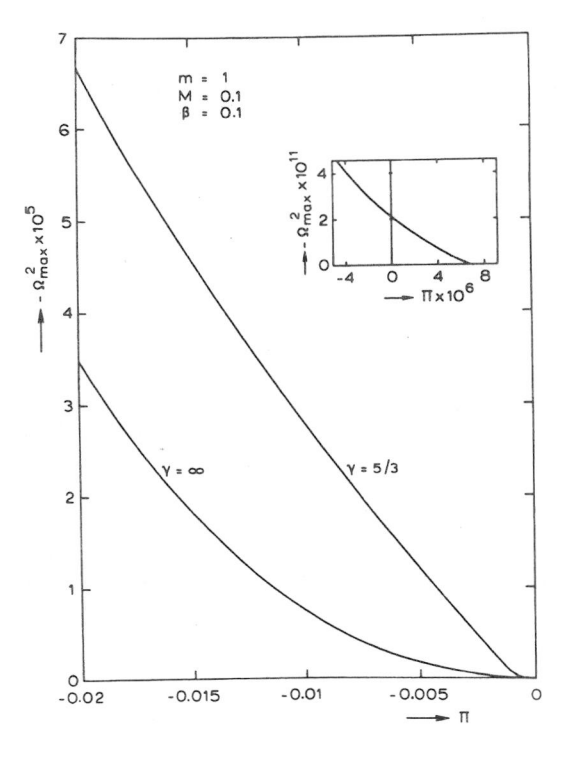

Fig. 9.16. Maximum growth rate as a function of the pressure gradient for interchanges and quasi-interchanges in a constant-pitch magnetic field. The inset shows the region of quasi-kinks. (From Goedbloed and Hagebeuk [86].)

the Lundquist field essentially occur for $B_\theta \sim B_z$, we cannot exploit the low-β tokamak approximation to simplify the Bessel function expressions.

(a) Solution of the marginal equation of motion Rather than exploiting Newcomb's equations, it is expedient to derive the marginal stability equations from the original expressions (6.29) and (6.30) for the force operator. The pressure terms disappear, since $\nabla p = 0$ and $\nabla \cdot \boldsymbol{\xi} = 0$ at marginal stability, so that the marginal equation of motion becomes

$$\mathbf{F}(\boldsymbol{\xi}) = -\mathbf{B} \times (\nabla \times \mathbf{Q}) + \mathbf{j} \times \mathbf{Q} = -\mathbf{B} \times (\nabla \times \mathbf{Q} - \alpha \mathbf{Q}) = 0. \quad (9.141)$$

Hence, the magnetic field perturbation $\mathbf{Q} = \nabla \times (\boldsymbol{\xi} \times \mathbf{B})$ satisfies the differential equation

$$\nabla \times \mathbf{Q} - \alpha \mathbf{Q} = \lambda \mathbf{B}, \quad (9.142)$$

where $\lambda(r)$ is a first order quantity (with suppressed phase factor $\exp i(m\theta + kz)$). To determine λ, we take the divergence of Eq. (9.142), which only gives a contribution from the RHS,

$$\mathbf{B} \cdot \nabla \lambda = iF\lambda = 0 \quad \Rightarrow \quad \lambda = u(r)\delta(r - r_s), \qquad (9.143)$$

when there is a singularity $F = 0$ at $r = r_s$. Clearly, the quantity λ represents a skin current induced by the perturbation at the singularity, where the strength $u(r)$ is to be determined yet.

We solve Eq. (9.142) by eliminating Q_r and Q_θ from the first two components,

$$Q_r = \frac{i}{(\alpha^2 - k^2)r}\left[kr(Q'_z + B_\theta\lambda) + m\alpha Q_z\right],$$
$$Q_\theta = \frac{-1}{(\alpha^2 - k^2)r}\left[\alpha r(Q'_z + B_\theta\lambda) + km Q_z\right], \qquad (9.144)$$

and substituting them in the third component:

$$\frac{1}{r}\frac{d}{dr}\left(r\frac{dQ_z}{dr}\right) + \left(\alpha^2 - k^2 - \frac{m^2}{r^2}\right)Q_z = -B_\theta\lambda' - 2\alpha B_z\lambda. \qquad (9.145)$$

(Here, we have used the force-free field condition $j_z = (1/r)(rB_\theta)' = \alpha B_z$ and cancelled a term $F\lambda = 0$.) The solution of the homogeneous equation that is regular at $r = 0$ is a Bessel function of the first kind:

$$\frac{1}{r}\frac{d}{dr}\left(r\frac{d\phi}{dr}\right) + \left(\alpha^2 - k^2 - \frac{m^2}{r^2}\right)\phi = 0 \quad \Rightarrow \quad \phi = C J_m(\sqrt{\alpha^2 - k^2}\,r), \qquad (9.146)$$

where we may restrict the analysis to $\alpha^2 > k^2$ since the boundary condition at $r = a$ (to be discussed below) requires oscillatory solutions. A solution of the inhomogeneous equation (9.145) can then be constructed that is only different from zero on the interval between the axis ($r = 0$) and the first singularity ($r = r_s$):

$$Q_z = \phi(r)H(r_s - r) = C J_m(\sqrt{\alpha^2 - k^2}\,r)H(r_s - r). \qquad (9.147)$$

Substituting this expression back into Eq. (9.145), and eliminating the homogeneous contributions, yields *two* conditions from the factors multiplying δ' and δ:

$$\phi\delta' + \left(2\phi' + \frac{\phi}{r}\right)\delta = B_\theta u\delta' + (B_\theta u' + 2\alpha B_z u)\delta \quad \Rightarrow \quad \begin{cases} u = \dfrac{\phi}{B_\theta} \\ B_\theta\phi' - \alpha B_z\phi = 0. \end{cases} \qquad (9.148)$$

The first one is the relation between u and ϕ we were looking for. However, we obtain an additional condition, a kind of internal boundary condition, to be satisfied by ϕ in order for the discontinuous solution (9.147) to be acceptable.

Since these conditions fix everything in the problem, let us consider them in detail. From $\mathbf{j}_1 = \nabla \times \mathbf{Q} = \alpha \mathbf{Q} + \lambda \mathbf{B}$, the skin current at the singularity is now determined:

$$\mathbf{j}_1^\star = \int_{r_s^-}^{r_s^+} \mathbf{j}_1 \, dr = \int_{r_s^-}^{r_s^+} u \delta(r - r_s) \mathbf{B} \, dr = \left[(\mathbf{B}/B_\theta) \phi \right]_{r_s}. \qquad (9.149)$$

According to Eq. (4.167) of Section 4.5.2, this skin current also fixes the two tangential components of $\mathbf{Q}$:

$$\mathbf{j}_1^\star = \mathbf{n} \times [\![\mathbf{Q}]\!] \quad \Rightarrow \quad \begin{cases} j_{1\theta}^\star = -[\![Q_z]\!] = Q_z(r_s^-) = \phi \big|_{r_s} \\ j_{1z}^\star = [\![Q_\theta]\!] = -Q_\theta(r_s^-) = \left[(B_z/B_\theta) \phi \right]_{r_s}. \end{cases} \qquad (9.150)$$

Evaluating everything at r_s^-, where $F \equiv m B_\theta / r + k B_z = 0$, this yields:

$$B_z j_{1\theta}^\star - B_\theta j_{1z}^\star = B_z Q_z + B_\theta Q_\theta = \frac{r B_z}{m} \left(\frac{m}{r} Q_z - k Q_\theta \right) \stackrel{(9.142)}{=} -i \frac{\alpha r B_z}{m} Q_r$$

$$\stackrel{(9.144)}{=} \frac{\alpha B_z}{m(\alpha^2 - k^2)} (kr\phi' + m\alpha \phi) = -\frac{\alpha}{\alpha^2 - k^2} (B_\theta \phi' - \alpha B_z \phi) = 0. \qquad (9.151)$$

Hence, the internal boundary condition (9.148)(b) turns out to be equivalent to the requirement that the perturbed radial magnetic field Q_r has to vanish at the singularity, i.e. we have recovered the condition that the displacement $\xi_r = -iQ_r/F$ has to be 'small' there: the circle is closed.

Inserting the expression (9.144)(a) for Q_r, with the solution (9.147) for Q_z, in the expression for ξ_r, and transforming the Bessel function derivatives, finally yields the explicit solution of the marginal equation of motion:

$$\xi_r = \frac{C\alpha}{B_0(\alpha^2 - k^2)} \frac{m(1 - k/\alpha) J_m(\beta r) + (k/\alpha) \beta r J_{m-1}(\beta r)}{m J_1(\alpha r) + kr J_0(\alpha r)} H(r_s - r)$$

$$\equiv \frac{C\alpha}{B_0(\alpha^2 - k^2)} \frac{\widetilde{Q}(m, k/\alpha, \alpha r)}{\widetilde{F}(m, k/\alpha, \alpha r)} H(r_s - r), \quad \text{where } \beta \equiv \sqrt{\alpha^2 - k^2}. \qquad (9.152)$$

Clearly, it would have been extremely difficult to produce this solution directly from Newcomb's equations (9.105)–(9.107). More importantly, we now have the physical reason of how 'small' solutions come about: the skin currents effectively

(b) Stability analysis and calculation of the growth rates According to Newcomb's stability theorem, or the MHD oscillation theorem, the expression (9.152) is to be studied with respect to the zeros of both ξ_r (i.e. Q_r) and F. Since both functions are oscillatory in the relevant parameter domain, this involves a rather subtle analysis, that was carried out first by Voslamber and Callebaut [241]. Roughly speaking, for instability, the function $\widetilde{Q}$ in the numerator should oscillate faster than the function $\widetilde{F}$ in the denominator, since the zeros of the latter function (in addition to the origin $r = 0$ and the wall position $r = a$) delimit the independent sub-intervals. The first transition to instability is found when $F \sim k + \mu m = 0$ somewhere (e.g. at the position indicated by the first dotted line in the top panel of Fig. 9.18) and Q_r also vanishes there, so that it may be chosen to vanish identically to the right of that point. Hence, at that transition (where $\omega^2 = 0$), a kind of block-function displacement is obtained for ξ_r.

For most choices of the parameters m, k/α and αr, the expression (9.152) is not a marginal solution at all. Whereas it satisfies the boundary condition at the origin, $(r\xi_r)|_{r=0} = 0$, only for very specific parameters, the outer boundary condition is satisfied as well, i.e. $\xi(a) = 0$ when there is no singularity, or $\xi(r_s)$ is 'small' (i.e. $Q_r(r_s) = 0$) when there is a singularity $r_s < a$. (Vice versa, if the expression in the numerator $\neq 0$ at the wall or at the singularity and one insists on satisfying the outer boundary condition, one needs to replace that expression by a linear combination of the J_ms and N_ms (Bessel functions of the second order, also called Neumann functions) that has the required property. Of course, the resulting function ξ_r will not be a marginal solution either (it will be irregular at $r = 0$), but we know how to draw conclusions from its oscillatory behaviour: if it has no other zero on $(0, r_s)$ the sub-interval is stable, if there are zeros it is unstable. Clearly, there is no advantage to this approach, so that we will stick to the simpler expression (9.152).) Hence, for an arbitrary choice of parameters, satisfaction of both the inner and the outer boundary condition will not be obtained and Eq. (9.152) is just an auxiliary expression from which we may draw the proper conclusions with respect to stability.

Recall from Section 9.4.1 that only the $m = 0$, $k \to 0$ and $|m| = 1$ modes have to be investigated to determine stability of a particular configuration. The $m = 0$, $k \to 0$ modes are stable since $\xi_r \sim J_1(\alpha r)/J_0(\alpha r)$ in that case, and the first zero of J_0 is smaller than the first zero of J_1. Hence, only the $|m| = 1$ modes need to be investigated. These are the only unstable modes since their instability is due to 'fine-tuning' of the oscillations of $\widetilde{Q}$ and $\widetilde{F}$, which is lost for higher values of $|m|$, so that the $|m| \geq 2$ modes are stable, even if the $|m| = 1$ modes are unstable.

To compute the stability of the $m = 1$ modes, we define the parameters $\kappa \equiv k/\alpha$ and $x \equiv \alpha r$, and construct the following curves in the x–κ parameter plane:

$$\begin{aligned} \widetilde{Q}(m = 1, \kappa, x) = 0 \quad &\Rightarrow \quad x = x_0(\kappa) \quad \text{(zeros)}, \\ \widetilde{F}(m = 1, \kappa, x) = 0 \quad &\Rightarrow \quad x = x_s(\kappa) \quad \text{(singularities)}. \end{aligned} \qquad (9.153)$$

These curves are very close, they only intersect or touch at three marginally stable points $x^{(i)}$, $\kappa^{(i)}$ ($i = 1, 2, 3$), defined below, and they just leave two tiny patches of the x–κ plane where $x_0 < x_s$. For values of κ in those patches, the critical wall parameter for marginal stability is $x_a \equiv \alpha a = x_0$, and the plasma is unstable for all $x_a > x_0$ (including $x_a > x_s$ since skin-current stabilization does not work when $x_0 < x_s$). Consequently, the $m = 1$ modes are unstable in two strips of the x_a–κ plane:

$$\begin{aligned} 0.272 > \kappa > 0, \quad & x_a > x^+(\kappa) \quad \left[\text{with } x^{(1)} < x^+(\kappa) < x^{(2)}\right], \\ 0 > \kappa > -0.237, \quad & x_a > x^-(\kappa) \quad \left[\text{with } x^{(2)} < x^-(\kappa) < x^{(3)}\right], \end{aligned} \qquad (9.154)$$

where the three limiting values are given by

$$\begin{aligned} x^{(1)} &\approx 3.176, & x^{(2)} &\approx 3.832, & x^{(3)} &\approx 4.744, \\ \kappa^{(1)} &\approx 0.272, & \kappa^{(2)} &= 0, & \kappa^{(3)} &= -0.237. \end{aligned} \qquad (9.155)$$

At the central point, $x^{(2)} = j_{11} \approx 3.832$ (the first zero of J_1), $\kappa^{(2)} = 0$, the functions $\widetilde{Q}$ and $\widetilde{F}$ coincide (so that ξ_r becomes a step function) indicating marginal stability for $\alpha a \geq 3.832$ and $k = 0$.

Thus, stability of the Lundquist field depends on the two parameters αa and k/α, and the unstable region with respect to the $m = 1$ modes falls within the strip $\alpha a > 3.176$, $0.271 > k/\alpha > 0.271 - 0.237$. Instability with respect to the $m = -1$ modes is governed by similar conditions with the sign of k reversed. The result is quite reasonable: $m = \pm 1$ kink modes are long-wavelength, current driven, instabilities which occur when the total current ($\sim \alpha a$) is large enough and when the longitudinal wavelength ($\sim k^{-1}$) is large compared to a typical length scale of the radial inhomogeneity ($\sim \alpha^{-1}$).

To compute the growth rates of the instabilities of the Lundquist field by means of the generalized Hain–Lüst equation is significantly simpler than the stability analysis (granted that a computer program for the solution of this equation has been written), since the equation is non-singular for $\omega^2 < 0$ so that the shooting method (Section 7.5.1) can be applied. The results of such computations are shown in Figs. 9.17 and 9.18. In the first figure, the growth rate of the $m = 1$ kink mode

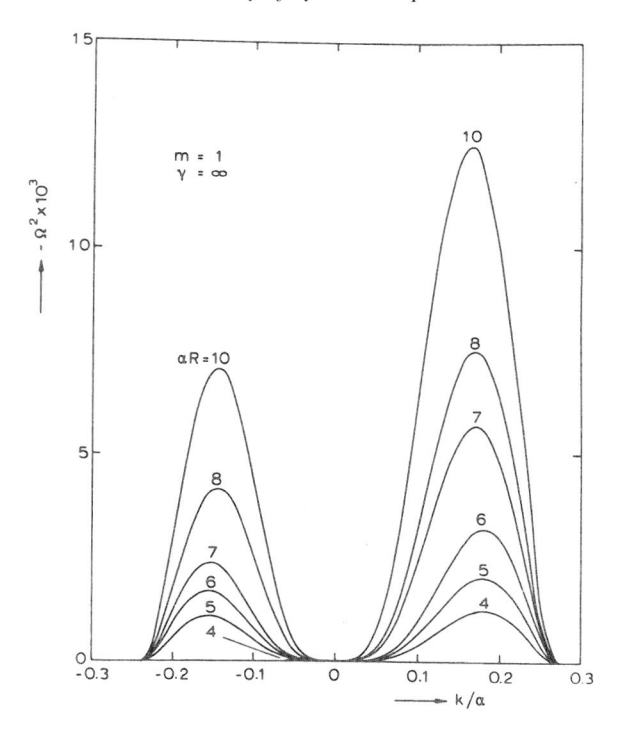

Fig. 9.17. Normalized growth rate, $-\Omega^2 \equiv -(\rho_0 R^2/B_0^2)\omega^2$, of the $m = 1$ kink mode in a Lundquist field as a function of k/α for several values of αR. (In this figure, the wall position is indicated by R instead of a.) (From Goedbloed and Hagebeuk [86].)

is shown for the unstable range of the parameters αa and k/α, found by Voslamber and Callebaut. Note the typical asymmetry of the growth rate with respect to the wave number k. The top panel of Fig. 9.18 shows the radial dependence of the inverse pitch μ of the magnetic field lines for a highly unstable Lundquist field (when the direction of the field has turned around several times), and two singular positions $k + \mu m = 0$ for a particular choice of the wave numbers. The bottom panel of this figure shows the eigenfunction of the unstable kink mode for that case, consisting of two approximately rigid displacements of different amplitude in the two independent sub-intervals of the marginal stability analysis. Note that the jumps of the marginal solution (the trial function of the energy principle) have disappeared for the finite value of $-\omega^2$ so that the actual eigenfunction is much smoother and, hence, more realistic than the trial function. With respect to the numerics, this implies that the computation of the growth rates requires significantly fewer grid points for the same accuracy of the result.

In conclusion, the stability of the force-free magnetic field configuration with respect to kink modes is determined by the competition between the destabilizing

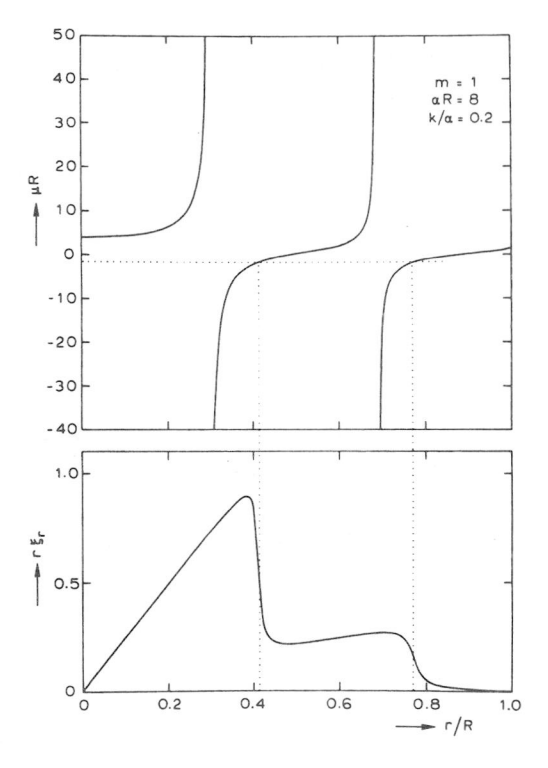

Fig. 9.18. Eigenfunction of the kink mode in a Lundquist field, $\alpha R = 8, k/\alpha = 0.2$. (In this figure, the wall position is indicated by R instead of a.) Top: inverse pitch of the field lines, indicating the radial positions of the singularity $k + \mu m = 0$ (dotted lines). Bottom: the $m = 1$ eigenfunction has smoothed out the jumps of the marginal mode at the singularities. (From Goedbloed and Hagebeuk [86].)

factor of the force-free equilibrium currents and the stabilizing factor of the skin-current perturbations at the singularities. This turned out to produce a very delicate balance in the stability analysis, where the force-free current destabilization just dominates, but only in a very tiny part of parameter space. Of course, this does not imply that the instabilities are hard to realize (nature immediately finds the route to the lowest energy state), but it does imply that the growth rates of the instabilities are significantly lower than they would be in the absence of skin-current stabilization.

9.4.4 Stability of the 'straight tokamak'

In this section, we present the energy principle counterpart of the normal-mode analysis of Section 9.2 for a cylindrical plasma surrounded by vacuum. We depart

9.4 Stability of cylindrical plasmas

from the extended energy principle of Section 6.6.3 with $W = W^p + W^s + W^v$, where the expressions for the plasma energy W^p, the surface energy W^s, and the vacuum energy W^v are given by Eqs. (6.157)–(6.159), respectively. The variables ξ and $\hat{Q}$ are subject to the boundary conditions (6.161)–(6.163). This general formalism is applied to the straight tokamak, where a toroidal configuration is represented by a periodic cylinder of finite length $L = 2\pi R_0$, so that the volume element becomes $dV = 4\pi^2 R_0 dr$.

Whereas some concepts of the preceding section on force-free fields will return (like singular trial functions and skin-current stabilization), the tokamak configuration is essentially different in two ways: (1) the cylinder has a finite length, so that only integer values of the longitudinal wave number $n = k/R_0$ need to be considered; (2) the q-profile is mainly increasing with radius and has a magnitude of order unity, in contrast to the Lundquist field, where $q \equiv (\mu R_0)^{-1}$ is radially decreasing (even going through zero) and $q_0 = 2\epsilon(\alpha a)^{-1} \ll 1$. These two factors make the stability properties completely different.

(a) Energy expression for a general cylindrical plasma–vacuum configuration
In the expressions of Newcomb's minimized form (9.104) of the plasma energy, a boundary term has been dropped (since only internal modes were considered in Section 9.4.1) that needs to be restored when we also permit external modes:

$$W^p = 2\pi^2 R_0 \int_0^a \left(f_0 \xi'^2 + g_0 \xi^2 \right) dr - 2\pi^2 R_0 \left[\frac{m^2 B_\theta^2 - k^2 r^2 B_z^2}{m^2 + k^2 r^2} \xi^2 \right]_{r=a}. \tag{9.156}$$

As in Section 6.6.4, Eq. (6.201), inserting back the minimizing solution of the Euler–Lagrange equation (9.105) into W^p yields:

$$W^p = 2\pi^2 R_0 \left[\frac{r^3 F^2}{m^2 + k^2 r^2} \frac{\xi'}{\xi} - \frac{m^2 B_\theta^2 - k^2 r^2 B_z^2}{m^2 + k^2 r^2} \right]_{r=a} \xi^2(a), \tag{9.157}$$

where the logarithmic derivative $(\xi'/\xi)|_{r=a}$ contains now all the necessary information of the plasma interval. The expression for the surface energy becomes

$$W^s = -2\pi^2 R_0 \left[\hat{B}_\theta^2 - B_\theta^2 \right]_{r=a} \xi^2(a). \tag{9.158}$$

As in Section 6.6.4(b), the vacuum energy is obtained by eliminating the variable $\hat{Q}_z$ and rearranging the resulting expression to get one positive definite term involving $\hat{Q}_\theta$:

$$W^v = 2\pi^2 R_0 \int_a^b \left[|\hat{Q}_r|^2 + |\hat{Q}_\theta|^2 + \frac{1}{k^2 r^2} \left|(r\hat{Q}_r)' + im\hat{Q}_\theta\right|^2 \right] r\, dr$$

$$= 2\pi^2 R_0 \int_a^b \left[\frac{1}{m^2 + k^2 r^2} \left|(r\hat{Q}_r)'\right|^2 + |\hat{Q}_r|^2 \right.$$

$$\left. + \frac{m^2 + k^2 r^2}{k^2 r^2} \left| i\hat{Q}_\theta + \frac{m}{m^2 + k^2 r^2} (r\hat{Q}_r)' \right|^2 \right] r\, dr. \tag{9.159}$$

Upon minimization, the last term vanishes and the first terms yield the Euler–Lagrange equation (9.56) with the Bessel function derivative solution (9.59), derived in Section 9.2.2. Inserting this solution again back into W^v, and expressing $\hat{Q}_r$ in ξ_r through the boundary condition (9.51), gives

$$W^v = -2\pi^2 R_0 \left[\frac{r}{m^2 + k^2 r^2} (r\hat{Q}_r)(r\hat{Q}_r)' \right]_{r=a}$$

$$= -2\pi^2 R_0 \left[\frac{r^3 \hat{F}^2}{m^2 + k^2 r^2} \frac{(r\hat{Q}_r)'}{r\hat{Q}_r} \right]_{r=a} \xi^2(a). \tag{9.160}$$

Collecting terms provides the general expression for the energy of a cylindrical plasma–vacuum configuration:

$$W = 2\pi^2 R_0 \left[\frac{r^3 F^2}{m^2 + k^2 r^2} \frac{\xi'}{\xi} - \frac{m^2 B_\theta^2 - k^2 r^2 B_z^2}{m^2 + k^2 r^2} + B_\theta^2 - \hat{B}_\theta^2 \right.$$

$$\left. - \frac{r^3 \hat{F}^2}{m^2 + k^2 r^2} \frac{(r\hat{Q}_r)'}{r\hat{Q}_r} \right]_{r=a} \xi^2(a), \tag{9.161}$$

where the logarithmic derivative for ξ is determined by solving the Euler–Lagrange equation (9.105), and the logarithmic derivative for $r\hat{Q}_r$ follows from Eq. (9.59):

$$\left[\frac{(r\hat{Q}_r)'}{\hat{Q}_r} \right]_{r=a} = \frac{m^2 + k^2 a^2}{ka} \frac{I_m(ka) K_m'(kb) - K_m(ka) I_m'(kb)}{I_m'(ka) K_m'(kb) - K_m'(ka) I_m'(kb)}$$

$$\left(\approx -|m| \frac{1 + (a/b)^{2|m|}}{1 - (a/b)^{2|m|}} \right). \tag{9.162}$$

The approximation in brackets on the RHS holds for long-wavelength perturbations ($k^2 a^2 \sim k^2 b^2 \ll m^2 \sim 1$), appropriate for the tokamak problem considered below.

(b) Energy expression for the 'straight tokamak' We now construct the leading order expression for the energy in the low-β tokamak ordering. This ordering

9.4 Stability of cylindrical plasmas

continues with genuine toroidal contributions in the next orders, see Volume 2. In the leading order (straight cylindrical) contribution, only the safety factor q enters:

$$q \approx \frac{1}{\mu R_0} = \frac{r B_z}{R_0 B_\theta} \approx \frac{B_0}{R_0} \frac{r}{B_\theta} \sim 1, \qquad (9.163)$$

the longitudinal field $B_z \approx B_0$, and pressure effects are negligible:

$$\beta \equiv 2p_0/B_0^2 \sim \epsilon^2 \ll 1. \qquad (9.164)$$

Hence, to leading order in the low-β tokamak ordering, we may forget about the cylindrical equilibrium equation (9.3), and just arbitrarily specify either one of the profiles for $q(r)$, $B_\theta(r)$, or the current density

$$j_z(r) = \frac{1}{r}(r B_\theta)' = \frac{B_0}{R_0} \frac{1}{r}\left(\frac{r^2}{q}\right)'. \qquad (9.165)$$

Next, as in Eq. (9.81), we replace the longitudinal wave number k by the toroidal mode number n, $k \equiv n/R_0$, and assume both, toroidal and poloidal, mode numbers to be of order unity:

$$m \sim n \sim 1 \quad \Rightarrow \quad m^2 + k^2 r^2 \approx m^2, \qquad F \approx \frac{B_0}{R_0}\left(n + \frac{m}{q}\right). \qquad (9.166)$$

The expression (9.161) for W then simplifies to:

$$W = \frac{2\pi^2 a^2 B_0^2}{R_0} \left[\left(\frac{n}{m} + \frac{1}{q}\right)^2 \frac{r\xi'}{\xi} + \frac{n^2}{m^2} - \frac{1}{\hat{q}^2} - \left(\frac{n}{m} + \frac{1}{\hat{q}}\right)^2 \frac{(r\hat{Q}_r)'}{\hat{Q}_r} \right]_{r=a} \xi^2(a). \qquad (9.167)$$

To complete this expression, we need to solve the Euler–Lagrange equation (9.105) for ξ with the approximations (9.163)–(9.166):

$$\frac{d}{dr}\left[r^3 \left(\frac{n}{m} + \frac{1}{q}\right)^2 \frac{d\xi}{dr} \right] - (m^2 - 1)r\left(\frac{n}{m} + \frac{1}{q}\right)^2 \xi = 0, \qquad (9.168)$$

where $q(r)$ needs to be specified. The equation for $r\hat{Q}_r$ simplifies from the Bessel equation (9.56) to an elementary differential equation that can be solved directly:

$$\frac{d}{dr}\left[r\frac{d}{dr}(r\hat{Q}_r) \right] - m^2 \hat{Q}_r = 0 \quad \Rightarrow \quad r\hat{Q}_r = C\left[(r/b)^{|m|} - (r/b)^{-|m|}\right], \qquad (9.169)$$

giving the approximate form of Eq. (9.162) for the logarithmic derivative. Restricting the analysis to configurations without surface currents now, so that $q_a = \hat{q}_a$, the expression for W becomes:

$$W = \frac{2\pi^2 a^2 B_0^2}{R_0}\left[\left(\frac{n}{m}+\frac{1}{q}\right)^2 \frac{r\xi'}{\xi} + \frac{n^2}{m^2} - \frac{1}{q^2}\right.$$
$$\left. + \left(\frac{n}{m}+\frac{1}{q}\right)^2 |m|\frac{1+(a/b)^{2|m|}}{1-(a/b)^{2|m|}}\right]_{r=a} \xi^2(a). \tag{9.170}$$

Clearly, stability only depends on the profile $q(r)$, the relative wall position b/a and the mode numbers n and m.

For a flat current profile, $j_z = $ const, also $q = $ const so that Eq. (9.168) reduces to an elementary equation that is easily solved:

$$(r^3\xi')' - (m^2-1)r\xi = 0 \quad\Rightarrow\quad \xi = r^{|m|-1} \quad\Rightarrow\quad \left(\frac{r\xi'}{\xi}\right)_{r=a} = |m|-1. \tag{9.171}$$

Then,

$$W = \frac{4\pi^2 a^2 B_0^2}{R_0} \frac{(nq+m)[nq+m-\text{sg}(m)(1-(a/b)^{2|m|})]}{|m|q^2(1-(a/b)^{2|m|})}, \tag{9.172}$$

in agreement with the earlier derived expression (9.99) for the growth rates of external kink modes in a constant-pitch magnetic field. Considering negative values of m only (for $n > 0$, instability only occurs when $m < 0$, and vice versa for $n < 0$ and $m > 0$), the plasma becomes unstable in the ranges

$$\frac{|m|-1+(a/b)^{2|m|}}{n} < q < \frac{|m|}{n}, \tag{9.173}$$

that is everywhere when the wall is at infinity. With the wall at a finite distance, stable windows for q are found, but only for single values of m and n. When all values of m and n are considered, no stable window of finite size remains: a flat current profile is unacceptable in a tokamak.

(c) Internal and external kink modes of the 'straight tokamak' It remains to solve Eq. (9.168) for realistic choices of $q(r)$ and to show that stable operating windows in parameter space can be found. This problem has been solved by Shafranov [207] in a very satisfactory manner, as we will see. The current profile is now assumed to be peaked in the centre ($r = 0$) and to fall off to zero at the plasma boundary ($r = a$), so that the q-profile is monotonically increasing:

$$q_0 \le q(r) \le q_a \quad (0 \le r \le a). \tag{9.174}$$

We already know from Section 9.3.2(b), Eq. (9.91), that there is a simple (geometrical) cure for the external $|m| = 1$ kink instability (which is independent of the shape of the current profile), viz. the Kruskal–Shafranov limit:

$$q_a \equiv \frac{2\pi a^2 B_0}{R_0 I_z} > 1. \tag{9.175}$$

With this limitation of the total plasma current I_z, we have to determine whether a reasonable current density profile $j_z(r)$ can be found for which the other modes are stable as well. 'Reasonable' here means: a current profile that can be experimentally realized and maintained on the time scale needed for fusion (i.e. many orders of magnitude longer than the characteristic time scale of these ideal MHD instabilities!). This leads to further restrictions on the value of q in the centre, q_0, and on the shape of the q-profile.

First, consider the $n = 1$, $|m| = 1$ *internal kink mode*. If $q_0 < 1$ and $q_a > 1$, there is a singular point in the plasma at $r = r_s$, where $q(r_s) = 1$. According to Newcomb's theorem, at such a point, the 'small' solution may jump so that we get the following solution of the Euler–Lagrange equation (9.168):

$$\xi = CH(r_s - r). \tag{9.176}$$

Inside the cylinder of radius $r = r_s$ the plasma is displaced rigidly with amplitude C, at $r = r_s$ the displacement jumps to zero (with a concomitant skin current), outside $r = r_s$ the displacement vanishes identically. Consequently, the full expression (9.170) for W vanishes: *the internal kink mode is marginally stable in the 'straight tokamak' limit*. At this point, one should realize that the expression (9.170) for W is actually only the leading order expression of the low-β-tokamak ordering, giving a growth rate $-\omega^2 \sim \epsilon^2$ when it is negative; see, e.g., Eq. (9.99). Exploiting the subscript 2 for this order, the result is a typical one for singular expansions, viz. that the leading order expression vanishes, $W_2 = 0$, so that the question immediately becomes: what is the sign of the next order, W_4? To answer that question, genuine toroidal contributions need to be calculated, which is beyond the present cylindrical analysis. However, the result of the toroidal calculation can be expressed simply as a condition on the q-profile: $W_4 > 0$ if $q_0 > 1$. Accidentally, the same condition is also obtained from a toroidal expansion of the Mercier criterion [153] for interchange modes in a low-β tokamak with confined pressure profile ($p' < 0$), typically involving mode numbers $|m|, |n| \gg 1$. Hence,

$$q_0 > 1 \tag{9.177}$$

is a condition for stability of the internal kink mode as well as the interchange modes in a low-β tokamak.

Large-amplitude sawtooth oscillations (oscillations of the central electron temperature with a period of 10–100 msec and leading to periodic loss of plasma confinement in the central region) occur in a tokamak when $q_0 < 1$. Hence, it was generally believed that the value of q_0 can never be smaller than 1 in a tokamak, until it was finally measured with the necessary precision by Soltwisch [209, 210] and found to be in the range of 0.73–0.78 in a particular tokamak (TEXTOR) and since then in many other tokamaks as well! (Precision measurement of the current profile is still very much needed for laboratory plasmas, and even more so for astrophysical plasmas, in order for MHD spectroscopy (Section 7.2.4) to become a mature method to determine the internal characteristics of plasmas.) Clearly, internal $|m| = 1$ kink modes (strictly limiting the central value of q to $q_0 = 1$) should not be considered as a similar threat to tokamak confinement as the external $|m| = 1$ kink modes (limiting the boundary value of q to the Kruskal–Shafranov limit $q_a = 1$). However, since the present section is concerned with the leading order ideal MHD stability of the 'straight tokamak', we will maintain the condition (9.177) for consistency of the analysis. Satisfying it, for an increasing q-profile, the Kruskal–Shafranov limit (9.175) is automatically satisfied as well.

Next, consider the *external* $|m| \geq 2$ *kink modes*. To determine the marginal stability boundaries, one needs to solve the differential equation (9.168) for ξ and subject it to the boundary condition

$$\left(\frac{r\xi'}{\xi}\right)_{r=a} = S_0 \equiv \frac{m - nq}{m + nq} - |m|\frac{1 + (a/b)^{2|m|}}{1 - (a/b)^{2|m|}}, \qquad (9.178)$$

obtained from Eq. (9.170) by putting $W = 0$. This boundary condition brings in the destabilizing free-boundary motion of the plasma–vacuum interface, as well as the stabilizing reaction of image currents induced in the conducting wall by the perturbations. For a flat current distribution, this boundary value problem yields the unstable windows (9.173): a kind of worst case scenario. To avoid these instabilities, one needs to shape the q-profile, i.e. the current density profile.

For definiteness, we now assume current profile distributions and associated q-profiles, as investigated by Wesson [243]:

$$j_z = j_0(1 - r^2/a^2)^\nu \quad \Rightarrow \quad q = q_0 \frac{(\nu + 1) r^2/a^2}{1 - (1 - r^2/a^2)^{\nu+1}}, \quad q_0 = \frac{2B_0}{R_0 j_0}.$$
$$(9.179)$$

This provides the necessary minimum number of parameters to fix q_0 and the overall shape of the q-profile (shear of the magnetic field), q_a/q_0, expressed by the parameter ν:

$$q_a/q_0 = \nu + 1. \qquad (9.180)$$

As illustrated in the rightmost part of Fig. 9.19, the parameter ν conveniently ranges from $\nu = 0$ (flat current), through $\nu = 1$ (parabolic current), to $\nu > 1$, corresponding to current distributions with an ever smaller gradient at the plasma

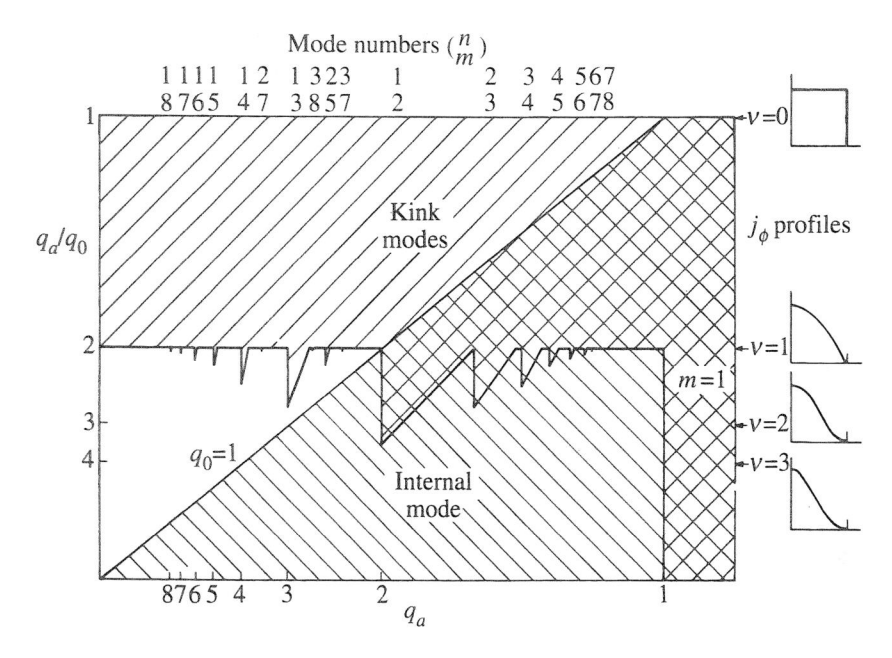

Fig. 9.19. Stability diagram for kink modes for the current distribution $j = j_0[1 - (r/a)^2]^\nu$, without a conducting wall. The vertical axis measures the peaking of the current as given by q_a/q_0 ($= \nu + 1$), and the horizontal coordinate is proportional to $1/q_a$ and therefore the total current. (In this figure, the definition of q is chosen such that the instabilities have positive m and n.) The $n = m = 1$ internal kink modes are unstable in the hatched area below the diagonal $q_0 = 1$. The $m = 1$ external kink modes are unstable for $q_a < 1$, and the $m \geq 2$ external kink modes are unstable in the upper hatched area. (From Wesson [243].)

boundary. The latter property guarantees stability with respect to the higher-$|m|$ kink modes. These current profiles also produce a smooth transition from the q-profile in the plasma to the $\hat{q}$-profile in the vacuum,

$$\hat{q}(r) = q_a(r/a)^2 \qquad (a \leq r \leq b), \tag{9.181}$$

following from Eq. (9.165) with $j_z = 0$.

The expressions (9.179) and (9.181) for the q-profiles permit one to precisely locate the position of the *rational magnetic surfaces* for each pair of mode numbers (m, n),

$$\text{in the plasma:} \quad m + nq(r_s) = 0 \quad (0 \leq r_s \leq a), \tag{9.182}$$

$$\text{in the vacuum:} \quad m + n\hat{q}(\hat{r}_s) = 0 \quad (a < \hat{r}_s \leq b). \tag{9.183}$$

Pairs (m, n) satisfying the condition (9.182), corresponding to a rational surface in the plasma, give rise to the interchange singularity that has been central to most of the stability analysis of the previous sections. In fact, the internal kink mode stability criterion (9.177) is just the condition for the absence of such a singularity for the $n = 1$, $m = -1$ internal kink mode in a plasma with an increasing q-profile. Because of the enormous simplification due to the low-β tokamak approximation, this was the only internal kink mode singularity we had to worry about. On the other hand, pairs (m, n) satisfying the condition (9.183), corresponding to a rational surface in the vacuum, do not give rise to a singularity at all: nothing prevents vacuum magnetic field lines from breaking and rejoining. (One would have to bring in the 'ghost plasma' of Section 9.2.2(b) to permit the induction of skin currents on the rational magnetic surfaces of the outer region (which would stabilize the external kink modes, as illustrated in Fig. 9.10).) It is a direct consequence of the stability criterion (9.173) (assuming that it is, in fact, the worst case scenario, which is confirmed by the numerical results) that potentially unstable pairs (m, n) with respect to external kink modes necessarily correspond to a rational surface in the vacuum:

$$q_a < |m|/n \equiv \hat{q}(\hat{r}_s). \tag{9.184}$$

Hence, external kink modes are due to the absence of electrical conductivity of the outer region, called 'vacuum'. In a tokamak, such a region is produced by a limiter scraping off the outer plasma layers.

For the analytically minded, a trace of the influence of the interchange singularity $m + nq = 0$ is still to be found in the differential equation (9.168) and the first term of the expression (9.178) for S_0. The latter term is positive (destabilizing) and can be made large (not infinite) by choosing $m + nq$ small (but always $\neq 0$) so that there is a 'virtual singularity' located in the vacuum, but close to the plasma boundary. By means of analytic continuation of the plasma equations, one can make a Frobenius expansion around that point, compute the indices, and construct a solution where the large part, involving terms $\ln(r_s - r)$ and $(r_s - r)^{-1}$, still dominates at the plasma boundary when $r_s - a$ can be made small enough. Consequently, the logarithmic derivative $(r\xi'/\xi)_{r=a}$ is obtained as a local quantity, depending on derivatives of the current density or the q-profile at the plasma boundary only. For the equilibrium (9.179), the results of this analysis are as follows:

(1) if $j_z(a) \neq 0$ the plasma is kink unstable for all $|m|$ (this involves extension of the current density profile with a pedestal, showing that current peaking is not enough);
(2) if $\nu \leq 1$ the plasma is unstable for any wall position (since the local analysis indicates instability);
(3) if $\nu > 1$ the plasma is only stable for $b < b_{\text{crit}}$, where b_{crit} is to be determined by the global boundary value problem (in other words: the local analysis fails).

9.4 Stability of cylindrical plasmas

This analysis is tedious and requires severe testing of the limits of validity. It is clear that the borders of the kingdom where singularity reigns have been reached here: analysis is superior to numerics in the presence of singularities, in the absence of singularities the opposite holds.

In contrast, the numerical solution of the boundary value problem (9.168), (9.178) for external $|m| \geq 2$ kink modes is nearly trivial:

(1) specify a monotonically increasing profile $q(r)$, a relative wall position b/a, and a mode pair (m, n) satisfying the condition (9.184);
(2) integrate the differential equation (9.168) for ξ from the magnetic axis, where $\xi(0) = 0$, to the plasma boundary, determine the logarithmic derivative $L \equiv (r\xi'/\xi)_{r=a}$, and compare it with the quantity S_0 defined in Eq. (9.178):

$$\text{if } L < S_0 \Rightarrow \text{unstable}, \quad \text{if } L = S_0 \Rightarrow \text{marginal}, \quad \text{if } L > S_0 \Rightarrow \text{stable}; \quad (9.185)$$

(3) repeat this process for all pertinent mode pairs (m, n), iterating on the values of the parameters q_0 and q_a until $L = S_0$.

The differential equation for ξ is not only non-singular, but also admits no oscillatory solutions (the coefficients are positive definite). Hence, numerical integration is extremely fast and accurate. The only issue is the proper bookkeeping of the mode number pairs. Also, once the numerical scheme is established, extension with the computation of the growth rate of the instabilities is completely straightforward. This we leave as an exercise for the reader (Exercises [9.5] and [9.6]).

The stability diagram of Fig. 9.19 is obtained by such a numerical procedure applied to an equilibrium with the $q(r)$-profile (9.179). Without the internal kink mode condition $q_0 > 1$, there is a fairly wide region of parameters where the $|m| \geq 2$ external kink modes are stable, roughly corresponding to a current profile with $q_a/q_0 \sim 3$ ($\nu \sim 2$) and $q_a > 2$. This confirms the earlier conclusion of Shafranov [207] that complete stability of the 'straight tokamak' may be obtained if the radial current density profile is sufficiently peaked on axis. Imposing the condition $q_0 > 1$ as well, the stable region becomes significantly smaller, although an operating region remains (the white, roughly triangular, region), which is, however, rather hard to control experimentally. It has already been indicated that relaxation of the internal kink mode condition may not be disastrous, whereas relaxation of the condition for the higher $|m|$ external kink modes (excluding $|m| = 2$ and 3) may just give rise to what is called 'enhanced MHD activity'. Also, stabilization by the external wall has not been taken into account in Fig. 9.19. This will slow down the ideal MHD modes to the resistive skin time of the wall, where feedback stabilization techniques may be applied.

492   *Cylindrical plasmas*

To assess the implications of these results for tokamak operation, one should realize that a number of important effects have not been taken into account here: (1) the presence of *plasma flow* in the edge region, (2) the *finite conductivity* of that plasma, (3) the influence of β, i.e. of the *toroidicity*. These are topics, necessarily involving advanced numerical solution techniques, to be addressed in Volume 2 on *Advanced Magnetohydrodynamics*.

9.5 Literature and exercises

Notes on literature

MHD spectral theory of cylindrical plasmas:

– Hain & Lüst [102], in 'Zur Stabilität zylinder-symmetrischer Plasmakonfigurationen mit Volumenströmen', derive an ordinary differential equation to determine the growth rates of MHD instabilities in diffuse linear pinch configurations.

– Goedbloed [81](II), in 'Stabilization of magnetohydrodynamic instabilities by force-free magnetic fields – linear pinch', rederives the (apparently forgotten) Hain–Lüst equation, generalizes it to $\gamma \neq 1$, and applies it to stabilization of external kink modes by a force-free magnetic field in the outer region.

– Grad [99], in 'Magnetofluid-dynamic spectrum and low shear stability', puts the subject in the context of spectral theory, points out the presence of four types of singularities (unfortunately associating a continuous spectrum with each of them), and applies the theory to demonstrate stability of a large class of low-shear systems.

– Appert, Gruber, & Vaclavik [9], in 'Continuous spectra of a cylindrical magnetohydrodynamic equilibrium', derive the equivalent system of first order differential equations and demonstrate that the $D = 0$ singularities are apparent.

MHD stability of cylindrical plasmas:

– Newcomb [164], in 'Hydromagnetic stability of a diffuse linear pinch', presents the classical treatise of the stability of cylindrical plasmas, with a careful exposition of the techniques needed from variational analysis.

– Voslamber & Callebaut [241], in 'Stability of force-free magnetic fields', present a beautiful example of the subtleties of MHD stability theory applied to the Lundquist field.

– Shafranov [207], in 'Hydrodynamic stability of a current-carrying pinch in a strong longitudinal field', demonstrates stability of low-β tokamaks with respect to all kink modes for realistic current distributions.

– Robinson [195], in 'High-β diffuse pinch configurations', analyses the different diffuse pinch configurations from the point of view of maximizing β, demonstrating the favourable properties of the shear profile (opposite to that of the tokamak) of the reversed field pinch.

– Goedbloed & Hagebeuk [86] numerically solve the generalized Hain–Lüst equation to obtain the growth rates of instabilities of the Lundquist and constant-pitch magnetic fields; Goedbloed & Sakanaka [89] and Sakanaka & Goedbloed [199] continue to construct the different classes of σ-stable diffuse linear pinch configurations this way.

- Freidberg, *Ideal Magnetohydrodynamics* [72], Chapter 9 on stability of one-dimensional configurations, presents the MHD stability theory of cylindrical plasmas, with applications to a variety of experimental fusion devices.
- Wesson, *Tokamaks* [244], Chapter 6 on MHD stability, contains the essential elements (with the simple diagrams that are his hallmark) of the review paper [243] on 'Magnetohydrodynamic stability of tokamaks'.

Exercises

[9.1] *Cylindrical force-free magnetic fields*

Force-free magnetic fields, $\mathbf{j} = \alpha \mathbf{B}$, with cylindrical symmetry can be fixed in several different ways, viz. by prescribing the function $\alpha(r)$, or $\mu(r)$, or $B^2(r)$. Each of these quantities has a physical meaning: α is the ratio of the current and the magnetic field, μ is the inverse pitch of the field lines, and B^2 is the magnetic energy density (all disregarding constant factors). Derive the differential equations for B_z and B_θ for the three cases where either $\alpha(r)$, or $\mu(r)$, or $B^2(r)$ is prescribed. Comment on the conditions that have to be imposed for physical reality and on the advantages and disadvantages of the three prescriptions. Construct as many explicit solutions as you can for either of those cases. (They are useful as explicit equilibria when you want to check stability calculations.) Also, using the equilibrium relations, find relations between the three quantities.

[9.2] *Newcomb's stability equations*

Derive Newcomb's Euler–Lagrange equation, $(f_0 \xi')' - g_0 \xi = 0$, from the generalized Hain–Lüst equation, $(P\chi')' - Q\chi = 0$, by inserting the value $\omega^2 = 0$ in the latter and transforming from the variable ξ to $\chi \equiv r\xi$. Pay attention to how the derivative term is transformed (recall that it is associated with the perturbed total pressure).

- Is marginal stability analysis equivalent to the variational analysis associated with the energy principle? Comment on the role of singularities.
- Derive the one-dimensional form of the energy principle for a cylindrical plasma with length L and radius a from the Euler–Lagrange equation (i.e. the other way around with respect to the usual order) and get rid of the boundary terms.
- Expand about a point where $F = mB_\theta/r + kB_z = 0$ and derive Suydam's local stability criterion from the indicial equation.
- What type of solution is associated with the violation of Suydam's criterion? How is it related to a genuine eigenfunction?

[9.3] *WKB solution of the generalized Hain–Lüst equation*

Consider a cylindrical plasma with weak inhomogeneity so that the generalized Hain–Lüst equation, $(P\xi')' - Q\chi = 0$, can be solved by means of the WKB method. Writing

$$\chi(r) = p(r) \exp\left[i \int q(r)\, dr\right],$$

the expressions $p(r)$ and $q(r)$ are determined by requiring that the solution be correct to leading order in the inhomogeneity. This yields

$$p \approx (-PQ)^{-1/4}, \qquad q \approx (-Q/P)^{1/2},$$

where we have to demand that $|qL| \gg 1$ in order for the WKB approximation to be valid. Here, L is the scale length for the inhomogeneities. This yields a local dispersion equation relating ω^2 and the local radial wave number q,

$$q^2 = -Q/P,$$

which is a quintic in ω^2. (No, this has nothing to do with the fact that the discrete spectrum consists of five sub-spectra.) This equation may be solved in the neighbourhood of the Alfvén and slow continua, when these are sufficiently far apart. Show that this gives cluster spectra of the form

$$\rho\omega^2 \approx F^2 - \frac{A}{q^2}, \qquad \rho\omega^2 \approx \left(\frac{\gamma p}{\gamma p + B^2}\right)^2 F^2 - \frac{S}{q^2}.$$

Determine the expressions A and S. Compare them with the exact expressions for the cluster conditions of Section 9.3.3 and comment. Discuss the validity of the WKB approximation.

[9.4] *Instabilities of shearless magnetic fields*

In Section 9.4.2(b) the pressure-driven instabilities of plasmas with shearless magnetic fields were analysed. Using the same techniques, investigate the residual instabilities that occur when the pressure gradient is much smaller, $\Pi \equiv a^2 p'/(2r^2 B^2) \sim \epsilon^4$. The stability threshold for these modes can be found from the condition that there should be no real values of $k_\parallel$ where marginal stability occurs. Show that this gives the following local criterion for high-m modes:

$$p' > \frac{2B_\theta^4}{rm^2 B^2}.$$

For $m \to \infty$, this criterion transforms into Suydam's criterion for shearless magnetic fields. For low-m modes, the criterion is much more stringent, in particular when B_θ is large. Derive the dispersion equation for these modes, called quasi-kink modes. Show that they are unstable for values of $k_\parallel$ on only one side of $k_\parallel = 0$. Derive the expression for the maximum growth rate and estimate the time scale.

[9.5]* *Marginal stability of external kink modes in a tokamak*

Write a computer program constructing the Wesson diagram of Fig. 9.19 from the boundary value problem (9.168), (9.178) outlined in Section 9.4.4. Use the criterion $L = S_0$ of Eq. (9.185) to determine the stable regions. Figure out how to iterate on the parameters such that rapid convergence to marginal stability is obtained.

- When you have obtained agreement with the figure, extend the program with finite wall positions, $b/a = 1.1, 1.2, 1.5, 2.0$. Comment on the results.
- Run the program for $\nu = 0.5$ and try to stabilize by moving the wall in. What happens? Plot the solution ξ.

[9.6]* *Growth rates of external kink modes in a tokamak*

Assuming $\rho = \rho_0 = \text{const}$, and normalizing the eigenvalue $\bar{\omega}^2 \equiv (\rho_0 R_0^2/B_0^2)\omega^2$, derive the equation of motion for a 'straight tokamak' from the low frequency limit (9.46) of the generalized Hain–Lüst equation:

$$\frac{d}{dr}\left[r^3\{(n+m/q)^2 - \bar{\omega}^2\}\frac{d\xi}{dr}\right] - (m^2-1)r\{(n+m/q)^2 - \bar{\omega}^2\}\xi = 0.$$

Derive the associated plasma–vacuum boundary condition from Eq. (9.60):

$$\left(\frac{r\xi'}{\xi}\right)_{r=a} = S \equiv \frac{1}{(n+m/q)^2 - \bar{\omega}^2}\left[\bar{\omega}^2 - n^2 + m^2/q^2 - (n+m/q)^2|m|\frac{1+(a/b)^{2|m|}}{1-(a/b)^{2|m|}}\right].$$

- Noting that this boundary value problem only requires some additional terms $\bar{\omega}^2 < 0$, modify the marginal stability program of the previous exercise to calculate the growth rates of the external kink modes.
- Pick a reasonable cutoff value for $\sigma^2 \equiv -\bar{\omega}^2$ and construct the σ-stability contour diagram for that value. What do you conclude from the result?

10

Initial value problem and wave damping*

10.1 Implications of the continuous spectrum★

We now embark on the solution of the initial value problem (IVP) as outlined in Section 6.3.2. In general, the IVP in ideal MHD arises when one wishes to solve the equation of motion (6.23), which we now write as

$$\rho^{-1}\mathbf{F}(\boldsymbol{\xi}) - \frac{\partial^2 \boldsymbol{\xi}}{\partial t^2} = 0, \tag{10.1}$$

for arbitrary initial values $\boldsymbol{\xi}_i$,

$$\boldsymbol{\xi}(\mathbf{r}; t=0) = \boldsymbol{\xi}_i(\mathbf{r}). \tag{10.2}$$

Obviously, the spatial boundary value problem (BVP) must be solved simultaneously, but we have sufficiently studied that part of the problem for inhomogeneous plane slab and cylindrical plasmas in Chapters 7 and 9. In the present chapter, we wish to investigate how such plasmas evolve in time. What is the role of the intricate MHD spectra derived in the previous chapters? How do the discrete and continuous parts of those spectra enter the temporal description of an arbitrarily excited plasma?

The IVP for macroscopic plasmas is a natural analogue of the IVP for microscopic plasmas first (correctly) analysed in 1946 by Landau [136], who derived the surprising result that electrostatic waves ('plasma oscillations'), in the absence of any dissipation mechanism, are damped (Section 2.3.3). The mechanism is called Landau damping and, since it is associated with a microscopic description in terms of the velocity space distribution function, it was (and still frequently is) thought that such dissipationless damping processes are restricted to the microscopic picture. However, in 1971 Sedláček [206] showed that a macroscopic description of plasma oscillations of a cold plasma also leads to dissipationless damping, due to inhomogeneity in ordinary space. (Note that Landau damping is due to the variation of the distribution function, i.e. to inhomogeneity in velocity space). Next, in 1973 Tataronis and Grossmann [225] showed that ideal MHD waves actually also exhibit damping, due to spatial inhomogeneities of the plasma.

The bottom line appears to be that, *in conservative systems with a continuous spectrum* (whether it concerns a quantum mechanical system of scattering particles, microscopic plasma oscillations of a collection of charged particles, or macroscopic Alfvén waves of a confined plasma), *damping of the initial perturbations occurs through redistribution over the different improper continuum modes*. In plasmas described by MHD, the dissipationless development, on an ideal MHD time scale, of very localized structures builds up large spatial gradients of the macroscopic variables that will enormously enhance the genuine dissipation rate associated with, for example, the resistivity (Section 4.3.1). We will call the ideal MHD part of this evolution 'quasi-dissipation'. Its practical consequences for resonant wave absorption and heating, in particular for astrophysical plasmas, will be discussed in Chapter 11. The latter chapter deals with the implications of the continuous spectrum for systems that are actively excited (i.e. model III of Section 4.6.1). In the present chapter, we will deal with the implications of the continuous spectrum for IVP and wave damping for passive systems, left to themselves.

10.2 Initial value problem★

To solve the equation of motion (10.1) with the initial data (10.2), one may exploit the techniques of forward and inverse Laplace transformation that were already introduced in Section 6.3, of which we will repeat the necessary equations for convenience of reading. The forward Laplace transformation to the complex ω-plane introduces the initial data in the equations, whereas the inverse Laplace transformation back to the time domain then has to deal with the special values of ω that belong to the spectrum. In this manner, the contributions of the different parts of the spectrum become manifest. The forward Laplace transformation

$$\hat{\boldsymbol{\xi}}(\mathbf{r}; \omega) \equiv \int_0^\infty \boldsymbol{\xi}(\mathbf{r}; t)\, e^{i\omega t}\, dt \tag{10.3}$$

transforms the homogeneous equation (10.1) into the inhomogeneous equation

$$(\rho^{-1}\mathbf{F} + \omega^2 \mathbf{I}) \cdot \hat{\boldsymbol{\xi}}(\mathbf{r}; \omega) = i\omega \boldsymbol{\xi}_i(\mathbf{r}) - \dot{\boldsymbol{\xi}}_i(\mathbf{r}) \equiv i\omega \mathbf{X}, \tag{10.4}$$

where the initial displacement $\boldsymbol{\xi}_i$ and the initial velocity $\dot{\boldsymbol{\xi}}_i$ are absorbed in the definition of the vector $\mathbf{X}$. As compared with Eq. (6.72) of Section 6.3.2, we have now introduced the more appropriate notation of matrix operators $\mathbf{F}$ for the force and **I** for the identity:

$$\mathbf{F} \cdot \hat{\boldsymbol{\xi}} \equiv \mathbf{F}(\hat{\boldsymbol{\xi}}), \qquad \mathbf{I} \cdot \hat{\boldsymbol{\xi}} \equiv \mathbf{I}(\hat{\boldsymbol{\xi}}) = \hat{\boldsymbol{\xi}}. \tag{10.5}$$

The formal solution of the inhomogeneous equation (10.4) involves the inversion of the differential operator on the left hand side:

$$\hat{\boldsymbol{\xi}}(\mathbf{r}; \omega) = (\rho^{-1}\mathbf{F} + \omega^2 \mathbf{I})^{-1} \cdot i\omega \mathbf{X}(\mathbf{r}; \omega). \tag{10.6}$$

Clearly, the construction of the inverse operator $(\rho^{-1}\mathbf{F} + \omega^2 \mathbf{I})^{-1}$ will be one of the major tasks of this section. This operator is *the resolvent operator* for this problem, which is the most compact expression of the different spectral alternatives, as we saw in Section 6.3.1. Not surprisingly, since the resolvent operator is the inverse of a differential operator, it will turn out to be an integral operator involving Green's functions. To obtain the actual solution of the IVP, the inverse Laplace transformation is applied to Eq. (10.6):

$$\begin{aligned}
\boldsymbol{\xi}(\mathbf{r}; t) &= \frac{1}{2\pi} \int_{iv_0-\infty}^{iv_0+\infty} \hat{\boldsymbol{\xi}}(\mathbf{r}; \omega) \, e^{-i\omega t} \, d\omega \\
&= \frac{1}{2\pi} \int_C (\rho^{-1}\mathbf{F} + \omega^2)^{-1} \cdot i\omega \mathbf{X}(\mathbf{r}; \omega) e^{-i\omega t} \, d\omega,
\end{aligned} \tag{10.7}$$

as we already saw in Section 6.3.2. Here, judicious deformation of the contour C in the complex ω-plane (Fig. 6.11) will reveal the characteristic temporal behaviour of the different contributions of the spectrum.

10.2.1 Reduction to a one-dimensional representation★

Let us now specify the plasma to be a plane gravitating slab, as considered in Chapter 7. (We follow the highly delayed publication of Goedbloed [85], based on an unpublished memorandum of 1973, which contained the construction of the resolvent operator in the apparently singular ranges $D = 0$ for the analogous case of cylindrical geometry.) The direction of inhomogeneity is represented by the x-coordinate and the dependence on the symmetry coordinates y and z may be eliminated by considering Fourier modes $\hat{f}_{k_y k_z}(x; \omega) \exp i(k_y y + k_z z)$, where we omit the subscripts k_y and k_z and suppress the exponential factor from now on. Since the dynamics is strongly guided by the magnetic surfaces ($x = $ const) and the magnetic field lines in those surfaces, it is again expedient to exploit the field line projection:

$$\hat{\boldsymbol{\xi}} = \hat{\xi} \mathbf{e}_x - i\hat{\eta} \mathbf{e}_\perp - i\hat{\zeta} \mathbf{e}_\parallel, \qquad \mathbf{X} \equiv X \mathbf{e}_x - iY \mathbf{e}_\perp - iZ \mathbf{e}_\parallel, \tag{10.8}$$

10.2 Initial value problem*

where

$$\hat{\xi} \equiv \hat{\xi}_x, \qquad X \equiv \xi_i + (i/\omega)\dot{\xi}_i,$$
$$\hat{\eta} \equiv i(B_z\hat{\xi}_y - B_y\hat{\xi}_z)/B, \qquad Y \equiv \eta_i + (i/\omega)\dot{\eta}_i, \qquad (10.9)$$
$$\hat{\zeta} \equiv i(B_y\hat{\xi}_y + B_z\hat{\xi}_z)/B, \qquad Z \equiv \zeta_i + (i/\omega)\dot{\zeta}_i.$$

Exploiting the same reductions as used in Section 7.3.2 for the derivation of the wave equation (7.89), this leads to the following representation of the inhomogeneous problem (10.4):

$$\begin{pmatrix} \frac{d}{dx}(\gamma p + B^2)\frac{d}{dx} - f^2 B^2 + \rho\omega^2 & \frac{d}{dx}g(\gamma p + B^2) + g\rho\hat{g} & \frac{d}{dx}f\gamma p + f\rho\hat{g} \\ -g(\gamma p + B^2)\frac{d}{dx} + g\rho\hat{g} & -g^2(\gamma p + B^2) - f^2 B^2 + \rho\omega^2 & -gf\gamma p \\ -f\gamma p\frac{d}{dx} + f\rho\hat{g} & -fg\gamma p & -f^2\gamma p + \rho\omega^2 \end{pmatrix} \begin{pmatrix} \hat{\xi} \\ \hat{\eta} \\ \hat{\zeta} \end{pmatrix}$$

$$= i\rho\omega \begin{pmatrix} X \\ Y \\ Z \end{pmatrix}. \qquad (10.10)$$

Our next task is the inversion of Eq. (10.10), producing the explicit solution of $\hat{\xi}$ as a function of $\mathbf{X}$ as formally expressed by Eq. (10.6).

The elimination of the tangential components $\hat{\eta}$ and $\hat{\zeta}$ in terms of $\hat{\xi}$ and the initial data takes the following form:

$$\hat{\eta} = S\hat{\xi}' + K\hat{\xi} + i\rho\omega(HY + IZ),$$
$$\hat{\zeta} = A\hat{\xi}' + L\hat{\xi} + i\rho\omega(IY + JZ), \qquad (10.11)$$

where the expressions S, A, K, L correspond to Eqs. (7.90) for the homogeneous problem:

$$S \equiv \frac{g(b^2 + c^2)}{D}(\omega^2 - \omega_S^2), \qquad K \equiv -\frac{g\hat{g}\omega^2}{D},$$
$$A \equiv \frac{fc^2}{D}(\omega^2 - \omega_A^2), \qquad L \equiv -\frac{f\hat{g}(\omega^2 - k_0^2 b^2)}{D}, \qquad (10.12)$$

and new expressions H, I, J enter with the initial data:

$$H \equiv \frac{\omega^2 - f^2 c^2}{\rho D}, \qquad I \equiv \frac{fgc^2}{\rho D},$$

$$J \equiv \frac{\omega^2 - f^2 b^2 - g^2(b^2 + c^2)}{\rho D}. \qquad (10.13)$$

Here, S and A introduce the slow and Alfvén continuum frequencies in the formulation:

$$\omega_S^2(x) \equiv f^2 \frac{b^2 c^2}{b^2 + c^2}, \qquad \omega_A^2(x) \equiv f^2 b^2. \qquad (10.14)$$

The elimination of the tangential components involves the determinant D of the four lower right corner elements of the matrix of Eq. (10.10):

$$D(x) \equiv \omega^4 - k_0^2(b^2 + c^2)\omega^2 + k_0^2 f^2 b^2 c^2 = (\omega^2 - \omega_{s0}^2)(\omega^2 - \omega_{f0}^2), \qquad (10.15)$$

where the notation of the local frequencies,

$$\omega_{s0,f0}^2 \equiv \tfrac{1}{2} k_0^2 (b^2 + c^2) \left[1 \pm \sqrt{1 - \frac{4 f^2 b^2 c^2}{k_0^2 (b^2 + c^2)^2}} \right], \qquad (10.16)$$

indicates a relationship to the slow and fast wave motion. This brings the issue of the nature of the apparent $D = 0$ singularities into focus. By now, we know that those frequency ranges do not constitute continuous spectra, in contrast to the genuine slow and Alfvén continua, $\{\omega_S^2(x)\}$ and $\{\omega_A^2(x)\}$. The latter continua manifest themselves in the numerators of the expressions (10.12) for S and A, multiplying the derivative $\hat{\xi}'$ in the expressions (10.11) for $\hat{\eta}$ and $\hat{\zeta}$. This is more decisive for the local dynamics than the apparent $D = 0$ singularities, which all turn out to cancel in the final analysis.

Substitution of $\hat{\eta}$ and $\hat{\zeta}$ into the normal component of the equation of motion (10.10) leads to an inhomogeneous differential equation for $\hat{\xi}$:

$$(P\hat{\xi}')' - Q\hat{\xi} = R \equiv i\omega \left\{ \rho X - [\rho(SY + AZ)]' + \rho(KY + LZ) \right\}, \qquad (10.17)$$

where

$$P \equiv \frac{N}{D}, \qquad N(x) \equiv \rho(b^2 + c^2)(\omega^2 - \omega_A^2)(\omega^2 - \omega_S^2),$$

$$Q \equiv -\left[\rho(\omega^2 - f^2 b^2) + \rho' \hat{g} - k_0^2 \rho \hat{g}^2 \frac{\omega^2 - f^2 b^2}{D} - \left\{ \rho \hat{g} \frac{\omega^2(\omega^2 - f^2 b^2)}{D} \right\}' \right].$$

$$(10.18)$$

Eq. (10.17) is solved by the integral

$$\hat{\xi}(x) = \int_{x_1}^{x_2} G(x, x') R(x') \, dx', \qquad (10.19)$$

where $G(x, x')$ is the *Green's function* satisfying the equation

$$\frac{d}{dx}\left[P(x)\frac{dG(x,x')}{dx}\right] - Q(x)\,G(x,x') = \delta(x-x'), \qquad (10.20)$$

subject to the boundary conditions

$$G(x=x_1, x') = G(x=x_2, x') = 0,$$

$$[\![G]\!]_{x=x'} = 0, \qquad \left[\!\!\left[P\frac{\partial G}{\partial x} \right]\!\!\right]_{x=x'} = 1. \qquad (10.21)$$

The double brackets indicate possible jumps: $[\![f]\!] \equiv \lim_{\epsilon \to 0}[f(x=x'+\epsilon) - f(x=x'-\epsilon)]$. (See Fig. 10.2(a) for an illustration of what the Green's function might look like.)

For simplicity, we here consider a slab $[x_1, x_2]$, assuming that either the inhomogeneity of the equilibrium is not too strong or that the slab is thin enough that the continua $\{\omega_S^2\}$ and $\{\omega_A^2\}$, where $N = 0$, and the ranges $\{\omega_{s0}^2\}$ and $\{\omega_{f0}^2\}$, where $D = 0$, do not overlap. This is possible since the frequencies are well ordered when considered for fixed position x:

$$0 \leq \omega_S^2 \leq \omega_{s0}^2 \leq \omega_A^2 \leq \omega_{f0}^2 \leq \omega_F^2 = \infty. \qquad (10.22)$$

Overlapping could hardly be avoided in a cylinder when the origin is included since $\omega_{s0}^2 \to \omega_S^2$ and $\omega_{f0}^2 \to \infty$ there. This does not happen in a plane slab. Hence, $\{\omega_{f0}^2\}$ is extremely dependent on the choice of coordinates, which is another clear indication that the $D = 0$ singularities cannot represent continuous spectra. However, this is a detail that is not important for our present purpose.

The Green's function is built from solutions of the homogeneous equations

$$(PU_i')' - QU_i = 0 \qquad (i = 1, 2), \qquad (10.23)$$

producing a left component $U_1(x)$ satisfying the left boundary condition $U_1(x_1) = 0$ and a right component $U_2(x)$ satisfying the right boundary condition $U_2(x_2) = 0$. The formal solution of Eq. (10.20) then reads:

$$G(x, x'; \omega^2) = \frac{\Gamma(x, x'; \omega^2)}{\Delta(\omega^2)}, \qquad (10.24)$$

where

$$\Gamma(x, x'; \omega^2) \equiv U_1(x; \omega^2)U_2(x'; \omega^2)H(x'-x) + U_1(x'; \omega^2)U_2(x; \omega^2)H(x-x'),$$

$$\Delta(\omega^2) \equiv P(x; \omega^2)\left[U_1(x; \omega^2)U_2'(x; \omega^2) - U_1'(x; \omega^2)U_2(x; \omega^2)\right]. \qquad (10.25)$$

For the study of the initial value problem with respect to the response of the Alfvén and slow continua $N(\omega^2) = 0$, the consideration of the logarithmic singularities of the components $U_{1,2}(x)$ is imperative. They lead to corresponding branch cuts of Γ and Δ when considered as a function of complex ω, due to contributions like $\ln(\omega^2 - \omega_A^2(x))$. The analysis of these singularities is fully analogous to that of cold plasma oscillations of an inhomogeneous plasma which has been given in great detail by Sedláček [206]. His analysis explicitly shows that the zeros of the *conjunct* $\Delta(\omega^2)$ constitute the discrete spectrum whereas the branch cuts of Γ and Δ constitute the continuous spectrum. The corresponding analysis for Alfvén waves has been given by Tataronis [224].

In Sections 10.3 and 10.4, we will consider explicit solutions of the Green's function, and their implications for the spectrum, for a highly simplified model. In the present section, we keep the analysis as general as possible in order to stay close to the original goal of the paper [85] on which this section is based. This is to construct the full three-dimensional response for MHD waves in the frequency range $D(\omega^2) = 0$ and to show that no singularities of G occur there other than possible poles $\Delta(\omega^2) = 0$ corresponding to discrete modes. Expansion of the coefficients P and Q of Eq. (10.20) in terms of the distance $s \equiv x - x_0$ from an apparent singularity $D(x_0) = 0$ satisfies the special property (J. M. Greene, unpublished, 1974)

$$P = \frac{P_0}{s} + \cdots, \quad Q = \frac{Q_0}{s^2} + \frac{Q_1}{s} + \cdots \quad \Rightarrow \quad Q_0^2 - P_0 Q_1 = 0, \quad (10.26)$$

already encountered in Eq. (7.143). This guarantees absence of logarithmic singularities so that the two independent solutions close to the apparent singularity are both regular:

$$U_s = s^2 + \cdots, \quad U_\ell = 1 + \alpha s + \cdots,$$

$$\alpha \equiv -Q_0/P_0 = \left(\frac{\hat{g} \omega^2}{(b^2 + c^2)(\omega^2 - \omega_S^2)} \right)_0. \quad (10.27)$$

Hence, in the mentioned frequency range, there are no logarithmic singularities, no branch cuts, but an almost forbidding number of occurrences of a vanishing denominator D in all of the coefficients (10.12)–(10.13). Next, we will show that this is not only just an apparent obstacle, but even a necessary element in the response to the genuine continua.

10.2.2 Restoring the three-dimensional picture*

Having solved for the normal response $\hat{\xi}(x)$, expressed by Eq. (10.19) with the Green's function $G(x, x')$ given by Eqs. (10.24)–(10.25), it remains to construct the tangential response $\hat{\eta}(x)$, $\hat{\zeta}(x)$ according to Eqs. (10.11). This part of the analysis is usually underexposed since it does not require the solution of additional differential equations but just involves the substitution of the normal solutions obtained. However, this part is really the most significant in this case. We first complete the response (10.19) for $\hat{\xi}(x)$ by partially integrating the derivative term on the RHS of Eq. (10.17):

$$\hat{\xi} = i\omega \int \left\{ GX + \left[\left(S\frac{\partial}{\partial x'} + K \right) G \right] Y + \left[\left(A\frac{\partial}{\partial x'} + L \right) G \right] Z \right\} \rho(x')\, dx'. \tag{10.28}$$

Here, the functions in the integrand are considered as functions of x' and the square brackets indicate that the derivative $\partial/\partial x'$ is to be taken on $G(x, x')$ only. This expression reveals two special normal operators, already encountered in Eq. (10.11) for the tangential components, but now they act on the Green's function itself:

$$\tilde{S}(x) \equiv S(x)\frac{\partial}{\partial x} + K(x),$$

$$\tilde{A}(x) \equiv A(x)\frac{\partial}{\partial x} + L(x). \tag{10.29}$$

Eq. (10.28) provides the complete normal response $\hat{\xi}$ in terms of the initial data X, Y, Z. In turn, substitution of this expression in Eqs. (10.11) provides the response of the tangential components $\hat{\eta}$ and $\hat{\zeta}$ in terms of the initial data. The latter involves the operation of $\tilde{S}$ and $\tilde{A}$ on the Green's function twice in succession.

We now compose *the full three-dimensional response*:

$$\hat{\xi}(\mathbf{r}; \omega^2) = i\omega \int dx'\, \rho(x')\, \mathbf{G}(\mathbf{r}, \mathbf{r}'; \omega^2) \cdot \mathbf{X}(\mathbf{r}'; \omega), \tag{10.30}$$

where we recall that we have suppressed the Fourier factors providing the dependence on the coordinates y and z and, consequently, also the corresponding factors in $\mathbf{G} \sim \delta(y - y')\delta(z - z')$ since integration over those coordinates is trivial. In terms of the field line projection, this response may be written as

$$(\hat{\xi}(x), \hat{\eta}(x), \hat{\zeta}(x))^T = i\omega \int dx'\, \rho(x')\, \mathbf{G}(\mathbf{r}, \mathbf{r}'; \omega^2) \cdot (X(x'), Y(x'), Z(x'))^T, \tag{10.31}$$

where *the Green's dyadic* $\mathbf{G}$ is given by

$$\mathbf{G} = \begin{pmatrix} 1 & \tilde{S}(x') & \tilde{A}(x') \\ \tilde{S}(x) & \tilde{S}(x)\tilde{S}(x') & \tilde{S}(x)\tilde{A}(x') \\ \tilde{A}(x) & \tilde{A}(x)\tilde{S}(x') & \tilde{A}(x)\tilde{A}(x') \end{pmatrix} G(x, x') + \begin{pmatrix} 0 & 0 & 0 \\ 0 & H(x) & I(x) \\ 0 & I(x) & J(x) \end{pmatrix} \delta(x - x'), \tag{10.32}$$

with the one-dimensional Green's function $G(x, x')$ given by Eqs. (10.24)–(10.25). This provides the three-dimensional response we are looking for but, since all constituent functions $S, A, K, L, H, I, J \sim 1/D$, the representation is full of apparent singularities. However, we should not lose sight of the essential feature here, viz. that the operators $\tilde{S}$ and $\tilde{A}$ separately 'kill' the localized response associated with large normal derivatives $(\partial/\partial x \to \infty)$, for either one of the two tangential components, leaving the perpendicular component $\hat{\eta}$ as the dominant response for the Alfvén frequencies ω_A and the parallel component $\hat{\zeta}$ as the dominant response for the slow frequencies ω_S.

With the representation (10.32), we have the key to open up the structure of the Green's dyadic **G** to finally produce the complete three-dimensional response in a form that is free of apparent singularities and also exhibits the genuine ones. To that end, we study **G** in the neighbourhood of the point $x = x_0$ (and $x' = x_0'$) where $D(x) = 0$, of course, exploiting the fact that the constituent functions $U_{1,2}(x)$ of the one-dimensional Green's function $G(x, x')$ are regular there, according to Eqs. (10.27). We now transform to a three-dimensional picture exploiting solution vectors $\mathbf{U}_{1,2}(x)$ with components $(U_{1,2}, V_{1,2}, W_{1,2})$, where the tangential components satisfy the homogeneous counterparts of Eqs. (10.11):

$$V_{1,2}(x) \equiv \tilde{S}(x)U_{1,2},$$
$$W_{1,2}(x) \equiv \tilde{A}(x)U_{1,2}. \tag{10.33}$$

In the neighbourhood of $x = x_0$ we then have

$$U = C_s U_s + C_\ell U_\ell = C_s(s^2 + \cdots) + C_\ell(1 + \alpha s + \cdots),$$
$$V \equiv \tilde{S}U = C_s(2Ss + Ks^2 + \cdots) + C_\ell(S\alpha + K + \cdots),$$
$$W \equiv \tilde{A}U = C_s(2As + Ls^2 + \cdots) + C_\ell(A\alpha + L + \cdots), \tag{10.34}$$

where the leading order terms are all finite due to the equalities

$$(K/S)_0 = (L/A)_0 = -\alpha. \tag{10.35}$$

Hence, the operators $\tilde{S}$ and $\tilde{A}$ produce the finite expressions (10.34) for $V_{1,2}$ and $W_{1,2}$, which may then be used to define the three-dimensional generalization of the function $\Gamma(x, x'; \omega^2)$ defined in Eq. (10.25):

$$\Gamma_{UV}(x, x'; \omega^2) \equiv U_1(x; \omega^2)V_2(x'; \omega^2)H(x' - x)$$
$$+ V_1(x'; \omega^2)U_2(x; \omega^2)H(x - x'), \tag{10.36}$$

and analogous definitions for Γ_{UW}, Γ_{VU}, etc. Note that $\Gamma_{UU} \equiv \Gamma$ as defined by Eq. (10.25).

10.2 Initial value problem★

With the definitions (10.36) of the Γs, the different linear contributions to the Green's dyadic **G** given in Eq. (10.32) may be decomposed as follows:

$$\tilde{S}(x)\Gamma(x,x') = \left[\tilde{S}U_1(x)\right]U_2(x')H(x'-x) + U_1(x')\left[\tilde{S}U_2(x)\right]H(x-x')$$

$$\equiv V_1(x)U_2(x')H(x'-x) + U_1(x')V_2(x)H(x-x') \equiv \Gamma_{VU}(x,x'),$$

$$\tilde{S}(x')\Gamma(x,x') \equiv \Gamma_{UV}(x,x'), \quad \text{etc.} \tag{10.37}$$

Here, the square brackets are put around the first factor only because operation of the normal derivatives on the Heaviside functions does not produce additional terms since G is continuous according to the boundary conditions (10.21)(a). However, the boundary conditions (10.21)(b) show that the derivative of G is not continuous, so that the quadratic contributions to the Green's dyadic produce decompositions with additional δ-functions:

$$\tilde{S}(x)\tilde{S}(x')\Gamma(x,x') = \left[\tilde{S}U_1(x)\right]\left[\tilde{S}U_2(x')\right]H(x'-x) + \left[\tilde{S}U_1(x')\right]$$

$$\times \left[\tilde{S}U_2(x)\right]H(x-x') - S^2\left(U_1'U_2 - U_1U_2'\right)\delta(x-x')$$

$$\equiv \Gamma_{VV}(x,x') - (S^2/P)\Delta\,\delta(x-x'),$$

$$\tilde{S}(x)\tilde{A}(x')\Gamma(x,x') = \Gamma_{VW}(x,x') - (SA/P)\Delta\,\delta(x-x'),$$

$$\tilde{A}(x)\tilde{S}(x')\Gamma(x,x') = \Gamma_{WV}(x,x') - (SA/P)\Delta\,\delta(x-x'),$$

$$\tilde{A}(x)\tilde{A}(x')\Gamma(x,x') = \Gamma_{WW}(x,x') - (A^2/P)\Delta\,\delta(x-x'). \tag{10.38}$$

The δ-function contributions combine with those already present in the second matrix of Eq. (10.32) according to the equalities

$$-\frac{S^2}{P} + H = \frac{1}{\rho(\omega^2 - \omega_A^2)}, \quad -\frac{SA}{P} + I = 0, \quad -\frac{A^2}{P} + J = \frac{1}{\rho(\omega^2 - \omega_S^2)}. \tag{10.39}$$

Now, insert the expressions (10.37)–(10.39) into Eq. (10.32) for **G** and watch the miracle happen:

$$\mathbf{G}(x,x';\omega^2) = \frac{1}{\Delta(\omega^2)}\begin{pmatrix} \Gamma_{UU} & \Gamma_{UV} & \Gamma_{UW} \\ \Gamma_{VU} & \Gamma_{VV} & \Gamma_{VW} \\ \Gamma_{WU} & \Gamma_{WV} & \Gamma_{WW} \end{pmatrix}$$

$$+ \begin{pmatrix} 0 & 0 & 0 \\ 0 & \dfrac{\delta(x-x')}{\rho(\omega^2-\omega_A^2)} & 0 \\ 0 & 0 & \dfrac{\delta(x-x')}{\rho(\omega^2-\omega_S^2)} \end{pmatrix}. \tag{10.40}$$

Since the Γs defined in Eq. (10.36) are all finite for $D = 0$, this representation is manifestly free of apparent singularities. More importantly, in addition to the implicit presence in the branch cuts of the Γs and Δ, the genuine Alfvén and slow singularities $N = 0$ are now distinctly present on the diagonal. Starting from this dyadic form of **G**, the resolution of the identity, the expansion in eigenfunctions, and the question of completeness of the MHD spectrum can be established, in analogy with the analysis of Sedláček [206].

We have now completed the construction of the resolvent operator $(\mathbf{F} + \rho\omega^2 \mathbf{I})^{-1}$ connecting the Laplace transformed variable $\hat{\boldsymbol{\xi}}$ with the initial data **X**,

$$\hat{\boldsymbol{\xi}} = (\mathbf{F} + \rho\omega^2 \mathbf{I})^{-1} \cdot i\rho\omega \mathbf{X}, \tag{10.41}$$

explicitly given by Eq. (10.30) involving the Green's dyadic **G** as given by Eq. (10.40). The solution of the initial value problem involves the inverse Laplace transform:

$$\begin{aligned}
\boldsymbol{\xi}(\mathbf{r}; t) &= \frac{1}{2\pi} \int_{i\nu_0 - \infty}^{i\nu_0 + \infty} \hat{\boldsymbol{\xi}}(\mathbf{r}; \omega) \, e^{-i\omega t} \, d\omega \\
&= \frac{1}{2\pi} \int_C d\omega \, e^{-i\omega t} \int_{x_1}^{x_2} dx' \rho(x') \, \mathbf{G}(\mathbf{r}, \mathbf{r}'; \omega^2) \cdot \left[i\omega \boldsymbol{\xi}_i(x') - \dot{\boldsymbol{\xi}}_i(x') \right].
\end{aligned} \tag{10.42}$$

The integration contour of Figs. 6.10 and 6.11 is to be placed above the largest point eigenvalue $i\nu_{\max}$ (i.e. $\nu_0 > \nu_{\max}$) of **F**. Upon deformation of this contour *branch cuts of the Γs and* Δ, corresponding to the Alfvén and slow *continua* $\{\pm\omega_A\}$ and $\{\pm\omega_S\}$ and *zeros of* Δ, corresponding to the different *discrete modes*, are encountered.

The explicit evaluation of the integrals in Eq. (10.42) requires the specification of an equilibrium and initial data. For example, equilibria with a steep gradient in the Alfvén frequency exhibit damping of Alfvén waves [225] which may be described by deforming the integration contour off the principal Riemann sheet across the branch cuts to another Riemann sheet where poles corresponding to quasi-modes may be encountered (see Section 10.3). Thus the great example of Landau's prescription of handling the poles in plasma kinetic theory [136] can be applied to ideal MHD theory as well. Preparation of special initial data, singling out the δ-functions of Eq. (10.40), leads to improper Alfvén and slow modes where a single magnetic surface oscillates with the frequency ω_A or ω_S. This is the analogue of the Van Kampen modes [237] for ideal MHD. Consequently, the phenomena of damping and singular oscillation are exclusively associated with the genuine $N = 0$ singularities, corresponding to *local* perturbations (large 'wave numbers') which rapidly vary in the direction of inhomogeneity.

Once more: the $D = 0$ singularities are apparent, not genuine, and correspond in an average sense to the *global* slow and fast discrete modes in the turning point frequency ranges $\{\omega_{s0}\}$ and $\{\omega_{f0}\}$. They just complicate the analysis of the discrete modes with respect to their monotonicity properties (as described by the oscillation theorem of Section 7.4.4) since they usually overlap with the genuine $N = 0$ continua. However, there is no place for additional continua besides the slow and Alfvén continua, except for the cluster point $\omega_F^2 = \infty$ which provides the asymptotic behaviour of the localized fast modes for large 'wave numbers'. In that sense, there are *three continua* [82] in ideal MHD. They correspond to the slow, Alfvén and fast degrees of freedom. This structure is already present in homogeneous plasmas. Inhomogeneity, in e.g. cylindrical and toroidal geometry, extends this structure by quite a number of additional important features, but it does not change the fundamental number of degrees of freedom: *the threefold ideal MHD spectrum is complete!*

10.3 Damping of Alfvén waves★

In principle, the initial value problem has now been solved. However, this solution consists of the simultaneous evolution of all the MHD modes. In order not to get lost by all formal generalities, let us now concentrate on the important features. To that end, we make some simplifying assumptions to the effect that the three sub-spectra become widely separated. We may then study the separate influence of one sub-spectrum, in this case the Alfvén continuum.

For the study of the Alfvén continuum by itself, we may ignore the gravitational acceleration terms. We also assume the density to be constant and the magnetic field **B** to be *uni-directional*, so that the functions f and g become constant wave numbers:

$$f = k_\parallel, \qquad g = k_\perp. \tag{10.43}$$

Next, we consider a *low-β* plasma ($\beta \equiv 2p/B^2$), so that

$$c^2 \ll b^2. \tag{10.44}$$

This assumption separates the slow from the Alfvén modes:

$$\omega_S^2 \approx k_\parallel^2 c^2 \approx \omega_{s0}^2 \ll \omega_A^2 = k_\parallel^2 b^2. \tag{10.45}$$

In order to separate off the influence of the fast modes as well (Fig. 10.1) we concentrate our study on *nearly perpendicular propagation*:

$$k_\parallel \ll k_\perp \approx k_0, \tag{10.46}$$

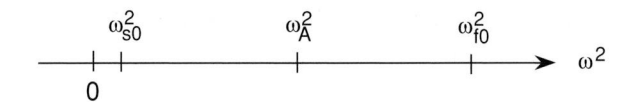

Fig. 10.1. Separation of the three frequency ranges.

so that
$$\omega_A^2 = k_\parallel^2 b^2 \ll \omega_{f0}^2 \approx k_0^2 b^2 . \tag{10.47}$$

Under these conditions, there is no parallel motion to leading order, $\hat{\zeta} = Z = 0$, so that Eq. (10.10) simplifies to

$$\begin{pmatrix} \dfrac{d}{dx} b^2 \dfrac{d}{dx} - k_\parallel^2 b^2 + \omega^2 & \dfrac{d}{dx} k_\perp b^2 \\ -k_\perp b^2 \dfrac{d}{dx} & -k_\perp^2 b^2 - k_\parallel^2 b^2 + \omega^2 \end{pmatrix} \begin{pmatrix} \hat{\xi} \\ \hat{\eta} \end{pmatrix} = i\omega \begin{pmatrix} X \\ Y \end{pmatrix} . \tag{10.48}$$

Only transverse motion needs to be studied. In this equation we have kept terms of unequal order in $k_\parallel$ and $k_\perp$ because large terms cancel upon elimination of $\hat{\eta}$. After elimination we keep terms of comparable order only, resulting in the following equations:

$$-\dfrac{1}{k_0^2}\left[(\omega^2 - \omega_A^2)\hat{\xi}'\right]' + (\omega^2 - \omega_A^2)\hat{\xi} = i\omega\left(X + \dfrac{1}{k_0}Y'\right), \tag{10.49}$$

$$\hat{\eta} = -\dfrac{1}{k_0}\hat{\xi}' - i\omega\dfrac{Y}{k_0^2 b^2} , \tag{10.50}$$

where all equilibrium variations are expressed by the Alfvén frequency,
$$\omega_A^2 = \omega_A^2(x) = k_\parallel^2 b^2(x) . \tag{10.51}$$

Introducing the short-hand notation
$$P(x; \omega^2) \equiv -\rho(\omega^2 - \omega_A^2)/k_0^2 ,$$
$$Q(x; \omega^2) \equiv -\rho(\omega^2 - \omega_A^2) , \tag{10.52}$$
$$R(x; \omega^2) \equiv i\rho\omega\left(X + Y'/k_0\right) ,$$

the inhomogeneous second order differential equation (10.49) may be written as
$$(P\hat{\xi}')' - Q\hat{\xi} = R . \tag{10.53}$$

Of course, the basic equation is of the same form as the general equation (10.17) so that the solution $\hat{\xi}(x)$ may be represented by the integral (10.19) involving a

10.3 Damping of Alfvén waves*

Green's function $G(x, x')$ which is a solution of the differential equation (10.20) subject to the boundary conditions (10.21).

10.3.1 Green's function*

The inhomogeneous equation (10.20) allows for a unique solution for the Green's function (Fig. 10.2(a)) when the homogeneous equation does not have a non-trivial solution (*Fredholm alternative*). Proper and improper solutions of the homogeneous equation (Figs. 10.2(b) and (c)) occur for values of ω^2 inside the spectrum, which is confined to the real ω^2-axis, so that we certainly have a unique Green's function for complex values of ω on the Laplace contour. The procedure is then to construct the Green's function for complex values of ω^2 where existence is guaranteed and to deform the contour in such a way that the spectrum is approached.

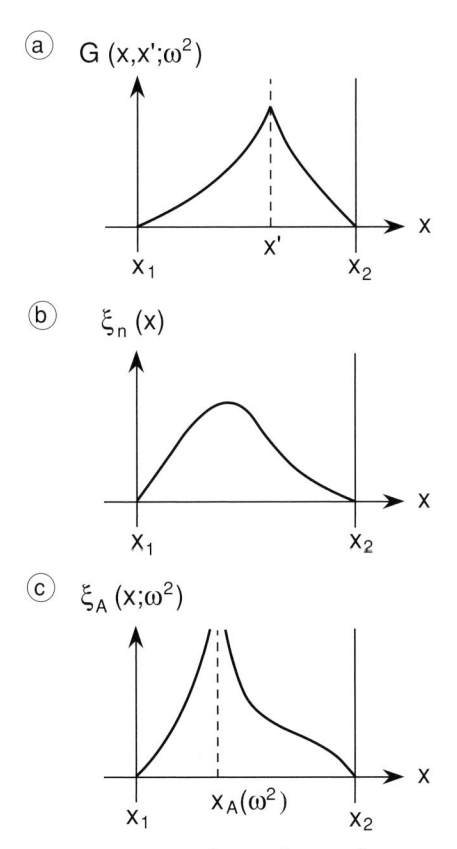

Fig. 10.2. (a) Green's function for $\omega^2 \notin \{\omega_A^2(x), \omega_n^2\}$; (b) proper eigenfunction for $\omega^2 = \omega_n^2$; (c) improper eigenfunction for $\omega^2 \in \{\omega_A^2(x)\}$.

As in Section 10.2.1, the symmetric expression for $G(x, x'; \omega^2)$ is found in terms of the fundamental solutions $U_1(x; \omega^2)$ and $U_2(x; \omega^2)$ of the homogeneous equation satisfying the left and right boundary conditions, respectively:

$$(PU_1')' - QU_1 = 0, \qquad U_1(x_1) = 0,$$
$$(PU_2')' - QU_2 = 0, \qquad U_2(x_2) = 0. \qquad (10.54)$$

In terms of these functions one finds for the Green's function:

$$G(x, x'; \omega^2) = \frac{\Gamma(x, x'; \omega^2)}{\Delta(\omega^2)}, \qquad (10.55)$$

where

$$\Gamma(x, x'; \omega^2) \equiv U_1(x_<; \omega^2) U_2(x_>; \omega^2) \qquad (10.56)$$
$$= U_1(x; \omega^2) U_2(x'; \omega^2) H(x' - x) + U_1(x'; \omega^2) U_2(x; \omega^2) H(x - x'),$$
$$\Delta(\omega^2) \equiv P(x; \omega^2) \left[U_1(x; \omega^2) U_2'(x; \omega^2) - U_1'(x; \omega^2) U_2(x; \omega^2) \right]. \qquad (10.57)$$

Here, we have introduced the notation

$$x_< \equiv \inf(x, x'), \qquad x_> \equiv \sup(x, x'). \qquad (10.58)$$

The expression inside the square brackets in the definition of Δ is recognized as the Wronskian. By means of Eqs. (10.54) one proves

$$\frac{\partial \Delta}{\partial x} = P'(U_1 U_2' - U_1' U_2) + P(U_1 U_2'' - U_1'' U_2)$$
$$= U_1(PU_2')' - U_2(PU_1')' = QU_1 U_2 - QU_2 U_1 = 0, \qquad (10.59)$$

so that $\Delta \neq \Delta(x)$. For eigenfunctions, the solution of the homogeneous equation satisfies both left and right boundary conditions, so that $U_1 = U_2$. In that case $\Delta(\omega^2) = 0$. For that reason, $\Delta(\omega^2)$ is called the *dispersion function*.

Let us again specify the profile $\omega_A^2 = \omega_A^2(x)$ to be monotonically increasing on the interval (x_1, x_2), as in Section 7.4, and construct the inverse profile $x_A = x_A(\omega^2)$. For example, for a simple linear profile (Fig. 10.3) the explicit functions would read:

$$\omega_A^2(x) = \omega_0^2(x) + \omega_A^{2\,\prime}(x - x_0), \qquad \omega_0^2 \equiv \tfrac{1}{2}(\omega_{A1}^2 + \omega_{A2}^2),$$
$$x_A(\omega^2) = x_0 + (\omega^2 - \omega_0^2)/\omega_A^{2\,\prime}, \qquad x_0 \equiv \tfrac{1}{2}(x_1 + x_2). \qquad (10.60)$$

In Section 7.4 we expanded around the singularity $x = x_A(\omega^2)$ of Eq. (10.60) in terms of the variable $s = x - x_A(\omega^2)$. Here, ω^2 is complex so that the corresponding singularity of Eq. (10.54) occurs in the complex z-plane for $z = z_A(\omega^2)$ (see Fig. 10.4) where $z_A(\omega^2)$ is the analytic continuation of $x_A(\omega^2)$. For the linear

10.3 Damping of Alfvén waves*

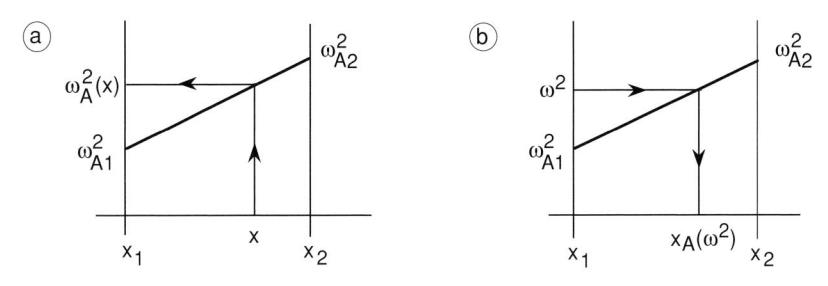

Fig. 10.3. Inversion of $\omega_A^2 = \omega_A^2(x)$ for linear profiles.

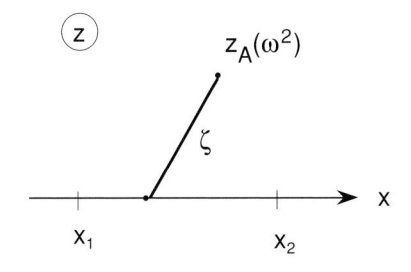

Fig. 10.4. Analytic continuation of $x_A(\omega^2)$.

profile the explicit expression for $z_A(\omega^2)$ would be

$$z_A(\omega^2) = x_0 + (\omega^2 - \omega_0^2)/\omega_A^{2\,\prime}. \tag{10.61}$$

Introducing a complex variable ζ replacing s,

$$\zeta = \zeta(x;\omega^2) \equiv x - z_A(\omega^2), \tag{10.62}$$

the solutions U_1 and U_2 of the equations (10.54) may be expressed as a linear combination of the functions

$$\begin{cases} u(\zeta) \\ u(\zeta)\ln\zeta + v(\zeta), \end{cases} \tag{10.63}$$

where $u(\zeta)$ and $v(\zeta)$ are the analytic continuations of the functions $u(s)$ and $v(s)$ introduced in Eq. (7.140), which may be written as a power series in ζ: $u = a + b\zeta + \cdots$, and similarly for v. Hence,

$$U_1(\zeta) = \left[\ln\frac{\zeta(x;\omega^2)}{\zeta_1(\omega^2)} - \frac{v_1(\omega^2)}{u_1(\omega^2)}\right] u(\zeta;\omega^2) + v(\zeta;\omega^2),$$

$$U_2(\zeta) = \left[\ln\frac{\zeta(x;\omega^2)}{\zeta_2(\omega^2)} - \frac{v_2(\omega^2)}{u_2(\omega^2)}\right] u(\zeta;\omega^2) + v(\zeta;\omega^2). \tag{10.64}$$

Substituting these expressions into Eq. (10.55) provides us with the formal solution of the Green's function:

$$G(x, x'; \omega^2) = \left\{ \left[\ln \frac{\omega^2 - \omega_A^2(x_<)}{\omega^2 - \omega_{A1}^2} - \frac{v_1(\omega^2)}{u_1(\omega^2)} \right] u(x_<; \omega^2) + v(x_<; \omega^2) \right\}$$

$$\times \left\{ \left[\ln \frac{\omega^2 - \omega_A^2(x_>)}{\omega^2 - \omega_{A2}^2} - \frac{v_2(\omega^2)}{u_2(\omega^2)} \right] u(x_>; \omega^2) + v(x_>; \omega^2) \right\}$$

$$\Bigg/ \left[\ln \frac{\omega^2 - \omega_{A2}^2}{\omega^2 - \omega_{A1}^2} - \frac{v_1(\omega^2)}{u_1(\omega^2)} + \frac{v_2(\omega^2)}{u_2(\omega^2)} \right]. \qquad (10.65)$$

Here, the logarithmic expression in terms of ζ has been converted into the more transparent form in terms of $\omega^2 - \omega_A^2(x)$ by means of the relation

$$\zeta = x - z_A(\omega^2) = -(\omega^2 - \omega_A^2(x))/\omega_A^{2'}, \qquad (10.66)$$

which is, strictly speaking, only valid for the linear profile. However, for an arbitrary monotonically increasing profile Eq. (10.65) is also valid if the functions u and v are redefined such that the expressions for the basic solutions are written as

$$\begin{cases} u(\omega^2 - \omega_A^2(x)) \\ u(\omega^2 - \omega_A^2(x)) \ln(\omega^2 - \omega_A^2(x)) + v(x; \omega^2) \end{cases} \qquad (10.67)$$

instead of Eq. (10.63). Clearly, for the derivation of the expression (10.65) of the Green's function no other property has been used than the fact that $\omega_A^2(x)$ is a monotonic function and that the slow continuum is far away so that we are dealing with only one singularity at a time.

10.3.2 Spectral cuts★

For the completion of the initial value problem we now need to study the behaviour of the Green's function when ω approaches the spectrum. We have already seen that the zeros of the denominator $\Delta(\omega^2)$ represent the discrete spectrum. The continuous spectrum arises as a result of the multi-valuedness of the logarithmic terms appearing in both $\Gamma(x, x'; \omega^2)$ and $\Delta(\omega^2)$. In order to make these logarithmic terms single-valued one needs to cut the complex ω-plane along branch cuts that precisely correspond to the continuous spectra $\pm\{\omega_A(x)\}$, as we will see.

In order to make a logarithmic function $\ln z$ single-valued one may cut the z-plane along any curve starting at the branch point $z = 0$ and extending to ∞. Let us choose the negative real axis as a branch cut (Fig. 10.5). Along this branch

10.3 Damping of Alfvén waves*

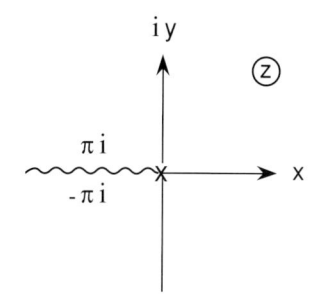

[Figure: coordinate system with iy axis, x axis, branch points at πi and -πi marked with wavy line branch cut, point z labeled]

Fig. 10.5. Branch cut for $\ln z$.

cut one may write

$$\lim_{y \to 0^\pm} \ln z = \ln |z| \pm \pi i \tag{10.68}$$

(on the principal, $n = 0$, Riemann sheet), where $+\pi i$ is the value immediately above the branch cut and $-\pi i$ immediately below. If one wishes to deform a contour across a branch cut one moves to another Riemann sheet of the logarithmic function. These sheets are labelled by n, and the logarithmic function increases by an amount $2\pi i$ every time one encircles the branch point and moves to the next Riemann sheet. Therefore, the general expression for the logarithmic function when approaching the real axis may be written as

$$\lim_{y \to 0^\pm} \ln z = \ln |x| \pm \pi i H(-x) + 2n\pi i, \tag{10.69}$$

where the jump of the Heaviside function occurs at the branch point. Accordingly, for complex values of $\omega = \operatorname{Re} \omega + i\nu$, one may write for a logarithmic expression of the type $\ln[(\omega^2 - \omega_\beta^2)/(\omega^2 - \omega_\alpha^2)]$ when approaching the real axis:

$$\lim_{\nu \to 0^\pm} \ln \frac{\omega^2 - \omega_\beta^2}{\omega^2 - \omega_\alpha^2} = \ln \left| \frac{\omega^2 - \omega_\beta^2}{\omega^2 - \omega_\alpha^2} \right|$$

$$\pm i\pi \Big[H(\omega - \omega_\alpha) - H(\omega - \omega_\beta) - H(\omega + \omega_\alpha) + H(\omega + \omega_\beta) \Big] + 2n\pi i. \tag{10.70}$$

Hence, assuming $\omega_\beta^2 > \omega_\alpha^2$, the branch cuts and jumps are as indicated in Fig. 10.6. Here, we have indicated how one moves from the principal sheet to the $n = 1$ and $n = -1$ sheets when crossing the branch cuts.

On the basis of the expression (10.67) we find that the function $\Gamma(x, x'; \omega^2)$ has branch points $\omega_{A<}^2 \equiv \omega_A^2(x_<)$, $\omega_{A>}^2 \equiv \omega_A^2(x_>)$, ω_{A1}^2, and ω_{A2}^2, whereas the

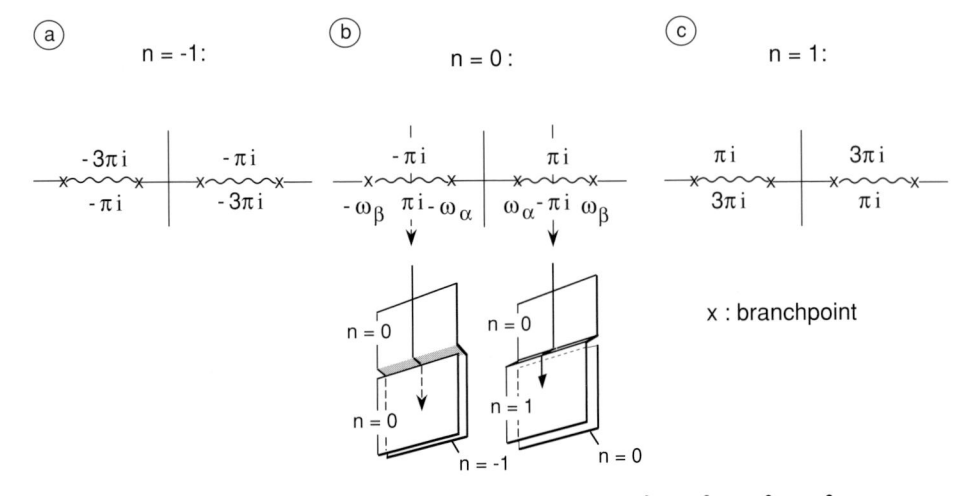

Fig. 10.6. Riemann sheets for the function $\ln[(\omega^2 - \omega_\beta^2)/(\omega^2 - \omega_\alpha^2)]$.

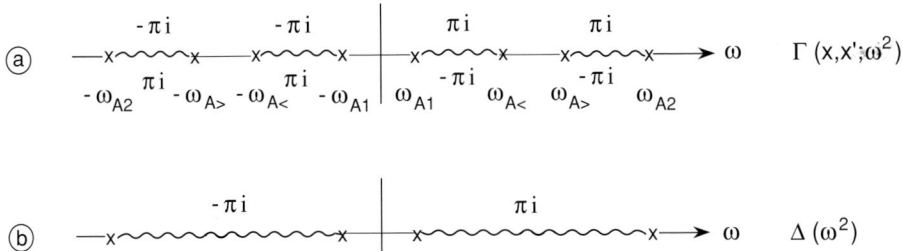

Fig. 10.7. Branch cuts for: (a) numerator, and (b) denominator of the Green's function.

function $\Delta(\omega^2)$ only has branch points at ω_{A1}^2 and ω_{A2}^2. One may connect these branch points as indicated in Fig. 10.7. For the Green's function $G = \Gamma/\Delta$ these branch points should be joined. One may do this by choosing the branch cuts for Δ differently, so that the Laplace contour C may be deformed to a contour C' as shown in Fig. 10.8 (see Sedláček [206]). This clearly shows that the contribution of the continuous spectrum is due to the jump in the logarithmic function along the branch cuts.

Let us now calculate the typical contributions of the spectral cuts to the solution of the initial value problem. Take special initial data: $\xi_i(x) \neq 0$, $\dot{\xi}_i(x) = \eta_i(x) = \dot{\eta}_i(x) = 0$. The solution of the initial value problem can then be written from the

10.3 Damping of Alfvén waves*

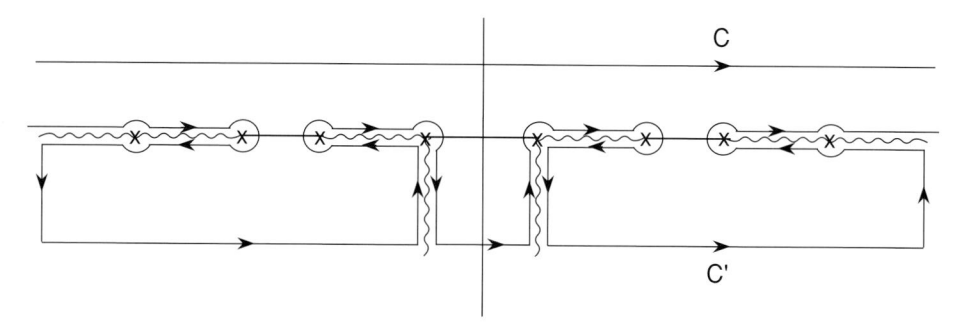

Fig. 10.8. Laplace contours for the Green's function $G(x, x'; \omega^2) \equiv \Gamma(x, x'; \omega^2)/\Delta(\omega^2)$.

Eqs. (10.1), (10.50) and (10.19) as:

$$\xi(x; t) = \frac{1}{2\pi} \int_C d\omega \frac{i\omega}{\Delta(\omega^2)} e^{-i\omega t} \int_{x_1}^{x_2} dx' \, \Gamma(x, x'; \omega^2) \xi_i(x'),$$

$$\eta(x; t) = -\frac{1}{k_0} \frac{\partial}{\partial x} \xi(x; t).$$

(10.71)

From Eq. (10.70) one then finds as the typical contribution from a jump of the logarithmic function at some real frequency ω_α:

$$\xi(t) \sim \int_C i\omega e^{-i\omega t} H(\omega - \omega_\alpha) \, d\omega = -\int_C \omega \frac{\partial}{\partial \omega}\left(\frac{e^{-i\omega t}}{t}\right) H(\omega - \omega_\alpha) \, d\omega$$

$$= \int_C \frac{e^{-i\omega t}}{t} H(\omega - \omega_\alpha) \, d\omega + \int_C \omega \frac{e^{-i\omega t}}{t} \delta(\omega - \omega_\alpha) \, d\omega. \quad (10.72)$$

Asymptotically, the first integral may be neglected because the rapidly oscillating integrand kills this contribution for large t. Thus, as shown by Tataronis [224], the asymptotic behaviour in time of the Alfvén continuum modes is given by:

$$\xi(t) \sim \frac{\omega_\alpha}{t} e^{-i\omega_\alpha t}, \qquad \eta(t) \sim -i \frac{\omega_\alpha \omega'_\alpha}{k_0} e^{-i\omega_\alpha t}. \quad (10.73)$$

Consequently, the continuous spectrum of Alfvén modes yields oscillatory normal components that are damped like t^{-1} and *undamped* oscillatory tangential components, perpendicular to the field lines, where each point oscillates with its own local Alfvén frequency. As time goes on, the factor $\exp(-i\omega_\alpha t)$ gives rise to an ever more fluctuating spatial structure of the motion, finally resulting in completely uncoordinated oscillations, which is called *phase mixing*. In this way, large spatial gradients are built up so that, eventually, dissipative effects lead to dissipation of the energy of the continuum modes and heating of the plasma (see Chapter 11).

10.4 Quasi-modes*

In contrast to the situation just described, another kind of motion exists that displays coherent oscillations. To exhibit this, let us start with a profile $\omega_A^2(x)$ that has a step discontinuity at some value of x, say in the middle of the slab at $x = x_0 \equiv \frac{1}{2}(x_1 + x_2)$ (Fig. 10.9). The singularities of the continuous spectrum $\omega_{A1}^2 \leq \omega^2 \leq \omega_{A2}^2$ are now all concentrated in the point $x = x_0$. This gives rise to a special mode which is called *a surface mode*. It may be found from the homogeneous equation corresponding to Eq. (10.49):

$$\frac{1}{k_0^2}\left[(\omega^2 - \omega_A^2)\xi'\right]' - (\omega^2 - \omega_A^2)\xi = 0, \qquad (10.74)$$

where $\omega_A^2(x) = \omega_{A1}^2 H(x_0 - x) + \omega_{A2}^2 H(x_0 - x)$. On the left and right intervals $x_1 \leq x < x_0$ and $x_0 \leq x < x_2$ this equation reduces to

$$\xi'' - k_0^2 \xi = 0, \qquad (10.75)$$

having the solutions $\exp(k_0 x)$ and $\exp(-k_0 x)$, when $\omega^2 \neq \omega_{A1}^2$ and $\omega^2 \neq \omega_{A2}^2$, respectively. The solution $\xi_1 = \sinh[k_0(x - x_1)]$ satisfying the left hand boundary condition may be combined with the solution $\xi_2 = \sinh[k_0(x_2 - x)]$ satisfying the right hand boundary condition to form a cusp-shaped perturbation which is an eigenfunction of the system (Fig. 10.10). That this is so may be seen by applying the proper boundary condition to join ξ_1 to ξ_2. This condition is found from Eq. (10.74) by integrating across the jump:

$$\int_{x_0^-}^{x_0^+} \left\{\left[\omega^2 - \omega_{A1}^2 H(x_0 - x) - \omega_{A2}^2 H(x - x_0)\right]\xi'\right\}' dx$$

$$= (\omega^2 - \omega_{A1}^2)\xi_1' - (\omega^2 - \omega_{A2}^2)\xi_2' = 0,$$

or

$$\left[\!\left[(\omega^2 - \omega_A^2)\xi'\right]\!\right]_{x=x_0} = 0. \qquad (10.76)$$

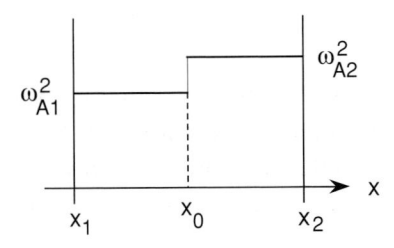

Fig. 10.9. Step discontinuity of the Alfvén frequency.

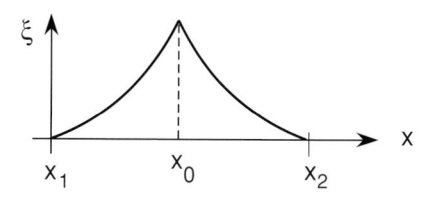

Fig. 10.10. Surface mode.

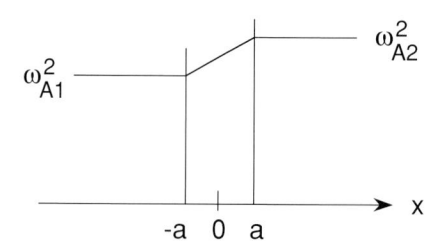

Fig. 10.11. Smoothing the discontinuity.

This condition is fulfilled for $\omega^2 = \omega_0^2 \equiv \frac{1}{2}(\omega_{A1}^2 + \omega_{A2}^2)$, which is the eigenfrequency of the cusped surface wave.

10.4.1 Dispersion equation*

Let us now remove the degeneracy of the step and introduce a genuine continuum by smoothing out the discontinuity (Fig. 10.11). This we do by replacing the step by a linearly increasing profile between $x = -a$ and $x = a$, where we have fixed $x_0 = 0$. For simplicity, we also take $x_1 \to -\infty$ and $x_2 \to +\infty$. The spectrum of the system then changes as shown in Fig. 10.12. Notice that for the stepped and the continuous profile there are also infinitely many discrete Alfvén modes with eigenfrequencies $\omega = \pm\omega_{A1}$ and $\omega = \pm\omega_{A2}$. These are localized on the left and the right homogeneous intervals, respectively (Fig. 10.13). That this is so may be seen from Eq. (10.74) by pulling out the factor $\omega^2 - \omega_A^2$ which is constant on the homogeneous intervals:

$$(\omega^2 - \omega_A^2)(\xi'' - k_0^2 \xi) = 0. \tag{10.77}$$

Hence, for $\omega^2 = \omega_{A1}^2$ on the left homogeneous interval ξ may be chosen arbitrarily. Each choice of this function is a proper Alfvén eigenfunction. Likewise, for $\omega^2 = \omega_{A2}^2$ on the right interval. However, here we wish to concentrate on the influence of the inhomogeneity. In particular, we want to see what happened to the surface wave by the introduction of the linear profile. Does the appearance of a continuous spectrum imply that all of a sudden the coherent oscillations of

Fig. 10.12. Change of the spectrum due to smoothing.

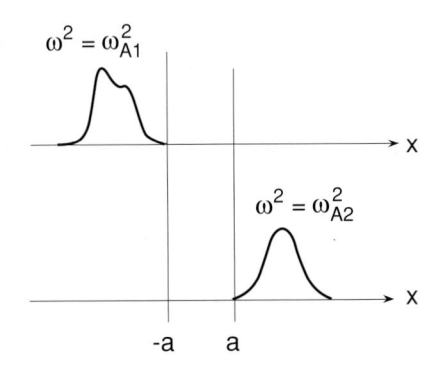

Fig. 10.13. Alfvén modes localized left and right of the inhomogeneity.

the surface wave have disappeared to make place for the kind of chaotic response expressed by Eq. (10.73)? This is hard to believe.

We already noticed that the discrete spectrum comes about from the poles of the Green's function, i.e. the zeros of the dispersion function $\Delta(\omega^2)$. Let us, therefore, study the expression $\Delta(\omega^2)$ for the present case. To that end, we need the explicit solutions U_1 and U_2 to the homogeneous equations (10.54) on the three intervals $(-\infty, -a)$, $(-a, a)$ and (a, ∞). The virtue of the choice of a linear profile on $(-a, a)$ is that the homogeneous equation for this interval may be written as

$$\frac{d}{d\zeta} \zeta \frac{dU_1}{d\zeta} - k_0^2 \zeta\, U_1 = 0, \qquad \zeta \equiv -2a\, \frac{\omega^2 - \omega_A^2(x)}{\omega_{A2}^2 - \omega_{A1}^2}, \qquad (10.78)$$

so that we obtain modified Bessel functions of complex argument as solutions:

$$U_1, U_2 = \begin{cases} I_0(k\zeta) = 1 + \tfrac{1}{4}(k_0\zeta)^2 + \cdots \\ K_0(k_0\zeta) = -(\ln \tfrac{1}{2}k_0\zeta + \gamma) I_0(k_0\zeta) + \tfrac{1}{4}(k_0\zeta)^2 + \cdots, \end{cases} \quad (10.79)$$

where $\gamma \approx 0.577$ is Euler's constant.

10.4 Quasi-modes*

Consequently, the following solutions are obtained:

$$U_1 = \begin{cases} e^{k_0 x} \\ A_1 I_0(k_0\zeta) + B_1 K_0(k_0\zeta), \\ C_1 e^{k_0 x} + D_1 e^{-k_0 x} \end{cases} \quad U_2 = \begin{cases} C_2 e^{k_0 x} + D_2 e^{-k_0 x} & (-\infty, -a) \\ A_2 I_0(k_0\zeta) + B_2 K_0(k_0\zeta) & (-a, a) \\ e^{-k_0 x} & (a, \infty). \end{cases}$$
(10.80)

The constants $A_{1,2}$, $B_{1,2}$, $C_{1,2}$ and $D_{1,2}$ are fixed by equating functions and first derivatives at the boundaries of the intervals. For the calculation of $\Delta(\omega^2)$ we actually only need to compute $A_{1,2}$ and $B_{1,2}$ because $\Delta(\omega^2)$ is independent of x, so that we may choose to evaluate it in the inhomogeneous layer. The solutions U_1 and U_2 on $(-a, a)$ read:

$$U_1 = k_0 \zeta_1 e^{-k_0 a} \left\{ \left[K_0(k_0\zeta_1) + K_1(k_0\zeta_1) \right] I_0(k_0\zeta) \right.$$
$$\left. - \left[I_0(k_0\zeta_1) - I_1(k_0\zeta_1) \right] K_0(k_0\zeta) \right\},$$

$$U_2 = -k_0 \zeta_2 e^{-k_0 a} \left\{ \left[K_0(k_0\zeta_2) + K_1(k_0\zeta_2) \right] I_0(k_0\zeta) \right.$$
$$\left. - \left[I_0(k_0\zeta_2) + I_1(k_0\zeta_2) \right] K_0(k_0\zeta) \right\}, \quad (10.81)$$

where

$$\zeta_{1,2} \equiv -2a \, (\omega^2 - \omega^2_{A1,2}) / (\omega^2_{A2} - \omega^2_{A1}). \quad (10.82)$$

Inserting these solutions into the dispersion function we find

$$\Delta = P(U_1 U_2' - U_1' U_2)$$
$$= C \zeta_1 \zeta_2 \left\{ \left[I_0(k_0\zeta_1) - I_1(k_0\zeta_1) \right] \left[K_0(k_0\zeta_2) - K_1(k_0\zeta_2) \right] \right.$$
$$\left. - \left[K_0(k_0\zeta_1) + K_1(k_0\zeta_1) \right] \left[I_0(k_0\zeta_2) + I_1(k_0\zeta_2) \right] \right\}, \quad (10.83)$$

where C is a constant that is not important for the present purpose. To obtain Eq. (10.83) we have used the property $z[I_0(z) K_1(z) + I_1(z) K_0(z)] = 1$.

The dispersion equation is given by

$$\Delta(\omega^2) = 0. \quad (10.84)$$

Note that the two trivial solutions $\zeta_1 = 0$ and $\zeta_2 = 0$ on the homogeneous intervals, corresponding to the two discrete eigenvalues $\omega^2 = \omega^2_{A1}$ and $\omega^2 = \omega^2_{A2}$, are *not* contained in this dispersion equation because the factors ζ_1 and ζ_2 cancel out against the singularities of the exponential Bessel functions.

10.4.2 Exponential damping*

Let us now investigate whether some more solutions exist, hopefully corresponding to the surface wave solution of the step function model. To that end we study a situation where the continuous profile is close to the step function model, i.e., a is considered to be small. Since the other intervals are infinite, the only scale to compare a with is the perpendicular wavelength k_0^{-1}. Hence, we assume $k_0 a \ll 1$ and expand Eq. (10.84) in orders of $k_0 a$. By means of the expansions (10.79) of the Bessel functions we find to leading order

$$\ln \frac{\zeta_2}{\zeta_1} + \frac{1}{k_0}\left(\frac{1}{\zeta_1} + \frac{1}{\zeta_2}\right) = 0, \tag{10.85}$$

or

$$\ln \frac{\omega^2 - \omega_{A2}^2}{\omega^2 - \omega_{A1}^2} - \frac{\omega_{A2}^2 - \omega_{A1}^2}{2 k_0 a}\left[\frac{1}{\omega^2 - \omega_{A1}^2} + \frac{1}{\omega^2 - \omega_{A2}^2}\right] = 0. \tag{10.86}$$

Let us now study this expression in the neighbourhood of the real axis so that $\nu \ll \omega$. We then have from Eq. (10.70) for ω in the range of the continua:

$$\ln \frac{\omega^2 - \omega_{A2}^2}{\omega^2 - \omega_{A1}^2} \approx \ln \left|\frac{\omega^2 - \omega_{A2}^2}{\omega^2 - \omega_{A1}^2}\right|$$

$$+ \mathrm{sg}(\omega)\mathrm{sg}(\nu)\,\pi \mathrm{i} + 2n\pi \mathrm{i} + 2\mathrm{i}\nu\omega \frac{\omega_{A2}^2 - \omega_{A1}^2}{(\omega^2 - \omega_{A1}^2)(\omega^2 - \omega_{A2}^2)}, \tag{10.87}$$

where the last term may be dropped again as it is small compared to the other imaginary contributions. This gives

$$\ln \left|\frac{\omega^2 - \omega_{A2}^2}{\omega^2 - \omega_{A1}^2}\right| - \frac{\omega_{A2}^2 - \omega_{A1}^2}{k_0 a} \frac{\omega^2 - \tfrac{1}{2}(\omega_{A1}^2 + \omega_{A2}^2)}{(\omega^2 - \omega_{A1}^2)(\omega^2 - \omega_{A2}^2)}$$

$$+ \mathrm{sg}(\omega)\mathrm{sg}(\nu)\,\pi \mathrm{i} + 2n\pi \mathrm{i} + \mathrm{i}\frac{\nu\omega(\omega_{A2}^2 - \omega_{A1}^2)}{k_0 a} \frac{(\omega^2 - \omega_{A1}^2)^2 + (\omega^2 - \omega_{A2}^2)^2}{(\omega^2 - \omega_{A1}^2)^2(\omega^2 - \omega_{A2}^2)^2} = 0. \tag{10.88}$$

The real and imaginary parts of this dispersion equation give the roots we are looking for:

$$\mathrm{Re}\,\omega = \pm\omega_0 \equiv \pm\sqrt{\tfrac{1}{2}(\omega_{A1}^2 + \omega_{A2}^2)},$$

$$\nu = \nu_0 \equiv -\frac{1}{8}\pi k_0 a \left[\mathrm{sg}(\nu)\mathrm{sg}(\omega) + 2n\right]\frac{\omega_{A2}^2 - \omega_{A1}^2}{\mathrm{sg}(\omega)\omega_0}. \tag{10.89}$$

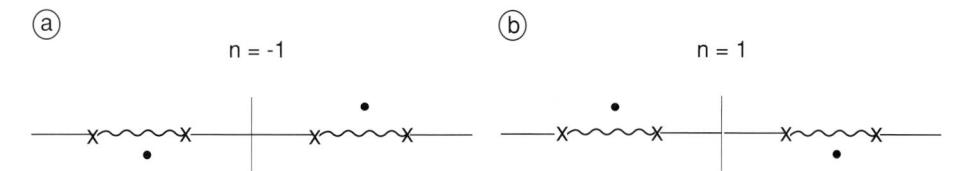

Fig. 10.14. Poles of the Green's function.

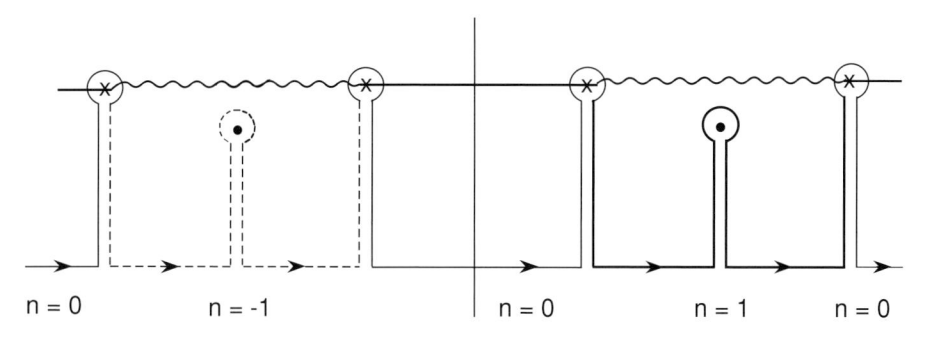

Fig. 10.15. Deformation of the Laplace contour.

This seems to give a satisfactory generalization of the surface mode as it reduces to $\omega = \omega_0$ for $a = 0$. If $a \neq 0$ a 'mode' is obtained which has a small imaginary part to the 'eigenfrequency'. We have put quotation marks here because we have proved already that in ideal MHD normal modes cannot have complex eigenvalues. On the other hand, we have obtained a genuine pole of the Green's function, which certainly will influence the response to the initial data.

For $n = 0$ the expression for ν_0 in Eq. (10.89) gives a contradiction, so that no solutions are found on the principal Riemann sheet, corresponding to the fact that complex eigenvalues do not exist in ideal MHD. For $n = 1$ and $n = -1$, however, we find two poles (see Fig. 10.14) with

$$|\nu_0| = \frac{1}{8}\pi k_0 a \, (\omega_{A2}^2 - \omega_{A1}^2)/\omega_0 . \tag{10.90}$$

We may now deform the Laplace contour across the branch cuts so that the contributions of the complex poles on the neighbouring Riemann sheets are picked up (Fig. 10.15). Ignoring the contributions of the branch cuts corresponding to the continuous spectrum (and also the contribution of the branch points which are simultaneously poles corresponding to the degenerate Alfvén modes), we find asymptotically for large t for the contributions of these poles:

$$\xi(t) \sim \frac{1}{2\pi} \int d\omega \, \frac{i\omega}{\Delta(\omega^2)} e^{-i\omega t} \sim \frac{1}{2\pi} \int d\omega \, \frac{i\omega}{\omega - \omega_0} e^{-i\omega t}$$

$$= \omega_0 e^{-i\omega_0 t} \sim e^{-|\nu_0|t} e^{-i\omega_0 t} . \tag{10.91}$$

Likewise, $\eta(t) \sim e^{-|\nu_0|t} e^{-i\omega_0 t}$.

Hence, we have found a 'mode' that is exponentially damped. Since the pole is not on the principal branch of the Green's function, there is no contradiction with the general proof that complex eigenvalues do not occur for self-adjoint linear operators. On the other hand, it is clear that the present 'mode' of the plasma is of physical interest as it represents a coherent oscillation of the inhomogeneous system. In contrast to the chaotic response produced by the branch cuts of the continuous spectrum this 'mode' constitutes a very orderly motion. The plasma as a whole oscillates with a definite frequency that cannot be distinguished from a true eigenmode during times $\tau \ll \nu_0^{-1}$. 'Modes' like these occur in many branches of physics and, accordingly, they have received many different names, like *quasi-modes, collective modes, virtual eigenmodes, resonances*, etc. The damping is completely analogous to the well-known phenomenon of Landau damping in the Vlasov description of plasmas. Landau damping is due to inhomogeneity of the equilibrium in velocity space. Damping of Alfvén waves is due to inhomogeneity of the equilibrium in ordinary space.

The expression (10.89) for the frequency of a quasi-mode in a plasma–plasma interface configuration has been derived under the assumption that the density is constant (so that ω_A^2 variations are due to the magnetic field). Permitting a jump in the density, the expression for the real part of the quasi-mode becomes a weighted average of the Alfvén frequencies on both sides,

$$\text{Re}\,\omega = \pm\sqrt{\frac{\rho_1 \omega_{A1}^2 + \rho_2 \omega_{A2}^2}{\rho_1 + \rho_2}}. \tag{10.92}$$

In this form, the expression also describes the quasi-modes of a plasma–vacuum interface configuration (in the limit $\rho_2 \to 0$), which play an important role in resonant absorption processes (see Section 11.1).

10.4.3 Different kinds of quasi-modes*

The quasi-modes derived above originate from surface waves propagating along a thin transition region between two homogeneous plasmas. However, quasi-modes can originate from many different kinds of waves and the term is used for any 'discrete' mode with an oscillatory part of the frequency in the range of the continuous spectrum, so that it couples to the continuum modes resulting in damping.

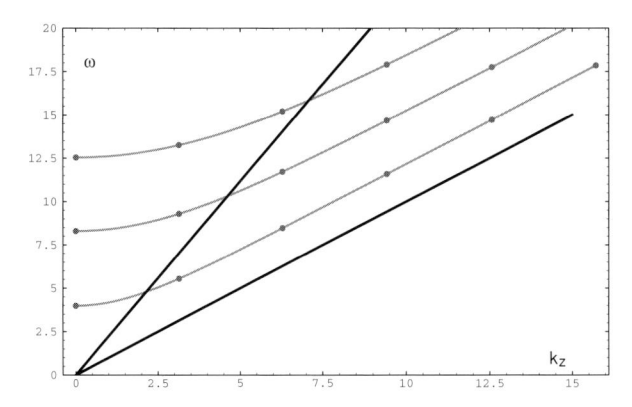

Fig. 10.16. Eigenfrequencies of the first three fast eigenmodes (grey lines) with upper and lower bound of the Alfvén continuum (black lines) as functions of k_z (for $L = a = 1$, and $k_y = 0$). (From De Groof et al. [61].)

In MHD, quasi-modes can be due to fast and slow magneto-sonic waves, discrete Alfvén waves, and different kinds of 'gap' modes in two-dimensional configurations. Even kink modes in a plasma–vacuum configuration with a wall may turn into quasi-modes when the wall is moved in and the kink mode moves into the Alfvén continuum, see Chance et al. [50].

To demonstrate how easily fast magneto-sonic waves can turn into quasi-modes, we consider a pressureless plasma slab with a uniform magnetic field $B_0 = B_{0z} = 1$. Assume that the slab has a finite width a in the x-direction, L in the z-direction, and is infinite in the y-direction. The wave number in the z-direction is then quantized, $k_z = n\pi/L$, and the eigenfrequencies of the first three fast magneto-sonic eigenmodes (0, 1 and 2 nodes in the x-direction) are shown in Fig. 10.16 as functions of k_z. The different grey lines connect fast modes with the same number of nodes in the x-direction. The figure also shows the upper and lower limits of the Alfvén continuum as functions of k_z. The density for this case is chosen as $\rho_0(x) = 0.6 + 0.4\cos(\pi/a\, x)$. Keeping k_y fixed while increasing k_z then results in ever more fast magneto-sonic modes 'swallowed' by the continuum. Since $k_y \neq 0$, the fast eigenmodes with an eigenfrequency within the range of the Alfvén continuum couple to the shear Alfvén continuum modes and become quasi-modes. For more realistic (larger) values of L, many more quasi-modes are present, see De Groof et al. [61].

10.5 Leaky modes*

Most of the plasmas we considered so far were isolated from their surroundings by either a perfectly conducting wall or a vacuum, or both. In the previous section we considered an inhomogeneous, thin, plasma layer surrounded on both sides

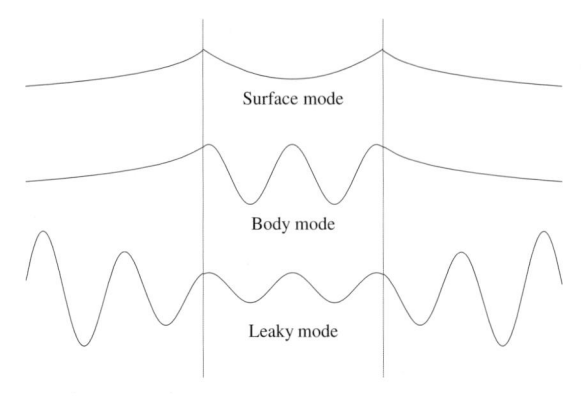

Fig. 10.17. Typical structures of surface waves, body waves and leaky waves inside a plasma, where the boundaries are represented by the two vertical lines. The leaky waves are defined completely by their outward propagating external behaviour. (Courtesy of A. De Groof.)

by homogeneous plasmas supporting wave modes, where we concentrated on the surface waves and the effect of the inhomogeneity on them. This configuration is generic for many applications in solar astrophysics and, more particular, coronal seismology.

In general, in solar astrophysics, one considers magnetic structures, such as loops and arcades, embedded in another plasma that may, or may not, be magnetized. The possible wave solutions in the magnetic structures are then classified on the basis of their spatial and temporal character inside and outside the magnetic structure (see Fig. 10.17). A wave that is propagating along the boundary of the magnetic structure, and shows exponentially damped behaviour both in the central and in the surrounding plasma, is called a *surface wave*. An oscillating wave solution that shows evanescent behaviour in the surrounding plasma is called a *body mode*. Both surface and body modes thus have a non-propagating character in the surrounding plasma and are therefore called *non-leaky modes*, as they do not leak out energy from the magnetic structure to the environment. Waves that have an outward propagating behaviour in the surrounding plasma are called *leaky waves*.

In this section we will derive leaky wave solutions in a simple configuration so that analytical methods can be exploited and the derivation is not too complicated. In Section 10.5.1, we discuss the equations to be solved and the boundary conditions to be imposed. Section 10.5.2 is devoted to the normal-mode analysis of this problem, where the eigenvalue problem is formulated and solved. In Section 10.5.3, the initial value problem is solved by means of the Laplace transform. The wave equation is Laplace transformed with respect to time to obtain a second order ODE with an RHS given by the initial conditions. Just like in

Section 10.2, the Green's function is constructed by joining two linearly independent solutions. The conjunct of these solutions is a function dependent only on the complex frequency. The Laplace transform of the solution is then expressed by means of this Green's function. In this initial value problem approach, the leaky modes correspond to the poles of the Green's function (the zeros of the conjunct). These poles are independent of the spatial coordinates and correspond to the eigenvalues found in Section 10.5.2, demonstrating that the two approaches of the problem are equivalent.

10.5.1 Model equations and boundary conditions*

Consider a uniform, pressureless, plasma slab confined in the x-direction between $x = \pm a$ and infinite in the y- and z-directions. In the x-direction, the plasma is surrounded on both sides by another uniform plasma that supports waves and extends up to $\pm\infty$. The Alfvén velocity inside the plasma slab, indicated by b, is assumed to be lower than the Alfvén velocity in the external plasma, indicated by b_e. This model problem can be regarded as a slab version of the uniform tube model for solar coronal loops studied by Cally [49]. Here, we consider a different parameter regime.

The profiles of the Alfvén frequency and the cutoff frequency (ω_{f0}) are illustrated in Fig. 10.18. In this figure, the frequencies of the surface mode and the first three 'fast' modes are also indicated. The lowest two of the latter modes are situated below the cutoff frequency in the external plasma so that these two modes will have an exponentially decreasing behaviour in the external plasma. Therefore, as will be shown, these modes correspond to 'body' modes and have real frequencies.

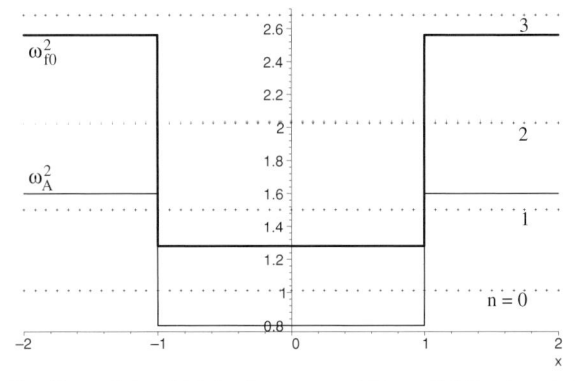

Fig. 10.18. Profiles of the Alfvén frequency (thin solid line) and the cutoff frequency ω_{f0} (thick solid line); $a = 1, b = 0.5, b_e = 1, k_y = 2, k_z = 1.6$. The horizontal dotted lines indicate the real parts of the frequencies of the surface mode (lowest) and the first three 'fast' modes.

The third mode, however, lies above the external cutoff frequency and thus has a propagating character in the external region. This is a 'leaky' mode with a complex frequency.

Consider normal-mode solutions of the form

$$\xi(\mathbf{r}, t) = \hat{\xi}(x)\, e^{i(k_y y + k_z z - \omega t)}. \tag{10.93}$$

In the pressureless, uniform, plasma slab considered here, Eq. (7.91) for $\xi \equiv \hat{\xi}_x$ reduces to a simple Helmholtz equation:

$$\xi'' + k_x^2 \xi = 0, \qquad \text{with} \quad k_x^2 \equiv \frac{\omega^2}{b^2} - k_0^2, \tag{10.94}$$

where $k_0^2 \equiv k_y^2 + k_z^2$. This equation also applies to the displacement in the external plasma, ξ_e, with k_x^2 replaced by $k_{xe}^2 \equiv \omega^2/b_e - k_0^2$. The general solution of these equations can be written as

$$\xi = \begin{cases} \alpha_L e^{ik_{xe}x} + \beta_L e^{-ik_{xe}x} & \text{for } x < -a, \\ \alpha e^{ik_x x} + \beta e^{-ik_x x} & \text{for } -a < x < a, \\ \alpha_R e^{ik_{xe}x} + \beta_R e^{-ik_{xe}x} & \text{for } a < x, \end{cases} \tag{10.95}$$

where the coefficients are determined by applying the boundary conditions.

Note that the results of Chapter 5 are recovered when perfectly conducting walls are put at $x = \pm a$. The boundary conditions $\xi(-a) = \xi(a) = 0$ then yield the dispersion relation $\sin(2ak_x) = 0$, so that

$$\omega_n^2 = (k_x^2 + k_0^2)\, b^2, \qquad k_x = \frac{n\pi}{2a} \quad (n = 0, \pm 1, \pm 2, \ldots), \tag{10.96}$$

gives the frequencies of the fast magneto-sonic modes of a finite homogeneous slab.

In the plasma–plasma case considered here, however, there is an external plasma and $k_{xe} \equiv (\omega^2/b_e - k_0^2)^{1/2}$ can be real or imaginary, corresponding to oscillatory or evanescent behaviour in the external region. The boundary conditions to be considered in this case are

$$[\![\xi]\!]_{x=-a} = 0, \quad [\![\xi]\!]_{x=a} = 0 \quad \text{and} \quad [\![\Pi]\!]_{x=-a} = 0, \quad [\![\Pi]\!]_{x=a} = 0, \tag{10.97}$$

where Π denotes the total perturbed pressure. These boundary conditions can be written in terms of the mechanical impedance, the ratio of the alternating force to

10.5 Leaky modes*

the alternating velocity:

$$\left[\frac{\Pi}{-i\omega\xi}\right]_{x=-a} = 0, \quad \left[\frac{\Pi}{-i\omega\xi}\right]_{x=a} = 0. \tag{10.98}$$

Imposing these equivalent boundary conditions is known as 'impedance matching'. As a matter of fact, the internal solution gives rise to the 'transmitted' or 'internal mechanical' impedance $Z_T \equiv \Pi/(-i\omega\xi)$. The external solution can be split into outgoing and incoming waves (see Keppens [121, 122]), each with a corresponding impedance:

$$Z_I \equiv \frac{\Pi_{in}}{-i\omega\xi_{in}}, \quad \text{and} \quad Z_O \equiv \frac{\Pi_{out}}{-i\omega\xi_{out}}. \tag{10.99}$$

The boundary conditions can be written in terms of these impedances as follows:

$$\frac{\xi_{out}}{\xi_{in}} = \frac{Z_I - Z_T}{Z_T - Z_O}. \tag{10.100}$$

To determine the possibly complex eigenfrequencies of the leaky modes, we have to consider the case of no incoming waves. The impedance criterion to be satisfied is then $Z_T = Z_O$. (In the next chapter, we have to impose the impedance criterion $Z_T = Z_I$ to find the frequencies that yield 100% absorption of the corresponding incoming wave, i.e. no outgoing waves.)

A similar treatment for the boundary conditions was carried out by Stenuit et al. [216] in the case of a cylindrical flux tube. The cylindrical geometry complicates the analysis considerably and equation (10.94) takes the form of a Bessel equation. The in- and outgoing wave solutions can then be expressed in terms of Hankel functions, where the issue of the boundary conditions becomes non-trivial. In order to determine what boundary conditions have to be imposed, one has to check the asymptotic behaviour of the Hankel functions and their contribution to the radial energy flux. In the slab geometry considered here this issue is trivial. The energy flux averaged over a period is defined as $S_x \equiv \frac{1}{2}\text{Re}(-\Pi^*i\omega\xi)$, where the asterisk denotes the complex conjugate. It is clear that, in the RHS plasma for instance, the solution $\alpha_R \exp(ik_{xe}x)$ yields a positive outward energy flux in the limit of infinitely large x, i.e. $\lim_{x\to\infty} S_x^{out} > 0$. On the other hand, the solution $\beta_R \exp(-ik_{xe}x)$ yields $\lim_{x\to\infty} S_x^{in} < 0$. Hence, to find the leaky modes, the incoming wave has to be rejected and one has to set $\beta_R = 0$. For non-leaky or body modes, the outgoing wave has to be rejected, i.e. $\alpha_R = 0$. Similarly, for the LHS plasma.

10.5.2 Normal-mode analysis★

To get the leaky modes, we set $\alpha_L = \beta_R = 0$ and define

$$F \equiv -\rho b \frac{\omega^2 - \omega_A^2}{\omega^2 - \omega_{f0}^2} i k_x, \qquad F_e \equiv -\rho_e b_e \frac{\omega^2 - \omega_A^2}{\omega^2 - \omega_{f0}^2} i k_{xe}. \qquad (10.101)$$

Impedance matching at $s = \pm a$ discussed in the previous section then yields a homogeneous system for the constants α and β:

$$\frac{F\left(\alpha e^{ik_x a} - \beta e^{-ik_x a}\right)}{\alpha e^{ik_x a} + \beta e^{-ik_x a}} = +F_e, \qquad \frac{F\left(\alpha e^{-ik_x a} - \beta e^{+ik_x a}\right)}{\alpha e^{-ik_x a} + \beta e^{+ik_x a}} = -F_e. \qquad (10.102)$$

This system has non-trivial solutions if

$$\frac{(F + F_e)^2}{(F - F_e)^2} = e^{4ik_x a}. \qquad (10.103)$$

This is the dispersion relation for the leaky modes we were looking for.

Let us consider the simple case with $k_y = k_z = 0$, so that $\omega_A = \omega_{f0} = 0$ and $k_x^2 = \omega^2/b^2$, while $k_{xe}^2 = \omega^2/b_e^2$. The dispersion relation can then be solved analytically:

$$\omega = -\frac{\pi b}{2a} n - \frac{ib}{2a} \log\left(\frac{\rho b + \rho_e b_e}{\rho b - \rho_e b_e}\right), \qquad (n = 0, \pm 1, \pm 2, \ldots). \qquad (10.104)$$

For $b = 0.5$, $b_e = 1$ and $a = 1$, this then yields the following leaky modes:

$$\omega = -\frac{\pi}{4} n - \frac{i}{4} \log(3) \qquad (n = 0, \pm 1, \pm 2, \ldots). \qquad (10.105)$$

A typical eigenfunction is shown in Fig. 10.19.

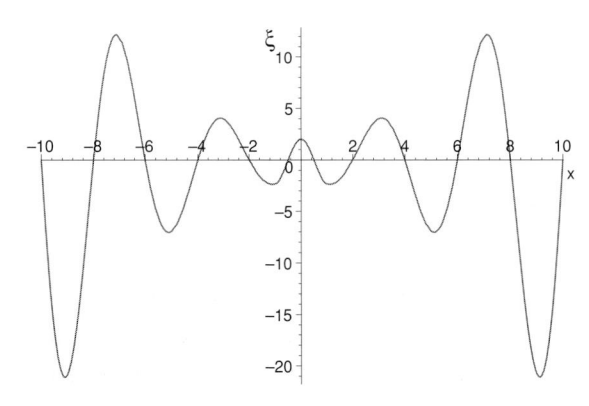

Fig. 10.19. The profile of ξ for the leaky mode with $n = 2$ and $b = 0.5$, $b_e = 1$, $a = 1$, $k_y = 0$, $k_z = 0$.

Before we proceed, let us also consider the non-leaky modes. As the slab plasma is uniform, no coupling with continuum modes is possible. For body modes, the external solution is non-propagating so that there is no damping mechanism and the corresponding eigenfrequencies must be purely real. Thus, $k_{xe}^2 = \omega^2/b_e^2 - k_0^2 \leq 0$. A similar analysis as above, but now ignoring the outgoing wave solutions in the external plasma, gives rise to the dispersion relation

$$1 = e^{4ik_x a}. \tag{10.106}$$

For $b = 0.5$, $b_e = 1$ and $a = 1$, this then yields the following body modes:

$$\omega = -\frac{\pi}{4}n \quad (n = 0, \pm 1, \pm 2, \ldots). \tag{10.107}$$

Notice that in the limit $\rho_e \to 0$ (plasma surrounded by a vacuum), i.e. $\omega_{f0e} \to \infty$ and $b_e \to \infty$ (while ω stays finite), no propagation is possible in the external region so that only body modes are found. Applying the limit to the dispersion relation found above indeed yields the dispersion relation for the body modes. In the opposite limit, $\rho_e \to \infty$ and $b_e \to 0$, we also only find body modes so that no outward propagation is possible.

10.5.3 Initial value problem approach★

In order to show how the modes found appear in the response of the slab to an initial perturbation, we now formulate the corresponding initial value problem which we will solve by means of a Laplace transform. As usual, the Laplace transform of the solution of the wave equation for positive times is defined by the formula

$$\hat{\xi}(\mathbf{r}; \omega) \equiv \int_0^\infty \xi(\mathbf{r}; t) e^{i\omega t} \, dt. \tag{10.108}$$

Applying this transformation to the wave equation (10.1) yields, after reduction to the one-dimensional representation in terms of ξ (see Section 10.2.1),

$$\hat{\xi}''(x; \omega) + k_x^2 \hat{\xi}(x; \omega) = \frac{i\omega}{b^2} \xi(x; 0) - \frac{1}{b^2} \dot{\xi}(x; 0), \tag{10.109}$$

which is a simplified form of Eq. (10.4) applied to the uniform, pressureless, slab plasma considered here. The solution of this ODE for $\hat{\xi}(x; \omega)$ can be obtained conveniently by means of a Green's function. According to the general theory, the latter may be expressed in terms of two linearly independent solutions $\hat{\xi}_1$ and $\hat{\xi}_2$ to the homogeneous equation, where the first satisfies the boundary condition at $x = -\infty$, and the second the boundary condition at $x = +\infty$. The two solutions read:

$$\hat{\xi}_1 = \begin{cases} e^{-ik_{xe}x} & \text{for } x < -a, \\ \alpha_1 e^{ik_x x} + \beta_1 e^{-ik_x x} & \text{for } -a < x < a, \\ \alpha_R e^{ik_{xe}x} + \beta_R e^{-ik_{xe}x} & \text{for } a < x, \end{cases} \quad (10.110)$$

and

$$\hat{\xi}_2 = \begin{cases} \alpha_L e^{ik_{xe}x} + \beta_L e^{-ik_{xe}x} & \text{for } x < -a, \\ \alpha_2 e^{ik_x x} + \beta_2 e^{-ik_x x} & \text{for } -a < x < a, \\ e^{ik_{xe}x} & \text{for } a < x. \end{cases} \quad (10.111)$$

According to the general theory, the conjunct $\hat{\xi}'_2 \hat{\xi}_1 - \hat{\xi}'_1 \hat{\xi}_2$ is independent of x. Hence, we determine the two solutions only in the region $-a < x < +a$, i.e. we determine the constants $\alpha_{1,2}$ and $\beta_{1,2}$ by applying the boundary conditions specified in Section 10.5.1. The zeros of this conjunct give rise to poles of the Green's function and determine the discrete spectrum of the problem. These poles are the solutions of the equation

$$\frac{(F + F_e)^2}{(F - F_e)^2} = e^{4ik_x a}, \quad (10.112)$$

which is exactly the same as the dispersion relation found in Section 10.5.2! Clearly, the solutions are also the same, so that both normal-mode and initial value problem approaches yield the same discrete spectrum.

10.6 Literature and exercises*

Notes on literature

Alfvén resonance: mode conversion:

– An alternative treatment of waves in inhomogeneous plasmas can be found in 'Waves in plasmas' by Stix [218]. Chapter 13 considers wave propagation through an inhomogeneous plasma. Starting from a WKB approach, the Alfvén resonance is discussed in terms of the singular-turning-point theory and mode conversion.

Continuous spectrum in cold plasma oscillations:

– Barston [16] elegantly solves the problem of cold plasma oscillations by means of a normal-mode analysis.
– Sedláček [206] analyses 'Electrostatic oscillations in cold inhomogeneous plasmas' and shows that inhomogeneities of the plasma lead to dissipationless damping. He applies both normal-mode analysis and the initial value problem approach and shows that the two are equivalent. However, since the normal-mode analysis may overlook collective modes, he concludes that it cannot replace the Laplace transform technique completely.

Continuous spectrum and damping in MHD:

- Tataronis & Grossmann [225] show that the continuous spectrum of ideal MHD, due to spatial inhomogeneities of the plasma, leads to dissipationless damping as a result of phase mixing (see further in Chapter 11).
- Chen & Hasegawa [54] extend the analysis of Tataronis & Grossmann [225] to compressible plasmas with magnetic shear (see further in Chapter 11).
- Goedbloed [85] constructs the resolvent operator, involved in the general solution of the initial value problem, for one-dimensional MHD problems (as presented in Section 10.2).

Leaky modes:

- Wilson [246] studies the eigenspectrum of a flux tube of finite width embedded in a non-magnetized medium and considers complex frequencies and leakage of energy into the surroundings. This leakage of wave energy is further developed by Spruit [215], in the thin flux tube approximation, and by Cally [49] for a tube with arbitrary radius. Cally sets up a classification scheme with seven types of non-leaky modes.

Exercises

[10.1] *Derivation of Eq. (10.92)*
Generalize the derivation in Section 10.4.2 by considering a density profile that is constant but different in each of the two plasmas to derive Eq. (10.92).

[10.2] *Continuum damping in cylindrical plasmas*
In Chapter 9 we studied the continuous MHD spectrum in cylindrical plasmas. Consider a uniform cylindrical plasma with radius 1 and Alfvén frequency ω_{Ai} separated from another uniform surrounding plasma, characterized by ω_{Ae}, by a thin transition region $1 - a \leq r \leq 1 + a$ in which the square of the local Alfvén frequency varies linearly from ω_{Ai}^2 to ω_{Ae}^2. Repeat the derivation of the quasi-mode in Section 10.4 in this set-up.

[10.3] *Quasi-modes in real plasmas*
Use the tables in Appendix B to get a rough estimate of the oscillatory frequency and the damping time scales of the quasi-mode resulting from a surface wave in laboratory plasmas and in solar coronal loops (cf. Eq. (10.89)).

[10.4] *Damping of quasi-modes*
The damping rate of ideal quasi-modes does not correspond to a heating rate of the plasma. Why not? What is the meaning of this damping rate then? Explain.

[10.5] *Understanding Fig. 10.16*
Consider a finite ($-a < x < +a$), homogeneous, magnetized slab plasma surrounded by another homogeneous plasma, also with uniform magnetic field (in the z-direction). Write a Maple or Mathematica work sheet to plot the internal and external Alfvén frequency, ω_{Ai} and ω_{Ae}, versus k_z (choose parameter values so that these two frequencies are different). Also plot the frequency ω_{f0}, both for the internal and external plasma. Now add plots of the frequencies of the first three fast magneto-sonic modes (notice that k_x is quantized). Play around with the parameters (k_y, internal and external magnetic field and density) and observe that the frequencies of the fast magneto-sonic modes lie between ω_{Ai} and ω_{Ae} for

large k_z. What is the fundamental difference of the figure you get from Fig. 10.16? How do you get Fig. 10.16 from here?

[10.6] *Leaky waves in cylindrical geometry*

Consider a uniform, cylindrical magnetic-flux tube embedded in a uniform, wave carrying plasma and repeat the derivation of the leaky waves in Section 10.5 for this configuration.

[10.7] *Resonant leaky modes*

Combine the derivations in Sections 10.4 and 10.5 and derive the dispersion relation for resonant leaky modes in a non-uniform plasma. (Consider a plane slab.)

11
Resonant absorption and wave heating

In Chapters 7 and 9, the MHD spectral analysis of an ideal plasma with inhomogeneities in one spatial direction led to singular second order differential equations for the plasma displacement in the direction of inhomogeneity: Eqs. (7.91) and (9.31). The two singularities of these equations give rise to two continuous parts of the MHD spectrum, as demonstrated in Section 7.4 for slab geometry and in Section 9.2.2 for cylindrical geometry. It was shown that the eigenfunctions corresponding to these Alfvén and slow magneto-sonic continua possess non-square integrable tangential components leading to extreme anisotropic behaviour. Clearly, this has a dramatic effect on the dynamical behaviour of inhomogeneous plasmas. In the present chapter, we will discuss the consequences of these continuous spectra for the dynamical response of an inhomogeneous plasma slab or cylinder to periodic, multi-periodic, or random external drivers. This will lead to the concepts of *resonant absorption* of waves and *phase mixing* of neighbouring magnetic field lines.

Resonant 'absorption' (or 'dissipation') and phase mixing are fundamental properties of MHD waves that are studied in many different plasma systems. These phenomena affect the dynamics of plasma perturbations significantly and often dominate the energy conversion and transport in inhomogeneous plasmas. Since they are basic to MHD wave heating and acceleration of plasmas, they deserve special attention. In fact, since all plasmas occurring in nature are – to a higher or lower degree – *inhomogeneous* and since waves can be excited easily in plasmas, *resonant absorption* and *phase mixing* frequently occur. These phenomena have been studied in the context of wave damping and heating for *controlled thermonuclear fusion experiments* (Tataronis and Grossmann [225], Chen and Hasegawa [54], Balet *et al*. [15], Poedts *et al*. [185], Vaclavik and Appert [235]) and for *solar and astrophysical plasmas* (Ionson [116], Kuperus *et al*. [133], Poedts *et al*. [177, 182], Goossens [93]), whereas resonant mode conversion is also studied in *magnetospheric physics* (Kivelson and Southwood [128], Zhu and Kivelson [249], Kivelson and Russell [127], Mann and Wright [151]).

In the solar context, resonant absorption and phase mixing play a dominant role in 'sunspot seismology' (Bogdan and Knölker [34]) and 'coronal seismology' (Nakariakov *et al.* [161]) in which the interactions of MHD waves with magnetic structures are exploited as *diagnostic tools*. The goal here is to deduce the values of plasma parameters from the comparison of theoretical results and observations of MHD waves whose characteristics (amplitudes, frequencies, etc.) are determined by the ambient plasma, as in helioseismology [97]. In confined laboratory plasmas, this method was called 'MHD spectroscopy' [88] (Section 7.2.4). Clearly, MHD spectroscopy and coronal seismology are complicated by the inhomogeneity of the studied plasmas, the nonlinearity of the dynamics, and the complex magnetic geometries.

In this volume, we focus on the basics of the resonant absorption and phase mixing mechanisms and only consider relatively simple geometries and physics. Nonlinear, two-dimensional, resistive and viscous effects, etc. will be discussed in the companion Volume 2 on *Advanced Magnetohydrodynamics*.

In this chapter, the physical mechanisms of resonant absorption and phase mixing are first explained in detail for a simple model configuration (Section 11.1). In Section 11.2 applications of this mechanism to wave damping and heating are discussed and illustrated for a generic cylindrical plasma. In Section 11.3, alternative configurations with different boundary conditions are presented together with their consequences for applications to solar and magnetospheric physics.

11.1 Ideal MHD theory of resonant absorption

11.1.1 Analytical solution of a simple model problem

Consider a semi-infinite plasma that occupies the half space $x < 0$ next to a vacuum in the half space $x > 0$. This is the plane slab version of model problem III introduced in Section 4.6.1. Let us assume, for simplicity, that there is a unidirectional magnetic field in both vacuum and plasma and choose the z-axis in the direction of the magnetic field. We also assume that all equilibrium quantities are constant for $x < x_1$ ($\equiv -a$) and only depend on x in a region $x_1 \leq x \leq x_2$ ($\equiv 0$). In the vacuum ($x > 0$), we assume the presence of a sheet current $\mathbf{j}_c^\star$ in a coil which represents the external driving source or 'antenna' at $x = c$ (see the sketch in Fig. 11.1, where the plasma inhomogeneity is indicated by the quantity ϵ that is still to be defined).

This configuration is a simplified version of the set-up studied in one of the first papers in which the mechanism of resonant 'absorption' was investigated in the framework of thermonuclear fusion [54], viz. as a possible supplementary heating mechanism to bring tokamak plasmas into the ignition regime. (In the original set-

11.1 Ideal MHD theory of resonant absorption

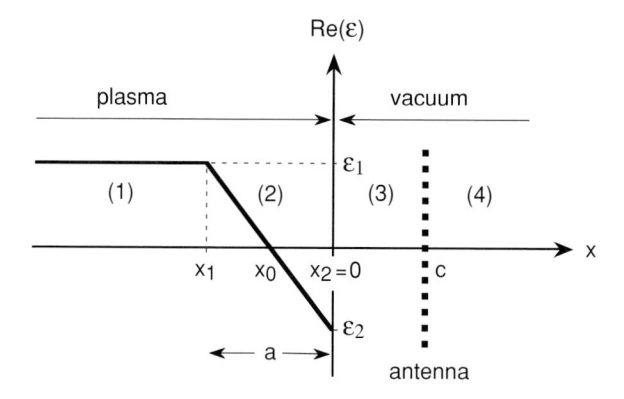

Fig. 11.1. Sketch of the configuration in which the resonant absorption of Alfvén wave energy is studied (adapted from Ref. [54]). The power source consists of an external 'antenna' with surface current $\mathbf{j}_c^*$ situated in the vacuum region next to the plasma. Here $\epsilon(x) \equiv \rho(x)[\omega^2 - \omega_A^2(x)]$.

up of Chen and Hasegawa the magnetic field is sheared. They generalized earlier results of Grossmann and Tataronis [225] who did not include magnetic shear and assumed incompressibility, so that the important role of fast magneto-sonic waves was not fully considered.) As a matter of fact, the singular behaviour of the excited waves results in the small length scales that are required for efficient dissipation of wave energy in highly conductive plasmas. The solution derived by Chen and Hasegawa will be discussed for the special case of a uni-directional magnetic field.

Since the equilibrium quantities are constant in time and do not depend on y and z, we assume linear perturbations of the form

$$\boldsymbol{\xi}(x, y, z; t) = \boldsymbol{\xi}(x)\, e^{i(k_y y + k_z z - \omega t)}, \qquad (11.1)$$

where $\omega = \omega_d - i\delta$, with ω_d the frequency of the external driver and δ a small positive constant ($0 \leq \delta \ll \omega_d$) that will be discussed below. We concentrate on Alfvén wave absorption and separate off the slow magneto-sonic waves by considering the low-β approximation ($p \ll \tfrac{1}{2}B^2$) of Eq. (7.89) of Section 7.3.2. Since the magnetic field is uni-directional, the parallel and perpendicular 'wave numbers' are constant: $f \equiv k_\parallel = k_z$ and $g \equiv k_\perp = k_y$. In the limit of vanishing gravitation, we then get a system of two differential equations:

$$\begin{pmatrix} \rho\omega^2 + \dfrac{d}{dx} B^2 \dfrac{d}{dx} - k_\parallel^2 B^2 & \dfrac{d}{dx} k_\perp B^2 \\[1em] -k_\perp B^2 \dfrac{d}{dx} & \rho\omega^2 - k_0^2 B^2 \end{pmatrix} \begin{pmatrix} \xi_x \\ i\xi_\perp \end{pmatrix} = 0, \qquad (11.2)$$

where $k_0^2 \equiv k_\perp^2 + k_\parallel^2$. This system describes Alfvén modes and fast magneto-sonic modes. (Note that we now exploit $\xi_x \equiv \xi$ and $i\xi_\perp \equiv \eta$ as variables since we will need the symbol η for the plasma resistivity in Section 11.1.2.) In general, i.e. for $k_\perp \neq 0$, these two modes are coupled since the equations (11.2) are coupled. This *coupling is of vital importance* for the mechanism of resonant absorption. We will see that the mechanism *does not work for* $k_\perp = 0$, at least not in this set-up with sideways excitation (see Section 11.2). The second equation is algebraic in $\xi_\perp$ so that it allows us to express $\xi_\perp$ in terms of ξ_x:

$$i\xi_\perp = \frac{k_\perp B^2}{\rho\omega^2 - k_0^2 B^2} \frac{d\xi_x}{dx}. \tag{11.3}$$

Upon substitution of this expression in the first part of Eq. (11.2) one obtains a second order ODE for the variable ξ_x:

$$\frac{d}{dx}\left(\frac{\rho\omega^2 - k_\parallel^2 B^2}{\rho\omega^2 - k_0^2 B^2} B^2 \frac{d\xi_x}{dx}\right) + (\rho\omega^2 - k_\parallel^2 B^2)\xi_x = 0, \tag{11.4}$$

which is the simplified form of Eq. (7.91) under the assumptions made.

Note that the Alfvén frequency,

$$\omega_A \equiv k_\parallel v_A, \qquad v_A \equiv B(x)/\sqrt{\rho(x)}, \tag{11.5}$$

is a function of x in the layer $-a \leq x \leq 0$ through the inhomogeneity of the magnetic field or the density, or both. These inhomogeneities may be considered to be arbitrary, although the equilibrium condition $(p + \tfrac{1}{2}B^2)' = 0$ would require the gradient of the magnetic pressure to be balanced by $-p'$, which should be small in general because of the low-β assumption. For definiteness, we will assume that the magnetic field is constant and that inhomogeneities are exclusively due to the density profile $\rho(x)$.

We now focus on nearly perpendicular propagation, i.e. $k_\parallel \ll k_\perp \approx k_0$, and very strong coupling (although the fast wave sub-spectrum is well separated from the Alfvén wave sub-spectrum in this case: $\omega_A^2 = k_\parallel^2 v_A^2 \ll \omega_{f0}^2 \approx k_0^2 v_A^2$). We also assume that the oscillatory part ω_d of the frequency ω lies in the range of the Alfvén continuum. Equation (11.4) then reduces to

$$\frac{d}{dx}\left(\rho[\omega^2 - \omega_A^2(x)]\frac{d\xi_x}{dx}\right) - k_\perp^2 \rho[\omega^2 - \omega_A^2(x)]\xi_x = 0, \tag{11.6}$$

which is equivalent to (10.53) since $k_0^2 \approx k_\perp^2$ and the right hand side term is zero in this case with an initial static equilibrium. (As a matter of fact, remark that (10.53) was derived under the same assumptions.) Defining a complex quantity $\epsilon(x) \equiv \rho(x)[\omega^2 - \omega_A^2(x)]$, we then get

$$\frac{d^2\xi_x}{dx^2} + \frac{1}{\epsilon}\frac{d\epsilon}{dx}\frac{d\xi_x}{dx} - k_\perp^2 \xi_x = 0. \tag{11.7}$$

11.1 Ideal MHD theory of resonant absorption

We now consider the simple case where $\text{Re}[\epsilon(x)]$ depends linearly on x in the inhomogeneous layer:

$$\text{Re}[\epsilon(x)] = \begin{cases} \epsilon_1 & (x < x_1) \\ \epsilon_1 + \dfrac{\epsilon_2 - \epsilon_1}{x_2 - x_1}(x - x_1) & (x_1 \leq x \leq x_2), \end{cases} \quad (11.8)$$

as displayed in Fig. 11.1. The zero of the function $\text{Re}[\epsilon(x)]$ corresponds to the resonant position $x = x_0 \equiv (\epsilon_1 x_2 - \epsilon_2 x_1)/(\epsilon_1 - \epsilon_2)$, where the driving frequency matches the local Alfvén frequency: $\omega_d = \omega_A(x_0)$. For this simple configuration, Eq. (11.7) can be solved analytically.

We can distinguish four regions in the set-up as sketched in Fig. 11.1. The plasma occupies regions (1) and (2), while the vacuum is situated in regions (3) and (4). The equations for the vacuum regions will be derived below. They can also be solved analytically. The main issue is then to connect the solutions across those regions. This 'simple' problem actually involves a mixture of all the laboratory plasma models introduced in Section 4.6.1, and elaborated for the linearized interface problems in Section 6.6.1. Going from left to right, subsequently, the following boundary conditions are to be imposed:

- At $x = x_l$ ($\to -\infty$): the plasma–rigid wall (model I) boundary conditions (4.169)–(4.170) for the plasma displacement at a conducting wall which, since it is actually absent in the present case, is equivalent to imposing a boundary condition at $-\infty$:

$$\xi_x \to 0 \quad (\text{for } x \to -\infty); \quad (11.9)$$

- At $x = x_1$ ($\equiv -a$): the plasma–plasma (model II*) boundary conditions (4.164)–(4.166) which, upon linearization about the *perturbed* boundary, yield the first and second interface conditions (6.144) and (6.147) derived in Section 6.6.1:

$$[\![\xi_x]\!] = 0 \quad (\text{at } x = x_1), \quad (11.10)$$

$$[\![\xi'_x]\!] = 0 \quad (\text{at } x = x_1); \quad (11.11)$$

- At $x = x_2$ ($\equiv 0$): the plasma–vacuum (model II) boundary conditions (4.175)–(4.176) which, upon linearization about the *perturbed* boundary, yield the first and second interface conditions (6.140) and (6.143) derived in Section 6.6.1:

$$ik_\| \hat{B} \xi_x = \hat{Q}_x \quad (\text{at } x = x_2), \quad (11.12)$$

$$Q_z = -B(\xi'_x + ik_\perp \xi_y) = \hat{Q}_z \quad (\text{at } x = x_2); \quad (11.13)$$

- At $x = c$: the antenna (model III) boundary conditions (4.179)–(4.180) for the jump of the perturbed magnetic field produced by the currents in the antenna:

$$[\![\hat{Q}_x]\!] = 0 \quad (\text{at } x = c), \quad (11.14)$$

$$\mathbf{n} \times [\![\hat{\mathbf{Q}}]\!] = \mathbf{j}_c^* \quad (\text{at } x = c); \quad (11.15)$$

- At $x = x_r$ ($\to \infty$): the vacuum–rigid wall (model II) boundary condition (4.173) for the perturbed magnetic field at a conducting wall which, since it is again absent in the present case, is equivalent to imposing a boundary condition at ∞:

$$\hat{Q}_x \to 0 \quad (\text{for } x \to \infty). \tag{11.16}$$

Now count: Eqs. (11.9)–(11.16) provide eight boundary conditions to determine the eight arbitrary amplitudes of the independent solutions of the second order differential equations for the plasma variable ξ_x and the vacuum variable $\hat{Q}_z$ in the four regions (1)–(4). Also note the difference between the present forced oscillation problem, where arbitrary values of ω_d and the amplitude of the antenna current $\mathbf{j}_c^*$ determine everything else, and the problem of free oscillations, where $\mathbf{j}_c^* = 0$ and ω becomes an eigenvalue (because eight boundary conditions is one too much in that case, since the amplitude of a linear eigenoscillation is arbitrary then).

(a) Plasma solution In the left plasma region (1), the coefficient ϵ is constant so that Eq. (11.7) reduces to a simple Helmholtz equation with the general solution

$$\xi_x^{(1)} = A_1 \, e^{k_\perp (x+a)} + B_1 \, e^{-k_\perp (x+a)}, \tag{11.17}$$

where $B_1 = 0$ because of the boundary condition (11.9), and A_1 will be determined by one of the boundary conditions (11.10), (11.11). This specific form has been chosen to simplify the implementation of the latter conditions.

In the right plasma region (2), the coefficient function ϵ varies linearly with x. In terms of the normalized variable

$$X \equiv \frac{k_\perp \epsilon}{d\epsilon/dx}, \quad \text{so that } \operatorname{Re}(X) = k_\perp (x - x_0), \tag{11.18}$$

Eq. (11.7) reduces to the zeroth order modified Bessel equation:

$$\frac{d^2 \xi_x}{dX^2} + \frac{1}{X} \frac{d\xi_x}{dX} - \xi_x = 0. \tag{11.19}$$

Hence, the solution for ξ_x in region (2) can be written as a linear combination of the modified Bessel functions of the first and second kind [1]:

$$\xi_x^{(2)} = A_1 \left[A_2 I_0(X) + B_2 K_0(X) \right], \tag{11.20}$$

where multiplication with A_1 again serves the purpose of simplifying implementation of the boundary conditions at $x = x_1$.

▷ **Bessel function identities.** The following identities are needed here:

$$I_0' = I_1, \qquad I_1' = I_0 - X^{-1} I_1,$$

$$K_0' = -K_1, \qquad K_1' = -K_0 - X^{-1} K_1,$$

$$I_0 K_1 + K_0 I_1 = X^{-1}, \tag{11.21}$$

where the primes denote differentiation with respect to the argument X. ◁

11.1 Ideal MHD theory of resonant absorption

The boundary conditions (11.10) and (11.11) at $x = x_1$ then yield

$$A_2 I_0(X_1) + B_2 K_0(X_1) = 1,$$
$$A_2 I_1(X_1) - B_2 K_1(X_1) = 1,$$

where the Bessel functions are evaluated for the value $X_1 \equiv X(x = x_1)$ at the left boundary of interval (2). From this system, the constants A_2 and B_2 can be determined easily:

$$A_2 = X_1 \Big[K_0(X_1) + K_1(X_1) \Big], \qquad B_2 = -X_1 \Big[I_0(X_1) - I_1(X_1) \Big], \quad (11.22)$$

so that ξ_x is determined in the whole plasma region up to a scale factor A_1. This scale factor depends on the current density in the external antenna and follows from linking the plasma solution to the vacuum solution, which will be determined now.

(b) Vacuum solution The vacuum magnetic field perturbations in regions (3) and (4) are determined by the simpler equations $\nabla \times \hat{\mathbf{Q}} = 0$ and $\nabla \cdot \hat{\mathbf{Q}} = 0$. The first condition gives

$$\hat{Q}_y = \frac{k_\perp}{k_\parallel} \hat{Q}_z, \qquad \hat{Q}_x = -\frac{i}{k_\parallel} \frac{d \hat{Q}_z}{dx}, \quad (11.23)$$

which upon substitution in the second condition yields a Helmholtz equation for $\hat{Q}_z$:

$$\frac{d^2 \hat{Q}_z}{dx^2} - k_0^2 \hat{Q}_z = 0. \quad (11.24)$$

Hence, the solutions in regions (3) and (4) can be written in the convenient form

$$\hat{Q}_z^{(3)} = A_3 e^{-k_0(x-c)} + B_3 e^{k_0(x-c)}, \qquad \hat{Q}_z^{(4)} = A_4 e^{-k_0(x-c)} + B_4 e^{k_0(x-c)}, \quad (11.25)$$

where the boundary condition (11.16) implies that $B_4 = 0$.

Now comes the important part. The spatial and temporal dependence of the dynamics is forced onto the system by means of the antenna surface current

$$\mathbf{j}_c^\star = \mathbf{J}^\star e^{i(k_y y + k_z z - \omega t)}, \quad (11.26)$$

where $\mathbf{J}^\star$ is the amplitude of $\mathbf{j}_c^\star$ with components $J_y^\star$ and $J_z^\star$, having the dimension of current per unit length. Since $\nabla \cdot \mathbf{j}_c^\star = 0$, the wave vector $\mathbf{k}_0 \equiv (0, k_\perp, k_\parallel)$ will be perpendicular to the antenna current: $\mathbf{k}_0 \cdot \mathbf{J}^\star = 0$. Hence, four of the five constants $J_y^\star, J_z^\star, k_\perp, k_\parallel, \omega_d$ can be freely chosen, corresponding to a possible construction of the antenna as a series of wires with a current per unit length of amplitude $|\mathbf{J}^\star|$ that varies harmonically from wire to wire, with a spatial constant k_0 and with the same temporal constant ω_d, whereas the inclination with respect to the z-axis determines $J_y^\star/J_z^\star = -k_\parallel/k_\perp$. The experimenter appears to have complete control over everything! Not quite though: the external antenna excites the 'quasi-mode' in the plasma that is related to the plasma–vacuum surface mode (see Section 10.4). This collective mode has a frequency $\omega = \omega_0 + i\nu_0$, where the parameter ν_0 contains the information of the continuous spectrum of Alfvén frequencies (11.5) in the inhomogeneous layer (2). Such easy-to-excite quasi-modes will turn out to determine the efficiency of the resonant absorption mechanism.

The jump of the vacuum magnetic field components at $x = c$ is now determined by the boundary conditions (11.14) and (11.15) giving two independent relations,

$$[[\hat{Q}_x]] = \hat{Q}_x^{(4)}\big|_{x=c} - \hat{Q}_x^{(3)}\big|_{x=c} = i(k_0/k_\parallel)(A_4 - A_3 + B_3) = 0, \quad (11.27)$$

$$[[\hat{Q}_z]] = \hat{Q}_z^{(4)}\big|_{x=c} - \hat{Q}_z^{(3)}\big|_{x=c} = A_4 - A_3 - B_3 = -J_y^\star, \quad (11.28)$$

whereas the third relation is not independent, $[[\hat{Q}_y]] = (k_\perp/k_\parallel)[[\hat{Q}_z]]$, in agreement with Eq. (11.23)(a). With these two conditions, the constants A_3 and B_3 can be expressed in terms of A_4 and the current in the antenna coil:

$$A_3 = A_4 + \tfrac{1}{2}J_y^\star, \qquad B_3 = \tfrac{1}{2}J_y^\star, \quad (11.29)$$

so that the magnetic field perturbation in regions (3) and (4) becomes

$$\hat{Q}_z^{(3)} = A_4 e^{-k_0(x-c)} + J_y^\star \cosh[k_0(x-c)], \qquad \hat{Q}_z^{(4)} = A_4 e^{-k_0(x-c)}. \quad (11.30)$$

Hence, the vacuum solution consists of a *driven part*, confined to region (3), that depends on the current in the external coil, and an *induced part* with scale factor A_4 that depends on the plasma response, which is still unknown at this point.

(c) Linking the vacuum solution to the plasma solution We applied six of the eight boundary conditions determining the plasma and vacuum solutions up to the scale factors A_1 and A_4. Since we are dealing with a driven problem, there is a *unique solution*. In order to determine it, we only have to find expressions for these scale factors. They follow in a natural way from linking the vacuum solution to the plasma solution by means of the two boundary conditions (11.12) and (11.13) at the plasma–vacuum interface:

$$k_\parallel B \xi_x^{(2)}\big|_{x=x_2} = -\frac{1}{k_\parallel}\frac{d\hat{Q}_z^{(3)}}{dx}\bigg|_{x=x_2}, \quad (11.31)$$

$$\frac{B\epsilon}{k_\perp^2 B^2 - \epsilon}\frac{d\xi_x^{(2)}}{dx}\bigg|_{x=x_2} = \hat{Q}_z^{(3)}\big|_{x=x_2}. \quad (11.32)$$

By substituting the derived expressions $\xi_x^{(2)}$ and $\hat{Q}_z^{(3)}$ in these equations, one obtains two equations for A_1 and A_4:

$$CA_1 - e^{k_0c}A_4 = J_y^\star \sinh(k_0c),$$

$$DA_1 - e^{k_0c}A_4 = J_y^\star \cosh(k_0c),$$

where

$$C \equiv (k_\parallel^2/k_0)B\Big[A_2 I_0(X_2) + B_2 K_0(X_2)\Big],$$

$$D \equiv \frac{k_\perp B \epsilon_2}{k_\perp^2 B^2 - \epsilon_2}\Big[A_2 I_1(X_2) - B_2 K_1(X_2)\Big], \quad (11.33)$$

involving the Bessel functions evaluated for the value $X_2 \equiv X(x = x_2)$ at the right boundary of interval (2), and the constants A_2 and B_2 determined in Eqs. (11.22). The straightforward solution is:

$$A_4 = \Big[C\cosh(k_0c) - D\sinh(k_0c)\Big]A_1, \quad A_1 = -J_y^\star \frac{e^{-k_0c}}{C-D}, \quad (11.34)$$

so that all constants are proportional to $J_y^\star$.

11.1.2 Role of the singularity

Hence, the solution appears to be determined completely in the plasma as well as in the vacuum: eight boundary conditions determine eight constants A_i, B_i ($i = 1, \ldots, 4$). Clearly, the whole solution is proportional to the current in the external coil ($J_y^\star$) and vanishes when there is no external driving current. Concerning the dependence on the driving frequency ω_d: when ω_d^2 is outside the range of the Alfvén continuum $\{\omega_A^2(x)|x_1 \leq x \leq x_2\}$, we could assume $\delta = 0$ and all constants to be real so that the obtained solution would be adequate. When ω_d^2 is in the continuum range, Chen and Hasegawa conveniently avoided the singularity at x_0 where $\omega_d = \omega_A(x_0)$ by considering *a complex $\omega = \omega_d + i\delta$ with $\delta = 0^+$*. The role of the 'artificial' damping ([235, 185]) is to mimic real resistive (or viscous) dissipation while keeping the analysis tractable. Below it will be shown that the actual energy 'absorption' thus obtained *corresponds to real damping when dissipation is taken into account*. This may appear strange since in *ideal* MHD there can be no dissipation or heating. Hence, the early ideal MHD studies introduced the term resonant 'absorption' rather than resonant 'heating' or 'dissipation'. *The point is, however, that the actual energy absorption (or dissipation) rate does not depend* *on the value δ (nor on the resistivity or viscosity).* It is the logarithmic singularity that causes the wave phase mixing and, hence, the dissipation of the energy of the excited wave. Mathematically, this process is reversible in ideal MHD but real dissipation destroys the reversibility and converts the dissipated energy into thermal energy.

Looking at the above solution in terms of modified Bessel functions one may wonder what happened to the logarithmic singularity of ξ_x that was shown to be characteristic for Alfvén continuum modes in Section 7.4. Of course, this singularity is still present here. Indeed, close to the singularity we have $|\epsilon| \ll 1$ and thus $|X| \ll 1$. The modified Bessel functions of complex argument can be expanded in this limit and to leading order we have $I_0(X) \approx 1$ and $K_0(X) \approx \ln(\frac{1}{2}X)$. In Fig. 11.2 the solution in region (2) has been plotted for a choice of parameters that yields a singularity at $x = 0.5$. The artificial damping $\delta = -0.001$ prevents unlimited amplitudes and gives rise to nearly singular behaviour. However, the ideal singularity is still clearly recognizable with the logarithmic behaviour near $x = x_0$ in the imaginary part of ξ_x, the jump in the real part of ξ_x, and the corresponding

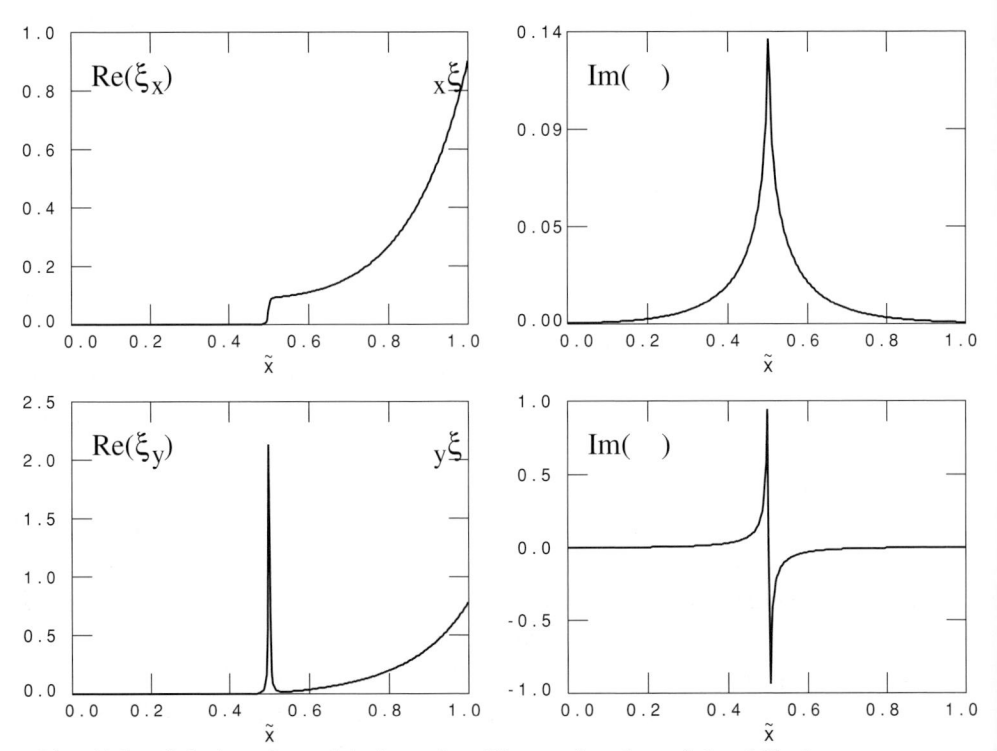

Fig. 11.2. Solutions ξ_x and ξ_y in region (2) as a function of the shifted coordinate $\tilde{x} \equiv x/a + 1$ for $a = B = 1$, $\omega_d = k_\| = J_y^\star = 1$, and a density such that $\text{Re}(\epsilon_1) = 1$ and $\text{Re}(\epsilon_2) = -1$ so that $\tilde{x}_0 = 0.5$; $\delta = -0.001$, $k_\perp = 5$.

$(x - x_0)^{-1}$ and $\delta(x - x_0)$ behaviour of the tangential perpendicular component ξ_y. The jump is due to the analytic continuation of the logarithm through the branch cut (see Section 10.3.2) and the size of this jump determines the energy absorption rate, as shown below.

The physical interpretation of Fig. 11.2 is not trivial. Remark that we have exploited the widely-used 'complex notation' throughout this book. This is done in the interest of mathematical simplicity when calculating with harmonic functions exploiting, instead of sines and cosines, exponential functions with imaginary argument. Physical quantities, however, are *real*. The rule for using the exponential functions thus is that the harmonic function is obtained by taking the real part at any convenient point in the calculation. With $A = A_r + iA_i$ the real part of $Ae^{i\omega t}$ is $A_r \cos \omega t - A_i \sin \omega t$. Hence, $Ae^{i\omega t}$ represents a harmonic containing a component of amplitude A_r in phase with $\cos \omega t$ and a component of amplitude A_i in phase with $\sin \omega t$ (and thus in quadrature with $\cos \omega t$).

Hence, the temporal behaviour of the solution shown in Fig. 11.2 shows at the beginning of each oscillation period (multiples of $2\pi/\omega$) the real parts (left) and half a period later the imaginary parts (right). At any other moment a combination of both solutions is obtained. Clearly, the different magnetic surfaces (or field lines) are 'out of phase' in this steady state solution. This is due to the 'phase mixing' mechanism, as we will see below.

Energy 'absorption rate' Remark that in the complex notation, used throughout this book, nonlinear operations such as multiplying together two harmonic functions require special attention since the real part of, e.g., $e^{i\omega t} e^{i\omega t}$ is $\cos 2\omega t$ which is not identical to $\cos \omega t \cos \omega t$. The *time average* of two harmonic quantities, which are the real parts of $Ae^{i\omega t}$ and $Be^{i\omega t}$, is half the product of A and the complex conjugate of B, i.e. $\frac{1}{2}AB^*$. The latter result is useful when discussing energy equations in linear MHD, the only place(s) in this book where nonlinear operations of harmonic functions occur.

With the solution obtained in the previous section, the energy 'absorption rate' resulting from the resonant plasma dynamics can be determined explicitly for the simple configuration considered. The easiest way to do this is by considering the Poynting vector $\mathbf{S} = \frac{1}{2} \mathbf{E}^* \times \mathbf{Q}$. Here we exploit the simplified notation $\mathbf{Q}$ for $\mathbf{B}_1$, introduced in Eq. (6.25), while $\mathbf{E}$ denotes the perturbed electric field in this chapter (the subscript 1 is dropped to simplify the notation). The energy absorption rate (averaged over a driving period) in the steady state can then be written as

$$\frac{dW}{dt} = L_y L_z \, \text{Re}[\mathbf{S}(x_2) - \mathbf{S}(x_1)] \cdot \mathbf{e}_x , \quad (11.35)$$

with L_y and L_z the size of the plasma in the y- and z-direction, respectively. Since $\mathbf{E} = i\omega \boldsymbol{\xi} \times \mathbf{B}$, we have

$$E_\| = 0, \quad \text{and} \quad E_\perp = -i\omega B \xi_x. \tag{11.36}$$

Upon substituting this result in Eq. (11.35), we obtain

$$\begin{aligned}\frac{dW}{dt} &= \frac{\omega_d}{2} L_y L_z \, \text{Im} \left[B Q_\| \xi_x^* \right]_{x_1}^{x_2}, \\ &= \frac{\omega_d}{2} L_y L_z \, \text{Im} \left[\frac{B^2 \epsilon}{k_\perp^2 B^2 - \epsilon} \frac{d\xi_x}{dx} \xi_x^* \right]_{x_1}^{x_2}, \end{aligned} \tag{11.37}$$

by means of Eq. (11.3). Since we solved the complete boundary value problem for ξ_x we can and will compute the energy absorption rate.

First remark, however, that most of the absorption will take place near the singular layer at $x = x_0$ where $\epsilon(x) \approx d\epsilon_r/dx_{x_0}(x - x_0) + i\epsilon_i(x_0)$ (with $\epsilon_r = \text{Re}(\epsilon)$ and $\epsilon_i = \text{Im}(\epsilon)$). Thus, since $\epsilon(x) \approx 0$ near x_0, Eq. (11.7) reduces to

$$\frac{d^2 \xi_x}{dx^2} + \frac{1}{x - x_0 + i\delta'} \frac{d\xi_x}{dx} - k_\perp^2 \xi_x = 0, \tag{11.38}$$

where $\delta' = \dfrac{\epsilon_i(x_0)}{(d\epsilon_r/dx)_{x_0}}$. In Chapter 10 we noticed that in the limit $\delta' \to 0$ this equation has a logarithmic singularity at $x = x_0$ so that

$$\xi_x = C \ln(x - x_0 + i\delta'). \tag{11.39}$$

Consider now $x_1 = x_0 - \varepsilon$ and $x_2 = x_0 + \varepsilon$ with $\varepsilon \gg |\delta'|$ in Eq. (11.37) in order to evaluate the energy absorption rate near the singularity (denoted as dW_1/dt). In the limit $\varepsilon \to 0$ we then obtain

$$\frac{dW_1}{dt} = \frac{\omega_d}{2} \pi L_y L_z \frac{|C|^2}{k_\perp^2} \left. \frac{d\epsilon_r}{dx} \right|_{x_0}. \tag{11.40}$$

Hence, due to the logarithmic singularity ($C \neq 0$) energy is absorbed collisionlessly by the plasma. As time progresses, the energy supplied by the driver accumulates in an ever diminishing plasma layer around $x = x_0$, where the frequency of the driver matches the local Alfvén frequency. In this singular point the fields grow unbounded as time proceeds and the external driver continues to pump energy in to the plasma. Since we determined the plasma solution analytically in the previous section, we can calculate, in principle, the actual energy absorption rate and express $|C|^2$ in terms of the external driving source.

From Eq. (11.37) it is clear that the average energy absorption rate depends on the driving frequency. The original idea of Chen and Hasegawa was to exploit the

11.1 Ideal MHD theory of resonant absorption

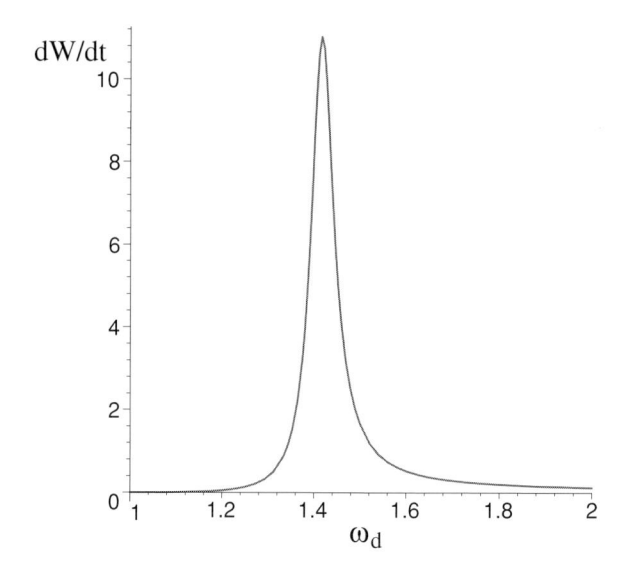

Fig. 11.3. Scan of the average energy absorption rate dW/dt in the steady state versus the driving frequency for $1 \leq \omega_d \leq 2$ for $\delta = 0.001$, $a = 0.005$, $c = 0.1$, $k_\parallel = 1$, $k_\perp = 10$, $\rho_1 = 1$, $\rho_2 = 0$ and $J_y^\star = 1$.

surface mode located at the plasma–vacuum surface. In the previous chapter it was shown that this mode transforms to a weakly-damped quasi-mode for an inhomogeneous plasma. The surface mode character should guarantee a good plasma–driver coupling while the coupling with the ideal Alfvén continuum modes should guarantee a good plasma heating through the resonant behaviour near the singularity. In Fig. 11.3 the average energy absorption rate, calculated from the above solution, is plotted versus the driving frequency. The peak in the power absorption profile is due to the presence of the quasi-mode. As a matter of fact, the oscillatory part of the frequency of this mode is given by

$$\omega_0 = \sqrt{2}\,\omega_{A1}, \qquad (11.41)$$

as is clear from Eq. (10.92) with $\rho_2 = 0$ (vacuum) and for a uniform magnetic field. For Fig. 11.3 the normalizing current in the antenna was fixed to 1 for all driving frequencies. As expected, driving at the oscillatory part of the quasi-mode frequency yields a much larger power absorption than driving at other continuum frequencies. The quasi-mode is the natural oscillation mode of the system and, hence, easy to excite. Due to the coupling of this mode to the singular shear Alfvén continuum, the energy pumped into the quasi-mode is efficiently absorbed (or 'dissipated' in non-ideal MHD). The latter remark brings us to an important issue, viz. the efficiency of the resonant absorption or heating mechanism.

Efficiency and role of the 'quasi-modes' In the literature, the efficiency of the resonant absorption mechanism has been expressed in terms of the coupling factor C, the fractional absorption f_a, and the antenna impedance. Clearly, the definition of these quantities is closely related to the energetics of the driven system. The power emitted by the antenna is given by

$$P_{\text{ant}}(\omega_d) = \tfrac{1}{2} \int_{\hat{V}} \mathbf{j}_c^* \cdot \hat{\mathbf{E}}^* \, d\hat{V}, \tag{11.42}$$

where $\hat{V}$ denotes the volume of the vacuum region. The real part of this power corresponds to the energy absorbed by the plasma, i.e. the quantity dW/dt defined in Eq. (11.35). The imaginary part, however, corresponds to the energy fluctuating in the system, i.e. the kinetic and potential energy of the plasma and the potential (magnetic) energy of the vacuum, as will be shown below. The coupling factor C depends on the driving frequency and is defined as:

$$C(\omega_d) \equiv \frac{\text{Im}(P_{\text{ant}})}{\text{Re}(P_{\text{ant}})}. \tag{11.43}$$

It gives the ratio of the amount of energy fluctuating in the system to the amount of energy converted into heat (in a dissipative system) which depends on the driving frequency (see e.g. [15], although these authors use the symbol Q to denote the coupling factor). It is closely related to the impedance of the driven system (for a real driving frequency it is proportional to it). This impedance can be measured at the antenna as $\text{Im}Q_z(c)/\text{Im}Q_x(c)$ which is $\sim \text{Im}A_4/\text{Re}A_4$ for the solution obtained in the previous subsection.

The fractional absorption f_a, on the other hand, is defined (by [100]) as

$$f_a(\omega_d) \equiv \frac{\text{Re}(P_{\text{ant}})}{|P_{\text{ant}}|}, \tag{11.44}$$

and thus compares the absorbed power to the total power of the disturbance in the system in the steady state. The functional relation of the two quantities is given by

$$f_a = \frac{1}{1 + C^2}. \tag{11.45}$$

Hence, *perfect* coupling, meaning that all the energy that enters the plasma remains inside the plasma and is (at least in the dissipative steady state) converted into heat,[1] corresponds to $C(\omega_d) = 0$ and $f_a(\omega_d) = 1$ (or 100%). No coupling at all corresponds to $C(\omega_d) \to \infty$ and $f_a(\omega_d) = 0$.

The presence of quasi-modes in the Alfvén continuum is crucial for the efficiency of the resonant absorption mechanism. The global character of the quasi-

---

[1] i.e. the Poynting vector at the plasma surface is pointing inwards all the time.

11.1 Ideal MHD theory of resonant absorption

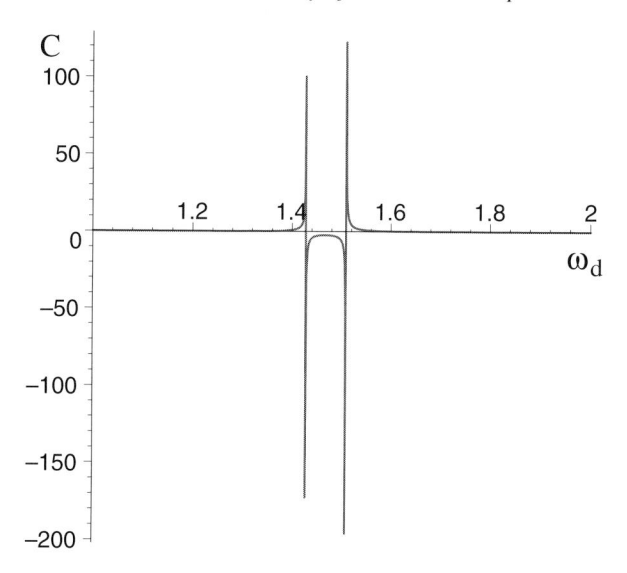

Fig. 11.4. Scan of the coupling factor C in the steady state versus the driving frequency for $\delta = 0.001$, $a = 0.005$, $c = 0.1$, $k_\parallel = 1$, $k_\perp = 10$, $\rho_1 = 1$, $\rho_2 = 0$, $J_y^* = 1$.

modes guarantees a good plasma–vacuum coupling, while the coupling of the quasi-mode with the singular Alfvén continuum modes guarantees efficient dissipation. The good plasma–vacuum coupling at $\omega_d = \sqrt{2}$ is confirmed in Fig. 11.4 which displays the coupling factor versus the driving frequency. Fig. 11.4, however, reveals a second driving frequency that yields excellent plasma–vacuum coupling. This is due to the presence of the antenna, which modifies the eigenvalue problem to be solved. As a matter of fact, besides the plasma–vacuum surface mode calculated in the previous chapter, there is now also a mode that 'fits' perfectly in the plasma–vacuum–antenna system.

This second mode can easily be derived. Consider the plasma to be completely uniform ($a = 0$). The energy flux is given by the Poynting vector which is proportional to $\hat{Q}_x \hat{Q}_z$ in the vacuum region. In the eigenvalue problem, i.e. without the driving current in the antenna, this energy flux must be continuous everywhere, in particular at the antenna: $[\![\hat{Q}_x \hat{Q}_z]\!]_c = 0$. Since $\hat{Q}_x$ is continuous everywhere, this means that the boundary condition at the antenna can be written as

$$\hat{Q}_x \Big|_c [\![\hat{Q}_z]\!]_c = 0. \tag{11.46}$$

For the surface mode we required both $\hat{Q}_x$ and $\hat{Q}_z$ to be continuous at the antenna position $x = c$. There is, however, a second possibility to satisfy the above boundary condition, viz. by requiring that $\hat{Q}_x$ vanishes at $x = c$. This yields $A_3 = B_3$ and $A_4 = 0$ which means that the induced solution vanishes at $x = c$. Linking the

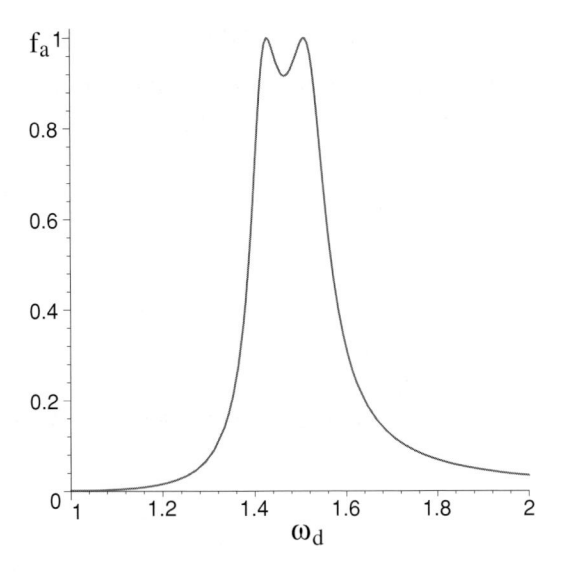

Fig. 11.5. Scan of the fractional absorption f_a in the steady state versus the driving frequency for $\delta = 0.001$, $a = 0.005$, $c = 0.1$, $k_\parallel = 1$, $k_\perp = 10$, $\rho_1 = 1$, $\rho_2 = 0$, $J_y^* = 1$.

vacuum solution to the plasma solution as before then yields an eigenvalue problem the solution of which is given by

$$\omega^2 = \omega_{A1}^2 \frac{1 + \tanh(k_0 c)}{k_\parallel^2/k_\perp^2 + \tanh(k_0 c)}. \qquad (11.47)$$

When the antenna is placed far away from the plasma ($c \to +\infty$), $\tanh(k_0 c) \approx 1$ and the frequency of this mode matches the frequency of the plasma–vacuum surface mode, i.e. $\omega = \sqrt{2}\,\omega_{A1}$. For the parameters chosen in Fig. 11.4, however, one finds $\omega \approx 1.52$. When there is a transition layer ($a \neq 0$) this mode also couples to the continuum modes and its frequency becomes complex. When the system is driven externally, the oscillatory part of this frequency yields perfect coupling, as can be concluded from Figs. 11.4 and 11.5.

Figure 11.5 displays the dependence of the fractional absorption on the driving frequency which quantifies the efficiency of the resonant absorption mechanism in a different way (see above). It shows that, in this case, both optimal driving frequencies 1.41 and 1.52 yield total or perfect absorption, i.e. $f_a = 1$ (or 100%). This means that, in the steady state, all the energy that enters the plasma stays in the plasma. For other driving frequencies the plasma–driver coupling is not perfect.

11.1.3 Resonant 'absorption' versus resonant 'dissipation'

Clearly, *dissipative effects* will prevent unlimited growth and will cause a real, dissipative, plasma to reach a *stationary state* after a finite time. In this stationary state all physical quantities oscillate harmonically with a constant amplitude (in time) and with the frequency imposed by the external source. The power supplied by the external source is exactly balanced by the energy dissipation rate in the plasma. In the above ideal MHD solution obtained in Section 11.1.1 this steady state was simulated in ideal MHD by considering artificial damping in the form of an imaginary part of the external driving frequency.

This is simpler than solving the dissipative MHD equations which are more complicated. The visco-resistive MHD equivalent of Eq. (11.2), for example, reads

$$\begin{pmatrix} \rho\omega^2 + \dfrac{d}{dx} B^2 \dfrac{d}{dx} - k_\parallel^2 B^2 - i\omega\rho(\eta+\nu)\dfrac{d^2}{dx^2}\dfrac{d}{dx} k_\perp B^2 \\ -k_\perp B^2 \dfrac{d}{dx} \rho\omega^2 - k^2 B^2 - i\omega\rho(\eta+\nu)\dfrac{d^2}{dx^2} \end{pmatrix} \begin{pmatrix} \xi_x \\ i\xi_\perp \end{pmatrix} = 0, \tag{11.48}$$

where the scalar viscosity ν and the electrical resistivity η are only retained in terms with derivatives in the x-direction. The reason for doing so is that these dissipative effects are extremely small in hot plasmas (e.g. in the solar corona or in tokamaks) and the dissipative terms only contribute in combination with x-derivatives which become extremely large in the neighbourhood of ideal singularities, as we have seen in the above ideal solution. The visco-resistive terms have the same effect as the artificial damping term in the previous sections, viz. to remove the singularity from the equations and, hence, to keep the solution finite. The fact that the dissipative effects are extremely small in the hot plasmas of interest results in nearly-singular behaviour in the neighbourhood of the ideal MHD resonance positions in a very similar way to that in the above solution. We will see examples of this below when some numerical solutions of the dissipative MHD equations are discussed.[2]

Remark on the resistive MHD solution and energetics Kappraff and Tataronis [120] considered a similar configuration of a planar sheet pinch but with sheet currents on both sides of the plasma and with finite plasma resistivity. These

---

[2] Analytical solutions of Eqs. (11.48) have been derived but these will not be discussed here since the analysis becomes quite involved. Moreover, for non-trivial equilibria the solutions are only partly analytical and the help of a computer is required after all.

authors demonstrated that the high absorption rate obtained in ideal MHD indeed corresponds to an effective heating rate in highly conductive tokamak plasmas. Their analysis is quite involved, including an asymptotic matching procedure of the solution in the resistive layer with the ideal MHD solution outside the layer and an inverse Laplace transformation of the resistive layer solution. The analysis reveals two time periods which characterize the energy absorption. Initially the energy accumulates at the singular layer, just like in ideal MHD, up to a critical time t_h which scales as $\eta^{-1/3}$. In our notation their solution reads

$$\xi_\perp(t) \cong \begin{cases} N(\omega_d)\, t\, \exp(-i\omega_d t) & (\text{for } t < t_h) \\ N(\omega_d)\, t_h\, \exp(-i\omega_d t) & (\text{for } t > t_h), \end{cases} \quad (11.49)$$

with $N(\omega_d)$ a multiplicative constant, that depends on the nature of the equilibrium and the source but is independent of η, and with

$$t_h = \left(\frac{24\mu_0}{\omega_d^2 \eta}\right)^{1/3} \left(\frac{2B_1'(x_0)}{B} - \frac{\rho'(x_0)}{\rho}\right)^{-2/3}. \quad (11.50)$$

Hence, for $t > t_h$ the growing solution in the resistive layer has reached a saturation amplitude and the absorbed energy is transferred to heat through Ohmic dissipation.

Sakurai et al. [200] and Goossens et al. [96] followed a similar approach but for cylindrical geometry. These authors used asymptotic expansions of solutions of the dissipative MHD equations to derive jump conditions or connection formulas that are used to integrate the equations through the near-singularities at the resonance points. This method was used in many papers on resonant absorption later on.

In resistive MHD, the plasma heating is treated consistently. This makes it possible to get a more detailed picture of the energetics in the driven system which can be derived from the linearized MHD equations. Below, we will denote the perturbed current density by $\mathbf{j}_1$ (here we need the subscript 1 on $\mathbf{j}_1$ because there is also $\mathbf{j}_0$). The perturbed electric field is still denoted by $\mathbf{E}$. Combining Maxwell's law and Faraday's law, for example, we easily get

$$-\nabla \cdot (\mathbf{E}^* \times \mathbf{Q}) = \mathbf{j}_1 \cdot \mathbf{E}^* + \mathbf{Q} \cdot \frac{\partial \mathbf{Q}^*}{\partial t}, \quad (11.51)$$

which means that the inflow of electromagnetic energy (LHS term) produces electrical energy for the plasma (first RHS term) and gives a rise in the magnetic energy (second RHS term). In turn, the electrical energy given to the plasma by the

11.1 Ideal MHD theory of resonant absorption

disturbance of the electromagnetic field may be rewritten:

$$\mathbf{j}_1 \cdot \mathbf{E}^* = \eta |\mathbf{j}_1|^2 + \mathbf{v}^* \cdot \mathbf{j}_1 \times \mathbf{B}_0. \tag{11.52}$$

Thus, the electrical energy appears as heat by Ohmic dissipation (first RHS term) plus work done by the $\mathbf{j}_1 \times \mathbf{B}_0$-force. Furthermore, the scalar product of $\mathbf{v}^*$ with the equation of motion yields the so-called mechanical energy equation:

$$\rho_0 \mathbf{v}^* \cdot \frac{\partial \mathbf{v}}{\partial t} = -\mathbf{v}^* \cdot \nabla p_1 + \mathbf{v}^* \cdot \mathbf{j}_0 \times \mathbf{Q} + \mathbf{v}^* \cdot \mathbf{j}_1 \times \mathbf{B}_0. \tag{11.53}$$

The above three equations may be combined and integrated over the plasma volume V to give the *resistive energy balance*:

$$-\tfrac{1}{2}\int_V \nabla \cdot (\mathbf{E}^* \times \mathbf{Q})\, dV = \underbrace{\tfrac{1}{2}\int_V \rho_0 \mathbf{v}^* \cdot \frac{\partial \mathbf{v}}{\partial t}\, dV}_{\dot{K}}$$

$$+ \underbrace{\tfrac{1}{2}\int_V \mathbf{v}^* \cdot \nabla p_1 - \mathbf{v}^* \cdot \mathbf{j}_0 \times \mathbf{Q} + \mathbf{Q} \cdot \frac{\partial \mathbf{Q}^*}{\partial t}\, dV}_{\dot{W}_p} + \underbrace{\tfrac{1}{2}\int_V \eta |\mathbf{j}_1|^2\, dV}_{\dot{D}}. \tag{11.54}$$

This means that the inflow of electromagnetic energy (LHS term) in the plasma region yields a rate of change of the kinetic energy of the plasma ($\dot{K}$), a rate of change of the potential energy of the plasma ($\dot{W}_p$) and a heating rate due to Ohmic dissipation ($\dot{D}$). Of course, the LHS term in the above equation is related to the power emitted by the antenna. For $\eta \neq 0$ the tangential components of the magnetic field perturbation are continuous at the plasma–vacuum boundary, and with the help of Gauss' theorem this means that the LHS term can be written as

$$-\tfrac{1}{2}\int_V \nabla \cdot (\mathbf{E}^* \times \mathbf{Q}) = \tfrac{1}{2}\int_{\hat{V}} \hat{\mathbf{j}}_c^\star \cdot \hat{\mathbf{E}}^* d\hat{V} = \underbrace{-\tfrac{1}{2}\int_{\hat{V}} \hat{\mathbf{j}}_c^\star \cdot \hat{\mathbf{E}}^*}_{P_{\text{ant}}} - \underbrace{\tfrac{1}{2}\int_{\hat{V}} \hat{\mathbf{Q}} \cdot \frac{\partial \hat{\mathbf{Q}}^*}{\partial t}}_{\dot{W}_v},$$

$$\tag{11.55}$$

i.e., the inflow of electromagnetic energy in the plasma equals the outflow of electromagnetic energy out of the vacuum region, which is the power emitted by the antenna (P_{ant}) minus the rate of change of the vacuum magnetic energy ($\dot{W}_v$). Hence, combining Eq. (11.54) and Eq. (11.55) yields

$$P_{\text{ant}} = \dot{K} + \dot{W}_p + \dot{W}_v + \dot{D}, \tag{11.56}$$

i.e., the power emitted by the antenna produces a change of the kinetic and potential energy of the plasma and of the potential (magnetic) energy of the vacuum plus heat by Ohmic dissipation. It is also clear that, in the steady state, $\operatorname{Re} P_{\text{ant}} = \dot{D}$, which means that the energy supply by the external harmonic source is exactly balanced by the dissipation rate in the plasma.

For η sufficiently small, the Ohmic heating rate is very well approximated by the energy accumulation rate obtained in ideal MHD. Hence, Kappraff and Tataronis obtained the following expression for *the resonant dissipation rate*:

$$\int_V \eta J_1^2 \, dx = \left. \frac{dW}{dt} \right|_{\text{ideal MHD}}, \tag{11.57}$$

in the limit $\eta \to 0$.

The coupling factor and the fractional absorption express the efficiency of the resonant absorption mechanism in the steady state by comparing quantities averaged over a driving period. Hence, they do not contain any information on the time scales of the mechanism, which are nonetheless important. As a matter of fact, efficient absorption in the steady state means nothing if the steady state is only reached in an asymptotic way. The latter remark is related to the quality factor Q of the resonance which is defined as [178]:

$$Q = \frac{|\dot{K}| + |\dot{W}_p| + |\dot{W}_v|}{2\pi |\dot{D}|}, \tag{11.58}$$

i.e. the ratio of the total energy contained in the system to the Ohmic heating per driving cycle in the stationary state. Notice that the norm of each component of the total energy has been taken separately in the definition of Q. In perfect coupling cases these terms can be quite large, but $\dot{K}$ has the opposite sign of $\dot{W}_p$ and $\dot{W}_v$ and, therefore, these terms cancel out and $\operatorname{Im}(P_{\text{ant}})$ vanishes so that the fractional absorption becomes 1. In the quality factor, however, these terms do not cancel out and Q compares the amount of dissipation in one driving period to the amount of energy that has to be pumped into the system before the steady state is reached. It thus tells us something about the time scales of the process. Good quality resonances have little losses and thus high Q-values. This means that a lot of energy needs to be put into the system to reach the steady state. Hence, for efficient heating on short time scales, low Q-values are required. Good plasma–driver coupling does not guarantee a low-Q resonance, and vice versa.

Remark that the quality of LRC-circuits or resonant cavities is sometimes defined as

$$Q^* = \frac{\omega_0}{\Delta \omega}, \tag{11.59}$$

where ω_0 is the resonance frequency and $\Delta \omega$ denotes the width at half-maximum of the peak in the resonance curve which is obtained by plotting the power absorbed by the oscillator as a function of the driving frequency ω (see, e.g., [110]).

The superscript * is added to distinguish the two definitions. As a matter of fact, it can be shown that the definitions are equivalent in the simple case of a resonant LRC-circuit or a resonant cavity, at least when the damping is weak so that the resonance is sharp. Here, however, the two definitions are not equivalent [178].

11.2 Heating and wave damping in tokamaks and coronal magnetic loops

As already mentioned, resonant absorption of Alfvén waves was first studied in the context of controlled thermonuclear fusion research to provide an additional heating mechanism for tokamak plasmas. In 1978, Ionson proposed the same heating mechanism for solar coronal loops to explain the high temperature of the solar corona [116]. Since then, the efficiency of resonant absorption or 'Alfvén wave heating' has been studied extensively in this context too. In the present section, we briefly discuss these applications and the main results are illustrated for a simple cylindrical plasma.

11.2.1 Tokamaks

(a) Heating of tokamak plasmas In large tokamaks, Ohmic heating due to the plasma current of a few MA yields ion temperatures of a few keV (recall that 1 keV $\approx 1.1 \times 10^7$ K), typically 30–40% of the required 10^8 K for the fusion of deuterium and tritium (see Chapter 1). Hence, a considerable amount of supplementary heating is required.[3] In the early 1970s it was suggested that the Alfvén continuum resonances be exploited for this purpose [225]. Due to the low frequencies involved (a few MHz), it was relatively easy and cheap to build antennas that generate the required power at the required frequencies. Experiments, such as with the 'Tokamak Chauffage Alfvén' (TCA) and its successor the 'Tokamak à Configuration Variable' (TCV) in Lausanne [28], were done to check the theoretical predictions and it was demonstrated that the antennas can indeed couple electromagnetic energy to the plasma efficiently and that resonant absorption is an efficient heating mechanism for tokamak plasmas. However, it turned out that most of the resonances occur close to the plasma edge so that the experiments yielded edge heating instead of heating of the inner core [15]. The TCA tokamak is living its second life in Brazil as the TCA-BR, at present installed in the plasma laboratory of the Applied Physics Department of the University of São Paulo [7].

---

[3] An alternative would be to increase the plasma current, which requires a tokamak with a very large magnetic field. Such an experiment has been suggested in the IGNITOR project but it has not been built so far for various economic and political reasons [37].

Nowadays, similar heating mechanisms are used to deposit energy in the inner plasma layers of tokamaks but involving higher frequencies, such as ion cyclotron resonance heating (ICRH), with frequencies Ω_i in the range 20–120 MHz depending on the magnetic field and the ion species involved, lower hybrid resonance heating (LHRH) with frequencies between Ω_i and Ω_e (1–8 GHz), and electron cyclotron resonance heating (ECRH), with frequencies Ω_e in the range 50–200 GHz. These methods are known under the common name radio frequency (RF) heating. They are usually operated in conjunction, and supplemented with alternative heating methods like neutral beam injection.

(b) Damping of global Alfvén waves Additional plasma heating by neutral beam injection or ICRH at frequencies of the order of the MHD continuum frequencies generates energetic He nuclei flying through the plasma at supra-thermal speeds. These α-particles can then destabilize the global toroidicity induced Alfvén eigenmodes (so-called TAEs) by a particle–wave interaction which leads to the loss of them before their energy can be thermalized. Since the confinement of α-particles is essential for ignition, this particle–wave interaction has been studied in detail. TAEs are observed routinely in large tokamaks, such as for example TFTR, DIII-D, JET, JT-60 [232]. Experiments have shown that TAEs can indeed be destabilized by the energetic particles, but the threshold for instability turned out to be much higher than expected from theoretical estimates based on collisionless Landau damping (see Section 2.3.3). Cylindrical or slab geometry simplifications cannot be applied here since TAEs owe their existence to the toroidicity of the plasma, which results in 'gaps' in the continuous spectrum due to coupling of the poloidal wave numbers. Hence, the models are necessarily two-dimensional and will be discussed in detail in the chapter on waves and instabilities in tokamaks in the companion Volume 2. Numerical investigations showed that the global Alfvén modes also experience resonant damping, which might explain the increased instability threshold [185].

11.2.2 Coronal loops and arcades

The temperature increase above the photosphere of the Sun (Fig. 8.13) requires work which is, most probably, done by the motions of the convection zone. These motions produce much more energy than required to heat the upper solar atmosphere, which has a very low heat capacity due to its extremely low density. Hence, the only problems left are to explain (1) how this energy gets to the upper layers of the solar atmosphere, and (2) how it is transformed into heat in those layers. There is an additional problem associated with the inhomogeneity of the upper atmospheric layers which consist of many magnetic structures. All of these structures, with a variety of length scales, have to be heated. Many solar physicists think it is not likely that one single heating mechanism is responsible for the heating of all these magnetic structures but that, instead, several heating mechanisms operate. These may be more or less effective in coronal holes, short coronal loops, long coronal loops, etc. However, it is almost certain that the heating mechanisms that are responsible for the heating of the upper layers of the solar atmosphere are magnetic in nature.

The magnetic heating mechanisms that have been proposed, and that are still investigated, can be divided into two types [133] differing with respect to the time scales of the external driving source. The first type is the *wave heating mechanism*. Here, problem (1) is 'solved' by MHD waves that are generated by the upwelling convective motions in the photosphere and that propagate through the photosphere and the chromosphere to reach the corona. For the efficient dissipation of this wave energy in the upper atmospheric layers (problem (2)), small length scales have to be created in order to make the time scale for magnetic diffusion ($\tau_d \sim \mu_0 l^2/\eta$) reasonably short. Such small length scales could be produced, for example, by the transformation of the waves into shocks or by 'resonant absorption'. Clearly, the time scale of the driving source must be relatively short compared to the time a perturbation needs to cross the loop in order to generate waves in the loop. The second kind of magnetic heating mechanism involves *Joule dissipation of electric current sheets* along the coronal magnetic field lines and involves much longer perturbation time scales. These currents could be generated by slow twisting of the flux tubes by the photospheric convective motions (problem (1)) and the release of energy could take place in narrow current layers (problem (2)). In the following, we will concentrate on resonant heating.

11.2.3 Numerical analysis of resonant absorption

The efficiency of the resonant absorption mechanism has been investigated quantitatively in tokamak plasmas, coronal plasmas and magnetospheric plasmas. The coronal loops and arcades are usually modelled by a straight cylinder or even a slab, so that the curvature of the loops along their length is not taken into account (see Fig. 11.6). As we have seen, in the early studies the geometry of tokamak plasmas was simplified in the same way. Later, however, numerical studies quantifying resonant absorption took into account the full geometry of the plasma, including the toroidal curvature and the non-circularity of the poloidal cross-section as, e.g., in Ref. [185]. Here, the main numerical results for cylindrical plasmas are briefly reviewed. To illustrate these results, we consider a generic cylindrical equilibrium (coronal loop or tokamak) in which the profile of the local Alfvén frequency $\omega_A(r) = (mB_\theta/r + kB_z)/\sqrt{\rho}$ is parabolic (Fig. 11.7). For $m = 2$ and $k_z = 0.05$,

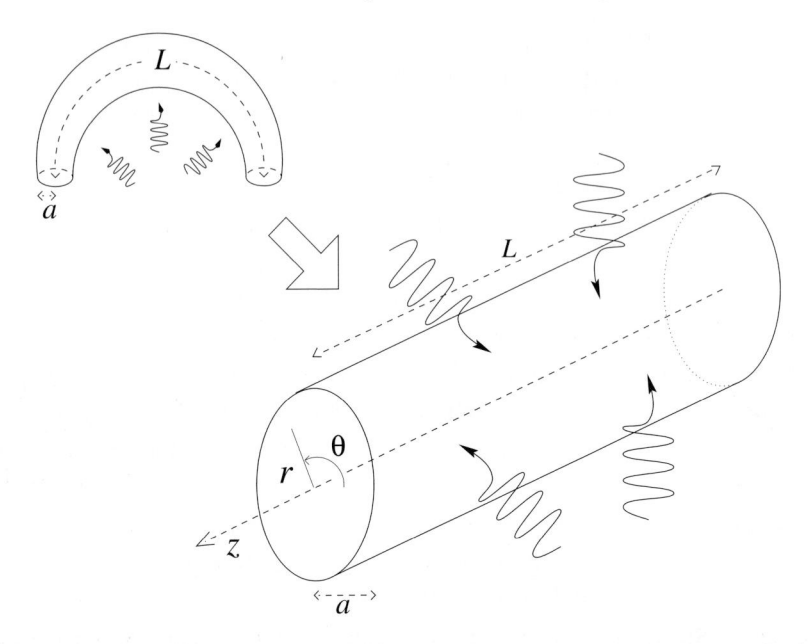

Fig. 11.6. Straightening a coronal magnetic loop with sideways wave excitation into a cylindrical model.

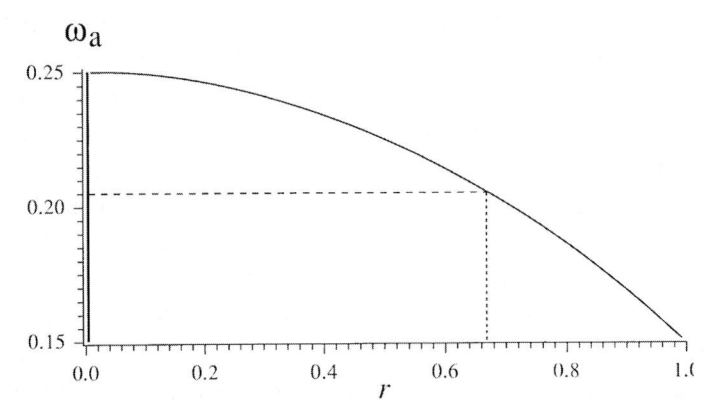

Fig. 11.7. Profile of the local Alfvén frequency resulting from a parabolic current density profile $j_z(r)$, constant magnetic field B_z, plasma density ρ, and wave numbers $m = 2$ and $k_z = 0.05$.

the ideal Alfvén continuum then ranges from $\omega_A(1) = 0.15$ to $\omega_A(0) = 0.25$ in dimensionless units (since $a = B_z = \rho = 1$).

(a) Temporal evolution This equilibrium is now excited periodically at a frequency $\omega_d = 0.205$ located in the range of the continuum. The ideal singular layer

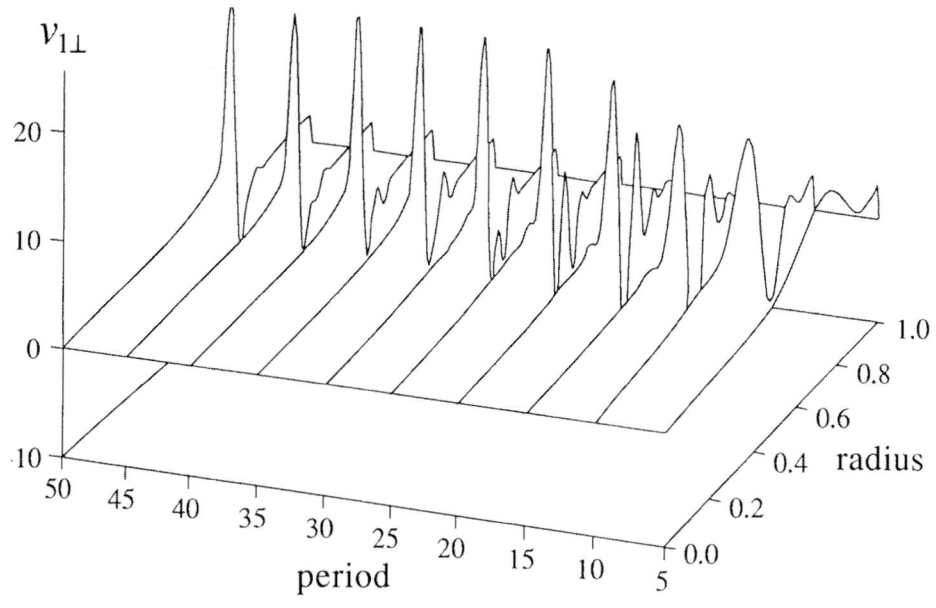

11.2 Heating and wave damping in tokamaks and loops

Fig. 11.8. Snapshots of $v_{1\perp}(r)$ for $\omega_d = 0.205$, $\eta = 10^{-6}$ (other parameters as in Fig. 11.7).

that corresponds to this oscillation frequency is located at $r = 0.671$, i.e. $\omega_d = \omega_A(0.671)$. In Fig. 11.8, snapshots of the radial profile of the velocity perturbation tangential to the magnetic surfaces and perpendicular to the magnetic field lines, $v_{1\perp} = (v_{1\theta} B_z - v_{1z} B_\theta)/B$, are plotted versus dimensionless time. The time step between the snapshots is five driving periods and the magnetic Reynolds number $R_m = 10^6$. In the initial phase, the relatively small plasma response is global and phase mixing only occurs as time progresses. The plasma response localizes in a gradually diminishing plasma layer around $r = r_0 = 0.671$, where large gradients build up. This localization and the growth of the oscillation amplitude in the resonant layer end after about 50 driving periods. From then on, the plasma oscillates purely harmonically, i.e. all the magnetic surfaces oscillate with constant amplitude and with the frequency ω_d of the external source. In other words, the system has reached a *stationary state*. The delta function $\delta(r - r_0)$ and singularity $(r - r_0)^{-1}$, characteristic for the perpendicular tangential component of the ideal Alfvén mode (see Eq. (7.154)), are still recognizable in this typical resistive solution.

(b) Energetics The time dependence of the energetics of the externally driven resistive plasma column is shown in Fig. 11.9. Here, the mean power P emitted by the external antenna, the mean rate of change of the kinetic energy $\dot{K}$ and of the potential energy $\dot{W}_p$ of the plasma, the mean Ohmic dissipation rate $\dot{D}$, and the

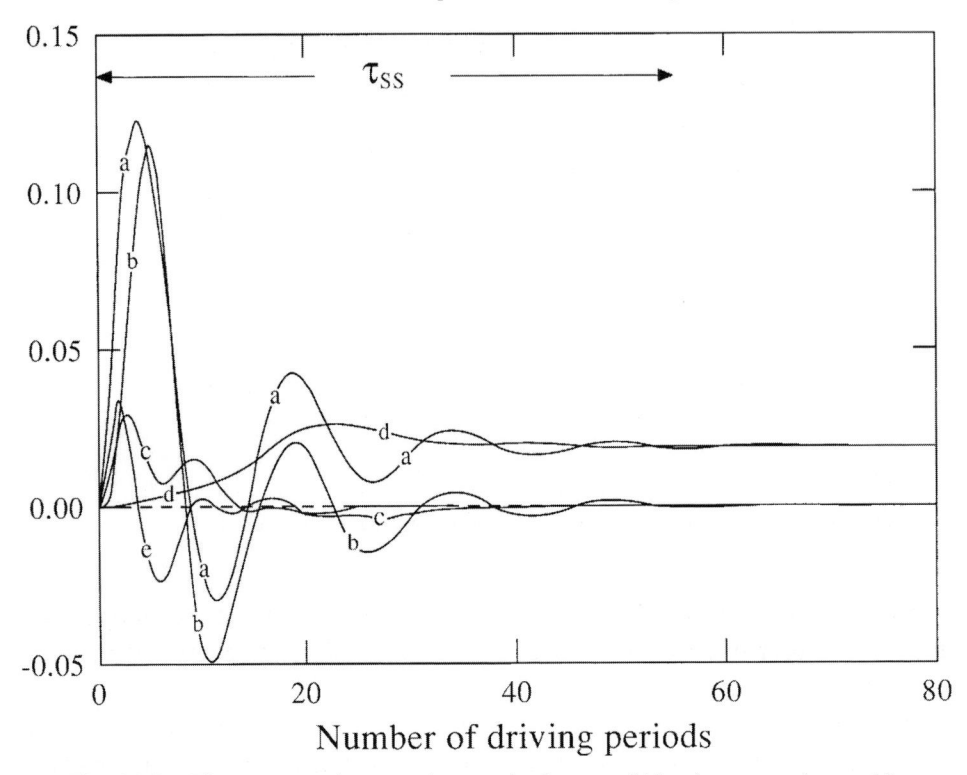

Fig. 11.9. Time-averaged power (curve a), change of kinetic energy (curve b), change of potential plasma energy (curve c), Ohmic dissipation rate (curve d) and change of vacuum magnetic energy (curve e) versus number of driving periods for $\omega_d = 0.205$ and $\eta = 10^{-6}$ (other parameters as in Fig. 11.7).

mean rate of change of the vacuum magnetic energy $\dot{W}_v$ are plotted versus the number of driving periods. These quantities are defined in Eqs. (11.54) and (11.55), and satisfy the relation (11.56). The initially oscillatory behaviour of these quantities is a consequence of the 'beats' that result from the initial excitation of the ideal quasi-mode.[4]

Initially, in the first 10 to 15 driving periods, the power supplied by the external source produces mainly a change of the kinetic and potential energy of the plasma and also, to a lesser degree, a change of the vacuum magnetic energy. In this phase, the Ohmic dissipation rate is very low. As time progresses, however, the rate of change of the kinetic and potential plasma energy and the rate of change of the vacuum magnetic energy gradually decrease and the Ohmic dissipation rate increases. Then, after about 55 driving periods the system attains a stationary state: the kinetic and potential plasma energy and the vacuum magnetic

---

[4] The beat frequency that results from $\sin(\omega_d t) + \sin(\omega_{qm} t)$ is $\frac{1}{2}(\omega_d - \omega_{qm})$.

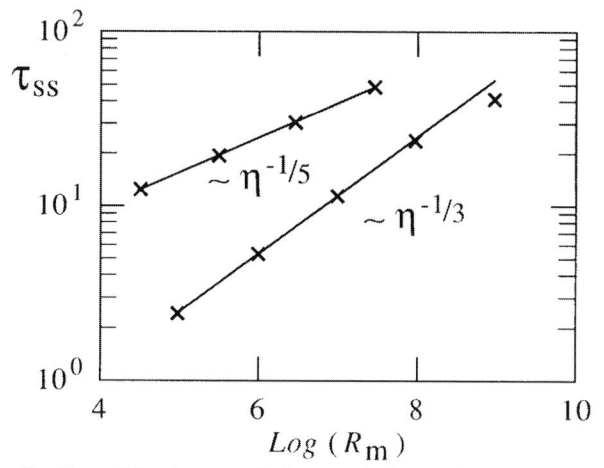

11.2 Heating and wave damping in tokamaks and loops

Fig. 11.10. Scaling of the time needed to reach the stationary state with the magnetic Reynolds number for typical coronal loop parameter values.

energy do not change any more ($\dot{K} = \dot{W}_p = \dot{W}_v = 0$) and the power supplied by the external source is exactly balanced by the Ohmic dissipation rate in the plasma ($P = \dot{D}$).

The time interval needed to reach the stationary state, τ_{ss}, depends on the plasma resistivity. We computed this dependence by means of numerical simulations keeping all parameters fixed except the plasma resistivity η. The time needed to reach the steady state is proportional to the cubic root of the magnetic Reynolds number. Hence, in terms of η, we get $\tau_{ss} \sim \eta^{-1/3}$ (see Fig. 11.10), in agreement with the analytical result of Kappraff and Tataronis [120] (see also Eq. (11.50)). However, the additional excitation of an ideal quasi-mode in the initial phase yields a different η-scaling, viz. $\tau_{ss} \sim \eta^{-1/5}$ (upper curve of Fig. 11.10).

(c) Discussion We have seen that the ideal quasi-modes play an important role in resonant absorption and affect both the efficiency and the time scales of this heating mechanism considerably. Excitation of a plasma at the frequency of such a weakly-damped quasi-mode yields 100% absorption, which means that all the energy that is supplied by the external source is absorbed by the plasma and, in the stationary state, converted into heat by Ohmic dissipation. Driving at an arbitrary frequency in the range of the ideal Alfvén continuum yields a less efficient plasma–driver coupling. This is a result of the fact that Alfvén waves *cannot be driven directly by a sideways external driver* as they propagate the energy mainly along the magnetic field lines. The contribution of the global mode is that it can transport the energy from the external driver *across the magnetic surfaces* to the inner plasma. The global mode has a frequency inside the continuous spectrum and

thus couples to an Alfvén wave, and it is this (nearly-singular) wave energy that is finally dissipated. However, for 'ordinary' continuum frequencies the fractional absorption can still be very high, depending on the equilibrium profiles. For typical coronal loop parameter values, driving at frequencies in the continuous spectrum often yields more than 90% 'absorption', i.e. more than 90% of the energy supplied by the external driver is actually dissipated and converted into heat. Driving at the foot points turns out to be even more efficient. This is due to the fact that in this way the Alfvén waves can be driven *directly* and there is *no need for a global mode to transport the energy from the external driver to the inner plasma* (see Section 11.3.1).

For *sideways driving*, it has been confirmed now that the basic time scale of resonant absorption depends on the proximity of the driving frequency to the quasi-mode frequency: for driving at the quasi-mode frequency, $\tau_{ss} \sim \eta^{-1/5}$, while for driving frequencies in the range of the ideal continuum but not close to the quasi-mode frequency, $\tau_{ss} \sim \eta^{-1/3}$. The results of the numerical simulations have been presented in dimensionless units. In active region loops on the Sun, the Alfvén continuum frequencies $\omega \approx 0.1$–1 typically correspond to 0.005–2 Hz. For realistic loop equilibria, τ_{ss} varies from a few minutes to a few hours, i.e. much shorter than the typical life time of coronal loops (1 day), even for realistic values of η. Hence, resonant absorption is very efficient for typical coronal loop values and a viable mechanism for heating of solar coronal loops. In tokamaks, the time scales for resonant absorption are much shorter. For a fully ionized tokamak plasma with an ion number density of 10^{20} m^{-3}, a small radius of 0.2 m, and a toroidal magnetic field of 2 T, the driving frequency $\omega = 0.205$ corresponds to a frequency of 1 MHz. Hence, for typical parameter values for small tokamaks, the time scale to reach the steady state is very short: for $R_m = 10^8$ the steady state is reached after about 150 driving periods in the simulation discussed above, i.e. after about 150 μs.

The cylindrical model considered here in the framework of linearized MHD is only a first approximation. For a final conclusion on the role of resonant absorption in coronal heating, more realistic simulations including effects of line-tying, curvature, foot point excitation, and nonlinearity need to be taken into account. Such simulations are being conducted at present. Line-tying has a drastic influence on the waves that appear in the loops and on the continuous spectrum. Also, for realistic magnetic Reynolds numbers the dynamics of the shear flow in the resonant layers turns out to be very nonlinear [170, 180, 186]. Nonlinear effects include background flows (leading to Doppler shifts of the continuum frequencies), nonlinear mode coupling, and Kelvin–Helmholtz instability of the resonant layers [170, 186]. Last but not least, a final conclusion also requires observational data on the power spectrum of the waves that are incident on the coronal loops.

11.3 Alternative excitation mechanisms

In the previous section we considered the simplest possible model to describe the mechanism of resonant absorption in a laboratory situation where the plasma is excited by a current in an external coil. In that case, the fast magneto-sonic wave mode plays the *crucial role of energy carrier*. As a matter of fact, shear Alfvén waves are unable to transfer the energy emitted by the external antenna from the antenna to the resonant layer since Alfvén waves can only transport energy *along* the magnetic field lines. Their group velocity is directed along the magnetic field, as discussed in Section 5.3. This is also clear from the system of equations (11.2) and the solution derived in the previous section: for $k_\perp = 0$ the Alfvén mode decouples completely from the fast magneto-sonic mode and the resonant absorption mechanism *simply does not work*. Indeed, the constants A_3 and B_3 are zero in that case (since $X = 0$ when $k_\perp = 0$) and the plasma does not respond to the external driver.

As mentioned in the introduction, the resonant absorption mechanism is intensively studied for many different situations. From a topological point of view these different physical set-ups can be classified into three configurations which are schematically displayed in Fig. 11.11. The first configuration, displayed in Fig. 11.11(a), is the plane slab version of the model problem III, discussed in Section 4.6. It corresponds to the sideways driven plasma discussed in the previous section. This configuration is used to study resonant absorption in tokamak devices, sideways driven coronal loops, and the Earth's magnetopause (Section 11.3.3). The second configuration (Fig. 11.11(b)) refers to the model

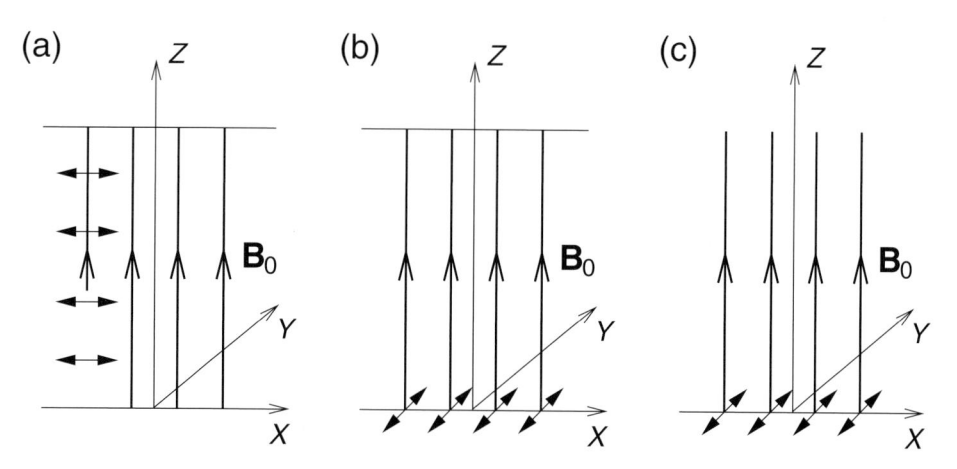

Fig. 11.11. Three different configurations corresponding to model problems III, IV and V of Section 4.6: (a) a sideways driven finite ('closed') loop, (b) the same system but driven at the 'foot points', and (c) an 'open' system without resonances but with phase mixing.

problem IV and is used to model the heating of solar coronal loops by foot point driving. From the physical point of view, this set-up is simpler than the previous one in the sense that the Alfvén waves can be excited directly, i.e. without the need of fast magneto-sonic wave modes, since the location of the energy source does not require the transfer of energy *through* magnetic flux surfaces here. Hence, this situation is mathematically tractable, even when dissipation is taken into account. It will be discussed in further detail below in Section 11.3.1. The first and second configurations have a finite length in the direction parallel to the magnetic field. The configuration can be periodic in this direction (as in a tokamak), or the ends of the field lines may be fixed (as in a coronal loop). The third configuration (Fig. 11.11(c)) corresponds to the plane slab version of model problem V of Section 4.6. It differs from the two previous ones by the fact that the field lines are 'open', i.e. very much longer than the wavelength in the parallel direction. It is representative for extremely long loops (as compared to the excited wavelengths) or coronal holes. In this situation, no resonances can occur but small length scales can still be created by phase mixing of the field lines. This process also occurs in resonant absorption and will be discussed in further detail in Section 11.3.2.

11.3.1 Foot point driving

Consider a simple slab model with a uniform magnetic field B in the z-direction and a density that is stratified in the x-direction only. Let us neglect the effect of plasma pressure and gravity for the time being and suppose the slab is bounded in the z-direction by two boundaries at $z = 0$ and $z = L$ (Fig. 11.12). This set-up corresponds to configuration (b) of Fig. 11.11 and could be a simple model for a solar coronal loop. With $L = 10^8$ m and an average Alfvén velocity $\bar{v}_A = 2 \times 10^6$ m s^{-1}, the characteristic time scale of such a loop would be $\tau_A = L/\bar{v}_A = 50$ s. As a result of the non-uniformity of the density, the Alfvén velocity and, hence, the local Alfvén frequency are functions of x. Thus, the ideal MHD spectrum of this simple model plasma contains an Alfvén continuum.

Let us assume that the above system is excited by a mono-periodic driver at one of the foot points of the loop, e.g. at $z = 0$, *with a frequency ω_d in the range of the continuous spectrum*. Then, as in the situation discussed in the previous section, a resonance will occur at $x = x_0$, where $\omega_d = \omega_A(x_0) = k_z B/[\rho(x_0)]^{1/2}$ (see Fig. 11.12). Hence, in the stationary state a solution of the form

$$\xi_y(x, z, t) = A(x)\, e^{i(k_z z - \omega_d t)}, \qquad (11.60)$$

is obtained. (We here ignore the fact that, in general, a Fourier series is required in the z-direction in order to satisfy the line-tying boundary conditions at $z = 0$ and $z = L$ [87, 103], according to Eq. (4.189) for model IV of Section 4.6.3.)

11.3 Alternative excitation mechanisms

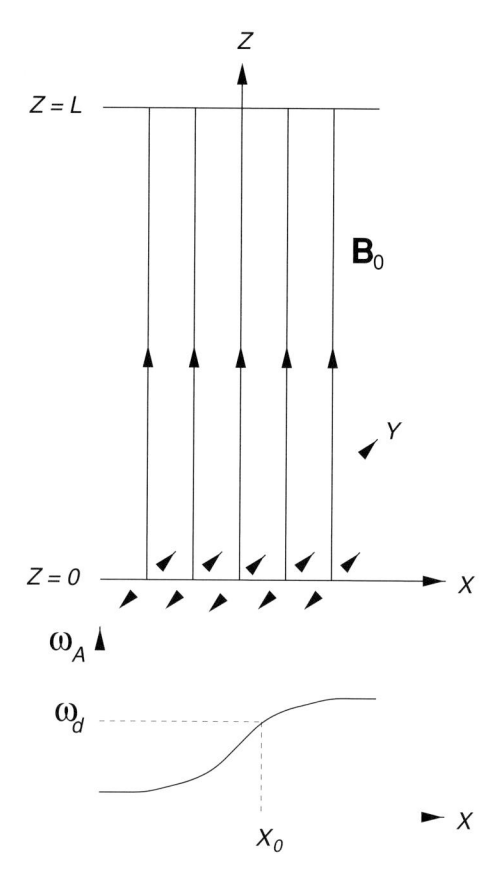

Fig. 11.12. Simple slab model for a coronal loop, ignoring curvature effects of the field lines and magnetic shear effects but taking into account 'line-tying', i.e. 'anchoring' of the 'foot points' of the magnetic field lines in the dense photosphere.

There are two important remarks to be made at this point. First of all, notice that the driving frequency is *real* now. This is because we will include viscosity and resistivity so that there is no need to include artificial damping to circumvent the singularity in the equations. Secondly, we simplified the model by considering perturbations which are polarized in the y-direction. The divergence-free condition for the perturbed magnetic field then reduces to $k_y \hat{Q}_y = 0$ so that $k_\perp = k_y = 0$ and the physical quantities do not depend on the y-coordinate. (Notice that, as a result of this assumption, the perturbations become automatically incompressible, $\nabla \cdot \mathbf{v}_1 = ik_y v_{1y} = 0$, so that the assumption $k_\parallel \ll k_\perp$ must be dropped. This gives an additional dissipative term $\sim k_z^2$ in Eq. (11.61) below.) We can then take dissipative effects into account while keeping the analysis tractable. As a matter of fact, the equations in the system (11.48) then decouple and the dynamic equation for the Alfvén waves can be studied separately. This decoupling of the Alfvén and fast wave modes means that in the present configuration the mechanism discussed in the previous section would not work. With the foot point driver, however, the fast wave modes are not required since the energy does not have to be transferred across the magnetic surfaces and the Alfvén waves are excited *directly* by foot point motions that are polarized in the tangential perpendicular (y) direction. As we will see below, considering $k_\perp = k_y \neq 0$, i.e. coupling between Alfvén and fast waves, affects the efficiency quite substantially, though, especially when random driving is considered.

The visco-resistive dynamic equation for the Alfvén wave then reads:

$$\frac{\partial^2 \xi_y}{\partial t^2} - (\eta + \nu)\left(\frac{\partial^2}{\partial x^2} - k_z^2\right)\frac{\partial \xi_y}{\partial t} + \omega_A^2(x)\,\xi_y = F(x)\,e^{i(k_z z - \omega_d t)}, \quad (11.61)$$

where the effect of an external harmonic driving term F with frequency ω_d is taken into account. The explicit form of $F(x)$ depends on details of the driver, i.e. the velocity profiles given on the boundaries at $z = 0$ and $z = L$. Equation (11.61) is the equation of a continuum of coupled forced oscillators which are *damped*.

▷ **Amplitude and phase at resonance.** The resonant behaviour described by Eq. (11.61) is similar to the well-known resonance of a driven oscillator described in many textbooks; see e.g. [39]. We can trace this comparison further by making a local analysis around a point $x = x_0$, i.e. a specific field line. This yields the classic equation for a forced and damped oscillator:

$$\frac{\partial^2 \xi_y}{\partial t^2} + (\eta + \nu)(k_x^2 + k_z^2)\frac{\partial \xi_y}{\partial t} + \omega_0^2 \xi_y = F_0\,e^{-i\omega_d t}, \quad (11.62)$$

where $F_0 \equiv F(x_0)$ and $\omega_0 \equiv \omega_A(x_0)$. Substitution of a solution of the form $\xi_y(x, z; t) = A_0 \exp\{i(k_x x + k_z z - \omega_d t)\}$ yields

$$A_0 = \frac{F_0}{(\omega_0^2 - \omega_d^2) - i\omega_d(\eta + \nu)(k_x^2 + k_z^2)}. \quad (11.63)$$

For $\eta = \nu = 0$, the amplitude A_0 becomes real (the plasma response is in phase or in opposite phase with the external driving force) and infinite when the driver frequency ω_d matches the 'natural' frequency ω_0 of the system. This is called 'resonance'. The behaviour of the field line resonator changes considerably if we include dissipation ($\eta \neq 0$ and/or $\nu \neq 0$). First of all, the absolute value of A_0 is reduced for every value of ω_d and remains finite for $\omega_d = \omega_0$. Next, the phase angle between the motion of the field line and the external driving force depends on the frequency,

$$\tan^{-1}\frac{\mathrm{Im}(A_0)}{\mathrm{Re}(A_0)} = \tan^{-1}\frac{\omega_d(\eta + \nu)(k_x^2 + k_z^2)}{(\omega_0^2 - \omega_d^2)}, \quad (11.64)$$

going through $\pi/2$ at the resonance. ◁

The characteristic time and length scales of the resonant absorption process can be estimated fairly easily. Let us assume that the system has been driven at a single frequency, viz. ω_d, for a long time so that it has reached a steady state. In this steady state, all physical quantities have a time behaviour of the form $e^{-i\omega_d t}$. Making the resonance condition explicit by using the linear Taylor expansion of $\omega_d^2 - \omega_A^2$ around the point x_0 where $\omega_d = \omega_A(x_0)$,

$$\omega_d^2 - \omega_A^2 = -(x - x_0)\frac{d\omega_A^2}{dx}\bigg|_{x=x_0}, \tag{11.65}$$

a simplified version of Eq. (11.61) is obtained which is valid only in the vicinity of the resonance position:

$$(x - x_0)\, 2\,\omega_A(x_0)\frac{d\omega_A}{dx}(x_0)\,\xi_y + i\omega_d(\eta + \nu)\frac{\partial^2 \xi_y}{\partial x^2} = F(x, z)\, e^{-i\omega_d t}. \tag{11.66}$$

Here, we also exploited the fact that $|k_z| = |\partial/\partial z| \ll |\partial/\partial x|$ in the steady state to drop the terms with k_z^2. The dissipation is only significant when the absolute value of the second, dissipative, term in the above equation becomes comparable with the first term. This comparison leads to an estimate of the characteristic length scale l_0 of the resonant absorption process:

$$l_0 \sim \left(\frac{\eta + \nu}{2\,|\omega_A'|_{x=x_0}}\right)^{1/3}, \tag{11.67}$$

where the prime denotes the derivative with respect to x. Upon substitution of this characteristic dissipation length scale in the expressions for the time scales of resistive diffusion, $\tau_{\text{res}} = l_0^2/\eta$, and viscous diffusion, $\tau_{\text{vis}} = l_0^2/\nu$, we find that the characteristic time scale for resonant absorption scales as

$$\tau_{\text{ra}} \sim (\nu + \eta)^{-1/3}\, |\omega_A'|^{-2/3}. \tag{11.68}$$

Notice that, if the profile of the local Alfvén frequency is not monotonic, there may exist two or more dissipation layers, even if the driver is mono-periodic. Also, a multi-periodic driver will create many dissipation layers in the plasma so that the whole plasma volume can be heated by resonant absorption.

11.3.2 Phase mixing

Let us again consider the 'closed' configurations discussed in the previous subsection. So far we assumed a mono-periodic driver. In a more realistic model for the solar corona, the driver has a broad spectrum and the time evolution of the wave amplitude is given by a superposition of the nearly-singular solutions discussed above. When each field line picks up its own characteristic frequency from the

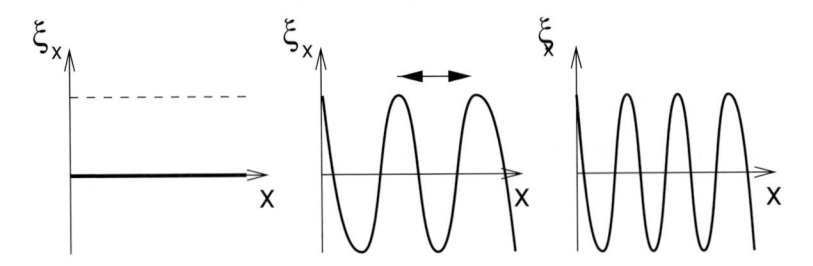

Fig. 11.13. 'Cascade' of energy to small length scales due to phase mixing: (a) in the initial state ($t = 0$ and/or $z = 0$) all field lines are in phase; (b) at $t = t_1 > 0$ phase mixing sets in; (c) at later times small length scales develop.

broad spectrum of the driver, we get a solution of the form

$$\xi_y(x, z, t) = A(x)\, e^{i[k_z z - \omega_A(x) t]}, \qquad (11.69)$$

which means that the field lines get more and more *out of phase* as time evolves, i.e. we get *phase mixing*. This yields an effective wave number

$$k_{x,\text{eff}} = \omega_A' t, \qquad (11.70)$$

which is proportional to t, meaning that the effective wavelength becomes smaller and smaller as time proceeds (see Fig. 11.13). This 'cascade' of the energy to ever smaller length scales continues until the effective wavelength becomes of the order of the length scale l_0 of Eq. (11.67) when the dissipative terms become important. Hence, we can define a phase mixing time τ_{mix} as the time at which $k_{x,\text{eff}} = 1/l_0$, i.e.

$$\tau_{\text{mix}} = \frac{1}{l_0 |\omega_A'|} \sim (\eta + \nu)^{-1/3} |\omega_A'|^{-2/3}. \qquad (11.71)$$

This phenomenon of phase mixing also occurs in the resonant absorption process discussed in the previous section. There too the field lines are initially in phase, but in the steady state shown in Fig. 11.2 the phases of the field lines are clearly not the same any more. In fact, the process of bringing neighbouring field lines out of phase is essential to get short length scales. Notice that we do not really need a resonance condition to get phase mixing. This process can also occur in coronal holes in which running waves along the 'open' field lines get phase mixed as time evolves. As a matter of fact, considering now the configuration displayed in Fig. 11.11(c) and assuming the foot points of the field lines are excited by a mono-periodic driver, we obtain *running wave* solutions of the form

$$\xi_y(x, z, t) \sim e^{i[k_z(x)z - \omega_d t]}, \tag{11.72}$$

where now, instead of the frequency, the *wave number depends on x*: $k_z(x) = \omega_d/v_A(x)$. Since the phase speed of the waves depends on x, the running waves phase mix as they propagate along the field lines. This results in an effective wave number in the x-direction which scales with z:

$$k_{x,\text{eff}} \sim \frac{1}{\xi_y} \frac{\partial \xi_y}{\partial x} \sim k'_z z. \tag{11.73}$$

Hence, large wave gradients are built up in the wave front as the waves propagate and z increases. These large gradients will appear at lower heights when the plasma is more inhomogeneous, i.e. when $k'_z(x)$ is larger. This means that we can define a phase mixing height z_{mix} as the height where $k_{x,\text{eff}} = 1/l_0$:

$$z_{\text{mix}} = \frac{1}{l_0 k'_z}. \tag{11.74}$$

Phase mixing of running Alfvén waves has been proposed as a possible means for explaining the heating of coronal holes and the acceleration of the solar wind [171]. The energy dissipation is now spread over the whole plasma volume even for a mono-periodic driver, in contrast to the resonant absorption case described in the previous section where it was limited to a narrow dissipative layer around the ideal resonance position in that case.

11.3.3 Applications to solar and magnetospheric plasmas

(a) Heating of line-tied loops and arcades As mentioned above, early loop heating models 'copied' the fusion set-up and considered a one-dimensional (periodic in two directions) flux tube excited *sideways* by incident fast magneto-sonic waves. Soon, however, it was realized that the foot points of the coronal loops are effectively 'anchored' ('line-tied') in the photosphere due to the high conductivity and the relatively high mass density of the latter. Moreover, the loops are also, perhaps mainly, excited at their foot points as a result of the 'anchoring' in the turbulent sub-photospheric plasma, and also due to the magnetic reconnection events that take place at the ends of the loops.

In Ref. [219] resonant heating or dissipation was simulated in an incompressible cylindrical line-tied plasma column excited at its foot points and it was remarked that the locations of the resonances differ from those in a periodic plasma. Later, these conclusions on the effect of line-tying were confirmed analytically [87, 103]: line-tying couples the Alfvén and fast magneto-sonic waves and also modifies the continuous spectrum drastically. For a cylindrical plasma with a sheared magnetic field, for example, the line-tied continuum is given by

$$\omega_A(r) = \frac{n\pi}{L} \frac{B_z(r)}{\sqrt{\rho(r)}}, \qquad (11.75)$$

i.e. *independent of the poloidal magnetic field component B_θ and the poloidal wave number*. For realistic coronal parameter values there are many quasi-modes in this line-tied continuum which positively affects the plasma–driver coupling [229], although they are, strictly speaking, not necessary here since the shear Alfvén waves can be driven *directly*.

Wave heating of line-tied loops has been studied extensively with linear and nonlinear methods, numerically and analytically, computing the steady state, the eigenvalues and the time evolution (see e.g. Refs. [181, 170, 172], and literature quoted there). Poedts and Boynton [179] showed with nonlinear MHD simulations that the heating can be very efficient (even without quasi-modes), and that it can easily compensate the radiative and conductive losses in a loop. Some studies even take into account the variation of the density [104] and magnetic field strength *along* the loops [22]. For realistic input power spectra the supplied energy is stored in the quasi-modes, which couple to Alfvén waves resulting in global heating of the loop on realistic time scales [229, 61, 62]. The outcome of all these studies is unanimous: *resonant dissipation is a viable heating mechanism* even when the heating rate drops due to 3D 'Kelvin–Helmholtz-like vortices' at the resonance layers [172].

However, it is important to include the coupling of the coronal parts of the loops with the transition region and chromosphere in order to take into account the effect of leakage [25] and tuning/detuning effects due to variation of the lengths of the loops. In a first attempt, Ofman *et al.* [173] tried this by using scaling laws in a 1D model to update the density due to the expected chromospheric evaporation. This yielded efficient heating concentrated in multiple resonant layers even for mono-periodic driving. Beliën *et al.* [20] even take into account the thermal structuring of the transition region and the higher chromosphere in a nonlinear numerical study. They note that only about 30% of the energy supplied by the driver takes the form of Alfvén waves. The bulk of this energy goes to slow magneto-sonic waves which are resonantly *dissipated in the foot points* and never reach the coronal part of the loops. This is actually compatible with the observed uniform temperature along the loops, which requires a more efficient heating of the foot points in order to compensate the larger radiation there due to the higher density. The results of Ref. [20], which considered mono-periodic driving only, should be extended with the consideration of more realistic drivers.

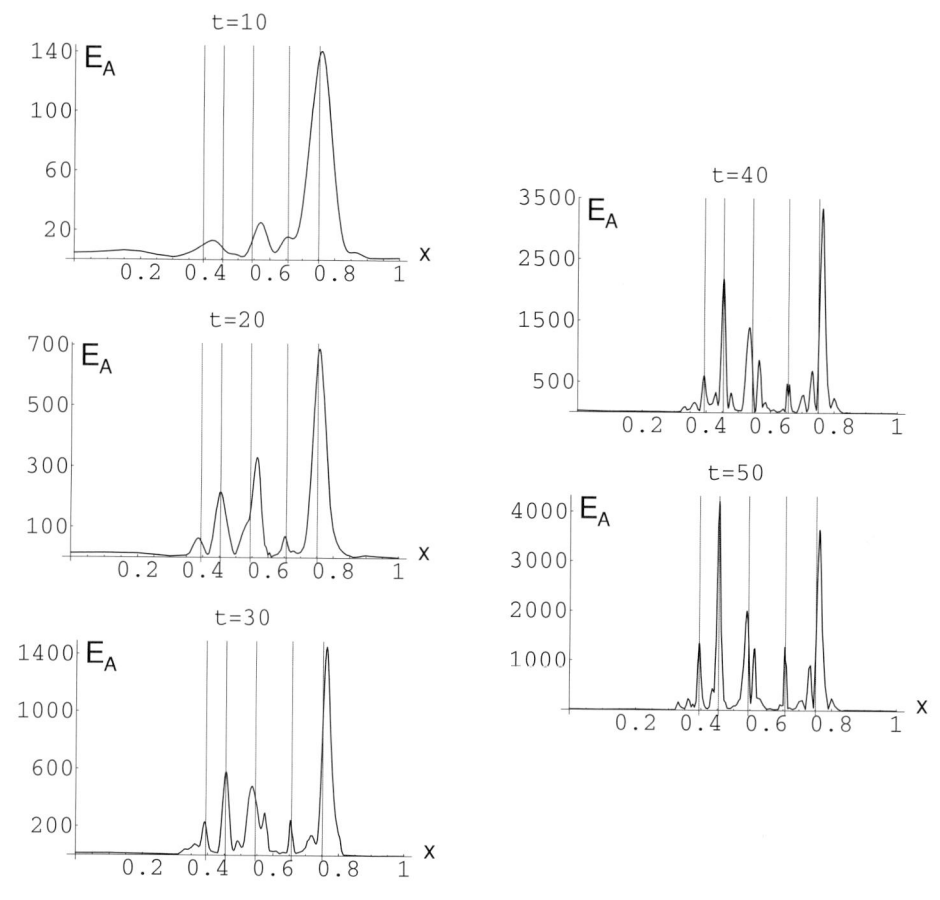

Fig. 11.14. Temporal evolution of the spatial distribution of the energy stored in Alfvén modes for a short loop with only five quasi-modes (from De Groof et al. [62]).

Such radial and azimuthal drivers have been considered in Refs. [61, 62], and reviewed in Ref. [94]. These linear MHD studies confirmed that for $k_y \neq 0$ the fast magneto-sonic waves and the Alfvén waves are coupled. The fast magneto-sonic body modes yield efficient coupling of the loops to the random (broadband) driver, and the coupling to the Alfvén waves yields efficient dissipation which is spread over the entire loop volume (Fig. 11.14).

(b) Absorption of acoustic waves by sunspots Since 1989, the resonant absorption mechanism has also been studied in the context of *p*-mode absorption by sunspots. Observations by Braun, Duvall and Labonte [42, 43] of high degree *p*-mode oscillations in regions around sunspots have revealed that sunspots act as strong absorbers of *p*-mode wave energy. Adopting a cylindrical coordinate system centred

on the sunspot, the amplitudes of the waves travelling inward and outward from the sunspot were determined and it was found that as much as 50% of the acoustic wave power can be lost. In a subsequent investigation, the same authors [44] explored the horizontal spatial distribution of high degree *p*-mode absorption in solar active regions. They found that the absorption reaches a maximum in the visible sunspot, but that it is not limited to the location of the visible spot and it is also associated with magnetic fields in the surrounding plage. Larger sunspots are observed to absorb even more *p*-mode energy. The discovery that sunspots are strong absorbers of acoustic wave energy opens up the new avenue of *sunspot seismology* [33], or *active region seismology* [228]: the effect of active regions on solar oscillations can be directly observed. The aim is to use observations of *p*-mode oscillations outside sunspots to derive information on the conditions inside the spots. This requires a basic theoretical model that describes the observed properties of *p*-mode oscillations, in particular the strong absorption.

As a model for a sunspot we adopt an ideal static cylindrically symmetric magnetic flux tube in which the equilibrium quantities are functions of the radial distance to the axis of symmetry only, similar to a model exploited by Lou [147]. At the boundary $r = a$ of the sunspot the total (plasma and magnetic) pressure has to be continuous. In the idealized equilibrium state considered the magnetic field in the sunspot gradually drops to zero with increasing radial distance r, and the non-uniform sunspot is surrounded by a non-magnetic and uniform plasma, so that for $r > a$ the equilibrium density, pressure and temperature are constant.

When the amplitude of the incoming wave is denoted by A^i and the amplitude of the outgoing wave by A^o the quantity we are interested in is the absorption coefficient

$$\alpha \equiv 1 - (A^o/A^i)^2. \tag{11.76}$$

Numerical simulations of the resonant absorption of incident sound waves for this simple sunspot model [95] reveal that the absorption coefficient α can indeed be of the order of 50% for typical sunspot parameters and for typical solar *p*-modes. This is illustrated in Figs. 11.15 and 11.16. In Fig. 11.15 the absorption coefficient is plotted versus the wave number k of the incident sound waves for two sunspot radii. In Fig. 11.16, α is plotted versus the sunspot radius for a fixed wave number. These results are for a spot with $B_z(0) = 0.2\,\text{T}\,(= 2000\,\text{gauss})$ and total photospheric pressure $p = 2.4 \times 10^2\,\text{N}\,\text{m}^{-2}\,(= 2.4 \times 10^4\,\text{Pa})$, as taken by Lou in a *viscous MHD model* [147] so that comparison is straightforward. In each figure, results are displayed for oscillations with $m = 1, 2, 3$, and 5 to see how the

11.3 Alternative excitation mechanisms

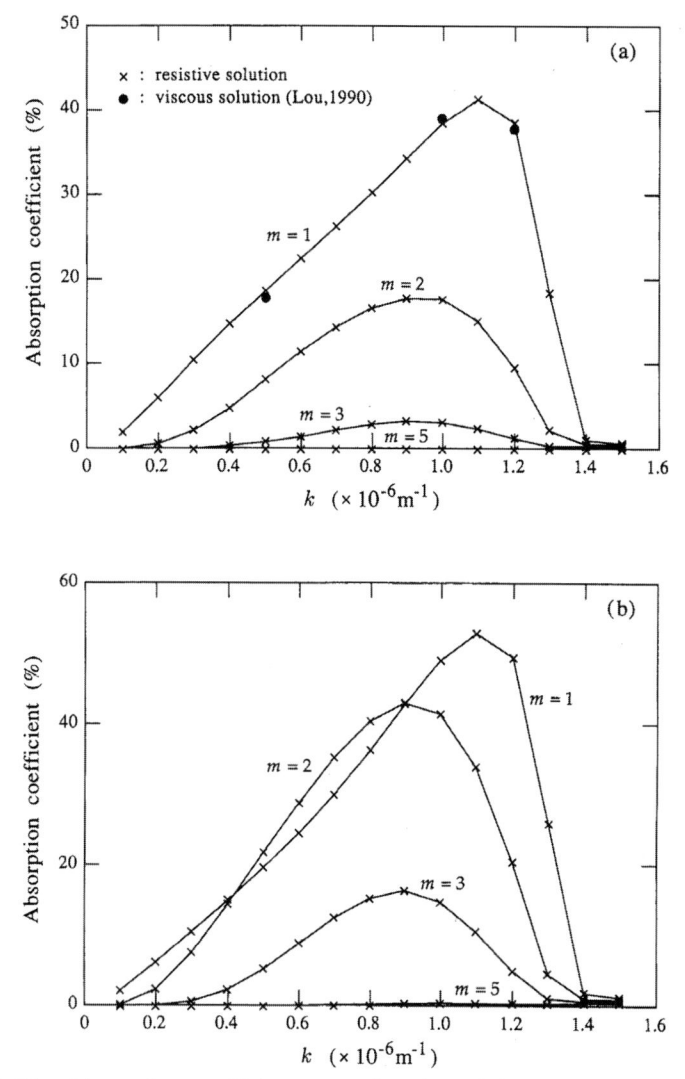

Fig. 11.15. Absorption coefficient α as a function of the wave number of acoustic oscillations of a sunspot with straight magnetic field and radius (a) $R = 4.2 \times 10^6$ m, and (b) $R = 6.3 \times 10^6$ m.

absorption coefficients vary as a function of azimuthal wave number. Since the magnetic field is straight, the absorption is independent of the sign of m so that it suffices to consider positive values only. Figures 11.15 and 11.16 apply to p-mode oscillations with a frequency $\omega = 0.02$ rad s^{-1}. For comparison, Lou's results for oscillations with $m = 1$ are also displayed.

From extensive parameter studies such as the ones discussed, the following conclusions can be drawn:

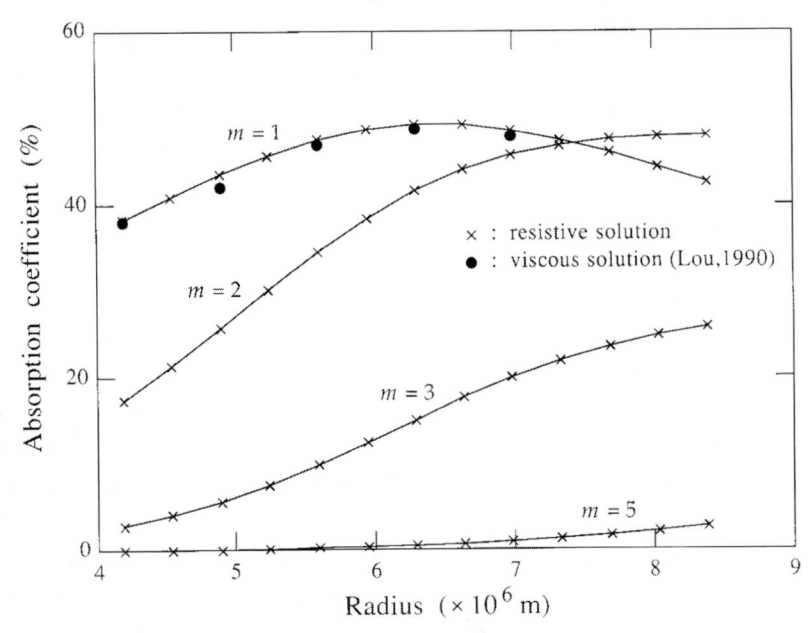

Fig. 11.16. Absorption coefficient α as a function of the radius of a sunspot with straight magnetic field, $B_z(0) = 0.2\,\text{T}$, $\omega = 0.02\,\text{rad s}^{-1}$, $k = 1 \times 10^{-6}\,\text{m}^{-1}$ and $m = 1, 2, 3, 5$. Crosses refer to resistive MHD, dots to viscous MHD [147].

- The results obtained in viscous MHD [147] are recovered in resistive MHD [95] showing that, in fact, the energy absorption rate is *independent of the actual dissipation mechanism*;
- Observed absorption coefficients of 50% and higher can easily be explained by resonant absorption;
- The variation of the absorption coefficient with the azimuthal wave number depends in a complicated way on the sunspot equilibrium and the characteristics of the incident acoustic mode (i.e. wavelength and angle of incidence);
- Resonant absorption of p-modes is more efficient in larger sunspots. (This does not conflict with the notion of 'scale independence' (see Chapter 4) since in these studies the mode number (wavelength) is fixed so that kR varies.) It is also more efficient in sunspots with twisted magnetic fields, in particular for higher azimuthal wave numbers.

The results discussed above are obtained for rather simple equilibrium models for sunspots. Clearly, for final conclusions numerical simulations are required based on more realistic sunspot models. These should take into account the flaring out of the field lines in the upper photospheric part (due to the decrease of the external pressure) and the stratification of the density with height (due to gravity). Also, there are other possible explanations for the p-mode absorption by sunspots which deserve further attention such as p-mode/s-mode conversion. In that mechanism, the incident p-modes couple to the slow magneto-sonic (s-)modes in the sunspot, in effect redirecting the energy flux downwards (or upwards) in the sunspot. Another explanation can be found in fibril models for sunspots. In such bundles of flux tubes multiple scattering resonances occur which can drastically damp the waves [123].

(c) Phase mixing of coronal loops and holes As already mentioned, phase mixing does not really need resonances to be effective. In 'open field' configurations, such as coronal holes, this mechanism also works. Phase mixing of coronal loops and holes was studied rigorously by many authors. It was realized that the flaring out of the magnetic field lines at the base of the coronal holes or long loops must affect the slope of the local Alfvén frequency substantially [186, 198, 64]. In the last reference, the authors considered a realistic two-dimensional configuration with diverging magnetic field and also took into account the plasma flow along the magnetic field lines. Their WKB solutions agree well with three-dimensional nonlinear MHD simulations. Strong damping is found in the layer where the velocity shear is concentrated. The total heat deposited in the coronal hole seems not to be affected by the vertical stratification, but the efficiency of the heating mechanism does depend on the geometry, the scale height, and the amplitude of the excited waves.

(d) Magnetospheric applications In 1861(!), Stewart observed ultra-low frequency (ULF) oscillations of magnetic fields on the surface of the Earth. These oscillations turned out to be caused by waves with periods ranging from seconds to minutes. They correspond to standing Alfvén waves with fixed ends in the ionosphere. In the 'box' model, the magnetic field lines are straightened and the natural frequencies are determined only by the length of the dipole magnetic flux tubes, the strength of the magnetic field, and the plasma density. Hence, for a given (known) magnetic geometry, the plasma density can be derived from the observed Alfvén wave frequencies [127]. This resonant field line model with the solar wind as sideways external driver is very successful in explaining many important properties of the ULF waves, e.g. the variation of the amplitude with latitude [127, 128].

11.4 Literature and exercises

Notes on literature

Resonant absorption in tokamaks:

– Tataronis & Grossmann [225] and Chen & Hasegawa [54] propose resonant absorption as a supplementary heating mechanism for tokamaks. Tataronis & Grossmann consider an incompressible plasma and show that MHD waves decay due to phase mixing. Chen & Hasegawa consider a compressible plasma slab with shear, using the surface wave for coupling since the fast magneto-sonic wave 'may produce undesirable effects'.

- Kappraff & Tataronis [120] analyse the time evolution of resonant absorption in linearized resistive MHD for a plane slab geometry. They confirm earlier ideal MHD results on time scales, efficiency and energetics of the heating process.
- Balet, Appert & Vaclavik [15] review the results of the TCA experiment in Lausanne where the theory of resonant absorption is tested. They also report the detection of a 'collective' mode as a 'natural oscillation mode of the system'.
- Poedts & Kerner [183] compute the resistive MHD spectrum and show that the ideal quasi-modes (important for excitation) correspond to resistive global eigenmodes. Poedts et al. [185] also analyse another application of resonant absorption, viz. damping of toroidicity-induced Alfvén eigenmodes ('gap' modes), further discussed in Volume 2.

Resonant absorption and phase mixing in coronal heating:

- Ionson [116] suggests resonant absorption as a heating mechanism for coronal loops, applying the model of Chen & Hasegawa [54] with parameter values for the solar corona.
- Heyvaerts & Priest [108] suggest a simple model for phase mixing as a mechanism to heat coronal loops as well as coronal holes.
- Kuperus, Ionson & Spicer [133] compare the different heating mechanisms proposed for the solar corona up to 1981, Goossens [93] reviews the resonant absorption models up to 1991, and Aschwanden [11] makes an evaluation for active regions based on observations by Yohkoh, SOHO and TRACE up to 2001.
- Grossmann & Smith [100] investigate the efficiency of resonant absorption for coronal heating with the power spectrum of driving photospheric oscillations as input.
- Hollweg & Yang [111], in 'Resonance absorption of compressible MHD waves at thin surfaces', calculate the damping of quasi-modes in finite loops.
- Most numerical results on resonant absorption are obtained with ideal MHD codes. Poedts, Goossens & Kerner [181, 182] solve the linearized resistive MHD equations with an accurate finite-element numerical code to obtain the resonant dissipation rate. An alternative, based on integration of the ideal MHD ODEs replacing the steep gradients in the dissipative layer by jump conditions, is developed by Sakurai et al. [200, 96].

Resonant absorption in sunspots:

- Thomas, Cram & Nye [228] propose the concept of sunspot seismology. They suggest that the response of the spots to forcing by the 5-minute oscillations from the surrounding photosphere 'may be used as a probe of sunspot structure below the surface'.
- Braun, Duvall & LaBonte [42, 43, 44] characterize the interaction of *p*-modes with sunspots by comparing the amplitudes of inward and outward propagating waves in an annular region around the sunspot or active region. They show that the loss, or absorption, of power of the incoming acoustic waves is robust enough to be measured.
- Hollweg [109] suggests that the observed *p*-mode absorption by sunspots is due to resonant absorption. His restrictions of plane slab and thin transition region are lifted by Lou [147], who solves the viscous MHD equations and obtains absorption

coefficients of 40–50%. These results are confirmed by Goossens & Poedts [95] in resistive MHD, demonstrating that the actual dissipation mechanism is of secondary importance.
– Bogdan & Braun [33] review the field of Active Region Seismology anno 1995.

Resonant absorption in magnetospheres:

– Kivelson & Southwood [128] discuss the role of global quasi-modes coupling to field line resonances in magnetospheric plasmas.
– Kivelson [126], in 'Pulsations and magnetohydrodynamic waves', explains the box model of the magnetosphere, and describes how wave disturbances at the magnetopause pump energy into the magnetospheric cavity and deposit it near magnetic shells where the conditions for transverse resonances are satisfied.

Exercises

[11.1] *Time average*
In Section 11.1.2 it was stated that the *time average* of two harmonic quantities, which are the real parts of $Ae^{i\omega t}$ and $Be^{i\omega t}$, is half the product of A and the complex conjugate of B, i.e. $\frac{1}{2}AB^*$. Check this statement.

[11.2] *Radial excitation*
In Section 11.3.1 we assumed that the perturbations were polarized in the tangential perpendicular (y) direction. Under which condition(s) can the Alfvén wave heating mechanism work with *radially* polarized foot point oscillations? Explain why.

[11.3] *Resonant absorption*
Write a Maple or Mathematica work sheet to reconstruct Fig. 11.5 and verify how the parameters a, c, $k_\parallel$, and $k_\perp$ affect the fractional absorption.

[11.4] *Time and length scales of resonant absorption*
Use the tables in Appendix B to get a rough estimate of the resonant heating time scales and the width of the resonant layers both in laboratory plasmas and in solar coronal loops (cf. Section 11.3.1).

[11.5] *Basic phase mixing result*
Consider Eq. (11.61) without the external driving term and assume that the gradients in the x-direction are much larger than the gradients in the z-direction ($\sim k_z$):

$$\frac{\partial^2 \xi_y}{\partial t^2} = v_A^2(x) \frac{\partial^2 \xi_y}{\partial z^2} + (\eta + \nu) \frac{\partial^2}{\partial x^2} \frac{\partial \xi_y}{\partial t}.$$

Heyvaerts and Priest [108] assume in their analysis a solution to this equation of the form

$$\xi_y \sim \xi_y(x, z) \exp i[\Omega t - k(x)z].$$

Derive the solution for $\xi_y(x, z)$ obtained by Heyvaerts and Priest, viz.

$$\xi_y(x, z) = \xi_y(x, 0) \exp\left[-\frac{(k(x)z)^3}{6R_T}\right], \qquad \text{where } R_T = \frac{\Omega}{\eta + \nu}\left[\frac{d}{dx}\log k(x)\right]^{-2},$$

which is valid under the condition of weak damping and strong phase mixing:

$$\frac{1}{k}\frac{\partial}{\partial z} \ll 1, \quad \text{and} \quad \frac{z}{k}\frac{\partial k}{\partial x} \gg 1,$$

and thus only for large z.

[11.6] *Phase mixing time and length scales*

Use the tables in Appendix B to get a rough estimate of the phase mixing time scales in solar coronal loops and the phase mixing height z_{mix} in coronal holes.

Appendix A

Vectors and coordinates

A.1 Vector identities

A list of the most frequently exploited identities:

$$\mathbf{a} \cdot (\mathbf{b} \times \mathbf{c}) = \mathbf{c} \cdot (\mathbf{a} \times \mathbf{b}) = \mathbf{b} \cdot (\mathbf{c} \times \mathbf{a}), \tag{A.1}$$

$$\mathbf{a} \times (\mathbf{b} \times \mathbf{c}) = \mathbf{a} \cdot \mathbf{c}\,\mathbf{b} - \mathbf{a} \cdot \mathbf{b}\,\mathbf{c}, \qquad (\mathbf{a} \times \mathbf{b}) \times \mathbf{c} = \mathbf{a} \cdot \mathbf{c}\,\mathbf{b} - \mathbf{b} \cdot \mathbf{c}\,\mathbf{a}, \tag{A.2}$$

$$\nabla \times \nabla \Phi = 0, \tag{A.3}$$

$$\nabla \cdot (\nabla \times \mathbf{a}) = 0, \tag{A.4}$$

$$\nabla \times (\nabla \times \mathbf{a}) = \nabla \nabla \cdot \mathbf{a} - \nabla^2 \mathbf{a}, \tag{A.5}$$

$$\nabla \cdot (\Phi \mathbf{a}) = \mathbf{a} \cdot \nabla \Phi + \Phi \nabla \cdot \mathbf{a}, \tag{A.6}$$

$$\nabla \times (\Phi \mathbf{a}) = \nabla \Phi \times \mathbf{a} + \Phi \nabla \times \mathbf{a}, \tag{A.7}$$

$$\mathbf{a} \times (\nabla \times \mathbf{b}) = (\nabla \mathbf{b}) \cdot \mathbf{a} - \mathbf{a} \cdot \nabla \mathbf{b}, \tag{A.8}$$

$$(\mathbf{a} \times \nabla) \times \mathbf{b} = (\nabla \mathbf{b}) \cdot \mathbf{a} - \mathbf{a} \nabla \cdot \mathbf{b}, \tag{A.9}$$

$$\nabla(\mathbf{a} \cdot \mathbf{b}) = (\nabla \mathbf{a}) \cdot \mathbf{b} + (\nabla \mathbf{b}) \cdot \mathbf{a}$$
$$= \mathbf{a} \cdot \nabla \mathbf{b} + \mathbf{b} \cdot \nabla \mathbf{a} + \mathbf{a} \times (\nabla \times \mathbf{b}) + \mathbf{b} \times (\nabla \times \mathbf{a}), \tag{A.10}$$

$$\nabla \cdot (\mathbf{a}\mathbf{b}) = \mathbf{a} \cdot \nabla \mathbf{b} + \mathbf{b} \nabla \cdot \mathbf{a}, \tag{A.11}$$

$$\nabla \cdot (\mathbf{a} \times \mathbf{b}) = \mathbf{b} \cdot \nabla \times \mathbf{a} - \mathbf{a} \cdot \nabla \times \mathbf{b}, \tag{A.12}$$

$$\nabla \times (\mathbf{a} \times \mathbf{b}) = \nabla \cdot (\mathbf{b}\mathbf{a} - \mathbf{a}\mathbf{b}) = \mathbf{a} \nabla \cdot \mathbf{b} + \mathbf{b} \cdot \nabla \mathbf{a} - \mathbf{b} \nabla \cdot \mathbf{a} - \mathbf{a} \cdot \nabla \mathbf{b}, \tag{A.13}$$

$$\iiint \nabla \cdot \mathbf{a}\, d\tau = \oiint \mathbf{a} \cdot \mathbf{n}\, d\sigma \quad \text{(Gauss)}, \tag{A.14}$$

$$\mathbf{a} \to \mathbf{a} \times \mathbf{c}(\text{onst}) \quad \Rightarrow \quad \iiint \nabla \times \mathbf{a}\, d\tau = \oiint \mathbf{n} \times \mathbf{a}\, d\sigma, \tag{A.15}$$

$$\mathbf{a} \to \Phi \mathbf{c}(\text{onst}) \quad \Rightarrow \quad \iiint \nabla \Phi\, d\tau = \oiint \Phi \mathbf{n}\, d\sigma, \tag{A.16}$$

$$\mathbf{a} \to \Phi \nabla \Psi - \Psi \nabla \Phi \Rightarrow$$

$$\iiint (\Phi \nabla^2 \Psi - \Psi \nabla^2 \Phi)\, d\tau = \oiint (\Phi \nabla \Psi - \Psi \nabla \Phi) \cdot \mathbf{n}\, d\sigma \quad \text{(Green)}, \tag{A.17}$$

$$\iint (\nabla \times \mathbf{a}) \cdot \mathbf{n}\, d\sigma = \oint \mathbf{a} \cdot d\mathbf{l} \quad \text{(Stokes)}, \tag{A.18}$$

$$\mathbf{a} \to \mathbf{a} \times \mathbf{c}(\text{onst}) \quad \Rightarrow \quad \iint (\mathbf{n} \times \nabla) \times \mathbf{a}\, d\sigma = \oint d\mathbf{l} \times \mathbf{a}, \tag{A.19}$$

$$\mathbf{a} \to \Phi \mathbf{c}(\text{onst}) \quad \Rightarrow \quad \iint \mathbf{n} \times \nabla \Phi\, d\sigma = \oint \Phi\, d\mathbf{l}. \tag{A.20}$$

A.2 Vector expressions in orthogonal coordinates

Considering the position vector as a function of orthogonal coordinates x_i,

$$\mathbf{r} = \mathbf{r}(x_1, x_2, x_3) \quad \Longleftrightarrow \quad \begin{cases} x = x(x_1, x_2, x_3) \\ y = y(x_1, x_2, x_3) \\ z = z(x_1, x_2, x_3), \end{cases} \tag{A.21}$$

the following geometric quantities are generated:

$$h_i \equiv \left|\partial \mathbf{r}/\partial x_i\right| \qquad \text{(scale factors)}, \tag{A.22}$$

$$\mathbf{e}_i \equiv (1/h_i)\, \partial \mathbf{r}/\partial x_i, \quad \mathbf{e}_i \cdot \mathbf{e}_j = \delta_{ij} \quad \text{(dimensionless unit vectors)}, \tag{A.23}$$

$$d\ell = \sqrt{\sum_i (h_i dx_i)^2} \qquad \text{(line element)}, \tag{A.24}$$

$$d\tau = h_1 h_2 h_3\, dx_1 dx_2 dx_3 \qquad \text{(volume element)}. \tag{A.25}$$

Vector representation:

$$\mathbf{V} = \sum_i \hat{V}_i\, \mathbf{e}_i \quad (\hat{V}_i - \text{physical components, same dimension as } \mathbf{V}). \tag{A.26}$$

Products:

$$\mathbf{A} \cdot \mathbf{B} = \sum_i \hat{A}_i \hat{B}_i \quad \text{(inner product)}, \quad (A.27)$$

$$\mathbf{A} \times \mathbf{B} = \sum_i \sum_j \sum_k \epsilon_{ijk} \hat{A}_j \hat{B}_j \mathbf{e}_i \quad \text{(vector product)}, \quad (A.28)$$

$$\epsilon_{ijk} \equiv \begin{cases} 1 & \text{if } ijk \text{ even permutation of 123} \\ -1 & \text{if } ijk \text{ odd permutation of 123} \\ 0 & \text{otherwise} \end{cases} \quad (A.29)$$

$$\text{(permutation symbol)},$$

$$\epsilon_{ijk} \epsilon_{ilm} = \delta_{jl} \delta_{km} - \delta_{jm} \delta_{kl}. \quad (A.30)$$

Differential operators:

$$\nabla \psi = \sum \frac{1}{h_i} \frac{\partial \psi}{\partial x_i} \mathbf{e}_i, \quad (A.31)$$

$$\nabla^2 \psi = \frac{1}{h_1 h_2 h_3} \left[\frac{\partial}{\partial x_1} \left(\frac{h_2 h_3}{h_1} \frac{\partial \psi}{\partial x_1} \right) + \frac{\partial}{\partial x_2} \left(\frac{h_1 h_3}{h_2} \frac{\partial \psi}{\partial x_2} \right) + \frac{\partial}{\partial x_3} \left(\frac{h_1 h_2}{h_3} \frac{\partial \psi}{\partial x_3} \right) \right], \quad (A.32)$$

$$\nabla \cdot \mathbf{A} = \frac{1}{h_1 h_2 h_3} \left[\frac{\partial}{\partial x_1} (h_2 h_3 \hat{A}_1) + \frac{\partial}{\partial x_2} (h_1 h_3 \hat{A}_2) + \frac{\partial}{\partial x_3} (h_1 h_2 \hat{A}_3) \right], \quad (A.33)$$

$$\nabla \times \mathbf{A} = \frac{1}{h_2 h_3} \left[\frac{\partial}{\partial x_2} (h_3 \hat{A}_3) - \frac{\partial}{\partial x_3} (h_2 \hat{A}_2) \right] \mathbf{e}_1$$

$$+ \frac{1}{h_1 h_3} \left[\frac{\partial}{\partial x_3} (h_1 \hat{A}_1) - \frac{\partial}{\partial x_1} (h_3 \hat{A}_3) \right] \mathbf{e}_2$$

$$+ \frac{1}{h_1 h_2} \left[\frac{\partial}{\partial x_1} (h_2 \hat{A}_2) - \frac{\partial}{\partial x_2} (h_1 \hat{A}_1) \right] \mathbf{e}_3. \quad (A.34)$$

Derivatives of the unit vectors:

$$\frac{\partial \mathbf{e}_1}{\partial x_1} = -\frac{1}{h_2} \frac{\partial h_1}{\partial x_2} \mathbf{e}_2 - \frac{1}{h_3} \frac{\partial h_1}{\partial x_3} \mathbf{e}_3, \quad \frac{\partial \mathbf{e}_2}{\partial x_1} = \frac{1}{h_2} \frac{\partial h_1}{\partial x_2} \mathbf{e}_1, \quad \frac{\partial \mathbf{e}_3}{\partial x_1} = \frac{1}{h_3} \frac{\partial h_1}{\partial x_3} \mathbf{e}_1,$$

$$\frac{\partial \mathbf{e}_1}{\partial x_2} = \frac{1}{h_1} \frac{\partial h_2}{\partial x_1} \mathbf{e}_2, \quad \frac{\partial \mathbf{e}_2}{\partial x_2} = -\frac{1}{h_1} \frac{\partial h_2}{\partial x_1} \mathbf{e}_1 - \frac{1}{h_3} \frac{\partial h_2}{\partial x_3} \mathbf{e}_3, \quad \frac{\partial \mathbf{e}_3}{\partial x_2} = \frac{1}{h_3} \frac{\partial h_2}{\partial x_3} \mathbf{e}_2,$$

$$\frac{\partial \mathbf{e}_1}{\partial x_3} = \frac{1}{h_1} \frac{\partial h_3}{\partial x_1} \mathbf{e}_3, \quad \frac{\partial \mathbf{e}_2}{\partial x_3} = \frac{1}{h_2} \frac{\partial h_3}{\partial x_2} \mathbf{e}_3, \quad \frac{\partial \mathbf{e}_3}{\partial x_3} = -\frac{1}{h_1} \frac{\partial h_3}{\partial x_1} \mathbf{e}_1 - \frac{1}{h_2} \frac{\partial h_3}{\partial x_2} \mathbf{e}_2.$$

$$(A.35)$$

Hence,

$$
\begin{aligned}
\mathbf{A} \cdot \nabla \mathbf{B} = &\left[\frac{\hat{A}_1}{h_1}\left(\frac{\partial \hat{B}_1}{\partial x_1} + \frac{\partial h_1}{\partial x_2}\frac{\hat{B}_2}{h_2} + \frac{\partial h_1}{\partial x_3}\frac{\hat{B}_3}{h_3} \right) + \frac{\hat{A}_2}{h_2}\left(\frac{\partial \hat{B}_1}{\partial x_2} - \frac{\partial h_2}{\partial x_1}\frac{\hat{B}_2}{h_1} \right) \right. \\
&\left. + \frac{\hat{A}_3}{h_3}\left(\frac{\partial \hat{B}_1}{\partial x_3} - \frac{\partial h_3}{\partial x_1}\frac{\hat{B}_3}{h_1} \right) \right] \mathbf{e}_1 \\
&+ \left[\frac{\hat{A}_1}{h_1}\left(\frac{\partial \hat{B}_2}{\partial x_1} - \frac{\partial h_1}{\partial x_2}\frac{\hat{B}_1}{h_2} \right) + \frac{\hat{A}_2}{h_2}\left(\frac{\partial \hat{B}_2}{\partial x_2} + \frac{\partial h_2}{\partial x_1}\frac{\hat{B}_1}{h_1} + \frac{\partial h_2}{\partial x_3}\frac{\hat{B}_3}{h_3} \right) \right. \\
&\left. + \frac{\hat{A}_3}{h_3}\left(\frac{\partial \hat{B}_2}{\partial x_3} - \frac{\partial h_3}{\partial x_2}\frac{\hat{B}_3}{h_2} \right) \right] \mathbf{e}_2 \\
&+ \left[\frac{\hat{A}_1}{h_1}\left(\frac{\partial \hat{B}_3}{\partial x_1} - \frac{\partial h_1}{\partial x_3}\frac{\hat{B}_1}{h_3} \right) + \frac{\hat{A}_2}{h_2}\left(\frac{\partial \hat{B}_3}{\partial x_2} - \frac{\partial h_2}{\partial x_3}\frac{\hat{B}_2}{h_3} \right) \right. \\
&\left. + \frac{\hat{A}_3}{h_3}\left(\frac{\partial \hat{B}_3}{\partial x_3} + \frac{\partial h_3}{\partial x_1}\frac{\hat{B}_1}{h_1} + \frac{\partial h_3}{\partial x_2}\frac{\hat{B}_2}{h_2} \right) \right] \mathbf{e}_3 .
\end{aligned}
\tag{A.36}
$$

▷ **Notation:** The awkward hat, used until here to avoid conflict with the covariant components of non-orthogonal coordinate systems, is dropped in the explicit expressions for the different coordinate systems below, by writing A_{x_i} instead of $\hat{A}_i$. ◁

A.2.1 Cartesian coordinates (x, y, z)

$$x \equiv x_1, \quad y \equiv x_2, \quad z \equiv x_3 \quad \Longrightarrow \quad h_1 = h_2 = h_3 = 1. \tag{A.37}$$

$$\nabla \psi = \frac{\partial \psi}{\partial x}\mathbf{e}_x + \frac{\partial \psi}{\partial y}\mathbf{e}_y + \frac{\partial \psi}{\partial z}\mathbf{e}_z, \tag{A.38}$$

$$\nabla^2 \psi = \frac{\partial^2 \psi}{\partial x^2} + \frac{\partial^2 \psi}{\partial y^2} + \frac{\partial^2 \psi}{\partial z^2}, \tag{A.39}$$

$$\nabla \cdot \mathbf{A} = \frac{\partial A_x}{\partial x} + \frac{\partial A_y}{\partial y} + \frac{\partial A_z}{\partial z}, \tag{A.40}$$

$$\nabla \times \mathbf{A} = \left(\frac{\partial A_z}{\partial y} - \frac{\partial A_y}{\partial z}\right)\mathbf{e}_x + \left(\frac{\partial A_x}{\partial z} - \frac{\partial A_z}{\partial x}\right)\mathbf{e}_y + \left(\frac{\partial A_y}{\partial x} - \frac{\partial A_x}{\partial y}\right)\mathbf{e}_z. \tag{A.41}$$

▷ **Note:** The vector identities of Section A.1, in particular the complicated ones involving vector products and curls, are most easily derived in Cartesian coordinates, exploiting Eqs. (A.31)–(A.36) with $h_i = 1$ (see, e.g., Goldston and Rutherford [92], p. 461). ◁

A.2.2 Cylinder coordinates (r, θ, z)
(see Fig. A.1)

Scale factors and unit vector derivatives:

$$\begin{cases} x = r\cos\theta \\ y = r\sin\theta \\ z = z \end{cases} \implies h_1 = 1, \quad h_2 = r, \quad h_3 = 1. \tag{A.42}$$

$$\frac{\partial \mathbf{e}_r}{\partial \theta} = \mathbf{e}_\theta, \quad \frac{\partial \mathbf{e}_\theta}{\partial \theta} = -\mathbf{e}_r \quad (\text{only} \neq 0 \text{ derivatives}). \tag{A.43}$$

Differential operators:

$$\nabla \psi = \frac{\partial \psi}{\partial r} \mathbf{e}_r + \frac{1}{r} \frac{\partial \psi}{\partial \theta} \mathbf{e}_\theta + \frac{\partial \psi}{\partial z} \mathbf{e}_z, \tag{A.44}$$

$$\nabla^2 \psi = \frac{1}{r} \frac{\partial}{\partial r}\left(r \frac{\partial \psi}{\partial r}\right) + \frac{1}{r^2} \frac{\partial^2 \psi}{\partial \theta^2} + \frac{\partial^2 \psi}{\partial z^2}, \tag{A.45}$$

$$\nabla \cdot \mathbf{A} = \frac{1}{r} \frac{\partial (rA_r)}{\partial r} + \frac{1}{r} \frac{\partial A_\theta}{\partial \theta} + \frac{\partial A_z}{\partial z}, \tag{A.46}$$

$$\nabla \times \mathbf{A} = \left(\frac{1}{r} \frac{\partial A_z}{\partial \theta} - \frac{\partial A_\theta}{\partial z}\right) \mathbf{e}_r \\
+ \left(\frac{\partial A_r}{\partial z} - \frac{\partial A_z}{\partial r}\right) \mathbf{e}_\theta + \left(\frac{1}{r} \frac{\partial (rA_\theta)}{\partial r} - \frac{1}{r} \frac{\partial A_r}{\partial \theta}\right) \mathbf{e}_z, \tag{A.47}$$

$$\nabla^2 \mathbf{A} = \left(\nabla^2 A_r - \frac{1}{r^2} A_r - \frac{2}{r^2} \frac{\partial A_\theta}{\partial \theta}\right) \mathbf{e}_r \\
+ \left(\nabla^2 A_\theta - \frac{1}{r^2} A_\theta + \frac{2}{r^2} \frac{\partial A_r}{\partial \theta}\right) \mathbf{e}_\theta + \nabla^2 A_z \mathbf{e}_z, \tag{A.48}$$

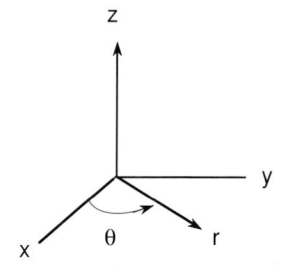

Fig. A.1.

$$\nabla \times \nabla \times \mathbf{A} = \left[-\frac{1}{r^2} \frac{\partial^2 A_r}{\partial \theta^2} - \frac{\partial^2 A_r}{\partial z^2} + \frac{1}{r^2} \frac{\partial^2 (rA_\theta)}{\partial \theta \, \partial r} + \frac{\partial^2 A_z}{\partial z \, \partial r} \right] \mathbf{e}_r$$
$$+ \left[\frac{\partial}{\partial r} \left(\frac{1}{r} \frac{\partial A_r}{\partial \theta} \right) - \frac{\partial}{\partial r} \left(\frac{1}{r} \frac{\partial (rA_\theta)}{\partial r} \right) - \frac{\partial^2 A_\theta}{\partial z^2} + \frac{1}{r} \frac{\partial^2 A_z}{\partial z \, \partial \theta} \right] \mathbf{e}_\theta$$
$$+ \left[\frac{1}{r} \frac{\partial}{\partial r} \left(r \frac{\partial A_r}{\partial z} \right) + \frac{1}{r} \frac{\partial^2 A_\theta}{\partial \theta \, \partial z} - \frac{1}{r} \frac{\partial}{\partial r} \left(r \frac{\partial A_z}{\partial r} \right) - \frac{1}{r^2} \frac{\partial^2 A_z}{\partial \theta^2} \right] \mathbf{e}_z,$$
$$\tag{A.49}$$

$$\mathbf{A} \cdot \nabla \mathbf{B} = \left[A_r \frac{\partial B_r}{\partial r} + \frac{A_\theta}{r} \left(\frac{\partial B_r}{\partial \theta} - B_\theta \right) + A_z \frac{\partial B_r}{\partial z} \right] \mathbf{e}_r$$
$$+ \left[A_r \frac{\partial B_\theta}{\partial r} + \frac{A_\theta}{r} \left(B_r + \frac{\partial B_\theta}{\partial \theta} \right) + A_z \frac{\partial B_\theta}{\partial z} \right] \mathbf{e}_\theta$$
$$+ \left[A_r \frac{\partial B_z}{\partial r} + \frac{A_\theta}{r} \frac{\partial B_z}{\partial \theta} + A_z \frac{\partial B_z}{\partial z} \right] \mathbf{e}_z. \tag{A.50}$$

A.2.3 Spherical coordinates (r, θ, ϕ)
(see Fig. A.2)

Scale factors and unit vector derivatives:

$$\begin{cases} x = R \cos \phi, & R = r \sin \theta \\ y = R \sin \phi \\ z = r \cos \theta \end{cases}$$
$$\implies h_1 = 1, \quad h_2 = r, \quad h_3 = r \sin \theta. \tag{A.51}$$

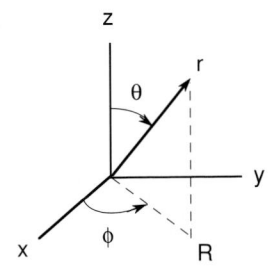

Fig. A.2.

$$\frac{\partial \mathbf{e}_r}{\partial \theta} = \mathbf{e}_\theta, \qquad \frac{\partial \mathbf{e}_\theta}{\partial \theta} = -\mathbf{e}_r,$$

$$\frac{\partial \mathbf{e}_r}{\partial \phi} = \sin\theta\, \mathbf{e}_\phi, \qquad \frac{\partial \mathbf{e}_\theta}{\partial \phi} = \cos\theta\, \mathbf{e}_\phi, \qquad \frac{\partial \mathbf{e}_\phi}{\partial \phi} = -\sin\theta\, \mathbf{e}_r - \cos\theta\, \mathbf{e}_\theta. \quad (A.52)$$

Differential operators:

$$\nabla \psi = \frac{\partial \psi}{\partial r}\mathbf{e}_r + \frac{1}{r}\frac{\partial \psi}{\partial \theta}\mathbf{e}_\theta + \frac{1}{r\sin\theta}\frac{\partial \psi}{\partial \phi}\mathbf{e}_\phi, \tag{A.53}$$

$$\nabla^2 \psi = \frac{1}{r^2}\frac{\partial}{\partial r}\left(r^2 \frac{\partial \psi}{\partial r}\right) + \frac{1}{r^2 \sin\theta}\frac{\partial}{\partial \theta}\left(\sin\theta \frac{\partial \psi}{\partial \theta}\right) + \frac{1}{r^2 \sin^2\theta}\frac{\partial^2 \psi}{\partial \phi^2}, \tag{A.54}$$

$$\nabla \cdot \mathbf{A} = \frac{1}{r^2}\frac{\partial}{\partial r}(r^2 A_r) + \frac{1}{r\sin\theta}\frac{\partial}{\partial \theta}(\sin\theta\, A_\theta) + \frac{1}{r\sin\theta}\frac{\partial A_\phi}{\partial \phi}, \tag{A.55}$$

$$\nabla \times \mathbf{A} = \frac{1}{r\sin\theta}\left[\frac{\partial}{\partial \theta}(\sin\theta\, A_\phi) - \frac{\partial A_\theta}{\partial \phi}\right]\mathbf{e}_r$$

$$+ \frac{1}{r}\left[\frac{1}{\sin\theta}\frac{\partial A_r}{\partial \phi} - \frac{\partial}{\partial r}(r A_\phi)\right]\mathbf{e}_\theta + \frac{1}{r}\left[\frac{\partial}{\partial r}(r A_\theta) - \frac{\partial A_r}{\partial \theta}\right]\mathbf{e}_\phi,$$

$$\tag{A.56}$$

$$\nabla^2 \mathbf{A} = \left[(\nabla^2 A_r) - \frac{2}{r^2}A_r - \frac{2}{r^2 \sin\theta}\frac{\partial}{\partial \theta}(\sin\theta\, A_\theta) - \frac{2}{r^2 \sin\theta}\frac{\partial A_\phi}{\partial \phi}\right]\mathbf{e}_r$$

$$+ \left[(\nabla^2 A_\theta) + \frac{2}{r^2}\frac{\partial A_r}{\partial \theta} - \frac{1}{r^2 \sin^2\theta}A_\theta - \frac{2\cos\theta}{r^2 \sin^2\theta}\frac{\partial A_\phi}{\partial \phi}\right]\mathbf{e}_\theta$$

$$+ \left[(\nabla^2 A_\phi) + \frac{2}{r^2 \sin\theta}\frac{\partial A_r}{\partial \phi} + \frac{2\cos\theta}{r^2 \sin^2\theta}\frac{\partial A_\theta}{\partial \phi} - \frac{1}{r^2 \sin^2\theta}A_\phi\right]\mathbf{e}_\phi,$$

$$\tag{A.57}$$

$$\nabla \times \nabla \times \mathbf{A} = \frac{1}{r^2 \sin\theta}\left[-\frac{\partial}{\partial \theta}\left(\sin\theta \frac{\partial A_r}{\partial \theta}\right) - \frac{1}{\sin\theta}\frac{\partial^2 A_r}{\partial \phi^2}\right.$$

$$\left. + \frac{\partial}{\partial \theta}\left(\sin\theta \frac{\partial(r A_\theta)}{\partial r}\right) + \frac{\partial^2(r A_\phi)}{\partial \phi\, \partial r}\right]\mathbf{e}_r$$

$$+ \frac{1}{r^2}\left[r\frac{\partial^2 A_r}{\partial r\, \partial \theta} - \frac{\partial}{\partial r}\left(r^2 \frac{\partial A_\theta}{\partial r}\right) - \frac{1}{\sin^2\theta}\frac{\partial^2 A_\theta}{\partial \phi^2}\right.$$

$$\left. + \frac{1}{\sin^2\theta}\frac{\partial^2(\sin\theta\, A_\phi)}{\partial \phi\, \partial \theta}\right]\mathbf{e}_\theta$$

$$+ \frac{1}{r^2} \left[\frac{r}{\sin\theta} \frac{\partial^2 A_r}{\partial r \partial \phi} + \frac{\partial}{\partial \theta}\left(\frac{1}{\sin\theta}\frac{\partial A_\theta}{\partial \phi}\right) \right.$$
$$\left. - \frac{\partial}{\partial r}\left(r^2 \frac{\partial A_\phi}{\partial r}\right) - \frac{\partial}{\partial \theta}\left(\frac{1}{\sin\theta}\frac{\partial(\sin\theta A_\phi)}{\partial \theta}\right) \right] \mathbf{e}_\phi,$$
(A.58)

$$\mathbf{A}\cdot\nabla\mathbf{B} = \left[A_r \frac{\partial B_r}{\partial r} + \frac{A_\theta}{r}\left(\frac{\partial B_r}{\partial \theta} - B_\theta\right) + \frac{A_\phi}{r}\left(\frac{1}{\sin\theta}\frac{\partial B_r}{\partial \phi} - B_\phi\right) \right]\mathbf{e}_r$$
$$+ \left[A_r \frac{\partial B_\theta}{\partial r} + \frac{A_\theta}{r}\left(B_r + \frac{\partial B_\theta}{\partial \theta}\right) + \frac{A_\phi}{r}\left(\frac{1}{\sin\theta}\frac{\partial B_\theta}{\partial \phi} - \cot\theta\, B_\phi\right) \right]\mathbf{e}_\theta$$
$$+ \left[A_r \frac{\partial B_\phi}{\partial r} + \frac{A_\theta}{r}\frac{\partial B_\phi}{\partial \theta} + \frac{A_\phi}{r}\left(B_r + \cot\theta\, B_\theta + \frac{1}{\sin\theta}\frac{\partial B_\phi}{\partial \phi}\right) \right]\mathbf{e}_\phi.$$
(A.59)

Appendix B

Tables of physical quantities

Table B.0.

Maxwell's equations

$$\frac{\partial \mathbf{B}}{\partial t} = -\nabla \times \mathbf{E} \qquad \text{Faraday}$$

$$\mathbf{j} = \frac{1}{\mu_0} \nabla \times \mathbf{B} - \epsilon_0 \frac{\partial \mathbf{E}}{\partial t} \qquad \text{'Ampère'}$$

$$\nabla \cdot \mathbf{B} = 0 \qquad \text{no magnetic monopoles}$$

$$\nabla \cdot \mathbf{E} = \frac{\tau}{\epsilon_0} \qquad \text{Poisson}$$

One cannot escape the feeling that these equations have an existence and intelligence of their own; that they are wiser than we are, wiser even than their discoverers; that we get more out of them than was originally put into them.

(Heinrich Hertz)

Table B.1.

Physical constants

Physical quantity	Symbol	Value	Units
charge of the electron	e	1.602×10^{-19}	C
mass of the electron	m_e	9.109×10^{-31}	kg
mass of the proton	m_p	1.673×10^{-27}	kg
permittivity of the vacuum	ϵ_0	8.854×10^{-12}	$\mathrm{F\,m^{-1}}$
permeability of the vacuum	μ_0	1.257×10^{-6} $(= 4\pi \times 10^{-7})$	$\mathrm{H\,m^{-1}}$
velocity of light	c	2.998×10^{8} $(= (\epsilon_0 \mu_0)^{-1/2})$	$\mathrm{m\,s^{-1}}$
Planck's constant	h	6.626×10^{-34}	J s
Boltzmann's constant	k	1.381×10^{-23}	$\mathrm{J\,K^{-1}}$
gravitational constant	G	6.671×10^{-11}	$\mathrm{m^3\,kg^{-1}\,s^{-2}}$

Table B.2.

Plasma parameters

Physical quantity	Definition	Expression*	Units
plasma frequency	$\omega_{p,e} \equiv \sqrt{\dfrac{ne^2}{\epsilon_0 m_e}}$	$= 56.6\sqrt{n}$	rad s^{-1}
Debye length	$\lambda_D \equiv \sqrt{\dfrac{\epsilon_0 kT}{ne^2}}$	$= 69.0\sqrt{\dfrac{T}{n}}$	m
electron cyclotron frequency	$\Omega_e \equiv \dfrac{eB}{m_e}$	$= 1.76 \times 10^{11}\, B$	rad s^{-1}
electron cyclotron radius	$R_e \equiv \dfrac{v_{\perp,e}}{\Omega_e}$	$= 2.21 \times 10^{-8}\, \dfrac{\sqrt{T}}{B}$	m
ion cyclotron frequency	$\Omega_i \equiv \dfrac{ZeB}{m_i}$	$= 9.58 \times 10^{7}\, \dfrac{Z}{A}\, B$	rad s^{-1}
ion cyclotron radius	$R_i \equiv \dfrac{v_{\perp,i}}{\Omega_i}$	$= 9.47 \times 10^{-7}\, \dfrac{\sqrt{A}}{Z}\, \dfrac{\sqrt{T}}{B}$	m
electron thermal speed	$v_{\text{th},e} \equiv \sqrt{\dfrac{2kT}{m_e}}$	$= 5.5 \times 10^{3}\, \sqrt{T}$	m s^{-1}
ion thermal speed	$v_{\text{th},i} \equiv \sqrt{\dfrac{2kT}{m_i}}$	$= 1.3 \times 10^{2}\, \dfrac{1}{\sqrt{A}}\, \sqrt{T}$	m s^{-1}
sound speed	$v_s \equiv \sqrt{\dfrac{\gamma p}{\rho}}$	$= 1.17 \times 10^{2}\, \sqrt{\dfrac{1+Z}{A}}\, \sqrt{T}$	m s^{-1}
Alfvén speed	$v_A \equiv \dfrac{B_0}{\sqrt{\mu_0 \rho}}$	$= 2.18 \times 10^{16}\, \sqrt{\dfrac{Z}{A}}\, \dfrac{B}{\sqrt{n}}$	m s^{-1}

*in $n\ (\equiv n_e \approx Zn_i)$, $T\ (\equiv T_e \approx T_i)$, B, with dimensions $[n] = $ m^{-3}, $[T] = $ K, $[B] = $ T;
A is the ion mass number (multiples of m_p), Z is the ion charge number (multiples of e).
$1\,\text{T} = 10^4\,\text{G}$,
$1\,\text{eV} = 1.60 \times 10^{-19}\,\text{J} \iff 1.16 \times 10^4\,\text{K}$.

Table B.3.

Orders of magnitude: kinetic theory

	Physical quantity	Symbol/definition	Tokamak	Coronal loop
⇒	particle density	n	10^{20} m^{-3}	10^{16} m^{-3}
⇒	magnetic field	B	3 T (30 kG)	0.03 T (300 G)
⇒	temperature	T	10^8 K ($\approx 10\,\text{keV}$)	10^6 K ($\approx 100\,\text{eV}$)
	electron th. speed	$v_{\text{th},e} \equiv \sqrt{2kT/m_e}$	5.9×10^7 m s^{-1}	5900 km s^{-1}
	ion thermal speed*	$v_{\text{th},i} \equiv \sqrt{2kT/m_i}$	1.4×10^6 m s^{-1}	140 km s^{-1}
	electron gyro freq.	$\Omega_e \equiv eB/m_e$	5.3×10^{11} rad s^{-1}	5.3×10^9 rad s^{-1}
		$\Omega_e/(2\pi)$	84 GHz	0.84 GHz
	electron gyro radius	$R_e \equiv v_{\perp,e}/\Omega_e$	0.1 mm	1 mm
	ion gyro freq.*	$\Omega_i \equiv ZeB/m_i$	2.9×10^8 rad s^{-1}	2.9×10^6 rad s^{-1}
		$\Omega_i/(2\pi)$	46 MHz	0.46 MHz
	ion gyro radius*	$R_i \equiv v_{\perp,i}/\Omega_i$	4.9 mm	4.9 cm
	plasma frequency	$\omega_{p,e} \equiv \sqrt{ne^2/(\epsilon_0 m_e)}$	5.7×10^{11} rad s^{-1}	5.7×10^9 rad s^{-1}
		$\omega_{p,e}/(2\pi)$	91 GHz	0.91 GHz
	electron skin depth	$\delta_e \equiv c/\omega_{p,e}$	0.53 mm	5.3 cm
	Debye length	$\lambda_D \equiv \sqrt{\epsilon_0 kT/(ne^2)}$	0.07 mm	0.7 mm

⇒ independent basic variables

*numbers for H plasma ($A = Z = 1$)

Table B.4.

Orders of magnitude: kinetic theory (cont'd)

		Solar wind	Magnetosphere	Your favourite plasma
⇒	n	$10^7 \, \text{m}^{-3}$	$10^{10} \, \text{m}^{-3}$	....
⇒	B	$6 \times 10^{-9} \, \text{T}$ (60 µG)	$3 \times 10^{-5} \, \text{T}$ (0.3 G)	....
⇒	T	$10^5 \, \text{K}$ ($\approx 10 \, \text{eV}$)	$10^4 \, \text{K}$ ($\approx 1 \, \text{eV}$)	....
	$v_{\text{th},e}$	$1700 \, \text{km s}^{-1}$	$590 \, \text{km s}^{-1}$	....
	$v_{\text{th},i}$	$41 \, \text{km s}^{-1}$	$14 \, \text{km s}^{-1}$	....
	Ω_e	$1.1 \times 10^3 \, \text{rad s}^{-1}$	$5.3 \times 10^6 \, \text{rad s}^{-1}$	....
	$\Omega_e/(2\pi)$	$0.17 \, \text{kHz}$	$0.84 \, \text{MHz}$	....
	R_e	$1.5 \, \text{km}$	$10 \, \text{cm}$	....
	Ω_i	$0.58 \, \text{rad s}^{-1}$	$2.9 \times 10^3 \, \text{rad s}^{-1}$	....
	$\Omega_i/(2\pi)$	$0.1 \, \text{Hz}$	$0.46 \, \text{kHz}$	....
	R_i	$71 \, \text{km}$	$4.9 \, \text{m}$	....
	$\omega_{p,e}$	$1.8 \times 10^5 \, \text{rad s}^{-1}$	$5.7 \times 10^6 \, \text{rad s}^{-1}$	....
	$\omega_{p,e}/(2\pi)$	$29 \, \text{kHz}$	$0.91 \, \text{MHz}$	....
	δ_e	$1.7 \, \text{km}$	$53 \, \text{m}$	....
	λ_D	$7 \, \text{m}$	$7 \, \text{cm}$	....

Table B.5.

Orders of magnitude: MHD

	Physical quantity	Symbol/definition	Tokamak	Coronal loop
⇒	width	a	1 m	10 000 km
⇒	length	L	20 m ($= 2\pi R$)	100 000 km
⇒	particle density	n	10^{20} m^{-3}	10^{16} m^{-3}
⇒	magnetic field	B	3 T (30 kG)	0.03 T (300 G)
⇒	temperature	T	10^8 K (≈ 10 keV)	10^6 K (≈ 100 eV)
	density*	$\rho \equiv (A/Z)nm_p$	1.7×10^{-7} kg m^{-3}	1.7×10^{-11} kg m^{-3}
	pressure	$p \equiv nkT$	1.4×10^5 N m^{-2}	0.14 N m^{-2}
	'plasma beta'	$\beta \equiv 2\mu_0 p/B^2$	0.04	0.0004
	sound speed*	$v_s \equiv \sqrt{\gamma p/\rho}$	1.2×10^6 m s^{-1}	120 km s^{-1}
	Alfvén speed*	$v_A \equiv B/\sqrt{\mu_0 \rho}$	6.5×10^6 m s^{-1}	6500 km s^{-1}
	Alfv. transit time	$\tau_A \equiv L/v_A$	3 µs	15 s
	times of interest	τ	seconds	days
	resist. decay time	$\tau_d \equiv \mu_0 a^2/\eta$	16 min	3.2×10^6 y
	Spitzer resistivity*	$\eta_\| = 65\, Z \ln\Lambda\, T_e^{-3/2}$	1.3×10^{-9} Ω m	1.2×10^{-6} Ω m
	magn. Reynolds nr.	$R_m \equiv \mu_0 a v_A/\eta_\|$	6.3×10^9	6.8×10^{13}

⇒ independent basic variables

*numbers for H plasma ($A = Z = 1$)

Table B.6.

Orders of magnitude: MHD (cont'd)

		Solar wind	Magnetosphere	Your favourite plasma
$\Rightarrow$	a	1.5×10^6 km	6×10^3 km	....
$\Rightarrow$	L	1.5×10^8 km	4×10^4 km	....
$\Rightarrow$	n	10^7 m^{-3}	10^{10} m^{-3}	....
$\Rightarrow$	B	6×10^{-9} T	3×10^{-5} T	....
$\Rightarrow$	T	10^5 K	10^4 K	....
	ρ	1.7×10^{-20} kg m^{-3}	1.7×10^{-17} kg m^{-3}	....
	p	1.4×10^{-11} N m^{-2}	1.4×10^{-9} N m^{-2}	....
	β	1	4×10^{-6}	....
	v_s	37 km s^{-1}	12 km s^{-1}	....
	v_A	41 km s^{-1}	6500 km s^{-1}	....
	τ_A	42 days	6 s	....
	τ	months	hours	....
	τ_d	1.7×10^9 y	1.2×10^3 y	....
	$\eta_\parallel$	5.2×10^{-5} Ω m	1.2×10^{-3} Ω m	....
	R_m	1.5×10^{12}	4.1×10^{10}	....

Table B.7.

Physical data of Sun and planets

	radius (km)	mass (kg)	rotation per. (d)	distance to Sun (AU)	orbital eccentr.	orbital per. (y)
Sun	7.0×10^5	2.0×10^{30}	25–27	–	–	–
Mercury	2.4×10^3	3.3×10^{23}	58.6	0.39	0.206	0.241
Venus	6.1×10^3	4.9×10^{24}	243.0	0.72	0.007	0.615
Earth	6.4×10^3	6.0×10^{24}	1.0	1.0 (def.)	0.017	1.0
Mars	3.4×10^3	6.4×10^{23}	1.03	1.52	0.093	1.88
Jupiter	7.1×10^4	1.9×10^{27}	0.42	5.20	0.048	11.9
Saturn	6.0×10^4	5.7×10^{26}	0.43	9.54	0.056	29.5
Uranus	2.6×10^4	8.7×10^{25}	0.72	19.2	0.047	84.0
Neptune	2.5×10^4	1.0×10^{26}	0.75	30.1	0.009	164.8
Pluto	1.8×10^3	1.0×10^{22}	6.4	39.4	0.249	248.6

$R_\odot = 6.96 \times 10^5$ km.

1 AU $= 1.496 \times 10^8$ km $= 215\, R_\odot$.

1 light-year $= 9.46 \times 10^{12}$ km $= 6.32 \times 10^4$ AU.

1 pc $= 3.09 \times 10^{13}$ km $= 3.26$ light-years $= 2.06 \times 10^5$ AU.

(Adapted from M. Zeilik and E. v. P. Smith, *Introductory Astronomy and Astrophysics* (Saunders College Publishing, 1987).)

Table B.8.

Planetary magnetic fields

	dipole moment (A m^2)	equatorial magn. field ($\times 10^{-4}$ T)	obliquity	tilt mag. axis with respect to rot. axis	size day-side magnetopause (planet. radii)
Mercury	4.0×10^{19}	0.003	0.0°	14.0°	1.4 (R_M)
Venus	$< 1.0 \times 10^{18}$	< 0.0003	177.4° [1]	–	1.1 (R_V)
Earth	8.1×10^{22}	0.31	23.5°	11.4°	10.4 (R_E)
Mars	2.3×10^{19}	0.0006	25.2°	–	?
Jupiter	1.5×10^{27}	4.3	3.1°	−9.6° [2]	65 (R_J)
Saturn	8.6×10^{25}	0.22	26.7°	−0° [2]	20 (R_S)
Uranus	3.9×10^{24}	0.23	97.9° [1]	−58.6° [2]	18 (R_U)
Neptune	2.0×10^{24}	0.13	28.8°	−46.8°	35 (R_N)
Pluto	?	?	65°	?	?

[1] retrograde

[2] dipole moment in the same direction as rotation vector.

(Adapted from S. K. Atreya, *Atmospheres and Ionospheres of the Outer Planets and Their Satellites* (Springer-Verlag, 1986);

L. J. Lanzerotti & C. Uberoi, 'The planets' magnetic environments', *Sky & Telescope*, Febr. 1989, 149;

G. Hunt & P. Moore, *Atlas of Uranus* (Cambridge University Press, 1989);

E. D. Miner, 'Voyager 2's encounter with the gas giants', *Phys. Today*, July 1990, 40–47.)

References

[1] M. Abramowitz and I. A. Stegun, *Handbook of Mathematical Functions with Formulas, Graphs, and Mathematical Tables* (Washington, D.C., National Bureau of Standards, 1964).

[2] A. Achterberg, 'Variational principle for relativistic magnetohydrodynamics', *Phys. Rev. A* **28** (1983), 2449–2458.

[3] J. A. Adam, 'A nonlinear eigenvalue problem in astrophysical magnetohydrodynamics: some properties of the spectrum', *J. Math. Phys.* **30** (1989), 744–756.

[4] A. I. Akhiezer, I. A. Akhiezer, P. V. Polovin, A. G. Sitenko and K. N. Stepanov, *Plasma Electrodynamics, Vol. 1, Linear Theory; Vol. 2, Non-linear Theory and Fluctuations* (Oxford, Pergamon Press, 1975).

[5] H. Alfvén, 'Existence of electromagnetic-hydrodynamic waves', *Nature* **3805** (1942), 405.

[6] H. Alfvén, *Cosmical Electrodynamics*, 1st edn (Oxford, Clarendon Press, 1950).

[7] G. Amarante-Segundo, A. G. Elfimov, D. W. Ross, R. M. O. Galvão and I. C. Nascimento, 'Calculations of wave excitation and dissipation in Tokamak Chauffage Alfvén wave heating experiment in Brazil', *Phys. Plasmas* **6** (1999), 2437–2442.

[8] A. M. Anile, *Relativistic Fluids and Magneto-Fluids* (Cambridge, Cambridge University Press, 1989).

[9] K. Appert, R. Gruber and J. Vaclavik, 'Continuous spectra of a cylindrical magnetohydrodynamic equilibrium', *Phys. Fluids* **17** (1974), 1471–1472.

[10] K. Appert, R. Gruber, F. Troyon and J. Vaclavik, 'Excitation of global eigenmodes of the Alfvén wave in tokamaks', *Plasma Phys.* **24** (1982), 1147–1159.

[11] M. J. Aschwanden, 'An evaluation of coronal heating models for active regions based on Yohkoh, Soho, and TRACE observations', *Astrophys. J.* **560** (2001), 1035–1044.

[12] W. A. Balbus and J. F. Hawley, 'Instability, turbulence, and enhanced transport in accretion disks', *Rev. Modern Physics* **70** (1998), 1–53.

[13] R. Balescu, 'Irreversible processes in ionised gases', *Phys. Fluids* **3** (1960), 52–63.

[14] R. Balescu, *Transport Processes in Plasmas; Vol. 1: Classical Transport Theory, Vol. 2: Neoclassical Transport* (Amsterdam, North Holland, 1988).

[15] B. Balet, K. Appert and J. Vaclavik, 'MHD studies on Alfvén wave heating of low-β plasmas', *Plasma Phys.* **24** (1982), 1005–1023.

[16] E. M. Barston, 'Electrostatic oscillations in inhomogeneous cold plasmas', *Ann. Phys. (New York)* **29** (1964), 282–303.

[17] G. K. Batchelor, *An Introduction to Fluid Dynamics* (Cambridge, Cambridge University Press, 1967).

[18] E. Battaner, *Astrophysical Fluid Dynamics* (Cambridge, Cambridge University Press, 1996).

[19] W. Baumjohann and R. A. Treumann, *Basic Space Plasma Physics* (London, Imperial College Press, 1996); R. A. Treumann and W. Baumjohann, *Advanced Space Plasma Physics* (London, Imperial College Press, 1997).

[20] A. J. C. Beliën, P. C. H. Martens and R. Keppens, 'Coronal heating by resonant absorption: the effects of chromospheric coupling', *Astrophys. J.* **526** (1999), 478–493.

[21] A. J. C. Beliën, S. Poedts and J. P. Goedbloed, 'Magnetohydrodynamic continua and stratification induced Alfvén eigenmodes in coronal magnetic loops', *Phys. Rev. Lett.* **76** (1996), 567–570.

[22] A. J. C. Beliën, S. Poedts and J. P. Goedbloed, 'Slow magnetosonic waves and instabilities in expanded flux tubes anchored in chromospheric/photospheric regions', in *The Corona and Solar Wind near Minimum Activity*, Proc. 5th SOHO Workshop, ESA SP-**404** (1997), 193–197.

[23] C. M. Bender and S. Orszag, *Advanced Mathematical Methods for Scientists and Engineers* (Tokyo, McGraw-Hill International Book Company, 1978).

[24] M. A. Berger, 'Rigorous new limits on magnetic helicity dissipation in the solar corona', *Geophys. Astrophys. Fluid Dynamics* **30** (1984), 79–104.

[25] D. Berghmans and P. De Bruyne, 'Coronal loop oscillations driven by footpoint motions: analytical results for a model problem', *Astrophys. J.* **453** (1995), 495–504.

[26] I. B. Bernstein, E. A. Frieman, M. D. Kruskal and R. M. Kulsrud, 'An energy principle for hydromagnetic stability problems', *Proc. Roy. Soc. (London)* **A244** (1958), 17–40.

[27] I. B. Bernstein, E. A. Frieman, M. D. Kruskal and R. M. Kulsrud, Appendix to Ref. [26], Project Matterhorn Report NYO-7896, Princeton (1957).

[28] G. Besson, A. de Chambrier, G. A. Collins, B. Joye, A. Lietti, J. B. Lister, J. M. Moret, S. Nowak, C. Simm and H. Weisen, 'A review of Alfvén wave heating', *Plasma Physics & Controlled Fusion* **28** (1986), 1291–1303.

[29] D. Biskamp, *Nonlinear Magnetohydrodynamics* (Cambridge, Cambridge University Press, 1993).

[30] D. Biskamp, *Magnetic Reconnection in Plasmas* (Cambridge, Cambridge University Press, 2000).

[31] J. A. Bittencourt, *Fundamentals of Plasma Physics* (New York, Pergamon Press, 1986); 2nd edn (Brazil, Foundation of the State of São Paulo for the Support of Research, 1995).

[32] A. A. Blank, K. O. Friedrichs and H. Grad, *Notes on Magnetohydrodynamics, V. Theory of Maxwell's Equations without Displacement Current* (New York, New York University, NYO-6486-V, 1957).

[33] T. J. Bogdan and D. C. Braun, 'Active region seismology', in *Helioseismology*, Proc. 4th SOHO Workshop, ESA SP-**376** (1995), 31–45.

[34] T. J. Bogdan and M. Knölker, 'Scattering of acoustic waves from a magnetic flux tube embedded in a radiating fluid', *Astrophys. J.* **369** (1991), 219–236.

[35] T. J. M. Boyd and J. J. Sanderson, *Plasma Dynamics* (London, Nelson, 1969).

[36] T. J. M. Boyd and J. J. Sanderson, *The Physics of Plasmas* (Cambridge, Cambridge University Press, 2003).
[37] C. M. Braams and P. E. Stott, *Nuclear Fusion: Half a Century of Magnetic Confinement Fusion Research* (Bristol, Institute of Physics Publishing, 2002).
[38] J. U. Brackbill and D. C. Barnes, 'The effect of non-zero $\nabla \cdot \mathbf{B}$ on the numerical solution of the magnetohydrodynamic equations', *J. Comp. Phys.* **35** (1980), 426–430.
[39] H. J. J. Braddick, *Vibrations, Waves, and Diffraction* (Maidenhead, McGraw-Hill Publishing Company Limited, 1965).
[40] S. I. Braginskii, 'On the modes of plasma oscillations in a magnetic field', *Sov. Phys. Dokl.* **2** (1957), 345–349.
[41] S. I. Braginskii, 'Transport processes in a plasma', in *Reviews of Plasma Physics, Vol. 1*, ed. M. A. Leontovich (New York, Consultants Bureau, 1965), pp. 205–311.
[42] D. C. Braun, T. L. Duvall and B. J. LaBonte, 'Acoustic absorption by sunspots', *Astrophys. J.* **319** (1987), L27–L31.
[43] D. C. Braun, T. L. Duvall and B. J. LaBonte, 'The absorption of high-degree p-mode oscillations in and around sunspots', *Astrophys. J.* **335** (1988), 1015–1025.
[44] D. C. Braun and T. L. Duvall, 'P-mode absorption in the giant active region of 10 March, 1989', *Solar Phys.* **129** (1990), 83–94.
[45] G. E. Brueckne, 'Space observations of solar corona', *Astronautics & Aeronautics* **10** (1972), 24.
[46] D. Burgess, 'Collisionless shocks', in *Introduction to Space Physics*, eds. Margaret G. Kivelson and Christopher T. Russell (New York, Cambridge University Press, 1995), pp. 129–163.
[47] L. F. Burlaga, L. Klein, N. R. Sheeley Jr., D. J. Michels, R. A. Howard, M. J. Koomen, R. Schwenn and H. Rosenbauer, 'A magnetic cloud and a coronal mass ejection', *Solar Phys.* **9** (1982), 1317–1320.
[48] F. H. Busse, 'Problems of planetary dynamo theory', in *Advances in Solar System Magnetohydrodynamics*, eds. E. R. Priest and A. W. Hood (Cambridge, Cambridge University Press, 1991), pp. 51–59.
[49] P. S. Cally, 'Leaky and non-leaky oscillations in magnetic-flux tubes', *Solar Phys.* **103** (1986), 277–298.
[50] M. S. Chance, J. M. Greene, R. C. Grimm and J. L. Johnson, 'Study of the MHD spectrum of an elliptic plasma column', *Nuclear Fusion* **17** (1977), 65–83.
[51] S. Chandrasekhar, *Hydrodynamic and Hydromagnetic Stability* (Oxford, Clarendon Press, 1961).
[52] S. Chapman and T. G. Cowling, *The Mathematical Theory of Non-uniform Gases*, 2nd edn (Cambridge, Cambridge University Press, 1970).
[53] F. C. Chen, *Introduction to Plasma Physics and Controlled Fusion, Vol. I: Plasma Physics*, 2nd edition (New York, Plenum Press, 1984).
[54] L. Chen and A. Hasegawa, 'Plasma heating by spatial resonance of Alfvén waves', *Phys. Fluids* **17** (1974), 1399–1403.
[55] A. R. Choudhuri, *The Physics of Fluids and Plasmas: an Introduction for Astrophysicists* (Cambridge, Cambridge University Press, 1998).
[56] J. Christensen-Dalsgaard, *Stellar Oscillations* (Aarhus, Lecture Notes, Astronomisk Institut, Aarhus Universitet, 1989).
[57] R. V. Churchill, *Complex Variables and Applications* (New York, McGraw-Hill Book Company, 1960).

[58] P. C. Clemmov and J. P. Dougherty, *Electrodynamics of Particles and Plasmas* (Reading, Addison-Wesley, 1969).

[59] R. Courant and K. O. Friedrichs, *Supersonic Flow and Shock Waves* (New York, Interscience Publishers, 1948).

[60] R. Courant and D. Hilbert, *Methods of Mathematical Physics II* (New York, Interscience, 1962).

[61] A. De Groof, W. J. Tirry and M. Goossens, 'Random driven fast waves in coronal loops – I. Without coupling to Alfvén waves', *Astron. Astrophys.* **335** (1998), 329–340; 'II. With coupling to Alfvén waves', *Astron. Astrophys.* **356** (2000), 724–734.

[62] A. De Groof and M. Goossens, 'Fast and Alfvén waves driven by azimuthal footpoint motions – I. Periodic driver', *Astron. Astrophys.* **386** (2002), 681–690; 'II. Random driver', *Astron. Astrophys.* **386** (2002), 691–698.

[63] P. J. Dellar, 'A note on magnetic monopoles and the one-dimensional MHD Riemann problem', *J. Comp. Physics* **172** (2001), 392–398.

[64] I. De Moortel, A. W. Hood, J. Ireland and T. D. Arber, 'Phase mixing of Alfvén waves in a stratified and open atmosphere', *Astron. Astrophys.* **346** (1999), 641–651.

[65] R. O. Dendy (ed.), *Plasma Physics: an Introductory Course* (Cambridge, Cambridge University Press, 1993).

[66] J. F. Denisse and J. L. Delcroix, *Théorie des Ondes dans les Plasmas* (Paris, Dunod, 1961). (Transl.: *Plasma Waves* (New York, John Wiley, Interscience, 1963).)

[67] P. A. M. Dirac, *The Principles of Quantum Mechanics*, 4th edition (London, Oxford University Press, 1958).

[68] A. J. H. Donné, A. L. Rogister, R. Koch and H. Soltwisch (eds.), *Proc. Second Carolus Magnus Summer School on Plasma Physics*, Aachen, 1995; *Transactions of Fusion Technology* **29** (1996), 1–432.

[69] P. V. Foukal, *Solar Astrophysics* (New York, John Wiley & Sons, 1990).

[70] J. Frank, A. King and D. Raine, *Accretion Power in Astrophysics*, 3rd edition (Cambridge, Cambridge University Press, 2002).

[71] J. P. Freidberg, 'Magnetohydrodynamic stability of a diffuse screw pinch', *Phys. Fluids* **13** (1970), 1812–1818.

[72] J. P. Freidberg, *Ideal Magnetohydrodynamics* (New York, Plenum Press, 1987).

[73] J. P. Freidberg and F. A. Haas, 'Kink instabilities in a high-β tokamak', *Phys. Fluids* **16** (1973), 1909–1916.

[74] B. Friedman, *Principles and Techniques of Applied Mathematics* (New York, Wiley & Sons, 1956)

[75] H. Friedman, *Sun and Earth* (New York, W. H. Freeman and Company, 1986).

[76] K. O. Friedrichs and H. Kranzer, *Notes on Magnetohydrodynamics, VIII. Non-linear wave motion* (New York, New York University, NYO-4686-VIII, 1958).

[77] A. A. Galeev and R. Z. Sagdeev, 'Theory of neo-classical diffusion', in *Reviews of Plasma Physics, Vol. 7*, ed. M. A. Leontovich (New York, Consultants Bureau, 1979), pp. 257–343.

[78] P. R. Garabedian, *Partial Differential Equations* (New York, John Wiley, 1964).

[79] G. A. Glatzmaier and P. H. Roberts, 'Three-dimensional self-consistent computer simulation of a geomagnetic field reversal', *Nature* **377** (1995), 203–209.

[80] S. K. Godunov, 'Symmetric form of the equations for magnetohydrodynamics', in *Numerical Methods for Mechanics of Continuous Media, Vol. 1*, 26–31 (1972) in

Russian; translation by T. Linde (Report of the Computer Centre of the Siberian branch of the USSR Academy of Sciences, 1972).

[81] J. P. Goedbloed, 'Stabilization of magnetohydrodynamic instabilities by force-free magnetic fields – I. Plane plasma layer', *Physica* **53** (1971), 412–444;
'II. Linear pinch', *Physica* **53** (1971), 501–534;
'III. Shearless magnetic fields', *Physica* **53** (1971), 535–570;
'IV. Boundary conditions for plasma–plasma interface', *Physica* **100C** (1980), 273–275.

[82] J. P. Goedbloed, 'Spectrum of ideal magnetohydrodynamics of axisymmetric toroidal systems', *Phys. Fluids* **18** (1975), 1258–1268.

[83] J. P. Goedbloed, *Lecture Notes on Ideal Magnetohydrodynamics*, Lectures at Universidade Estadual de Campinas, Brazil, 1978, Report Instituto de Física (1979); Rijnhuizen Report (1983) 83–145.

[84] J. P. Goedbloed, 'Plasma–vacuum interface problems in magnetohydrodynamics', *Physica* **12D** (1984), 107–132.

[85] J. P. Goedbloed, 'Once more: the continuous spectrum of ideal magnetohydrodynamics', *Phys. Plasmas* **5** (1998), 3143–3154.

[86] J. P. Goedbloed and H. J. L. Hagebeuk, 'Growth rates of instabilities of a diffuse linear pinch', *Phys. Fluids* **15** (1972), 1090–1101.

[87] J. P. Goedbloed and G. Halberstadt, 'Magnetohydrodynamic waves in coronal flux tubes', *Astron. Astrophys.* **281** (1994), 265–301.

[88] J. P. Goedbloed, G. T. A. Huysmans, H. A. Holties, W. Kerner and S. Poedts, 'MHD spectroscopy: free boundary modes (ELMs) and external excitation of TAE modes', *Plasma Physics & Controlled Fusion* **35** (1993), B277–292.

[89] J. P. Goedbloed and P. H. Sakanaka, 'New approach to magnetohydrodynamic stability. I. A practical stability concept', *Phys. Fluids* **17** (1974), 908–918.

[90] J. P. Goedbloed and J. W. A. Zwart, 'On the dynamics of the screw pinch', *Plasma Phys.* **17** (1975), 45–67.

[91] H. Goldstein, *Classical Mechanics*, 2nd edition (Reading, Addison Wesley, 1980).

[92] R. J. Goldston and P. H. Rutherford, *Introduction to Plasma Physics* (Bristol, Institute of Physics Publishing, 1995).

[93] M. Goossens, 'MHD waves and wave heating in nonuniform plasmas', in *Advances in Solar System Magnetohydrodynamics*, eds. E. R. Priest and A. W. Hood (Cambridge, Cambridge University Press, 1991), pp. 137–172.

[94] M. Goossens and A. De Groof, 'Resonant and phase-mixed magnetohydrodynamic waves in the solar atmosphere', *Phys. Plasmas* **8** (2001), 2371–2376.

[95] M. Goossens and S. Poedts, 'Linear resistive MHD computations of resonant absorption of acoustic oscillations in sunspots', *Astrophys. J.* **384** (1982), 348–360.

[96] M. Goossens, M. S. Ruderman and J. V. Hollweg, 'Dissipative MHD solutions for resonant Alfvén waves in 1-dimensional magnetic flux tubes', *Solar Phys.* **157** (1995), 75–102.

[97] D. Gough, 'Helioseismology: oscillations as a probe of the sun's interior', *Nature* **304** (1983), 689–690.

[98] H. Grad, *Notes on Magnetohydrodynamics, I. General Fluid Equations* (New York, New York University, NYO-6486-I, 1956).

[99] H. Grad, 'Magnetofluid-dynamic spectrum and low shear stability', *Proc. Natl. Acad. Sci. USA* **70** (1973), 3277–3281.

[100] W. Grossmann and R. A. Smith, 'Heating of solar coronal loops by resonant absorption of Alfvén waves', *Astrophys. J.* **332** (1988), 476–498.
[101] K. Hain, R. Lüst and A. Schlüter, 'Zur Stabilität eines Plasmas', *Z. Naturforsch.* **12a** (1957), 833–941.
[102] K. Hain and R. Lüst, 'Zur Stabilität zylinder-symmetrischer Plasmakonfigurationen mit Volumenströmen', *Z. Naturforsch.* **13a** (1958), 936–940.
[103] G. Halberstadt and J. P. Goedbloed, 'The continuous Alfvén spectrum of line-tied coronal loops', *Astron. Astrophys.* **280** (1993), 647–660.
[104] G. Halberstadt and J. P. Goedbloed, 'Alfvén-wave heating of coronal loops – photospheric excitation', *Astron. Astrophys.* **301** (1995), 559–576; 'Surface excitation revisited', *Astron. Astrophys.* **301** (1995), 577–592.
[105] J. Harvey, 'Helioseismology', *Phys. Today* **48** (1995), 32–38.
[106] A. Hasegawa and T. Sato, *Space Plasma Physics, Vol. I. Stationary Processes* (New York, Springer Verlag, 1989).
[107] R. D. Hazeltine and J. D. Meiss, *Plasma Confinement* (Redwood City, Addison Wesley Publishing Company, 1992).
[108] J. Heyvaerts and E. R. Priest, 'Coronal heating by phase-mixed shear Alfvén waves', *Astron. Astrophys.* **117** (1983), 220–234.
[109] J. V. Hollweg, 'Resonance-absorption of solar p-modes by sunspots', *Astrophys. J.* **335** (1988), 1005–1014.
[110] J. V. Hollweg, 'Resonant decay of global MHD modes at thick interfaces', *J. Geophys. Res.* **95** (1990), 2319–2324.
[111] J. V. Hollweg and G. Yang, 'Resonance-absorption of compressible MHD waves at thin surfaces', *J. Geophys. Res.* **93** (1988), 5423–5436.
[112] T. E. Holzer and E. Leer, 'Coronal hole structure and the high speed solar wind', in *The Corona and Solar Wind near Minimum Activity*, Proc. Fifth SOHO Workshop, ESA SP-**404** (1997), 65–74.
[113] A. J. Hundhausen, 'The solar wind', in *Introduction to Space Physics*, eds. Margaret G. Kivelson and Christopher T. Russell (New York, Cambridge University Press, 1995), pp. 91–128.
[114] S. Ichimaru, *Basic Principles of Plasma Physics; a Statistical Approach* (Reading, W. A. Benjamin Inc., 1973).
[115] E. L. Ince, *Ordinary Differential Equations* (New York, Dover Publications, 1956).
[116] J. A. Ionson, 'Resonant absorption of Alfvénic surface waves and the heating of solar coronal loops', *Astrophys. J.* **226** (1978), 650–673.
[117] J. D. Jackson, *Classical Electrodynamics*, 2nd edition (New York, John Wiley & Sons, 1975).
[118] P. Janhunen, 'A positive conservative method for magnetohydrodynamics based on HLL and Roe methods', *J. Comp. Physics* **160** (2000), 649–661.
[119] B. B. Kadomtsev, 'Hydromagnetic stability of a plasma', in *Reviews of Plasma Physics, Vol. 2*, ed. M. A. Leontovich (New York, Consultants Bureau, 1966), pp. 153–199.
[120] J. M. Kappraff and J. A. Tataronis, 'Resistive effects on Alfvén wave heating', *J. Plasma Physics*, **18** (1977), 209–226.
[121] R. Keppens, 'Flux tubes with a thin transition layer – scattering and absorption properties', *Solar Phys.* **161** (1995), 25–27.
[122] R. Keppens, 'Hot magnetic fibrils: the slow continuum revisited', *Astrophys. J.* **468** (1996), 907–920.

[123] R. Keppens, T. Bogdan and M. Goossens, 'Multiple-scattering and resonant absorption of p-modes by fibril sunspots', *Astrophys. J.* **436** (1994), 372–389.

[124] R. Keppens, F. Casse and J. P. Goedbloed, 'Waves and instabilities in accretion disks: magnetohydrodynamic spectroscopic analysis', *Astrophys. J.* **569** (2002), L121–L126.

[125] W. Kerner, 'Large-scale complex eigenvalue problems', *J. Comp. Physics* **85** (1989), 1–85.

[126] M. G. Kivelson, 'Pulsations and magnetohydrodynamic waves', in *Introduction to Space Physics*, eds. Margaret G. Kivelson and Christopher T. Russell (New York, Cambridge University Press, 1995), pp. 330–355.

[127] M. G. Kivelson and C. T. Russell (eds.), *Introduction to Space Physics* (New York, Cambridge University Press, Cambridge, 1995).

[128] M. G. Kivelson and D. J. Southwood, 'Coupling of global magnetospheric MHD eigenmodes to field line resonances', *J. Geophys. Res.* **91** (1986), 4345–4351.

[129] N. A. Krall and A. W. Trivelpiece, *Principles of Plasma Physics* (New York, McGraw-Hill, 1973).

[130] M. D. Kruskal and M. Schwarzschild, 'Some instabilities of a completely ionized plasma', *Proc. Roy. Soc. (London)* **A223** (1954), 348–360.

[131] M. D. Kruskal and J. L. Tuck, 'The instability of a pinched fluid with a longitudinal magnetic field', *Proc. Roy. Soc. (London)* **A245** (1958), 222–237.

[132] M. D. Kruskal, 'Asymptotology', in *Plasma Physics*, eds. B. B. Kadomtsev, M. N. Rosenbluth and W. B. Thompson (Vienna, IAEA, 1965), pp. 373–387.

[133] M. Kuperus, J. A. Ionson and D. S. Spicer, 'On the theory of coronal heating mechanisms', *Ann. Rev. Astron. Astrophys.* **19** (1971), 7–40.

[134] H. J. G. L. M. Lamers and J. P. Cassinelli, *Introduction to Stellar Winds* (Cambridge, Cambridge University Press, 1999).

[135] L. D. Landau, 'The transport equation in the case of Coulomb interactions', *J. Exp. Theor. Phys. USSR* **7** (1937), 203. (Transl.: *Phys. Z. Sowjet.* **10** (1936), 154; or *Collected Papers of L. D. Landau*, ed. D. ter Haar (Oxford, Pergamon Press, 1965) pp. 163–170.)

[136] L. D. Landau, 'On the vibrations of the electronic plasma', *J. Phys. USSR* **10** (1946), 25. (Transl.: *JETP* **16** (1946), 574; or *Collected Papers of L. D. Landau*, ed. D. ter Haar (Oxford, Pergamon Press, 1965) pp. 445–460.)

[137] L. D. Landau and E. M. Lifshitz, *Course of Theoretical Physics, Vol. 6: Fluid Mechanics*, 2nd edition (Oxford, Pergamon Press, 1987).

[138] L. D. Landau and E. M. Lifshitz, *Course of Theoretical Physics, Vol. 8: Electrodynamics of Continuous Media*, 2nd edition (Oxford, Pergamon Press, 1984), Chapter VIII.

[139] G. Laval, C. Mercier and R. M. Pellat, 'Necessity of the energy principle for magnetostatic stability', *Nuclear Fusion* **5** (1965), 156–158.

[140] J. D. Lawson, 'Some criteria for a power producing thermonuclear reactor', *Proc. Phys. Soc. (London)* **B70** (1957), 6–10.

[141] A. Lenard, 'On Bogoliubov's kinetic equation for a spatially homogeneous plasma', *Ann. Phys. (New York)* **10** (1960), 390–400.

[142] M. A. Leontovich (ed.), *Reviews of Plasma Physics, Vols. 1–5* (New York, Consultants Bureau, 1965–1970; translated from the Russian editions of 1963–1967).

[143] R. J. LeVeque, *Numerical Methods for Conservation Laws* (Berlin, Birkhäuser Verlag, 1990).

[144] R. J. LeVeque, D. Mihalas, E. A. Dorfi and E. Müller, *Computational Methods for Astrophysical Fluid Flow* (Berlin, Springer Verlag, 1998).

[145] A. Lichnerowicz, *Relativistic Hydrodynamics and Magnetohydrodynamics* (New York, W. A. Benjamin Inc., 1967).

[146] A. E. Lifschitz, *Magnetohydrodynamics and Spectral Theory* (Dordrecht, Kluwer Academic Publishers, 1989).

[147] Y.-Q. Lou, 'Viscous MHD modes and p-mode absorption by sunspots', *Astrophys. J.* **350** (1990), 452–462.

[148] B. C. Low, 'Magnetohydrodynamic processes in the solar corona: flares, coronal mass ejections, and magnetic helicity', *Phys. Plasmas* **1** (1994), 1684–1690.

[149] S. Lundquist, 'Magneto-hydrostatic fields', *Arkiv för Fysik* **2** (1951), 361–365.

[150] J. H. Malmberg and C. B. Wharton, 'Dispersion of electron plasma waves', *Phys. Rev. Lett.* **17** (1966), 175–178; J. H. Malmberg, C. B. Wharton, R. W. Gould and T. M. O'Neil, 'Observation of plasma wave echos', *Phys. Fluids* **11** (1968), 1147–1153.

[151] I. R. Mann and A. N. Wright, 'Coupling of magnetospheric cavity modes to field line resonances – a study of resonance widths', *J. Geophys. Res.* **100** (1995), 19 441–19 456.

[152] P. Martens and C. Zwaan, 'Origin and evolution of filament-prominence systems', *Astrophys. J.* **558** (2001), 872–887.

[153] C. Mercier, 'Un critère necessaire de stabilité hydromagnétique pour un plasma en symétrie de révolution', *Nuclear Fusion* **1** (1960), 47–53.

[154] L. Mestel, *Stellar Magnetism* (Oxford, Clarendon Press, 1999).

[155] F. Meyer, 'Untersuchung der Stabilität eines gravitierenden Plasmas in gekreuzten Magnetfeldern', *Z. Naturforsch.* **13a** (1958), 1016–1020.

[156] K. Miyamoto, *Plasma Physics for Nuclear Fusion*, revised edition (Cambridge, Mass., MIT Press, 1987).

[157] H. K. Moffatt, *Magnetic Field Generation in Electrically Conducting Fluids* (Cambridge, Cambridge University Press, 1978).

[158] D. C. Montgomery and D. A. Tidman, *Plasma Kinetic Theory* (New York, McGraw-Hill Book Company, 1964).

[159] P. M. Morse and H. Feshbach, *Methods of Theoretical Physics, Part I* (New York, McGraw-Hill, 1953).

[160] T. C. Mouschovias, 'The Parker instability in the interstellar medium', in *Solar and Astrophysical Magnetohydrodynamic Flows*, ed. K. C. Tsinganos (Dordrecht, Kluwer Academic Publishers, 1996), pp. 475–504.

[161] V. M. Nakariakov, L. Ofman, E. E. DeLuca, B. Roberts and J. M. Davila, 'TRACE observation of damped coronal loop oscillations: implications for coronal heating', *Science* **285** (1999), 862–864.

[162] Z. Nehari, *Conformal Mapping* (New York, Dover Publications, 1952).

[163] W. A. Newcomb, 'Motion of magnetic lines of force', *Ann. Phys. (New York)* **3** (1958), 347–385.

[164] W. A. Newcomb, 'Hydromagnetic stability of a diffuse linear pinch', *Ann. Phys. (New York)* **10** (1960), 232–267.

[165] W. A. Newcomb, 'Convective instability induced by gravity in a plasma with a frozen-in magnetic field', *Phys. Fluids* **4** (1961), 391–396.

[166] W. A. Newcomb, 'Lagrangian and Hamiltonian methods in magnetohydrodynamics', *Nucl. Fusion, 1962 Suppl.*, Part **2** (1962), 451–463.

[167] W. A. Newcomb, *Lecture Notes on Magnetohydrodynamics* (unpublished).

[168] K. Nishikawa and M. Wakatani, *Plasma Physics: Basic Theory with Fusion Applications*, 3rd edition (Berlin, Springer-Verlag, 2000).

[169] T. G. Northrop, *The Adiabatic Motion of Charged Particles* (New York, Interscience Publishers, 1963).

[170] L. Ofman and J. M. Davila, 'Nonlinear resonant absorption of Alfvén waves in three dimensions, scaling laws, and coronal heating', *J. Geophys. Res.* **100** (1995), 23 427–23 441.

[171] L. Ofman and J. M. Davila, 'Alfvén wave heating of coronal holes and the relation to the high-speed wind', *J. Geophys. Res.* **100** (1995), 23 413–23 425.

[172] L. Ofman and J. M. Davila, 'Nonlinear excitation of global modes and heating in randomly driven coronal loops', *Astrophys. J.* **456** (1996), L123–L126.

[173] L. Ofman, J. A. Klimchuk and J. M. Davila, 'A self-consistent model for the resonant heating of coronal loops: the effects of coupling with the chromosphere', *Astrophys. J.* **493** (1998), 474–479.

[174] E. N. Parker, 'Dynamics of the interplanetary gas and magnetic fields', *Astrophys. J.* **128** (1958), 664–676.

[175] E. N. Parker, 'The dynamical state of the interstellar gas and field', *Astrophys. J.* **145** (1966), 811–833.

[176] E. N. Parker, *Cosmical Magnetic Fields* (Oxford, Clarendon Press, 1979).

[177] S. Poedts, 'MHD waves and heating of the solar corona', in *Magnetic Coupling of the Solar Atmosphere*, Proc. Euroconference and IAU Colloquium 188, ESA SP-**505** (2002), 273–280.

[178] S. Poedts, A. J. C. Beliën and J. P. Goedbloed, 'On the quality of resonant absorption as a coronal loop heating mechanism', *Solar Phys.* **151** (1994), 271–304.

[179] S. Poedts and G. C. Boynton, 'Nonlinear MHD of footpoint driven coronal loops', *Astron. Astrophys.* **306** (1996), 610–620.

[180] S. Poedts and J. P. Goedbloed, 'Nonlinear wave heating of solar coronal loops', *Astron. Astrophys.* **321** (1997), 935–944.

[181] S. Poedts, M. Goossens and W. Kerner, 'Numerical simulation of coronal heating by resonant absorption of Alfvén waves', *Solar Phys.* **123** (1989), 83–115.

[182] S. Poedts, M. Goossens and W. Kerner, 'On the efficiency of coronal loop heating by resonant absorption', *Astrophys. J.* **360** (1990), 279–287.

[183] S. Poedts and W. Kerner, 'Ideal quasi-modes reviewed in resistive MHD', *Phys. Rev. Lett.* **66** (1991), 2871–2874.

[184] S. Poedts and W. Kerner, 'Time-scales and efficiency of resonant absorption in periodically driven plasmas', *J. Plasma Physics* **47** (1992), 139–162.

[185] S. Poedts, W. Kerner, J. P. Goedbloed, B. Keegan, G. T. A. Huysmans and E. Schwarz, 'Damping of global Alfvén waves in tokamaks due to resonant absorption', *Plasma Physics & Controlled Fusion* **34** (1992), 1397–1422.

[186] S. Poedts, G. Tóth, A. J. C. Beliën and J. P. Goedbloed, 'Nonlinear MHD simulations of wave dissipation in flux tubes', *Solar Phys.* **172** (1997), 45–52.

[187] R. V. Polovin and V. P. Demutskii, *Fundamentals of Magnetohydrodynamics* (New York, Consultants Bureau, 1990).

[188] K. G. Powell, 'An approximate Riemann solver for magnetohydrodynamics (that works in more than one dimension)', *ICASE Report No. 94-24*, Langley VA (1994).

[189] K. G. Powell, P. L. Roe, T. J. Linde, T. I. Gombosi and D. L. De Zeeuw, 'A solution-adaptive upwind scheme for ideal magnetohydrodynamics', *J. Comp. Phys.* **154** (1999), 284–309.

[190] E. R. Priest, *Solar Magnetohydrodynamics* (Dordrecht, Reidel, 1984).
[191] E. R. Priest and T. G. Forbes, *Magnetic Reconnection; MHD Theory and Applications* (Cambridge, Cambridge University Press, 2000).
[192] A. Reiman, 'Minimum energy state of a toroidal discharge', *Phys. Fluids* **23** (1980), 230–231.
[193] B. Roberts, 'Surface waves', in *Mechanisms of Chromospheric and Coronal Heating*, eds. P. Ulmschneider, E. R. Priest and R. Rosner (Berlin, Springer Verlag, 1991), pp. 494–507.
[194] P. H. Roberts, *An Introduction to Magnetohydrodynamics* (London, Longmans, Green and Co Ltd., 1967).
[195] D. C. Robinson, 'High-β diffuse pinch configurations', *Plasma Physics* **13** (1971), 439–462.
[196] M. N. Rosenbluth, W. M. MacDonald and D. L. Judd, 'Fokker–Planck equation for an inverse-square force', *Phys. Rev.* **107** (1957), 1–6.
[197] M. N. Rosenbluth, 'Stability and heating in the pinch effect', in *Proc. Second UN Intern. Conf. on Peaceful Uses of Atomic Energy* **31** (New York, Columbia University Press, 1959), 85–92.
[198] M. S. Ruderman, M. L. Goldstein, D. A. Roberts, A. Deane and L. Ofman, 'Alfvén wave phase mixing driven by velocity shear in two-dimensional open magnetic configurations', *J. Geophys. Res.* **104** (1999), 17 057–17 068.
[199] P. H. Sakanaka and J. P. Goedbloed, 'New approach to magnetohydrodynamic stability. II. Sigma-stable diffuse pinch configurations', *Phys. Fluids* **17** (1974), 919–929.
[200] T. Sakurai, M. Goossens and J. V. Hollweg, 'Resonant behaviour of MHD waves on magnetic-flux tubes. I. Connection formulas at the resonant surfaces', *Solar Phys.* **133** (1991), 227–245;
'II. Absorption of sound waves by sunspots', *Solar Phys.* **133** (1991), 247–262.
[201] M. Saunders, 'The earth's magnetosphere', in *Advances in Solar System Magnetohydrodynamics*, eds. E. R. Priest and A. W. Hood (Cambridge, Cambridge University Press, 1991), pp. 357–397.
[202] B. E. Schaeffer, J. R. King and C. P. Deliyannis, 'Superflares on ordinary solar-type stars', *Astrophys. J.* **529** (2000), 1026–1030.
[203] G. Schmidt, *Physics of High Temperature Plasmas*, 2nd edition (New York, Academic Press, 1979).
[204] C. J. Schrijver and C. Zwaan, *Solar and Stellar Magnetic Activity* (Cambridge, Cambridge University Press, 2000).
[205] L. Schwartz, *Théorie des Distributions* (Paris, Hermann et cie, 1950).
[206] Z. Sedláček, 'Electrostatic oscillations in cold inhomogeneous plasma', *J. Plasma Physics* **5** (1971), 239–263.
[207] V. D. Shafranov, 'Hydrodynamic stability of a current-carrying pinch in a strong longitudinal field', *Sov. Phys.–Tech. Phys.* **15** (1970), 175–183.
[208] V. L. Smirnov, *A Course of Higher Mathematics*, Vol. IV (Oxford, Pergamon Press, 1964).
[209] H. Soltwisch, 'Measurement of current-carrying changes during sawtooth activity in a tokamak by far-infrared polarimetry', *Rev. Sci. Instrum.* **59** (1992), 1599–1604.
[210] H. Soltwisch, 'Current density measurements in tokamak devices', *Plasma Physics & Controlled Fusion* **34** (1992), 1669–1698.
[211] G. O. Spies, 'Magnetohydrodynamic spectrum in a class of toroidal equilibria', *Phys. Fluids* **19** (1976), 427–437.

[212] G. O. Spies and J. A. Tataronis, 'The accumulation continua in ideal magnetohydrodynamics', *Phys. Plasmas* **10** (2003), 413–418.
[213] L. Spitzer, Jr, *Physics of Fully Ionized Gases* (New York, Interscience, 1962).
[214] L. Spitzer and R. Härm, 'Transport phenomena in a completely ionized gas', *Phys. Rev.* **89** (1953), 977–981.
[215] H. C. Spruit, 'Propagation speeds and acoustic damping of waves in magnetic flux tubes', *Solar Phys.* **75** (1982), 3–17.
[216] H. Stenuit, R. Keppens and M. Goossens, 'Eigenfrequencies and optimal driving frequencies of 1D non-uniform magnetic flux tubes', *Astron. Astrophys.* **331** (1998), 392–404.
[217] M. Stix, *The Sun, An Introduction*, 2nd edition (Berlin, Springer-Verlag, 2002).
[218] T. H. Stix, *Waves in Plasmas* (New York, American Institute of Physics, McGraw Hill, 1992).
[219] H. R. Strauss and W. S. Lawson, 'Computer-simulation of Alfvén resonance in a cylindrical, axially bounded flux tube', *Astrophys. J.* **346** (1989), 1035–1040.
[220] T. E. Stringer, 'Low-frequency waves in an unbounded plasma', *Plasma Physics (J. Nucl. Energy Part C)* **5** (1963), 89–107.
[221] P. A. Sturrock, *Plasma Physics: an Introduction to the Theory of Astrophysical, Geophysical, and Laboratory Plasmas* (Cambridge, Cambridge University Press, 1994).
[222] B. R. Suydam, 'Stability of a linear pinch', in *Proc. 2nd U.N. Intern. Conf. on Peaceful Uses of Atomic Energy* **31**, 157–159 (New York, Columbia University Press, 1959).
[223] D. G. Swanson, *Plasma Waves* (Boston, Academic Press, 1989).
[224] J. A. Tataronis, 'Energy absorption in the continuous spectrum of ideal MHD', *J. Plasma Phys.* **13** (1975), 87–105.
[225] J. A. Tataronis and W. Grossmann, 'Decay of MHD waves by phase mixing', *Z. Physik* **261** (1973), 203–216.
[226] R. J. Tayler, 'The influence of an axial magnetic field on the stability of a constricted gas discharge', *Proc. Roy. Soc. (London)* **B70** (1957), 1049–1063.
[227] J. B. Taylor, 'Relaxation of toroidal plasma and generation of reversed magnetic fields', *Phys. Rev. Lett.* **33** (1974), 1139–1141.
[228] J. H. Thomas, L. E. Cram and A. H. Nye, 'Five-minute oscillations as a probe of a sunspot structure', *Nature* **297** (1982), 485–487.
[229] W. J. Tirry and D. Berghmans, 'Wave heating of coronal loops driven by azimuthally polarised footpoint motions; 2. The time-dependent behaviour in ideal MHD', *Astron. Astrophys.* **325** (1997), 329–340.
[230] G. Tóth, 'The $\nabla \cdot \mathbf{B} = 0$ constraint in shock-capturing magnetohydrodynamics codes', *J. Comp. Phys.* **161** (2000), 1159–1170.
[231] B. A. Trubnikov, 'Particle interactions in a fully ionized plasma', in *Reviews of Plasma Physics, Vol. 1*, ed. M. A. Leontovich (New York, Consultants Bureau, 1965), pp. 105–204.
[232] A. D. Turnbull, E. J. Strait, W. W. Heidbrink, M. S. Chu, H. H. Duong, J. M. Greene, L. L. Lao, T. S. Taylor and S. J. Thompson, 'Global Alfvén modes. Theory and experiment', *Phys. Fluids* **B5** (1993), 2546–2553.
[233] C. Uberoi, 'Alfvén waves in inhomogeneous magnetic fields', *Phys. Fluids* **15** (1972), 1673–1675.
[234] W. Unno, Y. Osaki, H. Ando, H. Saio and H. Shibahashi, *Nonradial Oscillations of Stars*, 2nd edition (Tokyo, University of Tokyo Press, 1989).

[235] J. Vaclavik and K. Appert, 'Theory of plasma-heating by low-frequency waves – magnetic pumping and Alfvén resonance heating', *Nuclear Fusion* **31** (1991), 1945–1997.

[236] P. C. T. van der Laan, W. Schuurman, J. W. A. Zwart and J. P. Goedbloed, 'On the decay of the longitudinal current in toroidal screw pinches', in *Plasma Physics and Controlled Nuclear Fusion Research*, Proc. Fourth Intern. IAEA Conf, 17–23 June 1971, Madison, USA, Vol. **I** (Vienna, IAEA, 1971), 217–223.

[237] N. G. van Kampen, 'On the theory of stationary waves in plasmas', *Physica* **21** (1955), 949–963.

[238] N. G. van Kampen and B. U. Felderhof, *Theoretical Methods in Plasma Physics* (Amsterdam, North-Holland Publishing Company, 1967).

[239] A. A. Vlasov, 'The oscillation properties of an electron gas', *Zhur. Eksp. Teor. Fiz.* **8** (1938), 291–318; 'Vibrational properties, crystal structure, non-dissipated counterdirected currents and spontaneous origin of these properties in a "gas"', *J. Phys. USSR* **9** (1945), 25–40.

[240] J. Von Neumann, *Mathematical Foundations of Quantum Mechanics* (Princeton, Princeton University Press, 1955).

[241] D. Voslamber and D. K. Callebaut, 'Stability of force-free magnetic fields', *Phys. Rev.* **128** (1962), 2016–2021.

[242] A. A. Ware, 'Role of compressibility in the magnetohydrodynamic stability of the diffuse pinch discharge', *Phys. Rev. Lett.* **12** (1964), 439–441.

[243] J. A. Wesson, 'Magnetohydrodynamic stability of tokamaks', *Nuclear Fusion* **18** (1978), 87–132.

[244] J. Wesson, *Tokamaks*, 2nd edition (Oxford, Clarendon Press, 1996).

[245] R. B. White, *Theory of Toroidally Confined Plasmas*, 2nd edition (London, Imperial College Press, 2001).

[246] P. R. Wilson, 'Free and forced oscillations of a flux tube', *Astrophys. J.* **251** (1981), 756–767.

[247] L. Woltjer, 'A theorem on force-free magnetic fields', *Proc. Natl. Acad. Sci. USA* **44** (1958), 489–491.

[248] M. Zeilik and E. v. P. Smith, *Introductory Astronomy and Astrophysics* (Philadelphia, Saunders College Publishing, 1987).

[249] X. Zhu and M. G. Kivelson, 'Analytic formulation and quantitative solutions of the coupled ULF wave problem', *J. Geophys. Res.* **93** (1988), 8602–8612.

[250] H. Zirin, *Astrophysics of the Sun* (Cambridge, Cambridge University Press, 1988).

Index

Note: Italic page numbers indicate a main section on the subject.

active region seismology, 570
adiabatic invariant, 43
   first (transverse), 44
   second (longitudinal), 46
   third (magnetic flux), 46
Alfvén frequency, 73
Alfvén Mach number, 163
Alfvén velocity, 73
Alfvén waves, *71–74*, *199–200*, 208, 213
   damping, 507
   point disturbances, 208
   vectorial Alfvén velocity, 191
antenna impedance, 546
artificial damping, 541
astrophysical concepts
   absolute magnitude, 16
   apparent magnitude, 16
   basic postulate of astrophysics, 15
   Cowling approximation, 309
   evolution of stars, 15
   Hertzsprung–Russell diagram, 17
   luminosity, 304
   mean molecular weight, 303
   opacity, 304
   radiation pressure, 304
astrophysical objects
   accretion disc, 136
   active galactic nucleus (AGN), 137
   young stellar object (YSO), 137
astrophysical phenomena
   coronal mass ejection (CME), 19, 23
   dynamo, 19
   interplanetary magnetic field (IMF), 164
   magnetotail, 164
   solar eclipse, 19
   solar flare, 19, 150
   solar wind, 19, 164, 182
   space weather, 19
   spatial and temporal resolution, 20
   stellar oscillations, 309
   superflare, 403
   X-ray emitting stars, 19
astrophysical plasmas, *13–23*

body mode, 524, 526
boundary conditions, 140
   astrophysical models IV–VI, *180–182*
   laboratory models I–III, *174–178*
   line-tying, 141
   magnetic field, 237
   perfect wall, 143
   photosphere-corona, 141, 161, 181

celestial mechanics
   Kepler's laws, 13
   Newton's gravitational law, 15
characteristics, *213–227*
   Alfvén disturbances, 223
   characteristic directions, 218
   characteristic speed, 221
   degeneracy, 226
   entropy disturbances, 223
   initial value problem, *218–219*
   IVP in MHD, *220–223*
   magneto-acoustic disturbances, 223
   normal to space-part, 221
   ray surface, 224
   reciprocal normal surface, 224
   space-like boundaries, 219
   tangential discontinuities, 227
   time-like boundaries, 219
   weak discontinuities, 214, *221–226*
classical mechanics
   action variable, 44
   Hamiltonian mechanics, 15
classical transport coefficients, *99–104*
   Braginskii's expressions, 102
   electrical conductivity, 101
   electron and ion viscosities, 103
   electron thermal conductivity, 101
   heat transfer function, 103
   ion thermal conductivity, 103

Index 607

Spitzer resistivity, 101
thermo-electric coupling, 101
closed system, 151
collisionality, 65
collisions, *94–98*
   electron relaxation time, 95
   electron–electron collision frequency, 95
   electron–ion collision frequency, 95
   ion relaxation time, 96
   ion–ion collision frequency, 96
   temperature equilibration time, 97
complex notation, 543
computational MHD
   computing spectrum, 253
   computing stationary state, 253
   computing temporal evolution, 252
computational plasma physics, 395
conservation laws, *145–161*
continuous spectrum
   exponential damping, 520
   implications, 496
   relation with branch cuts, 502, 512
   resonant absorption, 541
   three continua, 507
coupling factor, 546
cyclotron motion, *34–37*
   frequency, 35
   radius, 36
   relativistic, 40
cylindrical plasmas
   Alfvén's model, 460
   apparent singularities, 444
   boundary condition at origin, 441
   boundary conditions, interface, *445–450*
   constant-pitch magnetic field, 459
   curvature magnetic field, 431
   diffuse linear pinch, 431
   dimensionless scaling, 433
   equilibrium, *431–438*
   field line projection, 438
   field line-bending, 457
   force-free magnetic fields, 434
   free-boundary modes, *455–459*
   generalized Hain–Lüst equation, 440, 444
   'ghost' plasma, 448, 458
   interface models, 436
   laser wake-field acceleration, 432
   magnetic flux tube, 431
   matrix eigenvalue problem, 440
   pressure, Eulerian, 443
   pressure, Lagrangian, 443
   singularities, 442
   spectral structure, *450–462*
   stability, *462–492*
   'straight tokamak', 435
   system of first order ODEs, 443
   tokamak approximation, 456
   wave equation, *438–450*
   waves in a θ-pinch, 454

discontinuities, *167–173*
   contact, 172
   jump conditions, 167
   Rankine–Hugoniot relations, 167
   shocks, 167
   tangential, 172
dispersion function, 510
dissipation
   artificial, 541
   resistivity, magnetic diffusivity, 165
   resonant, 552, 554, 567
   thermal conductivity, diffusivity, 10, 166
   viscosity, kinematic diffusivity, 166
dissipative and ideal fluids, *104–108*
Doppler shifted frequency, 188, 222
drift motion, *41–47*
   $\mathbf{B} \times \nabla \mathbf{B}$ drift, 43
   $\mathbf{E} \times \mathbf{B}$ drift, 42
dynamo
   Babcock model, 390
   coefficients α, β, 395
   Cowling's theorem, 394
   kinematic dynamo, 395
   magnetic buoyancy, 392
   magnetic diffusivity, 393
   solar dynamo, 390, 426
   Spitzer resistivity, 393
   turbulent magnetic diffusivity, 394

electrodynamics, 131
   Ampère's law, 133
   displacement current, 39, 133
   electromagnetic waves, 39
   Faraday's law, 142
   Maxwell's equations, *38–41*
   Ohm's law, 142
   Poisson's law, 133
   pre-Maxwell equations, 23, 39
electrostatic oscillations
   cold plasma, 356
energy
   absorption rate, 543, 544
   dissipation rate, 549
   flow, 148
   kinetic energy density, 149
   potential energy density, 149
energy principle, *261–263*
   normal modes and, 263
   proof of, 266
entropy waves, 189, *194–195*, 199
equilibrium, 230
   cylindrical, 233, see cylindrical plasmas
   inhomogeneous, 234
   lack of, 230
   magnetic surfaces, 233
   static, 74, 231, 233

fluid description, *65–79*
foot point driving, 562
force operator, *237–249*
   equation of motion, 237
   homogeneous plasma, 241
   self-adjointness, 238, *244–249*

force-free magnetic fields, 159
   constant pitch field, 160, 184, 185
   Lundquist field, 159
fractional absorption, 546
Fredholm alternative, 509

gauge transformation, 156
generalized eigenvalue problem, 193
global Alfvén eigenmode (GAE), 462
global conservation laws, 148
   energy, 151
   magnetic flux, 151
   mass, 151
   momentum, 151
gravitating fluid instabilities
   Brunt–Väisäläa frequency, 312
   convective cells, 313
   convective instabilities, *312–313*
   Schwarzschild criterion, 312
gravitating fluid slab
   boundary conditions, 311
   exponential stratification, 313
   HD wave equation, *309–312*
   planar stratification, 308
gravitating plasma instabilities
   energy principle, 366
   Euler–Lagrange equation, 368
   gravitational instabilities, *365–379*
   gravitational interchange, 344
   gravitational quasi-interchange, 344
   interchange point ($F = 0$), 334, 347, 371
   interchanges with shear, 371
   interchanges without shear, 376
   local interchange stability, 348, 373
   magnetic shear, 345, 372
   marginal equation of motion, 368
   Newcomb's procedure, 368
   Parker instability, 344
   Rayleigh–Taylor instability, 341, 374
   Suydam's criterion, 374
gravitating plasma slab
   Clebsch coordinates, 329
   derivation wave equation, *327–335*
   exponential stratification, 335
   field line projection, *328–333*
   first order differential equations, 334
   growth rate largest for $n = 1$, 343
   homogeneous wave problem, 323
   matrix wave equation, 330
   MHD wave equation, *322–345*
   one-dimensional representation, 498
   one-dimensional systems, 322
   second order differential equation, 332
   total pressure perturbation, 334
gravitation, 135
   external, 135
   internal, 135
gravito-acoustic waves, *313–317*
   acoustic cutoff frequency, 315

Brunt–Väisäläa frequency, 312
cavity modes, 311
dispersion equation, 314
effective 'wave number', 314
evanescence, 314
f-modes, 316
free-boundary modes, 316
g-modes, 314
Lamb frequency, 315
p-modes, 314
Sturmian, anti-Sturmian, 315
turning point frequencies, 314
gravito-MHD waves, *335–345*
   apparent crossing slow/Alfvén, 342
   Brunt–Väisäläa frequency N_B, 336
   Brunt–Väisäläa, magnetic N_m, 338
   dimensionless scaling, 338
   dispersion equation, 337
   oblique waves, 338
   parallel waves, 337
   perpendicular waves, 337
Green's dyadic, 503
Green's function, 509
   Laplace contour, 509, 514, 515, 521
   poles, 502, 518
   Riemann sheet, 513
   uniqueness, 509
guiding centre approximation, 37
gyro-motion, *see* cyclotron motion

helioseismology, 308, *317–322*, 534
   5 minute oscillations, 316
   analogy quantum mechanics, 320
   cavity modes, 320
   f-modes, 320
   Lamb frequency, 318
   g-modes, 320
   p-modes, 320
   power spectrum oscillations, 318
   radial wave equation, 318
   spherical geometry, 317
   systematics, 321
Hilbert space, *242–244*
   inner product, 242
   linear operator in, 242
   norm, 242
   self-adjoint operators, 244, 452
hydrodynamics
   convective instability, 307
   isentropic motion, 307
   Rayleigh–Taylor instability, 307
   Schwarzschild criterion, 307
   solar interior, 300
   wave equation of a gravitating slab, *309–312*

ideal fluids, 67
impedance, 527, 546
impedance matching, 527
incompressibility, 285
induction equation, 133, 142

inhomogeneity
  one-dimensional, *450–453*
initial data, 140
initial value problem, 497, 529
interchange instability, 44
interface plasmas, *274–296*
  boundary conditions, *276–280*
  first interface condition, 278, 279
  plasma–plasma (model II*), 279
  plasma–vacuum (model II), 277
  second interface condition, 278, 280
  self-adjointness, *280–282*
  variational principles, *283–285*
interface plasmas (nonlinear)
  energy conservation, *178–180*

kinematic expressions, *152–153*
  line element, 152
  surface element, 152
  volume element, 153
kinetic plasma theory, *47–65, 84–98*
  Balescu–Lenard collision integral, 88
  BBGKY hierarchy, 85
  Boltzmann equation, 49, *84–88*
  closure of kinetic equations, 94
  collisionless Boltzmann equation, 84
  collisions, 49
  definition of heat flow, 91
  definition of heat transfer, 92
  definition of stress tensor, 91
  definition of temperature, 91
  distribution function, 48, 84
  initial value problem, 61
  Landau collision integral, 49, 87
  Landau damping, *58–65*, 356
  local thermal equilibrium, 93
  Maxwell distribution, 52, 92
  moments of Boltzmann equation, 50, *88–90*
  phase mixing, 59
  phase space, 84
  Rosenbluth potentials, 87
  thermal fluctuations, 90
  thermal quantities, 52
  van Kampen modes, 59, 356
  Vlasov equation, 49, 84
  Vlasov–Poisson problem, 59
kink instability, 76
Kruskal–Shafranov condition, 78, 458

Lagrangian displacement vector, 235
Laplace contour, 509, 514, 515
  deformation, 521
Laplace transform
  contour and convergence, 254
  forward, 253, 497
  Green's function, 509
  inverse, 254, 498, 506
  leaky modes, 524
leaky mode, 523, 526
  initial value problem, 529

normal-mode analysis, 528
linearization
  MHD equations, *232–236*
  plasma oscillations, 55
linearized MHD
  counting boundary conditions, 238
  damped and overstable waves, 239
  Eulerian representation, 238
  initial value problem, *253–256*
  Lagrangian representation, 238
  stable waves and instabilities, 239
linked magnetic loops, 158
Liouville's theorem, 49
local conservation laws, 152
  energy, 154
  magnetic flux, 154
  mass, 153
  momentum, 153
Lorentz force, 70
loss cone, 45
low β plasma, 507, 535
Lundquist number, 163

macroscopic scales, 66
magnetic bottle, 43
magnetic confinement, 10
  θ-pinch, 10, 75
  cusp, 44
  magnetic mirror, 44
  optimization problem, 12
  screw pinch, 185
  spheromak, 4
  stellarator, 3
  tokamak, 3, 10
  z-pinch, 10, 75
magnetic field
  no spherical symmetry, 23
  shear, 158
  solenoidal condition, 156
magnetic field lines
  frozen-in, 155
  inverse pitch, 157, 433
  reconnection, 163
  safety factor (q), 435
  tearing, 163
  x-point separatrix, 164
magnetic flux, *140–144*
magnetic flux tube, 140
magnetic helicity, *155–161*
magnetic moment, 44
magnetic pressure, 77
magnetic Reynolds number, 71, 162
magnetic rigidity, 41
magnetic stress, 149
magnetic structures, 20
magnetic tension, 149
magnetic topology, 156
magneto-seismology accretion discs, 322
magneto-acoustic waves
  fast, *200–201*, 208, 213
  slow, *200–201*, 208, 213

magnetohydrodynamics (MHD), 7, 29
  electric field secondary, 29
  Ohm's law, 29
  spatial and temporal aspects, 186
  symmetric hyperbolic equations, 193, 219, 223
magnetosphere
  ring current, 46
magnetospheres, *415–425*
  bow shock, 422
  flux transfer event, 424
  geomagnetic storm, 413, 419
  Kelvin–Helmholtz instability, 424
  magnetic pressure, 420
  magnetopause, 422
  magnetospheric substorm, 413
  magnetotail, 424
  neutral sheet current, 424
  ultra low frequency wave, 424
Maxwell stress tensor, 149
Maxwell's equations, 38
MHD equations, 71, 133
  conservation form, *145–148*
  non-relativistic approximation, 133
  Ohm's law, 70
  scale independence, *138–139*
MHD spectral theory
  σ-stability, 371
  accumulation fast waves, 326
  accumulation slow waves, 326
  Alfvén and slow continua, *351–356*
  apparent singularities, 453
  body mode, 524, 526
  cluster point, potential, 347
  cluster spectra, *363–365*, 462, 467, 493
  clustering at edges of continua, 363
  complex indices, 373
  continuous spectrum, *345–365*, 452, 453, 496, 520
  $D = 0$ apparent singularities, 334, 335, 351
  damping of Alfvén waves, 356, 533, 534, 541
  degenerate Alfvén waves, 326, 454
  discrete spectrum, 452
  essential spectrum, 325, 341
  fast cluster point singularities, 356
  Frobenius expansion, 349
  HD & MHD, relation spectra, 341
  heating by Alfvén waves, 356, 533, 534, 541
  historical note, 356
  improper Alfvén eigenfunction, 353
  indicial equation, 349
  leaky mode, 523, 526
  MHD, initial value problem, 356
  $N = 0$ genuine singularities, 333
  non-singular ODE, 346
  normal dependence ω_A^2, 347
  number of nodes of eigenfunction, 325
  orthogonality of eigenfunctions, 362
  oscillation theorem for MHD wave equation, *360–362*
  oscillation theorems, *357–363*, 453
  quasi-mode, 516, 522, 540, 545, 546
  regular singularities, 348
  resolvent operator, 356
  role in temporal evolution, 496
  shooting method, 370
  singular differential equations, *345–351*
  small and large solutions, 350, 352
  small solution may jump, 353
  spectral structure, *345–365*, 460
  Sturm's oscillation theorem, 358
  Sturm's separation theorem, 358
  Sturm–Liouville system, 357, 451
  Sturmian, anti-Sturmian, 359, 462
  surface mode, 516, 525
  tangential components, non-square integrable, 354
  turning point frequencies, 326
  variational procedures, 453
MHD spectroscopy, *317–322*, 534
MHD wave equations, *190–204*
  3×3 representation, 198
  7×7 representation, 196
  8×8 representation, 192
  admitting monopoles, 197
  compressibility, 196
  counting variables, *186–198*
  dimensionless variables, 191
  dispersion equation, 199, 205
  gravitating plasma slab, 332
  inhomogeneous media, 198
  marginal solution, 194
  numerical $\nabla \cdot \mathbf{B}$ wave, *197*
  plane wave solutions, 192
  spurious eigenvalues, 194
  symmetric operator, 193
  velocity representation, *198–201*
  vorticity, 196
  wave vector projection, 196
MHD waves
  asymptotic properties, 212
  carrier wave, 206
  constructive interference, 206
  cusp velocity, 211
  dispersion diagrams, *202–204*
  eigenfrequency ordering, 202
  envelope wave fronts, 206
  Friedrichs diagrams, 208
  group diagrams, *205–213*
  group velocity, 205
  local propagation, 204
  low-β approximation, 210
  non-dispersive, 205
  orthogonal eigenfunctions, 202
  parallel propagation, 203
  parameter β, 204
  perpendicular propagation, 202
  phase diagrams, *205–213*
  phase velocity, 205
  relation to spectrum, 204
  relation to stability, 204
  return angle in group diagram, 211

mirror effect, 43
mirror ratio, 45
misnomers
    Larmor frequency, 36
    local field line coordinates, 329
    Larmor radius, 36
model problems, *173–182*

nature
    flaw in standard view of, 21
    fundamental forces, 21
nuclear fusion reactions, *4–6*, 18
    α-particle heating, 4
    Bremsstrahlung losses, 7
    CNO cycle, 6
    confinement time, 9
    core of the Sun, 5
    CTR, 3
    deuterium–deuterium reactions, 9
    deuterium–tritium reactions, 5
    heat transport losses, 7
    ignition condition, 8
    in stars, 18
    Lawson criterion, 8
    Li^6/Li^7 blanket, 4
    mass defect, 4
    product $n\tau_E T$, 9
    proton–proton chain, 6
    thermonuclear output power, 7

Ohmic dissipation, 70
one-fluid equations, *119–126*
    generalized Ohm's law, 121
    maximal ordering, *119–123*
    resistive and ideal, *124–126*
orbit theory, 37

partial differential equations
    Burgers' equation, 215
    Cauchy problem, 216
    domain of dependence, 219
    domain of influence, 219
    elliptic, 218
    hyperbolic, 218
    linear advection equation, 213
    parabolic, 218
    quasi-linear, 219
phase mixing, 533, 565
    coronal loops and holes, 573
    height, 567
    running waves, 567
    standing waves, 566
    time scale, 566
pinch effect, 75
planetary magnetism, *407–415*
    Earth, 407
    field reversal, 412
    geomagnetic dynamo, 409
    geomagnetic field, 408
    Jupiter, 407
    magnetic core spot, 410
    magnetic dipole, 408
    magnetic dipole moment, 408
    magnetic fields of planets, *413–415*
    non-dipolar field, 410
    periodicity of geomagnetic field, 411
    time scale of resistive decay, 411
plasma
    β, 44, 433
    Coulomb interaction, 25
    crude definition, 3
    Debye length, 25, 57
    Debye shielding, 26
    Debye sphere, 26
    Langmuir waves, 54
    macroscopic approach, *28–29*
    microscopic definition, *23–28*
    occurrence, 3, 22
    perfectly conducting fluid, 7
    plasma frequency, 56
    plasma oscillations, 54
    quasi charge-neutrality, 25
    Saha equation, 24
Poynting vector, 150, 151, 543
primitive variables, 148

quadratic forms in MHD, *256–263*
    linearized kinetic energy, 243
    linearized potential energy, 257
quality factor, 552
quantum mechanics
    de Broglie frequency, 40
    de Broglie wavelength, 40
    energy, 40
    momentum, 40
quasi-mode, 516, 522, 540, 545, 546

Rayleigh–Taylor instability, *287–296*
    growth rate, 294
    interchange instability, 287, 293
    magnetic shear, 287, 293
    Parker instability, 287
    wall stabilization, 287, 293
relativity
    energy, 40
    Lorentz transformation, 39, 42
    momentum, 40
    rest mass, 40
resistive energy balance, 551
resistive MHD equations, 69, *161*
    induction equation, 70
    time scale of resistive diffusion, 70
resolvent operator, 498
resonant absorption, 533, 534, 541
    absorption coefficient, 570
    applications, 553
    coupling factor, 546
    efficiency, 546
    energetics, 550, 557
    energy absorption rate, 543
    foot point driving, 562
    length scale, 565

resonant absorption (*cont.*)
  line-tied loops, 567
  magnetosphere, 573
  quality factor, 552
  role of quasi-modes, 546
  solar applications, 567
  solar loops, 554
  sunspots, 569
  temporal evolution, 556
  time scale, 565
  tokamak, 553
  total, 548
resonant damping, 554
resonant dissipation, 552, 554, 567
resonant dissipation rate, 552
resonant heating
  line-tied loops, 567
  solar loops, 554
  tokamaks, 553
Reynolds number, 71
Reynolds stress tensor, 149
Riemann sheet, 513
runaway electrons, 41

Schwarzschild radius, 137
Shafranov shift, 436
shocks
  collisionless, 167
  entropy condition, 170
  gas dynamics, 167
  MHD, 167
single particle motion, *34–47*
solar magnetism, *385–407*
  butterfly diagram, 388
  coronal heating problem, 406, 426
  coronal hole, 384, 403
  coronal loops, 401, 403
  coronal mass ejection, 403, 424
  granules, 388, 399
  heliosphere, 406
  helmet streamer, 404
  magnetograph (Babcock), 387
  Maunder minimum, 388
  neutral current sheet, 406
  penumbra, 397
  photospheric network, 388
  polar plume, 404
  polarities (p, f), 390
  prominence, 397, 401
  solar cycle, *387–395*
  solar dynamo, 390
  solar flares, 402, 426
  solar maxima/minima, 388
  sunspots, 387
  supergranules, 388, 399
  umbra, 397
  Zeeman splitting, 387, 396
solar wind
  critical point, 417
  interplanetary magnetic field, 418
  Parker model, 415
  solar breeze, 417

transonic flow, 418
sound waves, *186–190*
  compressible, 189
  in static media, 188
  longitudinal, 189
  sound velocity, 188
  wave equation, 187
space missions, 20
  Cluster, 415, 425, 426
  Skylab, 384, 403, 406
  SOHO, 384, 426
  Solar Orbiter, 426
  Ulysses, 426
  Voyager, 384, 415
space weather, 385, 426
specific heats (γ), ratio of, 53
spectral cut, 512
spectral theory
  alternatives, *250–256*
  analogy with quantum mechanics, 243, 271, 451
  approximate spectrum, 251
  compact operator, 251
  continuous spectrum, 251, 266
  discrete (point) spectrum, 252
  eigenvalue problem, 250
  Fredholm alternative, 251
  Heisenberg 'picture', 271
  ideal MHD spectrum, 238
  inhomogeneous equation, 250
  quadratic forms, 250
  resolvent operator, 252
  resolvent set, 252
  Schrödinger 'picture', 271
  self-adjoint operators, 244
  unbounded operator, 251
Spitzer resistivity, 54
stability
  σ-stability, *268*
  compressibility, 241
  constraints, 232, 236
  currents, 241
  exchange of stabilities, 265
  field line bending, 241
  gravity, 241
  homogeneous plasma, 258
  inhomogeneous plasma, 258
  internal and external modes, 273
  intuitive approach, *230–232*
  inverted glass of water, 232, 273
  marginal (neutral), 230, 240, 265
  marginal equation of motion, 240
  nonlinear, 232
  perturbation, 230
  pressure gradients, 241
stability of cylindrical plasmas
  σ-stability, *467–469*
  σ-stable configurations, 469
  constant-pitch field, *471–474*
  effective wall at singularity, 450, 458
  external kink mode, 455
  force-free magnetic fields, *475–482*
  general energy expression, 484

instabilities of a z-pinch, 469
kink modes in force-free fields, 480
Mercier criterion, 466
Newcomb's procedure, *463–467*
oscillation theorems, 462
pure interchanges, 460, 472
quasi-interchanges, 460, 472
skin current at singularity, 477
skin current model, 455
skin current perturbation, 450, 458
'small' solutions, 478
'straight tokamak', *482–492*
surface mode, 456
Suydam's criterion, *463–467*
stability of 'straight tokamak'
   $q_0 = 1$, 487
   energy expression, 486
   enhanced MHD activity, 491
   external kink modes, 488
   internal kink modes, 486
   low-β tokamak ordering, 472, 484
   rational magnetic surfaces, 489
   sawtooth oscillations, 487
   toroidal mode number (n), 456
   'virtual singularity', 490
   wall stabilization, 458
Sun
   chromosphere, 386, 395
   chromospheric spicules, 401
   convection zone, 301, *305–308*, 385
   convective stability, 307
   core, 301, 305, 385
   corona, 386
   coronagraph, 404
   differential rotation, 386, 387
   Doppler shift, 396
   dynamo, 308
   Fraunhofer lines, 395
   heliosphere, 386, 406, 418
   hydrodynamics of interior, *300–301*
   hydrostatic equilibrium, 303
   luminosity, 300
   photosphere, 386, 395
   radiative equilibrium, *301–305*
   radiative transport, 302
   radiative zone, 301, 385
   Schwarzschild criterion, 307
   solar constant, 300
   standard solar model, 300
   thermal conduction coefficient, 302
   thermonuclear energy, 302
   thermonuclear reactions, 300
   turbulent mixing, 307
sunspot seismology, 322, 397, 570
supersonic flow, 167
   Mach number, 167
surface current, 173
surface mode, 516, 525
surface vorticity, 173

thermodynamic variables, 134
   entropy, 134
   internal energy, 134
time derivative
   Eulerian, 132, 235
   Lagrangian, 132, 235
tokamak
   disruptions, 150
   resonant absorption, 553
   safety factor, 78, 157
transport theory, 53, 93
   Chapman–Enskog procedure, 93
   neo-classical transport, 98
   transport coefficients, 53
   turbulent transport, 98
two-fluid equations, 67, *98–118*
   electron skin depth, 69, 107
   heat flow, 67
   ideal, 108
   quasi charge-neutrality, 69
   ratio of masses over charges, 111
   resistive, 68, 108
   viscosity, 67

Universe
   big bang, 21
   plasmas everywhere, 20, 23

Van Allen belts, 44
variational analysis, 291
   Euler–Lagrange equations, 292
variational principles in MHD, *256–263*
   choice of norms, 262
   energy principle, 261
   extended σ-stability principle, 285
   extended energy principle, 285
   extended spectral principle, 283
   Hamilton's principle, 259
   interface extensions, *283–285*
   modified energy principle, 270
   Rayleigh–Ritz principle, 259
vector potential, 156
   for vacuum field, 276
viewpoints
   differential equations, 231
   energy and force, *230–232*
   variational quadratic forms, 231

wave packet shapes
   δ function, 206
   Gaussian, 206
waves in two-fluid plasmas, *108–118*
   cutoff frequencies, 116
   dispersion equation, 113
   high-frequency limits, 117
   MHD limit, 117
   resonance limits, 117
Wronskian, 510